Osswald / Hernández-Ortiz
Polymer Processing

Tim A. Osswald
Juan P. Hernández-Ortiz

Polymer Processing

Modeling and Simulation

HANSER
Hanser Publishers, Munich • Hanser Gardner Publications, Cincinnati

The Authors:
Prof. Dr. Tim A. Osswald, Department of Mechnical Engineering, University of Wisconsin-Madison, USA
Dr. Juan P. Hernández-Ortiz, Department of Chemical and Biological Engineering,
University of Wisconsin-Madison, USA

Distributed in the USA and in Canada by
Hanser Gardner Publications, Inc.
6915 Valley Avenue, Cincinnati, Ohio 45244-3029, USA
Fax: (513) 527-8801
Phone: (513) 527-8977 or 1-800-950-8977
www.hansergardner.com

Distributed in all other countries by
Carl Hanser Verlag
Postfach 86 04 20, 81631 München, Germany
Fax: +49 (89) 98 48 09
www.hanser.de

Library of Congress Cataloging-in-Publication Data

Osswald, Tim A.
Polymer processing : modeling and simulation / Tim A. Osswald, Juan P. Hernández-Oritz.-- 1st ed.
p. cm.
ISBN-13: 978-1-56990-398-8 (hardcover)
ISBN-10: 1-56990-398-0 (hardcover)
1. Polymers--Mathematical models. 2. Polymerization--Mathematical models.
I. Hernández-Oritz, Juan P. II. Title.
TP1087.O87 2006
668.901'5118--dc22

2006004981

Bibliografische Information Der Deutschen Bibliothek
Die Deutsche Bibliothek verzeichnet diese Publikation in der Deutschen Nationalbibliografie; detaillierte bibliografische Daten sind im Internet über <http://dnb.ddb.de> abrufbar.

ISBN-13: 978-3-446-40381-9
ISBN-10: 3-446-40381-7

Production Management: Oswald Immel
Coverconcept: Marc Müller-Bremer, Rebranding, München, Germany
Coverdesign: MCP • Susanne Kraus GbR, Holzkirchen, Germany
Printed and bound by Druckhaus "Thomas Müntzer" GmbH, Bad Langensalza, Germany

Lovingly dedicated to our maternal Grandfathers
Ernst Robert Georg Victor and Luis Guillermo Ortiz;
Two great men whose own careers in chemical
engineering influenced the paths we have taken

In gratitude to Professor R.B. Bird, the teacher and the pioneer who laid the groundwork for polymer processing – modeling and simulation

PREFACE

The groundwork for the fundamentals of polymer processing was laid out by Professor R. B. Bird, here at the University of Wisconsin-Madison, over 50 years ago. Almost half a century has past since the publication of Bird, Steward and Lightfoot's transport phenomena book. *Transport Phenomena* (1960) was followed by several books that specifically concentrate on polymer processing, such a the books by McKelvey (1962), Middleman (1977), Tadmor and Gogos (1979), and Agassant, Avenas, Sergent and Carreau (1991). These books have influenced generations of mechanical and chemical engineering students and practicing engineers. Much has changed in the plastics industry since the publication of McKelvey's 1962 *Polymer Processing* book. However, today as in 1962, the set-up and solution of processing problems is done using the fundamentals of transport phenomena.

What has changed in the last 50 years, is the complexity of the problems and how they are solved. While we still use traditional analytical, back-of-the-envelope solutions to model, understand and optimize polymer processes, we are increasingly using computers to numerically solve a growing number of realistic models. In 1990, Professor C.L. Tucker III, at the University of Illinois at Urbana-Champaign edited the book *Computer Simulation for Polymer Processes*. While this book has been out of print for many years, it is still the standard work for the graduate student learning computer modeling in polymer processing.

Since the publication of Tucker's book and the textbook by Agassant *et al.*, advances in the plastics industry have brought new challenges to the person modeling polymer processes. For example, parts have become increasingly thinner, requiring much higher injection pressures and shorter cooling times. Some plastic parts such as lenses and pats with microfeatures require much higher precision and are often dominated by three-dimensional flows.

The book we present here addresses traditional polymer processing as well as the emerging technologies associated with the 21st Century plastics industry, and combines the modeling aspects in *Transport Phenomena* and traditional polymer processing textbooks of the last few decades, with the simulation approach in *Computer Modeling for Polymer Processing*. This textbook is designed to provide a polymer processing background to engineering students and practicing engineers. This three-part textbook is written for a two-semester polymer processing series in mechanical and chemical engineering. The first and second part of the book are designed for the senior- to grad-level course, introducing polymer processing, and the third part is for a graduate course on simulation in polymer processing. Throughout the book, many applications are presented in form of examples and illustrations. These will also serve the practicing engineer as a guide when determining important parameters and factors during the design process or when optimizing a process.

Polymer Processing – Modeling and Simulation is based on lecture notes from intermediate and advanced polymer processing courses taught at the Department of Mechanical Engineering at the University of Wisconsin-Madison and a modeling and simulation in polymer processing course taught once a year to mechanical engineering students specializing in plastics technology at the University of Erlangen-Nurenburg, Germany. We are deeply indebted to the hundreds of students on both sides of the Atlantic who in the past few years endured our experimenting and trying out of new ideas and who contributed with questions, suggestions and criticisms.

The authors cannot acknowledge everyone who helped in one way or another in the preparation of this manuscript. We are grateful to the engineering faculty at the University of Wisconsin-Madison, and the University of Erlangen-Nurenberg for their support while developing the courses which gave the base for this book. In the Department of Mechanical Engineering at Wisconsin we are indebted to Professor Jeffrey Giacomin, for his suggestions and advise, and Professor Lih-Sheng Turng for letting us use his 3D mold filling results in Chapter 9. In the Department of Chemical and Biological Engineering in Madison we are grateful to Professors Juan dePablo and Michael Graham for JPH's financial support, and for allowing him to work on this project. We would like to thank Professor G.W. Ehrenstein, of the LKT-Erlangen, for extending the yearly invitation to teach the "Blockvorlesung" on *Modeling and Simulation in Polymer Processing*. The notes for that class, and the same class taught at the University of Wisconsin-Madison, presented the starting point for this textbook. We thank the following students who proofread, solved problems and gave suggestions: Javier Cruz, Mike Dattner, Erik Foltz, Yongho Jeon, Fritz Klaiber, Andrew Kotloski, Adam Kramschuster, Alejandro Londoño, Ivan López, Petar Ostojic, Sean Petzold, Brian Ralston, Alejandro Roldán and Himanshu Tiwari. We are grateful to Luz Mayed (Lumy) D. Nouguez for the superb job of drawing some of the figures. Maria del Pilar Noriega from the ICIPC and Whady F. Florez from the UPB, in Medellín, Colombia, are acknowledged for their contributions to Chapter 11. We are grateful to Dr. Christine Strohm and Oswald Immel of Hanser Publishers for their support throughout the development of this book. TAO thanks his wife, Diane Osswald, for as always serving as a sounding board from the beginning to the end of this project. JPH thanks his family for their continuing support.

TIM A. OSSWALD AND JUAN P. HERNANDEZ-ORTIZ

Madison, Wisconsin
Spring 2006

TABLE OF CONTENTS

PART III NUMERICAL TECHNIQUES

INTRODUCTION

Ignorance never settles a question.

—Benjamin Disraeli

The mechanical properties and the performance of a finished product are always the result of a sequence of events. Manufacturing of a plastic part begins with material choice in the early stages of part design. Processing follows this, at which time the material is not only shaped and formed, but the properties which control the performance of the product are set or frozen into place. During design and manufacturing of any plastic product one must always be aware that material, processing and design properties all go hand-in-hand and cannot be decoupled. This approach is often referred to as the five P's: polymer, processing, product, performance and post consumer life of the plastic product.

This book is primarily concerned with the first three P's. Chapters 1 and 2 of this book deal with the materials science of polymers, or the first P, and the rest of the book concerns itself with polymer processing. The performance of the product, which relates to the mechanical, electrical, optical, acoustic properties, to name a few, are not the focus of this book.

I.1 MODELING AND SIMULATION

A model of a process is a simplified physical or mathematical representation of that system, which is used to better understand the physical phenomena that exist within that process. A physical model is one where a simplified representation of that process is constructed,

Figure I.1: Photograph of the screw channel with nozzle and initial tracer ink position.

such as the screw extruder with a transparent barrel shown in Fig. I.1 [5]. The extruder in the photographs is a 6 inch diameter, 6D long constant channel depth screw pump that was built to demonstrate that a system where the screw rotates is equivalent to a system where the barrel is rotating. In addition, this physical model, which contained a Newtonian fluid (silicone oil), was used to test the accuracy of boundary element method simulations by comparing the deformation of tracer ink markings that were injected through various nozzles located at different locations in the screw channel.

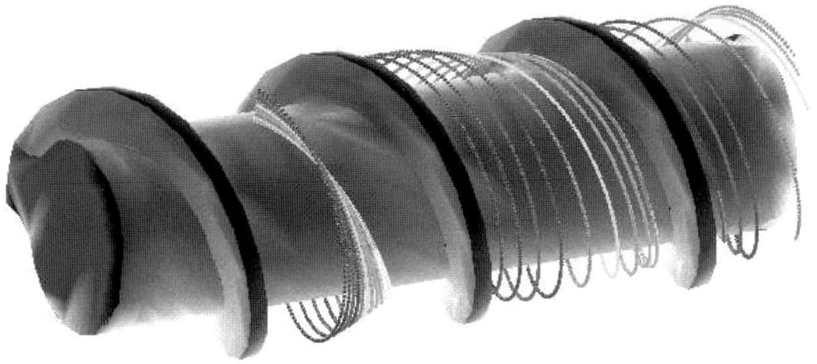

Figure I.2: BEM simulation results of the flow lines inside the screw channel of a single screw extruder.

Hence, the physical model of the screw pump served as a tool to understand the underlying physics of extrusion, as well as a means to validate mathematical models of polymer processes.

Physical models can be as complex as the actual system, except smaller in size. Such a model is called a pilot operation. Usually, such a system is built to experiment with different material formulations, screw geometries, processing conditions and many more, without having to use excessive quantities of material, energy and space. Once the desired results are achieved, or a specific invention has been realized on the pilot operation scale, it is important to scale it up to an industrial scale. Chapter 4 of this book presents how physical models can be used to understand and scale a specific process.

In lieu of a physical model it is often less expensive and time consuming to develop a mathematical model of the process. A mathematical model attempts to mimic the actual process with equations. The mathematical model is developed using material, energy and momentum balance equations, along with a series of assumptions that simplify the process sufficiently to be able to achieve a solution. Figure I.2 presents the flow lines in the metering section of a single screw extruder, computed using a mathematical model of the system, solved with the boundary element method (BEM), for a BEM representation shown in Fig. 11.25, composed of 373 surface elements and 1202 nodes [22, 5]. Here, although the geometry representation was accurate, the polymer melt was assumed to be a simple Newtonian fluid.

The more complex this mathematical model, the more accurately it represents the actual process. Eventually, the complexity is so high that we must resort to numerical simulation to model the process, or often the model is so complex that even numerical simulation fails to deliver a solution. Chapters 5 and 6 of this book address how mathematical models are used to represent polymer processes using analytical solutions. Chapters 7 to 11 present various numerical techniques used to solve more complex polymer processing models.

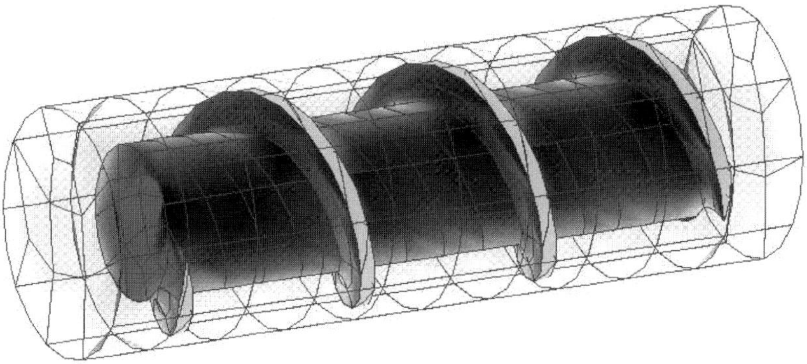

Figure I.3: BEM representation of the screw and barrel used to predict the results presented in Fig. I.2.

I.2 MODELING PHILOSOPHY

We model a polymer process or an event in order to better understand the system, to solve an existing problem or perhaps even improve the manufacturing process itself. Furthermore, a model can be used to optimize a given process or properties of the final product. In order to model or simulate a process we need to derive the equations that govern or represent the physical process. Before we solve the process' governing equations we must first simplify them by using a set of assumptions. These assumptions can be geometric simplifications, boundary conditions, initial conditions, physical assumptions, such as assuming isothermal systems or isotropic materials, as well as material models, such as Newtonian, elastic, visco-elastic, shear thinning, or others.

When modeling, it is good practice to break the analysis and solution process into set of standard steps that will facilitate a solution to the problem [1, 2, 4]. These steps are:

- Clearly define the ***scope*** of the problem and the goals you want to achieve,
- ***Sketch*** the system and define parameters such as dimensions and boundary conditions,
- Write down the general ***governing equations*** that govern the variables in the process, such as mass, energy and momentum balance equations,
- Introduce the ***constitutive equations*** that relate the problem's variables,
- State your ***assumptions*** and ***reduce the governing equations*** using these assumptions,
- ***Scale*** the variables and governing equations,
- ***Solve*** the equation and plot results.

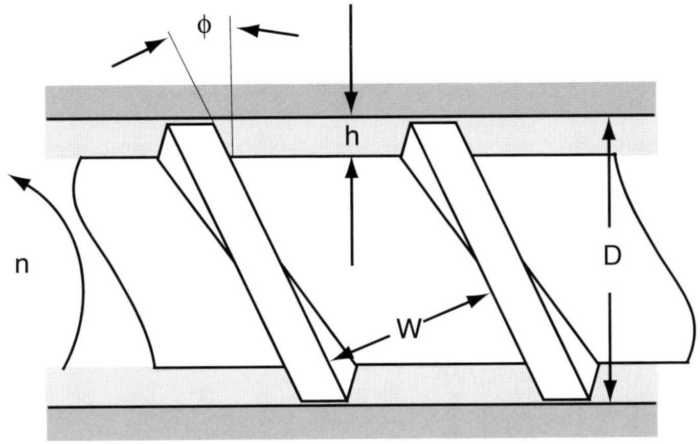

Figure I.4: Schematic diagram of a single screw mixing device.

■ **EXAMPLE 0.1.**

Physical and mathematical model of a single screw extruder mixer. To illustrate the concept of modeling, we will use a hypothetical small (pilot) screw extruder, like the one presented in Fig. I.4, and assume that it was successfully used to disperse solid agglomerates within a polymer melt. Two aspects are important when designing the process: the stresses required to disperse the solid agglomerates and controlling the viscous friction inside the melt to avoid overheating of the material. Both these aspects were satisfied in the pilot process, that had dimensions and process conditions given by:

- Geometric parameters - Diameter, D_1, channel depth, h_1, channel width, W_1 and helix angle, ϕ_1,
- Processing conditions - Heater temperature, T_1, and rotational speed of the screw, n_1,
- Material parameters - Viscosity, μ_1, and melting temperature, T_{m1}.

However, the pilot system is too small to be feasible, and must therefore be scaled up for production. We now begin the systematic solution of this problem, following the steps delineated above.

- ***Scope***
  The purpose of this analysis is to design an industrial size version of the pilot process, which achieves the same dispersive mixing without overheating the polymer melt. In order to simplify the solution we lay the helical geometry flat, a common way of analyzing single screw extruders.

- ***Sketch***

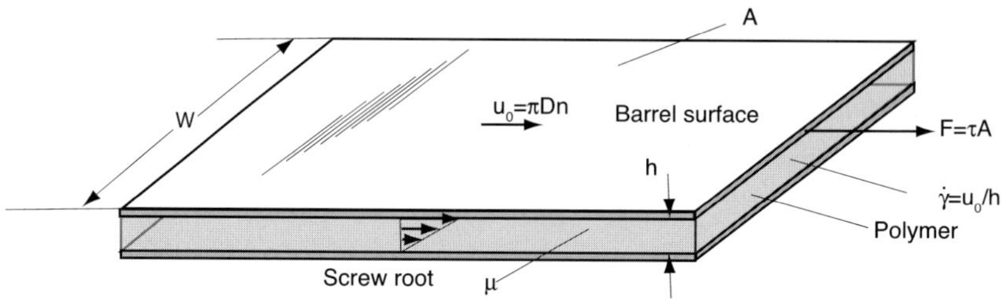

- ***Governing Equations***

The relative motion between the screw and the barrel is represented by the velocity

$$u_0^i = \pi D_i n_i \tag{I.1}$$

where D_i is the diameter of the screw and barrel and n_i the rotational speed of the screw in revolutions/second. The subscript i is 1 for the pilot process and 2 for the scaled-up industrial version of the process. For a screw pump system, the volumetric throughput is represented using

$$Q_i = \frac{u_0^i h_i W_i}{2} \cos\phi = \frac{\pi D_i n_i h_i W_i}{2} \cos\phi \tag{I.2}$$

The torque used to turn the screw, T, in Fig. I.4 is equivalent to the force used to move the plate in the model presented in the sketch, F, as

$$F_i = \frac{2T_i}{D_i} \tag{I.3}$$

Using the force we can compute the energy rate, per unit volume, that goes into the viscous polymer using

$$E_v^i = \frac{F_i u_0^i}{A_i h_i} \tag{I.4}$$

This viscous heating is conducted out from the polymer at a rate controlled by the thermal conductivity, k, with units W/m/K. The rate of heat per volume conducted out of the polymer can be estimated using

$$E_c^i = k\frac{\Delta T}{h_i^2} \tag{I.5}$$

where ΔT is a temperature difference characteristic of the process at hand given by the difference between the heater temperature and the melting temperature.

- ***Constitutive Equations***

The constitutive equation here is the relation between the shear stress, τ and the rate of deformation $\dot{\gamma}$. We can define the shear stress, τ_i, for system i using

$$\tau_i = \mu_i \dot{\gamma}_i = \mu_i \frac{u_0^i}{h_i} \tag{I.6}$$

- ***Assumptions and Reduction of Governing Equations***
  Since we are scaling the system with the same material, we can assume that the material parameters remain constant, and for simplicity, we assume that the heater temperature remains the same. In addition, we will fix our geometry to a standard square pitch screw (ϕ =17.65^o) and therefore, a channel width proportional to the diameter. Hence, the parameters to be determined are D_2, h_2 and n_2.

  Using the constitutive equation, we can also compute the force it takes to move the upper plate (barrel)

$$F_i = \tau A_i = \mu \frac{u_0^i}{h_i} \tag{I.7}$$

  This results in a viscous heating given by

$$E_v^i = \mu \left(\frac{u_0^i}{h_i} \right)^2 \tag{I.8}$$

  which, due to the high viscosity of polymers, is quite significant and often leads to excessive heating of the melt during processing.

- ***Scale***
  We can assess the amount of viscous heating if we compare it to the heat removed through conduction. To do this, we scale the viscous dissipation with respect to thermal conduction by taking the ratio of the viscous heating, E_v, to the conduction, E_c,

$$\frac{E_v^i}{E_c^i} = \frac{\mu u_0^{i2}}{k \Delta T} \tag{I.9}$$

  This ratio is often referred to as the Brinkman number, Br. When Br is large, the polymer may overheat during processing.

- ***Solve Problem***
  Since the important parameters for developing the pilot operation were the stress (to disperse the solid agglomerates) and the viscous dissipation (to avoid overheating), we need to maintain τ_i and the Brinkman number, Br, constant. If our scaling parameter is the diameter, we can say

$$D_2 = \mathcal{R} D_1 \tag{I.10}$$

  where $\mathcal{R}$ is the scaling factor. Hence, for a constant Brinkman number we must satisfy

$$n_2 = n_2 / \mathcal{R} \tag{I.11}$$

  which results in an industrial operation with the same viscous dissipation as the pilot process. Using this rotational speed we can now compute the required channel depth to maintain the same stress that led to dispersion. Therefore, for a constant stress, $\tau_2 = \tau_1$ we must satisfy,

$$h_2 = h_1 \tag{I.12}$$

  Although the above solution satisfies our requirements, it leads to a very small volumetric throughput. However, in industry there are various scaling rules that are used for extruder systems which compromise one or the other requirement. We cover this in more detail in Chapter 4 of this book.

I.3 NOTATION

There are many ways of writing equations that represent transport of mass, heat, and fluids trough a system, and the constitutive equations that model the behavior of the material under consideration. Within this book, tensor notation, Einstein notation, and the expanded differential form are considered. In the literature, many authors use their own variation of writing these equations. The notation commonly used in the polymer processing literature is used throughout this textbook. To familiarize the reader with the various notations, some common operations are presented in the following section.

The physical quantities commonly encountered in polymer processing are of three categories: *scalars*, such as temperature, pressure and time; *vectors*, such as velocity, momentum and force; and *tensors*, such as the stress, momentum flux and velocity gradient tensors. We will distinguish these quantities by the following notation,

$$\begin{aligned} T &\longrightarrow \text{scalar: italic} \\ \mathbf{u} = u_i &\longrightarrow \text{vector: boldface or one free subindex} \\ \boldsymbol{\tau} = \tau_{ij} &\longrightarrow \text{second-order tensor: boldface or two free subindices} \end{aligned}$$

The free subindices notation was introduced by Einstein and Lorentz and is commonly called the *Einstein notation*. This notation is a useful way to collapse the information when dealing with equations in *cartesian coordinates*, and it is equivalent to subindices used when writing computer code. The Einstein notation has some basic rules that are as follows,

- The subindices $i, j, k = 1, 2, 3$ and they represent the x, y and z Cartesian coordinates,
- Every free index represents an increase in the tensor order: one free index for vectors, u_i, two free indices for matrices (second order tensors), τ_{ij}, three free indices for third order tensors, ϵ_{ijk},
- Repeated subindices imply summation, $\tau_{ii} = \tau_{11} + \tau_{22} + \tau_{33}$,
- Comma implies differentiation, $u_{i,j} = \partial u_i/\partial x_j$.

The vector differential operator, ∇, is the most widely used vector and tensor differential operator for the balance equations. In Cartesian coordinates it is defined as

$$\begin{aligned} \nabla = \frac{\partial}{\partial x_j} &= \left(\frac{\partial}{\partial x}, \frac{\partial}{\partial y}, \frac{\partial}{\partial z}\right) \\ &= \left(\frac{\partial}{\partial x_1}, \frac{\partial}{\partial x_2}, \frac{\partial}{\partial x_3}\right) \end{aligned} \tag{I.13}$$

This operator will define the *gradient* of any scalar or vector quantity. For a scalar quantity it will produce a *vector gradient*

$$\nabla T = \frac{\partial T}{\partial x_i} = \left(\frac{\partial T}{\partial x_1}, \frac{\partial T}{\partial x_2}, \frac{\partial T}{\partial x_3}\right) \tag{I.14}$$

while for a vector it will produce a second-order tensor

$$\nabla \mathbf{u} = \frac{\partial u_i}{\partial x_j} = \begin{bmatrix} \frac{\partial u_x}{\partial x} & \frac{\partial u_y}{\partial x} & \frac{\partial u_z}{\partial x} \\ \frac{\partial u_x}{\partial y} & \frac{\partial u_y}{\partial y} & \frac{\partial u_z}{\partial y} \\ \frac{\partial u_x}{\partial z} & \frac{\partial u_y}{\partial z} & \frac{\partial u_z}{\partial z} \end{bmatrix} = \begin{bmatrix} \frac{\partial u_1}{\partial x_1} & \frac{\partial u_2}{\partial x_1} & \frac{\partial u_3}{\partial x_1} \\ \frac{\partial u_1}{\partial x_2} & \frac{\partial u_2}{\partial x_2} & \frac{\partial u_3}{\partial x_2} \\ \frac{\partial u_1}{\partial x_3} & \frac{\partial u_2}{\partial x_3} & \frac{\partial u_3}{\partial x_3} \end{bmatrix} \tag{I.15}$$

When the gradient operator is dotted with a vector or a tensor, the *divergence* of the vector or tensor is obtained. The divergence of a vector produces a scalar

$$\nabla \cdot \mathbf{u} = \frac{\partial u_i}{\partial x_i} = \frac{\partial u_x}{\partial x} + \frac{\partial u_y}{\partial y} + \frac{\partial u_z}{\partial z} = \frac{\partial u_1}{\partial x_1} + \frac{\partial u_2}{\partial x_2} + \frac{\partial u_3}{\partial x_3} \tag{I.16}$$

while for a second order tensor it produces the components of a vector as follows

$$\nabla \cdot \boldsymbol{\tau} = \frac{\partial \tau_{ij}}{\partial x_j} = \begin{pmatrix} \frac{\partial \tau_{xx}}{\partial x} + \frac{\partial \tau_{xy}}{\partial y} + \frac{\partial \tau_{xz}}{\partial z} \\ \frac{\partial \tau_{yx}}{\partial x} + \frac{\partial \tau_{yy}}{\partial y} + \frac{\partial \tau_{yz}}{\partial z} \\ \frac{\partial \tau_{zx}}{\partial x} + \frac{\partial \tau_{zy}}{\partial y} + \frac{\partial \tau_{zz}}{\partial z} \end{pmatrix} = \begin{pmatrix} \frac{\partial \tau_{11}}{\partial x_1} + \frac{\partial \tau_{12}}{\partial x_2} + \frac{\partial \tau_{13}}{\partial x_3} \\ \frac{\partial \tau_{21}}{\partial x_1} + \frac{\partial \tau_{22}}{\partial x_2} + \frac{\partial \tau_{23}}{\partial x_3} \\ \frac{\partial \tau_{31}}{\partial x_1} + \frac{\partial \tau_{32}}{\partial x_2} + \frac{\partial \tau_{33}}{\partial x_3} \end{pmatrix} \tag{I.17}$$

Finally, the *Laplacian* is defined by the divergence of the gradient. For a scalar quantity it is

$$\nabla \cdot \nabla T = \nabla^2 T = \frac{\partial^2 T}{\partial x_j \partial x_j} = \frac{\partial^2 T}{\partial x^2} + \frac{\partial^2 T}{\partial y^2} + \frac{\partial^2 T}{\partial z^2} = \frac{\partial^2 T}{\partial x_1^2} + \frac{\partial^2 T}{\partial x_2^2} + \frac{\partial^2 T}{\partial x_3^2} \tag{I.18}$$

while for a vector it is written as

$$\nabla \cdot \nabla \mathbf{u} = \nabla^2 u_i = \frac{\partial^2 u_i}{\partial x_j \partial x_j} = \begin{pmatrix} \frac{\partial^2 u_x}{\partial x^2} + \frac{\partial^2 u_x}{\partial y^2} + \frac{\partial^2 u_x}{\partial z^2} \\ \frac{\partial^2 u_y}{\partial x^2} + \frac{\partial^2 u_y}{\partial y^2} + \frac{\partial^2 u_y}{\partial z^2} \\ \frac{\partial^2 u_z}{\partial x^2} + \frac{\partial^2 u_z}{\partial y^2} + \frac{\partial^2 u_z}{\partial z^2} \end{pmatrix} = \begin{pmatrix} \frac{\partial^2 u_1}{\partial x_1^2} + \frac{\partial^2 u_1}{\partial x_2^2} + \frac{\partial^2 u_1}{\partial x_3^2} \\ \frac{\partial^2 u_2}{\partial x_1^2} + \frac{\partial^2 u_2}{\partial x_2^2} + \frac{\partial^2 u_2}{\partial x_3^2} \\ \frac{\partial^2 u_3}{\partial x_1^2} + \frac{\partial^2 u_3}{\partial x_2^2} + \frac{\partial^2 u_3}{\partial x_3^2} \end{pmatrix} \tag{I.19}$$

A very useful and particular case of the vector gradient is the *velocity vector gradient*, $\nabla \mathbf{u}$ shown in eqn. (I.15). With this tensor, two very useful tensors can be defined, the *strain rate tensor*

$$\dot{\boldsymbol{\gamma}} = \dot{\gamma}_{ij} = \nabla \mathbf{u} + (\nabla \mathbf{u})^T = \frac{\partial u_i}{\partial x_j} + \frac{\partial u_j}{\partial x_i} \tag{I.20}$$

which is a symmetric tensor. And the *vorticity tensor*

$$\boldsymbol{\omega} = \omega_{ij} = \nabla \mathbf{u} - (\nabla \mathbf{u})^T = \frac{\partial u_i}{\partial x_j} - \frac{\partial u_j}{\partial x_i} \tag{I.21}$$

which is an anti-symmetric tensor.

I.4 CONCLUDING REMARKS

This manuscript is concerned with modeling and simulation in polymer processing. We have divided the book into three parts: I. Background, II. Processing Fundamentals and III. Simulation in Polymer Processing. The background section introduces the student to polymer materials science (Chapter 1), to important material properties needed for modeling (Chapter 2) and gives an overview of polymer processing systems and equipment (Chapter 3). The second part introduces the student to modeling in polymer processing. The section covers dimensional analysis and scaling (Chapter 4), the balance equations with simple flow and heat transfer solutions in polymer processing (Chapter 5), and introduces many analytical solutions that can be used to analyze a whole variety of polymer processing techniques (Chapter 6). The third part of this book covers simulation in polymer processing. The section covers the various numerical simulation techniques, starting with numerical tools (Chapter 7), and covering the various numerical methods used to solve partial differential equations found in processing, such as the finite difference technique (Chapter 8), the finite element method (Chapter 9), the boundary element method (Chapter 10) and radial basis functions collocation method (Chapter 11).

REFERENCES

1. C.G. Baird and D.I. Collias. *Polymer Processing: Principles and Design.* John Wiley & Sons, New York, 1988.
2. J.A. Dantzig and C.T. Tucker III. *Modeling in Materials Processing.* Cambridge University Press, Cambridge, 2001.
3. P.J. Gramann. PhD thesis, University of Wisconsin-Madison, 1995.
4. C.T. Tucker III, editor. *Computer Modeling for Polymer Processing.* Hanser, Munich, 1989.
5. C. Rauwendaal, T.A. Osswald, G. Tellez, and P.J. Gramann. Flow analysis in screw extruders - effect of kinematic conditions. *International Polymer Processing*, 13(4):327–333, 1998.

PART I

BACKGROUND

CHAPTER 1

POLYMER MATERIALS SCIENCE

I just want to say one word to you, Ben.
Just one word - plastics.

—Advice given to the young graduate played by Dustin Hoffman in the 1967 movie *The Graduate*.

The material behavior of polymers is totally controlled by their molecular structure. In fact, this is true for all polymers; synthetically generated polymers as well as polymers found in nature (bio-polymers), such as natural rubber, ivory, amber, protein-based polymers or cellulose-based materials. To understand the basic aspects of material behavior and its relation to the molecular structure of polymers, in this chapter we attempt to introduce the fundamental concepts in a compact and simple way.

1.1 CHEMICAL STRUCTURE

As the word itself suggests, polymers are materials composed of molecules of very high molecular weight. These large molecules are generally referred to as *macromolecules*. Polymers are macromolecular structures that are generated synthetically or through natural processes. Historically, it has always been said that synthetic polymers, are generated through *addition* or *chain growth polymerization*, and *condensation* or *radical initiated polymerization*. In addition polymerization, the final molecule is a repeating sequence of blocks with a chemical formulae to those of the monomers. Condensation polymerization processes occur when the resulting polymers have fewer atoms than those present in the monomers from which they are generated. However, since many additional polymerization processes result in condensates, and various condensation polymerization processes are chain growth polymerization processes that resemble addition polymerization, today we rather break-down polymerization processes into *step* polymerization and *chain* polymerization. Table 1.1 shows a break-down of polymerization into step and chain polymerization, and presents examples for the various types of polymerization processes. Linear and non-linear step growth polymerization are processes in which the polymerization occurs with more than one molecular species. On the other hand, chain growth polymerization processes occur with monomers with a reactive end group. Chain growth polymerization processes include *free-radical* polymerization, *ionic* polymerization, *cationic* polymerization, *ring opening* polymerization, *Ziegler-Natta* polymerization and *Metallocene* catalysis polymerization. Free-radical polymerization is the most widely used polymerization process and it is used to polymerize monomers with the general structure $CH_2{=}CR_1R_2$. Here, the polymer molecules grow by addition of a monomer with a free-radical reactive site called an active site. A chain polymerization process can also take place when the active site has an ionic charge. When the active site is positively charged, the polymerization process is called a cationic polymerization, and when the active site is negatively charged it is called ionic polymerization. Finally, monomers with a cyclic or ring structure such as caprolactam can be polymerized using the ring-opening polymerization process. In the case of caprolactam, it is polymerized into polycaprolactam or polyamide 6.

The atomic composition of polymers encompasses primarily non-metallic elements such as carbon (C), hydrogen (H) and oxygen (O). In addition, recurrent elements are nitrogen (N), chlorine (Cl), fluoride (F) and sulfur (S). The so-called *semi-organic* polymers contain other non-metallic elements such as silicon (Si) in silicone or polysiloxane, as well as bor or beryllium (B). Although other elements can sometime be found in polymers, because of their very specific nature, we will not mention them here. The properties of the above elements lead to specific properties that are common of all polymers. These are:

- Polymers have very low electric conductance (i.e. they are electric insulators),
- Polymers have a very low thermal conductance (i.e. they are thermal insulators),
- Polymers have a very low density (between 0.8 and 2.2 g/cm^3),
- Polymers have a low thermal resistance and will easily irreversibly thermally degrade.

There are various ways that the monomers can arrange during polymerization; however, we can break them down into two general categories: *uncross-linked* and *cross-linked*. Furthermore, the uncross-linked polymers can be subdivided into linear and branched polymers. The most common example of uncross-linked polymers that present the various degrees of branching is polyethylene (PE).

Another important family of uncross-linked polymers are copolymers. Copolymers are polymeric materials with two or more monomer types in the same chain. A copolymer that is composed of two monomer types is referred to as a *bipolymer* (i.e., PS-HI), and one that is formed by three different monomer groups is called a *terpolymer* (i.e., ABS). Depending on how the different monomers are arranged in the polymer chain, one distinguishes between random, alternating, block, or graft copolymers, discussed later in this chapter.

Although thermoplastics can cross-link under specific conditions, such as gel formation when PE is exposed to high temperatures for prolonged periods of time, thermosets, and

Table 1.1: Polymerization Classification

Classification	Polymerization	Examples
Step Linear	Polycondensation	Polyamides Polycarbonate Polyesters Polyethers Polyimide Siloxanes
	Polyaddition	Polyureas Polyurethanes
Step Non-linear	Network Polymers	Epoxy resins Melamine Phenolic Polyurethanes Urea
Chain	Free radical	Polybutadiene Polyethylene (branched) Polyisoprene Polymethylmethacrylate Polyvinyl acetate Polystyrene
	Cationic	Polyethylene Polyisobutylene Polystyrene Vinyl esters
	Anionic	Polybutadiene Polyisoprene Polymethylmethacrylate Polystyrene
	Ring opening	Polyamide 6 Polycaprolactone Polyethylene oxide Polypropylene oxide
	Ziegler-Natta	Polyethylene Polypropylene Polyvinyl chloride Other vinyl polymers
	Metallocene	Polyethylene Polypropylene Polyvinyl chloride Other vinyl polymers

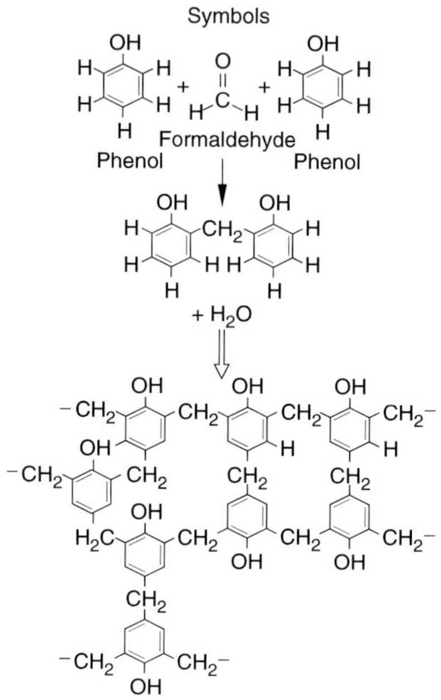

Figure 1.1: Symbolic representation of the condensation polymerization of phenol-formaldehyde.

some elastomers, are polymeric materials that have the ability to cross-link. The cross-linking causes the material to become heat resistant after it has solidified. The cross-linking usually is a result of the presence of double bonds that open, allowing the molecules to link with their neighbors. One of the oldest thermosetting polymers is phenol-formaldehyde, or phenolic. Figure 1.1 shows the chemical symbol representation of the reaction where the phenol molecules react with formaldehyde molecules to create a three-dimensional cross-linked network that is stiff and strong, leaving water as the by-product of this chemical reaction. This type of chemical reaction is a condensation polymerization.

With regard to the chemistry of polymerization processes, we will only introduce the topic superficially. A polymerization reaction is controlled by several conditions such as temperature, pressure, monomer concentration, as well as by structure-controlling additives such as catalysts, activators, accelerators, and inhibitors. There are various ways a polymerization process can take place such as schematically depicted in Fig. 1.1. There are numerous other types of reactions that are not mentioned here. When synthesizing some polymers there may be multiple ways of arriving at the finished product. For example, polyformaldehyde (POM) can be synthesized using all the reaction types presented in Table 1.1. On the other hand, polyamide 6 (PA6) is synthesized through various steps that are present in different types of reactions, such as polymerization and polycondenzation.

1.2 MOLECULAR WEIGHT

A polymeric material may consist of polymer chains of various lengths or repeat units. Hence, the molecular weight is determined by the average or mean molecular weight which

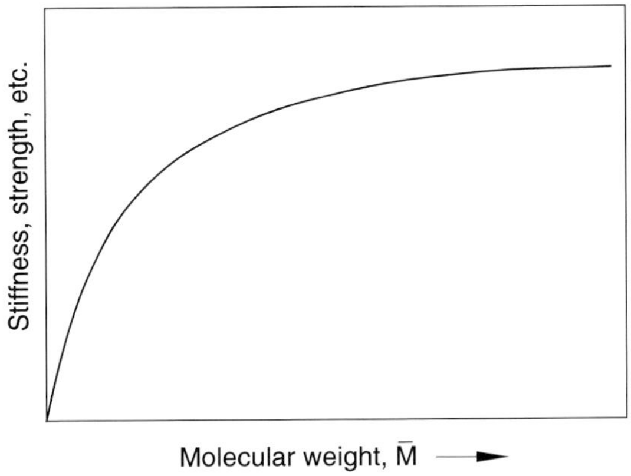

Figure 1.2: Influence of molecular weight on mechanical properties.

is defined by

$$\bar{M} = \frac{W}{N} \tag{1.1}$$

where W is the weight of the sample and N the number of moles in the sample.

The properties of polymeric material are strongly linked to the molecular weight of the polymer as shown schematically in Fig. 1.2. A polymer such as polystyrene is stiff and brittle at room temperature with a degree of polymerization, n, of 1,000. Polystyrene with a degree of polymerization of 10 is sticky and soft at room temperature.

Figure 1.3 shows the relation between molecular weight, temperature and properties of a typical polymeric material. The stiffness properties reach an asymptotic maximum, whereas the flow temperature increases with molecular weight. On the other hand, the degradation temperature steadily decreases with increasing molecular weight. Hence, it is necessary to find the molecular weight that renders ideal material properties for the finished polymer product, while having flow properties that make it easy to shape the material during the manufacturing process. It is important to mention that the temperature scale in Fig. 1.3 corresponds to a specific time scale, e.g., time required for a polymer molecule to flow through an injection molding runner system. If the time scale is reduced (e.g., by increasing the injection speed), the molecules have more difficulty sliding past each other. This would require a somewhat higher temperature to assure flow. In fact, at a specific temperature, a polymer melt may behave as a solid if the time scale is reduced sufficiently. Hence, for this new time scale the stiffness properties and flow temperature curves must be shifted upward on the temperature scale. A limiting factor is that the thermal degradation curve remains fixed, limiting processing conditions to remain above certain time scales. This relation between time, or time scale, and temperature is often referred to as *time-temperature superposition principle* and is discussed in detail in the literature [17].

With the exception of maybe some naturally occurring polymers, most polymers have a molecular weight distribution as shown in Fig. 1.4. We can define a number average, weight average, and viscosity average[1] for such a molecular weight distribution function. The *number average* is the first moment and the *weight average* the second moment of the

[1] There are other definitions of molecular weight which depend on the type of measuring technique.

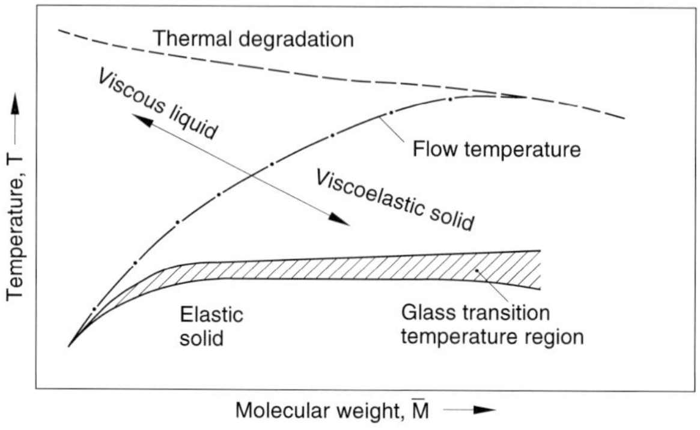

Figure 1.3: Diagram representing the relation between molecular weight, temperature and properties of a typical thermoplastic.

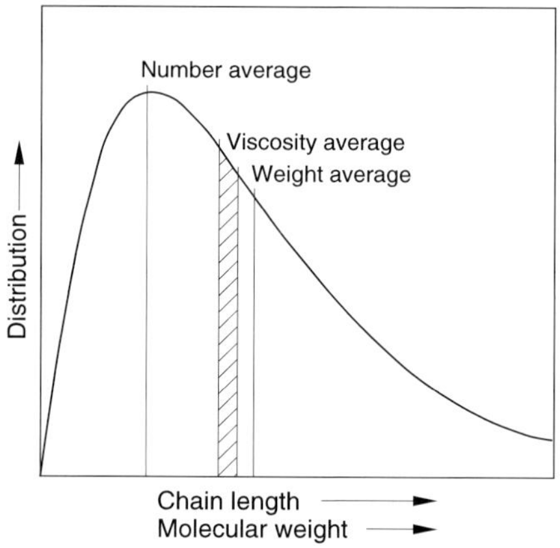

Figure 1.4: Molecular weight distribution of a typical thermoplastic.

distribution function. In terms of mechanics, this is equivalent to the center of gravity and the radius of gyration as first and second moments, respectively.

The *viscosity average* relates the molecular weight of a polymer to the measured viscosity as shown in Fig. 1.5.

Figure 1.5 [2] presents the viscosity of various undiluted polymers as a function of molecular weight. The figure shows how for all these polymers the viscosity goes from a linear (slope=1) to a power dependence (slope=3.4) at some critical molecular weight. The linear relation is sometimes referred to as *Staudinger's rule*[12] and applies for a perfectly *monodispersed polymer*, where the friction between the molecules increases proportionally to the molecule's length. The increased slope of 3.4 is due to molecular entanglement due to the long molecular chains. The *Mark-Houwink relation* is often used to represent this effect, and it is written as

$$\eta = k\bar{M}_v{}^{\alpha} \tag{1.2}$$

where η is the viscosity, $\bar{M}_v$ the *viscosity average molecular weight*, α the slope in the viscosity curve, and k a constant.

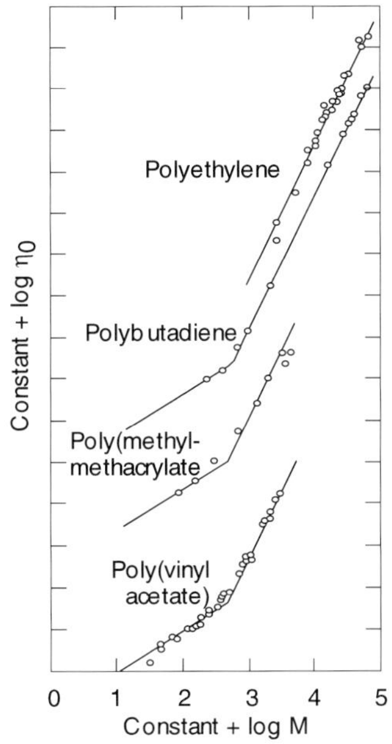

Figure 1.5: Zero shear rate viscosity for various polymers as a function of weight average molecular weight.

A measure of the broadness of a polymer's molecular weight distribution is the *polydispersity index* defined by

$$PI = \frac{\bar{M}_w}{\bar{M}_n} \tag{1.3}$$

Figure 1.6 [5] presents a plot of flexural strength versus *melt flow index*[2] for polystyrene samples with three different polydispersity indices. The figure shows that low polydispersity index-grade materials render higher-strength properties and flowability, or processing ease, than high polydispersity index grades.

Physically, the molecules can have rather large dimensions. For example, each repeat unit of a carbon backbone molecule, such as polyethylene, measures 0.252 nm in length. If completely stretched out, a high molecular weight molecule with, say 10,000 repeat units can measure over 2 μm in length. Figure 1.7 serves to illustrate the range in dimensions associated with polymers as well as which microscopic devices are used to capture the detail at various orders of magnitude. If we go from the atomic structure to the part geometry we easily travel between 0.1 nm and 1 mm, covering eight orders of magnitude.

■ **EXAMPLE 1.1.**

Polymer molecular weight and molecule size. You are asked to compute the maximum possible separation between the ends of a high density polyethylene molecule with an average molecular weight of 100,000.

[2]The melt flow index is the mass (grams) extruded through a capillary in a 10-minute period while applying a constant pressure. Increasing melt flow index signifies decreasing molecular weight.

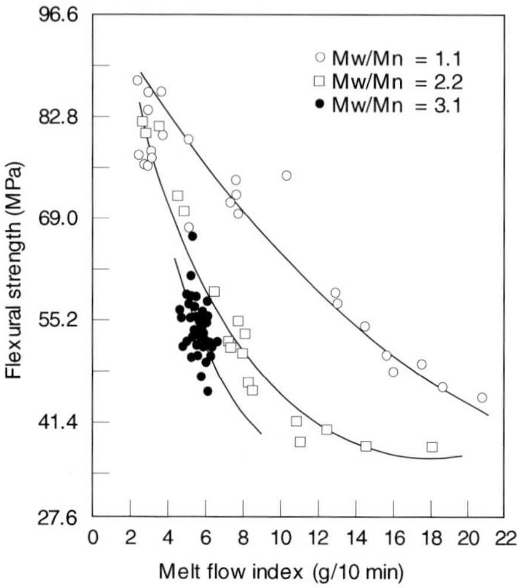

Figure 1.6: Effect of molecular weight on the strength-melt flow index interretationship of polystyrene for three polydispersity indices.

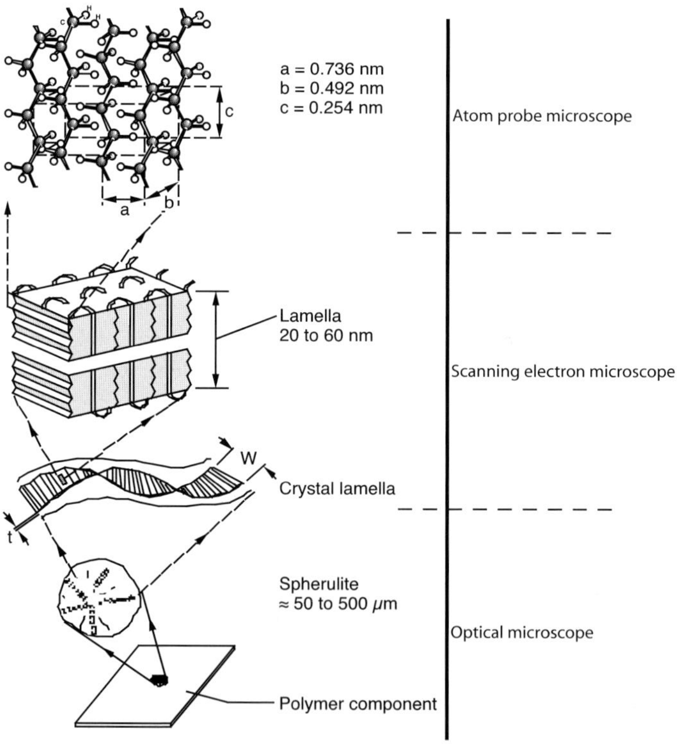

Figure 1.7: Schematic representation of the general molecular structure of semi-crystalline polymers and magnitudes as well as microscopic devices associated with such structures.

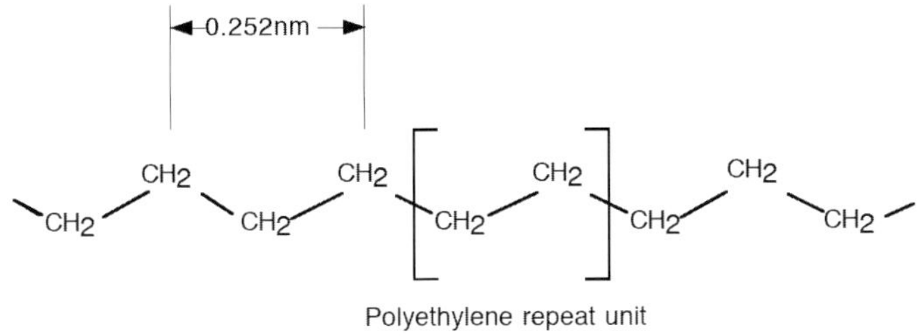

Figure 1.8: Schematic diagram of a polyethylene molecule.

The first task is to estimate the number of repeat units, n, in the polyethylene chain. Each repeat unit has 2 carbons and 4 hydrogen atoms. The molecular weight of carbon is 12 and that of hydrogen 1. Hence

$$MW/\text{repeat unit} = 2(12) + 4(1) = 28 \tag{1.4}$$

The number of repeat units is computed as

$$n = MW/(MW/\text{repeat unit}) = 100,000/28 = 3,571 \text{ units} \tag{1.5}$$

Using the diagram presented in Fig. 1.8 we can now estimate the length of the fully extended molecule using

$$\ell = 0.252 \text{ nm}(3,571) = 890 \text{ nm} = 0.89\ \mu\text{m} \tag{1.6}$$

1.3 CONFORMATION AND CONFIGURATION OF POLYMER MOLECULES

The conformation and configuration of the polymer molecules have a great influence on the properties of the polymer component. The conformation describes the preferential spatial positions of the atoms in a molecule. It is described by the polarity flexibility and regularity of the macromolecule. Typically, carbon atoms are tetravalent, which means that they are surrounded by four substituents in a symmetric tetrahedral geometry. The most common example is methane, CH_4, schematically depicted in Fig. 1.9. As the figure demonstrates, the tetrahedral geometry sets the bond angle at 109.5^o.

This angle is maintained between carbon atoms on the backbone of a polymer molecule, as shown in Fig. 1.10. As shown in the figure, each individual axis in the carbon backbone is free to rotate. The configuration gives the information about the distribution and spatial organization of the molecule.

During polymerization it is possible to place the X-groups on the carbon-carbon backbone in different directions. Thc order in which they are arranged is called the tacticity. The polymers with side groups placed randomly are called atactic. The polymers whose side groups are all on the same side are called isotactic, and those molecules with regularly alternating side groups are called syndiotactic. Figure 1.11 shows the three different tacticity cases for polypropylene. The tacticity in a polymer determines the degree of crystallinity that a polymer can reach. For example, a polypropylene with a high isotactic content will reach a high degree of crystallinity and as a result be stiff, strong and hard.

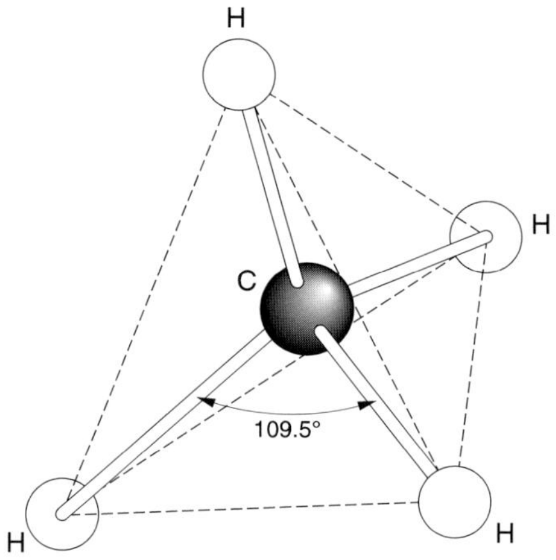

Figure 1.9: Schematic of a tetrahedron formed by methane.

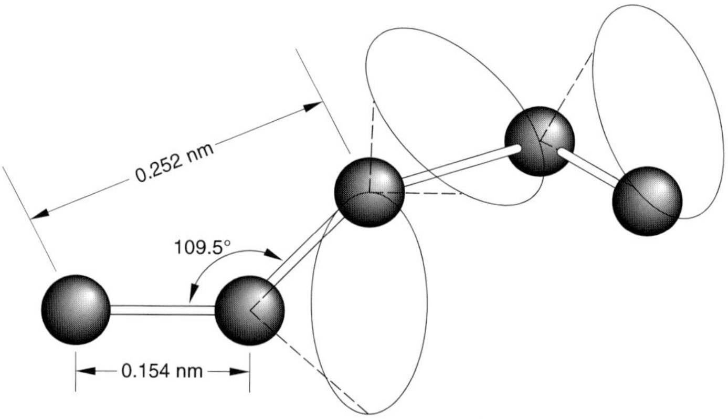

Figure 1.10: Random conformation of a polymer chain's carbon-carbon backbone.

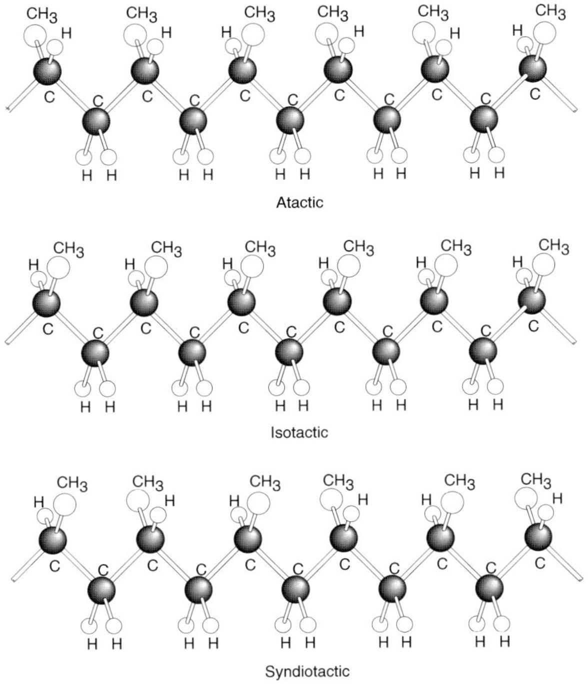

Figure 1.11: Different polypropylene structures.

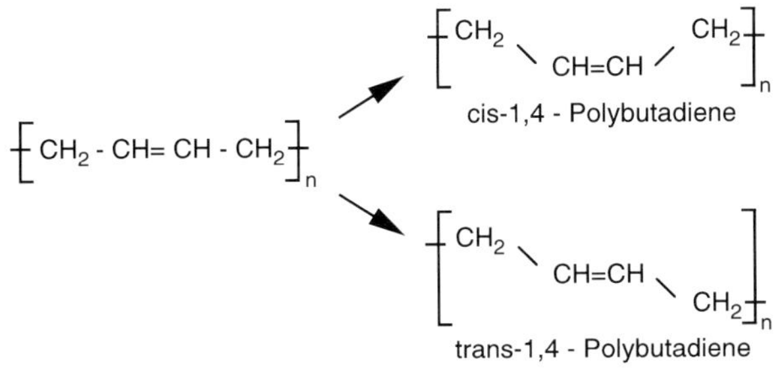

Figure 1.12: Symbolic representation of cis-1,4- and trans-1,4-polybutadiene molecules.

Another type of geometric arrangement arises with polymers that have a double bond between carbon atoms. Double bonds restrict the rotation of the carbon atoms about the backbone axis. These polymers are sometimes referred to as geometric isomers. The X-groups may be on the same side (cis-) or on opposite sides (trans-) of the chain as schematically shown for polybutadiene in Fig. 1.12. The arrangement in a cis-1,4-polybutadiene results in a very elastic rubbery material, whereas the structure of the trans-1,4-polybutadiene results in a leathery and tough material. Branching of the polymer chains also influences the final structure, crystallinity and properties of the polymeric material.

Figure 1.13 shows the molecular architecture of high density, low density and linear low density polyethylenes. The high density polyethylene has between 5 and 10 short branches every 1,000 carbon atoms. The low density material has the same number of branches as PE-HD; however, they are much longer and are themselves usually branched. The PE-LLD

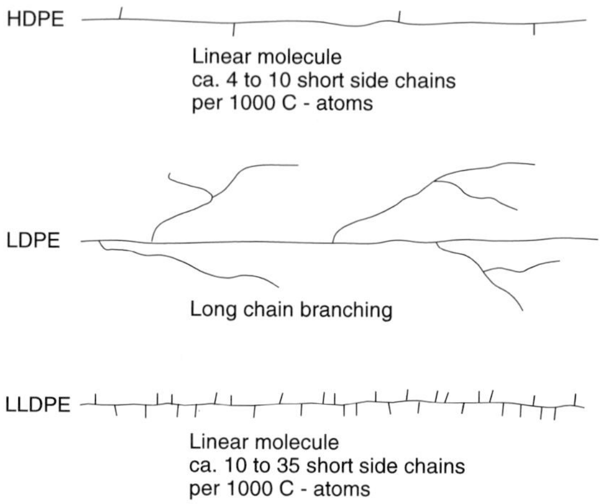

Figure 1.13: Schematic of the molecular structure of different polyethylenes.

has between 10 and 35 short chains every 1,000 carbon atoms. Polymer chains with fewer and shorter branches can crystallize with more ease, resulting in higher density.

The various intermolecular force, generally called Van der Waals forces, between macromolecules are of importance because of the size of the molecules. These forces are often the cause of the unique behavior of polymers. The so-called *dispersion forces*, the weakest of the intermolecular forces, are caused by the instantaneous dipoles that form as the charge in the molecules fluctuates. Very large molecules, such as ultra high molecular weight polyethylene can have significant dispersion forces. *Dipole-dipole forces* are those intermolecular forces that result from the attraction between polar groups. Hydrogen bonding intermolecular forces, the largest of them all, take place when a polymer molecule contains $-$OH or $-$NH groups. The degree of polarity within a polymer determines how strongly it is attracted to other molecules. If a polymer is composed of atoms with different electronegativity (EN) it has a high degree of polarity and it is usually called a *polar* molecule. A *non-polar* molecule is one that is composed of atoms with equal or similar electronegativity. For example, polyethylene, which is formed of carbon ($EN = 2.5$) and hydrogen ($EN = 2.1$) alone, is considered a non-polar material because $\Delta EN = 0.4$. An increase in polarity is to be expected when elements such as chlorine, fluorine, oxygen or nitrogen are present in a macromolecule. Table 1.2 presents the electronegativity of common elements found in polymers. The intramolecular forces affect almost every property that is important when processing a polymer, including the effect that low molecular weight additives, such as solvents, plasticizers and permeabilizers, as well as the miscibility of various polymers have when making blends.

1.4 MORPHOLOGICAL STRUCTURE

Morphology is the order or arrangement of the polymer structure. The possible *order* between a molecule or molecule segment and its neighbors can vary from a very ordered highly crystalline polymeric structure to an amorphous structure (i.e., a structure in greatest disorder or random). The possible range of order and disorder is clearly depicted on the left side of Fig. 1.14. For example, a purely amorphous polymer is formed only by the non-

Table 1.2: Electronegativity Number EN for Various Elements (After Pauling)

Element	EN
Flourine (F)	4.0
Oxygen (O)	3.5
Chlorine (Cl)	3.0
Nitrogen (N)	3.0
Carbon (C)	2.5
Sulfur (S)	2.5
Hydrogen (H)	2.1
Silicone (Si)	1.8
Zink (Zn)	1.6
Sodium (Na)	0.9

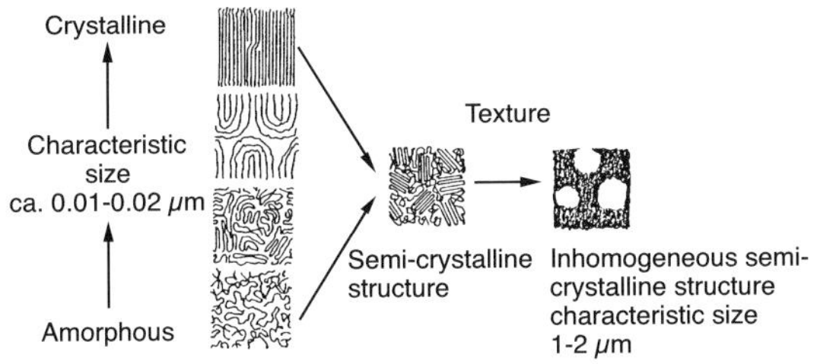

Figure 1.14: Schematic diagram of possible molecular structure which occur in thermoplastic polymers.

crystalline or *amorphous* chain structure, whereas the *semi-crystalline* polymer is formed by a combination of all the structures represented in Fig. 1.14.

The semi-crystalline arrangement that certain polymer molecules take during cooling is in great part due to intramolecular forces. As the temperature of a polymer melt is lowered, the free volume between the molecules is reduced, causing an increase in the intramolecular forces. As the free volume is reduced further, the intermolecular forces cause the molecules to arrange in a manner that brings them to a lower state of energy, as for example, the folded chain structure of polyethylene molecules shown in Fig. 1.7. This folded chain structure, which starts at a nucleus, grows into the spherulitic structure shown in Fig. 1.7 and in the middle of Fig. 1.14, an image that can be captured with an electron microscope. A macroscopic structure, shown in the right hand side of the figure, can be captured with an optical microscope. An optical microscope can capture the coarser macro-morphological structure such as the spherulites in semi-crystalline polymers. Figure 1.7, presented earlier, shows a schematic of the spherulitic structure of polyethylene with the various microscopic devices that can be used to observe different levels of the formed morphology. An amorphous polymer is defined as having a purely random structure. However, it is not quite clear if a *purely amorphous* polymer as such exists. Electron microscopic observations have shown amorphous polymers that are composed of relatively stiff chains, exhibit a certain degree of macromolecular structure and order, for example, globular regions or fibrilitic structures. Nevertheless, these types of amorphous polymers are still found to be optically isotropic. Even polymers with soft and flexible macromolecules, such as polyisoprene which was first considered to be random, sometimes show band-

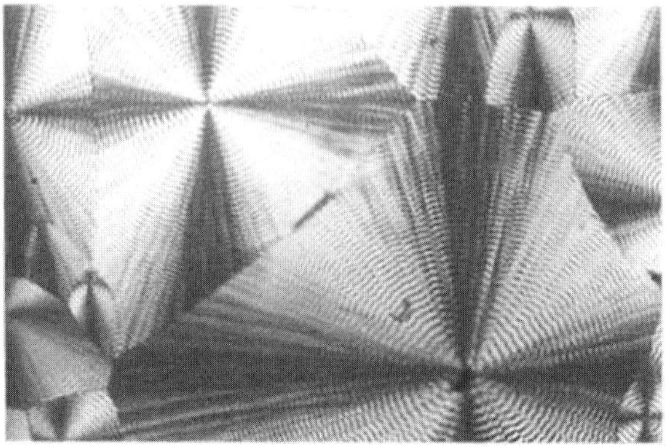

Figure 1.15: Polarized microscopic image of the spherulitic structure in polypropylene.

like and globular regions. These bundle-like structures are relatively weak and short-lived when the material experiences stresses. The shear-thinning viscosity effect of polymers sometimes is attributed to the breaking of such macromolecular structures.

Early on, before the existence of macromolecules had been recognized, the presence of highly crystalline structures had been suspected. Such structures were discovered when undercooling or when stretching cellulose and natural rubber. Later, it was found that a crystalline order also existed in synthetic macromolecular materials such as polyamides, polyethylenes, and polyvinyls. Because of the polymolecularity of macromolecular materials, a 100% degree of crystallization cannot be achieved. Hence, these polymers are referred to as semi-crystalline. It is common to assume that the semi-crystalline structures are formed by small regions of alignment or crystallites connected by random or amorphous polymer molecules.

With the use of electron microscopes and sophisticated optical microscopes the various existing crystalline structures are now well recognized. They can be listed as follows:

- *Single crystals*. These can form in solutions and help in the study of crystal formation. Here, plate-like crystals and sometimes whiskers are generated.

- *Spherulites*. As a polymer melt solidifies, several folded chain lamellae spherulites form which are up to 0.1 mm in diameter. A typical example of a spherulitic structure is shown in Fig. 1.15. The spherulitic growth in a polypropylene melt is shown in Fig. 1.16.

- *Deformed crystals*. If a semi-crystalline polymer is deformed while undergoing crystallization, oriented lamellae form instead of spherulites.

- *Shish-kebab*. In addition to spherulitic crystals, which are formed by plate- and ribbonlike structures, there are also shish-kebab crystals which are formed by circular plates and whiskers. Shish-kebab structures are generated when the melt undergoes a shear deformation during solidification. A typical example of a shish-kebab crystal is shown in Fig. 1.17.

The speed at which crystalline structures grow depends on the type of polymer and on the temperature conditions. Table 1.3 shows the maximum growth rate for common semi-crystalline thermoplastics as well the maximum achievable degree of crystallinity.

Figure 1.16: Development of the spherulitic structure in polypropylene. Images were taken at 30 seconds intervals.

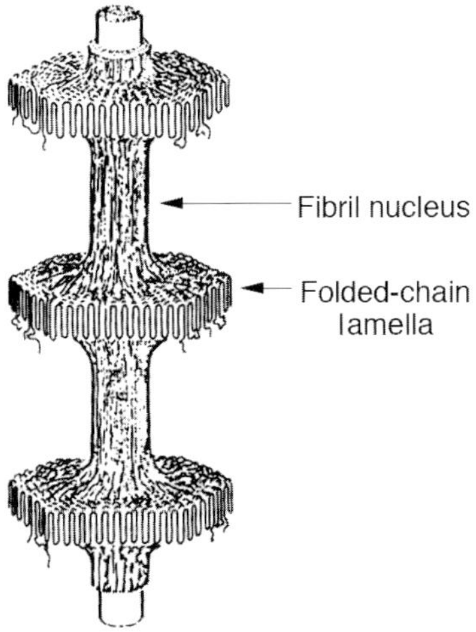

Figure 1.17: Model of the shish-kebab morphology.

Table 1.3: Maximum Crystalline Growth Rate and Maximum Degree of Crystallinity for Various Thermoplastics

Polymer	Growth rate(μ/min)	Maximum crystallinity(%)
Polyethylene	>1000	80
Polyamide 66	1000	70
Polyamide 6	200	35
Isotactic polypropylene	20	63
Polyethylene teraphthalate	7	50
Isotactic polystyrene	0.3	32
Polycarbonate	0.01	25

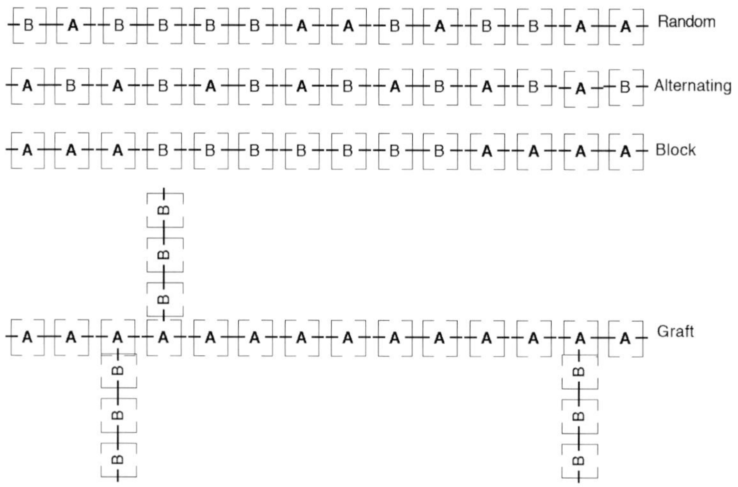

Figure 1.18: Schematic representation of different copolymers.

1.4.1 Copolymers and Polymer Blends

Copolymers are polymeric materials with two or more monomer types in the same chain. A copolymer that is composed of two monomer types is referred to as a bipolymer, and one that is formed by three different monomer groups is called a terpolymer. Depending on how the different monomers are arranged in the polymer chain, one distinguishes between random, alternating, block or graft copolymers. The four types of copolymers are schematically represented in Fig. 1.18.

A common example of a copolymer is an ethylene-propylene copolymer. Although both monomers would result in semi-crystalline polymers when polymerized individually, the melting temperature disappears in the randomly distributed copolymer with ratios between 35/65 and 65/35, resulting in an elastomeric material, as shown in Fig. 1.19. In fact, EPDM rubbers are continuously gaining acceptance in industry because of their resistance to weathering. On the other hand, the ethylene-propylene block copolymer maintains a melting temperature for all ethylene/propylene ratios, as shown in Fig. 1.20.

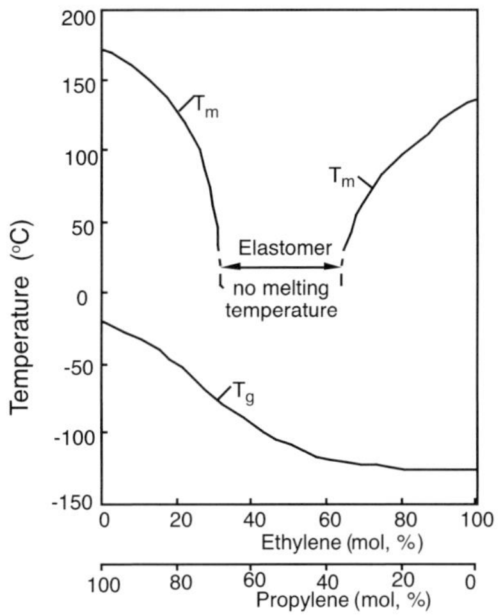

Figure 1.19: Melting and glass transition temperature for random ethylene-propylene copolymers.

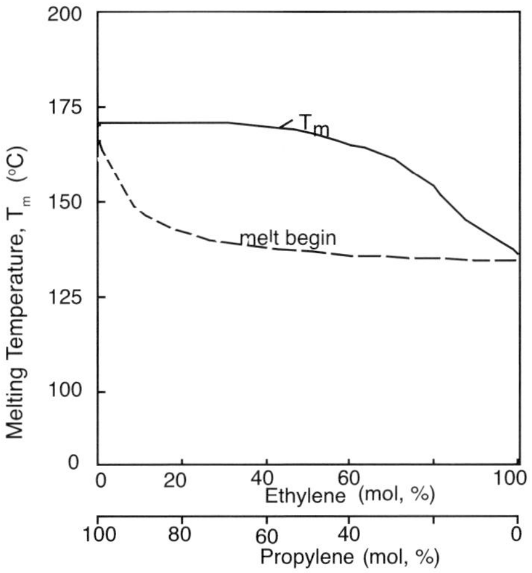

Figure 1.20: Melting temperature for ethylene-propylene block copolymers.

Another widely used copolymer is high impact polystyrene (PS-HI), which is formed by grafting polystyrene to polybutadiene. Again, if styrene and butadiene are randomly copolymerized, the resulting material is an elastomer called styrene-butadiene-rubber (SBR). Another classic example of copolymerization is the terpolymer acrylonitrile-butadiene-styrene (ABS). Polymer blends belong to another family of polymeric materials which are made by mixing or blending two or more polymers to enhance the physical properties of each individual component. Common polymer blends include PP-PC, PVC-ABS, PE-PTFE and PC-ABS.

1.5 THERMAL TRANSITIONS

A phase change or a thermal transition occurs with polymers when they undergo a significant change in material behavior. The phase change occurs as a result of either a reduction in material temperature or a chemical curing reaction. A thermoplastic polymer hardens as the temperature of the material is lowered below either the melting temperature for a semi-crystalline polymer, the glass transition temperature for an amorphous thermoplastic or the crystalline and glass transition temperatures in liquid crystalline polymers. A thermoplastic has the ability to soften again as the temperature of the material is raised above the solidification temperature. With thermoplastics the term *solidification* is often misused to describe the hardening of amorphous thermoplastics. On the other hand, the solidification of thermosets leads to cross-linking of molecules. The effects of cross-linkage are irreversible and result in a network that hinders the free movement of the polymer chains independent of the material temperature.

The solidification of most materials is defined at a discrete temperature, whereas amorphous polymers do not exhibit a sharp transition between the liquid and the solid states. Instead, an amorphous thermoplastic polymer vitrifies as the material temperature drops below the glass transition temperature, T_g. Due to their random structure, the characteristic size of the largest ordered region is on the order of a carbon-carbon bond. This dimension is much smaller than the wavelength of visible light and so generally makes amorphous thermoplastics transparent. Figure 1.21 shows the shear modulus, G', versus temperature for polystyrene, one of the most common amorphous thermoplastics. The figure shows two general regions: one where the modulus appears fairly constant, and one where the modulus drops significantly with increasing temperature. With decreasing temperatures, the material enters the glassy region where the slope of the modulus approaches zero. At high temperatures, the modulus is negligible and the material is soft enough to flow. Although there is not a clear transition between *solid* and *liquid*, the temperature at which the slope is highest is T_g. For the polystyrene in Fig. 1.21 the glass transition temperature is approximately 120°C. Although data are usually presented in the form shown in Fig. 1.21, it should be mentioned here that the curve shown in the figure was measured at a constant frequency.

If the frequency of the test is increased —reducing the time scale— the curve is shifted to the right, since higher temperatures are required to achieve movement of the molecules at the new frequency. This can be clearly seen for PVC in Fig. 1.22. A similar effect is observed when the molecular weight of the material is increased. The longer molecules have more difficulty sliding past each other, thus requiring higher temperatures to achieve *flow*.

The transition temperatures as well as flow behavior are significantly affected by the pressure one applies to the material. Higher pressures reduce the free volume between the

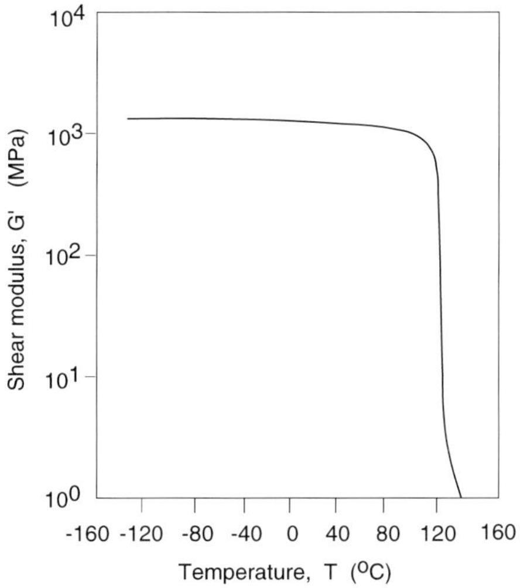

Figure 1.21: Shear modulus of polystyrene as a function of temperature.

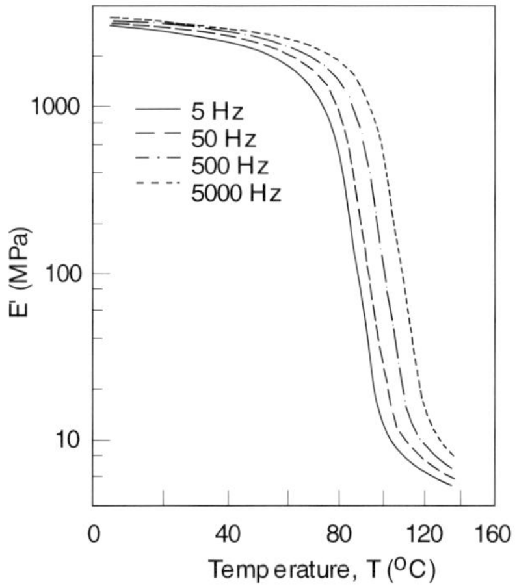

Figure 1.22: Modulus of polyvinyl chloride as a function of temprcaturc at various test frequencies.

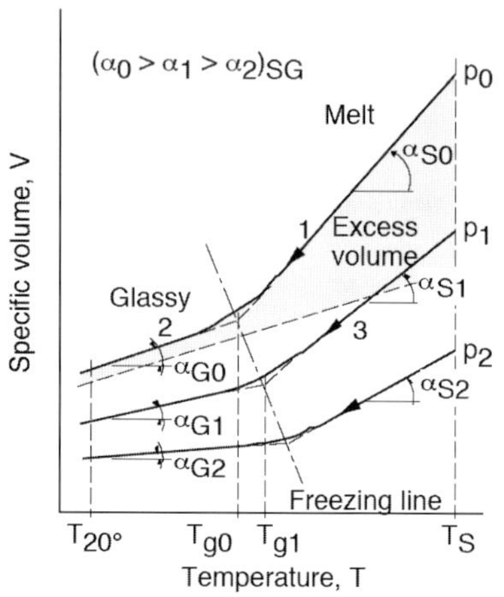

Figure 1.23: Schematic of a pvT diagram for amorphous thermoplastics.

molecules which restricts their movement. This requires higher temperatures to increase the free volume sufficiently to allow molecular movement. This is clearly depicted in Fig. 1.23, which schematically presents the pressure-volume-temperature (pvT) behavior of amorphous polymers.

Semi-crystalline thermoplastic polymers show more order than amorphous thermoplastics. The molecules align in an ordered crystalline form as shown for polyethylene in Fig. 1.24. The size of the crystals or spherulites is much larger than the wavelength of visible light, making semi-crystalline materials translucent and not transparent. However, the crystalline regions are very small, with molecular chains comprised of both crystalline and amorphous regions. The degree of crystallinity in a typical thermoplastic will vary from grade to grade as, for example, in polyethylene, where the degree of crystallinity depends on the branching and the cooling rate. Because of the existence of amorphous as well as crystalline regions, a semi-crystalline polymer has two distinct transition temperatures, the glass transition temperature, T_g, and the melting temperature, T_m.

Figure 1.25 shows the dynamic shear modulus versus temperature for a high density polyethylene, the most common semi-crystalline thermoplastic. Again, this curve presents data measured at one test frequency. The figure clearly shows two distinct transitions: one at about -110°C, the glass transition temperature, and another near 140°C, the melting temperature. Above the melting temperature, the shear modulus is negligible and the material will flow. Crystalline arrangement begins to develop as the temperature decreases below the melting point. Between the melting and glass transition temperatures, the material behaves as a leathery solid. As the temperature decreases below the glass transition temperature, the amorphous regions within the semi-crystalline structure solidify, forming a glassy, stiff, and in some cases brittle polymer.

Figure 1.26 summarizes the property behavior of amorphous, crystalline, and semi-crystalline materials using schematic diagrams of material properties plotted as functions of temperature. Again, pressures affect the transition temperatures as schematically depicted in Fig. 1.27 for a semi-crystalline polymer.

The transition regions in liquid crystalline polymers or mesogenic polymers is much more complex. These transitions are referred to as mesomorphic transitions, and occur when one goes from a crystal to a liquid crystal, from a liquid crystal to another liquid

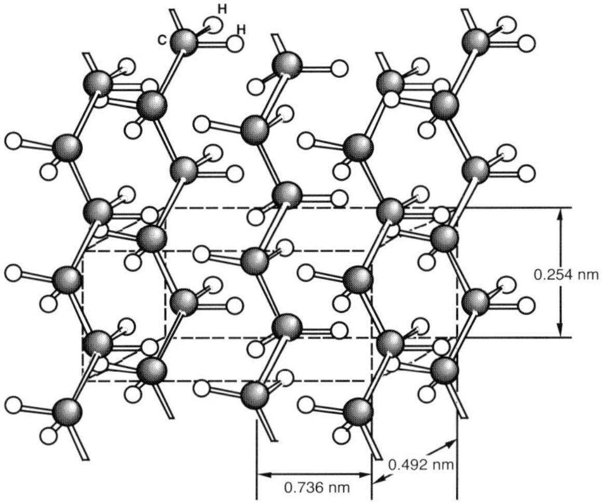

Figure 1.24: Schematic representation of the crystalline structure of polyethylene.

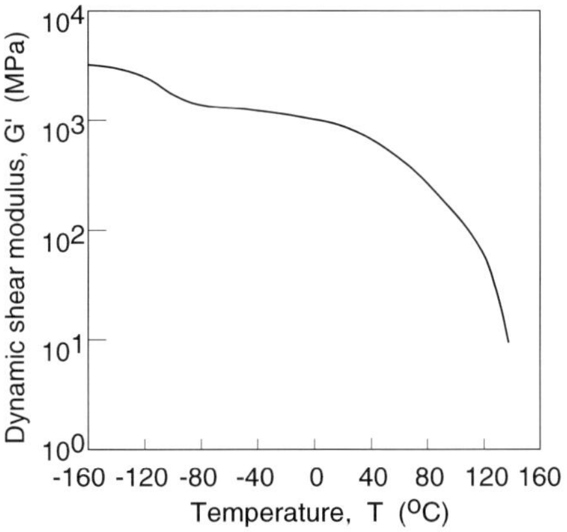

Figure 1.25: Shear modulus of high-density polyethylene as a function of temperature.

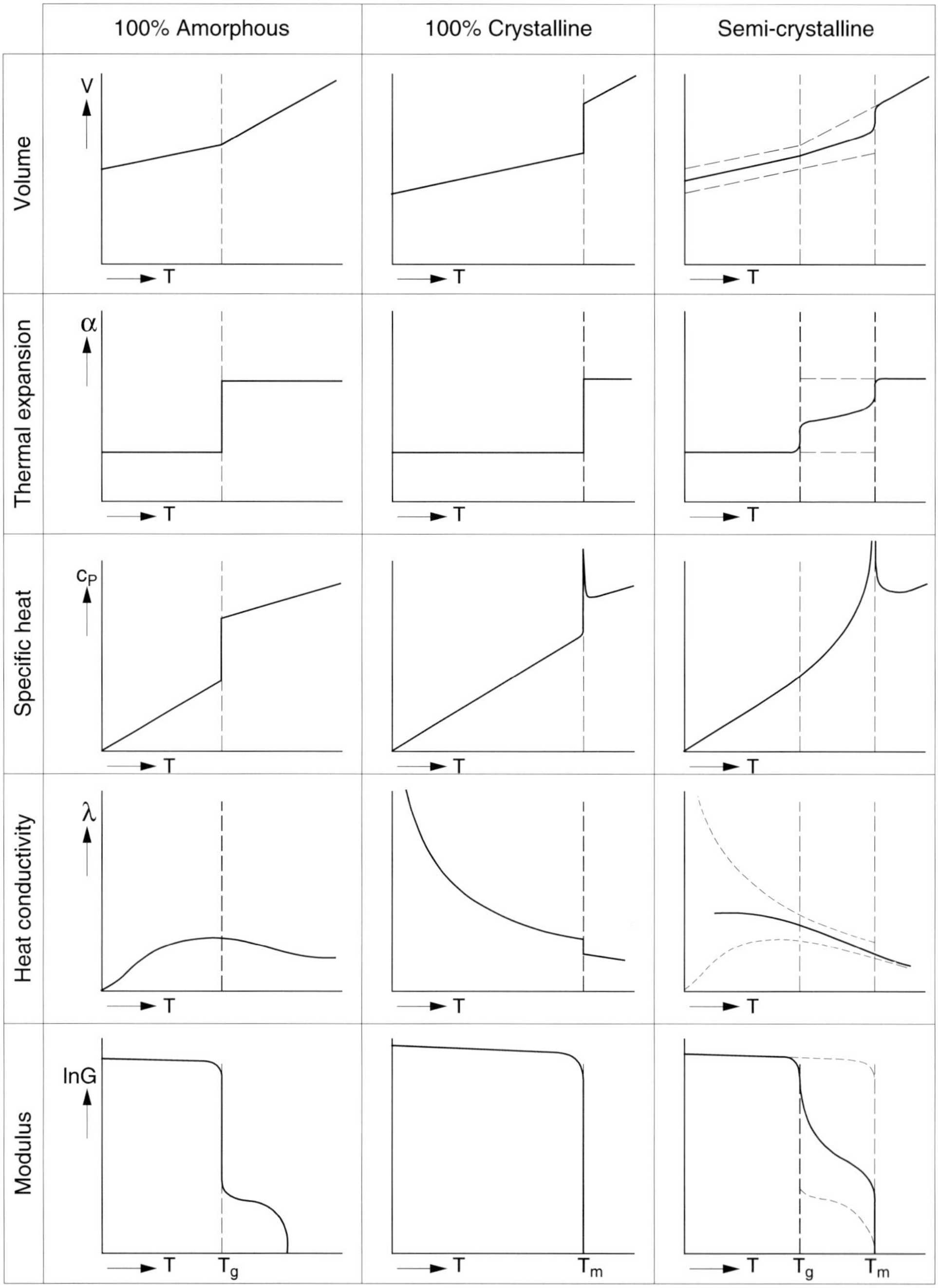

Figure 1.26: Schematic of the behavior of some polymer properties as a function of temperature for different thermoplastics [66].

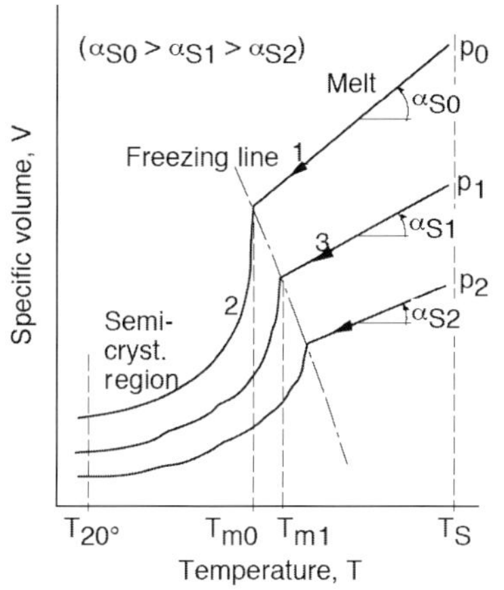

Figure 1.27: Schematic of a pvT diagram for semi-crystalline thermoplastics.

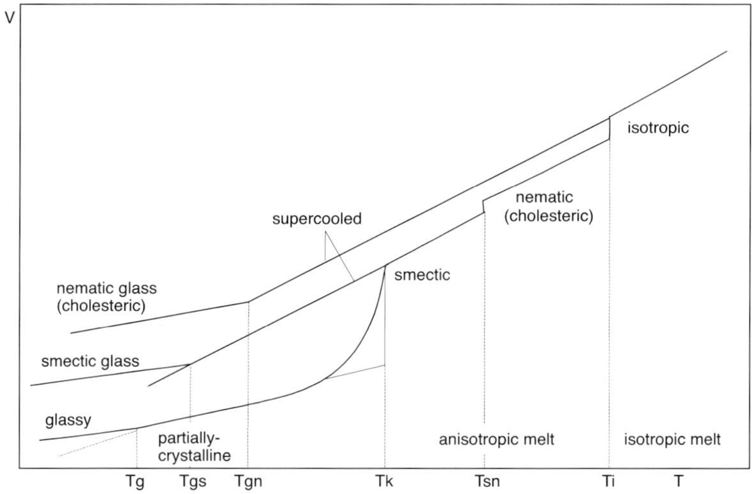

Figure 1.28: Schematic volume-temperature diagram for a liquid crystalline polymer [66].

crystal, and from a liquid crystal to and isotropic fluid. The volume-temperature diagram for a liquid crystalline polymer is presented in Fig. 1.28. The figure clearly depicts the various phases present in a liquid crystalline polymer. From a lower temperature to a high temperature, these are the glassy phase, partially crystalline phase, the smectic phase, the nematic or cholesteric phase and the isotropic phase. In the smectic phase the molecules have all distinct orientation and all their *centers of gravity* align with each other, giving them a highly organized structure. In the nematic phase the axes of the molecules are aligned, giving them a high degree of orientation, but where the *centers of gravity* of the molecules are not aligned. During cooling, both the nematic and the smectic phases can be maintained, leading to nematic glass and smectic glass, respectively.

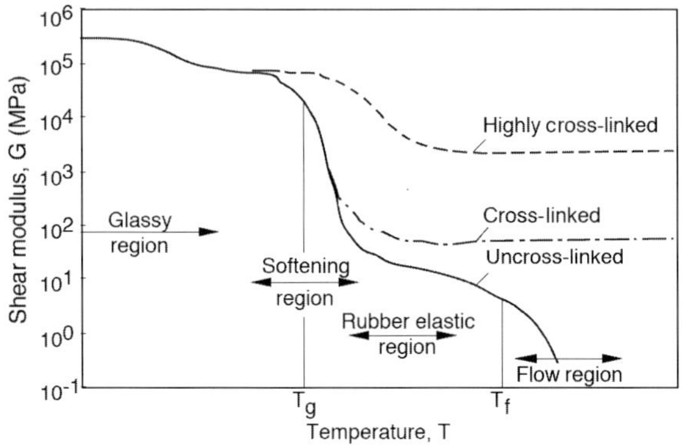

Figure 1.29: Shear modulus and behavior of cross-linked and uncross-linked polymers.

Cross-linked polymers, such as thermosets and elastomers, behave completely differently than their counterparts, thermoplastic polymers. In cross-linked systems, the mechanical behavior is also best reflected by the plot of the shear modulus versus temperature. Figure 1.29 compares the shear modulus between highly cross-liked, cross-linked, and uncross-linked polymers. The coarse cross-linked system, typical of elastomers, has a low modulus above the glass transition temperature. The glass transition temperature of these materials is usually below -50°C, so they are soft and flexible at room temperature. On the other hand, highly cross-linked systems, typical in thermosets, show a smaller decrease in stiffness as the material is raised above the glass transition temperature; the decrease in properties becomes smaller as the degree of cross-linking increases. With thermosetting polymers, strength remains fairly constant up to the thermal degradation temperature of the material.

1.6 VISCOELASTIC BEHAVIOR OF POLYMERS

Although polymers have their distinct transitions and may be considered liquid when above the glass transition or melting temperatures, or solid when below those temperatures, in reality they are neither liquid nor solid, but viscoelastic. In fact, at any temperature, a polymer can be either a liquid or a solid, depending on the time scale or speeds at which its molecules are being deformed. The most common technique of measuring and demonstrating this behavior is by performing a stress relaxation test and the time-temperature superposition principle.

1.6.1 Stress Relaxation

In a stress relaxation test, a polymer test specimen is deformed by a fixed amount, ϵ_0, and the stress required to hold that amount of deformation is recorded over time. This test is very cumbersome to perform, so the design engineer and the material scientist have tended to ignore it. In fact, several years ago, the standard relaxation test ASTM D2991 was dropped by ASTM. Rheologists and scientists, however, have been consistently using the stress relaxation test to interpret the viscoelastic behavior of polymers.

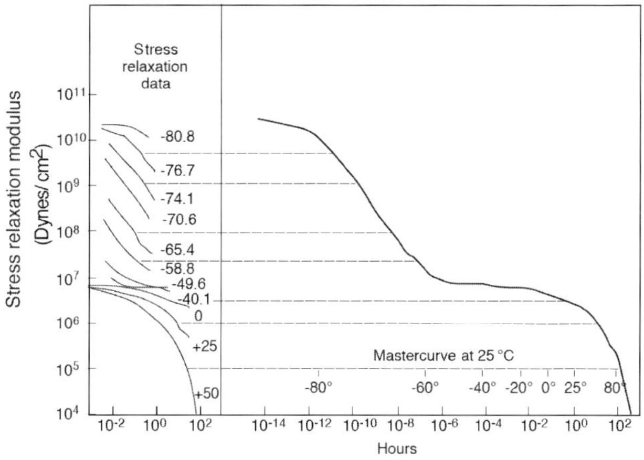

Figure 1.30: Relaxation modulus curves for polyisobutylene at various temperatures and corresponding master curve at 25°C.

Figure 1.30 [4] presents the stress relaxation modulus measured of polyisobutylene (chewing gum) at various temperatures. Here, the stress relaxation modulus is defined by

$$E_r(t) = \frac{\sigma(t)}{\epsilon_0} \tag{1.7}$$

where ϵ_0 is the applied strain and $\sigma(t)$ is the stress being measured. From the test results it is clear that stress relaxation is time and temperature dependent, especially around the glass transition temperature where the slope of the curve is maximal. In the case of the polyisobutylene shown in Fig. 1.30, the glass transition temperature is about -70°C. The measurements were completed in an experimental time window between a few seconds and one day. The tests performed at lower temperatures were used to record the initial relaxation, while the tests performed at higher temperatures only captured the end of relaxation of the rapidly decaying stresses.

It is well known that high temperatures lead to short molecular relaxation times and low temperatures lead to materials with long relaxation times. This is due to the fact that at low temperatures the free volume between the molecules is reduced, restricting or slowing down their movement. At high temperatures, the free volume is larger and the molecules can move with more ease. Hence, when changing temperature, the shape of creep or relaxation test results remain the same except that they are horizontally shifted to the left or right, which represent shorter or longer response times, respectively.

The same behavior is observed if the pressure is varied. As the pressure is increased, the free volume between the molecules is reduced, slowing down molecular movement. Here, an increase in pressure is equivalent to a decrease in temperature. In the melt state, the viscosity of a polymer increases with pressure. Figure 1.31 [7] is presented to illustrate the effect of pressure on stress relaxation.

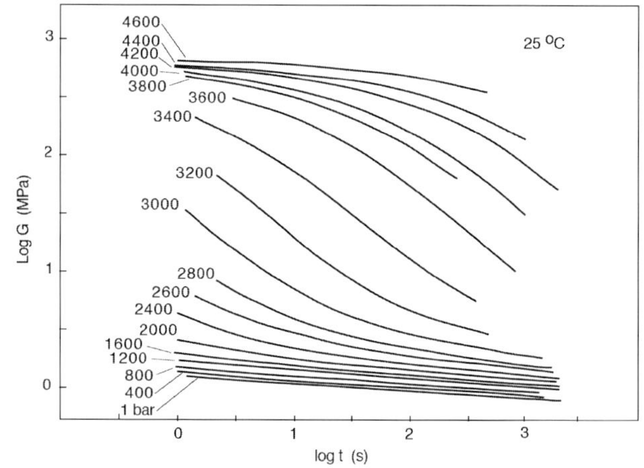

Figure 1.31: Shear relaxation modulus for a chlorosulfonated polyethylene at various pressures.

1.6.2 Time-Temperature Superposition (WLF-Equation)

The time-temperature equivalence seen in stress relaxation test results can be used to reduce data at various temperatures to one general master curve for a reference temperature, T_{ref}. To generate a master curve at the reference temperature, the curves shown in the left of Fig. 1.30 must be shifted horizontally, maintaining the reference curve stationary. Density changes are usually small and can be neglected, eliminating the need to perform tedious corrections. The master curve for the data in Fig. 1.30 is shown on the right side of the figure. Each curve was shifted horizontally until the ends of all the curves became superimposed. The amount that each curve was shifted can be plotted with respect to the temperature difference taken from the reference temperature. For the data in Fig. 1.30 the shift factor is shown in the plot in Fig. 1.32.

The amounts by which the curves where shifted are represented by

$$\log(t) - \log(t_{ref}) = \log\left(\frac{t}{t_{ref}}\right) = \log(a_T) \tag{1.8}$$

Although the results in Fig. 1.32 where shifted to a reference temperature of 298 K (25°C), Williams, Landel and Ferry [14] chose $T_{ref} = 243$K for

$$\log(a_T) = \frac{-8.86(T - T_{ref})}{101.6 + T - T_{ref}} \tag{1.9}$$

which holds for nearly all polymers if the chosen reference temperature is 45 K above the glass transition temperature. In general, the horizontal shift, $\log(a_T)$, between the relaxation responses at various temperatures to a reference temperature can be computed using the well known Williams-Landel-Ferry [14] (WLF) equation. The WLF equation is

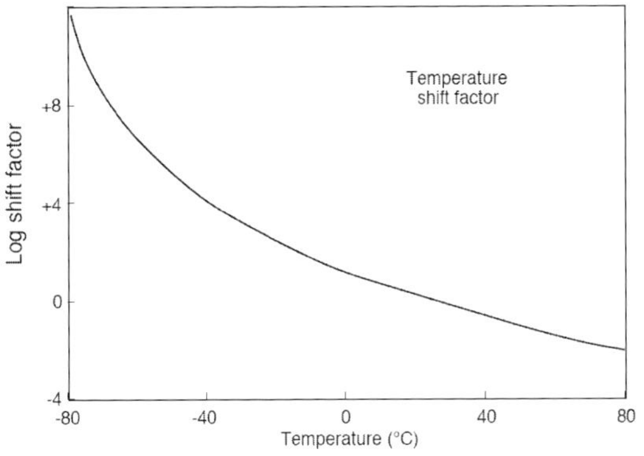

Figure 1.32: Plot of the shift factor as a function of temperature used to generate the master curve plotted in Fig. 1.30.

given by

$$\log a_T = \frac{C_1(T - T_{ref})}{C_2 + T - T_{ref}} \tag{1.10}$$

where C_1 and C_2 are material dependent constants. It has been shown that with the assumption $C_1 = 17.44$ and $C_2 = 51.6$, eqn. (1.10) fits well for a wide variety of polymers as long as the glass transition temperature is chosen as the reference temperature. These values for C_1 and C_2 are often referred to as universal constants. Often, the WLF equation must be adjusted until it fits the experimental data. Master curves of stress relaxation tests are important because the polymer's behavior can be traced over much greater periods of time than those determined experimentally.

■ **EXAMPLE 1.2.**

Stress relaxation master curve. For the poly-α-methylstyrene stress relaxation data in Fig. 1.33 [8], create a master creep curve at T_g (204°C). Identify the glassy, rubbery, viscous and viscoelastic regions of the master curve. Identify each region with a spring-dashpot diagram. Develop a plot of the shift factor, $\log(a_T)$ versus T, used to create your master curve $\log(a_T)$ is the horizontal distance that the curve at temperature T was slid to coincide with the master curve. What is the relaxation time of the polymer at the glass transition temperature?

The master creep curve for the above data is generated by sliding the individual relaxation curves horizontally until they match with their neighbors, using a fixed scale for a hypothetical curve at 204°C. Since the curve does not exist for the desired temperature, we can interpolate between 208.6°C and 199.4°C. The resulting master curve is presented in Fig.1.34. The amount each curve must be shifted from the master curve to its initial position is the shift factor, $\log(a_T)$. The graph also shows the spring-dashpot models and the shift factor for a couple of temperatures.

Figure 1.35 represents the shift factor versus temperature. The solid line indicates the shift factor predicted by the WLF equation. The relaxation time for the poly-α-

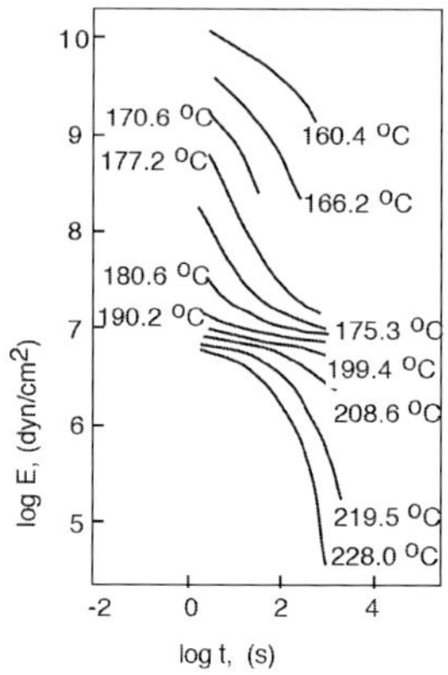

Figure 1.33: Stress relaxation data for poly-α-methylstyrene.

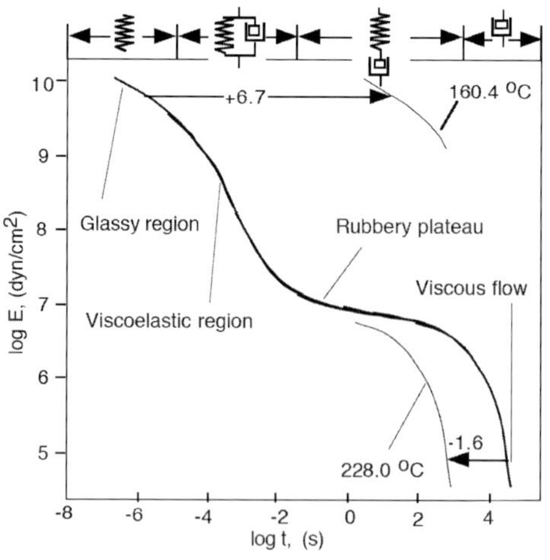

Figure 1.34: Master curve for poly-α-methylstyrene at 204°C.

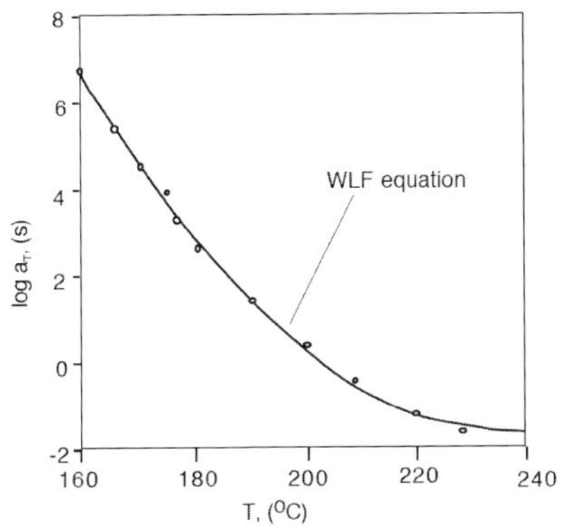

Figure 1.35: Shift factor and WLF curves for $T_{ref} = 204^oC$.

methylstyrene presented here is between 10^4 and $10^{4.5}$ s (8.8 h). The relaxation time for the remaining temperatures can be computed using the shift factor curve.

1.7 EXAMPLES OF COMMON POLYMERS

1.7.1 Thermoplastics

Examples of various thermoplastics are discussed in detail in the literature [6, 10] and can be found in commercial materials data banks [1]. Examples of the most common thermoplastic polymers, with a short summary, are given below. Ranges of typical processing conditions are also presented, grade dependent.

Polyacetal (POM). Polyacetal is a semi-crystalline polymer known for its high toughness, high stiffness and hardness. It is also highly sought after for its dimensional stability and its excellent electrical properties. It resists many solvents and is quite resistant to environmental stress cracking. Polyacetal has a low coefficient of friction. When injection molding polyacetal, the melt temperature should be between 200-210°C and the mold temperature should be above 90°C. Due to the flexibility and toughness of polyacetal it can be used in sport equipment, clips for toys and switch buttons.

Polyamide 66 (PA66). Polyamide 66 is a semi-crystalline polymer known for its hardness, stiffness, abrasion resistance and high heat deflection temperature. When injection molding PA66, the melt temperature should be between 260 and 320°C, and the mold temperature 80-90°C or above. The pellets must be dried before molding. Of the various polyamide polymers, this is the preferred material for molded parts that will be mechanically and thermally loaded. It is ideally suited for automotive and chemical applications such as gears, spools and housings. Its mechanical properties are significantly enhanced when reinforced with glass fiber.

Polyamide 6 (PA6). Polyamide 6 is a semi-crystalline polymer known for its hardness and toughness; however, with a toughness somewhat lower than PA66. When injection molding PA66, the melt temperature should be between 230 and 280°C, and the mold temperature 80-90°C or above. The pellets must be dried before molding. Low viscosity Polyamide 6 grades can be used to injection mold many thin-walled components. High viscosity grades can be used to injection mold various engineering components such as gears, bearings, seals, pump parts, cameras, telephones, etc. Its mechanical properties are significantly enhanced when reinforced with glass fiber.

Polycarbonate (PC). Polycarbonate is an amorphous thermoplastic known for its stiffness, toughness and hardness over a range from -150°C to 135°C. It is also known for its excellent optical properties and high surface gloss. When injection molding PC, the pellets must be dried before molding for 10 hours at about 130°C. The melt temperature should be between 280 and 320°C, and the mold temperature 85-120°C. Typical applications for injection molded polycarbonate parts are telephone housings, filter cups, lenses for glasses and optical equipment, camera housings, marine light covers, safety goggles, hockey masks, etc. Compact discs (CDs) are injection compression molded. Polycarbonate's mechanical properties can also be significantly enhanced when reinforced with glass fiber.

Polyethylene (PE). As was mentioned in previous sections, the basic properties of polyethylene depend on the molecular structure, such as degree of crystallinity, branching, degree of polymerization and molecular weight distribution. Due to all these factors, polyethylene can be a low density polyethylene (PE-LD), linear low density polyethylene (PE-LLD), high density polyethylene (PE-HD), ultra high molecular weight high density polyethylene (PE-HD-HMW), etc. When injection molding PE-LD, the melt temperature should be between 160 and 260°C, and the mold temperature 30-70°C, grade dependent. Injection temperatures for PE-HD are between 200-300°C and mold temperatures between 10 and 90°C. Typical applications for injection molding PE-LD parts are very flexible and tough components such as caps, lids and toys. Injection molding of PE-HD components include food containers, lids, toys, buckets, etc.

Polymethylmethacrylate (PMMA). Polymethylmethacrylate is an amorphous polymer known for its high stiffness, strength and hardness. PMMA is brittle but its toughness can be significantly increased when used in a copolymer. PMMA is also scratch resistant and it can have high surface gloss. When injection molding PMMA, the melt temperature should be between 210 and 240°C, and the mold temperature 50- 70°C. Typical applications for injection molded polymethylmethacrylate parts are automotive rear lights, drawing instruments, watch windows, lenses, jewlery, pipe fittings, etc.

Polypropylene (PP). Polypropylene is a semi-crystalline polymer known for its low density, and its somewhat higher stiffness and strength than PE-HD. However, PP has a lower toughness than PE-HD. Polypropylene homopolymer has a glass transition temperature of as high as -10°C, below which temperature it becomes brittle. However, when copolymerized with ethylene it becomes tough. Because of its flexibility and the large range of properties, including the ability to reinforce it with glass fiber, polypropylene is often used as a substitute for an engineering thermoplastic. When injection molding PP, the melt temperature should be between 250 and 270°C, and the mold temperature 40-100°C. Typical applications for injection molded polypropylene parts are housings for domestic appliances, kitchen utensils, storage boxes with integrated hinges (living hinges), toys, disposable syringes, food containers, etc.

Polystyrene (PS). Polystyrene is an amorphous polymer known for its high stiffness and hardness. PS is brittle but its toughness can be significantly increased when copolymerized with butadiene. PS is also known for its high dimensional stability, its clarity and it can have high surface gloss. When injection molding PS, the melt temperature should be between 180 and 280°C, and the mold temperature 10-40°C. Typical applications for injection molded polystyrene parts are pharmaceutical and cosmetic cases, radio and television housings, drawing instruments, clothes hangers, toys, etc.

Polyvinylchloride (PVC). Polyvinylchloride comes either unplasticized (PVC-U) or plasticized (PVC-P). Unplasticized PVC is known for its high strength, rigidity and hardness. However, PVC-U is also known for its low impact strength at low temperatures. In the plasticized form, the flexibility of PVC will vary over a wide range. Its toughness will be higher at low temperatures. When injection molding PVC-U pellets, the melt temperature should be between 180 and 210°C, and the mold temperature should be at least 30°C. For PVC-U powder the injection temperatures should be 10°C lower, and the mold temperatures at least 50°C. When injection molding PVC-P pellets, the melt temperature should be between 170 and 200°C, and the mold temperature should be at least 15°C. For PVC-P powder the injection temperatures should 5°C lower, and the mold temperatures at least 50°C. Typical applications for injection molded plasticized polyvinylchloride parts are shoe soles, sandals and some toys. Typical applications for injection molded unplasticized polyvinylchloride parts are pipefittings.

1.7.2 Thermosetting Polymers

Thermosetting polymers solidify by a chemical cure. Here, the long macromolecules crosslink during cure, resulting in a network. The original molecules can no longer slide past each other. These networks prevent *flow* even after re-heating. The high density of crosslinking between the molecules makes thermosetting materials stiff and brittle. The cross-linking causes the material to become resistant to heat after it has solidified. However, thermosets also exhibit glass transition temperatures which sometimes exceed thermal degradation temperatures.

The cross-linking usually is a result of the presence of double bonds that break, allowing the molecules to link with their neighbors. One of the oldest thermosetting polymers is phenolformaldehyde or phenolic. Figure 1.1 shows the chemical symbol representation of the reaction. The phenol molecules react with formaldehyde molecules to create a three-dimensional cross-linked network that is stiff and strong. The byproduct of this chemical reaction is water. Examples of the most common thermosetting polymers, with a short summary, are given below.

Phenol Formaldehyde (PF). Phenol formaldehyde is known for its high strength, stiffness, hardness and its low tendency to creep. It is also known for its high toughness, and depending on its reinforcement, it will also exhibit high toughness at low temperatures. PF also has a low coefficient of thermal expansion. Phenol formaldehyde can be compression molded, transfer molded and injection-compression molded. Typical applications for phenol formaldehyde include distributor caps, pulleys, pump components, handles for irons, etc. It should not be used in direct contact with food.

Unsaturated Polyester (UPE). Unsaturated polyester is known for its high strength, stiffness and hardness. It is also known for its dimensional stability, even when hot, making it ideal for under the hood applications. In most cases, UPE is found reinforced with glass fiber. Unsaturated polyester is processed by compression molding, injection molding, injection-compression molding and casting. Sheet molding compound (SMC) is used for compression molding and bulk molding compound is used for injection and injection-compression molding. Typical applications for fiber reinforced unsaturated polyester are automotive body panels, automotive valve covers and oil pans, breaker switch housings, electric motor parts, distributor caps, fans, bathroom sinks, bathtubs, etc.

Epoxy (EP). Epoxy resins are known for their high adhesion properties, high strength, and excellent electrical and dielectrical properties. They are also known for their low shrinkage, their high chemical resistance and their low susceptibility to stress crack formation. They are heat resistant up to their glass transition temperature (around 150-190°C) where they exhibit a significant reduction in stiffness. Typical applications for epoxy resins are switch parts, circuit breakers, housings, encapsulated circuits, etc.

Cross-linked Polyurethanes (PU). Cross-linked polyurethane is known for its high adhesion properties, high impact strength, rapid curing, low shrinkage and low cost. PU is also known for its wide variety of forms and applications. PU can be an elastomer, a flexible foam, a rigid foam, an integral foam, a lacquer, an adhesive, etc. Typical applications for cross-linked polyurethane are television and radio housings, copy and computer housings, ski and tennis racket composites, etc.

1.7.3 Elastomers

The rubber industry is one of the oldest industries. For many years now, rubber products have been found everywhere, from belts to seals, and from hoses to engine mounts. To design and manufacture such products, the rubber technologist must go through various procedures and steps, such as choice of materials and additives, choice of compounding equipment and vulcanization system, as well as testing procedures to evaluate the quality of the finished product. The choice of the base elastomer can be an overwhelming task, even for those with experience in rubber technology. There are hundreds of choices for rubber compounds and blends. In addition, there are hundreds of different additives for various tasks. Additives can be used for softening or plasticizing the rubber compound for easy processing. There are choices of additives that will protect the compound and the finished product from aging, ozone and fatigue, as well as various vulcanization additives that will accelerate or retard the curing process.

Natural Rubbers (NR). The chemical name for NR is polyisoprene, which is a homopolymer of isoprene. It has the cis-1,4 configuration. In addition, the polymer contains small amounts of non-rubber substances, notably fatty acids, proteins, and resinous materials that function as mild accelerators and activators for vulcanization. Raw materials for the production of NR must be derived from trees of the Hevea Brasiliensis species. NR is available in a variety of types and grades, including smoked sheets, air-dried sheets, and pale crepes.

Synthetic Polyisoprene Rubbers (IR). IR is a cis-1,4 polyisoprene synthetic natural rubber. However, it does not contain the non-rubber substances that are present in NR. One can differentiate between two basic types of synthetic polyisoprene by the polymerization catalyst system used. They are commonly referred to as *high* cis and *low* cis types. The *high* cis grades contain approximately 96-97% cis-1,4 polyisoprene.

Styrene-Butadiene Rubbers (SBR). Styrene-butadiene rubbers are produced by random copolymerization of styrene and butadiene. The higher the cis-1,4 content of BR, the lower its glass transition temperature T_g. Pure cis-1,4 BR grades have a T_g temperature of about -100^oC. Commercial grades with about 98% cis-1,4 content have a T_g temperature around -90^oC.

Acrylonitrile-Butadiene Rubbers (NBR). Acrylonitrile-butadiene rubbers (NBR), or simply nitrile rubbers, are copolymers of butadiene and acrylonitrile. They are available in five grades based on the acrylonitrile (ACN) content.

- Very low nitriles: typically 18-20% ACN
- Low nitriles: typically 26-29% ACN
- Medium nitriles: typically 33-35% ACN
- High nitriles: typically 38-40% ACM
- Very high nitriles: typically 45-48% ACN

The glass transition temperatures of polyacrylonitrile at +90°C and of polybutadiene at -90°C differ considerably; therefore, with an increasing amount of acrylonitrile in the polymer, the T_g temperature of NBR rises together with its brittleness temperature. The comonomer ratio is the single most important recipe variable for the production of acrylonitrile-butadiene rubbers.

Ethylene-Propylene Rubbers (EPM and EPDM). There are two types of ethylene-propylene rubbers:

- EPM: fully saturated copolymers of ethylene and propylene
- EPDM: terpolymers of ethylene, propylene, and a small percentage of a non-conjugated diene, which makes the side chains unsaturated.

There are three basic dienes used as the third monomer:

- 1,4 hexadiene (1,4 HD)
- Dicyclopentadiene (DCPD)
- 5-ethylidene norbornene (ENB)

The EPM rubbers, being completely saturated, require organic peroxides or radiation for vulcanization. The EPDM terpolymers can be vulcanized with peroxides, radiation, or sulfur.

Problems

1.1 Estimate the degree of polymerization of a polyethylene with an average molecular weight of 150,000. The molecular weight of an ethylene monomer is 28.

1.2 What is the maximum possible separation between the ends of a polystyrene molecule with a molecular weight of 160,000.

1.3 To enhance processability of a polymer why would you want to decrease its molecular weight?

1.4 Why would an uncrosslinked polybutadiene flow at room temperature?

1.5 Is it true that by decreasing the temperature of a polymer you can increase its relaxation time?

1.6 If you know the relaxation time of a polymer at one temperature, can you use the WLF equation to estimate the relaxation time of the same material at a different temperature? Explain.

1.7 What role does the cooling rate play in the morphological structure of semi-crystalline polymers?

1.8 Explain how cross-linking between the molecules affect the molecular mobility and elasticity of elastomers.

1.9 Increasing the molecular weight of a polymer increases its strength and stiffness, as well as its viscosity. Is too high of a viscosity a limiting factor when increasing the strength by increasing the molecular weight? Why?

1.10 Which broad class of thermoplastic polymers densifies the least during cooling and solidification from a melt state into a solid state? Why?

1.11 What class of polymers would you probably use to manufacture frying pan handles? Even though most polymers could not actually be used for this particular application, what single property do all polymers exhibit that would be considered advantageous in this particular application.

1.12 In terms of recycling, which material is easier to handle, thermosets or thermoplastics? Why?

1.13 You are to extrude a polystyrene tube at an average speed of 0.1 m/s. The relaxation time, λ, of the polystyrene, at the processing temperature, is 1 second. The die land length is 0.02m. Will elasticity play a significant role in your process.

1.14 Figure 1.36 presents some creep modulus data for polystyrene at various temperatures [11]. Create a master curve at 109.8°C by graphically sliding the curves at some temperatures horizontally until they line up.

- **a)** Identify the glassy, rubbery, and viscoelastic regions of the master curve.
- **b)** Develop a plot of the shift factor, $\log(a_T)$ versus T, used to create your master curve $\log(a_T)$ is the horizontal distance that the curve at temperature T was slid to coincide with the master curve. Compare your graphical result with the WLF equation. Note: The WLF equation is for a master curve at T_g (85°C for this PS), but your master curve is for 109.8°C, so be sure you make a fair comparison.

1.15 Figure 1.37 presents relaxation data for polycarbonate at various temperatures [8]. Create a master curve at 25°C by graphically sliding the curves at the various temperatures horizontally until they line up.

- **a)** Identify the glassy, rubbery, and viscoelastic regions of the master curve.
- **b)** Develop a plot of the shift factor, $\log(a_T)$ versus T, used to create your master curve. $\log(a_T)$ is the horizontal distance that the curve at temperature T was slid to coincide with the master curve.
- **c)** Compare your graphical result with the WLF equation. Note that the resulting master curve is far from the glass transition temperature of polycarbonate.

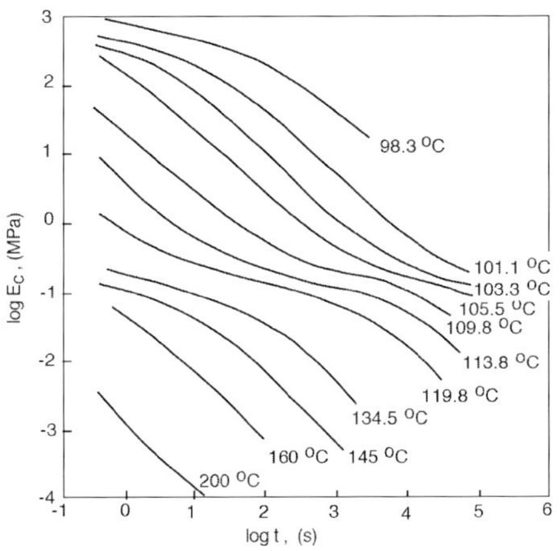

Figure 1.36: Creep modulus as a function of time for polystyrene.

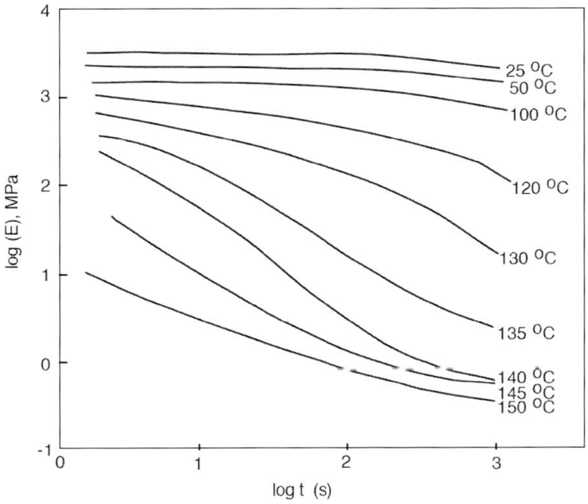

Figure 1.37: Relaxation modulus as a function of time for polycarbonate.

1.16 Figure 1.31 presents shear relaxation data for a chlorosulfonated polyethylene at various pressures. Create a master curve at 1 bar by graphically sliding the curves at the various pressures horizontally until they line up. On the same graph, draw the master curve at a pressure of 1,200 bar, a high pressure encountered during injection molding.

REFERENCES

1. Campus Group www.campusplastics.com 2005.
2. G.C. Berry and T.G. Fox. *Adv. Polymer Sci.*, 5:261, 1968.
3. P.J. Carreau, D.C.R. DeKee, and R.P. Chhabra. *Rheology of Polymeric Systems*. Hanser Publishers, Munich, 1997.
4. E. Castiff and A.V.J. Tobolsky. *Colloid Science*, 10:375, 1955.
5. M.L. Crowder, A.M. Ogale, E.R. Moore, and B.D. Dalke. *Polym. Eng. Sci.*, 34(19):1497, 1994.
6. H. Domininghaus. *Plastics for Engineers*. Hanser Publishers, 1991.
7. R.W. Fillers and N.W. Tschoegl. *Trans. Soc. Rheol.*, 21:51, 1977.
8. T. Fujimoto, M. Ozaki, and M. Nagasawa. *J. Polymer Sci.*, 2:6, 1968.
9. T. A. Osswald and G. Menges. *Material Science of Polymers for Engineers*. Hanser Publishers, Munich, 2nd edition, 2003.
10. T.A. Osswald, E. Baur, and E. Schmachtenberg. *Plastics Handbook*. Hanser Publishers, Munich, 2006.
11. D.J. Pazek. *J. Polym. Sci.*, A-2 6:621, 1968.
12. H. Staudinger and W. Huer. *Ber. der Deutschen Chem. Gesel.*, 63:222, 1930.
13. D.W. van Krevelen. *Properties of Polymers*. Elsevier, Amsterdam, 1990.
14. M.L. Williams, R.F. Landel, and J.D. Ferry. *J. Amer. Chem. Soc.*, 77:3701, 1955.

CHAPTER 2

PROCESSING PROPERTIES

Did you ever consider viscoelasticity?

—Arthur Lodge

2.1 THERMAL PROPERTIES

The heat flow through a material can be defined by Fourier's law of heat conduction. Fourier's law can be expressed as

$$q_x = -k_x \frac{\partial T}{\partial x} \tag{2.1}$$

where q_x is the energy transport per unit area in the x direction, k_x the thermal conductivity and $\partial T/\partial x$ the temperature gradient. At the onset of heating, the polymer responds solely as a heat sink, and the amount of energy per unit volume, Q, stored in the material before reaching steady state conditions can be approximated by

$$Q = \rho C_p \Delta T \tag{2.2}$$

where ρ is the density of the material, C_p the specific heat, and ΔT the change in temperature. The material properties found in eqns. (2.1) and (2.2) are often written as one single property,

Table 2.1: Thermal Properties for Selected Polymeric Materials

Polymer	Specific gravity	Specific heat kJ/kg/K	Thermal conduc. W/m/K	Coeff. therm. expan. μm/m/K	Thermal diffusivity $(m^2/s)10^{-7}$	Max temp. ^oC
ABS	1.04	1.47	0.3	90	1.7	70
CA	1.28	1.50	0.15	100	1.04	60
EP	1.9	-	0.23	70	-	130
PA66	1.14	1.67	0.24	90	1.01	90
PA66-30% glass	1.38	1.26	0.52	30	1.33	100
PC	1.15	1.26	0.2	65	1.47	125
PE-HD	0.95	2.3	0.63	120	1.57	55
PE-LD	0.92	2.3	.33	200	1.17	50
PET	1.37	1.05	0.24	90	-	110
PF	1.4	1.3	0.35	22	1.92	185
PMMA	1.18	1.47	0.2	70	1.09	50
POM	1.42	1.47	0.2	80	0.7	85
coPOM[a]	1.41	1.47	0.2	95	0.72	90
PP	0.905	1.95	0.24	100	0.65	100
PPO[b]	1.06	-	0.22	60	-	120
PS	1.05	1.34	0.15	80	0.6	50
PTFE	2.1	1.0	0.25	140	0.7	50
uPVC[c]	1.4	1.0	0.16	70	1.16	50
pPVC[d]	1.31	1.67	0.14	140	0.7	50
SAN	1.08	1.38	0.17	70	0.81	60
UPE	1.20	1.2	0.2	100	-	200
Steel	7.854	0.434	60	-	14.1	800

[a]Polyacetal copolymer; [b]Polyphenylene oxide copolymer; [c]Unplasticized PVC; [d]Plasticized PVC

namely the thermal diffusivity, α, which for an isotropic material is defined by

$$\alpha = \frac{k}{\rho C_p} \tag{2.3}$$

Typical values of thermal properties for selected polymers are shown in Table 6.1 [7, 17]. For comparison, the properties for stainless steel are also shown at the end of the list. It should be pointed out that the material properties of polymers are not constant and may vary with temperature, pressure or phase changes. This section will discuss each of these properties individually and present examples of some of the most widely used polymers and measurement techniques. For a more in-depth study of thermal properties of polymers the reader is encouraged to consult the literature [24, 46, 66].

2.1.1 Thermal Conductivity

When analyzing thermal processes, the thermal conductivity, k, is the most commonly used property that helps quantify the transport of heat through a material. By definition, energy is transported proportionally to the speed of sound. Accordingly, thermal conductivity

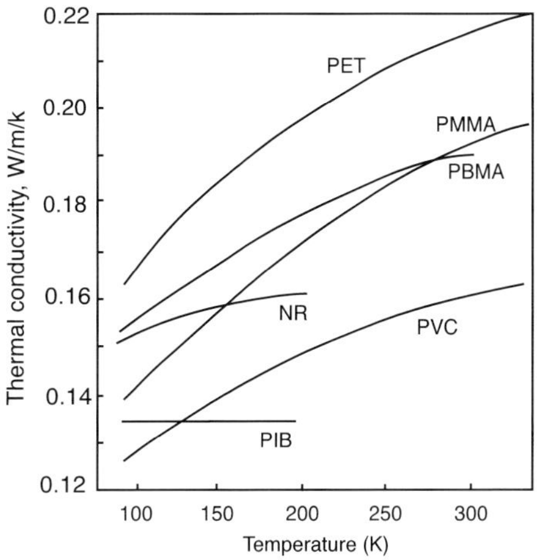

Figure 2.1: Thermal conductivity of various materials.

follows the relation

$$k \approx C_p \rho u l \tag{2.4}$$

where u is the speed of sound and l the molecular separation. Amorphous polymers show an increase in thermal conductivity with increasing temperature, up to the glass transition temperature, T_g. Above T_g, the thermal conductivity decreases with increasing temperature. Figure 2.1 [24] presents the thermal conductivity, below the glass transition temperature, for various amorphous thermoplastics as a function of temperature.

Due to the increase in density upon solidification of semi-crystalline thermoplastics, the thermal conductivity is higher in the solid state than in the melt. In the melt state, however, the thermal conductivity of semi-crystalline polymers reduces to that of amorphous polymers as can be seen in Fig. 2.2 [40].

Furthermore, it is not surprising that the thermal conductivity of melts increases with hydrostatic pressure. This effect is clearly shown in Fig. 2.3 [19]. As long as thermosets are unfilled, their thermal conductivity is very similar to amorphous thermoplastics. Anisotropy in thermoplastic polymers also plays a significant role in the thermal conductivity. Highly drawn semi-crystalline polymer samples can have a much higher thermal conductivity as a result of the orientation of the polymer chains in the direction of the draw.

For amorphous polymers, the increase in thermal conductivity in the direction of the draw is usually not higher than two. Figure 2.4 [24] presents the thermal conductivity in the directions parallel and perpendicular to the draw for high density polyethylene, polypropylene, and polymethyl methacrylate. A simple relation exists between the anisotropic and the isotropic thermal conductivity [39]. This relation is written as

$$\frac{1}{k_{||}} + \frac{2}{k_{\perp}} = \frac{3}{k} \tag{2.5}$$

where the subscripts $||$ and $\perp$ represent the directions parallel and perpendicular to the draw, respectively.

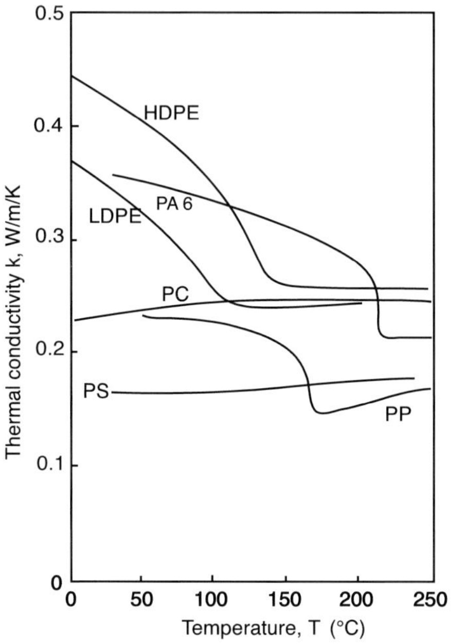

Figure 2.2: Thermal conductivity of various thermoplastics.

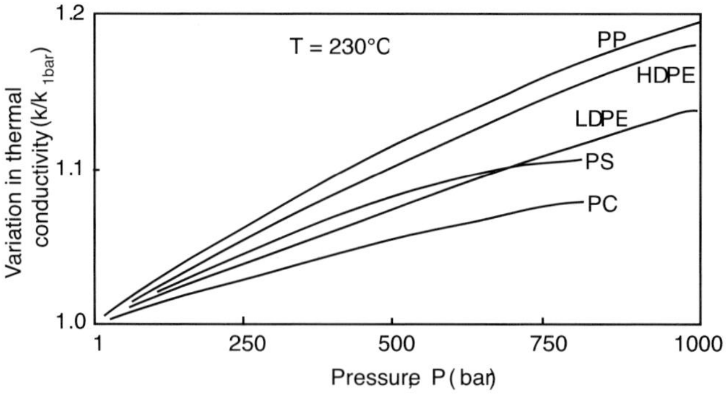

Figure 2.3: Influence of pressure on thermal conductivity of various thermoplastics.

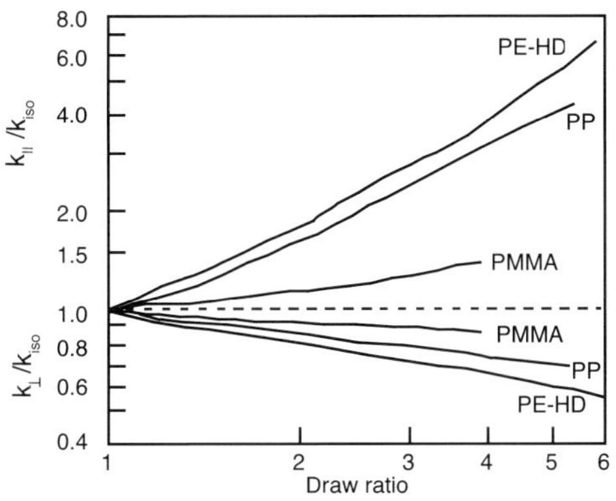

Figure 2.4: Thermal conductivity as a function of draw ratio in the directions perpendicular and parallel to the stretch for various oriented thermo-plastics.

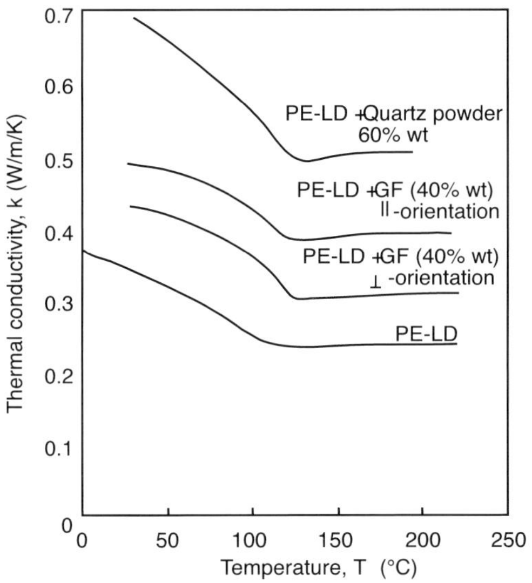

Figure 2.5: Influence of filler on the thermal conductivity of PE-LD.

The higher thermal conductivity of inorganic fillers increases the thermal conductivity of filled polymers. Nevertheless, a sharp decrease in thermal conductivity around the melting temperature of crystalline polymers can still be seen with filled materials. The effect of filler on thermal conductivity for PE-LD is shown in Fig. 2.5 [22]. This figure shows the effect of fiber orientation as well as the effect of quartz powder on the thermal conductivity of low density polyethylene.

Figure 2.6 demonstrates the influence of gas content on expanded or foamed polymers, and the influence of mineral content on filled polymers.

There are various models available to compute the thermal conductivity of foamed or filled plastics [39, 47, 51]. A rule of mixtures, suggested by Knappe [39], commonly used

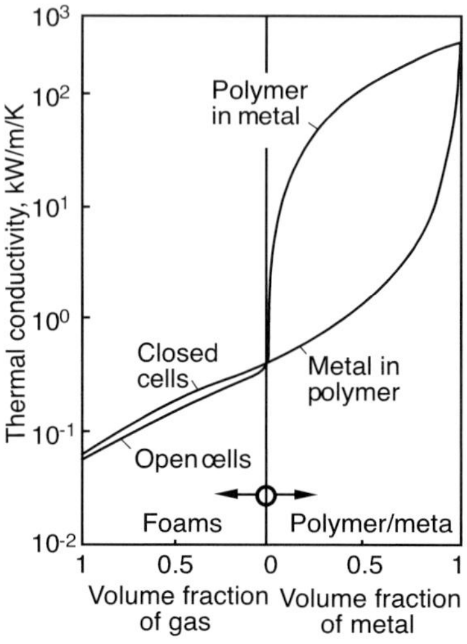

Figure 2.6: Thermal conductivity of plastics filled with glass or metal.

to compute thermal conductivity of composite materials is written as

$$k_c = \frac{2k_m + k_f - 2\phi_f(k_m - k_f)}{2k_m + k_f + \phi_f(k_{m-k_f})} k_m \tag{2.6}$$

where, ϕ_f is the volume fraction of filler, and k_m, k_f and k_c are the thermal conductivity of the matrix, filler and composite, respectively.

Figure 2.7 compares eqn. (2.6) with experimental data [2] for an epoxy filled with copper particles of various diameters. The figure also compares the data to the classic model given by Maxwell [47] which is written as

$$k_c = \left(1 + 3\phi_f \frac{\dfrac{k_f}{k_m} - 1}{\dfrac{k_f}{k_m} + 2} \right) k_m \tag{2.7}$$

In addition, a model derived by Meredith and Tobias [51] applies to a cubic array of spheres inside a matrix. Consequently, it cannot be used for volumetric concentration above 52% since the spheres will touch at that point. However, their model predicts the thermal conductivity very well up to 40% by volume of particle concentration. When mixing several materials the following variation of Knappe's model applies

$$k_c = \frac{1 - \sum_{i=1}^{n} 2\phi_i \dfrac{k_m - k_i}{2k_m + k_i}}{1 + \sum_{i=1}^{n} 2\phi_i \dfrac{k_m - k_i}{2k_m + k_i}} \tag{2.8}$$

where k_i is the thermal conductivity of the filler and ϕ_i its volume fraction. This relation is useful for glass fiber reinforced composites (FRC) with glass concentrations up to 50% by volume. This is also valid for FRC with unidirectional reinforcement. However, one must differentiate between the direction longitudinal to the fibers and that transverse to them.

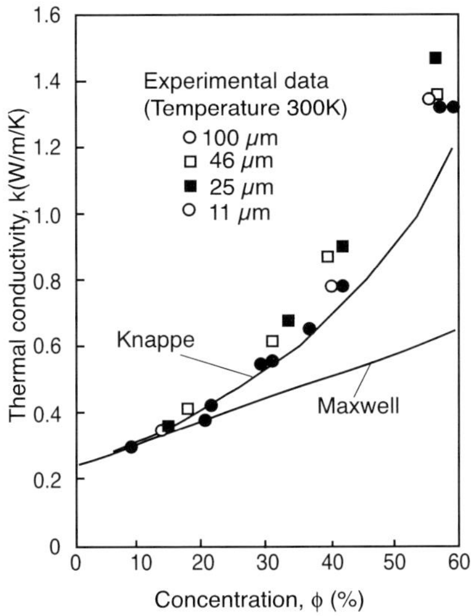

Figure 2.7: Thermal conductivity versus volume concentration of metallic particles of an epoxy resin. Solid lines represent predictions using Maxwell and Knappe models.

For high fiber content, one can approximate the thermal conductivity of the composite by the thermal conductivity of the fiber. The thermal conductivity can be measured using the standard tests ASTM C177 and DIN 52612. A new method currently being balloted (ASTM D20.30) is preferred by most people today.

2.1.2 Specific Heat

The specific heat, C, represents the energy required to change the temperature of a unit mass of material by one degree. It can be measured at either constant pressure, C_p, or constant volume, C_v. Since the specific heat at constant pressure includes the effect of volumetric change, it is larger than the specific heat at constant volume. However, the volume changes of a polymer with changing temperatures have a negligible effect on the specific heat. Hence, one can usually assume that the specific heat at constant volume or constant pressure are the same. It is usually true that specific heat only changes modestly in the range of practical processing and design temperatures of polymers. However, semi-crystalline thermoplastics display a discontinuity in the specific heat at the melting point of the crystallites. This jump or discontinuity in specific heat includes the heat that is required to melt the crystallites which is usually called the heat of fusion. Hence, specific heat is dependent on the degree of crystallinity. Values of heat of fusion for typical semi-crystalline polymers are shown in Table 2.2.

The chemical reaction that takes place during solidification of thermosets also leads to considerable thermal effects. In a hardened state, their thermal data are similar to the ones of amorphous thermoplastics. Figure 2.8 shows the specific heat graphs for the three polymer categories.

Table 2.2: Heat of Fusion of Various Thermoplastic Polymers [66]

Polymer	λ (kJ/kg)	T_m(^{o}C)
Polyamide 6	193-208	223
Polyamide 66	205	265
Polyethylene	268-300	141
Polypropylene	209-259	183
Polyvinyl chloride	181	285

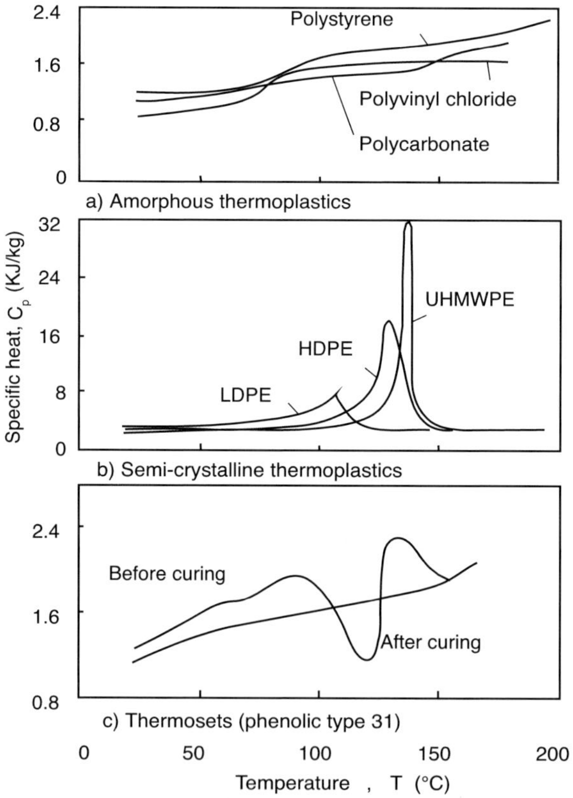

Figure 2.8: Specific heat curves for selected polymers of the three general polymer categories.

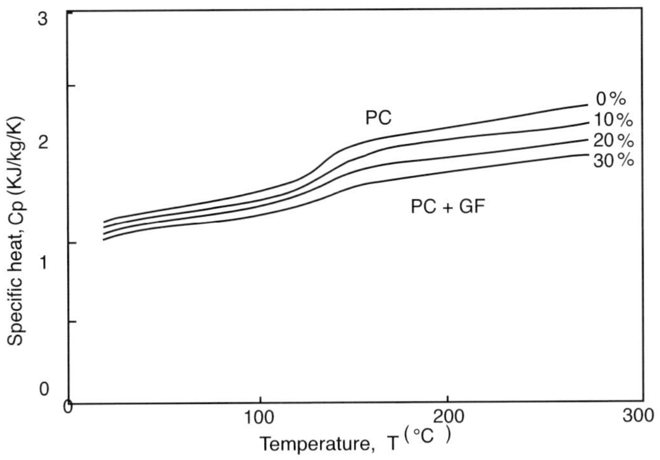

Figure 2.9: Generated specific heat curves for a filled and unfilled polycarbonate. Courtesy of Bayer AG, Germany.

For filled polymer systems with inorganic and powdery fillers, a rule of mixtures[1] can be written as

$$C_p(T) = (1 - \psi_f)C_{p_m}(T) + \psi_f C_{p_f}(T) \tag{2.9}$$

where ψ_f represents the weight fraction of the filler and C_{p_m} and C_{p_f} the specific heat of the polymer matrix and the filler, respectively. As an example of using eqn. (2.9), Fig. 2.9 shows a specific heat curve of an unfilled polycarbonate and its corresponding computed specific heat curves for 10%, 20%, and 30% glass fiber content. In most cases, temperature dependence of C_p on inorganic fillers is minimal and need not be taken into consideration. The specific heat of copolymers can be calculated using the mole fraction of the polymer components.

$$C_{p_{\text{copolymer}}} = \sigma_1 C_{p_1} + \sigma_2 C_{p_2} \tag{2.10}$$

where σ_1 and σ_2 are the mole fractions of the comonomer components and C_{p_1} and C_{p_1} the corresponding specific heats.

2.1.3 Density

The density or its reciprocal, the specific volume, is a commonly used property for polymeric materials. The specific volume is often plotted as a function of pressure and temperature in what is known as a pvT diagram. A typical pvT diagram for an unfilled and filled amorphous polymer is shown, using polycarbonate as an example, in Figs. 2.10 and 2.11 The two slopes in the curves represent the specific volume of the melt and of the glassy amorphous polycarbonate, separated by the glass transition temperature.

Figure 2.12 presents the pvT diagram for polyamide 66 as an example of a typical semi-crystalline polymer. Figure 2.13 shows the pvT diagram for polyamide 66 filled with 30%

[1] Valid up to 65% filler content by volume.

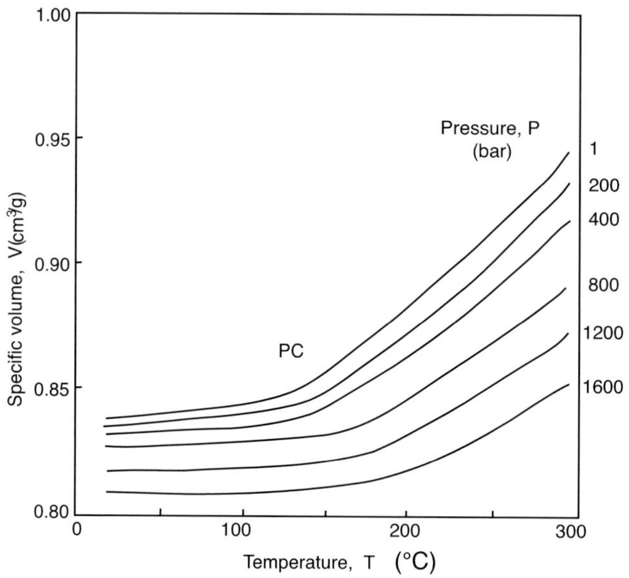

Figure 2.10: pvT diagram for a polycarbonate. Courtesy of Bayer AG, Germany.

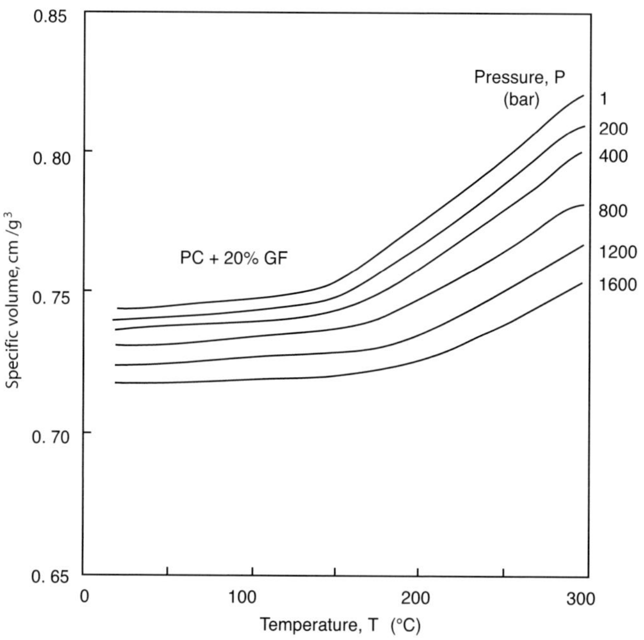

Figure 2.11: pvT diagram for a polycarbonate filled with 20% glass fiber. Courtesy of Bayer AG, Germany.

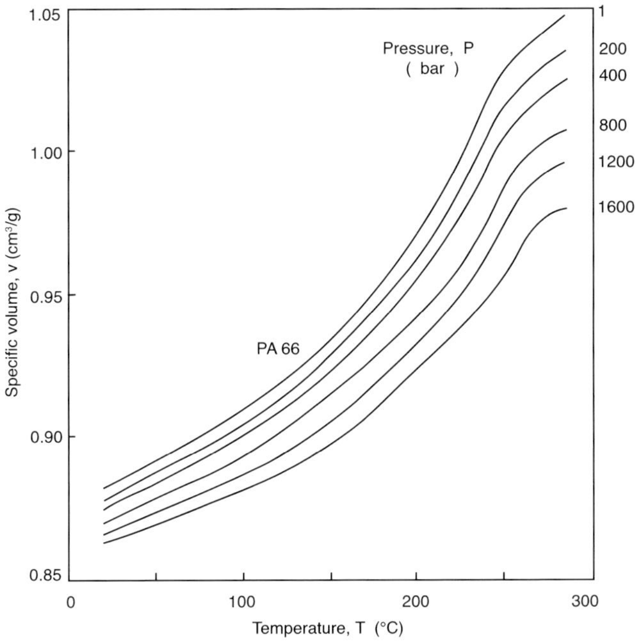

Figure 2.12: pvT diagram for a polyamide 66. Courtesy of Bayer AG, Germany.

glass fiber. The curves clearly show the melting temperature (i.e., $T_m \approx 250^o$C for the unfilled PA66 cooled at 1 bar, which marks the beginning of crystallization as the material cools). It should also come as no surprise that the glass transition temperatures are the same for the filled and unfilled materials.

When carrying out die flow calculations, the temperature dependence of the specific volume must often be dealt with analytically. At constant pressures, the density of pure polymers can be approximated by

$$\rho(T) = \rho_0 \frac{1}{1 + \alpha_t (T - T_0)} \tag{2.11}$$

where ρ_0 is the density at reference temperature, T_0, and α_t is the linear coefficient of thermal expansion. For amorphous polymers, eqn. (2.11) is valid only for the linear segments (i.e., below or above T_g), and for semi-crystalline polymers it is only valid for temperatures above T_m. The density of polymers filled with inorganic materials can be computed at any temperature using the following rule of mixtures

$$\rho_c(T) = \frac{\rho_m(T)\rho_f}{\psi \rho_m(T) + (1 - \psi)\rho_f} \tag{2.12}$$

where ρ_c, ρ_m and ρ_f are the densities of the composite, polymer and filler, respectively, and ψ is the weight fraction of filler.

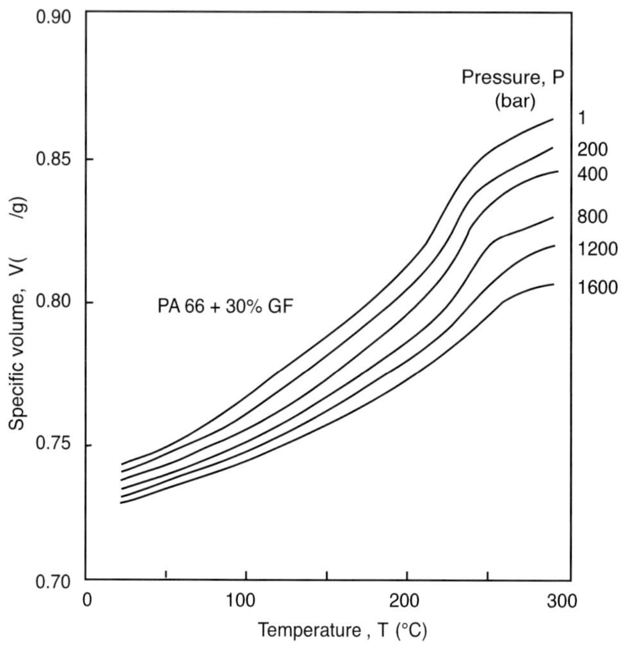

Figure 2.13: pvT diagram for a polyamide 66 filled with 30% glass fiber. Courtesy of Bayer AG, Germany.

A widely accepted form of modeling the density or specific volume is the Tait equation. It is often used to represent the pvT-behavior of polymers and it is represented as,

$$v(T,p) = v_0(T)\left[1 - C\ln\left(1 + \frac{p}{B(T)}\right)\right] + v_t(T,p) \tag{2.13}$$

where $C = 0.0894$. This equation of state is capable of describing both the liquid and solid regions by changing the constants in $v_0(T)$, $B(T)$ and $v_t(T,p)$, which are defined as,

$$v_0(T) = \begin{cases} b_{1,l} + b_{2,l}\bar{T} & \text{if } T > T_t(p) \\ b_{1,s} + b_{2,s}\bar{T} & \text{if } T < T_t(p) \end{cases} \tag{2.14}$$

$$B(T) = \begin{cases} b_{3,l}e^{-b_{4,l}\bar{T}} & \text{if } T > T_t(p) \\ b_{3,s}e^{-b_{4,s}\bar{T}} & \text{if } T < T_t(p) \end{cases} \tag{2.15}$$

and

$$v_t(T,p) = \begin{cases} 0 & \text{if } T > T_t(p) \\ b_7 e^{b_8\bar{T} - b_9 p} & \text{if } T < T_t(p) \end{cases} \tag{2.16}$$

where $\bar{T} = T - b_5$ and the transition temperature is assumed to be a linear function of pressure, i.e.,

$$T_f(p) = b_5 + b_6 p \tag{2.17}$$

Table 2.3 presents the constants for the Tait equation for given PC, PP and PS resins. Figures 2.14, 2.15 and 2.16 present the numerical pvT representation of the PC, PS and PP resins presented in Table 2.3.

Table 2.3: Tait equation constants for various materials based upon fitting data [13]

Material	PC	PP	PS
Grade	Lexan 101	PPN 1060	S3200
Manufacturer	G.E.	Hoechst	Hoechst
Cooling Rate (^{o}C/s)	8.0	5.4	4.0
$b_{1,l}$ (cm^3/g)	0.848	1.246	0.988
$b_{2,l}$ (cm^3/g/^oC)	5.28×10^{-4}	9.03×10^{-4}	6.10×10^{-4}
$b_{3,l}$ (dyne/cm^2)	2.37×10^{9}	9.28×10^{8}	1.15×10^{9}
$b_{4,l}$ (1/^{o}C)	5.54×10^{-3}	4.07×10^{-3}	3.66×10^{-3}
$b_{1,s}$ (cm^3/g)	0.848	1.160	0.988
$b_{2,s}$ (cm^3/g/^oC)	4.56×10^{-5}	3.57×10^{-4}	1.49×10^{-4}
$b_{3,s}$ (dyne/cm^2)	4.65×10^{9}	2.05×10^{9}	2.38×10^{9}
$b_{4,s}$ (1/^{o}C)	1.49×10^{-3}	2.49×10^{-3}	2.10×10^{-3}
b_5 (^{o}C)	149.0	123.0	112.0
b_6 (^{o}Cg/dyne)	3.2×10^{-8}	2.25×10^{-8}	7.8×10^{-8}
b_7 (cm^3/g)	0.0	0.087	0.0
b_8 (1/^{o}C)	–	5.37×10^{-1}	–
b_9 (cm^2/dyne)	–	1.26×10^{-8}	–

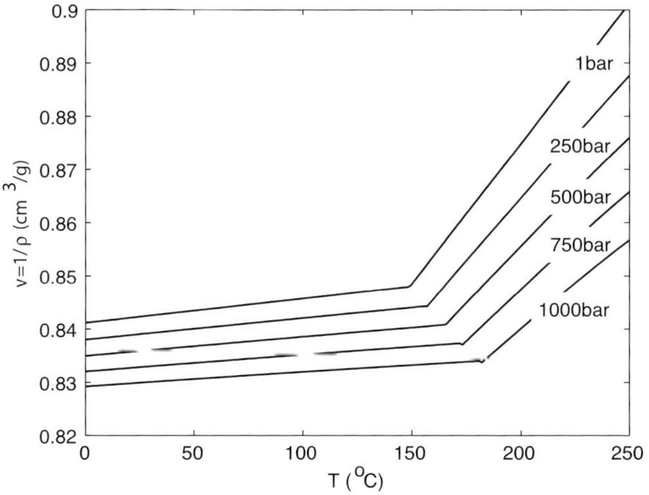

Figure 2.14: pvT diagram from the Tait equation for PC (Table 2.3).

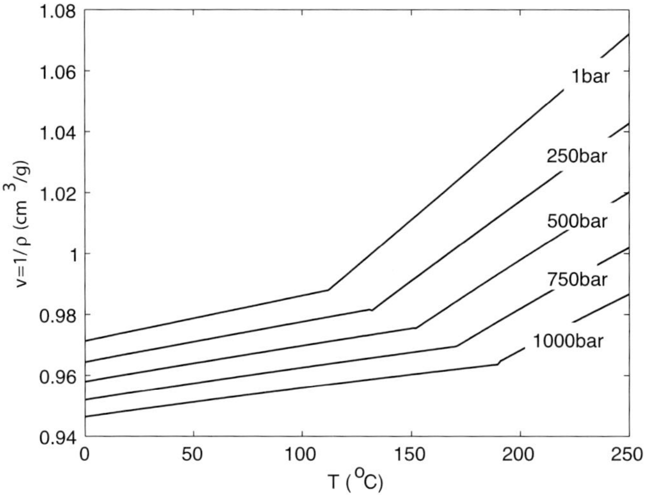

Figure 2.15: pvT diagram from the Tait equation for PS (Table 2.3).

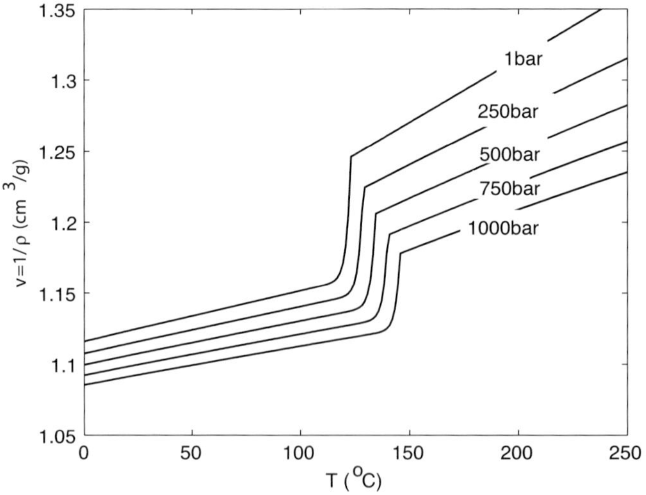

Figure 2.16: pvT diagram from the Tait equation for PP (Table 2.3).

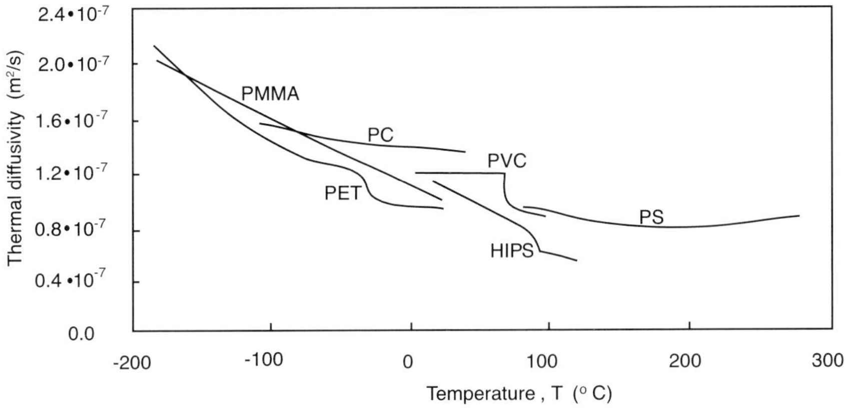

Figure 2.17: Thermal diffusivity as a function of temperature for various amorphous thermoplastics.

2.1.4 Thermal Diffusivity

Thermal diffusivity, defined in eqn. (2.3), is the material property that governs the process of thermal diffusion over time. The thermal diffusivity in amorphous thermoplastics decreases with temperature. A small jump is observed around the glass transition temperature due to the decrease in heat capacity at T_g. Figure 2.17 [24] presents the thermal diffusivity for selected amorphous thermoplastics.

A decrease in thermal diffusivity, with increasing temperature, is also observed in semi-crystalline thermoplastics. These materials show a minimum at the melting temperature as demonstrated in Fig. 2.18 [24] for a selected number of semi-crystalline thermoplastics. It has also been observed that the thermal diffusivity increases with increasing degree of crystallinity and that it depends on the rate of crystalline growth, hence, on the cooling speed.

2.1.5 Linear Coefficient of Thermal Expansion

The linear coefficient of thermal expansion is related to volume changes that occur in a polymer due to temperature variations and is well represented in the pvT diagram. For many materials, thermal expansion is related to the melting temperature of that material, demonstrated for some important polymers in Fig. 2.19.

Although the linear coefficient of thermal expansion varies with temperature, it can be considered constant within typical design and processing conditions. It is especially high for polyolefins, where it ranges from $1.5 \times 10^{-4}\mathrm{K}^{-1}$ to $2 \times 10^{-4}\mathrm{K}^{-1}$; however, fibers and other fillers significantly reduce thermal expansion. A rule of mixtures is sufficient to calculate the thermal expansion coefficient of polymers that are filled with powdery or small particles as well as with short fibers. In this case, the rule of mixtures is written as

$$\alpha_c = \alpha_p(1 - \phi_f) + \alpha_f\phi_f \tag{2.18}$$

where ϕ_f is the volume fraction of the filler, and α_c, α_p and α_f are coefficients for the composite, the polymer and the filler, respectively. In case of continuous fiber reinforcement, the rule of mixtures presented in eqn. (2.18) applies for the coefficient perpendicular

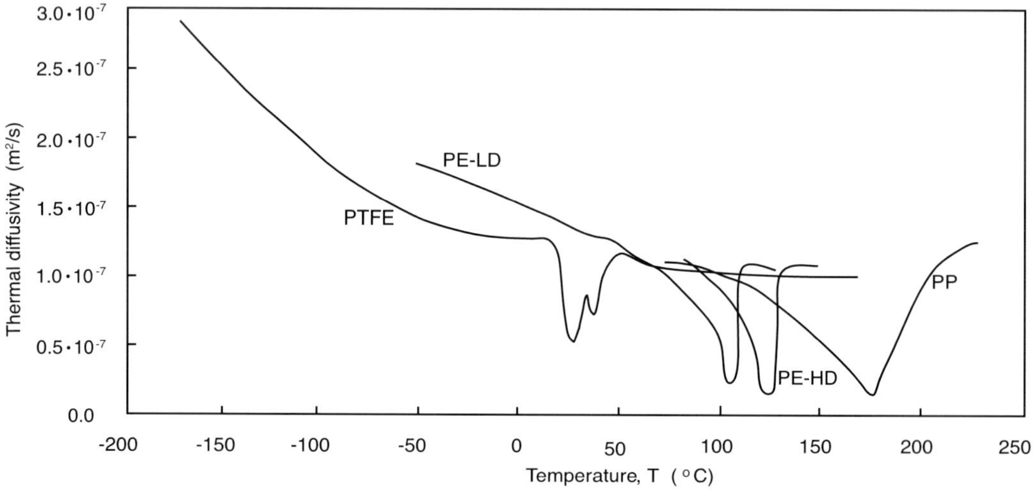

Figure 2.18: Thermal diffusivity as a function of temperature for various semi-crystalline thermoplastics.

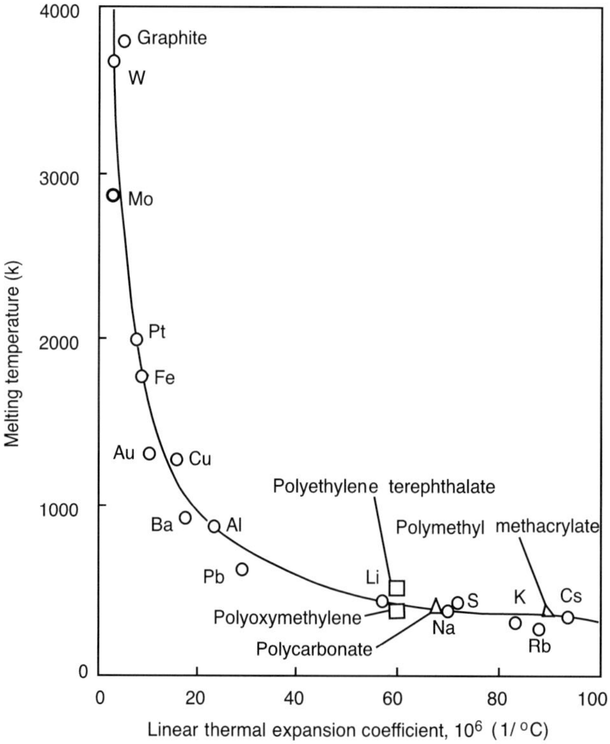

Figure 2.19: Relation between thermal expansion of some metals and plastics at 20°C and their melting temperature.

Table 2.4: Thermal Penetration of Various Thermoplastics[a]

Polymer	a_1	a_2
PE-HD	1.41	441.7
PE-LD	0.0836	615.1
PMMA	0.891	286.4
POM	0.674	699.6
PP	0.846	366.8
PS	0.909	188.9
PVC	0.649	257.8

[a] Coefficients to calculate thermal penetration using $b = a_1 T + a_2$ (W/s$^{1/2}$/m^2/K).

to the reinforcing fibers. In the fiber direction, however, the thermal expansion of the fibers determines the linear coefficient of thermal expansion of the composite. Extensive calculations are necessary to determine coefficients in layered laminated composites and in fiber reinforced polymers with varying fiber orientation distribution.

2.1.6 Thermal Penetration

In addition to thermal diffusivity, the thermal penetration number is of considerable practical interest. It is given by

$$b = \sqrt{kC_p\rho} \tag{2.19}$$

If the thermal penetration number is known, the contact temperature T_C, which results when two bodies A and B, which are at different temperatures, touch, can easily be computed using

$$T_C = \frac{b_A T_A + b_B T_B}{b_A + b_B} \tag{2.20}$$

where T_A and T_B are the temperatures of the touching bodies and b_A and b_B are the thermal penetrations for both materials. The contact temperature is very important for many objects in daily use (e.g., from the handles of heated objects or drinking cups made of plastic, to the heat insulation of space crafts). It is also very important for the calculation of temperatures in tools and molds during polymer processing. The constants used to compute temperature dependent thermal penetration numbers for common thermoplastics are given in Table 2.4 [11].

2.1.7 Measuring Thermal Data

Thanks to modern analytical instruments it is possible to measure thermal data with a high degree of accuracy. These data allow a good insight into chemical and manufacturing processes. Accurate thermal data or properties are necessary for everyday calculations and computer simulations of thermal processes. Such analyses are used to design polymer processing installations and to determine and optimize processing conditions. In the last twenty years, several physical thermal measuring devices have been developed to determine thermal data used to analyze processing and polymer component behavior.

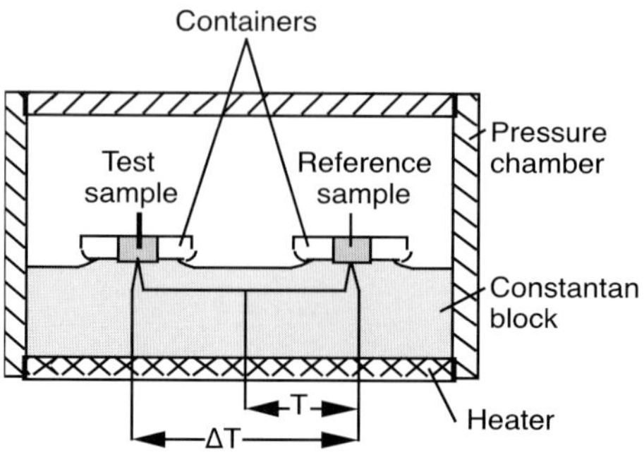

Figure 2.20: Schematic of a differential thermal analysis test.

Differential Thermal Analysis (DTA). The differential thermal analysis test serves to examine transitions and reactions which occur on the order between seconds and minutes, and involve a measurable energy differential of less than 0.04 J/g. Usually, the measuring is done dynamically (i.e., with linear temperature variations in time). However, in some cases isothermal measurements are also done. DTA is mainly used to determine the transition temperatures. The principle is shown schematically in Fig. 2.20. Here, the sample, S, and an inert substance, I, are placed in an oven that has the ability to raise its temperature linearly.

Two thermocouples that monitor the samples are connected opposite to one another such that no voltage is measured as long as S and I are at the same temperature:

$$\Delta T = T_S - T_I = 0 \tag{2.21}$$

However, if a transition or a reaction occurs in the sample at a temperature, T_C, then heat is consumed or released, in which case $\Delta T \neq 0$. This thermal disturbance in time can be recorded and used to interpret possible information about the reaction temperature, T_C, the heat of transition or reaction, ΔH, or simply about the existence of a transition or reaction.

Figure 2.21 presents the temperature history in a sample with an endothermic melting point (i.e., such as the one that occurs during melting of semi-crystalline polymers). The figure also shows the functions $\Delta T(T_I)$ and $\Delta T(T_S)$ which result from such a test. A comparison between Figs. 2.21 demonstrates that it is very important to record the sample temperature, T_S, to determine a transition temperature such as the melting or glass transition temperature.

Differential Scanning Calorimeter (DSC). The differential scanning calorimeter permits us to determine thermal transitions of polymers in a range of temperatures between -180 and +600°C. Unlike the DTA cell, in the DSC device, thermocouples are not placed directly inside the sample or the reference substance. Instead, they are embedded in the specimen holder or stage on which the sample and reference pans are placed; the thermocouples make contact with the containers from the outside. A schematic diagram of a differential scanning calorimeter is very similar to the one shown in Fig. 2.20. Materials that do not show or undergo transition or react in the measuring range (e.g., air, glass powder, etc.) are placed inside the reference container. For standardization, one generally uses mercury, tin, or zinc, whose properties are exactly known. In contrast to the DTA test, where samples larger than 10 g are needed, the DSC test requires samples that are in the mg range ($<$10 mg). Although DSC tests are less sensitive than the DTA tests, they are the most widely used tests for thermal analysis. In fact, DTA tests are rarely used in the polymer industry.

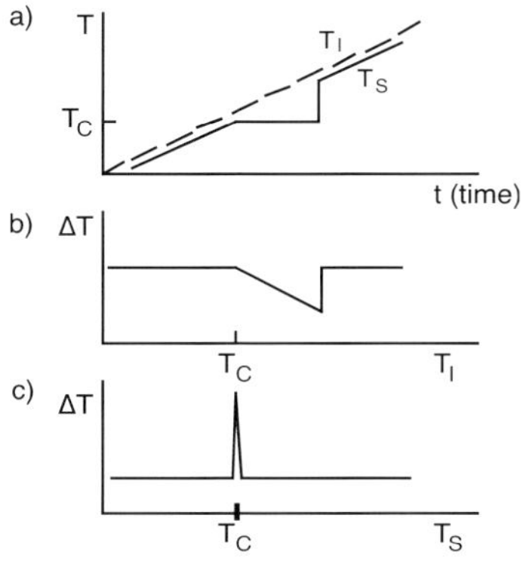

Figure 2.21: Temperature and temperature differences measured during melting of a semi-crystalline polymer sample.

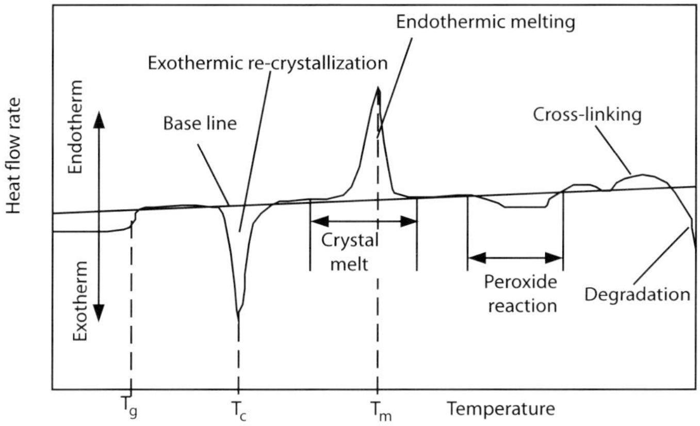

Figure 2.22: Typical DSC heat flow for a semi-crystalline polymer.

Figure 2.22 [66] shows a typical DSC curve measured using a partly crystalline polymer sample. In the figure, the area that is enclosed between the trend line and the base line is a direct measurement for the amount of heat, ΔH, needed for transition. In this case, the transition is melting and the area corresponds to the heat of fusion.

The degree of crystallinity, $\mathcal{X}$, is determined from the ratio of the heat of fusion of a polymer sample, ΔH_{SC}, and the enthalpy of fusion of a 100% crystalline sample ΔH_C.

$$\mathcal{X} = \frac{\Delta H_{SC}}{\Delta H_C} \tag{2.22}$$

In a DSC analysis of a semi-crystalline polymer, a jump in the specific heat curve, as shown in Fig. 2.22, becomes visible. The glass transition temperature, T_g, is determined at the inflection point of the specific heat curve. The release of residual stresses as a material's temperature is raised above the glass transition temperature is often observed in a DSC analysis.

Specific heat, C_p, is one of the many material properties that can be measured with DSC. During a DSC temperature sweep, the sample pan and the reference pan are maintained at the same temperature. This allows the measurement of the differential energy required

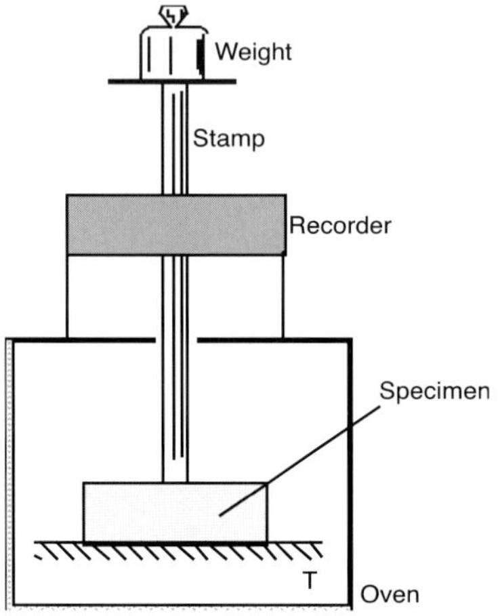

Figure 2.23: Schematic diagram of the thermomechanical analysis (TMA) device.

to maintain identical temperatures. The sample with the higher heat capacity will absorb a larger amount of heat, which is proportional to the difference between the heat capacity of the measuring sample and the reference sample. It is also possible to determine the purity of a polymer sample when additional peaks or curve shifts are detected in a DSC measurement.

Thermal degradation is generally accompanied by an exothermic reaction which may result from oxidation. Such a reaction can easily be detected in a DSC output. By further warming of the test sample, cross-linking may take place and, finally, chain breakage, as shown in Fig. 2.22.

An important aspect in DSC data interpretation is the finite heat flow resistance between the sample pan and the furnace surface. Recent studies by Janeschitz-Kriegl, Eder and co-workers [33, 70] have demonstrated that the heat transfer coefficient between the sample pan and furnace is of finite value, and cannot be disregarded when interpreting the data. In fact, with materials that have a low thermal conductivity, such as polymers, the finite heat transfer coefficient will significantly influence the temperature profiles of the samples.

Thermomechanical Analysis (TMA). Thermomechanical analysis (TMA) measures shape stability of a material at elevated temperatures by physically penetrating it with a metal rod. A schematic diagram of TMA equipment is shown in Fig. 2.23. In TMA, the test specimen's temperature is raised at a constant rate, the sample is placed inside the measuring device, and a rod with a specified weight is placed on top of it. To allow for measurements at low temperatures, the sample, oven, and rod can be cooled with liquid nitrogen.

Most instruments are so precise that they can be used to measure the melting temperature of the material and, by using linear dilatometry, to measure the thermal expansion coefficients. The thermal expansion coefficient can be measured using

$$\alpha_t \frac{1}{L_0}\frac{\Delta L}{\Delta T} \tag{2.23}$$

where L_0 is the initial dimension of the test specimen, ΔL the change in size and ΔT the temperature difference. For isotropic materials a common relation between the linear and

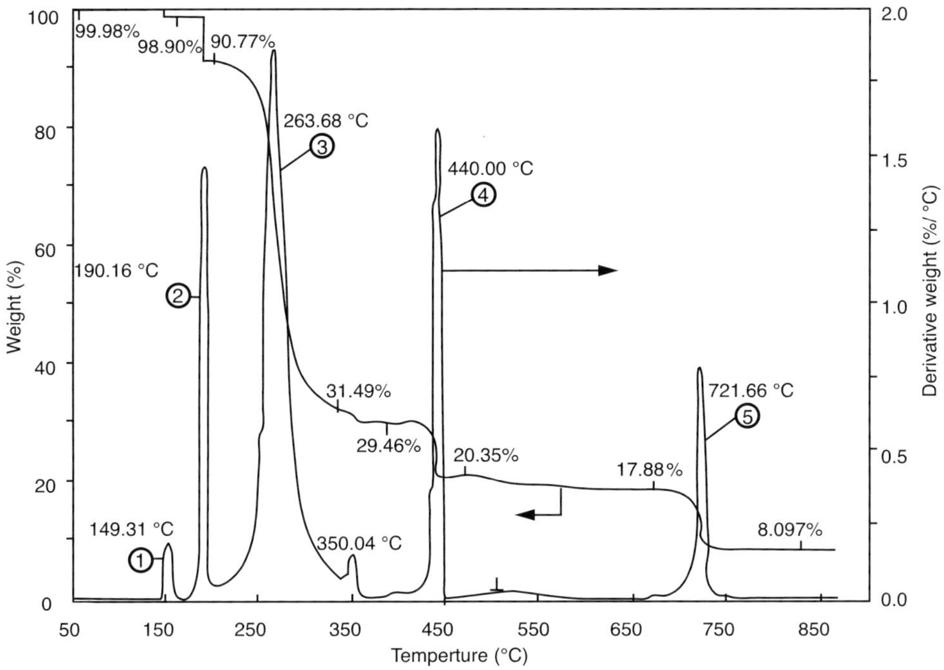

Figure 2.24: TGA analysis of a PVC fabric. (1) volatiles: humidity, monomers, solvents etc., (2) DOP plasticizer, (3) HCl formation, (4) carbon-carbon scission, and (5) CO_2 formation.

the volumetric thermal expansion coefficient can be used:

$$\gamma = 3\alpha_t \tag{2.24}$$

Thermogravimetry (TGA). A thermogravimetric analyzer can measure weight changes of less than 10 μg as a function of temperature and time. This measurement technique, typically used for thermal stability, works on the principle of a beam balance. The testing chamber can be heated (up to approximately 1,200^{o}C) and rinsed with gases (inert or reactive). Measurements are performed on isothermal reactions or at temperatures sweeps of less than 100 K/min. The maximum sample weight used in thermogravimetric analyses is 500 mg. Thermogravimetry is often used to identify the components in a blend or a compound based on the thermal stability of each component. Figure 2.24 shows results from a TGA analysis on a PVC fabric. The figure shows the transitions at which the various components of the compound decompose. The percent of the original sample weight is recorded along with the change of the weight with respect to temperature. Five transitions representing (1) the decomposition of volatile components, (2) decomposition of the DOP plasticizer, (3) formation of HCl, (4) carbon-carbon scission, and (5) the forming of CO_2, are clearly visible.

Density Measurements. One simple form of calculating the density of a polymer sample is to first weigh the sample immersed in water. Assuming the density of water to be 1.0g/cm^3, we can use the relation

$$\rho = \frac{m}{(m - m_i)(1\ \mathrm{cm}^3/\mathrm{g})} \tag{2.25}$$

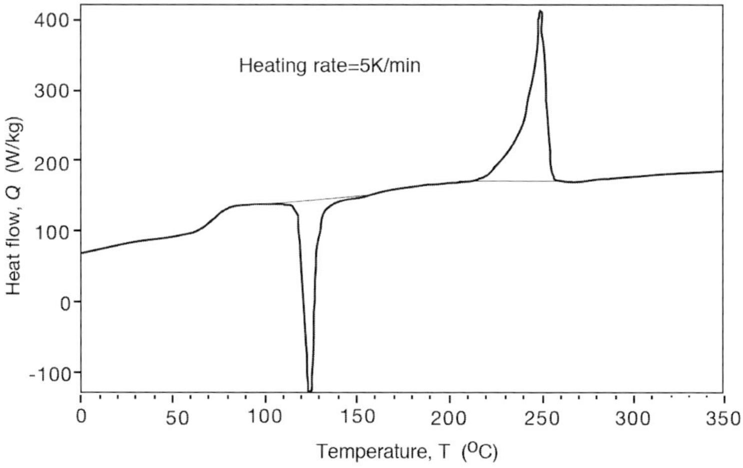

Figure 2.25: DSC scan of a PET bottle screw-top sample.

where m is mass of the specimen, m_i is the mass of the immersed specimen and $(m - m_i)$ is the mass of the displaced body of water.

Some common ways of determining density of polymeric materials are described by ASTM D792, ISO 1183, and DIN 53 479 test methods. Another common way of measuring density is the *through flow density meter.* Here, the density of water is changed to that of the polymer by adding ethanol until the plastic shavings are suspended in the solution. The density of the solution is then measured in a device that pumps the liquid through a U-pipe, where it is measured using ultrasound techniques. A density gradient technique is described by the standard ASTM D1505 test method.

■ **EXAMPLE 2.1.**

Degree of crystallinity for a PET bottle screw-top. A differential scanning calorimetry (DSC) test was performed on a 25 mg polyethylene terephthalate (PET) sample taken from the screw-top of a soda bottle. The test was performed using a heating rate of 5 K/min (5 K rise every minute). The DSC output is presented in Fig. 2.25.

From the curve estimate the glass transition temperature, T_g, the melting temperature, T_m, the crystallization temperature, T_c and the heat of fusion, λ, for this specific PET sample during the temperature ramp-up. If the heat of fusion for a hypothetically 100% crystalline PET sample is 137 kJ/kg [64], what was the degree of crystallinity in the PET bottle screw-top?

From the curve presented in Fig. 2.25 we can deduce that the glass transition temperature is around 72°C, the crystallization temperature at 125°C and the melting temperature at 250°C. Note that in all three cases there is a range of temperatures at which the transition occurs. To compute the heat of fusion during the ramp-up, we need to find the area between $\dot{Q}$ and the base-line for the endothermic deviation around the melting point, between 210 and 260°C. To do this we, must first transform the temperature scale to time by dividing it by the heating rate as

$$t = \frac{T}{5\ \text{K/min}} \frac{60\ \text{s}}{1\ \text{min}} \tag{2.26}$$

Hence, 50°C becomes 600 s, 100°C becomes 1,200 s, etc. The integral equals 37.8 kJ/kg, which represents the heat of fusion of the sample during the temperature ramp-up. However, one must consider that this melting energy includes the extra crystallization that occurs between 108 and 155°C. The exothermic energy is computed by integrating the curve between those two temperatures in a transformed time scale. The integral equals 22.9 kJ/kg. We can also find the area under the curve by transforming the heat flow, $\dot{Q}$, to heat capacity, C_p, and integrating using the temperature scale instead of a time scale. Heat capacity can be computed using

$$C_p = \frac{\dot{Q}}{5 \text{ K/min}} \frac{60 \text{ s}}{1 \text{ min}} \tag{2.27}$$

The degree of crystallinity of the initial PET bottle screw-top can now easily be computed using

$$\mathcal{X} = \frac{37.87 \text{ kJ/kg} - 22.9 \text{ kJ/kg}}{137 \text{ kJ/kg}} = 0.109 \quad \text{or} \quad 10.9\% \tag{2.28}$$

2.2 CURING PROPERTIES

Both, thermosets and elastomeric materials undergo a reaction process during processing. They can be classified in two general processing categories: heat activated cure and mixing activated cure thermosets. However, no matter which category a reactive polymer belongs to, its curing reaction can be described by the reaction between two chemical groups denoted by A and B which link two segments of a polymer chain. The reaction can be followed by tracing the concentration of unreacted As or Bs, C_A or C_B. If the initial concentration of A and B is defined as C_{A_0} and C_{B_0}, the degree of cure can be described by

$$c = \frac{C_{A_0} - C_{B_0}}{C_{A_0}} \tag{2.29}$$

The degree of cure or conversion, c, equals zero when there has been no reaction and equals one when all As have reacted and the reaction is complete. However, it is impossible to monitor reacted and unreacted As and Bs during the curing reaction of a thermoset polymer. It is known that the exothermic heat released during curing can be used to monitor the conversion, c. When a small sample of an unreacted thermoset polymer is placed in a differential scanning calorimeter (DSC) the sample will release a certain amount of heat, Q_T. This occurs because every cross-link that forms during a reaction releases a small amount of energy in the form of heat. For example, Fig. 2.26 shows the heat rate released during isothermal cure of a vinyl ester at various temperatures.

Using the exothermic heat as a measure of cure, the degree of cure can be defined by the following relation

$$c = \frac{Q}{Q_T} \tag{2.30}$$

where Q is the heat released up to an arbitrary time t, and is defined by

$$Q = \int_0^t \dot{Q} dt \tag{2.31}$$

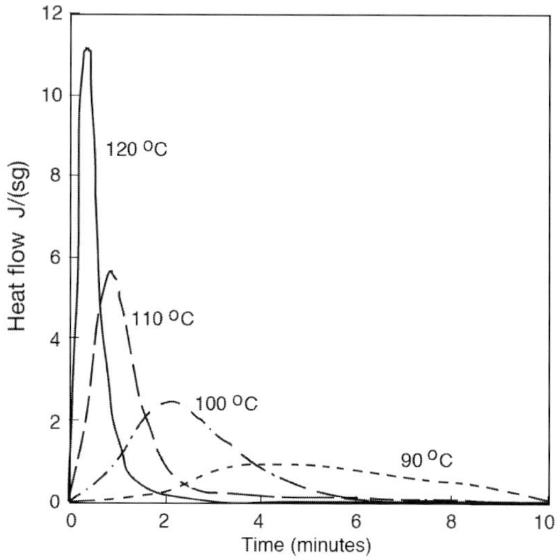

Figure 2.26: DSC scans of the isothermal curing reaction of vinyl ester at various temperatures.

DSC data is commonly fitted to semi-empirical models that accurately describe the curing reaction. Hence, the rate of cure can be described by the exotherm, $\dot{Q}$, and the total heat released during the curing reaction, Q_T, as

$$\frac{dc}{dt} = \frac{\dot{Q}}{Q_T} \tag{2.32}$$

With the use of eqn. (2.32), it is now easy to take the DSC data and find the models that best describe the curing reaction.

During cure, thermoset resins exhibit three distinct phases; viscous liquid, gel, and solid. Each of these three stages is marked by dramatic changes in the thermomechanical properties of the resin. The transformation of a reactive thermosetting liquid to a glassy solid generally involves two distinct macroscopic transitions: molecular gelation and vitrification. Molecular gelation is defined as the time or temperature at which covalent bonds connect across the resin to form a three-dimensional network which gives rise to long range elastic behavior in the macroscopic fluid. This point is also referred to as the gel point, where $c = c_g$. As a thermosetting resin cures, the cross-linking begins to hinder molecular movement, leading to a rise in the glass transition temperature. Eventually, when T_g nears the processing temperature, the rate of curing reduces significantly, and becomes dominated by diffusion. At this point, the resin has reached its vitrification point. Figure 2.27, which presents the degree of cure as a function of time, illustrates how an epoxy resin reaches a maximum degree of cure at various processing temperatures.

The resin processed at 200°C reaches 100% cure because the glass transition temperature of fully cured epoxy is 190°C, less than the processing temperature. On the other hand, the sample processed at 180°C reaches 97% cure and the one processed at 160°C only reaches 87% cure. Figures 2.26 and 2.27 also illustrate how the curing reaction is accelerated as the processing temperature is increased. The curing reaction of thermally cured thermoset

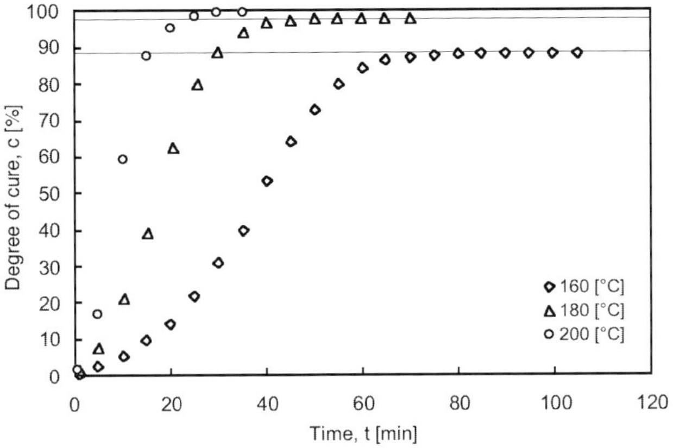

Figure 2.27: Degree of cure as a function time for an epoxy resin measured using isothermal DSC.

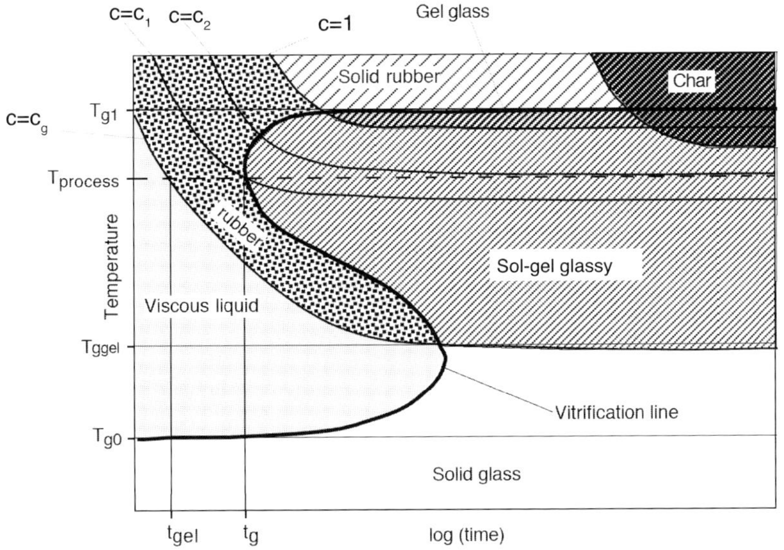

Figure 2.28: Time-temperature-transformation (TTT) diagram for a thermoset.

resins is not immediate, thus the blend can be stored in a refrigerator for a short period of time without having any significant curing reaction.

The behavior of curing thermosetting resins can be represented with the generalized time-temperature-transformation (TTT) cure diagram developed by Enns and Gillham[7]; it can be used to relate the material properties of thermosets as a function of time and the processing temperature as shown in Fig. 2.28.

The diagram presents various lines that represent constant degrees of cure. The curve labeled $c = c_g$ represents the gel point and $c = 1$ the fully cured resin. Both curves have their corresponding glass transition temperatures, T_{g_1} and $T_{g_{gel}}$, for the glass transition temperature of the fully cured resin and at its gel point, respectively. The glass transition temperature of the uncured resin, T_{g_0}, and an S-shaped curve labeled *vitrification line*, are also depicted. The vitrification line represents the boundary where the glass transition temperature becomes the processing temperature. Hence, to the left of the vitrification curve, the curing process is controlled by a very slow diffusion process. The TTT-diagram shows an arbitrary process temperature. The material being processed reaches the gel point at $t = t_{gel}$ and the vitrification line at $t = t_g$. At this point, the material has reached a degree of cure of c_1 and glass transition temperature of the resin is equal to the processing temperature. The material continues to cure very slowly (diffusion controlled) until it reaches a degree of cure just below c_2. There are also various regions labeled in the diagram. The one labeled *viscous liquid* is the one where the resin is found from the beginning of processing until the gel point has been reached. The flow and deformation that occurs during processing or shaping must occur within this region. The region labeled *char* must be avoided during processing, since at high processing temperatures the polymer will eventually undergo thermal degradation.

Many models exist that are used to predict the curing reaction during processing. All of them are of the form,

$$\frac{dc}{dt} = g(T, c) \tag{2.33}$$

The model that best represents the curing kinetics of thermosetting resins such as epoxy and unsaturated polyester, and as reflected in a TTT-diagram, is a diffusion modified Kamal-Sourour reaction model [9, 14, 13]. To model autocatalytic cure kinetics, the model can be applied as

$$\frac{dc}{dt} = (k_1 + k_2 c^m)(1 - c)^n \tag{2.34}$$

where m and n are reaction orders, and k_1 and k_2 are constants defined by

$$\frac{1}{k_i} = \frac{1}{k_i^c} + \frac{1}{k_D} \tag{2.35}$$

Here, k_i^c are Arrhenius overall rate constants defined by

$$k_i^c = a_i e^{-E_i/RT} \tag{2.36}$$

where a_1 and a_2 are fitting parameters, E_1 and E_2, activation energies and R the ideal gas constant. The constant k_D in eqn. (2.35) is the diffusion rate constant defined as

$$k_D = a_D e^{-E_D/RT} e^{-b/f} \tag{2.37}$$

where a_D and b are adjustable parameters, E_D is the activation energy of the diffusion process, and f is the equilibrium fractional free volume given by

$$f = 0.00048(T - T_g) + 0.025 \tag{2.38}$$

where T_g is the instantaneous glass transition temperature during cure. Equation (2.35) shows that the overall rate constant is governed at one extreme by the Arrhenius rate constant when $k_D \gg k_i^c$, which is the case prior to vitrification, and at the other extreme by the diffusion rate constant when $k_D \ll k_i^c$, which is the case well after vitrification. For a system exhibiting a unique one-to-one relationship between the glass transition temperature and conversion, DiBenedetto's equation [5] is one of the easiest approaches for stoichiometric ratios to express this relationship using only a single parameter as

$$T_g = T_{g0} + \frac{(T_{g1} - T_{g0})\lambda c}{1 - (1 - \lambda)c} \tag{2.39}$$

where λ is a structure dependent parameter theoretically equated to

$$\lambda = \frac{\Delta C_{p0}}{\Delta C_{p1}} \tag{2.40}$$

The values of ΔC_{p_o} and ΔC_{p_1} are the differences in the heat capacity between the glassy and rubbery state for the uncured resin and the fully cured network, respectively. However, the parameter λ can also be used as a fitting parameter.

Mixing activated cure materials such as polyurethanes will instantly start releasing exothermic heat after the mixture of its two components has occurred. The Castro-Macosko curing model, a second order reaction kinetic, accurately fits this behavior and is written as [44],

$$\frac{dc}{dt} = k_0 e^{-E/RT}(1 - c)^2 \tag{2.41}$$

In the above equations, the exponent of the $(1-c)^n$ term determines, which order reaction is being used to best fit the reaction kinetics. How these properties that describe the conversion or reaction are used, will be covered in more detail in Chapter 6 of this book.

2.3 RHEOLOGICAL PROPERTIES

Rheology is the field of science that studies fluid behavior during flow-induced deformation. From the variety of materials that rheologists study, polymers have been found to be the most interesting and complex. Polymer melts are shear thinning, viscoelastic, and their flow properties are temperature dependent. Viscosity is the most widely used material parameter when determining the behavior of polymers during processing. Since the majority of polymer processes are shear rate dominated, the viscosity of the melt is commonly measured using shear deformation measurement devices. However, there are polymer processes, such as blow molding, thermoforming, and fiber spinning, which are dominated by either elongational deformation or by a combination of shear and elongational deformation. In addition, some polymer melts exhibit significant elastic effects during deformation. Modeling and simulation of polymer flows will be briefly discussed in this chapter and are covered in more detail later in this book. For further reading on rheology of polymer melts, the reader should consult the literature [6, 8, 16, 45].

2.3.1 Flow Phenomena

There are three important phenomena seen is polymeric liquids that make them different from simple fluids: a non-Newtonian viscosity, normal stresses in shear flow, and elastic effects. All these effect are a result of the complex molecular structure of polymer macromolecules.

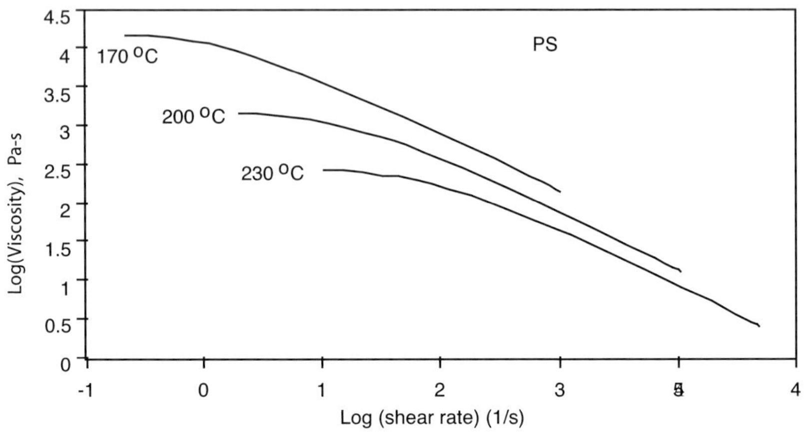

Figure 2.29: Viscosity curves for a polystyrene.

Non-Newtonian Viscosity. A Newtonian fluid is one where the deviatoric stresses that occur during deformation, $\boldsymbol{\tau}$, are directly proportional to the rate of deformation tensor, $\dot{\boldsymbol{\gamma}}$,

$$\boldsymbol{\tau} = \mu \dot{\boldsymbol{\gamma}} \tag{2.42}$$

The proportionality constant μ is the Newtonian viscosity, which is considered to be only dependent on temperature. However, the viscosity of most polymer melts is *shear thinning* in addition to being temperature dependent. The shear thinning effect is the reduction in viscosity at high rates of deformation. This phenomenon occurs because at high rates of deformation the molecules are stretched out and disentangled, enabling them to slide past each other with more ease, hence, lowering the bulk viscosity of the melt. Figure 2.29 shows the shear thinning behavior and temperature dependance of the viscosity of a specific general purpose polystyrene.

To take into consideration these non-Newtonian effects, it is common to use a viscosity which is a function of the strain rate and temperature to calculate the stress tensor in eqn. (2.42)

$$\boldsymbol{\tau} = \eta(\dot{\gamma}, T)\dot{\boldsymbol{\gamma}} \tag{2.43}$$

where η is the viscosity and $\dot{\gamma}$ the magnitude of strain rate or rate of deformation tensor defined by

$$\dot{\gamma} = \sqrt{\frac{1}{2}II} \tag{2.44}$$

where II is the second invariant of the strain rate tensor defined by

$$II = \sum_i \sum_j \dot{\gamma}_{ij}\dot{\gamma}_{ji} \tag{2.45}$$

The strain rate tensor components in eqn. (2.45) are defined by

$$\dot{\gamma}_{ij} = \frac{\partial u_i}{\partial x_j} + \frac{\partial u_j}{\partial x_i} \tag{2.46}$$

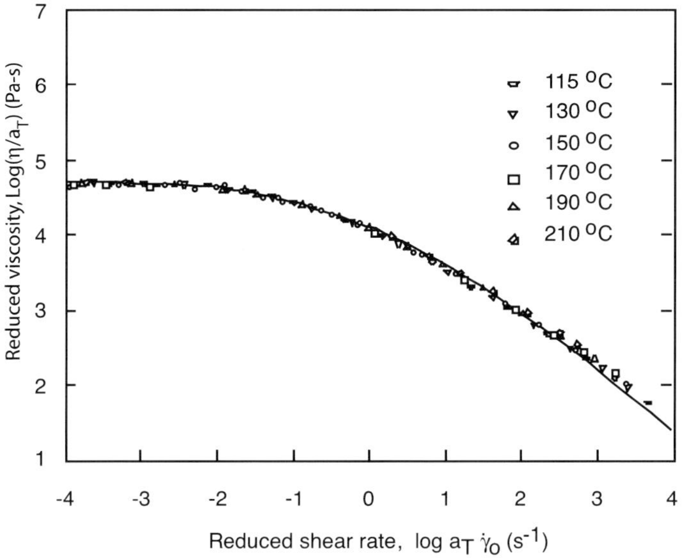

Figure 2.30: Reduced viscosity curve for a low density polyethylene at a reference temperature of 150°C.

The temperature dependence of the polymer's viscosity is normally factored out as

$$\eta(\dot{\gamma}, T) = f(T)\eta(\dot{\gamma}) \tag{2.47}$$

where for small variations in temperature, $f(T)$ can be approximated using an exponential function such as

$$f(T) = e^{-a(T-T_0)} \tag{2.48}$$

However, as mentioned in Chapter 1, a variation in temperature corresponds to a shift in the time scale. A shift commonly used for semi-crystalline polymers is the Arrhenius shift, which is written as

$$a_T(T) = \frac{\eta_0(T)}{\eta_0(T_0)} = e^{\frac{E_0}{R}\left(\frac{1}{T}-\frac{1}{T_0}\right)} \tag{2.49}$$

where E_0 is the activation energy, T_0 a reference temperature, and R the gas constant. Using this shift, one can translate viscosity curves measured at different temperatures to generate a master curve at a specific temperature. Figure 2.30 [41] presents the viscosity of a low density polyethylene with measured values shifted to a reference temperature of 150°C. For the shift in Fig. 2.30, an activation energy $E_0 = 54$ kJ/mol was used.

Several models that are used to represent the strain rate dependence of polymer melts are presented later in this chapter.

Normal Stresses in Shear Flow. The tendency of polymer molecules to *curl-up* while they are being stretched in shear flow results in normal stresses in the fluid. For example, shear flows exhibit a deviatoric stress defined by

$$\tau_{xy} = \eta(\dot{\gamma})\dot{\gamma}_{xy} \tag{2.50}$$

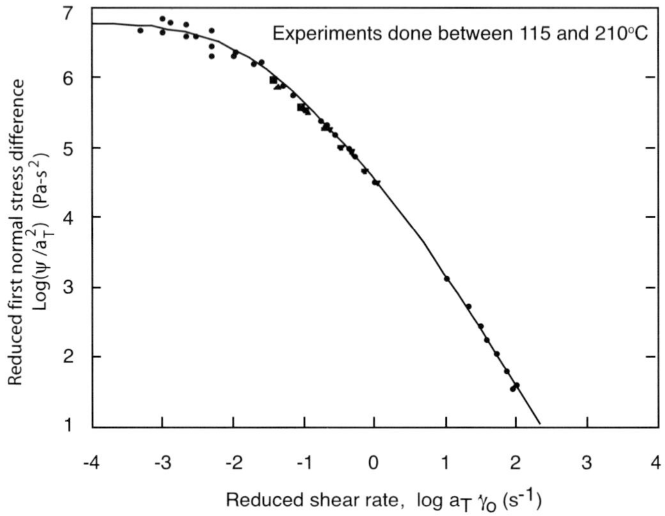

Figure 2.31: Reduced first normal stress difference coefficient for a low density polyethylene melt at a reference temperature of 150°C.

Measurable normal stress differences, $N_1 = \tau_{xx} - \tau_{yy}$ and $N_2 = \tau_{yy} - \tau_{zz}$ are referred to as the first and second normal stress differences. The first and second normal stress differences are material dependent and are defined by

$$N_1 = \tau_{xx} - \tau_{yy} = -\Psi_1 \dot{\gamma}_{xy}^2 \quad (2.51)$$

$$N_2 = \tau_{yy} - \tau_{zz} = -\Psi_2 \dot{\gamma}_{xy}^2 \quad (2.52)$$

The material functions, Ψ_1 and Ψ_2, are called the primary and secondary normal stress coefficients, and are also functions of the magnitude of the strain rate tensor and temperature. The first and second normal stress differences do not change in sign when the direction of the strain rate changes. This is reflected in eqns. (2.51) and (2.52). Figure 2.31 [41] presents the first normal stress difference coefficient for the low density polyethylene melt of Fig. 2.30 at a reference temperature of 150°C.

The second normal stress difference is difficult to measure and is often approximated by

$$\Psi_2 \approx -0.1\Psi_1 \quad (2.53)$$

Viscoelastic Memory Effects or Stress Relaxation. When a polymer melt is deformed, either by stretching, shearing or often by a combination of the above, the polymer molecules are stretched and untangled. In time, the molecules try to recover their initial shape, in essence getting used to their new state of deformation. If the deformation is maintained for a short period of time, the molecules may go back to their initial position, and the shape of the melt is fully restored to its initial shape. Here, it is said that the molecules *remembered* their initial position. However, if the shearing or stretching goes on for an extended period of time, the polymer cannot recover its starting shape, in essence *forgetting* the initial positions of the molecules. The time it takes for a molecule to fully relax and get used to its new state of deformation is referred to as the relaxation time, λ.

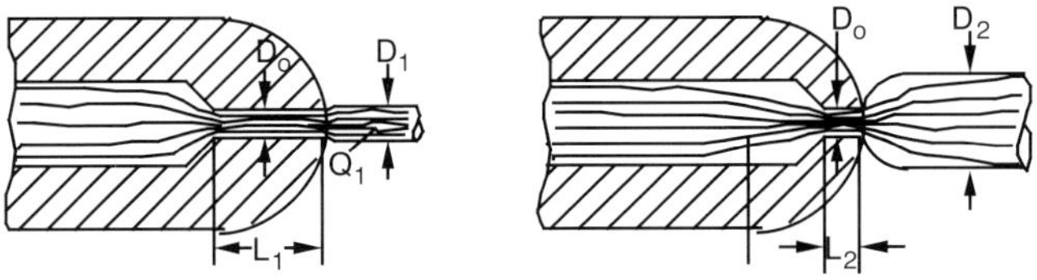

Figure 2.32: Schematic diagram of extrudate swell during extrusion.

A useful parameter often used to estimate the elastic effects during flow is the Deborah[2] number, De. The Deborah number is defined by

$$De = \frac{\lambda}{t_{\text{process}}} \tag{2.54}$$

where t_{process} is a characteristic process time. For example, in an extrusion die, a characteristic process time can be defined by the ratio of characteristic die dimension in the flow direction and average speed through the die. A Deborah number of zero represents a viscous fluid and a Deborah number of ∞ an elastic solid. As the Deborah number becomes > 1, the polymer does not have enough time to relax during the process, resulting in possible extrudate dimension deviations or irregularities such as extrudate swell, shark skin, or even melt fracture.

Although many factors affect the amount of extrudate swell, fluid *memory* and normal stress effects are the most significant ones. However, abrupt changes in boundary conditions, such as the separation point of the extrudate from the die, also play a role in the swelling or cross section reduction of the extrudate. In practice, the fluid memory contribution to die swell can be mitigated by lengthening the land length of the die. This is schematically depicted in Fig. 2.32 A long die land separates the polymer from the manifold for enough time to allow it to *forget* its past shape.

Waves in the extrudate may also appear as a result of high speeds during extrusion, where the polymer is not allowed to relax. This phenomenon is generally referred to as shark skin and is shown for a high density polyethylene in Fig. 2.33a [1]. It is possible to extrude at such high speeds that an intermittent separation of melt and inner die walls occurs as shown in Fig. 2.33b. This phenomenon is often referred to as the stick-slip effect or spurt flow and is attributed to high shear stresses between the polymer and the die wall. This phenomenon occurs when the shear stress is near the critical value of 0.1 MPa [30, 68, 67]. If the speed is further increased, a helical geometry is extruded as shown for a polypropylene extrudate in Fig. 2.33c. Eventually, the speeds are so high that a chaotic pattern develops, such as the one shown in Fig. 2.33d. This well known phenomenon is called melt fracture. The shark skin effect is frequently absent and spurt flow seems to occur only with linear polymers.

The critical shear stress has been reported to be independent of the melt temperature but to be inversely proportional to the weight average molecular weight [63, 68]. However, Vinogradov et al. [67] presented results that showed that the critical stress was independent of molecular weight except at low molecular weights. Dealy and co-workers [30], and Denn [17] give an extensive overview of various melt fracture phenomena which is recommended reading.

[2]From the Song of Deborah, Judges 5:5 - "The mountains flowed before the Lord." M. Rainer is credited for naming the Deborah number; Physics Today, 1, (1964).

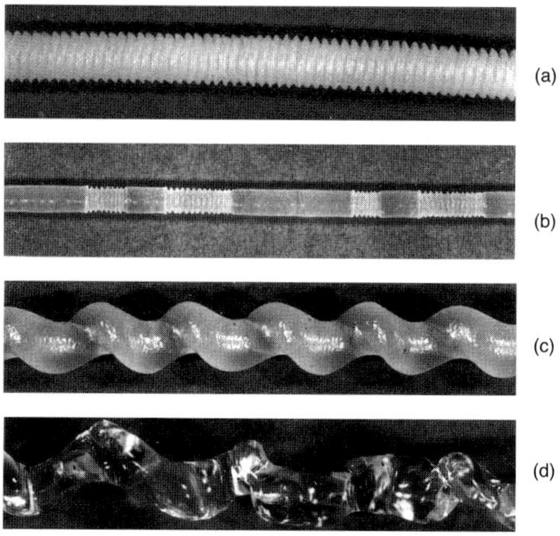

Figure 2.33: Various shapes of extrudates under melt fracture.

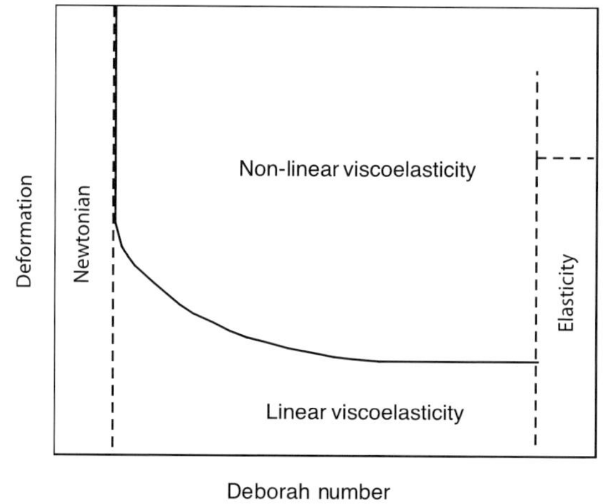

Figure 2.34: Schematic of Newtonian, elastic, linear, and non-linear viscoelastic regimes as a function of deformation and Deborah number during deformation of polymeric materials.

To summarize, the Deborah number and the size of the deformation imposed on the material during processing determine how the system can most accurately be modeled. Figure 2.34 [45] helps visualize the relation between time scale, deformation and applicable model. At small Deborah numbers, the polymer can be modeled as a Newtonian fluid, and at very high Deborah numbers the material can be modeled as a Hookean solid. In between, the viscoelastic region is divided in two: the linear viscoelastic region for small deformations, and the non-linear viscoelastic region for large deformations.

2.3.2 Viscous Flow Models

Strictly speaking, the viscosity η, measured with shear deformation viscometers, should not be used to represent the elongational terms located on the diagonal of the stress and strain rate tensors. Elongational flows are briefly discussed later in this chapter. A rheologist's

Table 2.5: Power Law and Consistency Indices for Common Thermoplastics

Polymer	m (Pa-s^n)	n	T (^{o}C)
High density polyethylene	2.0×10^4	0.41	180
Low density polyethylene	6.0×10^3	0.39	160
Polyamide 66	6.0×10^2	0.66	290
Polycarbonate	6.0×10^2	0.98	300
Polypropylene	7.5×10^3	0.38	200
Polystyrene	2.8×10^4	0.28	170
Polyvinyl chloride	1.7×10^4	0.26	180

task is to find the models that best fit the data for the viscosity represented in eqn. (2.43). Some of the models used by polymer processors on a day-to-day basis to represent the viscosity of industrial polymers are presented in this section.

The Power Law Model. The power law model proposed by Ostwald [57] and de Waale [15] is a simple model that accurately represents the shear thinning region in the viscosity versus strain rate curve but neglects the Newtonian plateau present at small strain rates. The power law model can be written as follows:

$$\eta = m(T)\dot{\gamma}^{n-1} \tag{2.55}$$

where m is referred to as the consistency index and n the power law index. The consistency index may include the temperature dependence of the viscosity such as represented in eqn. (2.48), and the power law index represents the shear thinning behavior of the polymer melt. It should be noted that the limits of this model are

$$\eta \longrightarrow 0 \text{ as } \dot{\gamma} \longrightarrow \infty$$

and

$$\eta \longrightarrow \infty \text{ as } \dot{\gamma} \longrightarrow 0$$

The infinite viscosity at zero strain rates leads to an erroneous result when there is a region of zero shear rate, such as at the center of a tube. This results in a predicted velocity distribution that is flatter at the center than the experimental profile, as will be explained in more detail in Chapter 5. In computer simulation of polymer flows, this problem is often overcome by using a truncated model such as

$$\eta = m_0(T)\dot{\gamma}^{n-1} \quad \text{for} \quad \dot{\gamma} > \dot{\gamma}_0 \tag{2.56}$$

and

$$\eta = m_0(T) \quad \text{for} \quad \dot{\gamma} \leqslant \dot{\gamma}_0 \tag{2.57}$$

where η_0 represents a zero-shear-rate viscosity ($\dot{\gamma_0}$). Table 2.5 presents a list of typical power law and consistency indices for common thermoplastics.

Table 2.6: Constants for Carreau-WLF (Amorphous) and Carreau-Arrhenius (Semi-Crystalline) Models for Various Common Thermoplastic

Polymer	k_1 Pa-s	k_2 s	k_3	k_4 ^{o}C	k_5 ^{o}C	T_0 ^{o}C	E_0 J/mol
High density polyethylene	24,198	1.38	0.60	-	-	200	22,272
Low density polyethylene	317	0.015	0.61	-	-	189	43,694
Polyamide 66	44	0.00059	0.40	-	-	300	123,058
Polycarbonate	305	0.00046	0.48	320	153	-	-
Polypropylene	1,386	0.091	0.68	-	-	220	427,198
Polystyrene	1,777	0.064	0.73	200	123	-	-
Polyvinyl chloride	1,786	0.054	0.73	185	88	-	-

The Bird-Carreau-Yasuda Model. A model that fits the whole range of strain rates was developed by Bird and Carreau [7] and Yasuda [72] and contains five parameters:

$$\frac{\eta - \eta_0}{\eta_0 - \eta_\infty} = [1 + |\lambda\dot{\gamma}|^a]^{(n-1)/a} \tag{2.58}$$

where η_0 is the zero-shear-rate viscosity, η_∞ is an infinite-shear-rate viscosity, λ is a time constant and n is the power law index. In the original Bird-Carreau model, the constant $a = 2$. In many cases, the infinite-shear-rate viscosity is negligible, reducing eqn. (2.58) to a three parameter model. Equation (2.58) was modified by Menges, Wortberg and Michaeli [50] to include a temperature dependence using a WLF relation. The modified model, which is used in commercial polymer data banks, is written as follows:

$$\eta = \frac{k_1 a_T}{[1 + k_2 \dot{\gamma} a_T]^{k_3}} \tag{2.59}$$

where the shift a_T applies well for amorphous thermoplastics and is written as

$$a_T = \frac{8.86(k_4 - k_5)}{101.6 + k_4 - k_5} - \frac{8.86(T - k_5)}{101.6 + T - k_5} \tag{2.60}$$

Table 2.6 presents constants for Carreau-WLF (amorphous) and Carreau-Arrhenius models (semi-crystalline) for various common thermoplastics. In addition to the temperature shift, Menges, Wortberg and Michaeli [50] measured a pressure dependence of the viscosity and proposed the following model, which includes both temperature and pressure viscosity shifts:

$$\log \eta(T, p) - \log \eta_0 = \frac{8.86(T - T_0)}{101.6 + T - T_0} - \frac{8.86(T - T_0 + 0.02p)}{101.6 + (T - T_0 + 0.02p)} \tag{2.61}$$

where p is in bar, and the constant 0.02 represents a 2°C shift per bar.

The Bingham Fluid. The Bingham fluid is an empirical model that represents the rheological behavior of materials that exhibit a *no flow* region below certain yield stresses, τ_Y, such as polymer emulsions and slurries. Since the material flows like a Newtonian liquid above the yield stress, the Bingham model can be represented by

$$\begin{aligned} \eta &= \infty \text{ and } \dot{\gamma} = 0 \text{ when } \tau \leqslant \tau_y \\ \eta &= \mu_0 + \frac{\tau_y}{\dot{\gamma}} \text{ when } \tau \geqslant \tau_y \end{aligned} \tag{2.62}$$

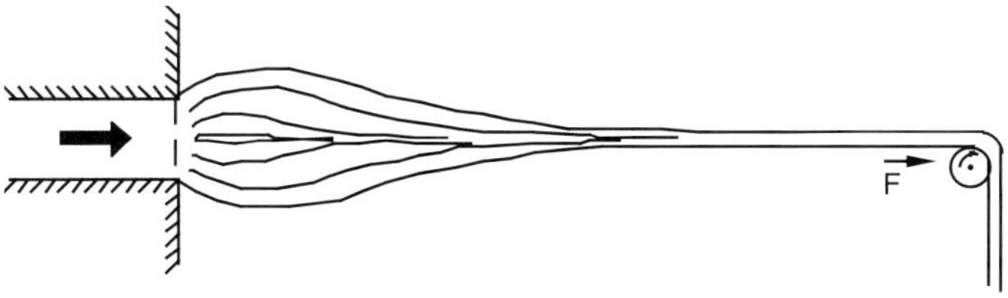

Figure 2.35: Schematic diagram of a fiber spinning process.

Here, τ is the magnitude of the deviatoric stress tensor and is computed in the same way as in eqn. (2.44).

Elongational Viscosity. In polymer processes such as fiber spinning, blow molding, thermoforming, foaming, certain extrusion die flows, and compression molding with specific processing conditions, the major mode of deformation is elongational. To illustrate elongational flows, consider the fiber spinning process shown in Fig. 2.35.

A simple elongational flow is developed as the filament is stretched with the following components of the rate of deformation:

$$\begin{aligned} \dot{\gamma}_{11} &= -\dot{\epsilon} \\ \dot{\gamma}_{22} &= -\dot{\epsilon} \\ \dot{\gamma}_{33} &= 2\dot{\epsilon} \end{aligned} \tag{2.63}$$

where $\dot{\epsilon}$ is the elongation rate, and the off-diagonal terms of $\dot{\gamma}_{ij}$ are all zero. The diagonal terms of the total stress tensor can be written as

$$\begin{aligned} \sigma_{11} &= -p - \eta\dot{\epsilon} \\ \sigma_{22} &= -p - \eta\dot{\epsilon} \\ \sigma_{33} &= -p + 2\eta\dot{\epsilon} \end{aligned} \tag{2.64}$$

Since the only outside forces acting on the fiber are in the axial or 3 direction, for the Newtonian case, σ_{11} and σ_{22} must be zero. Hence,

$$p = -\eta\dot{\epsilon} \tag{2.65}$$

and

$$\sigma_{33} = 3\eta\dot{\epsilon} = \bar{\eta}\dot{\epsilon} \tag{2.66}$$

which is known as elongational viscosity or Trouton viscosity [65]. This is analogous to elasticity where the following relation between elastic modulus, E, and shear modulus, G, can be written

$$\frac{E}{G} = 2(1 + \nu) \tag{2.67}$$

where ν is Poisson's ratio. For the incompressibility case, where $\nu = 0.5$, eqn. (2.67) reduces to

$$\frac{E}{G} = 3 \tag{2.68}$$

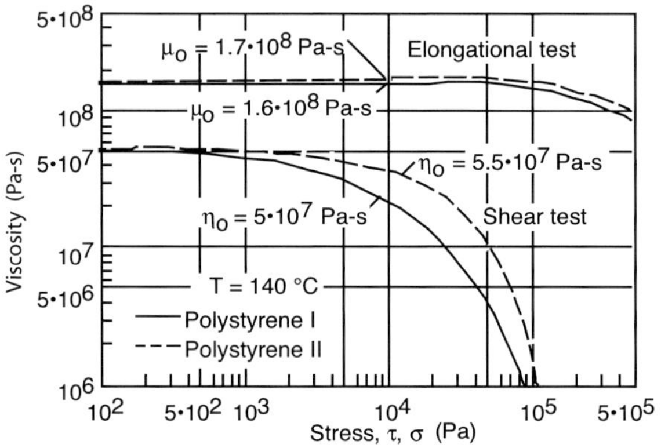

Figure 2.36: Shear and elongational viscosity curves for two types of polystyrene.

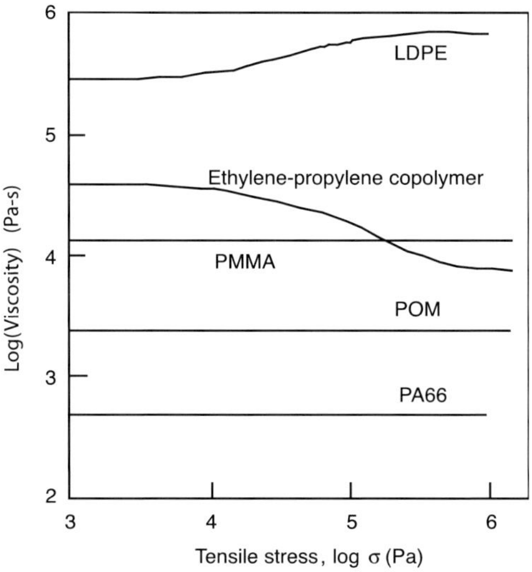

Figure 2.37: Elongational viscosity curves as a function of tensile stress for several thermoplastics.

Figure 2.36 [53] shows shear and elongational viscosities for two types of polystyrene. In the region of the Newtonian plateau, the limit of 3, shown in eqn. (2.66), is quite clear.

Figure 2.37 presents plots of elongational viscosities as a function of stress for various thermoplastics at common processing conditions. It should be emphasized that measuring elongational or extensional viscosity is an extremely difficult task. For example, in order to maintain a constant strain rate, the specimen must be deformed uniformly exponentially. In addition, a molten polymer must be tested completely submerged in a heated neutrally buoyant liquid at constant temperature.

Rheology of Curing Thermosets. A curing thermoset polymer has a conversion or cure dependent viscosity that increases as the molecular weight of the reacting polymer increases. For vinyl ester whose curing history is shown in Fig. 2.38 [29], the viscosity behaves as shown in Fig. 2.39 [29].

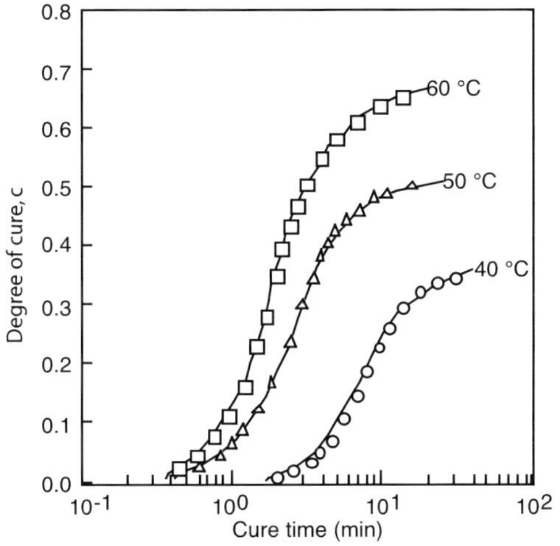

Figure 2.38: Degree of cure as a function of time for a vinyl ester at various isothermal cure temperatures.

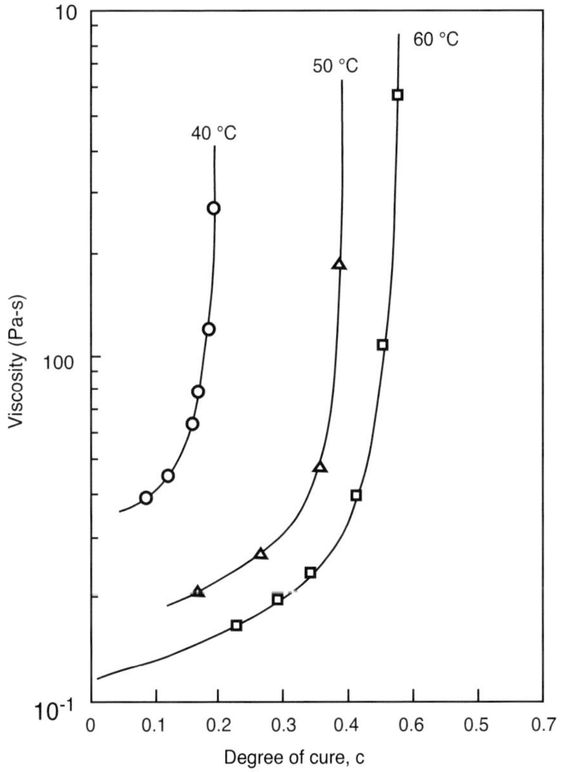

Figure 2.39: Viscosity as a function of degree of cure for a vinyl ester at various isothermal cure temperatures.

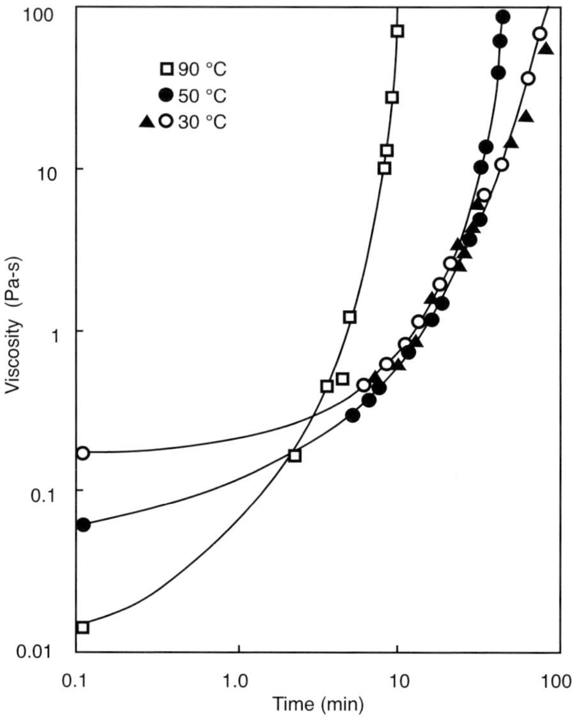

Figure 2.40: Viscosity as a function of time for a 47% MDI-BDO P(PO-EO) polyurethane at various isothermal cure temperatures.

Hence, a complete model for viscosity of a reacting polymer must contain the effects of strain rate, $\dot{\gamma}$, temperature, T, and degree of cure, c, such as

$$\eta = \eta(\dot{\gamma}, T, c) \tag{2.69}$$

There are no generalized models that include all these variables for thermosetting polymers. However, extensive work has been done on the viscosity of polyurethanes [9, 10] used in the reaction injection molding process. An empirical relation which models the viscosity of these mixing-activated polymers, given as a function of temperature and degree of cure, is written as

$$\eta = \eta_0 e^{E/RT} \left(\frac{c_g}{c_g - c} \right)^{c_1 + c_2 c} \tag{2.70}$$

where E is the activation energy of the polymer, R is the ideal gas constant, T is the temperature, c_g is the gel point[3], c the degree of cure, and c_1 and c_2 are constants that fit the experimental data. Figure 2.40 shows the viscosity as a function of time and temperature for a 47% MDI-BDO P(PO-EO) polyurethane.

Suspension Rheology. Particles suspended in a material, such as in filled or reinforced polymers, have a direct effect on the properties of the final article and on the viscosity during

[3] At the gel point, the cross-linking forms a closed network, at which point it is said that the molecular weight goes to infinity.

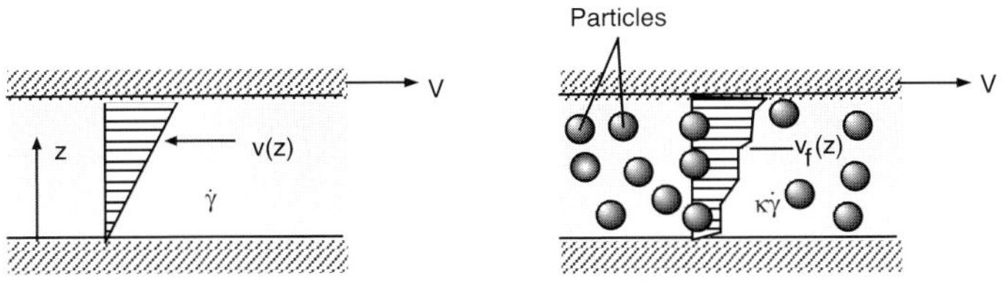

Figure 2.41: Schematic diagram of strain rate increase in a filled system.

processing. Numerous models have been proposed to estimate the viscosity of filled liquids [3, 15, 23, 25, 26]. Most models proposed are a power series of the form

$$\frac{\eta_f}{\eta} = 1 + a_1\phi + a_2\phi^2 + a_3\phi^3 + ... \tag{2.71}$$

The linear term in eqn. (2.71) represents the narrowing of the flow passage caused by the filler that is passively entrained by the fluid and sustains no deformation as shown in Fig. 2.41.

For instance, Einstein's model, which only includes the linear term with $a_1 = 2.5$, was derived based on a viscous dissipation balance. The quadratic term in the equation represents the first-order effects of interaction between the filler particles. Geisbüsch suggested a model with a yield stress and, where the strain rate of the melt increases by a factor κ as

$$\eta_f = \frac{\tau_0}{\dot{\gamma}} + \kappa\eta_0(\kappa\dot{\gamma}) \tag{2.72}$$

For high deformation stresses, which are typical in polymer processing, the yield stress in the filled polymer melt can be neglected. Figure 2.42 compares Geisbüsch's experimental data to eqn. (2.71) using the coefficients derived by Guth [25]. The data and Guth's model seem to agree well. A comprehensive survey on particulate suspensions was recently given by Gupta [25], and on short-fiber suspensions by Milliken and Powell [52].

2.3.3 Viscoelastic Constitutive Models

Viscoelasticity has already been introduced in Chapter 1, based on linear viscoelasticity. However, in polymer processing large deformations are imposed on the material, requiring the use of non-linear viscoelastic models. There are two types of general non-linear viscoelastic flow models: the differential type and the integral type.

Differential Viscoelastic Models. Differential models have traditionally been the choice for describing the viscoelastic behavior of polymers when simulating complex flow systems. Many differential viscoelastic models can be described by the general form

$$Y\boldsymbol{\tau} + \lambda_1\boldsymbol{\tau}_{(1)} + \lambda_2\{\dot{\boldsymbol{\gamma}}\cdot\boldsymbol{\tau} + \boldsymbol{\tau}\cdot\dot{\boldsymbol{\gamma}}\} + \lambda_3\{\boldsymbol{\tau}\cdot\boldsymbol{\tau}\} = \eta_0\left[\dot{\boldsymbol{\gamma}} + \lambda_4\boldsymbol{\gamma}_{(2)}\right] \tag{2.73}$$

where $\boldsymbol{\tau}_{(1)}$ is the first contravariant convected time derivative of the deviatoric stress tensor and represents rates of change with respect to a convected coordinate system that moves and deforms with the fluid. The convected derivative of the deviatoric stress tensor is defined as

$$\boldsymbol{\tau}_{(1)} = \frac{D\boldsymbol{\tau}}{Dt} - \left[(\nabla\mathbf{u})^T\cdot\boldsymbol{\tau} + \boldsymbol{\tau}\cdot(\nabla\mathbf{u})\right] \tag{2.74}$$

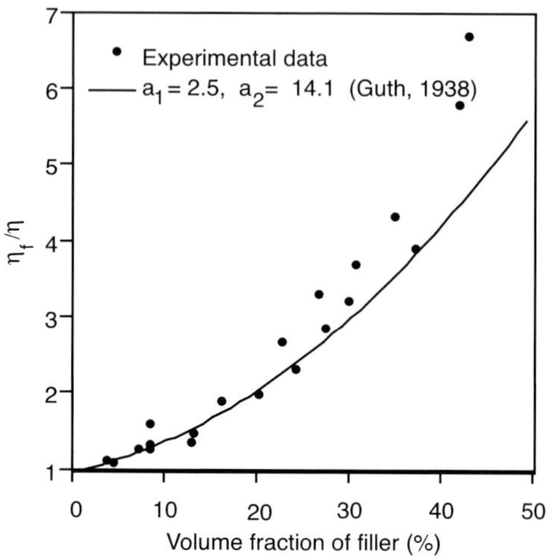

Figure 2.42: Viscosity increase as a function of volume fraction of filler for polystyrene and low density polyethylene containing spherical glass particles with diameters ranging between $36\mu m$ and $99.8\mu m$.

Table 2.7: Definition of Constants in eqn. (2.73)

Constitutive model	Y	λ_1	λ_2	λ_3	λ_4
Generalized Newtonian	1	0	0	0	0
Upper convected Maxwell	1	λ_1	0	0	0
Convected Jeffreys	1	λ_1	0	0	λ_4
White-Metzner	1	$\lambda_1(\dot{\gamma})$	0	0	0
Phan-Thien Tanner-1	$e^{(-\epsilon(\lambda/\eta_0)\mathrm{tr}\boldsymbol{\tau})}$	λ	$\frac{1}{2}\xi\lambda$	0	0
Phan-Thien Tanner-2	$1-\epsilon\left(\lambda/\eta_0\right)\mathrm{tr}\boldsymbol{\tau}$	λ	$\frac{1}{2}\xi\lambda$	0	0
Giesekus	1	λ_1	0	$-\left(\alpha\lambda_1/\eta_0\right)$	0

The constants in eqn. (2.73) are defined in Table 2.7 for various viscoelastic models commonly used to simulate polymer flows.

A review by Bird and Wiest [6] gives a more complete list of existing viscoelastic models. The upper convective model and the White-Metzner model are very similar with the exception that the White-Metzner model incorporates the strain rate effects of the relaxation time and the viscosity. Both models provide a first order approximation to flows, in which shear rate dependence and memory effects are important. However, both models predict zero second normal stress coefficients. The Giesekus model is molecular-based, non-linear in nature and describes the power law region for viscosity and both normal stress coefficients. The Phan-Thien Tanner models are based on network theory and give non-linear stresses. Both the Giesekus and Phan-Thien Tanner models have been successfully used to model complex flows.

■ **EXAMPLE 2.2.**

Shearing flows of the convected Jeffreys model. The convected Jeffreys model [6] or Oldroyd's B-fluid [54] is given by,

$$\boldsymbol{\tau} + \lambda_1 \boldsymbol{\tau}_{(1)} = \eta_0 \left[\dot{\boldsymbol{\gamma}} + \lambda_2 \boldsymbol{\gamma}_{(2)} \right] \tag{2.75}$$

Here we have three parameters: η_0 the zero-shear-rate viscosity, λ_1 the relaxation time and λ_2 the retardation time. In the case of $\lambda_2 = 0$ the model reduces to the convected Maxwell model, for $\lambda_1 = 0$ the model simplifies to a second-order fluid with a vanishing second normal stress coefficient [6], and for $\lambda_1 = \lambda_2$ the model reduces to a Newtonian fluid with viscosity η_0. If we impose a shear flow,

$$\frac{\partial u_x}{\partial y} = \dot{\gamma}_{yx}(t) \tag{2.76}$$

the constitutive equation (eqn. (2.75)) will be in tensor form (Table 2.8),

$$\begin{bmatrix} \tau_{xx} & \tau_{yx} & 0 \\ \tau_{yx} & \tau_{yy} & 0 \\ 0 & 0 & \tau_{zz} \end{bmatrix} + \lambda_1 \frac{d}{dt} \begin{bmatrix} \tau_{xx} & \tau_{yx} & 0 \\ \tau_{yx} & \tau_{yy} & 0 \\ 0 & 0 & \tau_{zz} \end{bmatrix} - \lambda_1 \dot{\gamma}_{yx} \begin{bmatrix} 2\tau_{yx} & \tau_{yy} & 0 \\ \tau_{yy} & 0 & 0 \\ 0 & 0 & 0 \end{bmatrix}$$
$$= \eta_0 \left\{ \dot{\gamma}_{yx} \begin{bmatrix} 0 & 1 & 0 \\ 1 & 0 & 0 \\ 0 & 0 & 0 \end{bmatrix} + \lambda_2 \frac{d\dot{\gamma}_{yx}}{dt} \begin{bmatrix} 0 & 1 & 0 \\ 1 & 0 & 0 \\ 0 & 0 & 0 \end{bmatrix} - 2\lambda_2 \dot{\gamma}_{yx}^2 \begin{bmatrix} 1 & 0 & 0 \\ 0 & 0 & 0 \\ 0 & 0 & 0 \end{bmatrix} \right\} \tag{2.77}$$

From this equations we can obtain the following set of partial differential equations,

$$\begin{aligned} \left(1 + \lambda_1 \frac{d}{dt}\right) \tau_{xx} - 2\tau_{yx}\lambda_1 \dot{\gamma}_{yx}(t) &= -2\eta_0 \lambda_2 \dot{\gamma}_{yx}^2(t) \\ \left(1 + \lambda_1 \frac{d}{dt}\right) \tau_{yy} &= 0 \\ \left(1 + \lambda_1 \frac{d}{dt}\right) \tau_{zz} &= 0 \\ \left(1 + \lambda_1 \frac{d}{dt}\right) \tau_{yx} - \tau_{yy}\lambda_1 \dot{\gamma}_{yx}(t) &= \eta_0 \left(1 + \lambda_2 \frac{d}{dt}\right) \dot{\gamma}_{yx}(t) \end{aligned} \tag{2.78}$$

which indicates that the normal stresses τ_{yy} and τ_{zz} are zero for any time-dependent shearing flow.

For *steady shear flow* these differential equations are simplified to give,

$$\begin{aligned}\tau_{yx} &= \eta_0\dot{\gamma}_{yx}\\ \tau_{xx}-\tau_{yy} &= 2\eta_0(\lambda_1-\lambda_2)\dot{\gamma}_{yx}^2\\ \tau_{yy}-\tau_{zz} &= 0\end{aligned} \tag{2.79}$$

and we obtain the following viscometric functions,

$$\begin{aligned}\eta &= \eta_0\\ \Psi_1 &= 2\eta_0(\lambda_1-\lambda_2)\\ \Psi_2 &= 0\end{aligned} \tag{2.80}$$

Indicating that the convected Jeffreys model gives a constant viscosity and first normal stress coefficient, while the second normal stress coefficient is zero.

For a *small amplitude oscillatory shearing flow*, the strain is defined as,

$$\gamma_{yx}(t) = \int_0^t \dot{\gamma}_0 \cos wt' dt' = \gamma_0 \sin wt \tag{2.81}$$

where $\gamma_0 = \dot{\gamma}_0/w$. The differential equation for the shear stress will be,

$$\left(1+\frac{d}{dt}\right)\tau_{yx} = \eta_0\gamma_0 w\left(\cos wt - \lambda_2 w \sin wt\right) \tag{2.82}$$

Seeking a steady periodic solution, the right hand side suggest that the solution should be [6],

$$\tau_{yx} = A\cos wt + B\sin wt \tag{2.83}$$

which, after replacing it into the original equation, we obtain,

$$\begin{aligned}A &= \eta_0\left(\frac{1+\lambda_1\lambda_2 w^2}{1+\lambda_1^2 w^2}\right)\gamma_0 w = \eta'(w)\gamma_0 w\\ B &= \eta_0\left(\frac{(\lambda_1-\lambda_2)\gamma_0 w^2}{1+\lambda_1^2 w^2}\right)\gamma_0 w = \eta''(w)\gamma_0 w\end{aligned} \tag{2.84}$$

■ EXAMPLE 2.3.

Steady shearfree flow for the White-Metzner model. This model is a nonlinear model which modifies the convected Maxwell model by including the dependence on $\dot{\gamma}$ in the viscosity, i.e.,

$$\boldsymbol{\tau} + \lambda_1(\dot{\gamma})\boldsymbol{\tau}_{(1)} = \eta(\dot{\gamma})\dot{\boldsymbol{\gamma}} \tag{2.85}$$

where $\dot{\gamma} = \sqrt{1/2\dot{\boldsymbol{\gamma}}:\dot{\boldsymbol{\gamma}}}$. For a shearfree flow we have that (Table 2.9),

$$\dot{\boldsymbol{\gamma}} = \dot{\epsilon}(t)\begin{bmatrix}-(1+b) & 0 & 0\\ 0 & (1-b) & 0\\ 0 & 0 & 2\end{bmatrix} \tag{2.86}$$

Table 2.8: Shearing Flow Tensors $\mathbf{u} = (\dot{\gamma}_{yx}(t)y, 0, 0)$ [6]

$\nabla\mathbf{u}$	$\dot{\gamma}_{yx}(t)\begin{bmatrix}0 & 0 & 0\\ 1 & 0 & 0\\ 0 & 0 & 0\end{bmatrix}$
$\dot{\boldsymbol{\gamma}} = \boldsymbol{\gamma}^{(1)} = \boldsymbol{\gamma}_{(1)}$	$\dot{\gamma}_{yx}(t)\begin{bmatrix}0 & 1 & 0\\ 1 & 0 & 0\\ 0 & 0 & 0\end{bmatrix}$
$\boldsymbol{\gamma}^{(2)}$	$\dfrac{\partial\dot{\gamma}_{yx}}{\partial t}\begin{bmatrix}0 & 1 & 0\\ 1 & 0 & 0\\ 0 & 0 & 0\end{bmatrix} + \dot{\gamma}_{yx}^2\begin{bmatrix}0 & 0 & 0\\ 0 & 2 & 0\\ 0 & 0 & 0\end{bmatrix}$
$\boldsymbol{\gamma}_{(2)}$	$\dfrac{\partial\dot{\gamma}_{yx}}{\partial t}\begin{bmatrix}0 & 1 & 0\\ 1 & 0 & 0\\ 0 & 0 & 0\end{bmatrix} - \dot{\gamma}_{yx}^2\begin{bmatrix}2 & 0 & 0\\ 0 & 0 & 0\\ 0 & 0 & 0\end{bmatrix}$
$\boldsymbol{\tau} = \boldsymbol{\tau}^{(0)} = \boldsymbol{\tau}_{(0)}$	$\begin{bmatrix}\tau_{xx} & \tau_{yx} & 0\\ \tau_{yx} & \tau_{yy} & 0\\ 0 & 0 & \tau_{zz}\end{bmatrix}$
$\boldsymbol{\tau}^{(1)}$	$\dfrac{\partial}{\partial t}\begin{bmatrix}\tau_{xx} & \tau_{yx} & 0\\ \tau_{yx} & \tau_{yy} & 0\\ 0 & 0 & \tau_{zz}\end{bmatrix} + \dot{\gamma}_{yx}\begin{bmatrix}0 & \tau_{xx} & 0\\ \tau_{xx} & 2\tau_{yx} & 0\\ 0 & 0 & 0\end{bmatrix}$
$\boldsymbol{\tau}_{(1)}$	$\dfrac{\partial}{\partial t}\begin{bmatrix}\tau_{xx} & \tau_{yx} & 0\\ \tau_{yx} & \tau_{yy} & 0\\ 0 & 0 & \tau_{zz}\end{bmatrix} - \dot{\gamma}_{yx}\begin{bmatrix}2\tau_{yx} & \tau_{yy} & 0\\ \tau_{yy} & 0 & 0\\ 0 & 0 & 0\end{bmatrix}$

for a steady flow $\dot{\epsilon}(t) = \dot{\epsilon}_0$ and

$$\dot{\gamma} \cdot \dot{\gamma} = \dot{\epsilon}_0^2 \begin{bmatrix} (1+b)^2 & 0 & 0 \\ 0 & (1-b)^2 & 0 \\ 0 & 0 & 4 \end{bmatrix} \tag{2.87}$$

and we have

$$\frac{1}{2}\dot{\gamma} : \dot{\gamma} = \frac{1}{2}\text{tr}\dot{\gamma} \cdot \dot{\gamma} = \sqrt{3+b^2}|\dot{\epsilon}_0| \tag{2.88}$$

Here $0 \leq b \leq 1$ and $\dot{\epsilon}$ is the elongation rate. Several special shearfree flows are obtained for particular choices of b, i.e.,

$b = 0$ and $\dot{\epsilon} > 0$ Elongational flow

$b = 0$ and $\dot{\epsilon} < 0$ Biaxial stretching flow

$b = 1$ Plannar elongational flow

The tensor form of the constitutive equation is,

$$\begin{aligned} \begin{bmatrix} \tau_{xx} & 0 & 0 \\ 0 & \tau_{yy} & 0 \\ 0 & 0 & \tau_{zz} \end{bmatrix} - \lambda_1(\dot{\gamma}) \begin{bmatrix} -(1+b)\tau_{xx} & 0 & 0 \\ 0 & -(1-b)\tau_{yy} & 0 \\ 0 & 0 & 2\tau_{zz} \end{bmatrix} \dot{\epsilon}_0 \\ = \eta(\dot{\gamma}) \begin{bmatrix} -(1+b) & 0 & 0 \\ 0 & -(1-b) & 0 \\ 0 & 0 & 2 \end{bmatrix} \dot{\epsilon}_0 \end{aligned} \tag{2.89}$$

which will give us the following differential equations,

$$\begin{aligned} \tau_{xx}\left[1 + (1+b)\left(\frac{\eta}{G}\dot{\epsilon}_0\right)\dot{\epsilon}_0\right] &= -(1+b)\eta\dot{\epsilon}_0 \\ \tau_{yy}\left[1 + (1-b)\left(\frac{\eta}{G}\dot{\epsilon}_0\right)\dot{\epsilon}_0\right] &= -(1-b)\eta\dot{\epsilon}_0 \\ \tau_{zz}\left[1 - 2\left(\frac{\eta}{G}\dot{\epsilon}_0\right)\dot{\epsilon}_0\right] &= 2\eta\dot{\epsilon}_0 \end{aligned} \tag{2.90}$$

From these equations we get the elongational viscosities as [6],

$$\begin{aligned} \bar{\eta}_1 = \tau_{zz} - \tau_{xx} &= \frac{(3+b)\eta(\dot{\gamma})\dot{\epsilon}_0}{[1 + (1+b)(\eta/G)\dot{\epsilon}_0]\,[1 - 2(\eta/G)\dot{\epsilon}_0]} \\ \bar{\eta}_2 = \tau_{yy} - \tau_{xx} &= \frac{2b\eta(\dot{\gamma})\dot{\epsilon}_0}{[1 + (1+b)(\eta/G)\dot{\epsilon}_0]\,[1 + (1-b)(\eta/G)\dot{\epsilon}_0]} \end{aligned} \tag{2.91}$$

Integral viscoelastic models. Integral models with a memory function have been widely used to describe the viscoelastic behavior of polymers and to interpret their rheological measurements [37, 41, 43]. In general one can write the single integral model as

$$\boldsymbol{\tau} = \int_{-\infty}^{t} M(t-t')\mathbf{S}(t')dt' \tag{2.92}$$

Table 2.9: Shearfree Flow Tensors $\mathbf{u} = (-1/2(1+b)\dot{\epsilon}(t)x, -1/2(1+b)\dot{\epsilon}(t)y, \dot{\epsilon}(t)z)$ [6]

$\nabla \mathbf{u}$	$\dot{\epsilon}(t)\begin{bmatrix} -1/2(1+b) & 0 & 0 \\ 0 & -1/2(1-b) & 0 \\ 0 & 0 & 1 \end{bmatrix}$
$\dot{\boldsymbol{\gamma}} = \boldsymbol{\gamma}^{(1)} = \boldsymbol{\gamma}_{(1)}$	$\dot{\epsilon}(t)\begin{bmatrix} -(1+b) & 0 & 0 \\ 0 & -(1-b) & 0 \\ 0 & 0 & 2 \end{bmatrix}$
$\boldsymbol{\gamma}^{(2)}$	$\frac{\partial \dot{\epsilon}}{\partial t}\begin{bmatrix} -(1+b) & 0 & 0 \\ 0 & -(1-b) & 0 \\ 0 & 0 & 2 \end{bmatrix} + \dot{\epsilon}^2\begin{bmatrix} (1+b)^2 & 0 & 0 \\ 0 & (1-b)^2 & 0 \\ 0 & 0 & 4 \end{bmatrix}$
$\boldsymbol{\gamma}_{(2)}$	$\frac{\partial \dot{\epsilon}}{\partial t}\begin{bmatrix} -(1+b) & 0 & 0 \\ 0 & -(1-b) & 0 \\ 0 & 0 & 2 \end{bmatrix} - \dot{\epsilon}^2\begin{bmatrix} (1+b)^2 & 0 & 0 \\ 0 & (1-b)^2 & 0 \\ 0 & 0 & 4 \end{bmatrix}$
$\boldsymbol{\tau} = \boldsymbol{\tau}^{(0)} = \boldsymbol{\tau}_{(0)}$	$\begin{bmatrix} \tau_{xx} & 0 & 0 \\ 0 & \tau_{yy} & 0 \\ 0 & 0 & \tau_{zz} \end{bmatrix}$
$\boldsymbol{\tau}^{(1)}$	$\frac{\partial}{\partial t}\begin{bmatrix} \tau_{xx} & 0 & 0 \\ 0 & \tau_{yy} & 0 \\ 0 & 0 & \tau_{zz} \end{bmatrix} + \dot{\epsilon}\begin{bmatrix} -(1+b)\tau_{xx} & 0 & 0 \\ 0 & -(1-b)\tau_{yy} & 0 \\ 0 & 0 & 2\tau_{zz} \end{bmatrix}$
$\boldsymbol{\tau}_{(1)}$	$\frac{\partial}{\partial t}\begin{bmatrix} \tau_{xx} & 0 & 0 \\ 0 & \tau_{yy} & 0 \\ 0 & 0 & \tau_{zz} \end{bmatrix} - \dot{\epsilon}\begin{bmatrix} -(1+b)\tau_{xx} & 0 & 0 \\ 0 & -(1-b)\tau_{yy} & 0 \\ 0 & 0 & 2\tau_{zz} \end{bmatrix}$

Table 2.10: Definition of Constants in eqn. (2.73)

Constitutive model	ϕ_1	ϕ_2
Lodge rubber-like liquid	1	0
K-BKZ[a]	$\frac{\partial W}{\partial I_1}$	$\frac{\partial W}{\partial I_2}$
Wagner[b]	$e^{\beta}\sqrt{\alpha I_1 + (1-\alpha) I_2 - 3}$	0
Papanastasiou-Scriven-Macosko[c]	$\frac{\alpha}{(\alpha - 3) + \beta I_1 + (1-\beta) I_2}$	0

[a] $W(I_1, I_2)$ represents a potential function which can be derived from empiricisms or molecular theory; [b] Wagner's model is a special form of the K-BKZ model; [c] The Papanastasiou-Scriven-Macosko model is also a special form of the K-BKZ model

where $M(t-t')$ is a memory function and $\mathbf{S}(t')$ a deformation dependent tensor defined by

$$\mathbf{S}(t') = \phi_1(I_1, I_2)\gamma_{[0]} + \phi_2(I_1, I_2)\gamma^{[0]} \tag{2.93}$$

where I_1 and I_2 are the first invariants of the Cauchy and Finger strain tensors, respectively.
Table 2.10 [4, 36, 42, 69] defines the constants ϕ_1 and ϕ_2 for various models. In eqn. (2.93), $\gamma_{[0]}$ and $\gamma^{[0]}$ are the finite strain tensors given by

$$\begin{aligned} \gamma_{[0]} &= \mathbf{\Delta}^t \cdot \mathbf{\Delta} - \boldsymbol{\delta} \\ \gamma^{[0]} &= \boldsymbol{\delta} - \mathbf{E} \cdot \mathbf{E}^t \end{aligned} \tag{2.94}$$

The terms Δ_{ij} and E_{ij} are displacement gradient tensors[4] defined by

$$\begin{aligned} \Delta_{ij} &= \frac{\partial x_i'(x, t, t')}{\partial x_j} \\ E_{ij} &= \frac{\partial x_i(x', t', t)}{\partial x_j'} \end{aligned} \tag{2.95}$$

where the components Δ_{ij} measure the displacement of a particle at past time t' relative to its position at present time t, and the terms E_{ij} measure the material displacements at time t relative to the positions at time t'.

A memory function $M(t-t')$, which is often applied and which leads to commonly used constitutive equations, is written as

$$M(t-t') = \sum_{k=1}^{n} \frac{\eta_k}{\lambda_k^2} e^{\left(-\frac{t-t'}{\lambda_k}\right)} \tag{2.96}$$

[4] Another combination of the displacement gradient tensors which are often used are the Cauchy strain tensor and the Finger strain tensor defined by $\mathbf{B}^{-1} = \mathbf{\Delta}^t \mathbf{\Delta}$ and $\mathbf{B} = \mathbf{E}\mathbf{E}^t$, respectively.

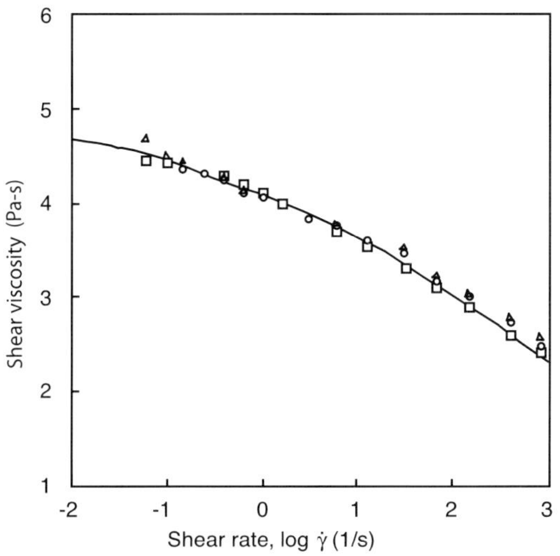

Figure 2.43: Measured and predicted shear viscosity for various high density poly-ethylene resins at 170°C.

where λ_k and η_k are relaxation times and viscosity coefficients at the reference temperature T_{ref}, respectively. Once a memory function has been specified one can calculate several material functions using [6]

$$\begin{aligned}\eta(\dot{\gamma}) &= \int_0^\infty M(s)s(\phi_1+\phi_2)ds \\ \psi_1(\dot{\gamma}) &= \int_0^\infty M(s)s^2(\phi_1+\phi_2)ds \\ \psi_2(\dot{\gamma}) &= \int_0^\infty M(s)s^2(\phi_2)ds\end{aligned} \tag{2.97}$$

For example, Figs. 2.43 and 2.44 present the measured [55] viscosity and first normal stress difference data, respectively, for three blow molding grade high density polyethylenes along with a fit obtained from the Papanastasiou-Scriven-Macosko [59] form of the K-BKZ equation. A memory function with a relaxation spectrum of 8 relaxation times was used.

The coefficients used to fit the data are summarized in Table 2.11 [43]. The viscosity and first normal stress coefficient data presented in Figs. 2.30 and 2.31 where fitted with the Wagner form of the K-BKZ equation [41].

■ **EXAMPLE 2.4.**

Shear flow for a Lodge rubber-liquid. If we consider the flow field,

$$\begin{aligned}u_x &= \dot{\gamma}_{yx}(t)y \\ u_y &= u_z = 0\end{aligned} \tag{2.98}$$

and we are seeking an expression of the stress tensor of a Lodge rubber-liquid, we start from the integral form of the stress tensor

$$\boldsymbol{\tau} = \int_{-\infty}^{t} M(t-t')\mathbf{S}(t')dt' \tag{2.99}$$

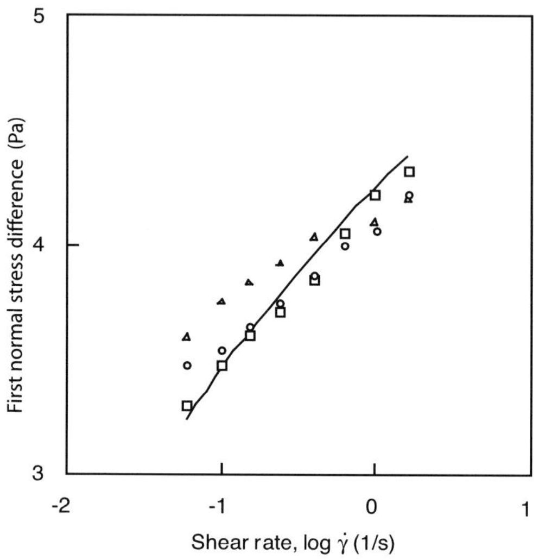

Figure 2.44: Measured and predicted first normal stress difference for various high density polyethylene resins at 170°C.

Table 2.11: Material Parameter Values in eqn. (2.96) for Fitting Data of High Density Polyethylene Melts at 170^oC

k	$\lambda_k(s)$	$\eta_k(Pa-s)$
1	0.0001	52
2	0.001	148
3	0.01	916
4	0.1	4,210
5	1	8,800
6	10	21,200
7	100	21,000
8	1,000	600

which for this type of materials reduces to

$$\boldsymbol{\tau} = \int_{-\infty}^{t} M(t-t')\boldsymbol{\gamma}_{[0]}dt' \tag{2.100}$$

For a simple shear flow

$$\boldsymbol{\gamma}_{[0]} = \begin{bmatrix} \gamma_{yx}^2 & \gamma_{yx} & 0 \\ \gamma_{yx} & 0 & 0 \\ 0 & 0 & 0 \end{bmatrix} \tag{2.101}$$

Thus, the components of the stress tensor are reduced to

$$\begin{aligned} \tau_{yx}(t) &= \int_{-\infty}^{t} M(t-t')\gamma(t,t')dt' \\ \tau_{xx}(t) - \tau_{yy}(t) &= \int_{-\infty}^{t} M(t-t')\gamma^2(t,t')dt' \\ \tau_{yy}(t) - \tau_{zz}(t) &= 0 \end{aligned} \tag{2.102}$$

where

$$\gamma_{yx}(t,t') = \int_{t}^{t'} \dot{\gamma}_{yx}(t'')dt'' \tag{2.103}$$

is the strain from t to t'.

For *steady shear flow* the strain is reduced to $\gamma_{yx}(t,t') = -\dot{\gamma}(t-t')$, where $\dot{\gamma}$ is the constant shear rate. This results in [6]

$$\begin{aligned} \tau_{yx}(t) &= \left(\int_0^s M(s)sds\right)\dot{\gamma} \\ \tau_{xx}(t) - \tau_{yy}(t) &= \left(\int_0^s M(s)s^2ds\right)\dot{\gamma}^2 \\ \tau_{yy}(t) - \tau_{zz}(t) &= 0 \end{aligned} \tag{2.104}$$

and the material functions will be,

$$\begin{aligned} \eta &= \sum_k^n \eta_k \\ \Psi_1 &= 2\sum_k^n \eta_k\lambda_k \\ \Psi_2 &= 0 \end{aligned} \tag{2.105}$$

2.3.4 Rheometry

In industry there are various ways to qualify and quantify the properties of the polymer melt. The techniques range from simple analyses for checking the consistency of the material at certain conditions, to more complex measurements to evaluate viscosity, and normal stress differences. This section includes three such techniques, to give the reader a general idea of current measuring techniques.

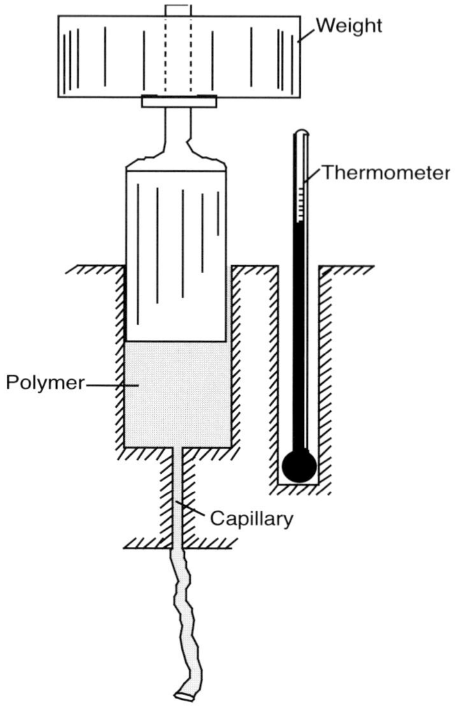

Figure 2.45: Schematic diagram of an extrusion plastometer used to measure the melt flow index.

The melt flow indexer. The melt flow indexer is often used in industry to characterize a polymer melt and as a simple and quick means of quality control. It takes a single point measurement using standard testing conditions specific to each polymer class on a ram type extruder or extrusion plastometer as shown in Fig. 2.45.

The standard procedure for testing the flow rate of thermoplastics using a extrusion plastometer is described in the ASTM D1238 test. During the test, a sample is heated in the barrel and extruded from a short cylindrical die using a piston actuated by a weight. The weight of the polymer in grams extruded during the 10-minute test is the melt flow index (MFI) of the polymer.

The capillary viscometer. The most common and simplest device for measuring viscosity is the capillary viscometer. Its main component is a straight tube or capillary, and it was first used to measure the viscosity of water by Hagen [28] and Poiseuille [60]. A capillary rheometer has a pressure driven flow for which the velocity gradient or strain rate and also the shear rate will be maximum at the wall and zero at the center of the flow, making it a non-homogeneous flow.

Since pressure driven viscometers employ non-homogeneous flows, they can only measure steady shear functions such as viscosity, $\eta(\dot{\gamma})$. However, they are widely used because they are relatively inexpensive to build and simple to operate. Despite their simplicity, long capillary viscometers give the most accurate viscosity data available. Another major advantage is that the capillary rheometer has no free surfaces in the test region, unlike other types of rheometers such as the cone and plate rheometers, which we will discuss in the next section. When the strain rate dependent viscosity of polymer melts is measured, capillary rheometers may provide the only satisfactory method of obtaining such data at shear rates

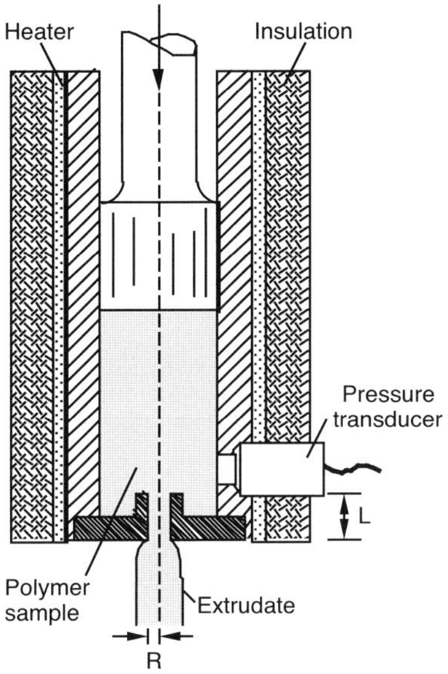

Figure 2.46: Schematic diagram of a capillary viscometer.

$>10\ s^{-1}$. This is important for processes with higher rates of deformation such as mixing, extrusion, and injection molding. Because its design is basic and it only needs a pressure head at its entrance, capillary rheometers can easily be attached to the end of a screw- or ram-type extruder for on-line measurements. This makes the capillary viscometer an efficient tool for industry.

The basic features of the capillary rheometer are shown in Fig. 2.46. A capillary tube of a specified radius, R, and length, L, is connected to the bottom of a reservoir. Pressure drop and flow rate through this tube are used to determine the viscosity. This will be covered in detail in Chapter 5.

The cone-plate rheometer. The cone-plate rheometer is often used when measuring the viscosity and the primary and secondary normal stress coefficient functions as a function of shear rate and temperature. The geometry of a cone-plate rheometer is shown in Fig. 2.47. Since the angle Θ_0 is very small, typically $< 5^o$, the shear rate can be considered constant throughout the material confined within the cone and plate. Although it is also possible to determine the secondary stress coefficient function from the normal stress distribution across the plate, it is very difficult to get accurate data.

The Couette rheometer. Another rheometer commonly used in industry is the concentric cylinder or Couette flow rheometer schematically depicted in Fig. 2.48. The torque, T, and rotational speed, Ω , can easily be measured. The torque is related to the shear stress that acts on the inner cylinder wall and the rate of deformation in that region is related to the rotational speed. The type of flow present in a Couette device is analyzed in detail in Chapter 5.

The major sources of error in a concentric cylinder rheometer are the end-effects. One way of minimizing these effects is by providing a large gap between the inner cylinder end and the bottom of the closed end of the outer cylinder.

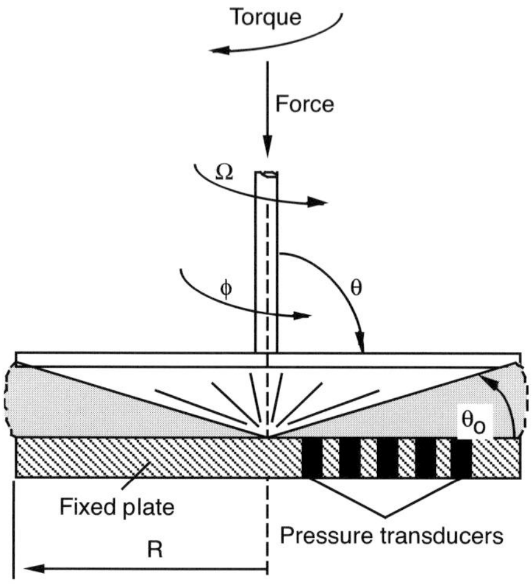

Figure 2.47: Schematic diagram of a cone-plate rheometer.

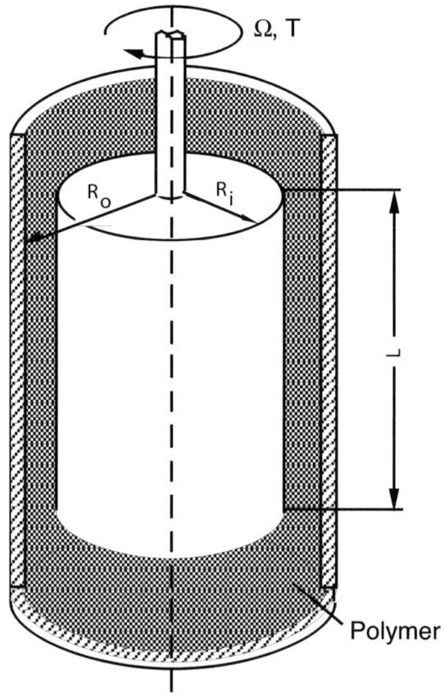

Figure 2.48: Schematic diagram of a Couette rheometer.

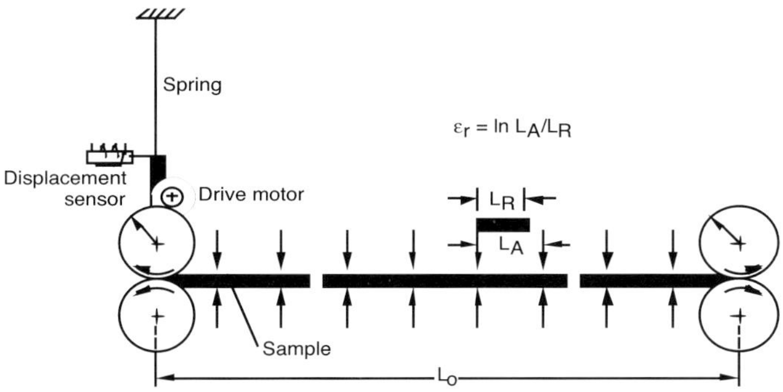

Figure 2.49: Schematic diagram of an extensional rheometer.

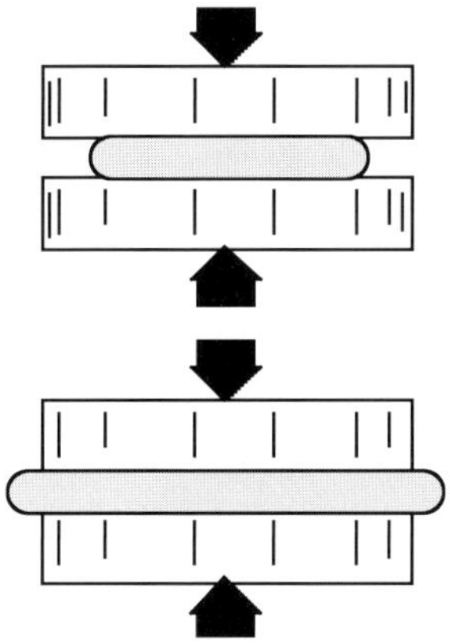

Figure 2.50: Schematic diagram of squeezing flow.

Extensional rheometry. It should be emphasized that the shear behavior of polymers measured with the equipment described in the previous sections cannot be used to deduce the extensional behavior of polymer melts. Extensional rheometry is the least understood field of rheology. The simplest way to measure extensional viscosities is to stretch a polymer rod held at elevated temperatures at a speed that maintains a constant strain rate as the rod reduces its cross-sectional area. The viscosity can easily be computed as the ratio of instantaneous axial stress to elongational strain rate. The biggest problem when trying to perform this measurement is to grab the rod at its ends as it is pulled apart. The most common way to grab the specimen is with toothed rotary clamps to maintain a constant specimen length [48]. A schematic of Meissner's extensional rheometer incorporating rotary clamps is shown in Fig. 2.49 [48].

Another set-up that can be used to measure extensional properties without clamping problems and without generating orientation during the measurement is the lubricating squeezing flow [12], which generates an equibiaxial deformation. A schematic of this apparatus is shown in Fig. 2.50.

It is clear from the apparatus description in Fig. 2.49 that carrying out tests to measure extensional rheometry is a very difficult task. One of the major problems arises because of the fact that, unlike shear tests, it is not possible to achieve steady state condition with elongational rheometry tests. This is simply because the cross-sectional area of the test

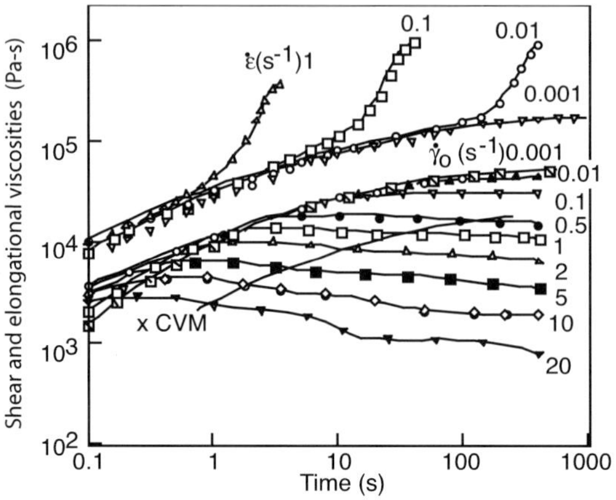

Figure 2.51: Development of elongational and shear viscosities during deformation for polyethylene samples.

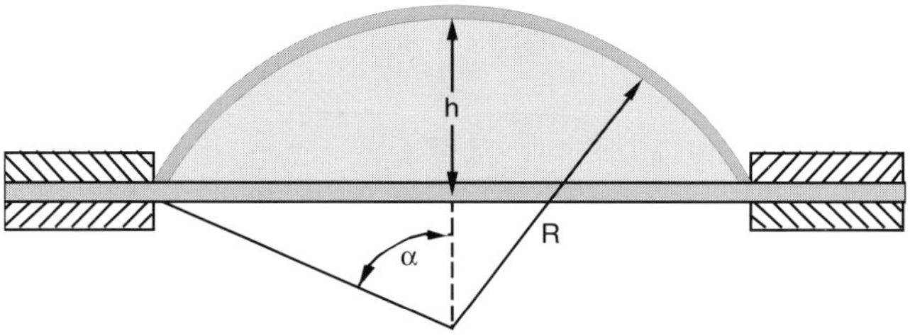

Figure 2.52: Schematic diagram of sheet inflation.

specimen is constantly diminishing. Figure 2.51 [48] shows this effect by comparing shear and elongational rheometry data on polyethylene.

Finally, another equibiaxial deformation test is carried out by blowing a bubble and measuring the pressure required to blow the bubble and the size of the bubble during the test, as schematically depicted in Fig. 2.52. This test has been successfully used to measure extensional properties of polymer membranes for blow molding and thermoforming applications. Here, a sheet is clamped between two plates with circular holes and a pressure differential is introduced to deform it. The pressure applied and deformation of the sheet are monitored over time and related to extensional properties of the material.

2.3.5 Surface Tension

Surface tension plays a significant role in the deformation of polymers during flow, especially in dispersive mixing of polymer blends. Surface tension, σ_S, between two materials appears as a result of different intermolecular interactions. In a liquid-liquid system, surface tension manifests itself as a force that tends to maintain the surface between the two materials to a minimum. Thus, the equilibrium shape of a droplet inside a matrix, which is at rest, is a sphere. When three phases touch, such as liquid, gas, and solid, we get different contact angles depending on the surface tension between the three phases.

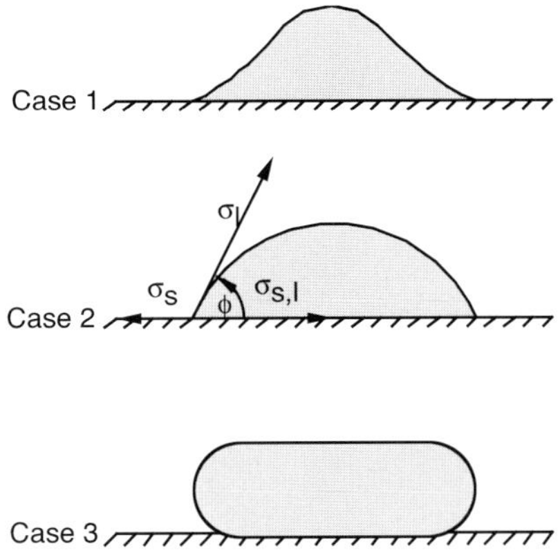

Figure 2.53: Schematic diagram of contact between liquids and solids with various surface tension effects.

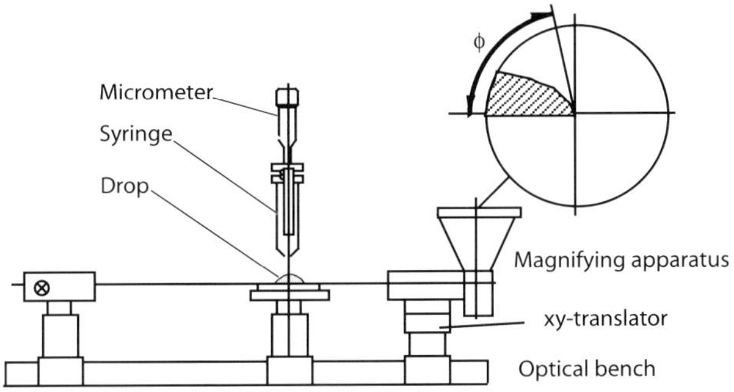

Figure 2.54: Schematic diagram of apparatus to measure contact angle between liquids and solids.

Figure 2.53 schematically depicts three different cases. In case 1, the liquid perfectly wets the surface with a continuous spread, leading to a wetting angle of zero. Case 2, with moderate surface tension effects, shows a liquid that has a tendency to flow over the surface with a contact angle between zero and $\pi/2$. Case 3, with a high surface tension effect, is where the liquid does not wet the surface which results in a contact angle greater than $\pi/2$. In Fig. 2.53, σ_S denotes the surface tension between the gas and the solid, σ_l the surface tension between the liquid and the gas, and σ_{sl} the surface tension between the solid and liquid. Using geometry one can write

$$\cos\theta = \frac{\sigma_s - \sigma_{sl}}{\sigma_l} \tag{2.106}$$

The wetting angle can be measured using simple techniques such as a projector, as shown schematically in Fig. 2.54. This technique, originally developed by Zisman [73], can be used in the ASTM D2578 standard test. Here, droplets of known surface tension, σ_l are applied to a film. The measured values of $\cos\phi$ are plotted as a function of surface tension, σ_l, as shown in Fig. 2.55, and extrapolated to find the critical surface tension, σ_c, required for wetting.

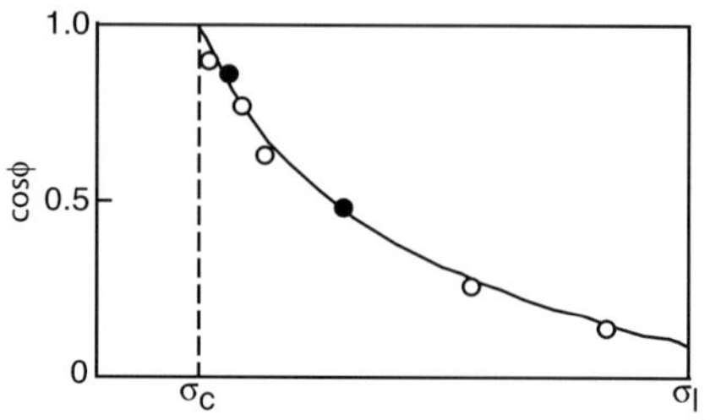

Figure 2.55: Contact angle as a function of surface tension.

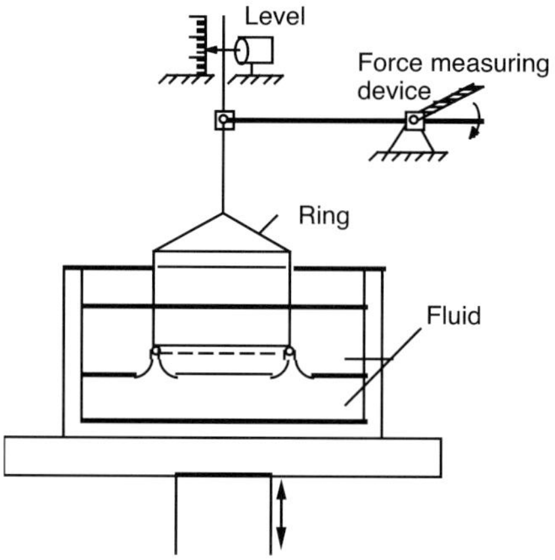

Figure 2.56: Schematic diagram of a tensiometer used to measure surface tension of liquids.

Table 2.12: Typical Surface Tension Values of Selected Polymers at 180°C

Polymer	σ_s (N/m)	$\partial\sigma_s/\partial T$ (N/m/K)
Polyamide resins (290°C)	0.0290	-
Polyethylene (linear)	0.0265	-5.7×10^{-5}
Polyethylene teraphthalate (290°C)	0.027	-
Polyisobutylene	0.0234	-6×10^{-5}
Polymethyl methacrylate	0.0289	-7.6×10^{-5}
Polypropylene	0.0208	-5.8×10^{-5}
Polystyrene	0.0292	-7.2×10^{-5}
Polytetrafluoroethylene	0.0094	-6.2×10^{-5}

For liquids of low viscosity, a useful measurement technique is the tensiometer, schematically represented in Fig. 2.56. Here, the surface tension is related to the force it takes to pull a platinum ring from a solution. Surface tension for selected polymers are listed in Table 2.12 [71], for some solvents in Table 2.13 [58] and between polymer-polymer systems in Table 2.14 [71].

Table 2.13: Surface Tension for Several Solvents

Solvent	σ_S(N/m)
n-Hexane	0.0184
Formamide	0.0582
Glycerin	0.0634
Water	0.0728

Table 2.14: Surface Tension Between Polymers

Polymers	σ_s (N/m)	$\partial\sigma_s/\partial T$ (N/m/K)	T (°C)
PE-PP	0.0011	-	140
PE-PS	0.0051	2.0×10^{-5}	180
PE-PMMA	0.0090	1.8×10^{-5}	180
PP-PS	0.0051	-	140
PS-PMMA	0.0016	1.3×10^{-5}	140

Furthermore, Hildebrand and Scott [32] found a relationship between the solubility parameter, δ, and surface tension, σ_S, for polar and non-polar liquids. Their relationship can be written as [66]

$$\sigma_S = 0.24\delta^{2.33}V^{0.33} \tag{2.107}$$

where V is the molar volume of the material. The molar volume is defined by

$$V = \frac{M}{\rho} \tag{2.108}$$

where M is the molar weight. It should be noted that the values in eqns. (2.107) and (2.108) must be expressed in cgs units.

There are many areas in polymer processing and in engineering design with polymers where surface tension plays a significant role. These are mixing of polymer blends, adhesion, treatment of surfaces to make them non-adhesive and sintering. During manufacturing, it is often necessary to coat and crosslink a surface with a liquid adhesive or bonding material. To enhance adhesion it is often necessary to raise surface tension by oxidizing the surface, by creating COOH-groups, using flames, etching or releasing electrical discharges. This is also the case when enhancing the adhesion properties of a surface before painting. On the other hand, it is often necessary to reduce adhesiveness of a surface such as required when releasing a product from the mold cavity or when coating a pan to give it nonstick properties. A material that is often used for this purpose is polytetrafluoroethylene (PTFE), mostly known by its tradename of teflon.

2.4 PERMEABILITY PROPERTIES

Because of their low density, polymers are relatively permeable by gases and liquids. A more in-depth knowledge of permeability is necessary when dealing with packaging applications and with corrosive protection coatings. The material transport of gases and liquids through polymers consists of various steps. They are:

- Absorption of the diffusing material at the interface of the polymer, a process also known as adsorption,
- *Diffusion*diffusion of the attacking medium through the polymer, and
- Delivery or secretion of the diffused material through the polymer interface, also known as desorption.

With polymeric materials these processes can occur only if the following rules are fulfilled:

- The molecules of the permeating materials are inert,
- The polymer represents a homogeneous continuum, and
- The polymer has no cracks or voids which channel the permeating material.

In practical cases, such conditions are often not present. Nevertheless, this chapter shall start with these *ideal cases,* since they allow for useful estimates and serve as learning tools for these processes.

2.4.1 Sorption

We talk about adsorption when environmental materials are deposited on the surface of solids. Interface forces retain colliding molecules for a certain time. Possible causes include Van der Waals' forces in the case of physical adsorption, chemical affinity (chemical sorption), or electrostatic forces. With polymers, we have to take into account all of these possibilities.

A gradient in concentration of the permeating substance inside the material results in a transport of that substance which we call molecular diffusion. The cause of molecular diffusion is the thermal motion of molecules that permit the foreign molecule to move along the concentration gradient using the intermolecular and intramolecular spaces. However, the possibility to migrate essentially depends on the size of the migrating molecule.

The rate of permeation for the case shown schematically in Fig. 2.57 is defined as the mass of penetrating gas or liquid that passes through a polymer membrane per unit time.

The rate of permeation, $\dot{m}$, can be defined using Fick's first law of diffusion as

$$\dot{m} = -DA\rho\frac{dc}{dx} \tag{2.109}$$

where D is defined as the *diffusion coefficient*, A is the area and ρ the density. If the diffusion coefficient is constant, eqn. (2.109) can be easily integrated to give

$$\dot{m} = -DA\rho\frac{c_1 - c_2}{L} \tag{2.110}$$

The equilibrium concentrations c_1 and c_2 can be calculated using the pressure, p, and the *sorption equilibrium parameter*, S:

$$c = Sp \tag{2.111}$$

which is often referred to as Henry's law.

The sorption equilibrium constant, also referred to as solubility constant, is almost the same for all polymer materials. However, it does depend largely on the type of gas and on the boiling, Tb, or critical temperatures, T_{cr}, of the gas, such as shown in Fig. 2.58.

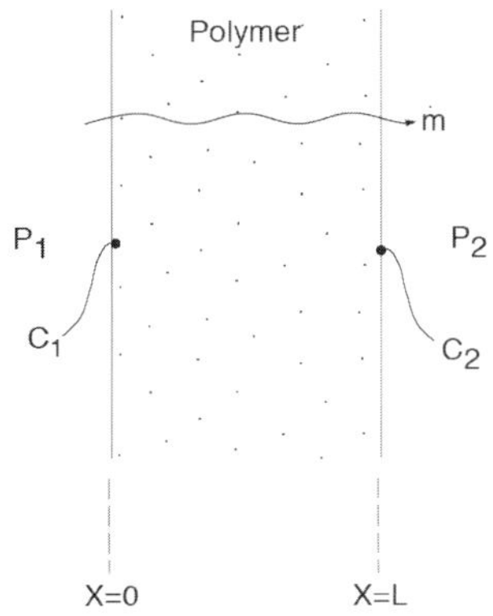

Figure 2.57: Schematic diagram of permeability through a film.

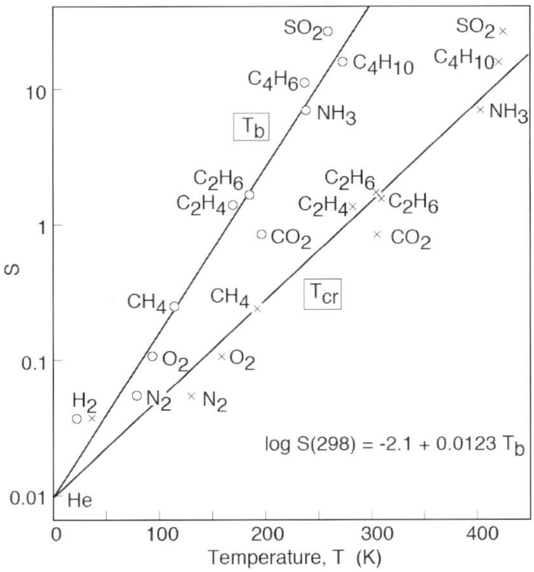

Figure 2.58: Solubility (cm^3/cm^3) of gas in natural rubber at 25°C and 1 bar as a function of the critical and the boiling temperatures.

Table 2.15: Permeability of Various Gases Through Several Polymer Films. Permeability units are in cm^3-mil/100in^2/24h/atm

Polymer	CO_2	O_2	H_2O
PET	12-20	5-10	2-4
OPET	6	3	1
PVC	4.75-40	8-15	2-3
PE-HD	300	100	0.5
PE-LD	-	425	1-1.5
PP	450	150	0.5
EVOH	0.05-0.4	0.05-0.2	1-5
PVDC	1	0.15	0.1

2.4.2 Diffusion and Permeation

Diffusion, however, is only one part of permeation. First, the permeating substance has to infiltrate the surface of the membrane; it has to be absorbed by the membrane. Similarly, the permeating substance has to be desorbed on the opposite side of the membrane. Combining eqn. (2.110) and (2.111), we can calculate the sorption equilibrium using

$$\dot{m} = -DS\rho A \frac{p_1 - p_2}{L} \tag{2.112}$$

where the product of the sorption equilibrium parameter and the diffusion coefficient is defined as the *permeability* of a material

$$P = -DS = \frac{\dot{m}L}{A\Delta p \rho} \tag{2.113}$$

Equation (2.113) does not take into account the influence of pressure on the permeability of the material and is only valid for dilute solutions. The Henry-Langmuir model takes into account the influence of pressure and works very well for amorphous thermoplastics. It is written as

$$P = -DS(1 + \frac{KR'}{1 + b\Delta p}) \tag{2.114}$$

where $K = c'_H b/S$, with c'_H being a saturation capacity constant and b an affinity coefficient. The constant R' represents the degree of mobility, where R'=0 for complete immobility and R'=1 for total mobility. Table 2.15 [62] presents permeability of various gases at room temperature through several polymer films.

In the case of multi-layered films commonly used as packaging material, we can calculate the permeation coefficient PC for the composite membrane using

$$\frac{1}{P_C} = \frac{1}{L_C} \sum_{i=1}^{n} \frac{L_i}{P_i} \tag{2.115}$$

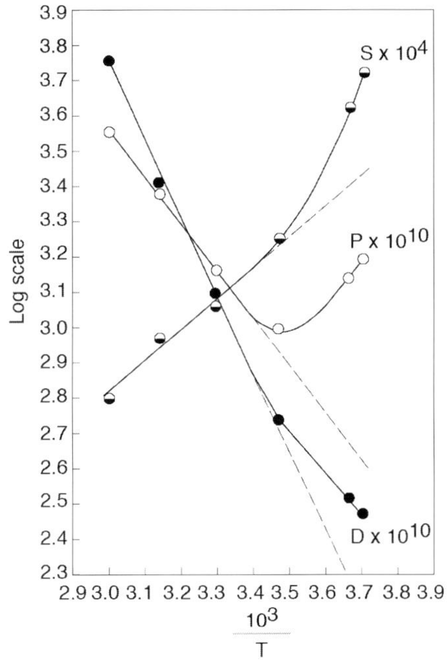

Figure 2.59: Sorption, diffusion, and permeability coefficients, as a function of temperature for polyethylene and methyl bromine at 600 mm of Hg.

Sorption, diffusion, and permeation are processes activated by heat and, as expected, follow an Arrhenius type behavior. Thus, we can write

$$\begin{aligned} S &= S_0 e^{-\Delta H_s/RT} \\ D &= D_0 e^{-E_D/RT} \\ P &= P_0 e^{-E_P/RT} \end{aligned} \tag{2.116}$$

where ΔH_S is the enthalpy of sorption, E_D and E_P are diffusion and permeation activation energies, R is the ideal gas constant, and T is the absolute temperature. The Arrhenius behavior of sorption, diffusion and permeability coefficients, as a function of temperature for polyethylene and methyl bromine at 600 mm of Hg are shown in Fig. 2.59 [61].

Figure 2.60 [38] presents the permeability of water vapor through several polymers as a function of temperature. It should be noted that permeability properties drastically change once the temperature exceeds the glass transition temperature. This is demonstrated in Table 2.16 [66], which presents Arrhenius constants for diffusion of selected polymers and CH_3OH.

The diffusion activation energy E_D depends on the temperature, the size of the gas molecule d, and the glass transition temperature of the polymer. This relationship is well represented in Fig. 2.61 [62] with the size of nitrogen molecules, d_{N_2} as a reference. Table 2.17 contains values of the effective cross section size of important gas molecules.

Using Fig. 2.61 with the values from Table 2.15 and using the equations presented in Table 2.18 the diffusion coefficient, D, for several polymers and gases can be calculated.

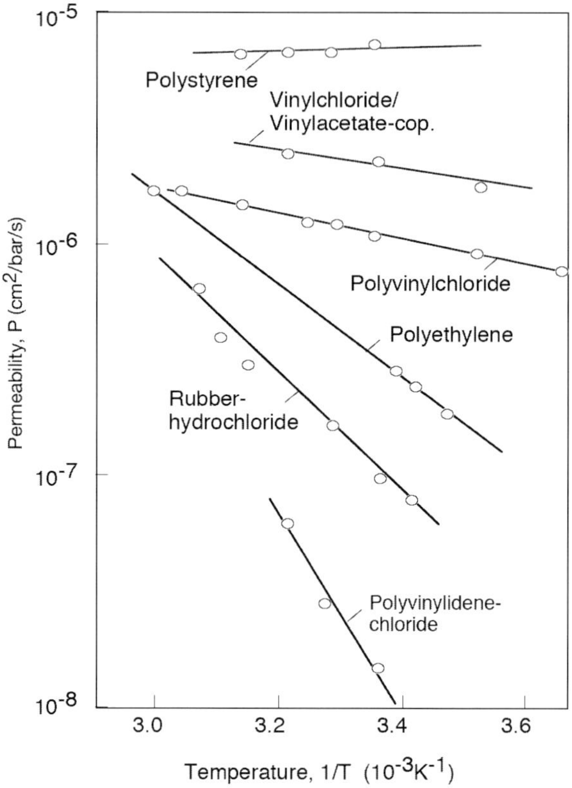

Figure 2.60: Permeability of water vapor as a function of temperature through various polymer films.

Table 2.16: Diffusion Constants Below and Above the Glass Transition Temperature

Polymer	T_g (°C)	$D_0(H_2O)$ (cm^2/s)		E_D (kcal/mol)	
		$T < T_g$	$T > T_g$	$T < T_g$	$T > T_g$
Polymethylmethacrylate	90	0.37	110	12.4	21.6
Polystyrene	88	0.33	37	9.7	17.5
Polyvinyl acetate	30	0.02	300	7.6	20.5

Table 2.17: Important Properties of Gases

Gas	d (nm)	V_{cr} (cm^3)	T_b (K)	T_{cr} (K)	dN_2/dx
He	0.255	58	4.3	5.3	0.67
H_2O	0.370	56	373	647	0.97
H_2	0.282	65	20	33	0.74
Ne	0.282	42	27	44.5	0.74
NH_3	0.290	72.5	240	406	0.76
O_2	0.347	74	90	55	0.91
Ar	0.354	75	87.5	151	0.93
CH_3OH	0.393	118	338	513	0.96
Kr	0.366	92	121	209	0.96
CO	0.369	93	82	133	0.97
CH_4	0.376	99.5	112	191	0.99
N_2	0.380	90	77	126	1.00
CO_2	0.380	94	195	304	1.00
Xe	0.405	119	164	290	1.06
SO_2	0.411	122	263	431	1.08
C_2H_4	0.416	124	175	283	1.09
CH_3Cl	0.418	143	249	416	1.10
C_2H_6	0.444	148	185	305	1.17
CH_2Cl_2	0.490	193	313	510	1.28
C_3H_8	0.512	200	231	370	1.34
C_6H_6	0.535	260	353	562	1.41

Table 2.18: Equations to Compute D Using Data from Table 2.15 and Table 2.16[a]

Elastomers	$\log D = \frac{E_D}{2.3R}\left(\frac{1}{T} - \frac{1}{T_R}\right) - 4$
Amorphous thermoplastics	$\log D = \frac{E_D}{2.3R}\left(\frac{1}{T} - \frac{1}{T_R}\right) - 5$
Semi-crystalline thermoplastics	$\log D = \left(\frac{E_D}{2.3R}\left(\frac{1}{T} - \frac{1}{T_R}\right) - 5\right)(1 - x)$

[a] $T_R = 435$K and $\mathcal{X}$ is the degree of crystallinity.

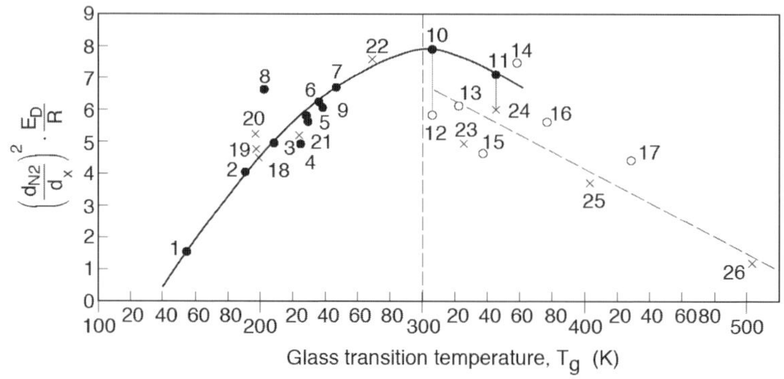

Figure 2.61: Graph to determine the diffusion activation energy E_D as a function of glass transition temperature and size of the gas molecule d_x, using the size of a nitrogen molecule, d_{N2}, as a reference. Rubbery polymers (•): 1 =Silicone rubber, 2 =Polybutadiene, 3 =Natural rubber, 4 =Butadiene/Acrylonitrile K 80/20, 5 =Butadiene/Acrylonitrile K 73/27, 6 =Butadiene/Acrylonitrile K 68/32, 7 =Butadiene/Acrylonitrile K 61/39, 8 =Butyl rubber, 9 =Polyurethane rubber, 10 =Polyvinyl acetate (r), 11 =Polyethylene terephthalate (r). Glassy polymers (*circ*): 12 =Polyvinyl acetate (g), 13 =Vinylchloride/vinyl acetate copolymer, 14 =Polyvinyl chloride, 15 =Polymethyl methacrylate, 16 =Polystyrene, 17 =Polycarbonate. Semi-crystalline polymers (×): 18 =High-density polyethylene, 19 =Low density polyethylene, 20 =Polymethylene oxide, 21 =Gutta percha, 22 =Polypropylene, 23 =Polychlorotrifluoroethylene, 24 =Polyethyleneterephthalate, 25 =Polytetraflourethylene, 26 =Poly(2,6-diphenylphenyleneoxide).

Table 2.18 also demonstrates that permeability properties are dependent on the degree of crystallinity. Figure 2.62 presents the permeability of polyethylene films of different densities as a function of temperature. Again, the Arrhenius relation becomes evident.

2.4.3 Measuring S, D, and P

The permeability P of a gas through a polymer can be measured directly by determining the transport of mass through a membrane per unit time. The sorption constant S can be measured by placing a saturated sample into an environment, which allows the sample to desorb and measure the loss of weight. As shown in Fig. 2.63, it is common to plot the ratio of concentration of absorbed substance $c(t)$ to saturation coefficient c_∞ with respect to the root of time.

The diffusion coefficient D is determined using sorption curves as the one shown in Fig. 2.63. Using the slope of the curve, a, we can compute the diffusion coefficient as

$$D = \frac{\pi}{16} L^2 a^2 \tag{2.117}$$

where L is the thickness of the membrane.

Another method uses the lag time, t_0, from the beginning of the permeation process until the equilibrium permeation has occurred, as shown in Fig. 2.64. Here, the diffusion coefficient is calculated using

$$D = \frac{L^2}{6t_0} \tag{2.118}$$

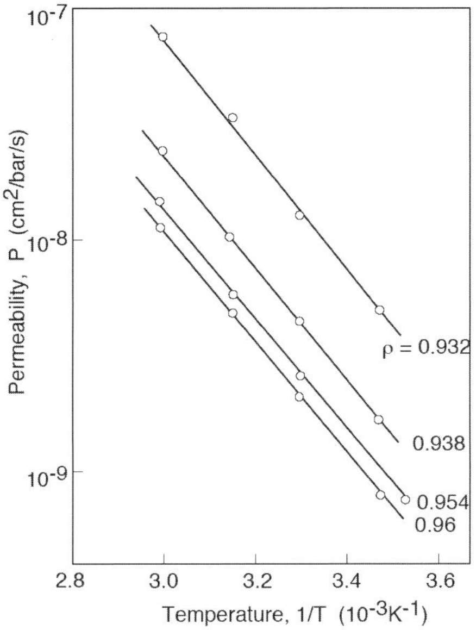

Figure 2.62: Permeation of nitrogen through polyethylene films of various densities.

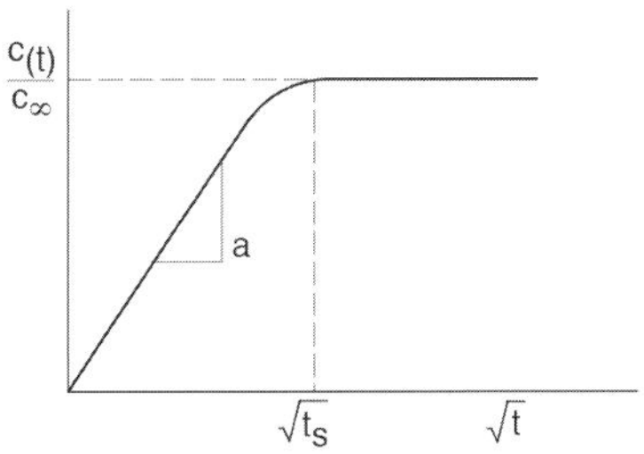

Figure 2.63: Schematic diagram of sorption as a function of time.

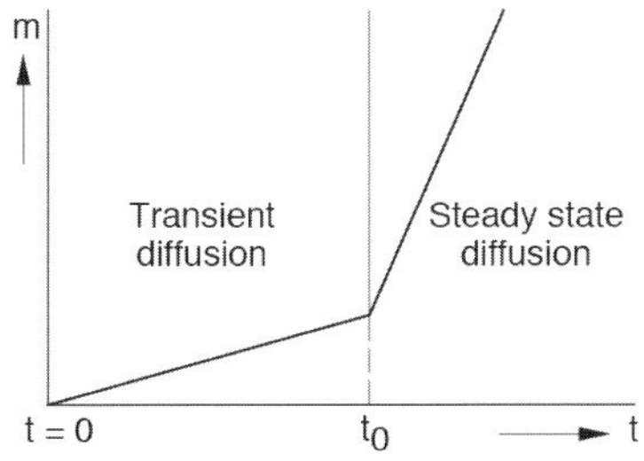

Figure 2.64: Schematic diagram of diffusion as a function of time.

The most important techniques used to determine gas permeability of polymers are the ISO 2556, DIN 53 380 and ASTM D 1434 standard tests.

2.4.4 Diffusion of Polymer Molecules and Self-Diffusion

The ability to infiltrate the surface of a host material decreases with molecular size. Molecules of $M > 5\times10^3$ can hardly diffuse through a porous-free membrane. Self-diffusion is when a molecule moves, say in the melt, during crystallization. Also, when bonding rubber, the so-called tack is explained by the self-diffusion of the molecules. The diffusion coefficient for self-diffusion is of the order of

$$D \sim \frac{T}{\eta} \tag{2.119}$$

where T is the temperature and η the viscosity of the melt.

2.5 FRICTION PROPERTIES

Friction is the resistance that two surfaces experience as they slide or try to slide past each other. Friction can be dry (i.e., direct surface-surface interaction) or lubricated, where the surfaces are separated by a thin film of a lubricating fluid.

The force that arises in a dry friction environment can be computed using Coulomb's law of friction as

$$F = \mu N \tag{2.120}$$

where F is the force in surface or sliding direction, N the normal force, and μ the coefficient of friction.

Coefficients of friction between several polymers and different surfaces are listed in Table 2.19 [49]. However, when dealing with polymers, the process of two surfaces sliding past each other is complicated by the fact that enormous amounts of frictional heat can be generated and stored near the surface due to the low thermal conductivity of the material. The analysis of friction between polymer surfaces is complicated further by environmental effects such as relative humidity and by the likeliness of a polymer surface to deform when stressed, such as shown in Fig. 2.65 [49]. The top two figures illustrate metal-metal friction, wheareas the bottom figures illustrate metal-polymer friction.

Temperature plays a significant role for the coefficient of friction μ as demonstrated in Fig. 2.66 for polyamide 66 and polyethylene. In the case of polyethylene, the friction first decreases with temperature. At 100°C, the friction increases because the polymer surface becomes tacky. The friction coefficient starts to drop as the melt temperature is approached. A similar behavior can be seen in the polyamide curve.

As mentioned earlier, temperature increases can be caused by the energy released by the frictional forces. A temperature increase in time, due to friction between surfaces of the same material, can be estimated using

$$\Delta T = \frac{2\dot{Q}\sqrt{t}}{\sqrt{\pi}\sqrt{k\rho C_p}} \tag{2.121}$$

where k is the thermal conductivity of the polymer, ρ the density, C_p the specific heat and the rate of energy created by the frictional forces, which can be computed using

$$\dot{Q} = Fu \tag{2.122}$$

Table 2.19: Coefficient of Friction for Various Polymers

Specimen	Partner			Velocity	(mm/s)		
		0.03	0.1	0.4	0.8	3.0	10.6
PP^i	PP^s	0.54	0.65	0.71	0.77	0.77	0.71
PA^i	PA^i	0.63	-	0.69	0.70	0.70	0.65
PP^s	PP^s	0.26	0.29	0.22	0.21	0.31	0.27
PA^m	PA^m	0.42	-	0.44	0.46	0.46	0.47
Steel	PP^s	0.24	0.26	0.27	0.29	0.30	0.31
Steel	PA^m	0.33	-	0.33	0.33	0.30	0.30
PP^s	Steel	0.33	0.34	0.37	0.37	0.38	0.38
PA^m	Steel	0.30	-	0.41	0.41	0.40	0.40

^{i}injection molded; ^{s}sandblasted; ^{m}machined

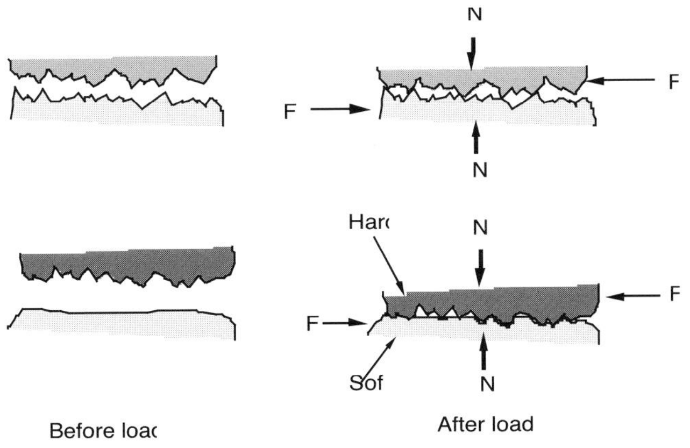

Figure 2.65: Effect of surface finish and hardness on frictional force build-up.

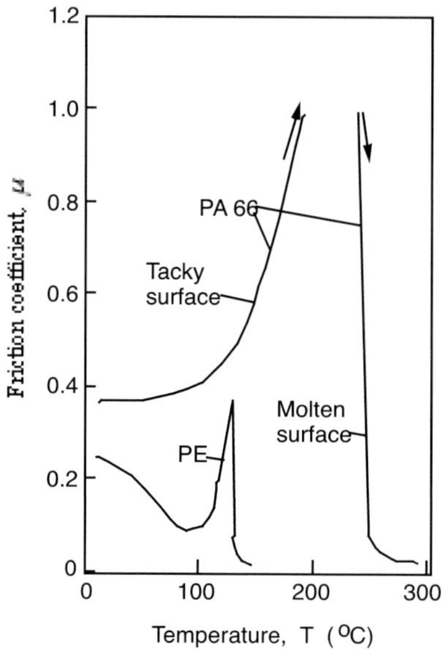

Figure 2.66: Temperature effect on coefficient of friction for a polyamide 66 and a high density polyethylene.

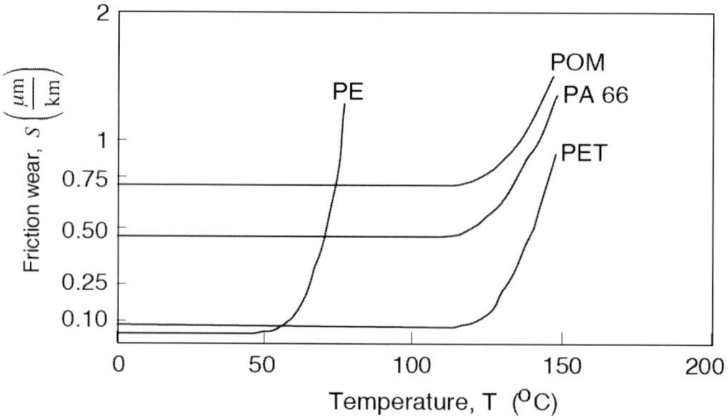

Figure 2.67: Wear as a function of temperature for various thermoplastics. Courtesy of BASF.

where u is speed between the sliding surfaces.

Wear is also affected by the temperature of the environment. Figure 2.67 shows how wear rates increase dramatically as the surface temperature of the polymer increases, causing it to become tacky.

Problems

2.1 Does the coefficient of linear expansion of a polymer increase or decrease upon the addition of glass fibers?

2.2 Plot T_g versus T_m for several polymers. What trend or relation do you observe?

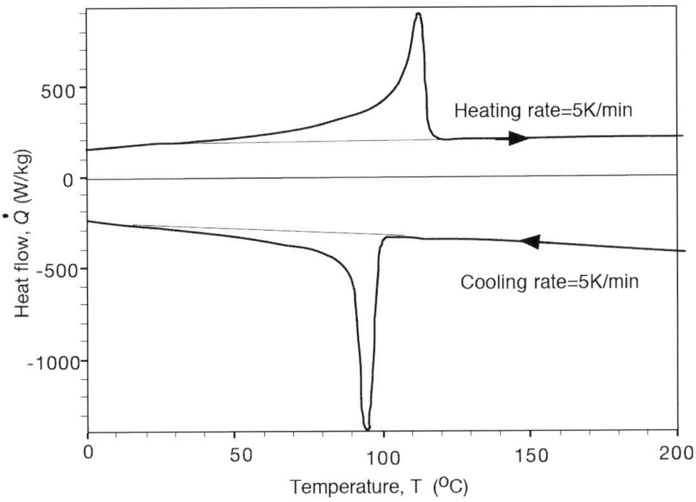

Figure 2.68: Heating and cooling DSC scans of a PE-LD sample.

2.3 In a soda bottle, how does the degree of crystallinity in the screw-top region compare to the degree of crystallinity in the wall? Explain.

2.4 A 5 K/min heating and 5 K/min cooling differential scanning calorimetry (DSC) test (Fig. 2.68) was performed on a 10.8 mg sample of PE-LD. What is the specific heat of the PE-LD just after melting, during heating, and just before crystallization during cooling. What is the degree of crystallinity of the initial and the final samples.

2.5 A differential scanning calorimetry (DSC) test was performed on an 11.4 mg polyethylene terephthalate (PET) sample using the standard ASTM D 3417 test method. The ASTM test calls for a temperature heating rate of 20°C/min (20°C rise every minute). The DSC output is presented in Fig. 2.69

a) From the curve, estimate the glass transition temperature, T_g, the melting temperature, T_m, the crystallization temperature, T_c, and the heat of fusion, λ, for this specific PET sample during the temperature ramp-up. Note that the heat flow scale has already been transformed to heat capacity. How do T_g and T_m compare to the "book values"?

b) If the heat of fusion for a hypothetically 100% crystalline PET is 137 kJ/kg, what was the degree of crystallinity of the original PET sample?

c) On the same graph below sketch a hypothetical DSC output for the original PET sample with a temperature heating rate that is too fast to allow any additional crystallization during heating.

2.6 Isothermal differential scanning calorimetry (DSC) tests were performed on three unsaturated polyester (UPE) samples at three different temperatures (100°C, 110°C, and 120°C). The output for the three DSC tests are presented in the Fig. 2.70. On the graph, label which curve is associated with which test temperature. From the curves in Fig. 2.70 estimate the total heat of reaction, Q_T.

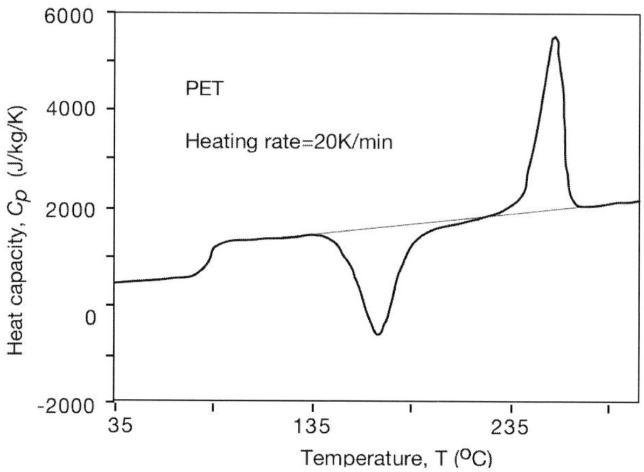

Figure 2.69: DSC scan of a PET sample. Courtesy of ICIPC, Medellín-Colombia.

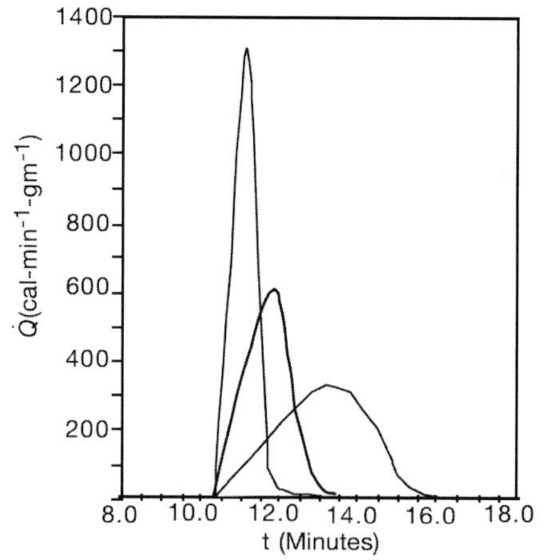

Figure 2.70: Isothermal DSC measurement of UPE samples. Courtesy of GenCorp Research, Akron-OH.

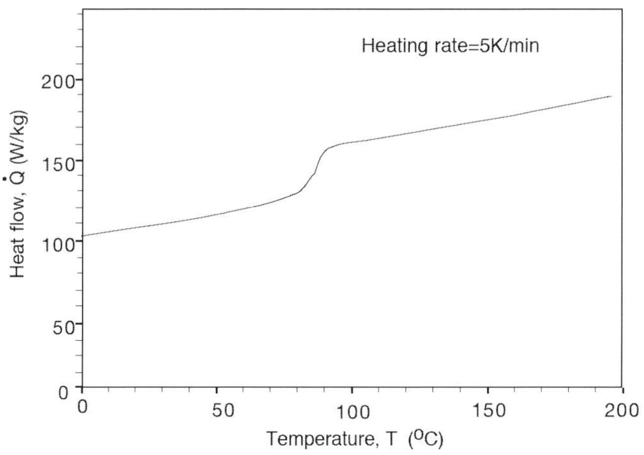

Figure 2.71: DSC scan of a PS sample.

2.7 A typical injection pack/hold pressure during injection molding of polyamide 66 components is 1,000 bar and the injection temperature is 280°C. The gate freezes shut when the average temperature inside the mold reaches 225°C.

a) Draw the process on the pvT-diagram given in Fig. 2.12.

b) What volume shrinkage should be taken into account when designing the mold? Note that the shrinkage is mostly taken up by a thickness reduction.

2.8 A differential scanning calorimetry (DSC) test (Fig. 2.71) was performed on an 18.3 mg sample of polystyrene.

a) What is the glass transition temperature of the sample?

b) Determine the specific heat of this PS just before the glass transition temperature has been reached.

c) What is C_p just after T_g?

d) Why is the heat larger as the temperature increases?

2.9 Sketch the pvT diagrams for a semi-crystalline polymer with a high and a low cooling rate.

2.10 In example 2.2 we obtained that for steady shearing flows the viscometric functions for this constitutive equation are defined by

$$\begin{aligned}\eta &= \eta_0 \\ \Psi_1 &= 2\eta_0(\lambda_1 - \lambda_2) \\ \Psi_2 &= 0\end{aligned} \tag{2.123}$$

What are your comments about this resutls? How this fuctions compare with experimental observations?

2.11 Develop expressions for the elongational viscosities $\bar{\eta}_1$ and $\bar{\eta}_2$ for steady shearfree flows of a convected Jeffreys model. Comment how this expression compares with experiments.

2.12 Develop expressions for the steady shear viscometric functions for the White-Metzner model.

2.13 Comment how the viscometric functions for the shear flow of a Lodge rubber-liquid develop in Example 2.4; compare with experimental observations.

2.14 Develop expressions for the elongational viscosities for the Lodge rubber-liquid in steady shearfree flow.

REFERENCES

1. J.F. Agassant, P. Avenas, J.-Ph. Sergent, and P.J. Carreau. *Polymer Processing - Principles and Modeling*. Hanser Publishers, Munich, 1991.
2. F.F.T Araujo and H.M. Rosenberg. *J. Phys. D*, 9:665, 1976.
3. G.K. Batchelor. *Annu. Rev. Fluid Mech.*, 6:227, 1974.
4. B. Bernstein, E. Kearsley, and L. Zappas. *Trans. Soc. Rheol.*, 7:391, 1963.
5. R.B. Bird, R.C. Armstrong, and O. Hassager. *Dynamics of Polymer Liquids: Fluid Mechanics*, volume 1. John Wiley & Sons, New York, 2nd edition, 1987.
6. R.B. Bird and J.M. Wiest. *Annu. Rev. Fluid Mech.*, 27:169, 1995.
7. P.J. Carreau. PhD thesis, University of Wisconsin-Madison, Madison, 1968.
8. P.J. Carreau, D.C.R. DeKee, and R.P. Chhabra. *Rheology of Polymeric Systems*. Hanser Publishers, Munich, 1997.
9. J.M. Castro and C.W. Macosko. *AIChE J.*, 28:250, 1982.
10. J.M. Castro, S.J. Perry, and C.W. Macosko. *Polym. Comm.*, 25:82, 1984.
11. I. Catic. PhD thesis, IKV, RWTH-Aachen, Germany, 1972.
12. Sh. Chatrei, C.W. Macosko, and H.H. Winter. *J. Rheol.*, 25:433, 1981.
13. H.H. Chiang. Simulation and verification of filling and post-filling stages of the injection-molding process. Technical Report 62, Cornell University, Ithaca, 1989.
14. R.J. Crawford. *Rotational Molding of Plastics*. Research Studies Press, Somerset, 1992.
15. A. de Waale. *Oil Color Chem. Assoc. J.*, 6:33, 1923.
16. J.M. Dealy and K.F. Wissbrun. *Melt Rheology and Its Role in Plastics Processing*. Van Nostrand, New York, 1990.
17. M.M. Denn. *Annu. Rev. Fluid Mech.*, 22:13, 1990.
18. A.T. DiBenedetto and L.E. Nielsen. *J. Macromol. Sci., Rev. Macromol. Chem.*, C3:69, 1969.
19. W. Dietz. *Kunststoffe*, 66(3):161, 1976.
20. A. Einstein. *Ann. Physik*, 19:549, 1906.
21. J.B. Enns and J.K. Gillman. Time-temperature-transformation (ttt) cure diagram: Modeling the cure behavior of thermosets. *J. Appl. Polym. Sci.*, 28:2567–2591, 1983.
22. F. Fischer. *Gummi-Asbest-Kunststoffe*, 32(12):922, 1979.
23. P. Geisbüsch. PhD thesis, IKV-RWTH-Aachen, Germany, 1980.
24. Y.K. Godovsky. *Thermophysical Properties of Polymers*. Springer-Verlag, Berlin, 1992.
25. R.K. Gupta. *Flow and rheology in polymer composites manufacturing*. Elsevier, Amsterdam, 1994.
26. E. Guth. *Phys. Rev.*, 53:321, 1938.
27. E. Guth and R. Simha. *Kolloid-Zeitschrift*, 74:266, 1936.

28. G.H.L. Hagen. *Annalen der Physik*, 46:423, 1839.

29. C.D. Han and K.W. Len. *J. Appl. Polym. Sci.*, 29:1879, 1984.

30. S.G. Hatzikiriakos and J.M. Dealy. In *SPE-ANTEC Tech. Papers*, volume 37, page 2311, 1991.

31. J.P. Hernandez-Ortiz and T.A. Osswald. A novel cure reaction model fitting technique based on dsc scans. *J. Polym. Eng.*, 25(1):23, 2005.

32. J. Hildebrand and R.L. Scott. *The solubility of non-electrolytes*. Reinhold Publishing Co., New York, 1949.

33. H. Janeschitz-Kriegl, H. Wippel, Ch. Paulik, and G. Eder. *Colloid Polym. Sci.*, 271(1107), 1993.

34. M.R. Kamal. *Polym. Eng. Sci.*, 14:231, 1979.

35. M.R. Kamal and S. Sourour. *Polym. Eng. Sci.*, 13:59, 1973.

36. A. Kaye. Non-newtonian flow in incompressible fluids. Technical Report CoA Note 134, The College of Aeronautics, Cranfield, 1962.

37. D.G. Kiriakidis and E. Mitsoulis. *Adv. Polym. Techn.*, 12:107, 1993.

38. W. Knappe. *VDI Berichte*, 68(29), 1963.

39. W. Knappe. *Adv. Polym. Sci.*, 7:477, 1971.

40. W. Knappe. *Kunststoffe*, 66(5):297, 1976.

41. H.M. Laun. *Rheol. Acta*, 17:1, 1978.

42. A.S. Lodge. *Elastic Liquids*. Academic Press, London, 1960.

43. X.-L. Luo. *J. Rheol.*, 33:1307, 1989.

44. C.W. Macosko. *RIM Fundamentals of reaction injection molding*. Hanser Publishers, Munich, 1989.

45. C.W. Macosko. *Rheology: Principles, Measuremenyts and Applications*. VCH, 1994.

46. V.B.F. Mathot. *Calorimetry and Thermal Analysis of Polymers*. Hanser Publishers, Munich, 1994.

47. J. C. Maxwell. *Electricity and Magnetism*. Clarendon Press, Oxford, 1873.

48. J. Meissner. *Rheol. Acta*, 10:230, 1971.

49. G. Menges. *Werkstoffkunde der Kunststoffe*. Hanser Publishers, Munich, 2 edition, 1984.

50. G. Menges, F. Wortberg, and W. Michaeli. *Kunststoffe*, 68:71, 1978.

51. R.E. Meredith and C.W. Tobias. *J. Appl. Phys.*, 31:1270, 1960.

52. W.J. Milliken and R.L. Powell. *Flow and rheology in polymer composites manufacturing*. Elsevier, Amsterdam, 1994.

53. H. Münstedt. *Rheol. Acta*, 14:1077, 1975.

54. J.G. Oldroyd. *Proc. Roy. Soc.*, A200:523, 1950.

55. N. Orbey and J.M. Dealy. *Polym. Eng. Sci.*, 24:511, 1984.

56. T. A. Osswald and G. Menges. *Material Science of Polymers for Engineers*. Hanser Publishers, Munich, 2nd edition, 2003.

57. W. Ostwald. *Kolloid-Z.*, 36:99, 1925.

58. D.K. Owens and R.C. Wendt. *J. Appl. Polym. Sci.*, 13:1741, 1969.

59. A.C. Papanastasiou, L.E. Scriven, and C.W. Macosko. *J. Rheol.*, 27:387, 1983.

60. L.J. Poiseuille. *Comptes Rendus*, 11:961, 1840.

61. C.E. Rogers. *Engineering Design for Plastics*. Krieger Publishing Company, Huntington, 1975.

62. D. Rosato and D.V. Rosato. *Blow Molding Handbook*. Hanser Publishers, Munich, 1989.

63. R.S. Spencer and R.D. Dillon. *J. Colloid. Sci.*, 3:167, 1947.

64. Z. Tadmor and C.G. Gogos. *Principles of Polymer Processing*. John Wiley & Sons, New York, 1979.

65. F.T. Trouton. *Proc. Roy. Soc. A*, 77, 1906.

66. D.W. van Krevelen. *Properties of Polymers*. Elsevier, Amsterdam, 1990.

67. G.V. Vinogradov, A.Y. Malkin, Y.G. Yanovskii, E.K. Borisenkova, B.V. Yarlykov, and G.V. Berenzhnaya. *J. Polym. Sci. A*, 10:1061, 1972.

68. J. Vlachopoulis and M. Alam. *Polym. Eng. Sci.*, 12:184, 1972.

69. M.H. Wagner. *Rheol. Acta*, 18:33, 1979.

70. C.H. Wu, G. Eder, and H. Janeschitz-Kriegl. *Colloid Polym. Sci.*, 271(1116), 1993.

71. S. Wu. *J. Macromol. Sci.*, 10:1, 1974.

72. K. Yasuda, R.C. Armstrong, and R.E. Cohen. *Rheol. Acta*, 20:163, 1981.

73. W.A. Zisman. *Ind. Eng. Chem.*, 55:19, 1963.

CHAPTER 3

POLYMER PROCESSES

There's a way to do it better - find it.

—Thomas A. Edison

Manufacturing of plastic parts can involve one or several of the following steps:

- Shaping operations - This involves transforming a polymer pellet, powder or resin into a final product or into a preform using extrusion or molding processes such as injection, compreession molding or rotomolding.
- Secondary shaping operation - Here a preform such as a parison or sheet is transformed into a final product using thermoforming or blow molding.
- Material removal - This type of operation involves material removal using machining operations, stamping, laser, drilling, etc.
- Joining operations - Here, two or more parts are assembled physically or by bonding or welding operations.

Most plastic parts are manufactured using shaping operations. Here, the material is deformed into its final shape at temperatures between room temperature and 350°C, using wear resistant tools, dies and molds. For example, an injection mold would allow making between 10^6 and 10^7 parts without much wear of the tool, justifying for the high cost of the molds utilized. One of the many advantages of polymer molding processes is the accuracy, sometimes with features down to the micrometer scale, with which one can shape the finished product without the need of trimming or material removal operations. For example, when making compact discs by an injection-compression molding process, it is possible to accurately produce features, that contain digital information smaller than 1 μm, in a disc with a thickness of less than 1 mm and a diameter of several centimeters. The cycle time to produce such a part can be less than 3 seconds.

In the past few years, we have seen trends where more complex manufacturing systems are developed that manufacture parts using various materials and components such as co-extrusion of multilayer films and sheets, multi-component injection molding, sandwiched parts or hollow products.

Thermoplastics and thermoplastic elastomers are shaped and formed by heating them above glass transition or melting temperatures and then freezing them into their final shape by lowering the temperature. At that point, the crystallization, molecular or fiber orientation and residual stress distributions are an integral feature of the final part, dominating the material properties and performance of the finished product. Similarly, thermosetting polymers and vulcanizing elastomers solidify by a chemical reaction that results in a cross-linked molecular structure. Here too, the filler or fiber orientation as well as the residual stresses are frozen into the finished structure after cross-linking.

This chapter is intended to give an introduction to the most important polymer processes[1].

3.1 EXTRUSION

During extrusion, a polymer melt is pumped through a shaping die and formed into a profile. This profile can be a plate, a film, a tube, or have any shape for its cross section. Ram-type extruders were first built by J. Bramah in 1797 to extrude seamless lead pipes. The first ram-type extruders for rubber were built by Henry Bewley and Richard Brooman in 1845. In 1846, a patent for cable coating was filed for trans-gutta-percha and cis-hevea rubber and the first insulated wire was laid across the Hudson River for the Morse Telegraph Company in 1849. The first screw extruder was patented by Mathew Gray in 1879 for the purpose of wire coating. However, the screw pump can be attributed to Archimedes, and the actual invention of the screw extruder in polymer processing by A.G. DeWolfe of the United States dates to the early 1860s. The first extrusion of thermoplastic polymers was done at the Paul Troester Maschinenfabrik in Hannover, Germany in 1935. Although ram and screw extruders are both used to pump highly viscous polymer melts through passages to generate specified profiles, they are based on different principles. The schematic in Fig. 3.1 shows under what principles ram extruders, screw extruders, and other pumping systems work.

The ram extruder is a positive displacement pump based on the pressure gradient term of the equation of motion. Here, as the volume is reduced, the fluid is displaced from one point to the other, resulting in a pressure rise. The gear pump, widely used in the polymer processing industry, also works on this principle. On the other hand, a screw extruder is a viscosity pump that works based on the pressure gradient term and the deformation of the fluid, represented as the divergence of the deviatoric stress tensor in Fig. 3.1. The centrifugal pump, based on the fluid inertia, and the Roman aqueduct, based on the potential energy of the fluid, are also represented in the figure and are typical of low viscosity liquids.

[1]For further reading in the area of extrusion and injection molding we recommend [9] and [21], respectively.

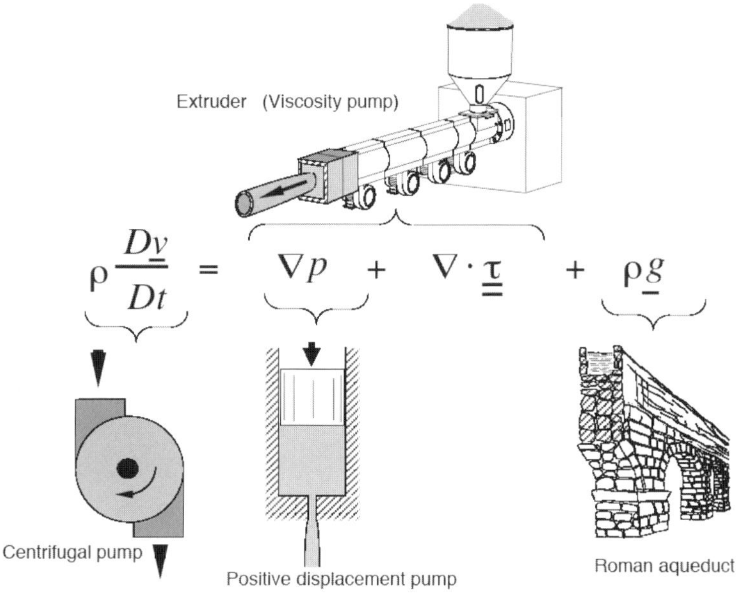

Figure 3.1: Schematic of pumping principles.

In today's polymer industry, the most commonly used extruder is the single screw extruder, schematically depicted in Fig. 3.2. A single screw extruder with a smooth inside barrel surface is called a conventional single screw extruder, with grooved feed zone it is called a grooved feed extruder. In some cases, an extruder can have a degassing zone, required to extract moisture, volatiles, and other gases that form during the extrusion process.

Another important class of extruders are the twin screw extruders, schematically depicted in Fig. 3.3. Twin screw extruders can have co-rotating or counter-rotating screws, and the screws can be intermeshing or non-intermeshing. Twin screw extruders are primarily employed as mixing and compounding devices, as well as polymerization reactors. The mixing aspects of single and twin screw extruders are detailed later in this chapter.

3.1.1 The Plasticating Extruder

The plasticating single screw extruder is the most common equipment in the polymer industry. It can be part of an injection molding unit and found in numerous other extrusion processes, including blow molding, film blowing, and wire coating. A schematic of a plasticating or three-zone single screw extruder, with its most important elements is given in Fig. 3.4.

Table 3.1 presents typical extruder dimensions and relationships common in single screw extruders, using the notation presented in Fig. 3.5.

The plasticating extruder can be divided into three main zones:

- The solids conveying zone
- The melting or transition zone

Table 3.1: Typical Extruder Dimensions and Relationships

L/D	Length to diameter ratio
	20 or less for feeding or melt extruders
	25 for blow molding, film blowing, and injection molding
	30 or higher for vented extruders or high output extruders
D	Standard diameter
US (inches)	0.75, 1.0, 1.5, 2, 2.5, 3.5, 4.5, 6, 8, 10, 12, 14, 16, 18, 20, and 24
Europe (mm)	20, 25, 30, 35, 40, 50, 60, 90, 120, 150, 200, 250, 300, 350, 400, 450, 500, and 600
ϕ	Helix angle
	17.65^o for a square pitch screw where $L_s = D$
	New trend: $0.8 < L_s/D < 1.2$
h	Channel depth in the metering section
	(0.05-0.07)D for D <30 mm
	(0.02-0.05)D for D >30 mm
β	Compression ratio
	$h_{feed} = \beta h$
	2 to 4
δ	Clearance between the screw flight and the barrel
	0.1 mm for D <30 mm
	0.15 mm for D >30 mm
N	Screw speed
	1-2 rev/s (60-120 rpm) for large extruders
	1-5 rev/s (60-300 rpm) for small extruders
V_b	Barrel velocity (relative to screw speed) $= \pi DN$
	0.5 m/s for most polymers
	0.2 m/s for unplasticized PVC
	1.0 m/s for PE-LD

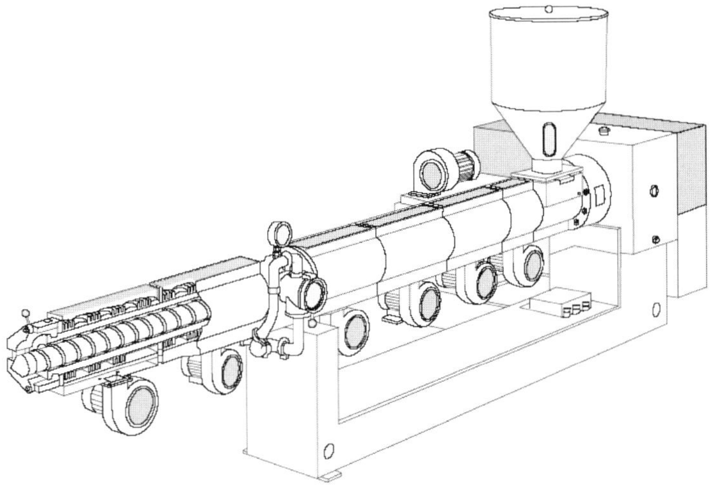

Figure 3.2: Schematic of a single screw extruder (Reifenhäuser).

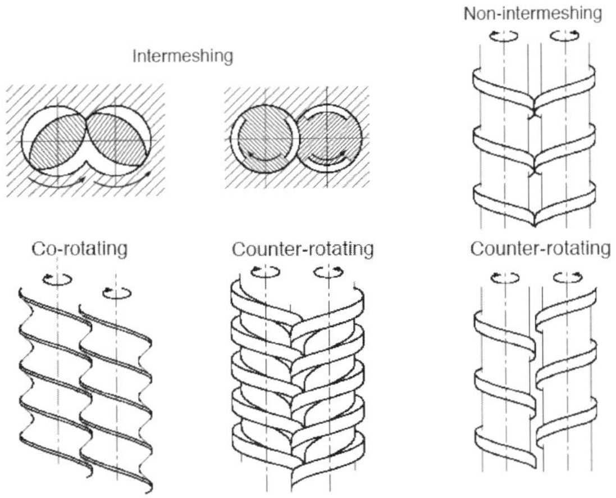

Figure 3.3: Schematic of different types of twin screw extruders.

- The metering or pumping zone

The tasks of a plasticating extruder are to:

- Transport the solid pellets or powder from the hopper to the screw channel
- Compact the pellets and move them down the channel

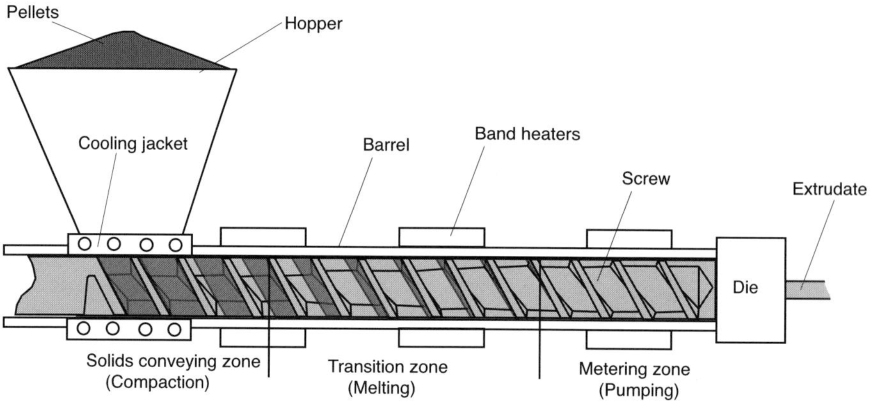

Figure 3.4: Schematic of a plasticating single screw extruder.

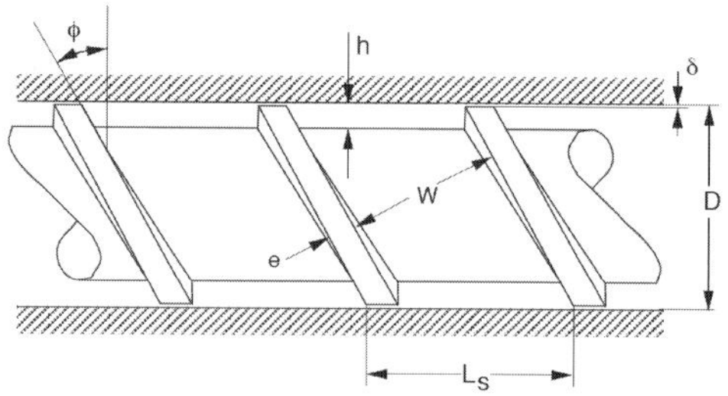

Figure 3.5: Schematic diagram of a screw section.

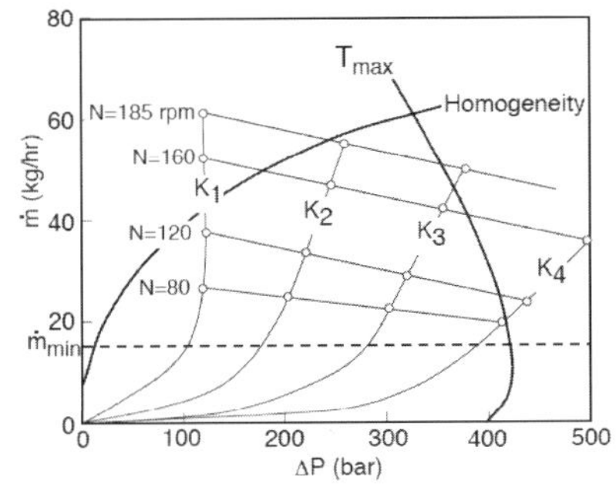

Figure 3.6: Screw and die characteristic curves for a 45 mm diameter extruder with an PE-LD.

- Melt the pellets
- Mix the polymer into a homogeneous melt
- Pump the melt through the die

The pumping capability and characteristic of an extruder can be represented with sets of die and screw characteristic curves. Figure 3.6 presents such curves for a conventional (smooth barrel) single screw extruder.

The die characteristic curves are labelled K_1, K_2, K_3, and K_4 in ascending order of die restriction. Here, K_1 represents a low resistance die such as for a thick plate, and K_4 represents a restrictive die, such as is used for film. The different screw characteristic curves represent different screw rotational speeds. In a screw characteristic curve, the point of maximum throughput and no pressure build-up is called the point of open discharge. This occurs when there is no die. The point of maximum pressure build-up and no throughput is called the point of closed discharge. This occurs when the extruder is plugged. Shown in Fig. 3.6 are also lines that represent critical aspects encountered during extrusion. The curve labeled T_{max} represents the conditions at which excessive temperatures are reached as a result of viscous heating. The feasibility line represents the throughput required to have an economically feasible system. The processing conditions to the right of the homogeneity line render a thermally and physically heterogeneous polymer melt.

The solids conveying zone. The task of the solids conveying zone is to move the polymer pellets or powders from the hopper to the screw channel. Once the material is in the screw channel, it is compacted and transported down the channel. The process to compact the pellets and to move them can only be accomplished if the friction at the barrel surface exceeds the friction at the screw surface. This can be visualized if one assumes the material inside the screw channel to be a nut sitting on a screw. As we rotate the screw without applying outside friction, the nut (polymer pellets) rotates with the screw without moving in the axial direction. As we apply outside forces (barrel friction), the rotational speed of the nut is less than the speed of the screw, causing it to slide in the axial direction. Virtually, the solid polymer is then "unscrewed" from the screw. To maintain a

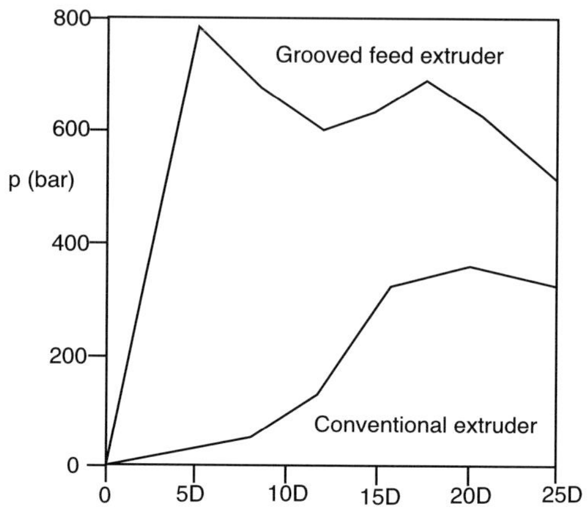

Figure 3.7: Typical conventional and grooved feed extruder pressure distributions in a 45 mm diameter extruder.

high coefficient of friction between the barrel and the polymer, the feed section of the barrel must be cooled, usually with cold water cooling lines. The frictional forces also result in a pressure rise in the feed section. This pressure compresses the solids bed which continues to travel down the channel as it melts in the transition zone. Figure 3.7 compares the pressure build-up in a conventional, smooth barrel extruder with that in a grooved feed extruder. In these extruders, most of the pressure required for pumping and mixing is generated in the metering section.

The simplest mechanism for ensuring high friction between the polymer and the barrel surface is grooving its surface in the axial direction [17, 16]. Extruders with a grooved feed section where developed by Menges and Predöhl [17, 16] in 1969, and are called grooved feed extruders. To avoid excessive pressures that can lead to barrel or screw failure, the length of the grooved barrel section must not exceed $3.5D$. A schematic diagram of the grooved section in a single screw extruder is presented in Fig. 3.8. The key factors that propelled the development and refinement of the grooved feed extruder were processing problems, excessive melt temperature, and reduced productivity caused by high viscosity and low coefficients of friction typical of high molecular weight polyethylenes and polypropylenes.

In a grooved feed extruder, the conveying and pressure build-up tasks are assigned to the feed section. The high pressures in the feed section (Fig. 3.7) lead to the main advantages over conventional systems. With grooved feed systems, higher productivity and higher melt flow stability and pressure invariance can be achieved. This is demonstrated with the screw characteristic curves in Fig. 3.9, which presents screw characteristic curves for a 45 mm diameter grooved feed extruder with comparable mixing sections and die openings as shown in Fig. 3.6.

The melting zone. The melting or transition zone is the portion of the extruder were the material melts. The length of this zone is a function of material properties, screw geometry, and processing conditions. During melting, the size of the solid bed shrinks as a melt pool forms at its side, as depicted in Fig. 3.10 which shows the polymer unwrapped from the screw channel.

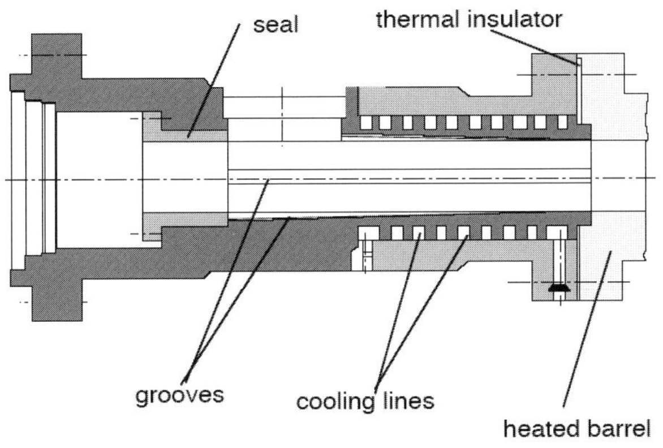

Figure 3.8: Schematic diagram of the grooved feed section of a single screw extruder.

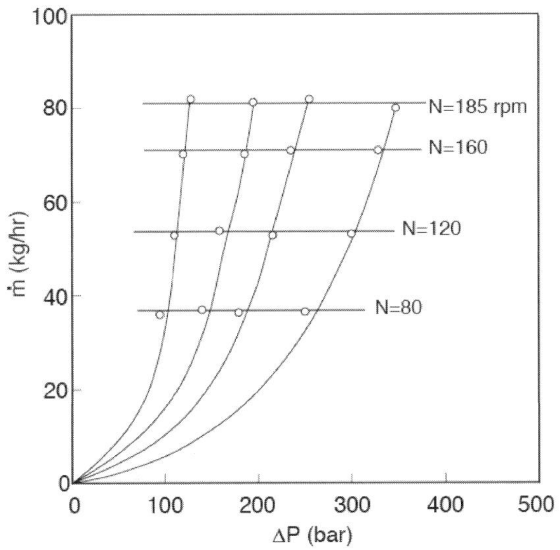

Figure 3.9: Screw and die characteristic curves for a grooved feed 45 mm diameter extruder for an PE-LD.

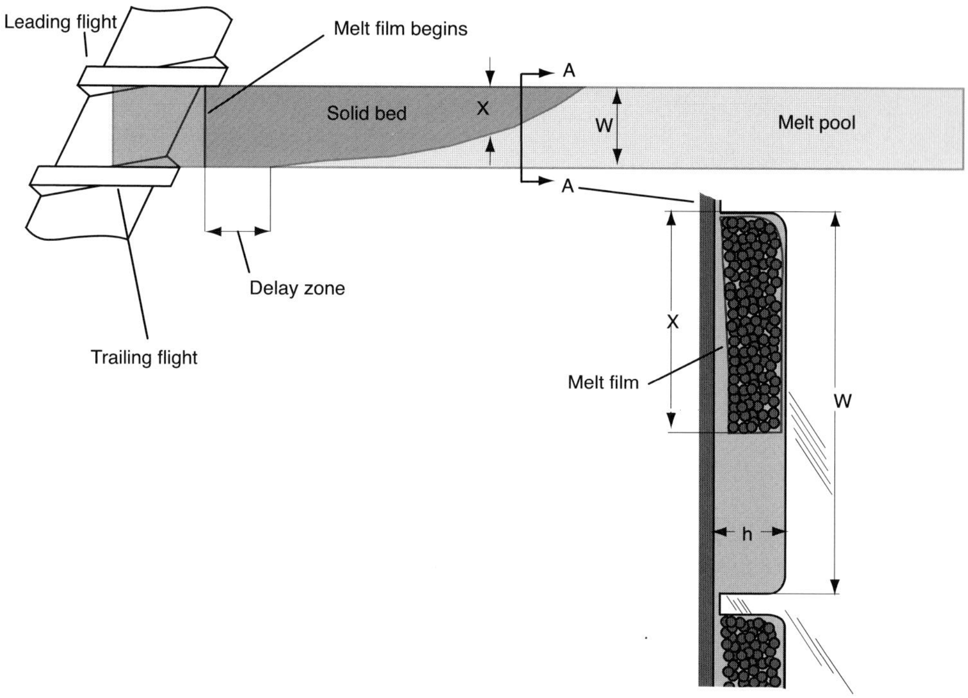

Figure 3.10: Solids bed in an unwrapped screw channel with a screw channel cross-section.

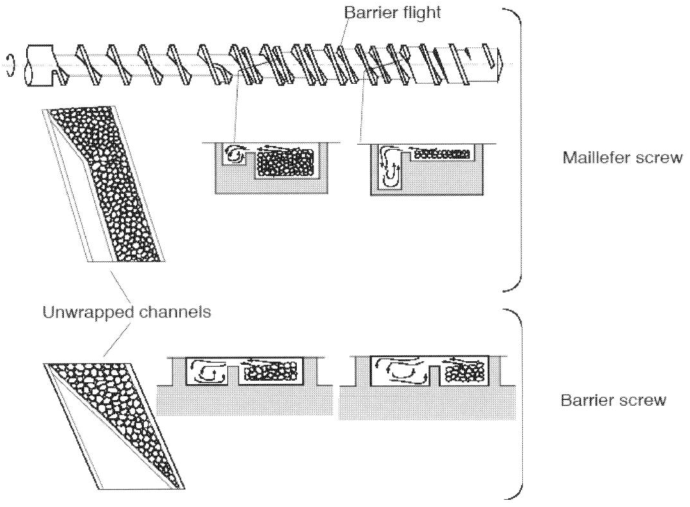

Figure 3.11: Schematic diagram of screws with different barrier flights.

Figure 3.10 also shows a cross section of the screw channel in the melting zone. The solid bed is pushed against the leading flight of the screw as freshly molten polymer is wiped from the melt film into the melt pool by the relative motion between the solids bed and the barrel surface. Knowing where the melt starts and ends is important when designing a screw for a specific application. The solid bed profile that develops during plastication remains one of the most important aspects of screw design.

From experiment to experiment there are always large variations in the experimental solids bed profiles. The variations in this section of the extruder are caused by slight variations in processing conditions and by the uncontrolled solids bed break up towards the end of melting. This effect can be eliminated by introducing a screw with a barrier flight that separates the solids bed from the melt pool. The Maillefer screw and barrier screw in Fig. 3.11 are commonly used for high quality and reproducibility. The Maillefer screw maintains a constant solids bed width, using most effectively the melting with melt-removal mechanism, while the barrier screw uses a constant channel depth with a gradually decreasing solids bed width.

The metering zone. The metering zone is the most important section in melt extruders and conventional single screw extruders that rely on it to generate pressures sufficient for pumping. In both the grooved barrel and the conventional extruder, the diameter of the screw determines the metering or pumping capacity of the extruder. Figure 3.12 presents typical normalized mass throughput as a function of screw diameter for both systems.

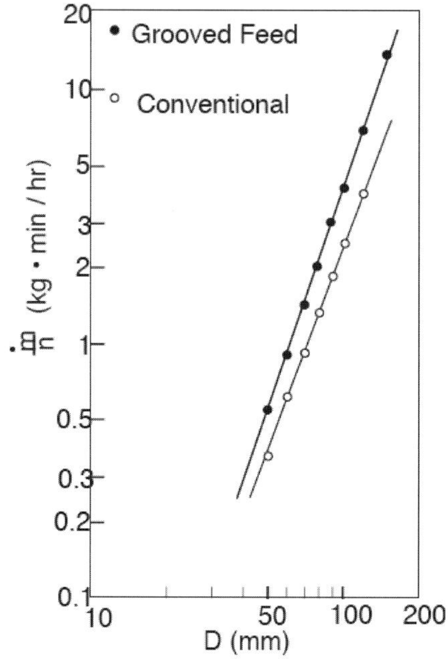

Figure 3.12: Throughput for conventional and grooved feed extruders.

3.1.2 Extrusion Dies

The extrusion die shapes the polymer melt into its final profile. It is located at the end of the extruder and used to extrude

- Flat films and sheets
- Pipes and tubular films for bags
- Filaments and strands
- Hollow profiles for window frames
- Open profiles

As shown in Fig. 3.13, depending on the functional needs of the product, several rules of thumb can be followed when designing an extruded plastic profile. These are:

- Avoid thick sections. Thick sections add to the material cost and increase sink marks caused by shrinkage.
- Minimize the number of hollow sections. Hollow sections add to die cost and make the die more difficult to clean.
- Generate profiles with constant wall thickness. Constant wall thickness in a profile makes it easier to control the thickness of the final profile and results in a more even crystallinity distribution in semi-crystalline profiles.

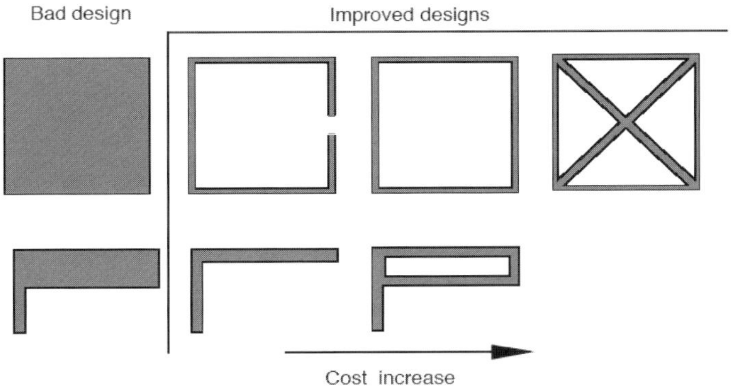

Figure 3.13: Extrusion profile design.

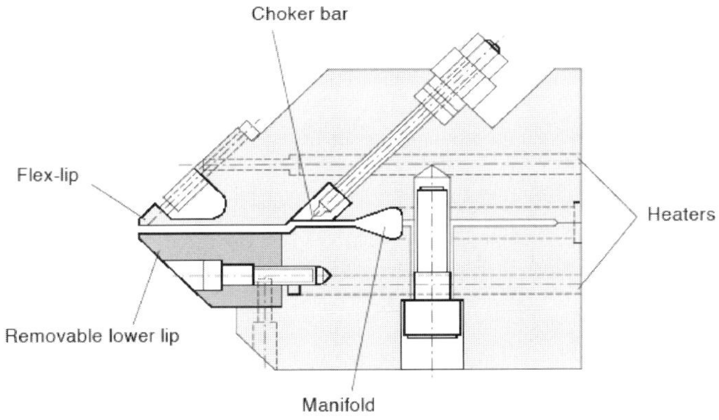

Figure 3.14: Cross-section of a coat-hanger die.

Sheeting dies. One of the most widely used extrusion dies is the coat-hanger sheeting die. A sheeting die, such as the one depicted in Fig. 3.14, is formed by the following elements:

- Manifold: evenly distributes the melt to the approach or land region
- Approach or land: carries the melt from the manifold to the die lips
- Die lips: perform the final shaping of the melt
- Flex lips: for fine tuning when generating a uniform profile

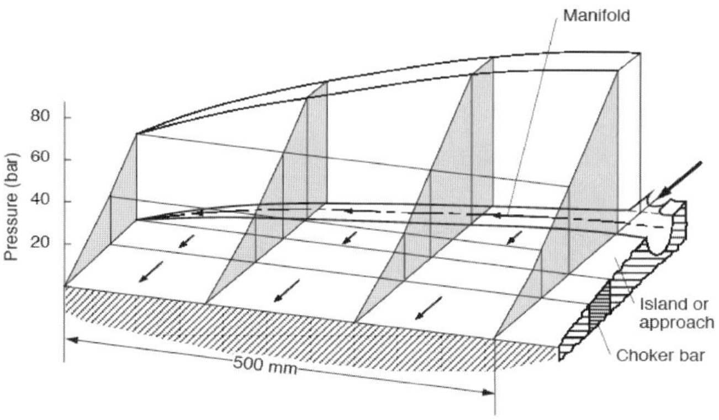

Figure 3.15: Pressure distribution in a coat-hanger die.

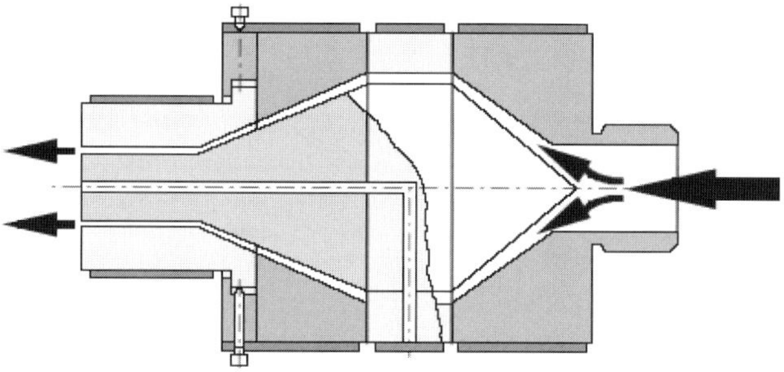

Figure 3.16: Schematic diagram of a spider leg tubing die.

To generate a uniform extrudate geometry at the die lips, the geometry of the manifold must be specified appropriately. Figure 3.15 presents the schematic of a coat-hanger die with a pressure distribution that corresponds to a die that renders a uniform extrudate. It is important to mention that the flow through the manifold and the approach zone depend on the non-Newtonian properties of the polymer extruded. Hence, a die designed for one material does not necessarily work for another.

Tubular dies. In a tubular die, the polymer melt exits through an annulus. These dies are used to extrude plastic pipes and tubular film. The film blowing operation is discussed in more detail later in this chapter. The simplest tubing die is the *spider die*, depicted in Fig. 3.16. Here, a symmetric mandrel is attached to the body of the die by several legs. The polymer must flow around the spider legs causing weld lines along the pipe or film. These weld lines, visible streaks along the extruded tube, are weaker regions.

To overcome weld line problems, the *cross-head tubing die* is often used. Here, the die design is similar to that of the coat-hanger die, but wrapped around a cylinder. This die is depicted in Fig. 3.17. Since the polymer melt must flow around the mandrel, the extruded

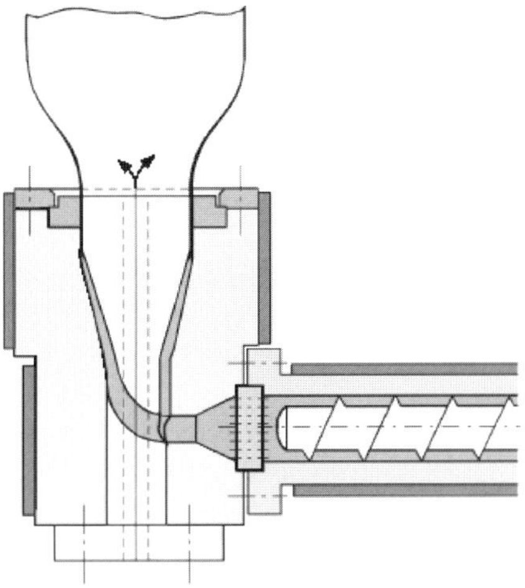

Figure 3.17: Schematic diagram of a cross-head tubing die used in film extrusion.

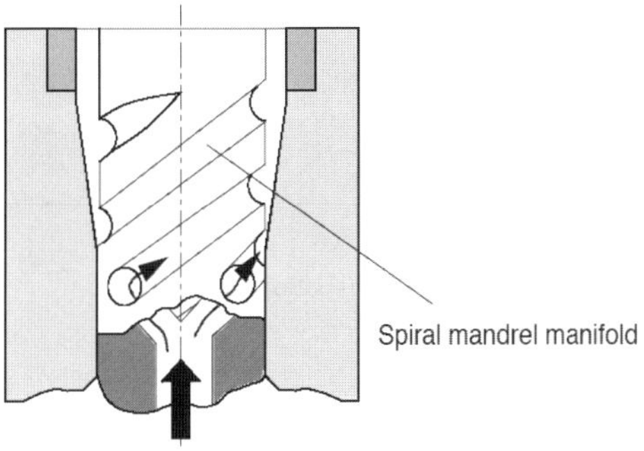

Figure 3.18: Schematic diagram of a spiral die.

tube exhibits one weld line. In addition, although the eccentricity of a mandrel can be controlled using adjustment screws, there is no flexibility to perform fine tuning such as in the coat-hanger die. This can result in tubes with uneven thickness distributions.

The *spiral die*, commonly used to extrude tubular blown films, eliminates weld line effects and produces a thermally and geometrically homogeneous extrudate. The polymer melt in a spiral die flows through several feed ports into independent spiral channels wrapped around the circumference of the mandrel. This type of die is schematically depicted in Fig. 3.18.

3.2 MIXING PROCESSES

Today, most processes involve some form of mixing. As discussed in the previous section, an integral part of a screw extruder is a mixing zone. In fact, most twin screw extruders are primarily used as mixing devices. Similarly, the plasticating unit of an injection molding

Table 3.2: Common Polymer Blends

Compatible polymer blends	
	Naturtal rubber and polybutadiene
	Polyamides (e.g., PA 6 and PA 66)
	Polyphenylene ether (PPE) and polystyrene
Partially incompatible polymer blends	
	Polyethylene and polyisobutylene
	Polyethylene and polypropylene (5% PE in PP)
	Polycarbonate and polyethylene teraphthalate
Incompatible polymer blends	
	Polystyrene/polyethylene
	Polyamide/polyethylene
	Polypropylene/polystyrene

machine often has a mixing zone. This is important because the quality of the finished product in almost all polymer processes depends in part on how well the material was mixed. Both the material properties and the formability of the compound into shaped parts are highly influenced by the mixing quality. Hence, a better understanding of the mixing process helps to optimize processing conditions and increase part quality.

The process of polymer blending or mixing is accomplished by distributing or dispersing a minor or secondary component within a major component serving as a matrix. The major component can be thought of as the continuous phase, and the minor components as distributed or dispersed phases in the form of droplets, filaments, or agglomerates. When creating a polymer blend, one must always keep in mind that the blend will probably be remelted in subsequent processing or shaping processes. For example, a rapidly cooled system, frozen as a homogenous mixture, can separate into phases because of coalescence when re-heated. For all practical purposes, such a blend is not processable. To avoid this problem, compatibilizers, which are macromolecules used to ensure compatibility in the boundary layers between the two phases, are common [26]..

The morphology development of polymer blends is determined by three competing mechanisms: distributive mixing, dispersive mixing, and coalescence. Figure 3.19 presents a model, proposed by Macosko and co-workers [26], that helps visualize the mechanisms governing morphology development in polymer blends. The process begins when a thin tape of polymer is melted away from the pellet. As the tape is stretched, surface tension causes it to rip and to form into threads. These threads stretch and reduce in radius, until surface tension becomes significant enough which leads to Rayleigh distrurbances. These cause the threads to break down into small droplets.

There are three general categories of mixtures that can be created:

- Homogeneous mixtures of compatible polymers,
- Single phase mixtures of partly incompatible polymers, and
- Multi-phase mixtures of incompatible polymers.

Table 3.2 lists examples of compatible, partially incompatible, and incompatible polymer blends.

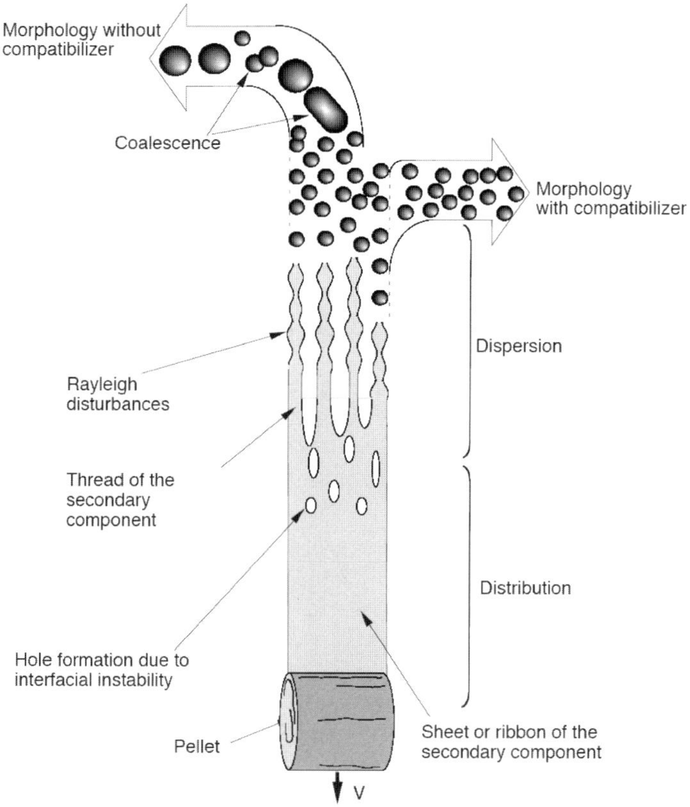

Figure 3.19: Mechanism for morphology development in polymer blends.

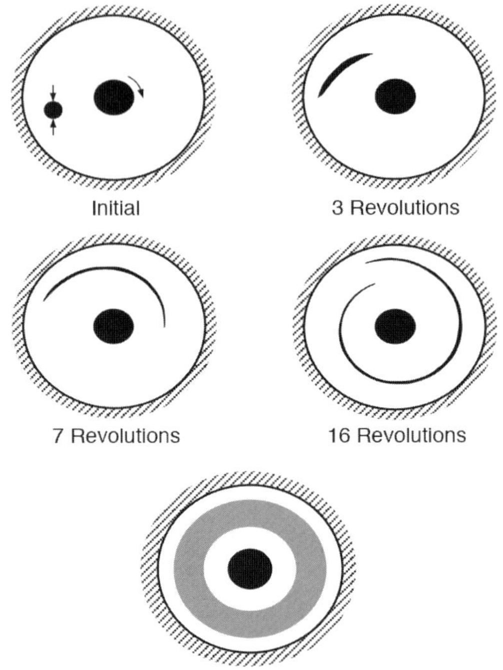

Figure 3.20: Experimental results of distributive mixing in Couette flow, and schematic of the final mixed system.

3.2.1 Distributive Mixing

Distributive mixing, or laminar mixing, of compatible liquids is usually characterized by the distribution of the droplet or secondary phase within the matrix. This distribution is achieved by imposing large strains on the system such that the interfacial area between the two or more phases increases and the local dimensions, or striation thicknesses, of the secondary phases decrease. This concept is shown schematically in Fig. 3.20 [23]. The figure shows a Couette flow device with the secondary component having an initial striation thickness of δ_0. As the inner cylinder rotates, the secondary component is distributed through the systems with constantly decreasing striation thickness; striation thickness depends on the strain rate of deformation which makes it a function of position.

Imposing large strains on the system is not always sufficient to achieve a homogeneous mixture. The type of mixing device, initial orientation and position of the two or more fluid components play a significant role in the quality of the mixture. For example, the mixing problem shown in Fig. 3.20 homogeneously distributes the melt within the region contained by the streamlines cut across by the initial secondary component. The final mixed system is shown in Fig. 3.20.

Figure 3.21 [19] shows another variation of initial orientation and arrangement of the secondary component. Here, the secondary phase cuts across all streamlines, which leads to a homogeneous mixture throughout the Couette device, under appropriate conditions.

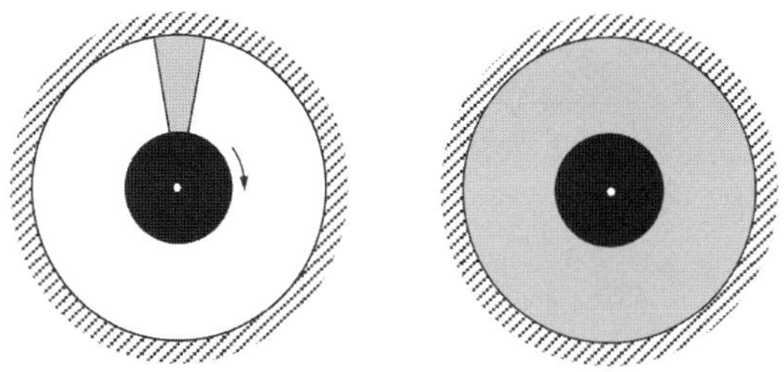

Figure 3.21: Schematic of distributive mixing in Couette flow.

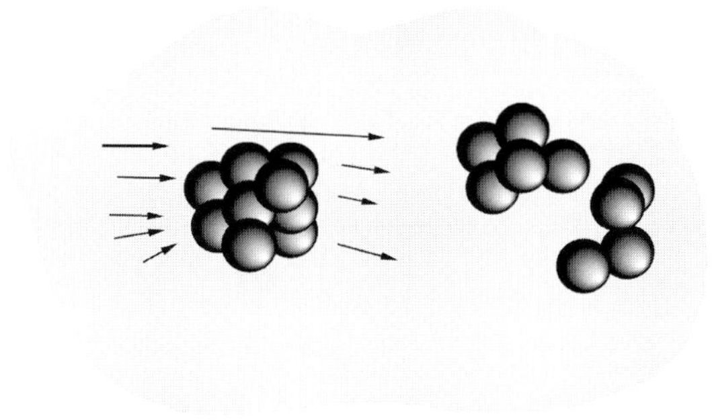

Figure 3.22: break up of particulate agglomerates during flow.

3.2.2 Dispersive Mixing

Dispersive mixing in polymer processing involves breaking a secondary immiscible fluid or an agglomerate of solid particles and dispersing them throughout the matrix. Here, the imposed strain is not as important as the imposed stress which causes the system to break up. Hence, the type of flow inside a mixer plays a significant role on the break up of solid particle clumps or fluid droplets when dispersing them throughout the matrix.

The most common example of dispersive mixing of particulate solid agglomerates is the dispersion and mixing of carbon black into a rubber compound. The dispersion of such a system is schematically represented in Fig. 3.22. However, the break up of particulate agglomerates is best explained using an ideal system of two small spherical particles that need to be separated and dispersed during a mixing process.

If the mixing device generates a simple shear flow, as shown in Fig. 3.23, the maximum separation forces that act on the particles as they travel on their streamline occur when they are oriented in a 45° position as they continuously rotate during flow. However, if the flow field generated by the mixing device is a pure elongational flow, such as shown in Fig. 3.24, the particles will always be oriented at 0°; the position of maximum force.

In general, droplets inside an incompatible matrix tend to stay or become spherical due to the natural tendencies of the drop to maintain the lowest possible surface-to-volume ratio.

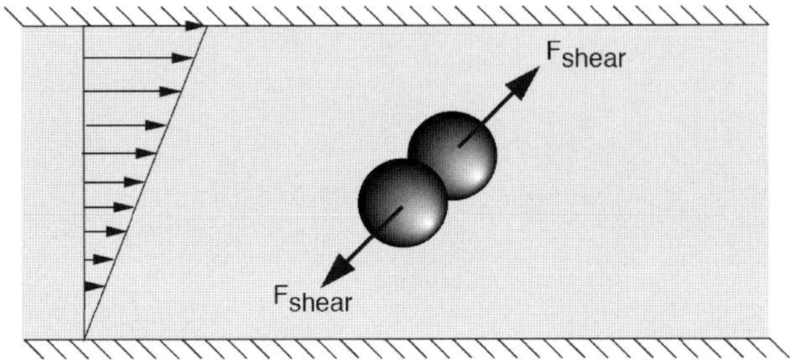

Figure 3.23: Force applied to a two particle agglomerate in a simple shear flow.

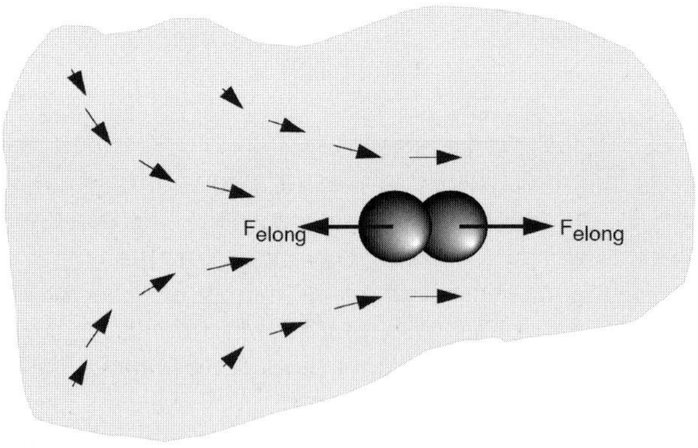

Figure 3.24: Force applied to a two particle agglomerate in an elongational flow.

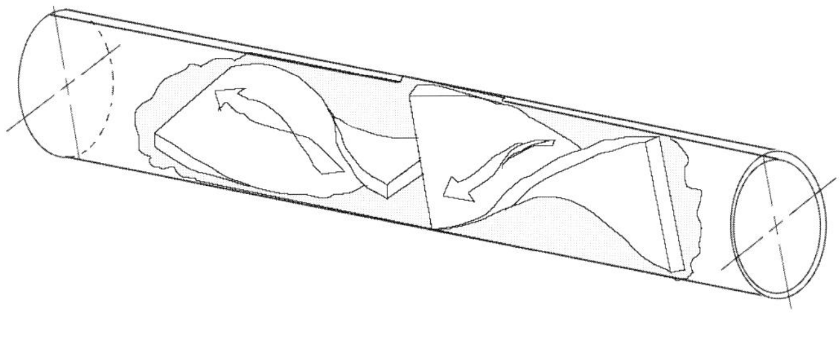

Figure 3.25: Schematic diagram of a Kenics static mixer.

However, a flow field within the mixer applies a stress on the droplets, causing them to deform. If this stress is high enough, it will eventually cause the drops to disperse. The droplets will disperse when the surface tension can no longer maintain their shape in the flow field and the filaments break up into smaller droplets. This phenomenon of dispersion and distribution continues to repeat itself until the deviatoric stresses of the flow field can no longer overcome the surface tension of the new droplets formed. As can be seen, the mechanism of fluid agglomerate break up is similar in nature to solid agglomerate break up in the sense that both rely on forces to disperse the particulates. Hence, elongation is also the preferred mode of deformation when breaking up fluid droplets and threads.

3.2.3 Mixing Devices

The final properties of a polymer component are heavily influenced by the blending or mixing process that takes place during processing or as a separate step in the manufacturing process. As mentioned earlier, when measuring the quality of mixing it is also necessary to evaluate the efficiency of mixing. For example, the amount of power required to achieve the highest mixing quality for a blend may be unrealistic or unachievable. This section presents some of the most commonly used mixing devices encountered in polymer processing.

In general, mixers can be classified in two categories: internal batch mixers and continuous mixers. Internal batch mixers, such as the Banbury type mixer, are the oldest type of mixing devices in polymer processing and are still widely used in the rubber compounding industry. Industry often also uses continuous mixers because they combine mixing in addition to their normal processing tasks. Typical examples are single and twin screw extruders that often have mixing heads or kneading blocks incorporated into their system.

Static mixers. Static mixers or motionless mixers are pressure-driven continuous mixing devices through which the melt is pumped, rotated, and divided, leading to effective mixing without the need for movable parts and mixing heads. One of the most commonly used static mixers is the twisted tape static mixer schematically shown in Fig. 3.25.

The polymer is sheared and then rotated by 90° by the dividing wall, the interfaces between the fluids increase. The interfaces are then re-oriented by 90° once the material enters a new section. The stretching-re-orientation sequence is repeated until the number of striations is so high that a seemingly homogeneous mixture is achieved. Figure 3.26 shows a sequence of cuts down a Kenics static mixer[2]. It can be seen that the number of striations

[2]Courtesy Chemineer, Inc., North Andover, Massachusetts.

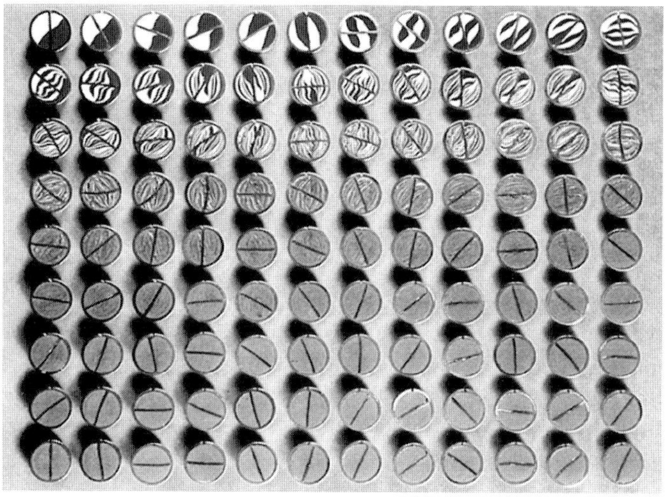

Figure 3.26: Experimental progression of the layering of colored resins in a Kenics static mixer.

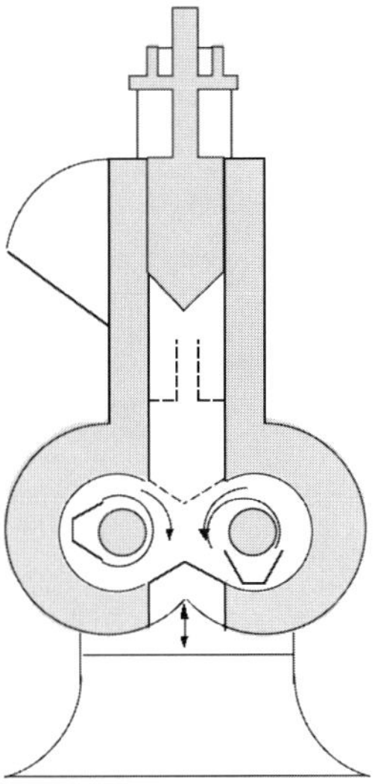

Figure 3.27: Schematic diagram of a Banbury type mixer.

increases from section to section by 2, 4, 8, 16, 32, etc., which can be computed using

$$N = 2^n \tag{3.1}$$

where N is the number of striations and n is the number of sections in the mixer.

Internal batch mixer. The internal batch or Banbury type mixer, schematically shown in Fig. 3.27, is perhaps the most commonly used internal batch mixer. Internal batch mixers are high-intensity mixers that generate complex shearing and elongational flows, which work especially well in the dispersion of solid particle agglomerates within polymer matrices. One of the most common applications for high-intensity internal batch mixing is the break up of carbon black agglomerates into rubber compounds.

The dispersion of agglomerates is strongly dependent on mixing time, rotor speed, temperature, and rotor blade geometry [3]. Figure 3.28 [6, 4] shows the fraction of undispersed carbon black as a function of time in a Banbury mixer at 77 rpm and 100°C. The broken line in the figure represents the fraction of particles smaller than 500 nm.

Mixing in single screw extruders. Mixing caused by the cross-channel flow component can be further enhanced by introducing pins in the flow channel. These pins can either sit on the screw as shown in Fig. 3.29 [9] or on the barrel as shown in Fig. 3.30 [15].

The extruder with the adjustable pins on the barrel is generally referred to as QSM-extruder[3]. In both cases, the pins disturb the flow by re-orienting the surfaces between fluids and by creating new surfaces by splitting the flow. Figure 3.31 shows the channel contents of a QSM-extruder[4]. The photograph demonstrates the re-orientation of the layers as the material flows past the pins. The pin type extruder is especially necessary for the mixing of high viscosity materials such as rubber compounds; thus, it is often called a cold

[3]QSM comes from the German words *Quer Strom Mischer* which translates into cross-flow mixing.
[4]Courtesy of the Paul Troester Maschinenfabrik, Hannover, Germany.

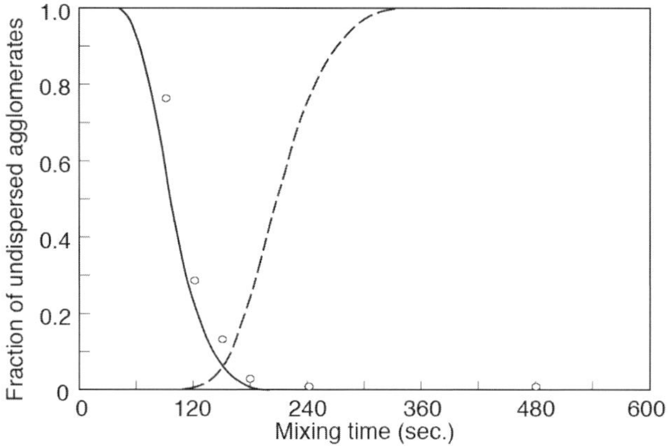

Figure 3.28: Fraction of undispersed carbon black, larger than 9 μm, as a function of mixing time inside a Banbury mixer. The open circles denote experimental results and the solid line a theoretical prediction. The broken line denotes the fraction of aggregates of size below 500 nm.

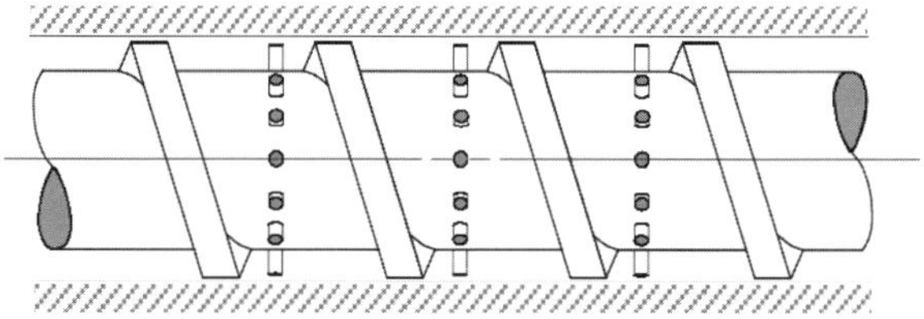

Figure 3.29: Pin mixing section on the screw of a single screw extruder.

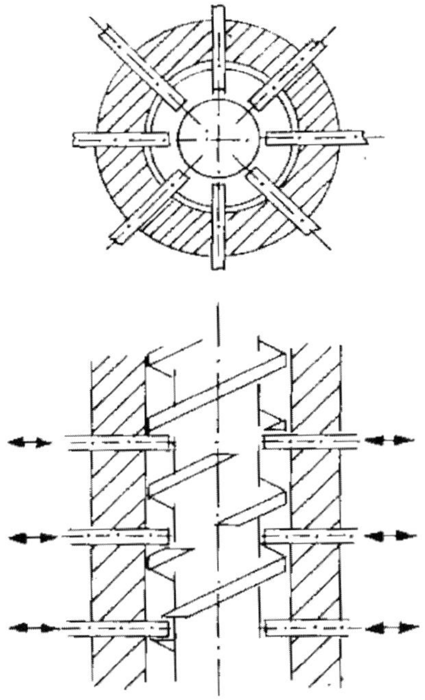

Figure 3.30: Pin barrel extruder.

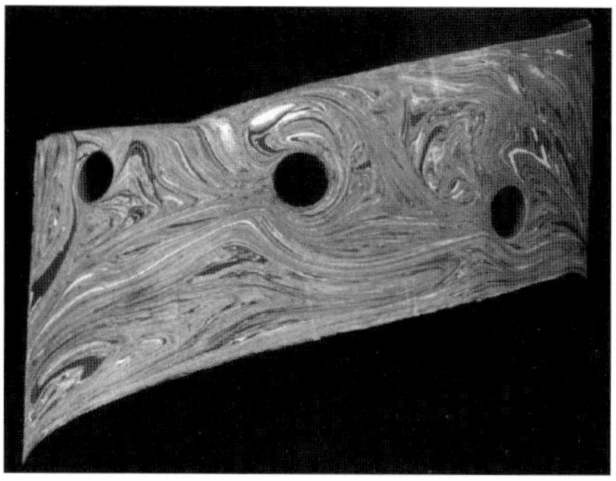

Figure 3.31: Photograph of the unwrapped channel contents of a pin barrel extruder.

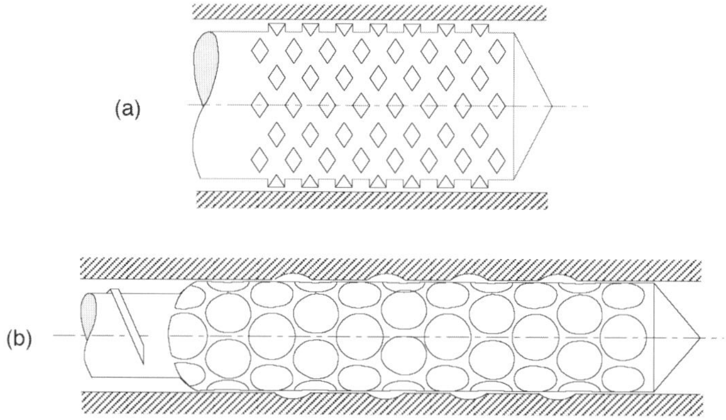

Figure 3.32: Distributive mixing sections: (a) Pineapple mixing section, (b) Cavity transfer mixing section.

feed rubber extruder. This machine is widely used in the production of rubber profiles of any shape and size.

For lower viscosity fluids, such as thermoplastic polymer melts, the mixing action caused by the cross-flow is often not sufficient to re-orient, distribute, and disperse the mixture, making it necessary to use special mixing sections. Re-orientation of the interfaces between primary and secondary fluids and distributive mixing can be induced by any disruption in the flow channel. Figure 3.32 [9] presents commonly used distributive mixing heads for single screw extruders. These mixing heads introduce several disruptions in the flow field, which have proven to perform well in mixing.

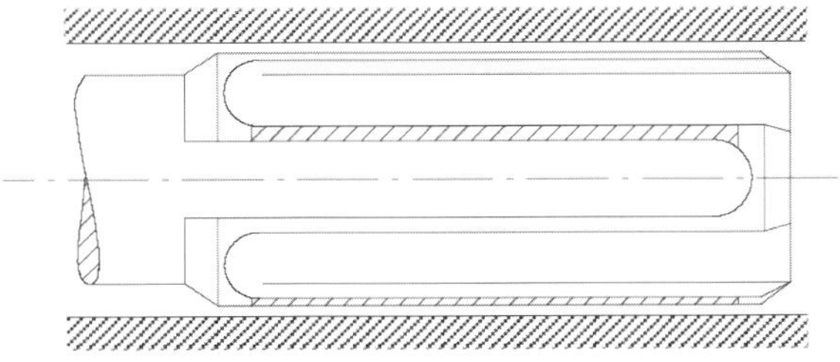

Figure 3.33: Maddock or Union Carbide mixing section.

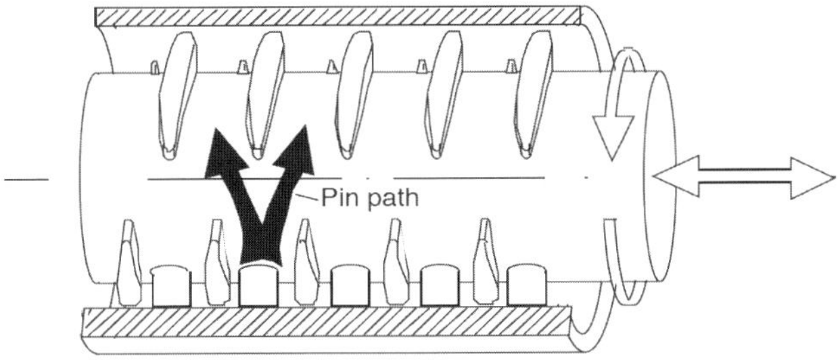

Figure 3.34: Schematic diagram of a cokneater.

As mentioned earlier, dispersive mixing is required when breaking down particle agglomerates or when surface tension effects exist between primary and secondary fluids in the mixture. To disperse such systems, the mixture must be subjected to large stresses. Barrier-type screws are often sufficient to apply high stresses to the polymer melt. However, more intensive mixing can be applied by using a mixing head. When using barrier-type screws or a mixing head as shown in Fig. 3.33 [9], the mixture is forced through narrow gaps, causing high stresses in the melt. It should be noted that dispersive as well as distributive mixing heads result in a resistance to the flow, which results in viscous heating and pressure losses during extrusion.

Cokneader. The cokneader is a single screw extruder with pins on the barrel and a screw that oscillates in the axial direction. Figure 3.34 shows a schematic diagram of a cokneader. The pins on the barrel practically wipe the entire surface of the screw, making it the only self-cleaning single-screw extruder. This results in a reduced residence time, which makes it appropriate for processing thermally sensitive materials. The pins on the barrel also disrupt the solid bed creating a dispersed melting [23] which improves the overall melting rate while reducing the overall temperature in the material.

A simplified analysis of a cokneader gives a number of striations per L/D of [24]

$$N_s = 2^{12} \tag{3.2}$$

which means that over a section of 4D the number of striations is $2^{12}(4) = 28^{13}$. A detailed discussion on the cokneader is given by Rauwendaal [24] and Elemans [9].

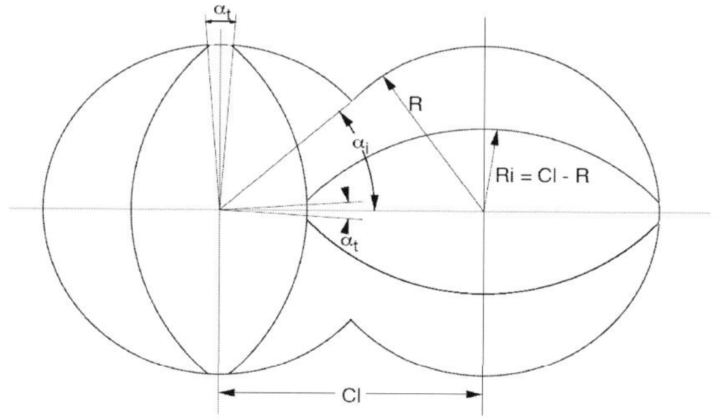

Figure 3.35: Geometry description of a double-flighted, co-rotating, self-cleaning twin screw extruder.

Twin screw extruders. In the past two decades, twin screw extruders have developed into the best available continuous mixing devices. In general, they can be classified into intermeshing or non-intermeshing, and co-rotating or counter-rotating twin screw extruders. The intermeshing twin screw extruders render a self-cleaning effect that evens out the residence time of the polymer in the extruder. The self-cleaning geometry for a co-rotating double flighted twin screw extruder is shown in Fig. 3.35. The main characteristic of this type of configuration is that the surfaces of the screws are sliding past each other, constantly removing the polymer that is stuck to the screw.

In the last two decades, the co-rotating twin screw extruder systems have established themselves as efficient continuous mixers, including reactive extrusion. In essence, the co-rotating systems have a high pumping efficiency caused by the double transport action of the two screws. Counter-rotating systems generate high stresses because of the calendering action between the screws, making them efficient machines to disperse pigments and lubricants[5].

Several studies have been performed to evaluate the mixing capabilities of twin screw extruders. Noteworthy are two studies performed by Lim and White [12, 13] that evaluated the morphology development in a 30.7 mm diameter screw co-rotating [28] and a 34 mm diameter screw counter-rotating [3] intermeshing twin screw extruder. In both studies they dry-mixed 75/25 blend of polyethylene and polyamide 6 pellets that were fed into the hopper at 15 kg/h. Small samples were taken along the axis of the extruder and evaluated using optical and electron microscopy.

The quality of the dispersion of the blend is assessed by the reduction of the characteristic size of the polyamide 6 phase. Figure 3.36 is a plot of the weight average and number average domain size of the polyamide 6 phase along the screw axis. The weight average phase size at the end of the extruder was measured to be 10 μm and the number average 6 μm.

[5]There seems to be considerable disagreement about co-versus counter-rotating twin screw extruders between different groups in the polymer processing industry and academic community.

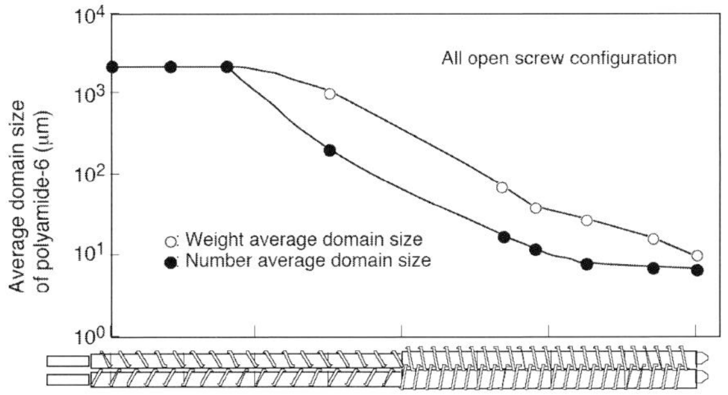

Figure 3.36: Number and weight average of polyamide 6 domain sizes along the screws for a counter-rotating twin screw extruder.

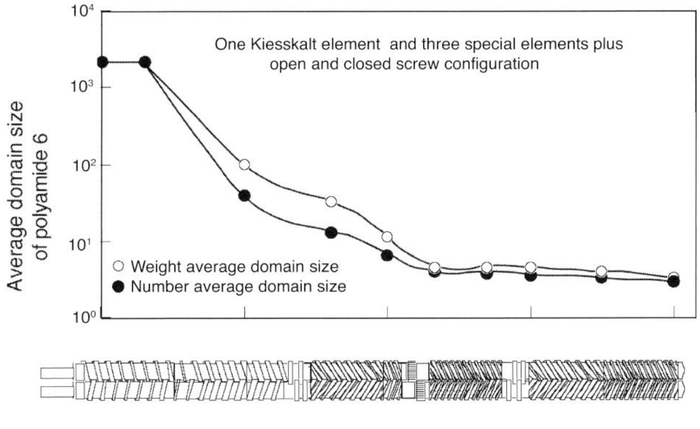

Figure 3.37: Number and weight average of polyamide 6 domain sizes along the screws for a counter-rotating twin screw extruder with special mixing elements.

By replacing sections of the screw with one kneading-pump element and three special mixing elements, the final weight average phase size was reduced to 2.2 μm and the number average to 1.8 μm, as shown in Fig. 3.37.

Using a co-rotating twin screw extruder with three kneading disk blocks, a final morphology with polyamide 6 weight average phase sizes of 2.6 μm was achieved. Figure 3.38 shows the morphology development along the axis of the screws. When comparing the outcome of both counter-rotating (Fig. 3.37) and co-rotating (Fig. 3.38), it is clear that both extruders achieve a similar final mixing quality. However, the counter-rotating extruder achieved the final morphology much earlier in the screw than the co-rotating twin screw extruder. A possible explanation for this is that the blend traveling through the counter-rotating configuration melted earlier than in the co-rotating geometry. In addition the phase

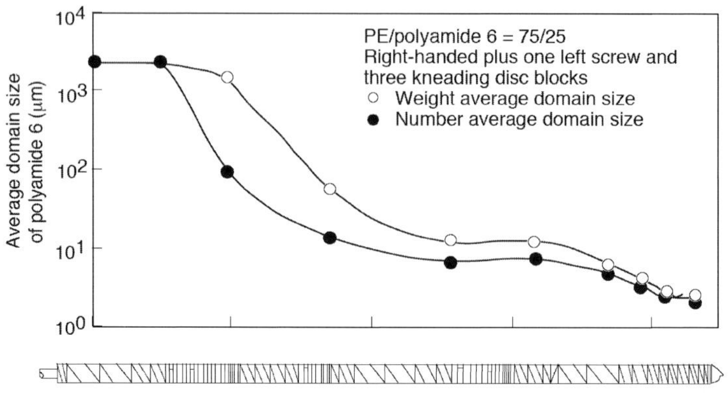

Figure 3.38: Number and weight average of polyamide 6 domain sizes along the screws for a co-rotating twin screw extruder with special mixing elements.

size was slightly smaller, possibly due to the calendering effect between the screws in the counter-rotating system.

3.3 INJECTION MOLDING

Injection molding is the most important process used to manufacture plastic products. Today, more than one-third of all thermoplastic materials are injection molded and more than half of all polymer processing equipment is for injection molding. The injection molding process is ideally suited to manufacture mass-produced parts of complex shapes requiring precise dimensions. The process goes back to 1872 when the Hyatt brothers patented their stuffing machine to inject cellulose into molds. However, today's injection molding machines are mainly related to the reciprocating screw injection molding machine patented in 1956. A modern injection molding machine with its most important elements is shown in Fig. 3.39. The components of the injection molding machine are the plasticating unit, clamping unit, and the mold.

Today, injection molding machines are classified by the following international convention[6]

MANUFACTURER T/P

where T is the clamping force in metric tons and P is defined as

$$P = \frac{V_{max} p_{max}}{1000} \tag{3.3}$$

where V_{max} is the maximum shot size in cm^3 and p_{max} is the maximum injection pressure in bar. The clamping forced T can be as low as 1metric ton for small machines, and as high as 11,000 tons.

[6]The old US convention uses MANUFACTURER T/V where T is the clamping force in British tons and V the shot size in ounces of polystyrene.

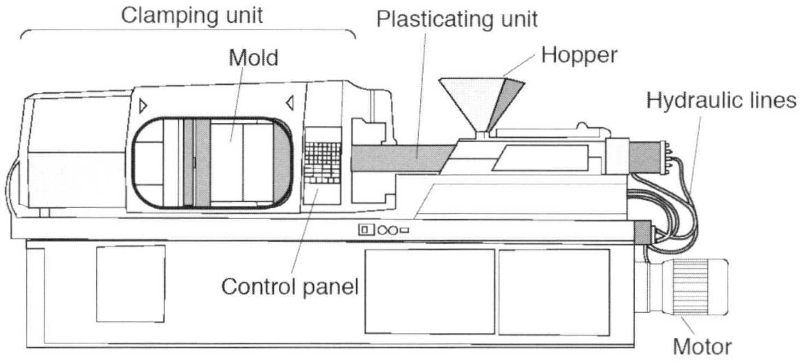

Figure 3.39: Schematic of an injection molding machine.

3.3.1 The Injection Molding Cycle

The sequence of events during the injection molding of a plastic part, as shown in Fig. 3.40, is called the injection molding cycle. The cycle begins when the mold closes, followed by the injection of the polymer into the mold cavity. Once the cavity is filled, a holding pressure is maintained to compensate for material shrinkage. In the next step, the screw turns, feeding the next shot to the front of the screw. This causes the screw to retract as the next shot is prepared. Once the part is sufficiently cool, the mold opens and the part is ejected.

Figure 3.41 presents the sequence of events during the injection molding cycle. The figure shows that the cycle time is dominated by the cooling of the part inside the mold cavity. The total cycle time can be calculated using

$$t_{\text{cycle}} = t_{\text{closing}} + t_{\text{cooling}} + t_{\text{ejection}} \tag{3.4}$$

where the closing and ejection times, t_{closing} and t_{ejection}, can last from a fraction of a second to a few seconds, depending on the size of the mold and machine. The cooling times, which dominate the process, depend on the maximum thickness of the part.

Using the average part temperature history and the cavity pressure history, the process can be followed and assessed using the pvT diagram as depicted in Fig. 3.42 [11, 18]. To follow the process on the pvT diagram, we must transfer both the temperature and the pressure at matching times. The diagram reveals four basic processes: an isothermal injection (0-1) with pressure rising to the holding pressure (1-2), an isobaric cooling process during the holding cycle (2-3), an isochoric cooling after the gate freezes with a pressure drop to atmospheric (3-4), and then isobaric cooling to room temperature (4-5).

The point on the pvT diagram at which the final isobaric cooling begins (4), controls the total part shrinkage. This point is influenced by the two main processing conditions –the melt temperature and the holding pressure as depicted in Fig. 3.43.

Here, the process in Fig. 3.42 is compared to one with a higher holding pressure. Of course, there is an infinite combination of conditions that render acceptable parts, bound by minimum and maximum temperatures and pressures. Figure 3.44 presents the molding diagram with all limiting conditions. The melt temperature is bound by a low temperature that results in a short shot or unfilled cavity and a high temperature that leads to material degradation. The hold pressure is bound by a low pressure that leads to excessive shrinkage or low part weight, and a high pressure that results in flash. Flash results when the cavity

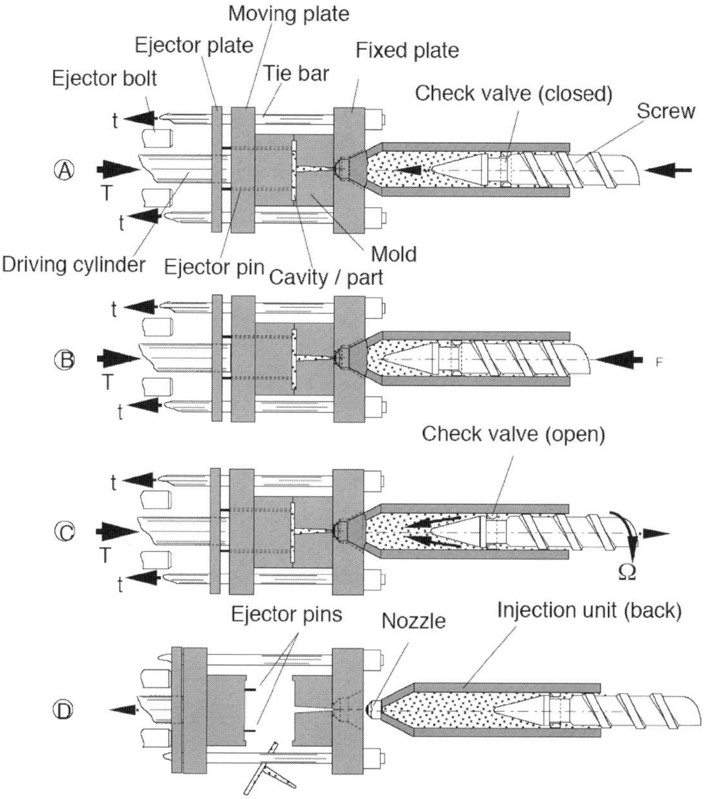

Figure 3.40: Sequence of events during an injection molding cycle.

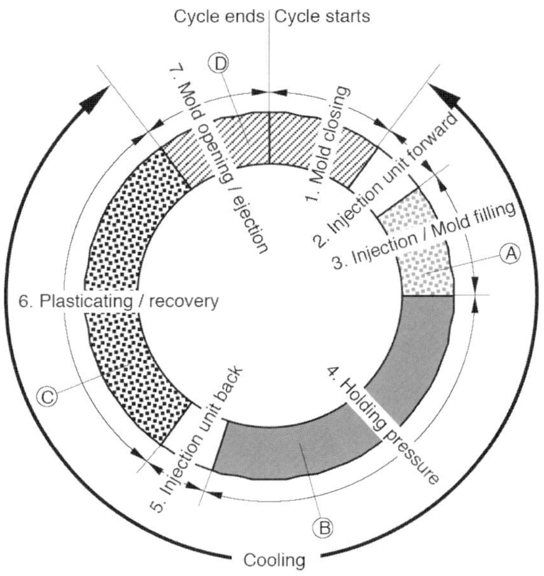

Figure 3.41: Injection molding cycle.

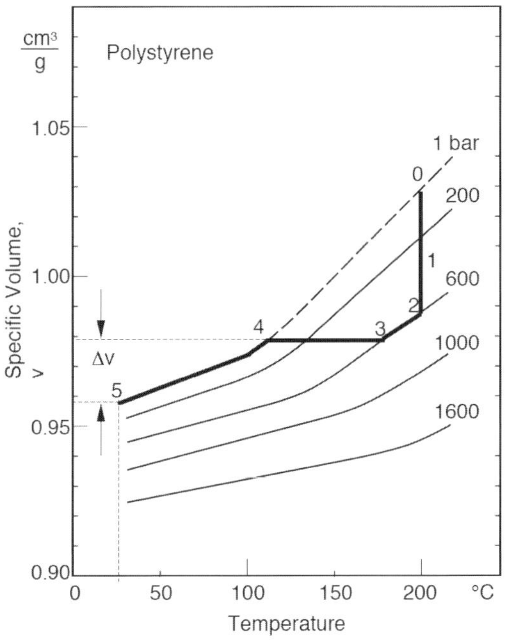

Figure 3.42: Trace of an injection molding cycle in a pvT diagram.

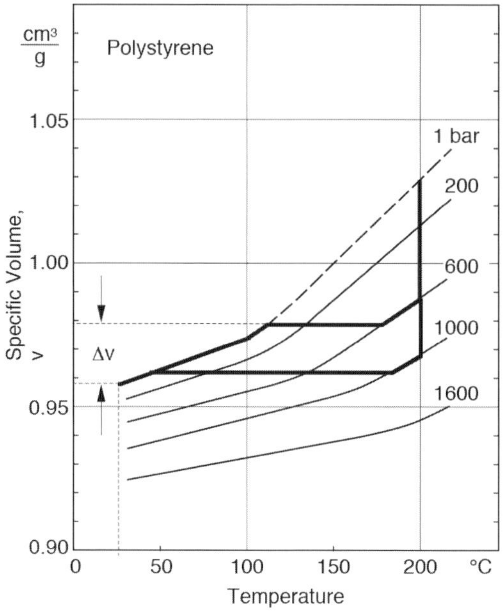

Figure 3.43: Trace of two different injection molding cycles in a pvT diagram.

pressure force exceeds the machine clamping force, leading to melt flow across the mold parting line. The holding pressure determines the corresponding clamping force required to size the injection molding machine. An experienced polymer processing engineer can usually determine which injection molding machine is appropriate for a specific application. For the untrained polymer processing engineer, finding this appropriate holding pressure and its corresponding mold clamping force can be difficult.

With difficulty one can control and predict the component's shape and residual stresses at room temperature. For example, sink marks in the final product are caused by material shrinkage during cooling, and residual stresses can lead to environmental stress cracking under certain conditions [17]. Warpage in the final product is often caused by processing conditions that lead to asymmetric residual stress distributions through the part thickness. The formation of residual stresses in injection molded parts is attributed to two major coupled factors: cooling and flow stresses. The first and most important is the residual stress formed as a result of rapid cooling which leads to large temperature variations.

3.3.2 The Injection Molding Machine

The plasticating and injection unit. A plasticating and injection unit is shown in Fig. 3.45. The major tasks of the plasticating unit are to melt the polymer, to accumulate the melt in the screw chamber, to inject the melt into the cavity, and to maintain the holding pressure during cooling.

The main elements of the plasticating unit follow:

- Hopper
- Screw

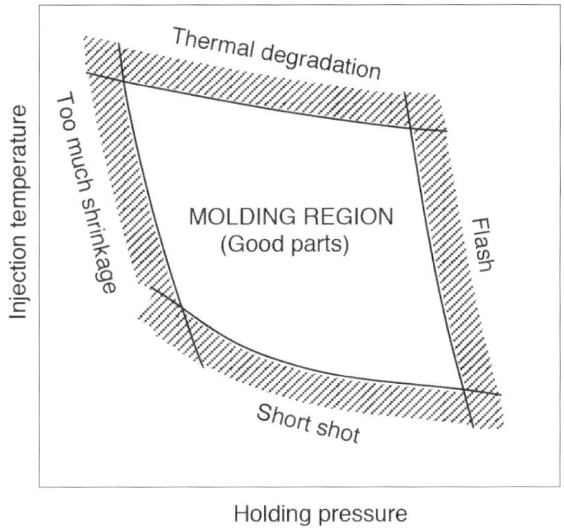

Figure 3.44: The molding diagram.

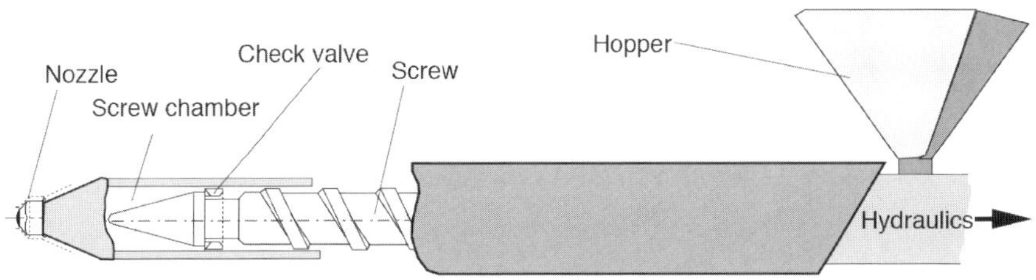

Figure 3.45: Schematic of a plasticating unit.

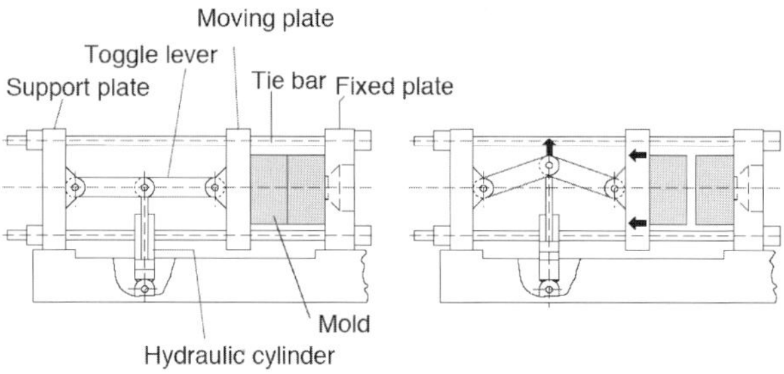

Figure 3.46: Clamping unit with a toggle mechanism.

- Heater bands
- Check valve
- Nozzle

The hopper, heating bands, and the screw are similar to a plasticating single screw extruder, except that the screw in an injection molding machine can slide back and forth to allow for melt accumulation and injection. This characteristic gives it the name reciprocating screw. For quality purposes, the maximum stroke in a reciprocating screw should be set shorter than $3D$. Although the most common screw used in injection molding machines is the three-zone plasticating screw, two-stage vented screws are often used to extract moisture and monomer gases just after the melting stage.

The check valve, or non-return valve, is at the end of the screw and enables it to work as a plunger during injection and packing without allowing polymer melt back to flow into the screw channel. A check valve and its function during operation is depicted in Fig. 3.40, and in Fig. 3.45. A high quality check valve allows less then 5% of the melt back into the screw channel during injection and packing. The nozzle is at the end of the plasticating unit and fits tightly against the sprue bushing during injection. The nozzle type is either open or shut-off. The open nozzle is the simplest, rendering the lowest pressure consumption.

The clamping unit. The job of a clamping unit in an injection molding machine is to open the mold, and to close it tightly to avoid flash during the filling and holding. Modern injection molding machines have two predominant clamping types: mechanical and hydraulic. Figure 3.46 presents a toggle mechanism in the open and closed mold positions. Although the toggle is essentially a mechanical device, it is actuated by a hydraulic cylinder. The advantage of using a toggle mechanism is that, as the mold approaches closure, the available closing force increases and the closing decelerates significantly. However, the toggle mechanism only transmits its maximum closing force when the system is fully extended.

Figure 3.47 presents a schematic of a hydraulic clamping unit in the open and closed positions. The advantages of the hydraulic system is that a maximum clamping force is attained at any mold closing position and that the system can take different mold sizes without major system adjustments.

The mold cavity. The central point in an injection molding machine is the mold. The mold distributes polymer melt into and throughout the cavities, shapes the part, cools the

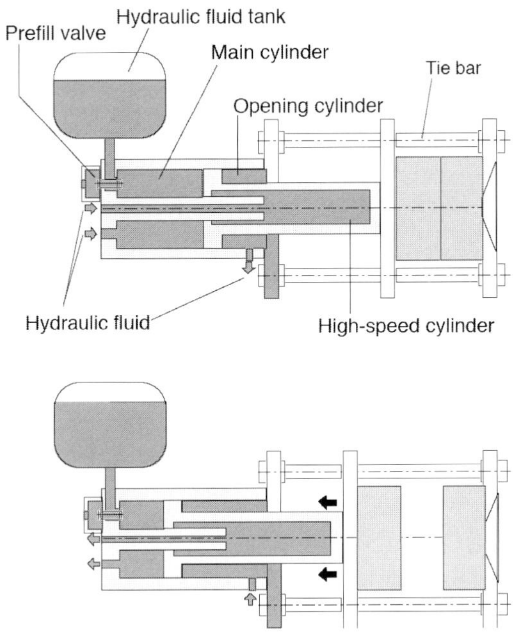

Figure 3.47: Hydraulic clamping unit.

melt and ejects the finished product. The mold is typically custom-made and consists of the following elements (see Fig. 3.48):

- Sprue and runner system
- Gate
- Mold cavity
- Cooling system (thermoplastics)
- Ejector system

During mold filling, the melt flows through the sprue and is distributed into the cavities by the runners, as seen in Fig. 3.49.

The runner system in Fig. 3.49(a) is symmetric, where all cavities fill at the same time causing the polymer to fill all cavities uniformly. The disadvantage of this balanced runner system is that the flow paths are long, leading to high material and pressure consumption. On the other hand, the asymmetric runner system shown in Fig. 3.49(b) leads to parts of different quality. Uniform filling of the mold cavities can also be achieved by varying runner diameters. There are two types of runner systems – cold and hot runners. Cold runners are ejected with the part and are trimmed after mold removal. The advantage of the cold runner is lower mold cost. The hot runner keeps the polymer at its melt temperature. The material stays in the runner system after ejection, and is injected into the cavity in the following cycle. There are two types of hot runner system: externally and internally

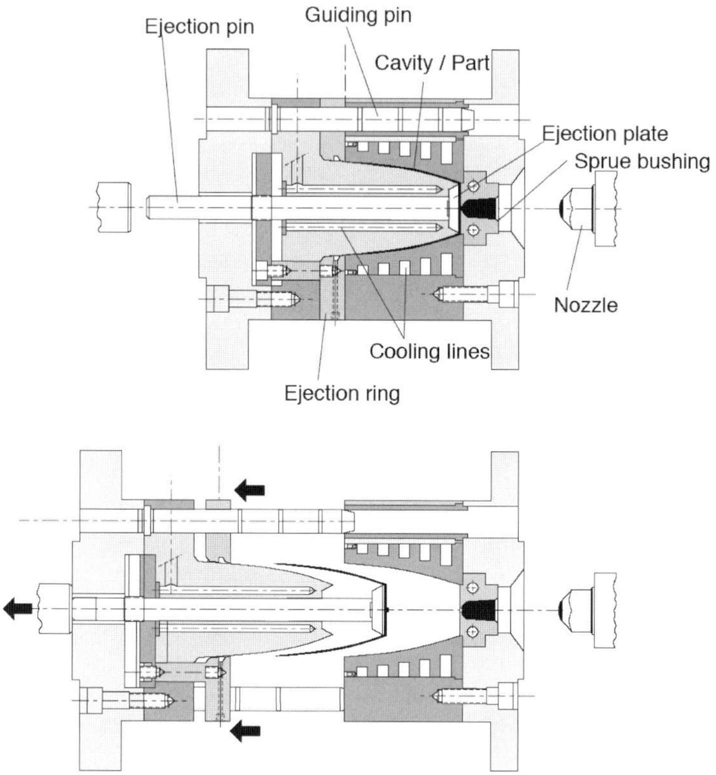

Figure 3.48: An injection mold.

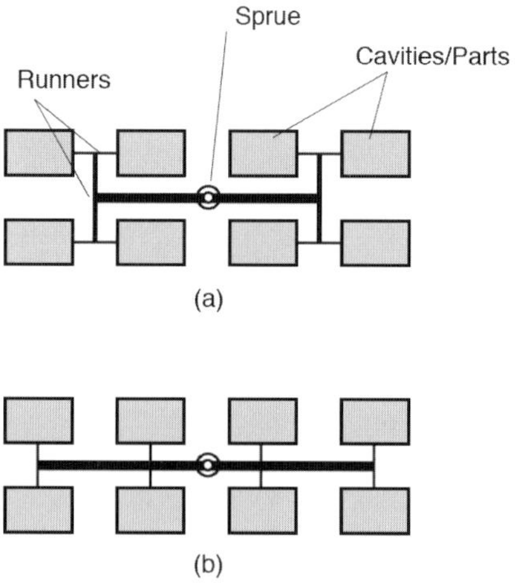

Figure 3.49: Schematic of different runner system arrangements.

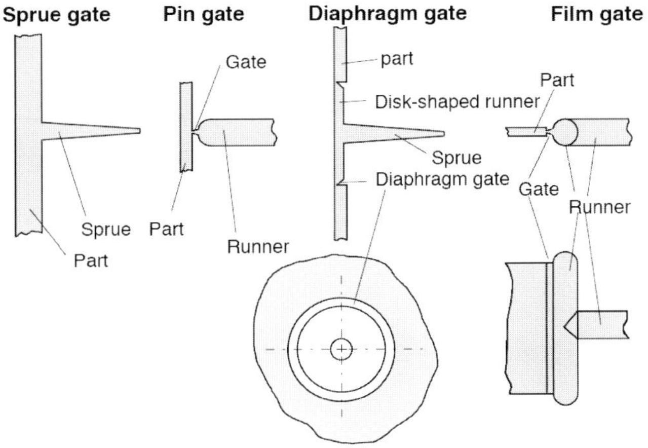

Figure 3.50: Schematic of different gating systems.

heated. The externally heated runners have a heating element surrounding the runner that keeps the polymer isothermal. The internally heated runners have a heating element running along the center of the runner, maintaining a polymer melt that is warmer at its center and possibly solidified along the outer runner surface. Although a hot runner system considerably increases mold cost, its advantages include elimination of trim and lower pressures for injection.

When large items are injection molded, the sprue sometimes serves as the gate, as shown in Fig. 3.50. The sprue must be subsequently trimmed, often requiring further surface finishing. On the other hand, a pin-type gate (Fig. 3.50) is a small orifice that connects the sprue or the runners to the mold cavity. The part is easily broken off from such a gate, leaving only a small mark that usually does not require finishing. Other types of gates, also shown in Fig. 3.50, are film gates, used to eliminate orientation, and disk or diaphragm gates for symmetric parts such as compact discs.

3.3.3 Related Injection Molding Processes

Although most injection molding processes are covered by the conventional process description discussed earlier in this chapter, there are several important molding variations including:

- Multi-color
- Multi-component
- Co-injection
- Gas-assisted

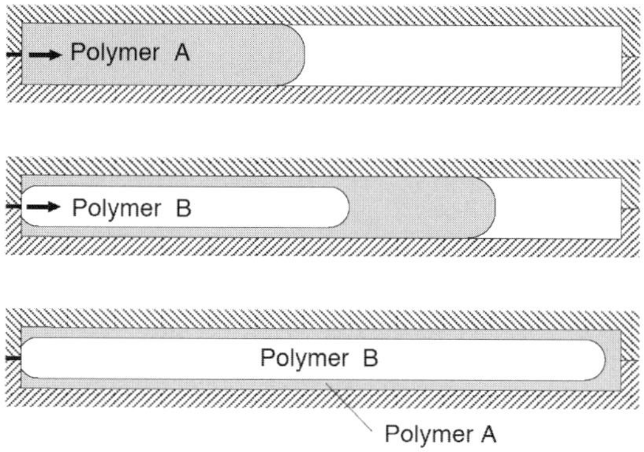

Figure 3.51: Schematic of the co-injection molding process.

- Injection-compression

Multi-component injection molding occurs when two or more polymers, or equal polymers of different color, are injected through different runner and gate systems at different stages during the molding process. Each component is injected using its own plasticating unit. The molds are often located on a turntable. Multi-color automotive stop lights are molded this way. In principle, the multi-component injection molding process is the same as the multi-color process. Here, either two incompatible materials are molded or one component is cooled sufficiently so that the two components do not adhere to each other. For example, to mold a ball and socket system, the socket of the linkage is molded first. The socket component is allowed to cool somewhat and the ball part is injected inside. This results in a perfectly movable system. This type of injection molding process is used to replace tedious assembling tasks and is becoming popular in countries where labor costs are high. In addition, today, a widely used application is the multi-component injection of a hard and a soft polymer such as polypropylene with a thermoplastic elastomer.

In contrast to multi-color and multi-component injection molding, co-injection molding uses the same gate and runner system. Here, the component that forms the outer skin of the part is injected first, followed by the core component. The core component displaces the first and a combination of the no-slip condition between polymer and mold and the freezing of the melt creates a sandwiched structure as depicted in Fig. 3.51.

In principle, the gas-assisted injection molding process is similar to co-injection molding. Here, the second or core component is nitrogen, which is injected through a needle into the polymer melt, blowing the melt out of the way and depositing it against the mold surfaces.

Injection-compression molding first injects the material into a partially opened mold, and then squeezes the material by closing the mold. Injection-compression molding is used for polymer products that require a high quality surface finish, such as compact discs and other optically demanding components because it practically eliminates tangential molecular orientation.

3.4 SECONDARY SHAPING

Secondary shaping operations such as extrusion blow molding, film blowing, and fiber spinning occur immediately after the extrusion profile emerges from the die. The thermo-

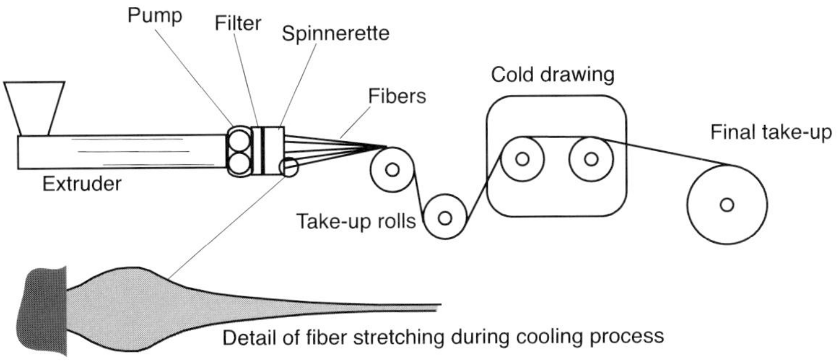

Figure 3.52: The fiber spinning process with detail of a stretching fiber during the cooling process.

forming process is performed on sheets or plates previously extruded and solidified. In general, secondary shaping operations consist of mechanical stretching or forming of a preformed cylinder, sheet, or membrane.

3.4.1 Fiber Spinning

Fiber spinning is used to manufacture synthetic fibers. During fiber spinning, a filament is continuously extruded through an orifice and stretched to diameters of 100 μm and smaller. The process is schematically depicted in Fig. 3.52. The molten polymer is first extruded through a filter or screen pack, to eliminate small contaminants. The melt is then extruded through a spinneret, a die composed of multiple orifices. A spinneret can have between one and 10,000 holes. The fibers are then drawn to their final diameter, solidified, and wound onto a spool. The solidification takes place either in a water bath or by forced convection. When the fiber solidifies in a water bath, the extrudate undergoes an adiabatic stretch before cooling begins in the bath. The forced convection cooling, which is more commonly used, leads to a non-isothermal spinning process. The drawing and cooling processes determine the morphology and mechanical properties of the final fiber. For example, ultra high molecular weight PE-HD fibers with high degrees of orientation in the axial direction can have the stiffness of steel with today's fiber spinning technology.

Of major concern during fiber spinning are the instabilities that arise during drawing, such as brittle fracture, Rayleigh disturbances, and draw resonance. Brittle fracture occurs when the elongational stress exceeds the melt strength of the drawn polymer melt. The instabilities caused by Rayleigh disturbances are like those causing filament break up during dispersive mixing, as discussed in Chapter 4. Draw resonance appears under certain conditions and manifests itself as periodic fluctuations that result in diameter oscillation.

3.4.2 Film Production

Cast film extrusion. In a cast film extrusion process, a thin film is extruded through a slit onto a chilled, highly polished, turning roll where it is quenched from one side. The speed of the roller controls the draw ratio and final film thickness. The film is then sent to a second roller for cooling of the other side. Finally, the film passes through a system of rollers and is wound onto a roll. A typical film casting process is depicted in Figs. 3.53

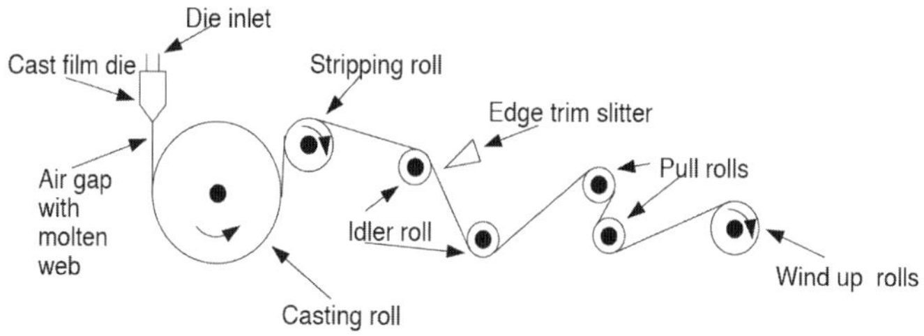

Figure 3.53: Schematic diagram of a film casting operation.

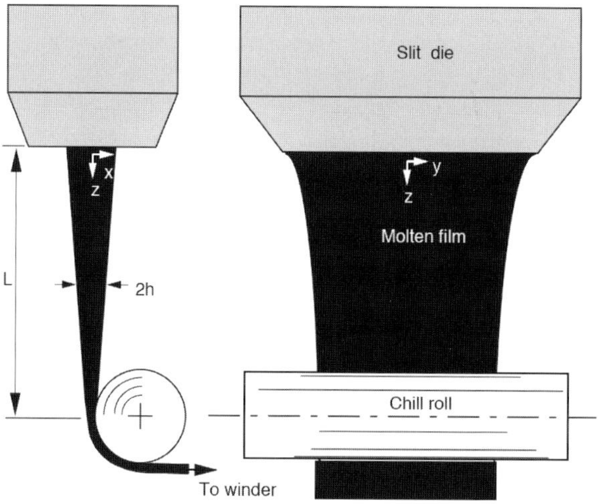

Figure 3.54: Film casting.

and 3.54. During the cast film extrusion process stability problems similar to those in fiber spinning encountered [2].

Film blowing. In film blowing, a tubular cross-section is extruded through an annular die, normally a spiral die, and is drawn and inflated until the freezing line is reached. Beyond this point, the stretching is practically negligible. The process is schematically depicted in Fig. 3.55 [14].

The advantage of film blowing over casting is that the induced biaxial stretching renders a stronger and less permeable film. Film blowing is mainly used with less expensive materials such as polyolefins. Polymers with lower viscosity such as PA and PET are better manufactured using the cast film process. The extruded tubular profile passes through one or two air rings to cool the material. The tube's interior is maintained at a certain pressure by blowing air into the tube through a small orifice in the die mandrel. The air is retained in the tubular film, or bubble, by collapsing the film well above its freeze-off point and tightly pinching it between rollers. The size of the tubular film is calibrated between the air ring and the collapsing rolls. The predecessor of the blow molding process was the blowing press developed by Hyatt and Burroughs in the 1860s to manufacture hollow celluloid articles.

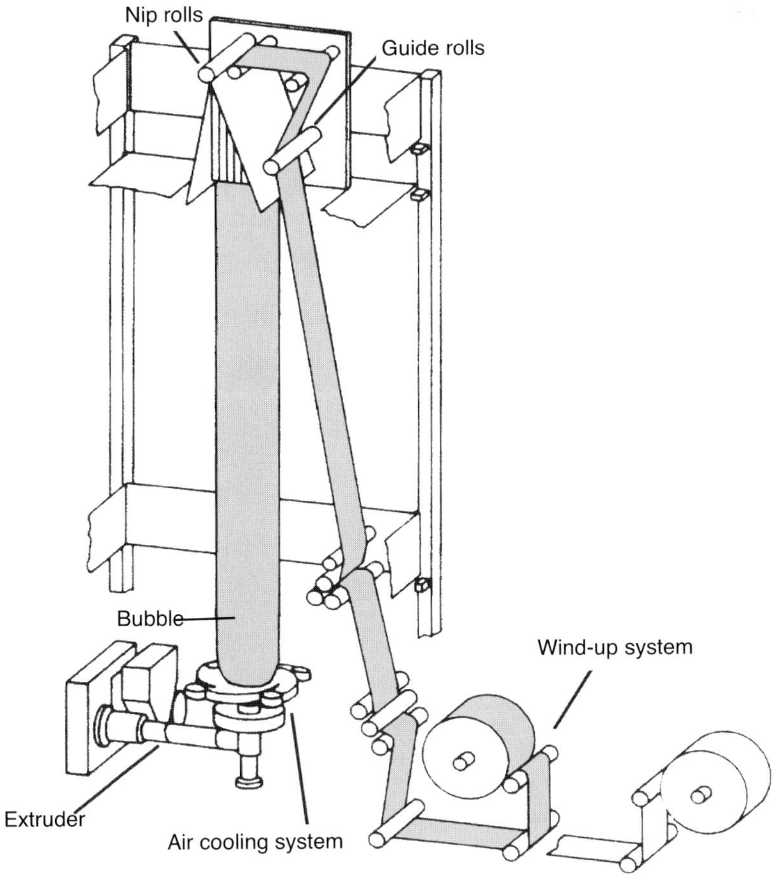

Figure 3.55: Schematic of a film blowing operation.

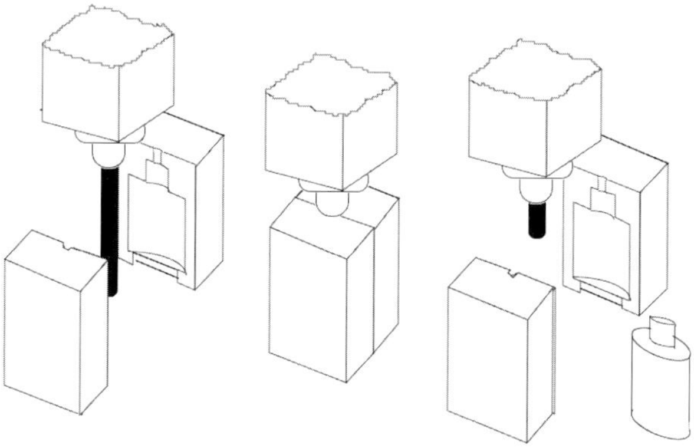

Figure 3.56: Schematic of the extrusion blow molding process.

Polystyrene was the first synthetic polymer used for blow molding during World War II and polyethylene was the first material to be implemented in commercial applications. Until the late 1950s, the main application for blow molding was the manufacture of PE-LD articles such as squeeze bottles. Blow molding produces hollow articles that do not require a homogeneous thickness distribution. Today, PE-HD, PE-LD, PP, PET, and PVC are the most common materials used for blow molding.

Extrusion blow molding. In extrusion blow molding, a parison or tubular profile is extruded and inflated into a cavity with a specified geometry. The blown article is held inside the cavity until it is sufficiently cool. Figure 3.56 [25] presents a schematic of the steps in blow molding.

During blow molding, one must generate the appropriate parison length such that the trim material is minimized. Another means of saving material is generating a parison of variable thickness, usually referred to as parison programming, such that an article with an evenly distributed wall thickness is achieved after stretching the material. An example of a programmed parison and finished bottle thickness distribution is presented in Fig. 3.57 [1].

A parison of variable thickness can be generated by moving the mandrel vertically during extrusion as shown in Fig. 3.58. A thinner wall not only results in material savings but also reduces the cycle time due to the shorter required cooling times.

As expected, the largest portion of the cycle time is the cooling of the blow molded container in the mold cavity. Most machines work with multiple molds in order to increase production. Rotary molds are often used in conjunction with vertical or horizontal rotating tables (Fig. 3.59 [14]).

Injection blow molding. Injection blow molding, depicted in Fig. 3.60 [25], begins by injection molding the parison onto a core and into a mold with finished bottle threads. The formed parison has a thickness distribution that leads to reduced thickness variations throughout the container.

Before blowing the parison into the cavity, it can be mechanically stretched to orient molecules axially, Fig. 3.61 [25]. The subsequent blowing operation introduces tangential orientation. A container with biaxial molecular orientation exhibits higher optical (clarity)

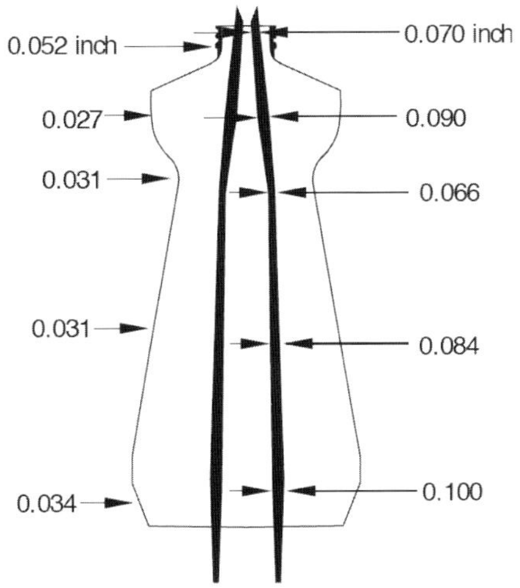

Figure 3.57: Wall thickness distribution in the parison and the bottle.

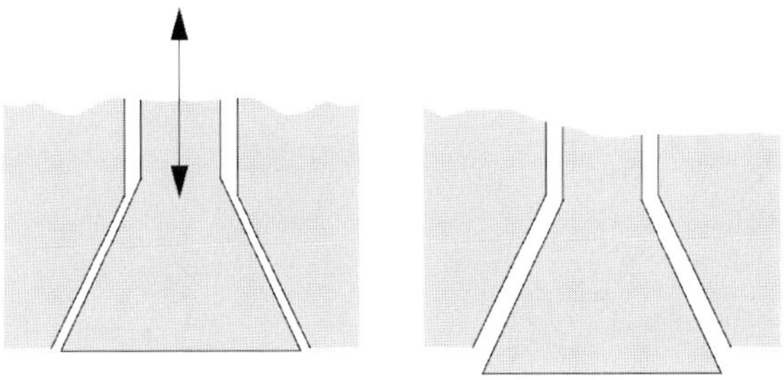

Figure 3.58: Moving mandrel used to generate a programmed parison.

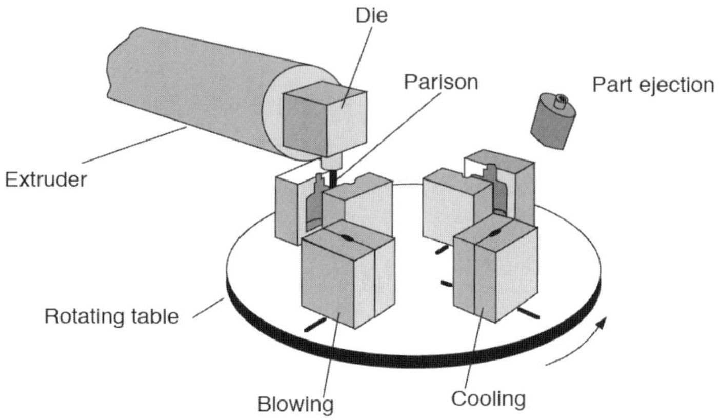

Figure 3.59: Schematic of an extrusion blow molder with a rotating table.

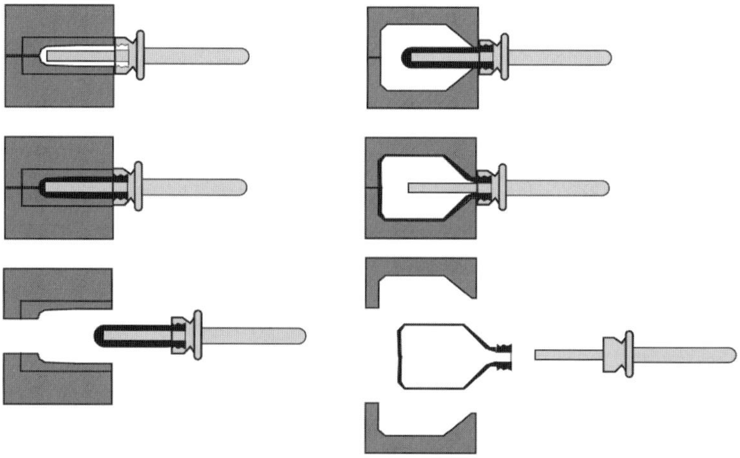

Figure 3.60: Injection blow molding.

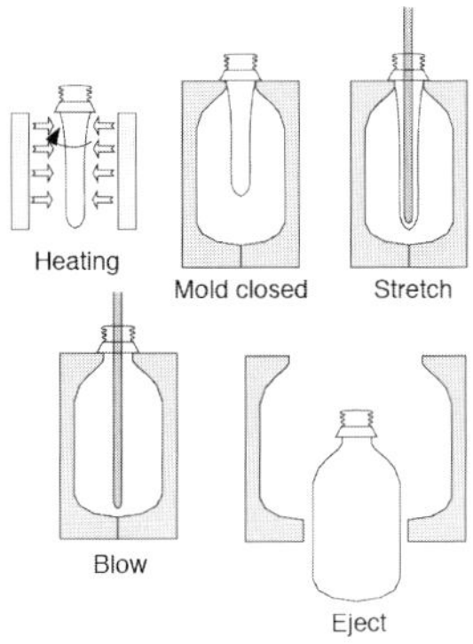

Figure 3.61: Stretch blow molding.

and mechanical properties and lower permeability. During injection blow molding one can go directly from injection to blowing or one can have a re-heating stage in-between.

The advantages of injection blow molding over extrusion blow molding are:

- Pinch-off and therefore post-mold trimming are eliminated
- Controlled container wall thickness
- Dimensional control of the neck and screw-top of bottles and containers

Disadvantages include higher initial mold cost, the need for both injection and blow molding units and lower volume production.

3.4.3 Thermoforming

Thermoforming is an important secondary shaping method of plastic film and sheet. Thermoforming consists of warming the plastic sheet and forming it into a cavity or over a tool using vacuum, air pressure, or mechanical means. During the 18th century, tortoiseshells and hooves were thermoformed into combs and other shapes. The process was refined during the mid-19th century to thermoform various cellulose nitrate articles. During World War II, thermoforming was used to manufacture acrylic aircraft cockpit enclosures, canopies, and windshields, as well as translucent covers for outdoor neon signs. During the 1950s, the process made an impact in the mass production of cups, blister packs, and other packaging commodities. Today, in addition to packaging, thermoforming is used to manufacture refrigerator liners, pick-up truck cargo box liners, shower stalls, bathtubs, as well as automotive trunk liners, glove compartments, and door panels.

A typical thermoforming process is presented in Fig. 3.62 [14]. The process begins by heating the plastic sheet slightly above the glass transition temperature for amorphous polymers, or slightly below the melting point for semi-crystalline materials. Although, both amorphous and semi-crystalline polymers are used for thermoforming, the process is most suitable for with amorphous polymers, because they have a wide rubbery temperature

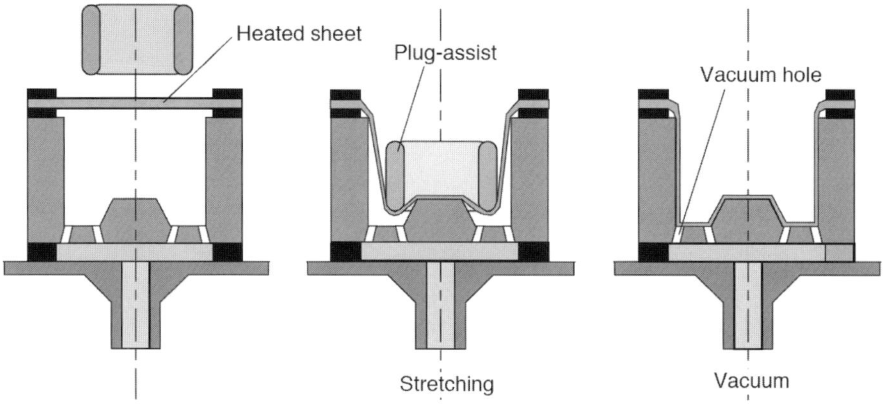

Figure 3.62: Plug-assist thermoforming using vacuum.

range above the glass transition temperature. At these temperatures, the polymer is easily shaped, but still has enough rigidity to hold the heated sheet without much sagging. Most semi-crystalline polymers lose their strength rapidly once the crystalline structure breaks up above the melting temperature.

The heating is achieved using radiative heaters and the temperature reached during heating must be high enough for sheet shaping, but low enough so the sheets do not droop into the heaters. One key requirement for successful thermoforming is to bring the sheet to a uniform forming temperature. The sheet is then shaped into the cavity over the tool. This can be accomplished in several ways. Most commonly, a vacuum sucks the sheet onto the tool, stretching the sheet until it contacts the tool surface. The main problem here is the irregular thickness distribution that arises throughout the part. Hence, the main concern of the process engineer is to optimize the system such that the differences in thickness throughout the part are minimized. This can be accomplished in many ways but most commonly by plug-assist. Here, as the plug pushes the sheet into the cavity, only the parts of the sheet not touching the plug-assist will stretch. Since the unstretched portions of the sheet must remain hot for subsequent stretching, the plug-assist is made of a low thermal conductivity material such as wood or hard rubber. The initial stretch is followed by a vacuum for final shaping. Once cooled, the product is removed. To reduce thickness variations in the product, the sheet can be pre-stretched by forming a bubble at the beginning of the process. This is schematically depicted in Fig. 3.63 [14]. The mold is raised into the bubble, or a plug-assist pushes the bubble into the cavity, and a vacuum finishes the process. One of the main reasons for the rapid growth and high volume of thermoformed products is that the tooling costs for a thermoforming mold are much lower than for injection molding.

3.5 CALENDERING

In a calender line, the polymer melt is transformed into films and sheets by squeezing it between pairs of co-rotating high-precision rollers. Calenders are also used to produce certain surface textures which may be required for different applications. Today, calendering lines are used to manufacture PVC sheet, floor covering, rubber sheet, and rubber tires. They are also used to texture or emboss surfaces. When producing PVC sheet and film, calender

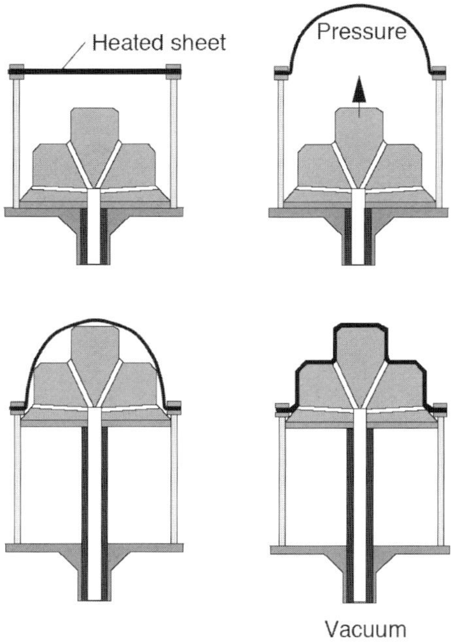

Figure 3.63: Reverse draw thermoforming with plug-assist and vacuum.

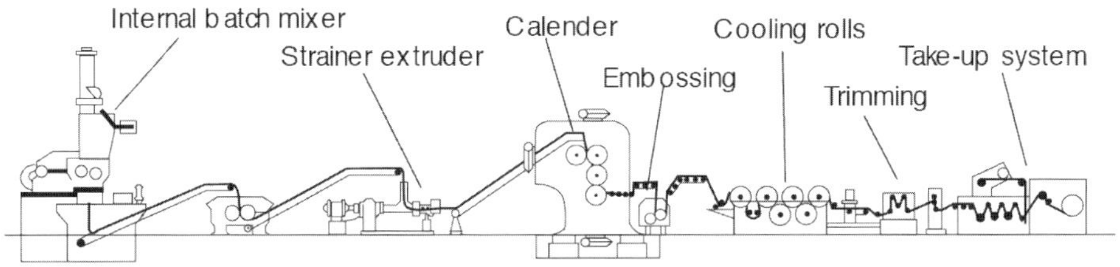

Figure 3.64: Schematic of a typical calendering process (Berstorff GmbH, Germany).

lines have a great advantage over extrusion processes because of the shorter residence times, resulting in a lower requirement for stabilizer. This can be cost effective since stabilizers are a major part of the overall expense of processing these polymers. Figure 3.64 [14] presents a typical calender line for manufacturing PVC sheet.

A typical system is composed of:

- Plasticating unit
- Calender
- Cooling unit
- Accumulator

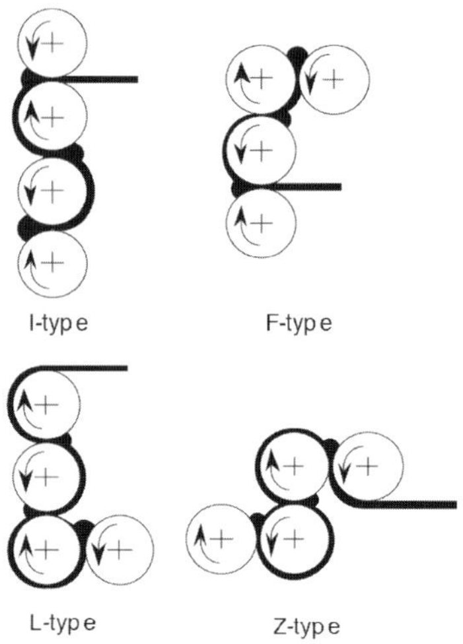

Figure 3.65: Calender arrangements.

- Wind-up station

In the plasticating unit, which is represented by the internal batch mixer and the strainer extruder, the material is melted and mixed and is fed in a continuous stream between the nip of the first two rolls. In another variation of the process, the mixing may take place elsewhere, and the material is simply reheated on the roll mill. Once the material is fed to the mill, the first pair of rolls controls the feeding rate, while subsequent rolls in the calender calibrate the sheet thickness. Most calender systems have four rolls as does the one in Fig. 3.64, which is an inverted L- or F-type system. Other typical roll arrangements are shown in Fig. 3.65. After passing through the main calender, the sheet can be passed through a secondary calendering operation for embossing. The sheet is then passed through a series of chilling rolls where it is cooled from both sides in an alternating fashion. After cooling, the film or sheet is wound.

One of the major concerns in a calendering system is generating a film or sheet with a uniform thickness distribution with tolerances as low as ± 0.005 mm. To achieve this, the dimensions of the rolls must be precise. It is also necessary to compensate for roll bowing resulting from high pressures in the nip region. Roll bowing is a structural problem that can be mitigated by placing the rolls in a slightly crossed pattern, rather than completely parallel, or by applying moments to the roll ends to counteract the separating forces in the nip region.

3.6 COATING

During coating, a liquid film is continuously deposited on a moving, flexible or rigid substrate. Coating is done on metal, paper, photographic films, audio and video tapes, and adhesive tapes. Typical coating processes include wire coating, dip coating, knife coating, roll coating, slide coating, and curtain coating.

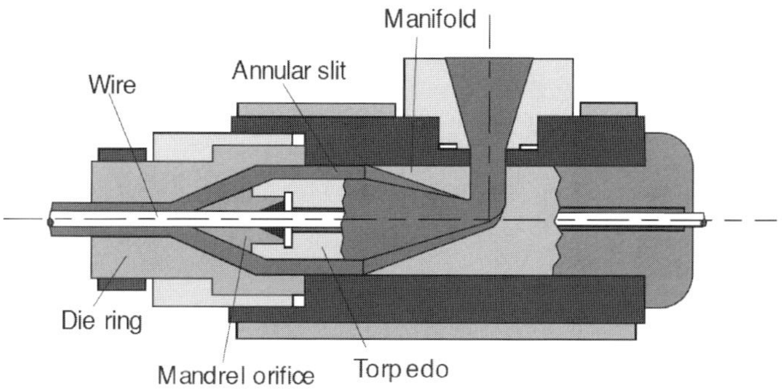

Figure 3.66: Schematic of the wire coating process.

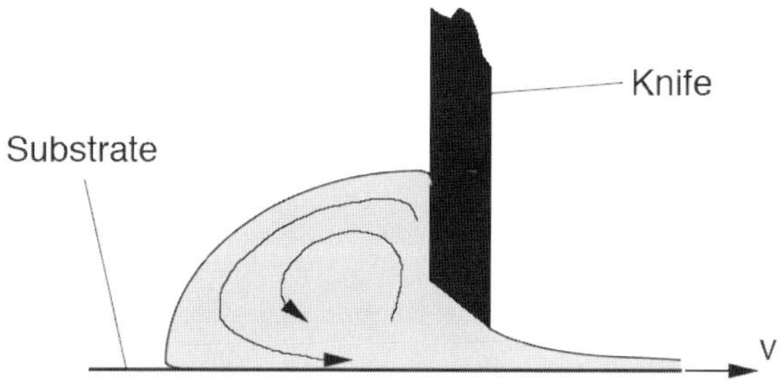

Figure 3.67: Schematic of the knife coating process.

In wire coating, a wire is continuously coated with a polymer melt by pulling the wire through an extrusion die. The polymer resin is deposited onto the wire using the drag flow generated by the moving wire and sometimes a pressure flow generated by the back pressure of the extruder. The process is schematically depicted in Fig. 3.66[7]. The second normal stress differences, generated by the high shear deformation in the die, help keep the wire centered in the annulus [29].

Dip coating is the simplest and oldest coating operation. Here, a substrate is continuously dipped into a fluid and withdrawn with one or both sides coated with the fluid. Dip coating can also be used to coat individual objects that are dipped and withdrawn from the fluid. The fluid viscosity and density and the speed and angle of the surface determine the coating thickness.

Knife coating, depicted in Fig. 3.67, consists of metering the coating material onto the substrate from a pool of material, using a fixed rigid or flexible knife. The knife can be normal to the substrate or angled and the bottom edge can be flat or tapered. The thickness of the coating is approximately half the gap between the knife edge and the moving substrate or web. A major advantage of a knife edge coating system is its simplicity and relatively low maintenance.

[7]Other wire coating operations extrude a tubular sleeve which adheres to the wire via stretching and vacuum. This is called tube coating.

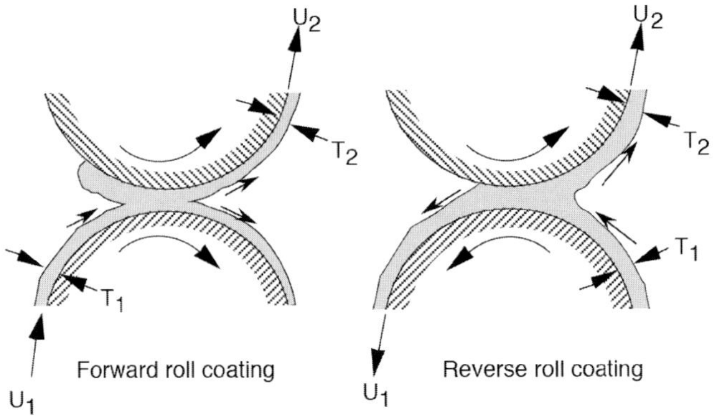

Figure 3.68: Schematic of forward and reversed roll coating processes.

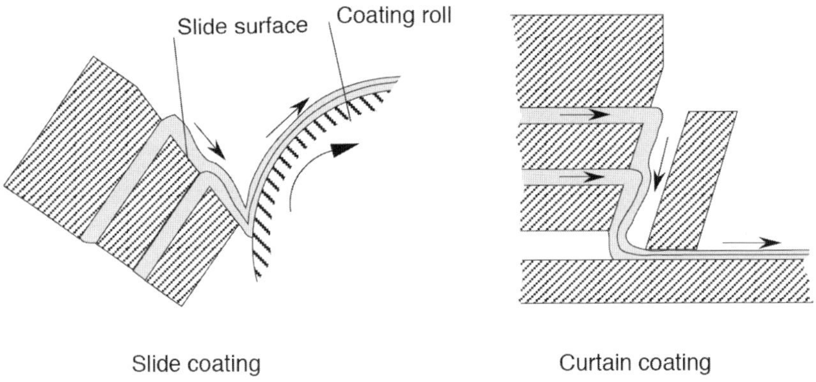

Figure 3.69: Schematic of slide and curtain coating.

Roll coating consists of passing a substrate and the coating simultaneously through the nip region between two rollers. The physics governing this process are similar to calendering, except that the fluid adheres to both the substrate and the opposing roll. The coating material is a low viscosity fluid, such as a polymer solution or paint and is picked up from a bath by the lower roll and applied to one side of the substrate. The thickness of the coating can be as low as a few μm and is controlled by the viscosity of the coating liquid and the nip dimension. This process can be configured as either forward roll coating for co-rotating rolls or reverse roll coating for counter-rotating rolls (Fig. 3.68). The reverse roll coating process delivers the most accurate coating thicknesses.

Slide coating and curtain coating, schematically depicted in Fig. 3.69, are commonly used to apply multi-layered coatings. However, curtain coating has also been widely used to apply single layers of coatings to cardboard sheet. In both methods, the coating fluid is pre-metered.

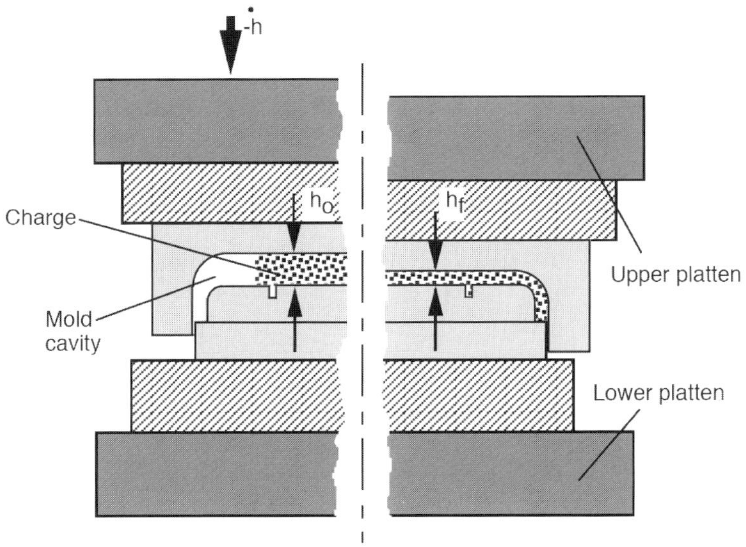

Figure 3.70: Schematic of the compression molding process.

3.7 COMPRESSION MOLDING

Compression molding is widely used in the automotive industry to produce parts that are large, thin, lightweight, strong, and stiff. It is also used in the household goods and electrical industries. Compression molded parts are formed by squeezing a charge, often glass fiber reinforced, inside a mold cavity, as depicted in Fig. 3.70.

The matrix can be either a thermoset or thermoplastic. The oldest and still widest used material for compression molded products is phenolic. The thermoset materials used to manufacture fiber reinforced compression molded articles is unsaturated polyester sheet or bulk, reinforced with glass fibers, known as sheet molding compound (SMC) or bulk molding compound (BMC). In SMC, the 25 mm long reinforcing fibers are randomly oriented in the plane of the sheet and make up for 20-30% of the molding compound's volume fraction.

A schematic diagram of an SMC production line is depicted in Fig. 3.71 [8]. When producing SMC, the chopped glass fibers are sandwiched between two carrier films previously coated with unsaturated polyester-filler matrix. A fiber reinforced thermoplastic charge is often called a glass mat reinforced thermoplastic (GMT) charge. The most common GMT matrix is polypropylene. More recently, long fiber reinforced themoplastics (LFT) have become common. Here, one squeezes sausage-shaped charges deposited on the mold by an extruder.

During processing of thermoset charges, the SMC blank is cut from a preformed roll and is placed between heated cavity surfaces. Generally, the mold is charged with 1 to 4 layers of SMC, each layer about 3 mm thick, which initially cover about half the mold cavity's surface. During molding, the initially randomly oriented glass fibers orient, leading to anisotropic properties in the finished product. When processing GMT charges, the preforms are cut and heated between radiative heaters. Once heated, they are placed inside a cooled mold that rapidly closes and squeezes the charges before they cool and solidify.

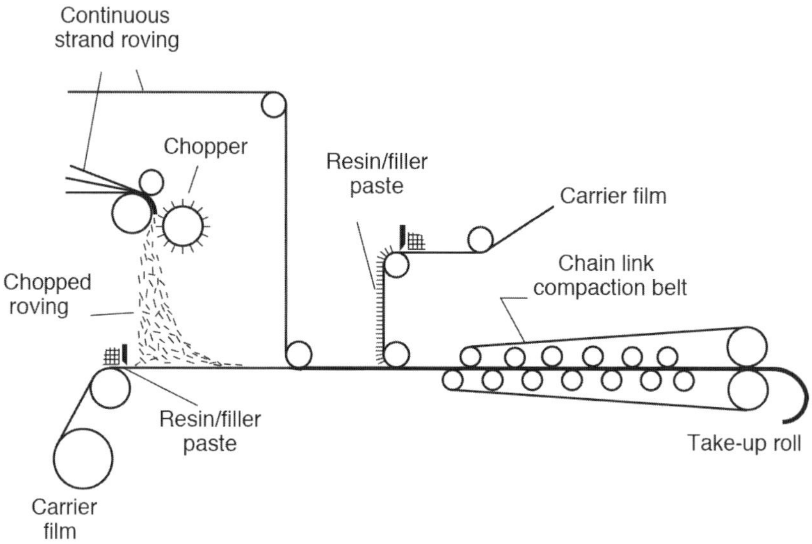

Figure 3.71: SMC production line.

One of the main advantages of the compression molding process is the low fiber attrition during processing. Here, relatively long fibers can flow in the melt without the fiber damage common during plastication and cavity filling during injection molding. An alternate process is injection-compression molding. Here, a charge is injected through a large gate followed by a compression cycle. The material used in the injection compression molding process is called bulk molding compound (BMC), which is reinforced with shorter fibers, generally 10 mm long, with an unsaturated polyester matrix. The main benefit of injection compression molding over compression molding is automation. The combination of injection and compression molding leads to lower degrees of fiber orientation and fiber attrition compared to injection molding.

3.8 FOAMING

In foam or a foamed polymer, a cellular or porous structure has been generated through the addition and reaction of physical or chemical blowing agents. The basic steps of foaming are cell nucleation, expansion or cell growth, and cell stabilization. Nucleation occurs when, at a given temperature and pressure, the solubility of a gas is reduced, leading to saturation, expelling the excess gas to form bubbles. Nucleating agents, such as powdered metal oxides, are used for initial bubble formation. The bubbles reach an equilibrium shape when their inside pressure balances their surface tension and surrounding pressures. The cells formed can be completely enclosed (closed cell) or can be interconnected (open cell).

In a physical foaming process a gas such as nitrogen or carbon dioxide is introduced into the polymer melt. Physical foaming also occurs after heating a melt that contains a low boiling point fluid, causing it to vaporize. For example, the heat-induced volatilization of low-boiling-point liquids, such as pentane and heptane, is used to produce polystyrene foams. Also, foaming occurs during volatilization from the exothermic reaction of gases produced during polymerization such as the production of carbon dioxide during the reac-

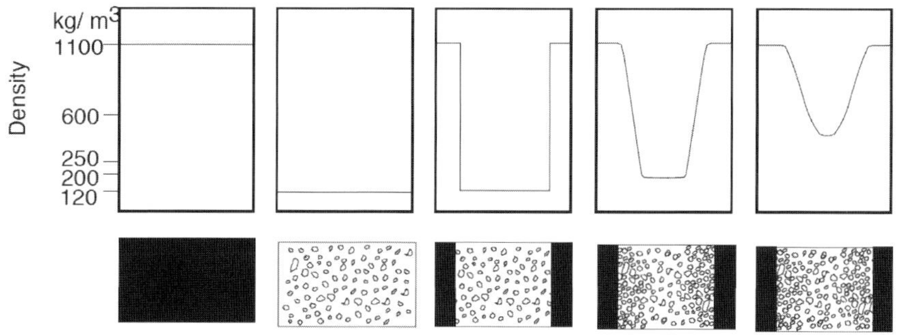

Figure 3.72: Schematic of various foam structures.

tion of isocyanate with water. Physical blowing agents are added to the plasticating zone of the extruder or molding machine. The most widely used physical blowing agent is nitrogen. Liquid blowing agents are often added to the polymer in the plasticating unit or the die. Chemical blowing agents are usually powders introduced in the hopper of the molding machine or extruder. Chemical foaming occurs when the blowing agent thermally decomposes, releasing large amounts of gas. The most widely used chemical blowing agent for polyolefin is azodicarbonamide. In mechanical foaming, a gas dissolved in a polymer expands upon reduction of the processing pressure. The foamed structures commonly generated are either homogeneous foams or integral foams.

Figure 3.72 [27] presents the various types of foams and their corresponding characteristic density distributions. In integral foam, the unfoamed skin surrounds the foamed inner core. This type of foam can be produced by injection molding and extrusion and it replaces the sandwiched structure also shown in Fig. 3.72.

Today, foams are of great commercial importance and are primarily used in packaging and as heat and noise insulating materials. Examples of foamed materials are polyurethane foams, expanded polystyrene (EPS) and expanded polypropylene particle foam (EPP). Polyurethane foam is perhaps the most common foaming material and is a typical example of a chemical foaming technique. Here, two low viscosity components, a polyol and an isocyanate, are mixed with a blowing agent such as pentane. When manufacturing semi-finished products, the mixture is deposited on a moving conveyor belt where it is allowed to rise, like a loaf of bread contained whithin shaped paper guides. The result is a continuous polyurethane block that can be used, among others, in the upholstery and matress industries. The basic material to produce expanded polystyrene products are small pearls produced by suspension styrene polymerization with 6-7% of pentane as a blowing agent. To process the pearls, they are placed in pre-expanding machines heated with steam until their temperature reaches 80 to 100°C. To enhance their expansion, the pearls are cooled in a vacuum and allowed to age and dry in ventilated storage silos before the shaping operation. Polystyrene foam is is used extensively in packaging, but its uses also extend to the construction industry as a thermal insulating material, as well as for shock absorption in children's safety seats and bicycle helmets. Expanded polypropylene particle foam is similar in to EPS but is characterized by its excellent impact absorption and chemical resistance. Its applications

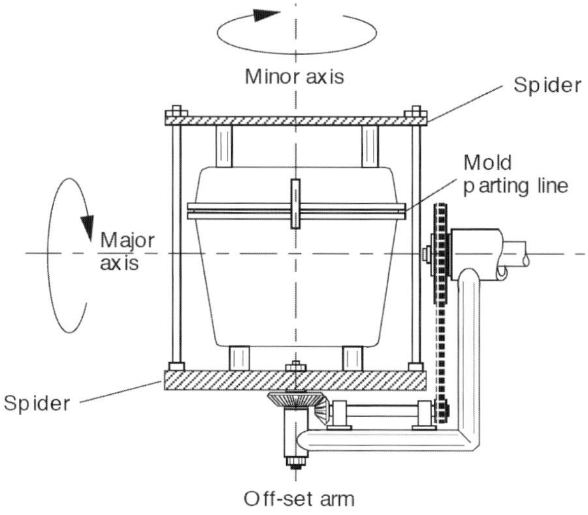

Figure 3.73: Schematic of the rotational molding process.

are primarely in the automotive industry as bumper cores, sun visors and knee cushions, to name a few.

3.9 ROTATIONAL MOLDING

Rotational molding is used to make hollow objects. In rotational molding, a carefully measured amount of powdered polymer, typically polyethylene, is placed in a mold. The mold is then closed and placed in an oven where the mold turns about two axes as the polymer melts, as depicted in Fig. 3.73.

During heating and melting, which occur at oven temperatures between 250 and 450°C, the polymer is deposited evenly on the mold's surface. To ensure uniform thickness, the axes of rotation should not coincide with the centroid of the molded product. The mold is then cooled and the solidified part is removed from the mold cavity. The parts can be as thick as 10 mm, and still be manufactured with relatively low residual stresses. The reduced residual stress and the controlled dimensional stability of the rotational molded product depend in great part on the cooling rate after the mold is removed from the oven. A mold that is cooled too fast yields warped parts. Usually, a mold is first cooled with air to start the cooling slowly, followed by a water spray for faster cooling. The main advantages of rotational molding over blow molding are the uniform part thickness and the low cost involved in manufacturing the mold. In addition, large parts such as play structures or kayaks can be manufactured more economically than with injection molding or blow molding. The main disadvantage of the process is the long cycle time for heating and cooling of the mold and polymer.

Figure 3.74 presents the air temperature inside the mold in a typical rotational molding cycle for polyethylene powders [7].

The process can be divided into six distinct phases:

1. Induction or initial air temperature rise

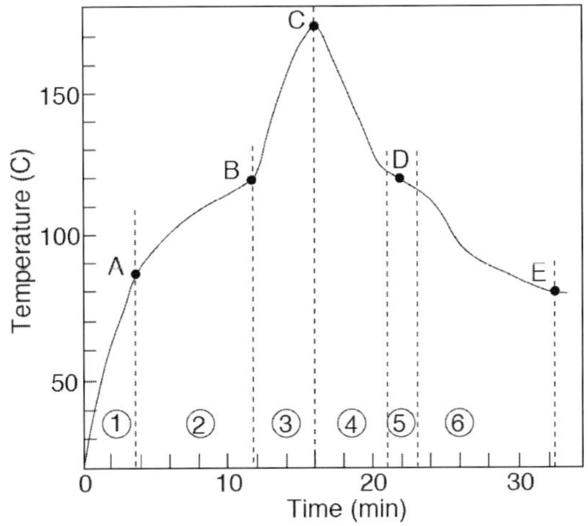

Figure 3.74: Typical air temperature in the mold while rotomolding polyethylene parts.

2. Melting and sintering
3. Bubble removal and densification
4. Pre-cooling
5. Crystallization of the polymer melt
6. Final cooling

The induction time can be significantly reduced by pre-heating the powder, and the bubble removal and cooling stage can be shortened by pressurizing the material inside the mold. The melting and sintering of the powder during rotational molding depends on the rheology and geometry of the particles. This phenomenon was studied in depth by Bellehumeur and Vlachopoulos [5].

REFERENCES

1. *Modern Plastics Encyclopedia*, volume 53. McGraw-Hill, New York, 1976.
2. N.R. Anturkar and A.J. Co. *J. non-Newtonian Fluid Mech.*, 28:287, 1988.
3. A. Biswas and T. A. Osswald. University of Wisconsin-Madison, 1994.
4. B.B. Boonstra and A.I. Medalia. *Rubber Age*, Mach-April 1963.
5. C.T. Callehumeur and J. Vlachopoulos. In *SPE ANTEC Tech. Pap.*, volume 56, 1998.
6. R.G. Cox. *J. Fluid Mech.*, 37(3):601, 1969.
7. R.J. Crawford. *Rotational Molding of Plastics*. Research Studies Press, Somerset, 1992.
8. D.L. Denton. The mechanical properties of an smc-r50 composite. Technical report, Owens-Corning Fiberglass Corporation, 1979.

9. P.H.M. Elemans. in: *Mixing and Compounding of Polymers*, I. Manas and Z. Tadmor (Eds.) Hanser Publishers, Munich, 1994.
10. P.J. Gramann, L. Stradins, and T. A. Osswald. *Intern. Polymer Processing*, 8:287, 1993.
11. J. Greener. *Polym. Eng. Sci*, 26:886, 1986.
12. S. Lim and J.L. White. *Intern. Polymer Processing*, 8:119, 1993.
13. S. Lim and J.L. White. *Intern. Polymer Processing*, 9:33, 1994.
14. G. Menges. *Einfuehrung in die Kunststoffverarbeitung*. Hanser Publishers, Munich, 1986.
15. G. Menges and E. Harms. Kautschuk und Gummi. *Kunststoffe*, 25:469, 1972.
16. G. Menges and W. Predoehl. *Plastverarbeiter*, 20:79, 1969.
17. G. Menges and W. Predoehl. *Plastverarbeiter*, 20:188, 1969.
18. W. Michaeli and M. Lauterbach. *Kunststoffe*, 79:852, 1989.
19. T. A. Osswald. *Polymer Processing Fundamentals*. Hanser Publishers, Munich, 1998.
20. T. A. Osswald and G. Menges. *Material Science of Polymers for Engineers*. Hanser Publishers, Munich, 2nd edition, 2003.
21. T. A. Osswald, L.S. Turng, and P.J. Gramann. (Eds.) *Injection Molding Handbook*. Hanser Publishers, Munich, 2001.
22. C. Rauwendaal. *Polymer Extrusion*. Hanser Publishers, Munich, 1990.
23. C. Rauwendaal. In *SPE ANTEC Tech. Pap.*, volume 39, page 2232, 1993.
24. C. Rauwendaal. *Mixing and Compounding of Polymers*, Hanser Publishers, Munich, 1994.
25. D.V. Rosato. *Blow Molding Handbook*. Hanser Publishers, Munich, 1989.
26. C.E. Scott and C.W. Macosko. *Polymer Bulletin*, 26:341, 1991.
27. F.A. Shutov. *Integral/Structural Polymer Foams*. Springer-Verlag, Berlin, 1986.
28. H.A. Stone and L.G. Leal. *J. Fluid Mech.*, 198:399, 1989.
29. Z. Tadmor and R.B. Bird. *Polym. Eng. Sci*, 14:124, 1973.

PART II

PROCESSING FUNDAMENTALS

CHAPTER 4

DIMENSIONAL ANALYSIS AND SCALING

A small leak can sink a great ship.

—Benjamin Franklin

Dimensional analysis is used by engineers to gain insight into a problem by allowing presentation of theoretical and experimental results in a compact manner. This is done by reducing the number of variables in a system by lumping them into meaningful dimensionless numbers. For example, if a flow system is dominated by the fluid's inertia as well as the viscous effects, it may best to present the results, i.e., pressure requirements, in terms of Reynolds number, which is the ratio of both effects. As one checks the order of these dimensionless numbers and compares them to one another, one can gain insight into what parameters, such as process conditions and material properties, are most important. Many researches also use dimensional analysis in theoretical studies. Often, dimensional analysis, in combination with experiments, results in fundamental relations that govern a process.

In polymer processing, as well as other manufacturing techniques or operations, one often works on a laboratory scale when developing new processes or materials, and when testing and optimizing a certain system. This laboratory operation, often referred to as a *pilot* plant, is a physical model of the actual or final system. How one goes from this laboratory model, that probably produces only a few cubic centimeters of material per hour, to the actual production process that can generate hundreds of kilograms per hour, is what is called *scale-up*. On some occasions, such as when trying to push the envelope in injection molding, where the thickness of the part is always being reduced, the term *scale-down* is also used. Since the methods mentioned in this chapter work for both, here we will simply call them scaling.

4.1 DIMENSIONAL ANALYSIS

Dimensional analysis, often referred to as the Π-theorem is based on the fact that every system that is governed by m physical quantities can be reduced to a set of $m - n$ mutually independent dimensionless groups, where n is the number of basic dimensions that are present in the physical quantities. The Π-theorem was introduced by Buckingham [1] in 1914 and is therefore known as the Buckingham Π-theorem. The Π-theorem is a procedure to determine dimensionless numbers from a list of variables or physical quantities that are related to a specific problem. This is best illustrated by an example problem.

Consider the classical problem of pressure drop during flow in a smooth straight pipe, ignoring the inlet effects. The first step is to list all possible variables or quantities that are related to the problem under consideration. In this case, we have:

- Target quantity: Pressure drop Δp
- Geometric variables: Pipe diameter D, and pipe length L
- Physical or material properties: the viscosity η, and the density ρ of the fluid
- Process variable: average fluid velocity u

Once we have defined all the physical quantities, also referred to as the relevance list, we write them with their respective dimensions in terms of mass M, length L, time T and temperature Θ, and in some cases force F, i.e.

$$\begin{array}{cccccc} \Delta p & D & L & \eta & \rho & u \\ \dfrac{M}{LT^2} & L & L & \dfrac{M}{LT} & \dfrac{M}{L^3} & \dfrac{L}{T} \end{array} \tag{4.1}$$

Table 4.1 presents various physical quantities with their respective dimensions in an $MLT\Theta$ system and in an $FLT\Theta$ system, respectively.

In this example there are $m = 6$ variables and since one only finds mass, length and time one can say that one has $n = 3$ dimensional quantities. Hence one can generate $m - n = 3$ dimensionless groups denoted by Π_1, Π_2 and Π_3. From the list above $n = 3$ repeating variables are selected. These variables can appear in all the dimensionless numbers. When selecting the repeating variables it is important that

- They are not dimensionless, and
- They must all have different units, i.e., one cannot choose both, the diameter and the length of the pipe, as repeating variables.

Here one can choose D, μ and ρ as the repeating variables. The first dimensionless group that was generated involves Δp. One can write the product of the repeating variables and Δp, where each of the repeating variables has an exponent that will render the whole product dimensionless ($M^0L^0T^0$),

$$\Pi_1 = \Delta p D^a u^b \rho^c = \left[\frac{M}{LT^2}\right]^1 [L]^a \left[\frac{L}{T}\right]^b \left[\frac{M}{L^3}\right]^c = M^0L^0T^0 \tag{4.2}$$

Table 4.1: Base and Secondary Quantities and Their Respective Dimensions According to the SI System

Quantity	Dimension	SI unit	Name
length (L, d)	L	m	meter
mass (m)	M	kg	kilogram
time (t)	T	s	second
temperature (T)	Θ	K	Kelvin
amount of substance	N	mol	mole
electric current	I	A	ampere
luminous intensity	J	cd	candela
area (A)	L^2	m^2	
volume (V)	L^3	m^3	
angular velocity (ω), shear rate $(\dot{\gamma})$, frequency	T^{-1}	s^{-1}	
velocity (u)	LT^{-1}	m/s	
acceleration $(a, \dot{u})$	LT^{-2}	m/s^2	
kinematic viscosity (ν), diffusivity (D)	L^2T^{-1}	m^2/s	
density (ρ)	ML^{-3}	kg/m^3	
surface tension (σ, σ_S)	MT^{-2}	kg/s^2	
force	MLT^{-2}	N	Newton
pressure (p), tension	$ML^{-1}T^{-2}$	Pa	Pascal
dynamic viscosity (μ, η)	$ML^{-1}T^{-1}$	Pa-s	
momentum	MLT^{-1}	kg-m/s	
angular momentum	ML^2T^{-1}	$kg\text{-}m^2/s$	
energy, work, torque	ML^2T^{-2}	J	Joule
power	ML^2T^{-3}	W	Watt
specific heat (C_p, C_V)	$L^2T^{-2}\Theta^{-1}$	J/kg/K	
conductivity (k)	$MLT^{-3}\Theta^{-1}$	W/m/K	
heat transfer coefficient (h)	$MT^{-3}\Theta^{-1}$	$W/m^2/K$	

For each dimensional quantity M, L and T we can write

$$\begin{aligned} M &\rightarrow 1 + c = 0 \\ L &\rightarrow -1 + a + b - 3c = 0 \\ T &\rightarrow -2 - b = 0 \end{aligned} \tag{4.3}$$

The three unknown exponents as $a = 0$, $b = -2$ and $c = -1$ can now be solved for

$$\Pi_1 = \frac{\Delta p}{u^2 \rho} \tag{4.4}$$

which is widely known as the *Euler* number, Eu. One can repeat the procedure for L so that

$$\Pi_2 = L D^a u^b \rho^c = L^1 \left[L\right]^a \left[\frac{L}{T}\right]^b \left[\frac{M}{L^3}\right]^c = M^0 L^0 T^0 \tag{4.5}$$

is dimensionless if $a = -1$, $b = 0$ and $c = 0$, resulting in

$$\Pi_2 = \frac{L}{D} \tag{4.6}$$

Similar, if we repeat this for η,

$$\Pi_3 = \eta D^a u^b \rho^c = \left[\frac{M}{LT}\right]^1 \left[L\right]^a \left[\frac{L}{T}\right]^b \left[\frac{M}{L^3}\right]^c = M^0 L^0 T^0 \tag{4.7}$$

is dimensionless when $a = -1$, $b = -1$ and $c = -1$, and

$$\Pi_3 = \frac{\eta}{D u \rho} \tag{4.8}$$

which is the inverse of the *Reynolds* number, Re. The above technique produces the relation

$$f\left(Eu, Re, \frac{L}{D}\right) = 0 \tag{4.9}$$

but cannot deduce the nature of this relation. The form of the function f can only be produced experimentally. Figure 4.1 presents results from such experiments performed by Stanton and Pannell [10, 14] where they plot $\lambda = 2EuD/L$ as a function of Re. This figure demonstrates the usefulness of dimensional analysis.

Tables 4.2 to 4.4 list several dimensionless numbers that are used in various areas of engineering. This list can be helpful in performing a dimensional analysis, to help interpret results that are sometimes difficult to discern from the variety of dimensionless numbers that can result during such an undertaking.

4.2 DIMENSIONAL ANALYSIS BY MATRIX TRANSFORMATION

The classic technique to determine dimensionless numbers, described above, is cumbersome to use in cases where the list of related physical quantities becomes large. Pawlowski [8] developed a matrix transformation technique that offers a systematic approach to the generation of Π-sets.

Table 4.2: Base and Secondary Quantities for Flow Problems

Name	**Symbol**	$\Pi-$**number**	**Remarks**
Archimedes	Ar	$\frac{g\Delta\rho l^3}{\rho\nu^2}$	$Ar = (\Delta\rho/\rho)Ga$
Bond	Bd	$\frac{\rho g l^2}{\sigma}$	$Bd = Ca/Fr$
Brinkman	Br	$\frac{u^2\eta}{k\Delta T}$	
Capillary	Ca	$\frac{\rho u^2 l}{\sigma}$	Also known as Weber
Deborah	De	$\lambda\dot{\gamma}$	λ: Relaxation time $\dot{\gamma}$: Shear rate
Eckert	Ec	$\frac{u^2}{C_p\Delta T}$	
Euler	Eu	$\frac{\Delta p}{\rho u^2}$	
Flow number	λ_F	$\frac{\dot{\gamma}}{\dot{\gamma}-\omega}$	ω: magnitude vorticity tensor
Froude	Fr	$\frac{u^2}{lq}$	
Galilei	Ga	$\frac{gl^3}{\nu_2}$	$Ga = Re^2/Fr$
Laplace	La	$\frac{\Delta pd}{\sigma}$	$La = EuCa$
Mach	Ma	u/u_s	u_s: velocity of sound
Newton	Ne	$\frac{F}{\rho u^2 l^2}$	F: force
Ohnesorge	Oh	$\frac{\eta}{\sqrt{\rho\sigma l}}$	$Oh = Ca^{1/2}/Re$
Reynolds	Re	$\frac{ul}{\nu}$	$Re = \frac{\rho ul}{\eta}$
Strouhal	Sr	$\frac{lf}{u}$	f: frequency
Weissenberg	We	$\frac{N_1}{\tau}$	N_1: Normal stress

Table 4.3: Base and Secondary Quantities for Heat and Mass Transfer Problems

Name	Symbol	$\Pi-$number	Remarks
Biot	Bi	$\frac{hl}{k}$	h: heat transfer coefficient k: thermal conductivity
Fourier	Fo	$\frac{\alpha t}{l^2}$	α: thermal diffusivity
Graetz	Gz	$\frac{ul}{\alpha}(d/l)$	$Gz = (d/l)Re_d Pr$
Grashof	Gr	$\frac{\beta \Delta T g l^3}{\nu^2}$	$Gr = \beta \Delta T Ga$ β: fluid expansion coefficient
Jakob	Ja	$\frac{C_p \Delta T_{sat}}{h_{fg}}$	h_{fg}: liquid-vapor enthalpy
Nahme-Griffith	Na	$a \Delta T Br$	a: Viscosity temperature dependence
Nusselt	Nu	$\frac{hl}{k_f}$	k_f: fluid thermal conductivity
Peclet	Pe	$\frac{ul}{\alpha}$	$Pe = RePr$
Prandtl	Pr	$\frac{\nu}{\alpha}$	
Rayleigh	Ra	$\frac{\beta \Delta T g l^3}{\alpha \nu}$	$Ra = GrPr$
Stanton	St	$\frac{h}{u \rho C_p}$	$St = Nu/Re/Pr$
Mass Biot	Bi_m	$\frac{h_m l}{D}$	h_m: mass transfer coefficient D: mass diffusivity
Bodenstein	Bo	$\frac{ul}{D_{ax}}$	D_{ax}: dispersion coefficient
Lewis	Le	$\frac{\alpha}{D}$	$Le = Sc/Pr$
Schmidt	Sc	$\frac{\nu}{D}$	

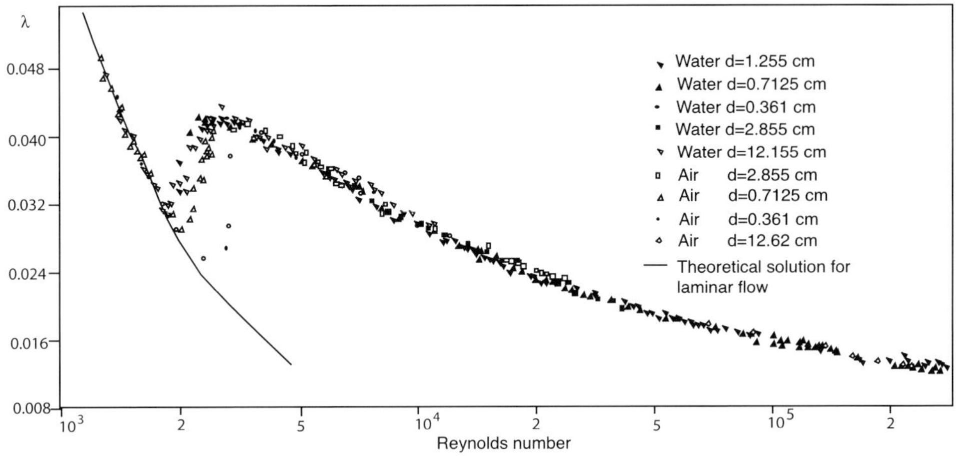

Figure 4.1: Pressure drop characteristic of a straight smooth tube.

Table 4.4: Base and Secondary Quantities for Problems with Reactions

Name	**Symbol**	$\Pi-$**number**	**Remarks**
Arrhenius	Arr	$\dfrac{E}{RT}$	R: Universal gas constant E: Energy of activation
Damköhler	Da	$\dfrac{c\Delta H_r}{\rho C_p T_0}$	
	Da_I	$k_1\tau$	k_1: Reaction rate constant τ: Mean residence time
	Da_{II}	k_1L^2/D	$Da_{II} = Da_I Bo$
	Da_{III}	$\dfrac{k_1\tau c\Delta H_r l^2}{kT_0}$	$Da_{III} = Da_I Da$
	Da_{IV}	$\dfrac{k_1 c\Delta H_r l^2}{kT_0}$	$Da_{IV} = Da_I Pe Da$
Hatta	Hat_1	$\dfrac{\sqrt{k_1 D}}{k_L}$	1st order reaction
	Hat_2	$\dfrac{\sqrt{k_2 c_2 D}}{k_L}$	2nd order reaction
Thiele modulus	Φ	$L\sqrt{k_1/D}$	$\Phi = \sqrt{Da_{II}}$

To demonstrate Pawlowski's matrix transformation technique, an example will be used in which a forced convection problem, where a fluid with a viscosity η, a density ρ, a specific heat C_p and a thermal conductivity k, is forced past a surface with a characteristic size D at an average speed u. The temperature difference between the fluid and the surface is described by $\Delta T = T_f - T_s$ and the resulting heat transfer coefficient is defined by h.

Again, the first step is to generate the relevance list. Here, the relevant list of physical quantities is:

- Geometric variable: D
- Process variables: u and ΔT
- Physical or material properties: η, ρ, C_p, and k
- Target quantity: h

The first step in generating the dimensionless variables is to set-up a dimensional matrix with the physical quantities and their respective units,

$$\begin{array}{ccccccccc} & D & u & \Delta T & \eta & \rho & C_p & k & h \\ M & 0 & 0 & 0 & 1 & 1 & 0 & 1 & 1 \\ L & 1 & 1 & 0 & -1 & -3 & 2 & 1 & 0 \\ T & 0 & -1 & 0 & -1 & 0 & -2 & -3 & -3 \\ \Theta & 0 & 0 & 1 & 0 & 0 & -1 & -1 & -1 \end{array} \tag{4.10}$$

The above dimensional matrix must be rearranged and divided into two parts, a square core matrix, which contains the dimensions pertaining to the repeating variables, and a residual matrix. Using the rules given in the previous section, the repeating variables are D, η, ρ and C_p and the dimensional matrix can be written as

$$\begin{array}{ccccccccc} & \eta & D & \rho & C_p & k & u & \Delta T & h \\ M & 1 & 0 & 1 & 0 & 1 & 0 & 0 & 1 \\ L & -1 & 1 & -3 & 2 & 1 & 1 & 0 & 0 \\ T & -1 & 0 & 0 & -2 & -3 & -1 & 0 & -3 \\ \Theta & 0 & 0 & 0 & -1 & -1 & 0 & 1 & -1 \\ & \multicolumn{4}{c}{\underbrace{\text{Core Matrix (4x4)}}} & \multicolumn{4}{c}{\underbrace{\text{Residual Matrix (4x4)}}} \end{array} \tag{4.11}$$

The next step is to transform the core matrix into a unity matrix. Hence, the order of the physical variables in the core matrix should be such that a minimum amount of linear transformations is required. Adding the M row to the L and T rows eliminates the non-zero term below the diagonal in the core matrix, i.e.,

$$\begin{array}{ccccccccc} & \eta & D & \rho & C_p & k & u & \Delta T & h \\ M & 1 & 0 & 1 & 0 & 1 & 0 & 0 & 1 \\ L+M & 0 & 1 & -2 & 2 & 2 & 1 & 0 & 1 \\ T+M & 0 & 0 & 1 & -2 & -2 & -1 & 0 & -2 \\ \Theta & 0 & 0 & 0 & -1 & -1 & 0 & 1 & -1 \end{array} \tag{4.12}$$

Performing the same operation with the upper portion of the core and residual matrices, for example, multiplying the $T + M$ row by 2 and add it to the $L + M$ row above, leads to a unity core matrix,

$$\begin{array}{c|cccc|cccc} & \eta & D & \rho & C_p & k & u & \Delta T & h \\ -T+2\Theta & 1 & 0 & 0 & 0 & 1 & 1 & 2 & 1 \\ L+3M+2T-2\Theta & 0 & 1 & 0 & 0 & 0 & -1 & -2 & -1 \\ T+M-2\Theta & 0 & 0 & 1 & 0 & 0 & -1 & -2 & 0 \\ -\Theta & 0 & 0 & 0 & 1 & 1 & 0 & -1 & 1 \end{array} \tag{4.13}$$

With the above matrix set the dimensionless numbers can be generated, in this case, 4 dimensionless groups, by placing the physical quantities in the residual matrix in the numerator and the quantities in the core matrix in the denominator with the coefficients in the residual matrix as their exponent. Hence,

$$\begin{aligned} \Pi_1 &= \frac{k}{\eta^1 D^0 \rho^0 C_p^1} = \frac{k}{\eta C_p} \\ \Pi_2 &= \frac{u}{\eta^1 D^{-1} \rho^{-1} C_p^0} = \frac{uD\rho}{\eta} \\ \Pi_3 &= \frac{\Delta T}{\eta^2 D^{-2} \rho^{-2} C_p^{-1}} = \frac{\Delta T D^2 \rho^2 C_p}{\eta^2} \\ \Pi_4 &= \frac{h}{\eta^1 D^{-1} \rho^0 C_p^1} = \frac{hD}{\eta C_p} \end{aligned} \tag{4.14}$$

If different repeating variables had been chosen in the core matrix, as for example η, D, u and ΔT, one would get

$$\begin{array}{c|cccc|cccc} & \eta & D & u & \Delta T & k & \rho & C_p & h \\ M & 1 & 0 & 0 & 0 & 1 & 1 & 0 & 1 \\ L & -1 & 1 & 1 & 0 & 1 & -3 & 2 & 0 \\ T & -1 & 0 & -1 & 0 & -3 & 0 & -2 & -3 \\ \Theta & 0 & 0 & 0 & 1 & -1 & 0 & -1 & -1 \end{array} \tag{4.15}$$

and after the matrix transformation

$$\begin{array}{c|cccc|cccc} & \eta & D & u & \Delta T & k & \rho & C_p & h \\ M & 1 & 0 & 0 & 0 & 1 & 1 & 0 & 1 \\ L+2M+T & 0 & 1 & 0 & 0 & 0 & -1 & 0 & -1 \\ -T-M & 0 & 0 & 1 & 0 & 2 & -1 & 2 & 2 \\ \Theta & 0 & 0 & 0 & 1 & -1 & 0 & -1 & -1 \end{array} \tag{4.16}$$

Here, the dimensionless numbers are

$$\begin{aligned} \Pi_1' &= \frac{k}{\eta^1 D^0 u^2 \Delta T^{-1}} = \frac{k\Delta T}{\eta u^2} \\ \Pi_2' &= \frac{\rho}{\eta^1 D^{-1} u^{-1} \Delta T^0} = \frac{uD\rho}{\eta} \\ \Pi_3' &= \frac{C_p}{\eta^0 D^0 u^2 \Delta T^{-1}} = \frac{\Delta T C_p}{u^2} \\ \Pi_4' &= \frac{h}{\eta^1 D^{-1} u^2 \Delta T^{-1}} = \frac{hD\Delta T}{\eta u^2} \end{aligned} \tag{4.17}$$

Taking the two Π-sets generates different dimensionless numbers that are more meaningful. For example, one such group of dimensionless numbers that can be generated is

$$
\begin{aligned}
Nu &= \frac{hD}{k} = \frac{\Pi_4}{\Pi_1} = \frac{\Pi'_4}{\Pi'_1} = \text{Nusselt Number} \\
Re &= \frac{\rho D u}{\eta} = \Pi_2 = \Pi'_2 = \text{Reynolds Number} \\
Br &= \frac{\eta u^2}{k \Delta T} = \frac{\Pi_2^2}{\Pi_1 \Pi_3} = \frac{1}{\Pi'_1} = \text{Brinkman Number} \\
Pr &= \frac{\eta C_p}{k} = \frac{1}{\Pi_1} = \frac{\Pi'_3}{\Pi'_1} = \text{Prandtl Number}
\end{aligned}
\tag{4.18}
$$

■ EXAMPLE 4.1.

Column buckling problem. Let us consider the classic column buckling problem depicted in Fig. 4.2.

General buckling in a slender column with a slenderness ratio, L/D, greater than 100, occurs when it is subjected to a critical compressive load. This load is much lower than the maximum load allowable for compressive yield. Although this problem can be easily solved using Euler's equation[1], which predicts the critical load applied to the slender column, it lends itself very well to illustrate dimensional analysis.

For this problem, the relevant physical quantities to be considered are:

- Target quantity: critical buckling load P_{cr}
- Geometric variables: area moment of inertia I and column length L
- Physical or material properties: Young's modulus E.

This choice of physical quantities reflects the experience of the authors; however, a different selection may also lead to satisfactory results. For example, for a column of circular cross-section, a geometric choice could have been diameter, D, instead of area moment of inertia, I. However, I is more general and works for every cross-sectional geometry.

Once the relevant parameters have been chosen, the dimensional matrix subdivided into core and residual matrix can be obtained. The core matrix is a 3×3 matrix, leaving a residual matrix of size 3×1. Since this will result in only one dimensionless number, the target value P_{cr} is left on the residual side, hence, choosing E, L and I as the repeating quantities

$$
\begin{array}{c|ccc|c}
 & E & L & I & P_{cr} \\
M & 1 & 0 & 0 & 1 \\
L & -1 & 1 & 4 & 1 \\
T & -2 & 0 & 0 & -2
\end{array}
\tag{4.19}
$$

It is clear that this problem does not have a solution as set-up since mass and time units only appear in one parameter of the repeating quantities. In order to solve this problem the number of equations can be reduced by reducing the dimensional

[1]The Euler column formula is stated as $P_{cr} = \pi^2 EI/L^2$ where P_{cr} is the critical buckling load.

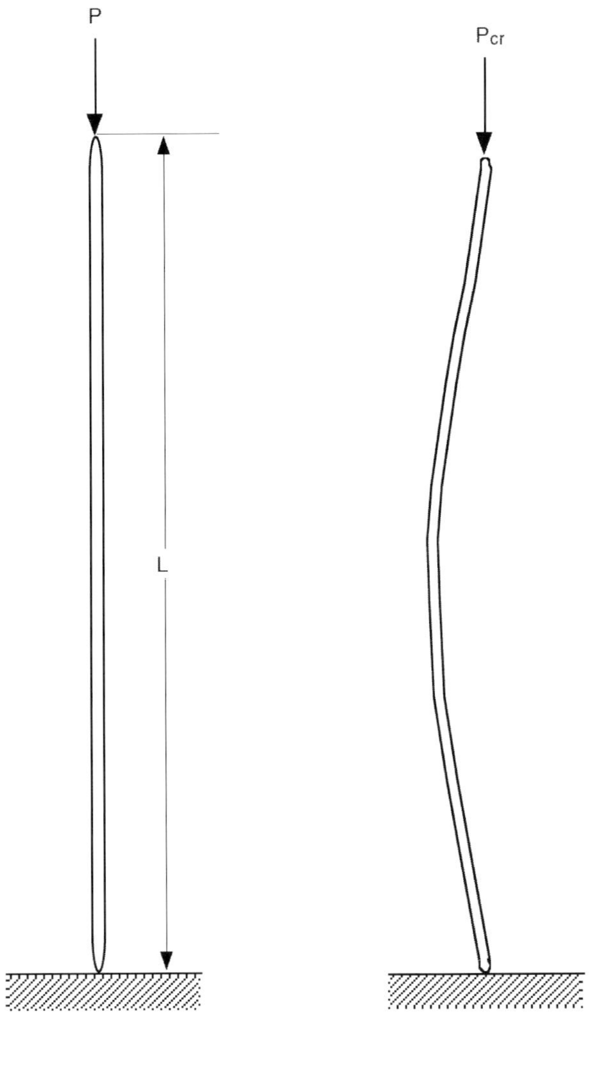

Figure 4.2: Schematic diagram of a column buckling case.

quantities from M, L and T to F and L. This reduces the core matrix to 2×2, with a 2×2 residual matrix

$$\begin{array}{c|cc|cc} & E & L & I & P_{cr} \\ F & 1 & 0 & 0 & 1 \\ L & -2 & 1 & 4 & 0 \end{array} \tag{4.20}$$

After transformation this gives

$$\begin{array}{c|cc|cc} & E & L & I & P_{cr} \\ F & 1 & 0 & 0 & 1 \\ L+2F & 0 & 1 & 4 & 2 \end{array} \tag{4.21}$$

which gives us $\Pi_1 = P_{cr}/L^2E$ and $\Pi_2 = I/L^4$ as the two dimensionless groups. Arranging the dimensional matrix with P_{cr} and L in the core matrix, the following dimensional matrix results

$$\begin{array}{c|cc|cc} & P_{cr} & L & I & E \\ F & 1 & 0 & 0 & 1 \\ L & 0 & 1 & 4 & -2 \end{array} \tag{4.22}$$

which does not need transformation and leads to $\Pi_1' = EL^2/P_{cr}$ and $\Pi_2' = L^4/I$, the inverse of the previous dimensionless numbers. Now a general relation between Π_1 and Π_2 can be written as

$$\Pi_1 = f(\Pi_2) \tag{4.23}$$

The relation between the resulting dimensionless groups can be found experimentally. For this problem, several experiments were performed using soldering rods of various materials and lengths and determining the critical buckling load for each case. A plot of Π_1 versus Π_2, shown in Fig. 4.3 for all the cases with a free-free end condition results in a straight line with a slope of π^2. Hence,

$$\Pi_1 = \pi^2 \Pi_2 \quad \text{or} \quad \frac{P_{cr}}{EL^2} = \pi^2 \frac{I}{L^4} \tag{4.24}$$

which is equivalent to Euler's column buckling formula

$$P_{cr} = \pi^2 \frac{EI}{L^2} \tag{4.25}$$

■ EXAMPLE 4.2.

Period of oscillation of small drops submerged in an incompatible fluid. Small drops are often submerged in incompatible fluids. For example, paint drops travel through air during the paint spraying process or polymer drops are carried inside a different polymer matrix during the mixing of polymer blends or simply during any polymer processing operation that involves polymer blends. As a drop is stressed and deforms during a given operation, it will oscillate due to the spring-effect given by surface tension (Fig. 4.4).

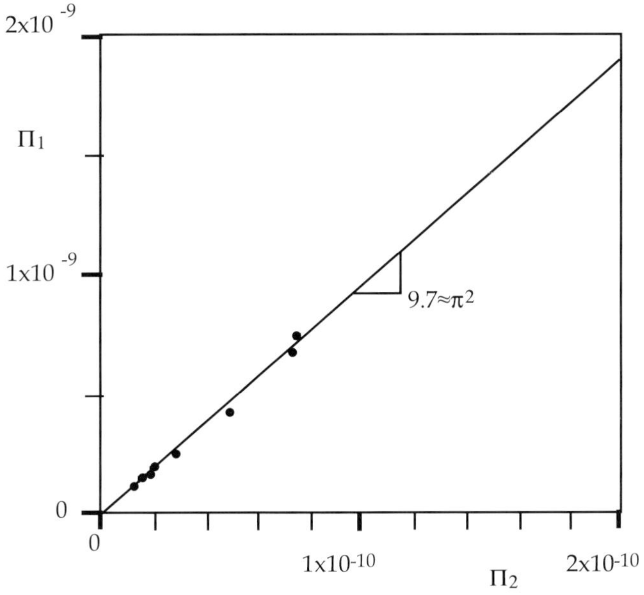

Figure 4.3: Results from a column buckling experiment.

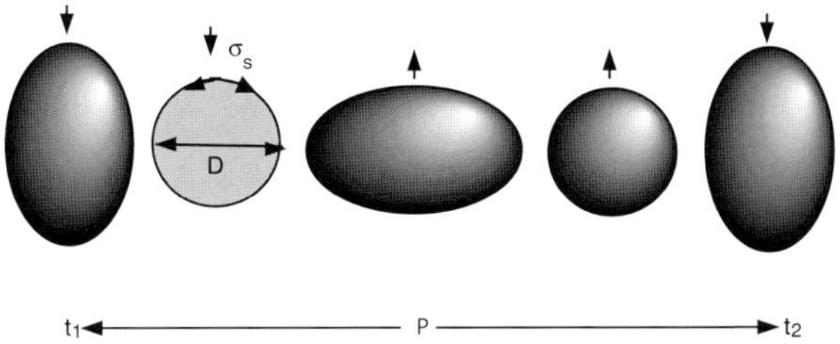

Figure 4.4: Schematic diagram of an oscillating drop.

During blending, for example, the period of oscillation is directly related to the time scale required to complete the dispersion or break up of the drop. Here, the relevant physical quantities chosen can be:

- Target quantity: period of oscillation P
- Geometric variables: diameter D
- Physical or material properties: surface tension σ_s and density ρ

Choosing σ_s, ρ and D as the terms in the core matrix, and arranging them such that the diagonal terms are populated we get

$$\begin{array}{cccc|c} & \rho & D & \sigma_s & P \\ M & 1 & 0 & 1 & 0 \\ L & -3 & 1 & 0 & 0 \\ T & 0 & 0 & -2 & 1 \end{array} \tag{4.26}$$

Applying the matrix transformation operations results in

$$\begin{array}{cccc|c} & \rho & D & \sigma_s & P \\ M+T/2 & 1 & 0 & 0 & 1/2 \\ L+3M+3T/2 & 0 & 1 & 0 & 3/2 \\ -T/2 & 0 & 0 & 1 & -1/2 \end{array} \tag{4.27}$$

resulting in $\Pi_1 = P\sigma_s^{1/2}/\rho^{1/2}D^{3/2}$. It can be shown numerically and experimentally that $\Pi_1 = K$ is a constant. Hence, the period of oscillation is $P = K\sqrt{\rho D^3/\sigma_s}$.

EXAMPLE 4.3.

Mixing time of two compatible fluids with the same density, viscosity and diffusivity. During mixing operations, it is often important to know when the blend can be considered homogeneous. In this example, consider t the time it takes for two compatible fluids of similar density and viscosity to be molecularly homogeneous [13]. Figure 4.5 depicts the set-up for this mixing operation.

Here, the relevant parameters can be

- Target quantity: mixing time t
- Process variables: rotational speed of the stirrer n, a tank with or without baffles
- Geometric variables: stirrer diameter d
- Physical or material properties: density ρ, diffusivity D and kinematic viscosity ν

The corresponding dimensional matrix can be written as

$$\begin{array}{ccccccc} & \rho & d & D & t & n & \nu \\ M & 1 & 0 & 0 & 0 & 0 & 0 \\ L & -3 & 1 & 2 & 0 & 0 & 2 \\ T & 0 & 0 & -1 & 1 & -1 & -1 \end{array} \tag{4.28}$$

From the dimensional matrix, it is clear that the mass unit appears only in the density term. Hence, density must be eliminated from the list along with the row corresponding to the mass unit, leaving a system with only 2 repeating parameters. In fact, the

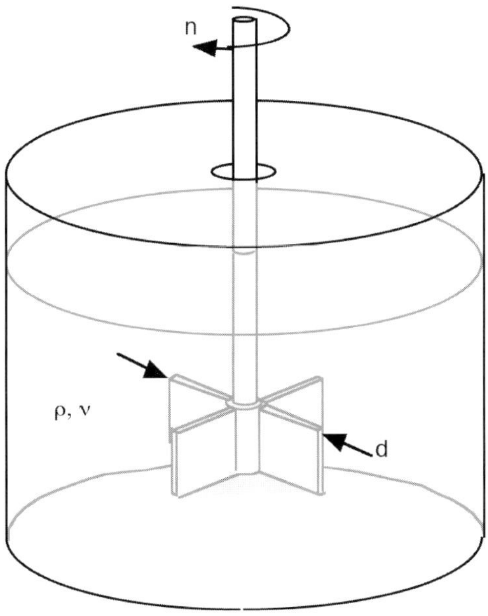

Figure 4.5: Schematic diagram of a stirring tank.

density is fully accounted for in the dynamic viscosity. Choosing ν and n as the repeating parameters results in

$$\begin{array}{c|cc|ccc} & \nu & n & d & D & t \\ L & 2 & 0 & 1 & 2 & 0 \\ T & -1 & -1 & 0 & -1 & 1 \end{array} \tag{4.29}$$

Matrix transformation results in

$$\begin{array}{c|cc|ccc} & \nu & n & d & D & t \\ L/2 & 1 & 0 & 1/2 & 1 & 0 \\ -T-L/2 & 0 & 1 & -1/2 & 0 & -1 \end{array} \tag{4.30}$$

Here, $\Pi_1 = dn^{1/2}/\nu^{1/2}$, $\Pi_2 = D/\nu$ and $\Pi_3 = nt$ are the resulting dimensionless groups. From these three dimensionless numbers can be deduced:

$$\begin{aligned} Re &= \frac{d^2 n}{\nu} = \Pi_1^2 = \text{Reynolds Number} \\ Sc &= \frac{\nu}{D} = \Pi_2^{-1} = \text{Schmidt Number} \\ \tau &= nt = \Pi_3 = \text{Dimensionless mixing time} \end{aligned} \tag{4.31}$$

The plot presented in Fig. 4.6 shows how the Reynolds number plays an effect on mixing time. The graph shows two sets of points, one for a mixing tank with baffles and the other for a mixing tank without baffles.

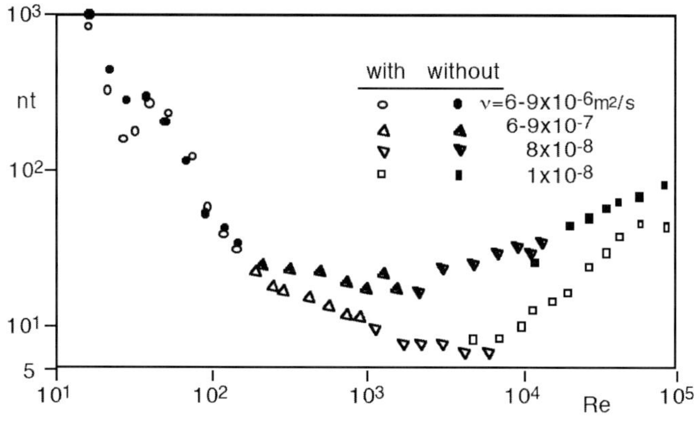

Figure 4.6: Dimensionless mixing time inside a stirring tank with and without baffles as a function of Reynolds number.

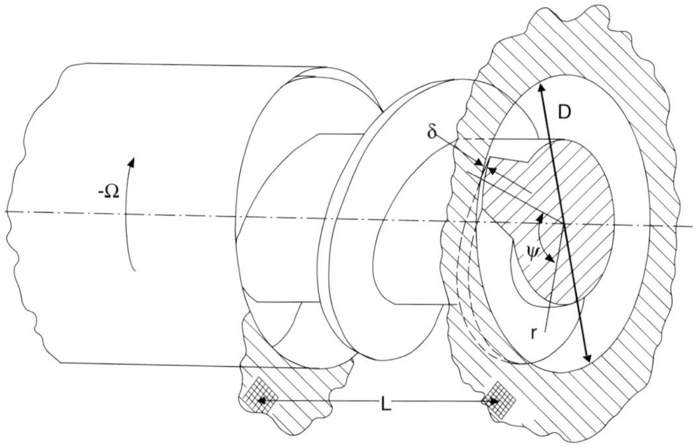

Figure 4.7: Schematic diagram of a screw geometry.

■ EXAMPLE 4.4.

Single screw extruder operating curves. The conveying characteristics of a single screw extruder can also be analyzed by use of dimensional analysis. Pawlowski [6, 7] used dimensional analysis and extensive experimental work to fully characterize the conveying and heat transfer characteristics of single screw extruders, schematically depicted in Fig. 4.7.

The relevant physical quantities that may be considered when characterizing a single screw extruder are:

- Target quantities: power consumption P, axial screw force F, pumping pressure Δp, temperature of the extrudate expressed in temperature difference $\Delta T = T - T_0$, and volumetric throughput Q

- Process variables: processing or heater temperature expressed in temperature difference $\Delta T_p = T_h - T_0$, and screw speed n
- Geometric variables: screw or inner barrel diameter D, axial screw length L, channel depth h, and clearance between the screw flight and the barrel δ
- Physical or material quantities: thermal conductivity k, density ρ, specific heat C_p, viscosity $\eta_0 = \eta(T_0)$, and viscosity temperature dependence a, from $\eta = \eta_0 e^{-a(T-T_0)}$

With this list of relevant parameters a dimensional matrix can be set up. Choosing η_0, D, n and ΔT as the repeating parameters the following dimensional matrix is set up

	η_0	D	n	ΔT_p	P	F	Δp	Q	C_p	ρ	k	a	h	L	δ	ΔT
M	1	0	0	0	1	1	1	0	0	1	1	0	0	0	0	0
L	−1	1	0	0	2	1	−1	3	2	−3	1	0	1	1	1	0
T	−1	0	−1	0	−3	−2	−2	−1	−2	0	−3	0	0	0	0	0
Θ	0	0	0	1	0	0	0	0	−1	0	−1	−1	0	0	0	1

(4.32)

which, after transformation, takes the following form

	η_0	D	n	ΔT_p	P	F	Δp	Q	C_p	ρ	k	a	h	L	δ	ΔT
M	1	0	0	0	1	1	1	0	0	1	1	0	0	0	0	0
$L+M$	0	1	0	0	3	2	0	3	2	−2	2	0	1	1	1	0
$-T-M$	0	0	1	0	2	1	1	1	2	−1	2	0	0	0	0	0
Θ	0	0	0	1	0	0	0	0	−1	0	−1	−1	0	0	0	1

(4.33)

Which contains the following dimensionless groups:

$$\Pi_1 = \frac{P}{\eta_0 D^3 n^2} \qquad \Pi_2 = \frac{F}{\eta_0 D^2 n} \qquad \Pi_3 = \frac{\Delta p}{\eta_0 n}$$

$$\Pi_4 = \frac{Q}{D^3 n} \qquad \Pi_5 = \frac{C_p \Delta T}{D^2 n^2} \qquad \Pi_6 = \frac{\rho D^2 n}{\eta_0} \tag{4.34}$$

$$\Pi_7 = \frac{k \Delta T}{\eta_0 D^2 n^2} \qquad \Pi_8 = a\Delta T \qquad \Pi_9 = \frac{h}{D}$$

$$\Pi_{10} = \frac{L}{D} \qquad \Pi_{11} = \frac{\delta}{D} \qquad \Pi_{12} = \frac{\Delta T}{\Delta T_p}$$

The first three and the last dimensionless groups are extruder operation characteristic values for power consumption, axial screw force, pumping pressure, and the extrudate temperature, respectively and depend on the process, material and geometry dimensionless groups. Π_4 is a dimensionless volumetric throughput and Π_6 the Reynolds number related to the rotational speed of the screw[2]. Π_9 through Π_{11} are geometry

[2]A single screw extruder has two Reynolds numbers. One, $Re_n = \rho D^2 n/\eta_0$, related to the rotational speed of the screw, and another, $Re_Q = Q\rho/\eta_0 D$, related to the mass throughput. The ratio of the two gives the dimensionless throughput $Re_Q/Re_n = Q/D^3 n$.

dependent dimensionless groups. The remaining are

$$\begin{aligned} Br &= \frac{\eta_0 D^2 n^2}{k \Delta T} = \frac{1}{\Pi_7} = \text{Brinkman Number} \\ Pr &= \frac{C_p \eta_0}{k} = \frac{\Pi_5}{\Pi_7} = \text{Prandtl Number} \\ Na &= a \Delta T Br = \frac{\Pi_8}{\Pi_7} = \text{Nahme-Griffith Number} \end{aligned} \tag{4.35}$$

The following relation between the dimensionless extruder operation curves and the other dimensionless groups can be expressed:

$$\frac{P}{\eta_0 D^3 n^2}, \frac{F}{\eta_0 D^2 n}, \frac{\Delta p}{\eta_0 n}, \frac{\Delta T}{\Delta T_p} = f\left(\frac{Q}{D^3 n}, Re_n, Br, Pr, Na, \frac{h}{D}, \frac{L}{D}, \frac{\delta}{D}\right) \tag{4.36}$$

The above equation can be simplified assuming a Newtonian isothermal problem. For such a case Pawlowski reduced the above equations to a set of characteristic functions that describe the conveying properties of a single screw extruder under isothermal and creeping flow ($Re < 100$) assumptions. These are written as

$$\begin{aligned} \Delta\widehat{p} = \frac{\Delta p D}{\eta_0 n L} &= f\left(\widehat{Q}\right) \\ \widehat{P} = \frac{P}{\eta_0 n^2 D^2 L} &= g\left(\widehat{Q}\right) \\ \widehat{F} = \frac{F}{\eta_0 n L D} &= h\left(\widehat{Q}\right) \end{aligned} \tag{4.37}$$

where $\Delta\hat{p}$, $\hat{P}$ and $\hat{F}$ are dimensionless pressure build-up, power consumption and axial screw force, respectively. These relationships are illustrated in the experimental measurements performed by Pawlowski [6, 7] and presented in Fig. 4.8.

From the experimental results eqns. (4.37) can be expressed as

$$\frac{1}{A_1}\widehat{Q} + \frac{1}{A_2}\Delta\widehat{p} = 1 \tag{4.38}$$

$$\frac{1}{B_1}\widehat{Q} + \frac{1}{B_2}\widehat{P} = 1 \tag{4.39}$$

$$\frac{1}{C_1}\widehat{Q} + \frac{1}{C_2}\widehat{F} = 1 \tag{4.40}$$

Equation (4.38) can be rewritten into the more familiar screw characteristic curve form as

$$\widehat{Q} = A_1 - \frac{A_1}{A_2}\Delta\widehat{p} \tag{4.41}$$

Figure 4.8 also presents an analytical solution for the screw characteristic curve of a single screw extruder with leakage flow effects. The discrepancies between analytical solution and experimental results arise due to the fact that the screw curvature, the flight angle and the fillet radii are not included in the analytical model. The analytical solution given by Tadmor and Klein [27] was used.

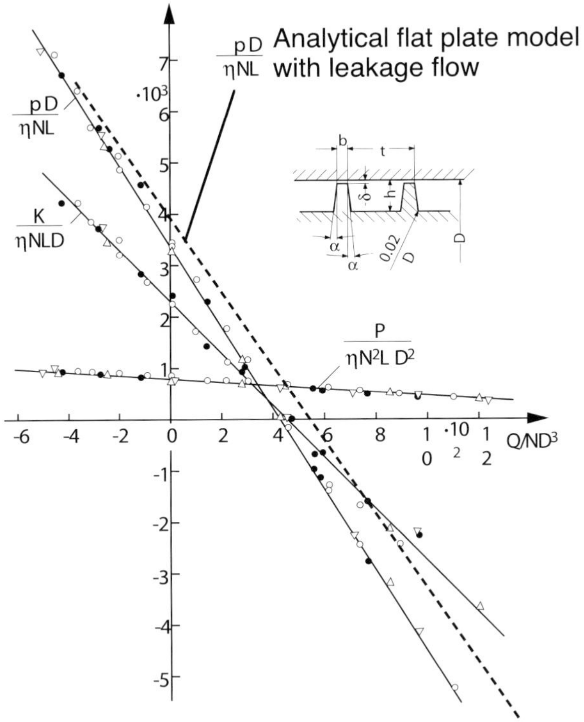

Figure 4.8: Throughput, power and axial force characteristic curves for a single screw extruder.

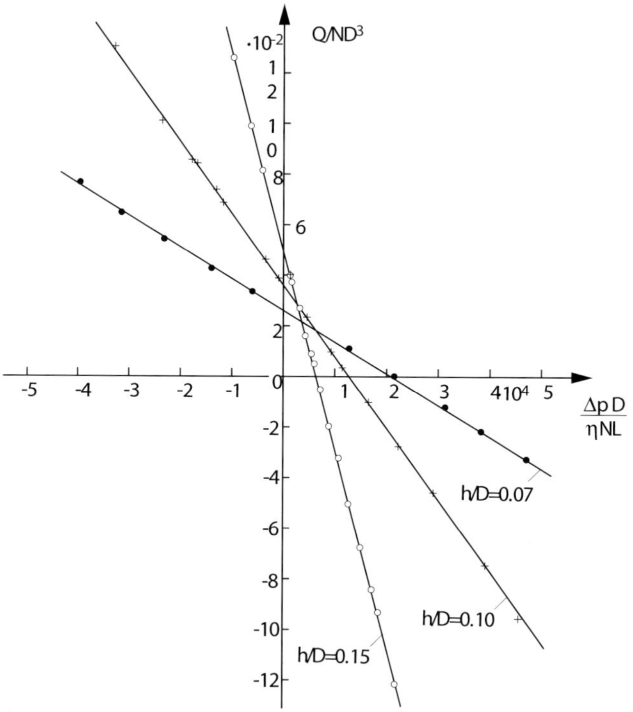

Figure 4.9: Screw characteristic curves as a function of channel depth.

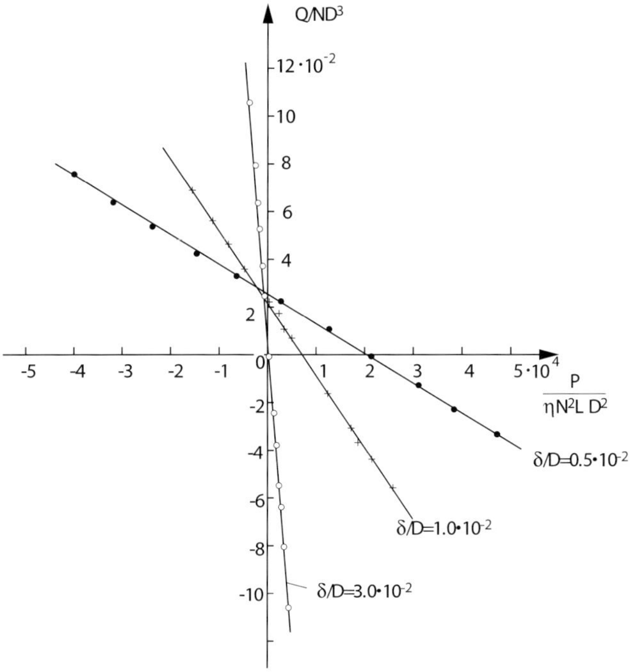

Figure 4.10: Screw characteristic curves as a function of flight clearance.

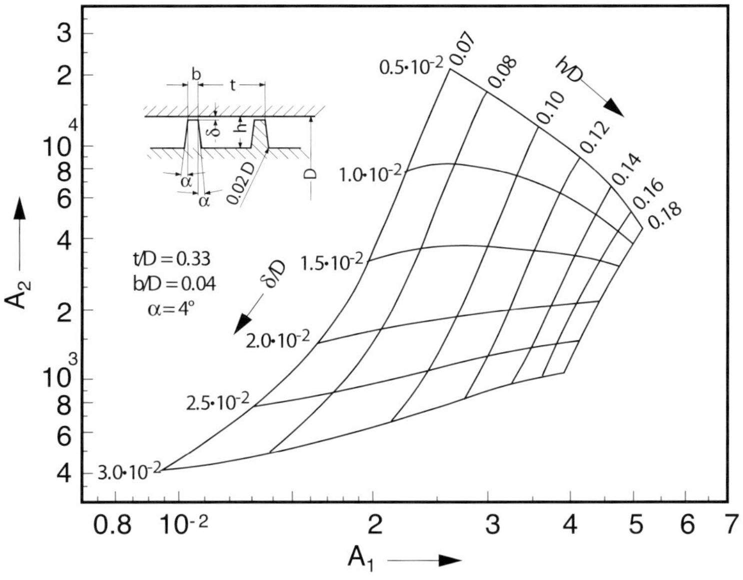

Figure 4.11: Nomogram summarizing the screw characteristic curves described in eqn. (4.38).

Through extensive experimental work Pawlowski was able to demonstrate the effect channel depth, h/D, and flight clearance, δ/D have on the screw characteristic curve. These are shown in Figs. 4.9 and 4.10, respectively. The nomogram presented in Fig. 4.11 summarizes the screw characteristic for a specific screw shape with different channel depths and flight clearances.

4.3 PROBLEMS WITH NON-LINEAR MATERIAL PROPERTIES

Most dimensional analyses deal with problems with linear material properties. However, in polymer processing, the viscosity is temperature as well as rate of deformation dependent. In addition, other properties are temperature and pressure dependent. In Example 4.4, one such non-linearity was introduced, namely the temperature dependance of the viscosity. In a similar way the rate of deformation dependence of the viscosity may also be introduced. Choosing the power-law model for the viscosity

$$\eta = m\dot{\gamma}^{\mathrm{n}-1} \tag{4.42}$$

where n here is the power-law index and in itself is a dimensionless number that represents the shear thinning (see Chapter 2) in a specific fluid.

In extrusion problems, the rate of deformation is directly proportional to the rotational speed of the screw. Hence, a characteristic viscosity can be defined as

$$\eta = mn^{\mathrm{n}-1} \tag{4.43}$$

For extrusion problems, Π_3 can be written as

$$\Pi_3 = \frac{\Delta p}{mn^{\mathrm{n}}} \tag{4.44}$$

In addition, the length of the extruder is directly proportional to the amount of pressure build-up in the pumping section of the extruder. Hence, to fully account for shear thinning as well as L/D of the extruder the operating curves data for an extruder can be significantly reduced. This is done by plotting Π_4 (throughput) as a function of Π_3/Π_{10} (pressure).

For the screw characteristic curves presented in Fig. 4.12 and in Fig. 4.13 for a conventional and grooved fed extruder, respectively, the reduced graphs are shown in Fig. 4.14. As can be seen here, each type of extruder can be represented with a single curve for a whole range of rotational speeds. It is to be noted that in this representation the effect of viscous dissipation was not included, which may explain why some of the points fall somewhat outside the fitted lines.

4.4 SCALING AND SIMILARITY

As pointed out at the beginning of the chapter, when designing a new polymer processing operation to produce a product or to blend or compound a new material, it is often desirable or necessary to work on a smaller scale such as a laboratory extruder, internal batch mixer, stirring tank, etc. The evolving model must then be scaled up or down to the actual operation.

When scaling a process, similarity between the various sizes and processes is sought. As a rule, a perfectly scalable prototype is one that is perfectly similar to its scaled system. A perfectly similar set of systems is one where all the dimensionless numbers or Π-groups

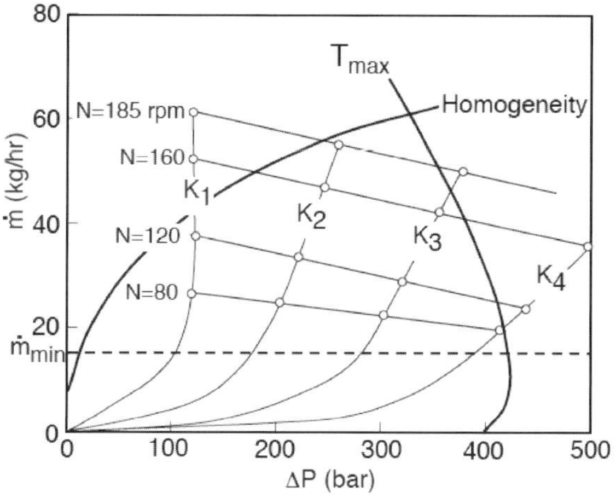

Figure 4.12: Screw and die characteristic curves for a 45 mm diameter extruder with an PE-LD.

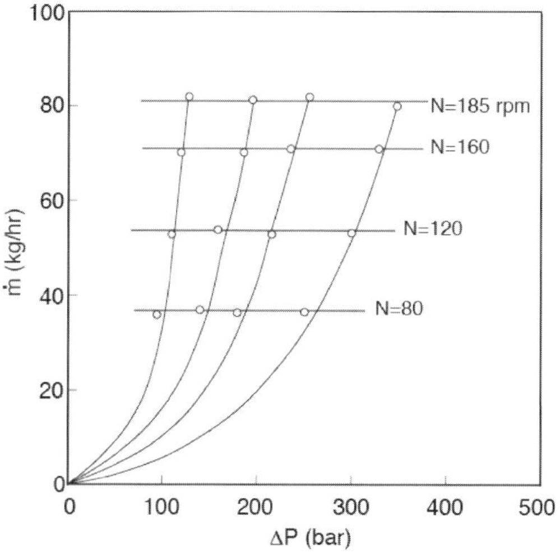

Figure 4.13: Screw and die characteristic curves for a grooved feed 45 mm diameter extruder with an PE-LD.

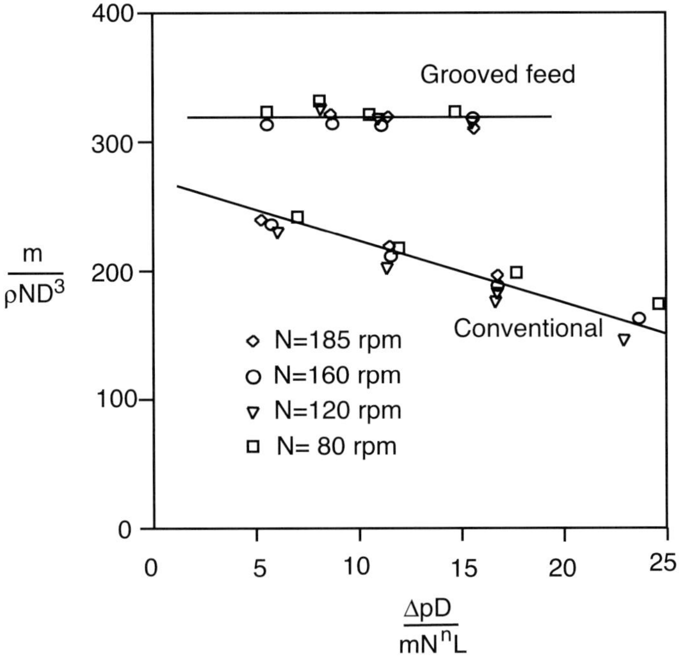

Figure 4.14: Reduced experimental screw characteristic curves for conventional and grooved feed single screw extruders.

produced have the same numerical value between each other. A rather simple system, for which this can be easily demonstrated, is the smooth pipe pressure drop experiments presented in Section 4.1. Here, the same dimensionless pressure drop, λ, versus Reynolds number curve was developed using pipes whose diameter varied from 3.61 mm to 126.2 mm and with viscosities of water and air that differ from each other by a factor of 100. Hence, if the system is a smooth pipe and the diameter were increased, the velocity determined would render the same Reynolds number and would then adjust the L/D to render the same λ. For example, if doubling the diameter of the original system that had a Reynolds number of 5 and therefore a $\lambda = 2.1$, would lead to a reduction of the speed by half, and an increase in the length of the pipe by a factor of 8 result in a perfectly similar system. In a similar way, scaling between two types of fluids with different viscosities could be achieved. In such as case, the speed or the diameter as well as the length of the pipe must be adjusted to have constant dimensionless numbers and therefore perfectly similar systems. Of course, not all systems are as straight forward as the smooth pipe flow system.

In most cases, scale-up by similarity is not always fully achieved. A process may be geometrically similar, but not thermally similar. Depending on the type of process involved, one or several kinds of similarities may be required. These may be geometric, kinematic, dynamic, thermal, kinetic or chemical similarities.

■ **EXAMPLE 4.5.**

Single screw extruder. Let us take the case of a single screw extruder section that works well when dispersing a liquid additive within a polymer matrix. The single screw extruder was already discussed in the previous section. However, the effect of surface tension, which is important in dispersive mixing, was not included in that analysis. Hence, if we also add surface tension as a relevant physical quantity, it would add one more column on the dimensional matrix. To find the additional dimensionless group associated with surface tension, σ_s, and size of the dispersed phase, R, two new columns to the matrix in eqn. (4.32) must be added resulting in:

$$\begin{array}{c|cccccc} & \eta_0 & D & n & \Delta T_p & \sigma_s & R \\ M & 1 & 0 & 0 & 0 & 1 & 0 \\ L & -1 & 1 & 0 & 0 & 0 & 1 \\ T & -1 & 0 & -1 & 0 & -2 & 0 \\ \Theta & 0 & 0 & 0 & 1 & 0 & 0 \end{array} \tag{4.45}$$

which, after transformation, results in

$$\begin{array}{c|cccccc} & \eta_0 & D & n & \Delta T_p & \sigma_s & R \\ M & 1 & 0 & 0 & 0 & 1 & 0 \\ L+M & 0 & 1 & 0 & 0 & 1 & 1 \\ -T-M & 0 & 0 & 1 & 0 & 1 & 0 \\ \Theta & 0 & 0 & 0 & 1 & 0 & 0 \end{array} \tag{4.46}$$

which results in $\Pi_{13} = \sigma_s/\eta_0 Dn$ and $\Pi_{14} = R/D$.

Combining these dimensionless along with Π_9 forms the well known capillary number as

$$Ca = \frac{\Pi_{14}}{\Pi_{13}\Pi_9} = \frac{\eta_0 DnR}{\sigma_s h} = \frac{\eta_0 \tau R}{\sigma_s} \tag{4.47}$$

In addition to the geometric parameters of this problem, of interest to us in this dispersive mixing process are the capillary number, Ca, which must be maintained constant in order to achieve the same amount of dispersion and the Brinkman number, Br, which must also be maintained constant so that the material is not overheated during mixing. Since our scaling factor, $\Re$, is determined by an increase in diameter,

$$D_{scaled} = D\Re \tag{4.48}$$

all our other dimensionless groups must be adjusted accordingly. Hence, if the Brinkman number is to remain constant, the rotational speed of the screw must be scaled by

$$n_{scaled} = \frac{n}{\Re} \tag{4.49}$$

Since the capillary number dominates the dispersion of the fluids, that dimensionless group must also be maintained constant. Since the rotational speed and the diameter have already been dealt with, the only remaining parameter in Ca is the channel depth, which must be maintained constant. Hence

$$h_{scaled} = h \tag{4.50}$$

which leads to a system that is economically unfeasible, since the material throughput increases proportionally to the increase in diameter, instead of the expected cubic relation. Hence, there is no geometric similarity between the model and the scaled process.

The scaling of extruders remains a very complex and controversial art. One form of scaling was proposed by Maddock [5] in 1959 and is still commonly used today. He suggested a constant shear rate within the extruder using

$$\begin{aligned} D_{scaled} &= D\Re \\ n_{scaled} &= \frac{n}{\sqrt{\Re}} \\ h_{scaled} &= h\sqrt{\Re} \end{aligned} \tag{4.51}$$

Scaling the Brinkman number using the above scaled parameters, gives

$$Br_{scaled} = Br\Re \tag{4.52}$$

which can potentially lead to viscous dissipation problems. To avoid these problems, Rauwendaal [9] suggests using,

$$\begin{aligned} D_{scaled} &= D\Re \\ n_{scaled} &= \frac{n}{\Re} \\ h_{scaled} &= h\sqrt{\Re} \end{aligned} \tag{4.53}$$

which gives,

$$Br_{scaled} = Br \tag{4.54}$$

It can be seen that neither of these leads to a perfectly similar scaled system.

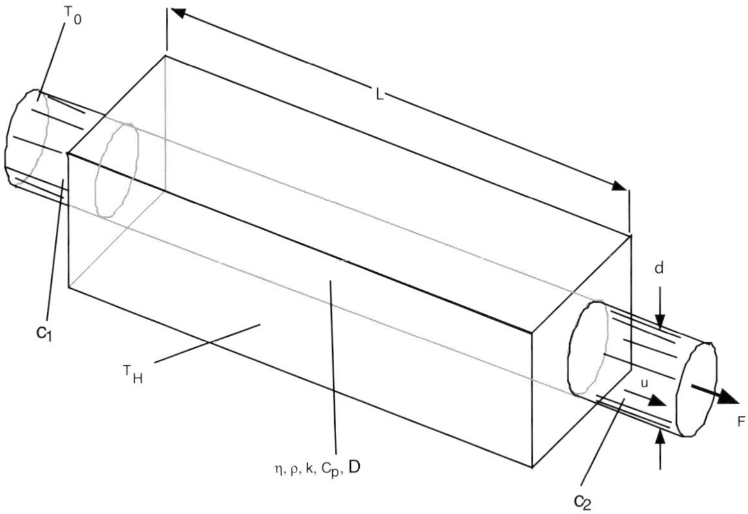

Figure 4.15: Schematic diagram of the pultrusion process.

■ **EXAMPLE 4.6.**

Curing reaction during the pultrusion process. The pultrusion process, as depicted in Fig. 4.15, involves a curing reaction as the fibers impregnated with the thermosetting resin are pulled through a heated die. As the fibers enter the die, the pressure increases causing the resin to fully impregnate the fibers, eliminating voids in the finished product. As the material advances through the die, the resin starts to cure, leading to an increase in viscosity and therefore a reduction of flow. The reaction will often lead to excessive temperature rises within the part, which in turn will lead to residual stresses, warpage and material degradation. This reaction process is very similar to the continuous chemical reaction process in tubular reactors studied by Damköhler [2].

In the set-up described by Fig. 4.15, the chosen physical quantities can be:

- Target quantity: maximum temperature inside the part due to exothermic reaction T_{max} $(\Delta T_{max} = T_{max} - T_0)$, pressure rise inside the die Δp, and pull force F,
- Geometric variables: length of the die L, and characteristic dimension across the die d,
- Process variables: pultrusion speed u, initial temperature T_0, die or heater temperature T_H $(\Delta T = T_H - T_0)$, inital degree of cure c_1, and final degree of cure c_2,
- Physical or material properties: viscosity η, density ρ, thermal conductivity k, specific heat C_p, molecular diffusion coefficient D, and permeability P.

During cure, thermosetting resins undergo a chemical reaction that follows laws of chemical thermodynamics and reaction kinetics. For simplicity it can be assumed as a second order reaction with a reaction rate defined by an Arrhenius relation as

$$\frac{dc}{dt} = \kappa_0 e^{-E/RT} c^2 \tag{4.55}$$

The chemical reaction is exothermic with a total heat of reaction per unit volume of Q_T. For an initial degree of cure of c_1, and a final degree of cure of c_2, the total heat of reaction inside the die is given by $(c_2 - c_1)\, Q_T$ or $\Delta c Q_T$. The complete relevance list is given by

$$(L, d, u, T_H, \Delta T_{max}, \Delta T, \eta, \rho, k, C_p, D, P, \Delta c Q_T, \kappa_0, E/R, \Delta p, F) \tag{4.56}$$

Following Damköhler's [2] analysis, M, L, T, Θ, and H can be used as dimensional quantities, where H are units of heat such as Joules or calories. Hence, five repeating variables must be picked. Next, ρ, L, κ_0, T_0 and $\Delta c Q_T$ are chosen as the repeating parameters. Eliminating L/d, $\Delta T_{max}/\Delta T$, $T_H/\Delta T$ and ER/T_H as obvious dimensionless groups reduces the dimensional matrix, which can now be written as

$$\begin{array}{cccccccccccccc}
 & \rho & L & \kappa_0 & T_H & \Delta c Q_T & u & \eta & D & C_p & k & P & \Delta p & F \\
M & 1 & 0 & 0 & 0 & 0 & 0 & 1 & 0 & -1 & 0 & 0 & 1 & 1 \\
L & -3 & 1 & 0 & 0 & -3 & 1 & -1 & 2 & 0 & -1 & 2 & -1 & 1 \\
T & 0 & 0 & -1 & 0 & 0 & -1 & -1 & -1 & 0 & -1 & 0 & -2 & -2 \\
\Theta & 0 & 0 & 0 & 1 & 0 & 0 & 0 & 0 & -1 & -1 & 0 & 0 & 0 \\
H & 0 & 0 & 0 & 0 & 1 & 0 & 0 & 0 & 1 & 1 & 0 & 0 & 0
\end{array} \tag{4.57}$$

which, after the transformation, becomes

$$\begin{array}{cccccccccccccc}
 & \rho & L & \kappa_0 & T_H & \Delta c Q_T & u & \eta & D & C_p & k & P & \Delta p & F \\
M & 1 & 0 & 0 & 0 & 0 & 0 & 1 & 0 & -1 & 0 & 0 & 1 & 1 \\
L+3M+3H & 0 & 1 & 0 & 0 & 0 & 1 & 2 & 2 & 0 & 2 & 2 & 2 & 4 \\
-T & 0 & 0 & 1 & 0 & 0 & 1 & 1 & 1 & 0 & 1 & 0 & 2 & -2 \\
\Theta & 0 & 0 & 0 & 1 & 0 & 0 & 0 & 0 & -1 & -1 & 0 & 0 & 0 \\
H & 0 & 0 & 0 & 0 & 1 & 0 & 0 & 0 & 1 & 1 & 0 & 0 & 0
\end{array} \tag{4.58}$$

Here the following dimensionless groups can be obtained:

$$\Pi_1 = \frac{u}{L\kappa_0} \qquad \Pi_2 = \frac{\eta}{\rho L^2 \kappa_0} \qquad \Pi_3 = \frac{D}{L^2 \kappa_0} \qquad \Pi_4 = \frac{C_p \rho T_H}{\Delta c Q_T}$$

$$\Pi_5 = \frac{k T_H}{L^2 \kappa_0 \Delta c Q_T} \qquad \Pi_6 = \frac{P}{L^2} \qquad \Pi_7 = \frac{\Delta p}{\rho L^2 \kappa_0^2} \qquad \Pi_8 = \frac{k \kappa_0^2}{\rho L^4}$$

The Π-groups presented above are formed by several known dimensionless numbers, such as Re, Pr and Sc. The first dimensionless group, Π_1 represents an inverse mean dimensionless residence time inside the die and can be also written as

$$\widehat{\tau} = \kappa_0 \tau \tag{4.59}$$

where $\tau = L/u$ is the mean residence time inside the pultrusion die. Π_4 is sometimes referred to as the Damköhler number, Da which is a reaction kinetic dimensionless number. Π_5 is the inverse of Damköhler IV number, Da^{IV}, which represents the ratio of the reaction heat generation to heat conduction. A large Da^{IV} is typical of a process during which a significant temperature increase occurs due to heat of reaction. Π_6 is a dimensionless permeability related to the fiber impregnation during the pultrusion process. Π_2 and Π_3 can be written as

$$\begin{aligned}\Pi_2 &= \frac{\eta}{\rho L^2 \kappa_0} = (\kappa_0 \tau Red/L)^{-1} \\ \Pi_3 &= \frac{D}{L^2 \kappa_0} = (\kappa_0 \tau ReScL/d)^{\ 1}\end{aligned} \tag{4.60}$$

Damköhler's original analysis [2] resulted in four dimensionless numbers which today are referred to as Damköhler numbers I through to IV and are given by

$$\begin{aligned}Da^I &= \frac{\kappa_0 L}{\eta} \\ Da^{II} &= \frac{\kappa_0 L^2}{D} \\ Da^{III} &= \frac{\kappa_0 \Delta c Q_T L}{\rho C_p T_0 u} = Da \kappa_0 \tau \\ Da^{IV} &= \frac{\kappa_0 \Delta c Q_T d^2}{k T_0} = Da \kappa_0 \tau RePr\end{aligned} \tag{4.61}$$

Note that Da^I is our dimensionless mean residence time and $Da^{II} = \Pi_3$. Da^{III} represents the ratio of the heat of reaction to the heat removed via convection. The ratio $Da^{IV}/Da^{III} = Ped/L = Gr$ represents the convection in the machine direction to the conduction through the thickness of the pultruded product.

Scaling a reactive system as the one described in this example is very complex. Let us assume that the engineer is scaling a system to a larger one, where similarity is maintained in the dimensionless group L/D, $L/D = idem$. In addition, the engineer is satisfied with the temperature build-up in the center of the part due to the heat of reaction. Hence, similarity must also be maintained in the dimensionless group $\Delta T_{max}/\Delta T$. For example, in this case the heater temperature, T_H, could be changed without changing the course of the reaction, knowing that the kinetic properties (κ_0) of our material cannot be changed, hence, maintaining $Da = idem$. Since the thicker part can lead to higher temperatures in the center (ΔT_{max}) due to difficulties in conducting the heat of reaction out of our system, it may be desirable to reduce the mean residence time by speeding up the pultruded product to the point that $\hat{\tau} \neq idem$.

In Damköhler's analysis, which applied to a continuous chemical reaction process in a tubular reactor, he solved these dilemmas by completely abandoning geometric similarity and fluid dynamic similarity. In other words, $L/D \neq idem$ and assuming that the Reynolds number is irrelevant in the scaling. Hence, his scale-up depends exclusively on thermal and reaction similarity. In our case it is even easier to see that the Reynolds number is very small and does not play a role in the process. By allowing to adjust L/D accordingly, there is more flexibility in the scaling problem.

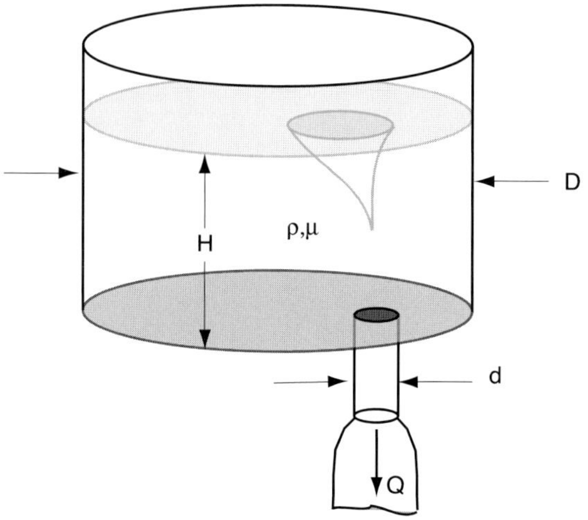

Figure 4.16: Schematic diagram of polymerization tank draining through a pipe on the bottom of the tank.

■ **EXAMPLE 4.7.**

Draining a polymerization tank. When draining a polymerization tank or a mixing vessel, a vortex forms, starting at the surface and moving toward the drainage pipe as schematically depicted in Fig. 4.16.

Eventually, as the level of the liquid inside the tank is low enough, the vortex is pulled into the pipe, entrapping air bubbles in the drained fluid. In most cases, air bubbles are not desired, and this situation should therefore be avoided. To better understand this situation, the engineer must build a model and determine the minimum fluid level height, H_{min}, to avoid air entrainment.

For this specific application a scale model of the system using a model fluid with a density, ρ_{model} =1,000 kg/m^3, and a Newtonian viscosity, $\mu_{\text{model}} = 1$ Pa-s, should be built. The full scale operation will have the following characteristics:

- Volumetric flow rate through the drain pipe - $Q = 60$ to 600 lt/min
- Tank dimensions - tank diameter of $D = 3$ m and drain diameter of $d = 30$ cm
- Fluid properties - Density of ρ =1,000 kg/m^3 and Newtonian viscosity of $\mu = 10$ Pa-s

A dimensional analysis of the system will result in four dimensionless numbers, Reynolds number, Froude number and geometric dimensionless parameters given by,

$$\begin{aligned} Re &= \frac{Q\rho}{D\mu} \\ Fr &= \frac{Q^2}{D^5 g} \\ \hat{H} &= \frac{H_{min}}{D} \\ \hat{d} &= \frac{d}{D} \end{aligned} \qquad (4.62)$$

respectively. If geometric similarity is assumed, the fourth dimensionless group can be eliminated, $\hat{d} \rightarrow idem$. If dynamic similarity is assumed

$$\begin{aligned} Re_{\text{model}} &= Re_{\text{tank}} \\ Fr_{\text{model}} &= Fr_{\text{tank}} \quad \text{and} \\ \hat{H}_{\text{model}} &= \hat{H}_{\text{tank}} \end{aligned} \tag{4.63}$$

Let us evaluate the Reynolds number and the Froude number for the actual process using $Q = 600$ lt/min ($Q = 0.01$ m^3/s),

$$Re_{\text{tank}} = \frac{0.01(1000)}{3(10)} = 0.333 \tag{4.64}$$

and

$$Fr_{\text{tank}} = \frac{(0.01)^2}{(3)^5(9.81)} = 4.2 \times 10^{-8} \tag{4.65}$$

The arbitrary decision to use a model with $D = 0.3$ m gives

$$\begin{aligned} Q_{\text{model}} &= \frac{D_{\text{model}}\mu_{\text{model}}Re_{\text{model}}}{\rho} \\ &= \frac{0.3(1)(0.333)}{1000} \\ &= 0.0001\text{m}^3/\text{s}(100\text{cm}^3/\text{s}) \end{aligned} \tag{4.66}$$

and

$$Fr_{\text{model}} = \frac{(0.0001)^2}{(0.3)^5(9.81)} = 4.2 \times 10^{-7} \tag{4.67}$$

It is noted here that the Froude number has changed and that dynamic similarity cannot be maintained if both, the model fluid viscosity and the model tank dimensions, are fixed because two unknowns (D and Q) are required to satisfy the two eqns. (4.64) and (4.65). Since gravity is a constant (9.81 m/s^2) and ρ/μ=1,000 s/m^2 is fixed for the model, obtaining that

$$\frac{Q_{\text{model}}}{D_{\text{model}}} = 0.000333 \tag{4.68}$$

and

$$Fr = \frac{Q^2_{\text{model}}}{D^5_{\text{model}}} = 4.1 \times 10^{-7} \tag{4.69}$$

Using the above equations yields a model tank diameter, $D = 0.647$ m, and drain flow rate, $Q = 0.000215$ m^3/s (215 cm^3/s). Similarly, using the lower flow rate of 60 lt/min, eqns. (4.70) and (4.71) become,

$$\frac{Q_{\text{model}}}{D_{\text{model}}} = 0.00000333 \tag{4.70}$$

and

$$Fr = \frac{Q^2_{\text{model}}}{D^5_{\text{model}}} = 4.1 \times 10^{-9} \tag{4.71}$$

which, as expected, results in the same model tank diameter $D = 0.647$ m, and a drain flow rate of $Q = 0.0000215$m^3/s (21.5 cm^3/s). Hence, performing the experiments in the range of 21.5 cm$^3/s < Q <$215 cm$^3/s$ is recommended. The observed minimum fluid level H^{model}_{min} can then be scaled to the actual H^{tank}_{min}, using eqn. (4.62) as

$$H^{\text{tank}}_{min} = \frac{D_{\text{tank}}}{D_{\text{model}}} H^{\text{model}}_{min} = 4.54 H^{\text{model}}_{min} \tag{4.72}$$

■ EXAMPLE 4.8.

Scaling approach for thin-wall injection molding. As material costs increase and weight and space restrictions are tightening, injection molders are facing the problem of having to inject thinner and thinner parts. This is especially true in the electronics industry where hand-held electronics housings now measure less than 0.6 mm in thickness. For example, there are cellphone battery housings that have a thickness of 0.18 mm (less than the thickness of three sheets of paper).

By definition, thin-wall injection molding processes are those with part thicknesses of less than 1 mm. They are produced using high injection pressures and velocities. Due to the high pressure requirement, modeling, with commercial computer programs has not been successful in predicting process physics during molding. García *et al.* [3] performed a dimensional analysis on injection molding and scaled-down the system by reducing the thickness of the part. In their analysis, they assumed a non-Newtonian, shear thinning and pressure dependent viscosity described by,

$$\eta = me^{-a\Delta T} e^{b\Delta p} \dot{\gamma}^{n-1} \tag{4.73}$$

which, when using a characteristic injection velocity, u, and mold thickness, h, can be written as,

$$\eta = me^{-a\Delta T} e^{b\Delta p} \left(\frac{2u}{h}\right)^{n-1} \tag{4.74}$$

where a and b represent the melt's sensitivity to temperature (ΔT) and pressure (Δp) changes, respectively. Furthermore, García *et al.* [3] assumed constant thermal conductivity, k, density, ρ and specific heat , C_p, and a mold length, L. Their dimensional analysis resulted in 5 dimensionless groups given by,

$$\begin{aligned}
\Pi_1 &= \frac{\Delta p L}{\eta u} \\
\Pi_2 &= \frac{u \rho L}{\eta} \\
\Pi_3 &= \frac{h}{L} \\
\Pi_4 &= \frac{k \Delta T}{\eta u^2} \\
\Pi_5 &= \frac{C_p \Delta T}{u^2}
\end{aligned} \tag{4.75}$$

where Π_2 is the Reynolds number, Π_4 is the inverse of the Brinkman number, Π_5/Π_4 is the Prandtl number and Π_1/Π_2 is the Euler number. Here, the thickness of the part is scaled down from h_1 to h_2 such that $h_2 = \Re h_1$. Obviously, in this problem geometric similarity is not maintained. In their analysis, to control viscous heating during molding, García *et al.* assumed that $Br \rightarrow idem$,

$$\frac{\eta_1 u_1^2}{k\Delta T_1} = \frac{\eta_2 u_2^2}{k\Delta T_2} \tag{4.76}$$

where $\Delta T_1 = \Delta T_2$. As the velocity, we take the equation for slit flow (see Chapter 5) where $\partial p/\partial z = \Delta p/L$,

$$u = \left(\frac{\Delta p h}{m e^{-a\Delta T} e^{b\Delta p} L}\right)^{1/n} \frac{h}{2\,(1/n + 2)} \tag{4.77}$$

which is used to compute u_1 and u_2 in eqn.(4.76) to become,

$$\frac{u_1}{u_2} = \frac{\Delta p_1^{1/n} h_1^{(1+n)/n} e^{b\Delta p_2/n}}{\Delta p_2^{1/n} h_2^{(1+n)/n} e^{b\Delta p_1/n}} \tag{4.78}$$

Combining eqns. (4.76) and (4.78) results in,

$$\Delta p_2^{1/n} \Re^{(-3n+1)(n^2+n)} = \Delta p_1^{1/n} e^{\left[\frac{(\Delta p_2 - \Delta p_1)(n+1)b + b}{n(n+1)}\right]} \tag{4.79}$$

García *et al.* plotted $\Delta p_2/\Delta p_1$ as a function of $\Re$ for various values of pressure dependence b. They used a power law index $n = 0.3$ and two typical low and high injection pressures of 40 MPa (400 bar) and 100 MPa (1000 bar), respectively. Figure 4.17 presents a plot of the pressure increase as a function of the down-scaled thickness, $\Re$. As expected, the pressure increases with a decrease in thickness and with an increase in the pressure dependence coefficient, b. The process that already begins with a high pressure requirement of 1000 bars,

Problems

4.1 A common way to produce small droplets is to let them drip slowly from a capillary. The the size of the droplet, D_d, depends on surface tension, σ_s, the diameter of the capillary, D, gravity, g, and density of the liquid. Perform a dimensional analysis to determine the dimensionless groups that govern this process.

- **a)** Set-up the dimensional matrix and rearrange it to form the core and residual matrices.
- **b)** Solve for the dimensionless numbers. Can you name these dimensionless numbers?
- **c)** Derive an expression that predicts drop size and relate this equation to the dimensionless numbers.

4.2 Perform a dimensional analysis on the polymerization tank draining problem of Example 4.7 and find the dimensionless groups used in the example.

4.3 Perform a dimensional analysis on the thin wall injection molding scaling problem of Example 4.8 and find the dimensionless groups used in the example.

- **a)** What other dimensionless numbers, if any, can be derived from this problem?
- **b)** What dimensionless number describes the cooling time in this problem?

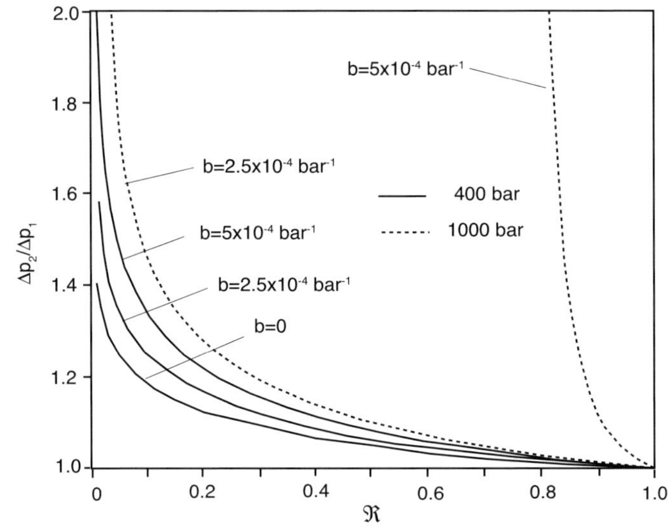

Figure 4.17: Pressure increase as a function of thickness reduction for various pressure dependence coefficients, b, and pressure requirements.

4.4 Perform a dimensional analysis on the heating of a plastic plate inside a convection oven. Your target value is the heating time to reach certain surface and center temperatures.

a) What are the process variables, geometric variables and the physical or material properties?

b) Setup the dimensional matrix and rearrange it to form the core and residual matrices.

c) Solve for the dimensionless numbers. Can you name these dimensionless numbers?

4.5 Go to your university engineering library and find the paper written by Grace in 1982 [4]. What is the relation between Grace's experiments and the dimensionless groups in Example 4.5? Attach a copy of the paper to the assignment.

4.6 During dispersion of incompatible fluids such as experienced when making polymer blends, a long thread of the dispersed phase breaks up due to Raleigh disturbances as shown in Fig. 4.18.

During this process, the relevant list of quantities is:

- Geometric variable: radius of the thread R and wave length of the disturbance, Λ
- Process variables: rate of deformation $\dot{\gamma}$ and type of flow[3], λ,
- Physical or material properties: viscosity of the continuous phase, η_c, of the dispersed phase, η_d, and surface tension, σ_s.

a) Perform a dimensional analysis and determine the relevant dimensionless groups.

[3]The flow parameter λ is often referred to as the flow number and is defined by: $\lambda = \frac{\dot{\gamma}}{\omega+\dot{\gamma}}$. When $\lambda = 0.5$ one has pure shear flow and when $\lambda = 1$ one has a stretching flow, which is very desirable for dispersion.

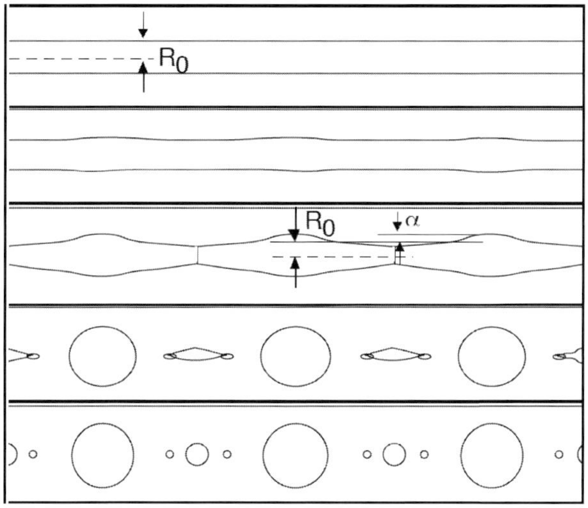

Figure 4.18: Disintegration of a Newtonian 0.35 mm diameter castor oil thread in a Newtonian silicone oil matrix. Redrawn from photographs taken every second.

b) Comment on the significance of each dimensionless group.

c) Go to your university engineering library and find the paper written by Grace in 1982 [4]. What is the relation between Grace's experiments and the dimensionless groups found in this problem? Attach a copy of the paper to the assignment.

4.7 You are asked to mix materials of different viscosities and densities in a mixing vessel or tank. Using two stirring tanks of 300 and 600 mm radius, viscosity ratios between 1 and 6,000 and differences in density between 0 and 0.3 g/cm^3, Zlokarnik [12] found a group of dimensionless numbers that were related as follows,

$$\frac{\sqrt{n\Theta Re}}{Ar^{1/3}+3} = 52$$

where, Ar is the Arquimedes number, Θ the mixing time and n the rotational speed of the stirrer.

a) Set-up the problem and find the dimensionless numbers found by Zlokarnik.

b) Comment on the significance of each dimensionless number. What do they mean?

c) Go to your university library and find Zlokarnik's paper. How much spread is there in the data that was used to determine the relation in the equation above? Attach a copy of the paper to the assignment.

REFERENCES

1. E. Buckingham. *Physical Review*, 4:345, 1914.
2. G. Damköhler. Einfluesse der Stroemung, Diffusion und des Waermeueberganges auf die Leistung von Reaktionsoefen. *Z.f. Elektrochemie*, 42:846, 1936.
3. S. Garcia, A. Roldán, J.P. Hernández-Ortiz, and T.A. Osswald. Dimensional analysis and scaling approach for thin wall injection molding. In *ANTEC 2004*, pages 3428–3422. SPE-ANTEC, 2004.
4. H.P. Grace. *Chem. Eng. Commun.*, 14:225, 1982.
5. B.H. Maddock. *SPE Journal*, 15:983, 1959.
6. J. Pawlowski. *Chemie-Ing. Techn.*, 38:1229, 1966.
7. J. Pawlowski. *Chemie-Ing. Techn.*, 39:1180, 1967.
8. J. Pawlowski. *Die Aehnlichkeitstheorie in der Physikalische-Technischen Forschung - Grundlagen und Anwendung*. Springer-Verlag, Berlin, 1971.
9. C. Rauwendaal. *Polymer Extrusion*. Hanser Publishers, Munich, 1990.
10. T.E. Stanton and J.R. Pannell. Similarity of motion in relation to the surface friction of fluids. *Phil. Trans. Roy. Soc. London*, 214:199, 1914.
11. Z. Tadmor and I. Klein. *Engineering Principles of Plasticating Extrusion*. Van Nostrand Reinhold Company, New York, 1970.
12. M. Zlokarnik. *Chemie-Ing.-Technik*, 38:357, 1966.
13. M. Zlokarnik. Eignung von ruehrem zum homogenisieren von fluessigkeitsgemischen. *Chem. Ing. Tech.*, 39(9/10):539, 1967.
14. M. Zlokarnik. *Scale-up in Chemical Engineering*. Wiley-VCH, Weinhein, 2002.

CHAPTER 5

TRANSPORT PHENOMENA IN POLYMER PROCESSING

So divinely is the world organized that every one of us, in our place and time, is in balance with everything else.

—Johann Wolfgang von Goethe

The field of transport phenomena is the basis of modeling in polymer processing. This chapter presents the derivation of the balance equations and combines them with constitutive models to allow modeling of polymer processes. The chapter also presents ways to simplify the complex equations in order to model basic systems such as flow in a tube or Hagen-Poiseulle flow, pressure flow between parallel plates, flow between two rotating concentric cylinders or Couette flow, and many more. These simple systems, or combinations of them, can be used to model actual systems in order to gain insight into the processes, and predict pressures, flow rates, rates of deformation, etc.

5.1 BALANCE EQUATIONS

When solving flow and heat transfer problems in polymer processing we must satisfy conservation of mass, forces or momentum and energy. Momentum and energy balances, in combination with material properties through constitutive relations, sometimes result in

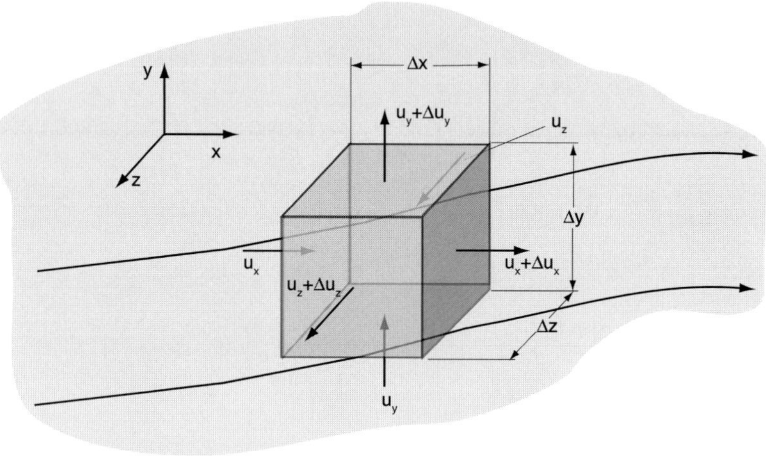

Figure 5.1: Differential frame immersed in a flow and fixed in space.

governing equations that are highly non-linear. This chapter presents the balance equations, making use of constitutive relations presented in Chapter 2 of this book.

5.1.1 The Mass Balance or Continuity Equation

The most basic aspect of modeling polymer processing is to satisfy the conservation of mass. When modeling the flow of polymers we can assume incompressibility[1], making a volume balance equivalent to a mass balance. The resulting equation is what is referred to as the *continuity equation*. In order to derive the continuity equation we place an imaginary wire frame of dimensions $\Delta x \times \Delta y \times \Delta z$ inside a flowing system as schematically depicted in Fig. 5.1.

Using the notation used in Fig. 5.1, we can perform a volumetric balance in and out of the differential element as,

$$\{\text{Volumetric flow rate}\}_{in} - \{\text{Volumetric flow rate}\}_{out} = 0 \tag{5.1}$$

or

$$\begin{aligned}&(u_z \Delta x \Delta y + u_y \Delta x \Delta z + u_x \Delta y \Delta z) - \\ &\qquad ([u_z + \Delta u_z]\Delta x \Delta y + [u_y + \Delta u_y]\Delta x \Delta z + [u_x + \Delta u_x]\Delta y \Delta z) = 0\end{aligned} \tag{5.2}$$

which results in,

$$-\Delta u_z \Delta x \Delta y - \Delta u_y \Delta x \Delta z - \Delta u_x \Delta y \Delta z = 0 \tag{5.3}$$

All balance equations are put in a more amenable form by dividing them by the differential volume, $\Delta x \Delta y \Delta z$, and putting them in terms per unite volume,

$$\frac{\Delta u_z}{\Delta z} + \frac{\Delta u_y}{\Delta y} + \frac{\Delta u_x}{\Delta x} = 0 \tag{5.4}$$

[1]From the pvT behavior of a polymer melt we know that, in principle, a polymer is not an incompressible fluid. However, the changes of volume with respect to pressure variations within a process are not significant enough to affect the flow field.

Table 5.1: Continuity Equation

Cartesian Coordinates (x, y, z):

$$\frac{\partial \rho}{\partial t} + \frac{\partial}{\partial x}(\rho u_x) + \frac{\partial}{\partial y}(\rho u_y) + \frac{\partial}{\partial z}(\rho u_z) = 0$$

Cylindrical Coordinates (r, θ, z):

$$\frac{\partial \rho}{\partial t} + \frac{1}{r}\frac{\partial}{\partial r}(\rho r u_r) + \frac{1}{r}\frac{\partial}{\partial \theta}(\rho u_\theta) + \frac{\partial}{\partial z}(\rho u_z) = 0$$

Spherical Coordinates (r, θ, ϕ):

$$\frac{\partial \rho}{\partial t} + \frac{1}{r^2}\frac{\partial}{\partial r}\left(\rho r^2 u_r\right) + \frac{1}{r \sin\theta}\frac{\partial}{\partial \theta}(\rho u_\theta \sin\theta) + \frac{1}{r \sin\theta}\frac{\partial}{\partial \phi}(\rho u_\phi) = 0$$

Letting the size of the differential element go to zero results in,

$$\frac{\partial u_z}{\partial z} + \frac{\partial u_y}{\partial y} + \frac{\partial u_x}{\partial x} = \frac{\partial u_i}{\partial x_i} = 0 \tag{5.5}$$

which states that the divergence of the velocity vector must equal zero when the mass or the volume is conserved. We can also write this equation as,

$$\nabla \cdot u = 0 \tag{5.6}$$

There are some aspects of polymer processing where the above forms of the continuity equation cannot be used, such as the flow of the nitrogen during gas assisted injection molding, the air inside the body during blow molding, the air inside the bubble during film blowing, and the gas inside a bubble during foaming. For all those cases, we have a compressible fluid with a variable density and the continuity equation must be written as,

$$\nabla \cdot (\rho u) = 0 \tag{5.7}$$

and for the transient case, we use,

$$\frac{\partial \rho}{\partial t} + \nabla \cdot (\rho u) = 0 \tag{5.8}$$

Table 5.1 presents the continuity equation in the Cartesian, cylindrical and spherical coordinate systems.

5.1.2 The Material or Substantial Derivative

It is possible to describe a flowing system from a fixed or moving observer point of view. A fixed observer, such as described in Fig. 5.2, feels the transient effects; a change in time before the system reaches steady state.

In a non-isothermal flow, a fixed observer feels

$$\frac{\partial u_x}{\partial t}, \frac{\partial u_y}{\partial t}, \frac{\partial u_z}{\partial t}, \frac{\partial T}{\partial t}, \text{ etc.}$$

Once the system reaches steady state, the fixed observer feels a constant velocity, temperature and other field variables.

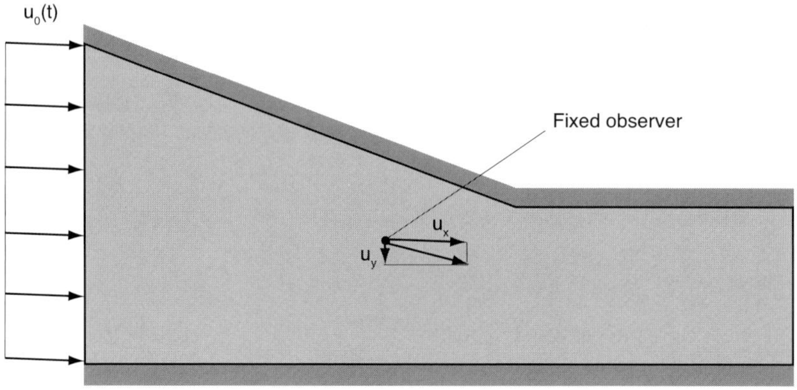

Figure 5.2: Flow system with a fixed obeserver.

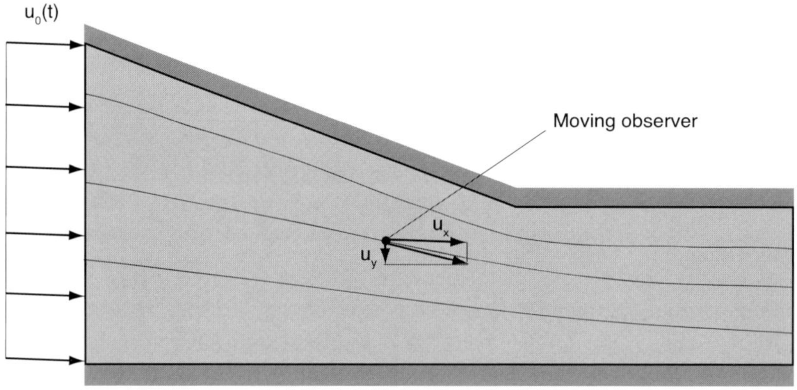

Figure 5.3: Flow system with an observer moving with a fluid particle on a given streamline.

On the other hand, a moving observer, such as the one shown in Fig. 5.3, not only feels the transient effects but also the changes that it undergoes as it travels through a gradient of velocity, temperature, concentration, etc.

The moving observer, described by a fluid particle, feels

$$\frac{\partial u_x}{\partial t} + u_x \frac{\partial u_x}{\partial x} + u_y \frac{\partial u_x}{\partial y} + u_z \frac{\partial u_x}{\partial z} = \frac{\partial u_i}{\partial t} + u_j \frac{\partial u_i}{\partial x_j} \tag{5.9}$$

as the change of u_x. Equation (5.6) is often written in short form as Du_x/Dt and is referred to as the *material derivative* or the *substantial derivative*.

5.1.3 The Momentum Balance or Equation of Motion

For a momentum balance, we take the same flow system as presented in Fig. 5.1, but instead of submerging an imaginary frame into the melt, we take an actual fluid element of dimensions $\Delta x \times \Delta y \times \Delta z$ (Fig. 5.4) and perform a force balance with the forces acting on its surfaces.

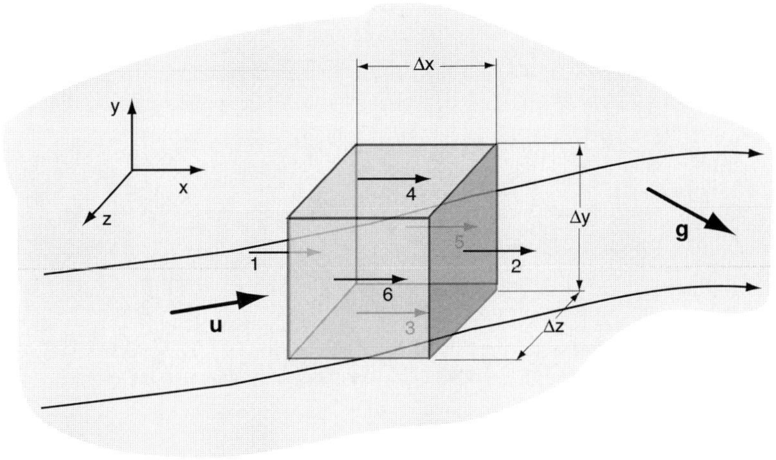

Figure 5.4: Differential fluid element traveling along its streamline x-direction forces that act on its surfaces.

The force balance can be written as

$$\sum \mathbf{f} = m\mathbf{a} \tag{5.10}$$

where the terms in the equation define force, $\mathbf{f}$, mass, m, and acceleration, $\mathbf{a}$, respectively. For simplicity, here we will only show the balance of forces in the x-direction. The balance in the y- and z-directions are left to the reader as a short exercise. The forces acting in the x-direction on a small fluid element are described in Fig. 5.4. Since the element in Fig. 5.4 is a fluid particle that moves with the flow, the change of its velocity components is described by the material derivative. Hence, the force balance in the x-direction is given by,

$$\sum \mathbf{f} = m\frac{Du_x}{Dt} \tag{5.11}$$

where $m = \rho \Delta x \Delta y \Delta z$. The following is a list of x forces that act on the surfaces of the differential fluid element:

1. $\sigma_{xx} \Delta y \Delta z$
2. $(\sigma_{xx} + \Delta\sigma_{xx}) \Delta y \Delta z$
3. $-\sigma_{yx} \Delta x \Delta z$
4. $(\sigma_{yx} + \Delta\sigma_{yx}) \Delta x \Delta z$
5. $-\sigma_{zx} \Delta x \Delta y$
6. $(\sigma_{zx} + \Delta\sigma_{zx}) \Delta x \Delta y$
7. $\rho g_x \Delta x \Delta y \Delta z$

Here, we used the mechanical engineering convention that takes as positive the forces that are pulling the element (tensile stresses) and negative the forces that push on the surface

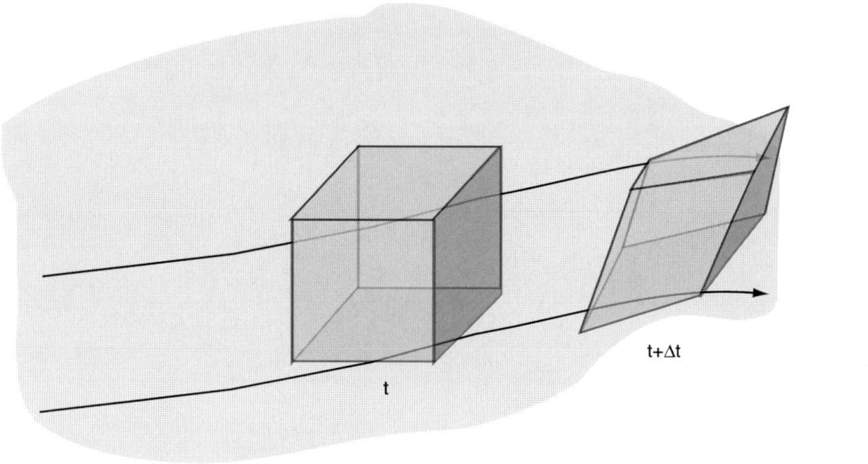

Figure 5.5: Effect of deviatoric stresses as the fluid element travels along its streamline.

(compressive stresses). In chemical engineering, the opposite is used since the stress field is regarded as a flux. A student may use any of the two conventions as long as she or he is consistent[2].

After adding the forces, dividing by the element's volume, and letting the volume go to zero, the force balance results in

$$\rho \frac{Du_x}{Dt} = \frac{\partial \sigma_{xx}}{\partial x} + \frac{\partial \sigma_{yx}}{\partial y} + \frac{\partial \sigma_{zx}}{\partial z} + \rho g_x \tag{5.12}$$

which for all three directions can be written as

$$\begin{aligned} \rho \frac{Du_i}{Dt} &= \frac{\partial \sigma_{ji}}{\partial x_j} + \rho g_i \\ \rho \frac{D\mathbf{u}}{Dt} &= \nabla \cdot \boldsymbol{\sigma} + \rho \mathbf{g} \end{aligned} \tag{5.13}$$

In fluid flow, however, it is necessary to split the total stress, σ_{ij}, into a deviatoric stress, τ_{ij}, and a hydrostatic stress, σ_H. The deviatoric stress is the one that leads to deformation (Fig. 5.5) and the hydrostatic stress is the one that is described by pressure (Fig. 5.6).

We can write,

$$\sigma_{ij} = \sigma_H \delta_{ij} + \tau_{ij} \tag{5.14}$$

where δ_{ij} is the Kronecker delta. As the above equation reveals, the hydrostatic stress can only act in the normal direction of a surface and it is equal in all three direction. Hence, we can write

$$\sigma_H = -p \tag{5.15}$$

[2]The chemical engineering convention has its roots at the University of Wisconsin-Madison. Professor C.L. Tucker III, the advisor of one of the authors at the University of Illinois at Urbana-Champaign, claimed that he could feel the stress tensor turn around on the windshield of his car every time he would drive across the Illinois-Wisconsin border.

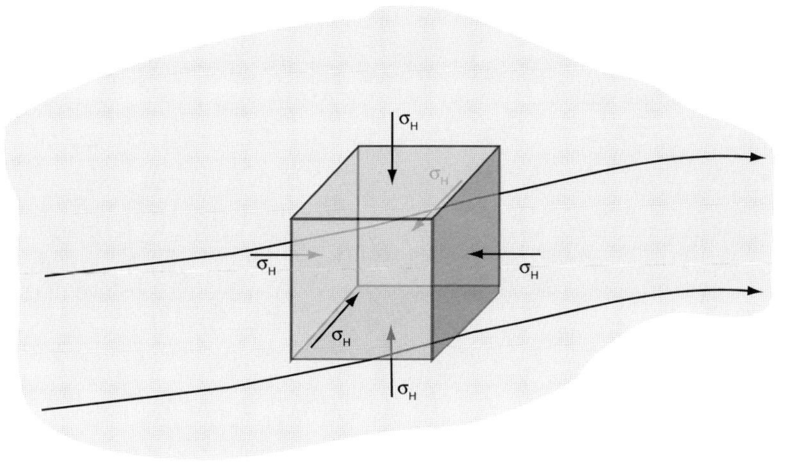

Figure 5.6: Hydrostatic stresses acting on a differential element.

where p defines the pressure. The negative pressure is due to the fact that a positive pressure causes a compressive stress. The total stress can be written as,

$$\sigma_{ij} = -p\delta_{ij} + \tau_{ij} \tag{5.16}$$

Using the definition of total stress given above, the momentum balance can now be written as,

$$\rho\frac{Du_i}{Dt} = -\frac{\partial p}{\partial x_i} + \frac{\partial \tau_{ji}}{\partial x_j} + \rho g_i \tag{5.17}$$

$$\rho\frac{D\mathbf{u}}{Dt} = -\nabla p + \nabla \cdot \boldsymbol{\tau} + \rho\mathbf{g} \tag{5.18}$$

Table 5.2 presents the momentum balance in terms of deviatoric stress in the Cartesian, cylindrical and spherical coordinate systems.

These forms of the equation of motion are commonly called *the Cauchy momentum equations*. For generalized Newtonian fluids we can define the terms of the deviatoric stress tensor as a function of a generalized Newtonian viscosity, η, and the components of the rate of deformation tensor, as described in Table 5.3.

In fluid mechanics, one common description of the deviatoric stress tensor is the Newtonian model given by,

$$\tau_{ij} = \mu\dot{\gamma}_{ij} \tag{5.19}$$

which reduces the Cauchy momentum equations to,

$$\begin{aligned} \rho\frac{Du_i}{Dt} &= -\frac{\partial p}{\partial x_i} + \mu\frac{\partial^2 u_i}{\partial x_j \partial x_j} + \rho g_i \\ \rho\frac{D\mathbf{u}}{Dt} &= -\nabla p + \mu\nabla^2\mathbf{u} + \rho\mathbf{g} \end{aligned} \tag{5.20}$$

which is often referred to as the *Navier-Stokes equations*. Table 5.4 presents the full form of the Navier-Stokes equations.

Table 5.2: Momentum Equation in terms of $\boldsymbol{\tau}$

Cartesian Coordinates (x, y, z):

$$\rho\left(\frac{\partial u_x}{\partial t}+u_x\frac{\partial u_x}{\partial x}+u_y\frac{\partial u_x}{\partial y}+u_z\frac{\partial u_x}{\partial z}\right)=-\frac{\partial p}{\partial x}+\left[\frac{\partial}{\partial x}\tau_{xx}+\frac{\partial}{\partial y}\tau_{yx}+\frac{\partial}{\partial z}\tau_{zx}\right]+\rho g_x$$

$$\rho\left(\frac{\partial u_y}{\partial t}+u_x\frac{\partial u_y}{\partial x}+u_y\frac{\partial u_y}{\partial y}+u_z\frac{\partial u_y}{\partial z}\right)=-\frac{\partial p}{\partial y}+\left[\frac{\partial}{\partial x}\tau_{xy}+\frac{\partial}{\partial y}\tau_{yy}+\frac{\partial}{\partial z}\tau_{zy}\right]+\rho g_y$$

$$\rho\left(\frac{\partial u_z}{\partial t}+u_x\frac{\partial u_z}{\partial x}+u_y\frac{\partial u_z}{\partial y}+u_z\frac{\partial u_z}{\partial z}\right)=-\frac{\partial p}{\partial z}+\left[\frac{\partial}{\partial x}\tau_{xz}+\frac{\partial}{\partial y}\tau_{yz}+\frac{\partial}{\partial z}\tau_{zz}\right]+\rho g_z$$

Cylindrical Coordinates (r, θ, z):

$$\rho\left(\frac{\partial u_r}{\partial t}+u_r\frac{\partial u_r}{\partial r}+\frac{u_\theta}{r}\frac{\partial u_r}{\partial \theta}+u_z\frac{\partial u_r}{\partial z}-\frac{u_\theta^2}{r}\right)=$$
$$-\frac{\partial p}{\partial r}+\left[\frac{1}{r}\frac{\partial}{\partial r}\left(r\tau_{rr}\right)\frac{1}{r}\frac{\partial}{\partial \theta}\left(\tau_{\theta r}\right)+\frac{\partial}{\partial z}\left(\tau_{zr}\right)-\frac{\tau_{\theta\theta}}{r}\right]+\rho g_r$$

$$\rho\left(\frac{\partial u_\theta}{\partial t}+u_r\frac{\partial u_\theta}{\partial r}+\frac{u_\theta}{r}\frac{\partial u_\theta}{\partial \theta}+u_z\frac{\partial u_\theta}{\partial z}+\frac{u_r u_\theta}{r}\right)=$$
$$-\frac{1}{r}\frac{\partial p}{\partial \theta}+\left[\frac{1}{r^2}\frac{\partial}{\partial r}\left(r^2\tau_{r\theta}\right)\frac{1}{r}\frac{\partial}{\partial \theta}\left(\tau_{\theta\theta}\right)+\frac{\partial}{\partial z}\left(\tau_{z\theta}\right)+\frac{\tau_{\theta r}-\tau_{r\theta}}{r}\right]+\rho g_\theta$$

$$\rho\left(\frac{\partial u_z}{\partial t}+u_r\frac{\partial u_z}{\partial r}+\frac{u_\theta}{r}\frac{\partial u_z}{\partial \theta}+u_z\frac{\partial u_z}{\partial z}\right)=$$
$$-\frac{\partial p}{\partial z}+\left[\frac{1}{r}\frac{\partial}{\partial r}\left(r\tau_{rz}\right)\frac{1}{r}\frac{\partial}{\partial \theta}\left(\tau_{\theta z}\right)+\frac{\partial}{\partial z}\left(\tau_{zz}\right)\right]+\rho g_z$$

Spherical Coordinates (r, θ, ϕ):

$$\rho\left(\frac{\partial u_r}{\partial t}+u_r\frac{\partial u_r}{\partial r}+\frac{u_\theta}{r}\frac{\partial u_r}{\partial \theta}+\frac{u_\phi}{r\sin\theta}\frac{\partial u_r}{\partial \phi}-\frac{u_\theta^2+u_\phi^2}{r}\right)=-\frac{\partial p}{\partial r}+$$
$$\left[\frac{1}{r^2}\frac{\partial}{\partial r}\left(r^2\tau_{rr}\right)+\frac{1}{r\sin\theta}\frac{\partial}{\partial \theta}\left(\tau_{\theta r}\sin\theta\right)+\frac{1}{r\sin\theta}\frac{\partial}{\partial \phi}\left(\tau_{\phi r}\right)-\frac{\tau_{\theta\theta}+\tau_{\phi\phi}}{r}\right]+\rho g_r$$

$$\rho\left(\frac{\partial u_\theta}{\partial t}+u_r\frac{\partial u_\theta}{\partial r}+\frac{u_\theta}{r}\frac{\partial u_\theta}{\partial \theta}+\frac{u_\phi}{r\sin\theta}\frac{\partial u_\theta}{\partial \phi}-\frac{u_r u_\theta-u_\phi^2\cot\theta}{r}\right)=-\frac{1}{r}\frac{\partial p}{\partial \theta}+$$
$$\left[\frac{1}{r^3}\frac{\partial}{\partial r}\left(r^3\tau_{r\theta}\right)+\frac{1}{r\sin\theta}\frac{\partial}{\partial \theta}\left(\tau_{\theta\theta}\sin\theta\right)+\frac{1}{r\sin\theta}\frac{\partial}{\partial \phi}\left(\tau_{\phi\theta}\right)+\frac{\left(\tau_{\theta r}-\tau_{r\theta}\right)-\tau_{\phi\phi}\cot\theta}{r}\right]+\rho g_\theta$$

$$\rho\left(\frac{\partial u_\phi}{\partial t}+u_r\frac{\partial u_\phi}{\partial r}+\frac{u_\theta}{r}\frac{\partial u_\phi}{\partial \theta}+\frac{u_\phi}{r\sin\theta}\frac{\partial u_\phi}{\partial \phi}+\frac{u_\phi u_r+u_\theta u_\phi\cot\theta}{r}\right)=-\frac{1}{r\sin\theta}\frac{\partial p}{\partial \phi}+$$
$$\left[\frac{1}{r^3}\frac{\partial}{\partial r}\left(r^3\tau_{r\phi}\right)+\frac{1}{r\sin\theta}\frac{\partial}{\partial \theta}\left(\tau_{\theta\phi}\sin\theta\right)+\frac{1}{r\sin\theta}\frac{\partial}{\partial \phi}\left(\tau_{\phi\phi}\right)+\frac{\left(\tau_{\phi r}-\tau_{r\phi}\right)+\tau_{\phi\theta}\cot\theta}{r}\right]+\rho g$$

Table 5.3: Stress Tensor: Generalized Newtonian Fluid

Cartesian Coordinates (x, y, z):

$$\tau_{xx} = 2\eta \frac{\partial u_x}{\partial x} \qquad \tau_{yy} = 2\eta \frac{\partial u_y}{\partial y}$$

$$\tau_{zz} = 2\eta \frac{\partial u_z}{\partial z} \qquad \tau_{xy} = \tau_{yx} = \eta \left(\frac{\partial u_x}{\partial y} + \frac{\partial u_y}{\partial x} \right)$$

$$\tau_{yz} = \tau_{zy} = \eta \left(\frac{\partial u_z}{\partial y} + \frac{\partial u_y}{\partial z} \right) \qquad \tau_{xz} = \tau_{zx} = \eta \left(\frac{\partial u_x}{\partial z} + \frac{\partial u_z}{\partial x} \right)$$

Cylindrical Coordinates (r, θ, z):

$$\tau_{rr} = 2\eta \frac{\partial u_r}{\partial r} \qquad \tau_{\theta\theta} = 2\eta \left(\frac{1}{r}\frac{\partial u_\theta}{\partial \theta} + \frac{u_r}{r} \right)$$

$$\tau_{zz} = 2\eta \frac{\partial u_z}{\partial z} \qquad \tau_{r\theta} = \tau_{\theta r} = \eta \left(r\frac{\partial}{\partial r}\left(\frac{u_\theta}{r}\right) + \frac{1}{r}\frac{\partial u_r}{\partial \theta} \right)$$

$$\tau_{\theta z} = \tau_{z\theta} = \eta \left(\frac{1}{r}\frac{\partial u_z}{\partial \theta} + \frac{\partial u_\theta}{\partial z} \right) \qquad \tau_{zr} = \tau_{rz} = \eta \left(\frac{\partial u_r}{\partial z} + \frac{\partial u_z}{\partial r} \right)$$

Spherical Coordinates (r, θ, ϕ):

$$\tau_{rr} = 2\eta \frac{\partial u_r}{\partial r} \qquad \tau_{\theta\theta} = 2\eta \left(\frac{1}{r}\frac{\partial u_\theta}{\partial \theta} + \frac{u_r}{r} \right)$$

$$\tau_{\phi\phi} = 2\eta \left(\frac{1}{r\sin\theta}\frac{\partial u_\phi}{\partial \phi} + \frac{u_r + u_\theta \cot\theta}{r} \right) \qquad \tau_{r\theta} = \tau_{\theta r} = \eta \left(r\frac{\partial}{\partial r}\left(\frac{u_\theta}{r}\right) + \frac{1}{r}\frac{\partial u_r}{\partial \theta} \right)$$

$$\tau_{\theta\phi} = \tau_{\phi\theta} = \eta \left(\frac{\sin\theta}{r}\frac{\partial}{\partial \theta}\left(\frac{u_\phi}{\sin\theta}\right) + \frac{1}{r\sin\theta}\frac{\partial u_\theta}{\partial \phi} \right) \qquad \tau_{\phi r} = \tau_{r\phi} = \eta \left(\frac{1}{r\sin\theta}\frac{\partial u_r}{\partial \phi} + r\frac{\partial}{\partial r}\left(\frac{u_\phi}{r}\right) \right)$$

Table 5.4: Navier-Stokes Equations

Cartesian Coordinates (x, y, z):

$$\rho\left(\frac{\partial u_x}{\partial t}+u_x\frac{\partial u_x}{\partial x}+u_y\frac{\partial u_x}{\partial y}+u_z\frac{\partial u_x}{\partial z}\right)=-\frac{\partial p}{\partial x}+\mu\left[\frac{\partial^2 u_x}{\partial x^2}+\frac{\partial^2 u_x}{\partial y^2}+\frac{\partial^2 u_x}{\partial z^2}\right]+\rho g_x$$

$$\rho\left(\frac{\partial u_y}{\partial t}+u_x\frac{\partial u_y}{\partial x}+u_y\frac{\partial u_y}{\partial y}+u_z\frac{\partial u_y}{\partial z}\right)=-\frac{\partial p}{\partial y}+\mu\left[\frac{\partial^2 u_y}{\partial x^2}+\frac{\partial^2 u_y}{\partial y^2}+\frac{\partial^2 u_y}{\partial z^2}\right]+\rho g_y$$

$$\rho\left(\frac{\partial u_z}{\partial t}+u_x\frac{\partial u_z}{\partial x}+u_y\frac{\partial u_z}{\partial y}+u_z\frac{\partial u_z}{\partial z}\right)=-\frac{\partial p}{\partial z}+\mu\left[\frac{\partial^2 u_z}{\partial x^2}+\frac{\partial^2 u_z}{\partial y^2}+\frac{\partial^2 u_z}{\partial z^2}\right]+\rho g_z$$

Cylindrical Coordinates (r, θ, z):

$$\rho\left(\frac{\partial u_r}{\partial t}+u_r\frac{\partial u_r}{\partial r}+\frac{u_\theta}{r}\frac{\partial u_r}{\partial \theta}+u_z\frac{\partial u_r}{\partial z}-\frac{u_\theta^2}{r}\right)=$$
$$-\frac{\partial p}{\partial r}+\mu\left[\frac{\partial}{\partial r}\left(\frac{1}{r}\frac{\partial}{\partial r}(ru_r)\right)+\frac{1}{r^2}\frac{\partial^2 u_r}{\partial \theta^2}+\frac{\partial^2 u_r}{\partial z^2}-\frac{2}{r^2}\frac{\partial u_\theta}{\partial \theta}\right]+\rho g_r$$

$$\rho\left(\frac{\partial u_\theta}{\partial t}+u_r\frac{\partial u_\theta}{\partial r}+\frac{u_\theta}{r}\frac{\partial u_\theta}{\partial \theta}+u_z\frac{\partial u_\theta}{\partial z}+\frac{u_r u_\theta}{r}\right)=$$
$$-\frac{1}{r}\frac{\partial p}{\partial \theta}+\mu\left[\frac{\partial}{\partial r}\left(\frac{1}{r}\frac{\partial}{\partial r}(ru_\theta)\right)+\frac{1}{r^2}\frac{\partial^2 u_\theta}{\partial \theta^2}+\frac{\partial^2 u_\theta}{\partial z^2}+\frac{2}{r^2}\frac{\partial u_r}{\partial \theta}\right]+\rho g_\theta$$

$$\rho\left(\frac{\partial u_z}{\partial t}+u_r\frac{\partial u_z}{\partial r}+\frac{u_\theta}{r}\frac{\partial u_z}{\partial \theta}+u_z\frac{\partial u_z}{\partial z}\right)=$$
$$-\frac{\partial p}{\partial z}+\left[\frac{\partial}{\partial r}\left(\frac{1}{r}\frac{\partial}{\partial r}(ru_z)\right)+\frac{1}{r^2}\frac{\partial^2 u_z}{\partial \theta^2}+\frac{\partial^2 u_z}{\partial z^2}\right]+\rho g_z$$

Spherical Coordinates (r, θ, ϕ):

$$\rho\left(\frac{\partial u_r}{\partial t}+u_r\frac{\partial u_r}{\partial r}+\frac{u_\theta}{r}\frac{\partial u_r}{\partial \theta}+\frac{u_\phi}{r\sin\theta}\frac{\partial u_r}{\partial \phi}-\frac{u_\theta^2+u_\phi^2}{r}\right)=-\frac{\partial p}{\partial r}+$$
$$\mu\left[\frac{1}{r^2}\frac{\partial^2}{\partial r^2}(r^2u_r)+\frac{1}{r^2\sin\theta}\frac{\partial}{\partial \theta}\left(\sin\theta\frac{\partial u_r}{\partial \theta}\right)+\frac{1}{r^2\sin^2\theta}\frac{\partial^2 u_r}{\partial \phi^2}\right]+\rho g_r$$

$$\rho\left(\frac{\partial u_\theta}{\partial t}+u_r\frac{\partial u_\theta}{\partial r}+\frac{u_\theta}{r}\frac{\partial u_\theta}{\partial \theta}+\frac{u_\phi}{r\sin\theta}\frac{\partial u_\theta}{\partial \phi}-\frac{u_r u_\theta-u_\phi^2\cot\theta}{r}\right)=-\frac{1}{r}\frac{\partial p}{\partial \theta}+$$
$$\mu\left[\frac{1}{r^2}\frac{\partial}{\partial r}\left(r^2\frac{\partial u_\theta}{\partial r}\right)+\frac{1}{r^2}\frac{\partial}{\partial \theta}\left(\frac{1}{\sin\theta}\frac{\partial}{\partial \theta}(u_\theta\sin\theta)\right)+\frac{1}{r^2\sin^2\theta}\frac{\partial^2 u_\theta}{\partial \phi^2}+\right.$$
$$\left.\frac{2}{r^2}\frac{\partial u_r}{\partial \theta}-\frac{2\cot\theta}{r^2\sin\theta}\frac{\partial u_\phi}{\partial \phi}\right]+\rho g_\theta$$

$$\rho\left(\frac{\partial u_\phi}{\partial t}+u_r\frac{\partial u_\phi}{\partial r}+\frac{u_\theta}{r}\frac{\partial u_\phi}{\partial \theta}+\frac{u_\phi}{r\sin\theta}\frac{\partial u_\phi}{\partial \phi}+\frac{u_\phi u_r+u_\theta u_\phi\cot\theta}{r}\right)=-\frac{1}{r\sin\theta}\frac{\partial p}{\partial \phi}+$$
$$\mu\left[\frac{1}{r^2}\frac{\partial}{\partial r}\left(r^2\frac{\partial u_\phi}{\partial r}\right)+\frac{1}{r^2}\frac{\partial}{\partial \theta}\left(\frac{1}{\sin\theta}\frac{\partial}{\partial \theta}(u_\phi\sin\theta)\right)+\frac{1}{r^2\sin^2\theta}\frac{\partial^2 u_\phi}{\partial \phi^2}+\right.$$
$$\left.\frac{2}{r^2\sin\theta}\frac{\partial u_r}{\partial \phi}+\frac{2\cot\theta}{r^2\sin\theta}\frac{\partial u_\theta}{\partial \phi}\right]+\rho g_\phi$$

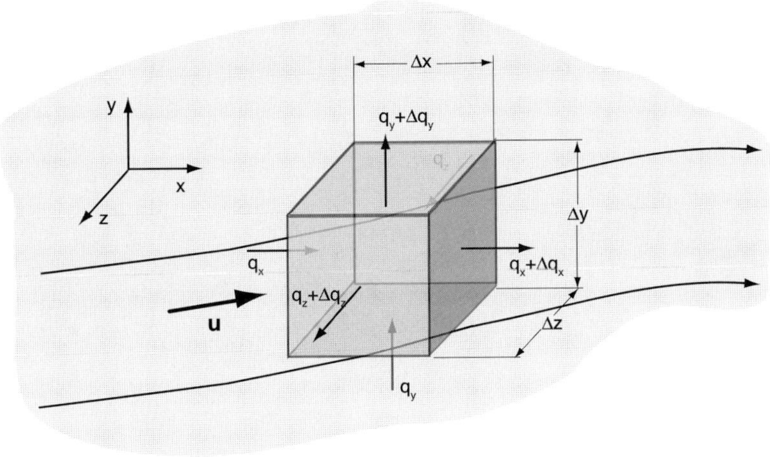

Figure 5.7: Heat flux across a differential fluid element during flow.

With a few exceptions one can say that a flowing polymer melt does not follow the model presented in eqn. (5.20). To properly model the flow of a polymer we must take into account the effects of rate of deformation, temperature and often time, making the partial differential equations that govern a system non-linear.

5.1.4 The Energy Balance or Equation of Energy

An energy balance around a moving fluid element, as shown in Fig. 5.5, can be written as,

$$\rho C_p \frac{DT}{Dt} = -\frac{\Delta q_x}{\Delta x} - \frac{\Delta q_y}{\Delta y} - \frac{\Delta q_z}{\Delta z} + \dot{Q} + \dot{Q}_{\text{viscous heating}} \tag{5.21}$$

where the left hand term represents the transient and convective effects and the right hand the conduction terms, arbitrary heat source ($\dot{Q}$), and viscous dissipation ($\dot{Q}_{\text{viscous heating}}$). Using Fourier's law for heat conduction

$$q_i = -k_i \frac{\partial T}{\partial x_i} \tag{5.22}$$

and assuming an isotropic material, $k_x = k_y = k_z = k$, we can write

$$\rho C_p \frac{DT}{Dt} = k \left(\frac{\partial^2 T}{\partial x^2} + \frac{\partial^2 T}{\partial y^2} + \frac{\partial^2 T}{\partial z^2} \right) + \dot{Q} + \dot{Q}_{\text{viscous heating}} \tag{5.23}$$

As an illustration, we will derive the viscous dissipation terms in the energy balance using a simple shear flow system such as the one shown in Fig. 5.6.

Here, the stresses within the system can be calculated using

$$\tau_{yx} = \mu \frac{\partial u_x}{\partial y} \tag{5.24}$$

which in terms of the parameters depicted in Fig. 5.6, such as force, F, area, A, gap height, h and plate speed, u_0, can be written as

$$\frac{F}{A} = \mu \frac{u_0}{h} \tag{5.25}$$

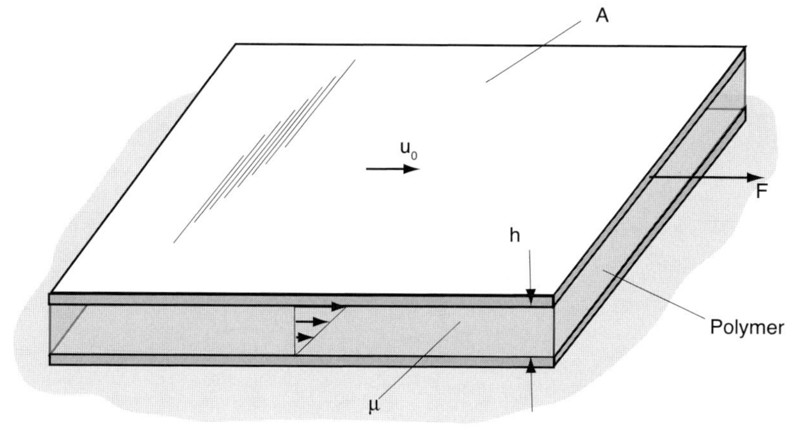

Figure 5.8: Schematic of a simple shear flow system used to illustrate viscous dissipation terms in the energy balance.

In the system, the *rate of energy input* is given by

$$Fu_0 = \mu \frac{u_0}{h} A u_0 \tag{5.26}$$

and the rate of energy input per unit volume is represented by

$$\frac{Fu_0}{Ah} = \mu \left(\frac{u_0}{h} \right) \left(\frac{u_0}{h} \right) \tag{5.27}$$

or

$$Q_{\text{viscous heating}} = \mu \left(\frac{\partial u_x}{\partial y} \right) \left(\frac{\partial u_x}{\partial y} \right) \tag{5.28}$$

From the above equation, we can deduce that for a Newtonian fluid the general term for viscous dissipation is given by $\mu(\dot{\boldsymbol{\gamma}} : \dot{\boldsymbol{\gamma}})$, where

$$\dot{\boldsymbol{\gamma}} : \dot{\boldsymbol{\gamma}} = \sum_{i=1}^{3} \sum_{j=1}^{3} \dot{\gamma}_{ij} \dot{\gamma}_{ji} \tag{5.29}$$

and for a non-Newtonian material, the viscous heating is written as $\boldsymbol{\tau} : \dot{\boldsymbol{\gamma}}$. Hence, the energy balance becomes,

$$\begin{aligned} \rho C_p \frac{\partial T}{\partial t} + \rho C_p u_j \frac{\partial T}{\partial x_j} &= \frac{\partial}{\partial x_j} k \frac{\partial T}{\partial x_j} + \tau_{ij} \dot{\gamma}_{ji} + \dot{Q} \\ \rho C_p \frac{\partial T}{\partial t} + \rho C_p \mathbf{u} \cdot \nabla T &= \nabla \cdot k \nabla T + \boldsymbol{\tau} : \dot{\boldsymbol{\gamma}} + \dot{Q} \end{aligned} \tag{5.30}$$

Table 5.5 presents the complete energy equation in the Cartesian, cylindrical and spherical coordinate systems. Table 5.6 defines the viscous dissipation terms for an incompressible Newtonian fluid.

Table 5.5: Energy Equation for a Newtonian Fluid with Constant Properties

Cartesian Coordinates (x, y, z):

$$\rho C_p \left(\frac{\partial T}{\partial t} + u_x \frac{\partial T}{\partial x} + u_y \frac{\partial T}{\partial y} + u_z \frac{\partial T}{\partial z} \right) = k \left[\frac{\partial^2 T}{\partial x^2} + \frac{\partial^2 T}{\partial y^2} + \frac{\partial^2 T}{\partial z^2} \right] + \mu \Phi_v$$

Cylindrical Coordinates (r, θ, z):

$$\rho C_p \left(\frac{\partial T}{\partial t} + u_r \frac{\partial T}{\partial r} + \frac{u_\theta}{r} \frac{\partial T}{\partial \theta} + u_z \frac{\partial T}{\partial z} \right) = k \left[\frac{1}{r} \frac{\partial}{\partial r} \left(r \frac{\partial T}{\partial r} \right) + \frac{1}{r^2} \frac{\partial^2 T}{\partial \theta^2} + \frac{\partial^2 T}{\partial z^2} \right] + \mu \Phi_v$$

Spherical Coordinates (r, θ, ϕ):

$$\rho C_p \left(\frac{\partial T}{\partial t} + u_r \frac{\partial T}{\partial r} + \frac{u_\theta}{r} \frac{\partial T}{\partial \theta} + \frac{u_\phi}{r \sin\theta} \frac{\partial T}{\partial \phi} \right) =$$

$$k \left[\frac{1}{r^2} \frac{\partial}{\partial r} \left(r^2 \frac{\partial T}{\partial r} \right) + \frac{1}{r^2 \sin\theta} \frac{\partial}{\partial \theta} \left(\sin\theta \frac{\partial T}{\partial \theta} \right) + \frac{1}{r^2 \sin^2\theta} \frac{\partial^2 T}{\partial \phi^2} \right] + \mu \Phi_v$$

Table 5.6: Viscous Dissipation Function Φ_v for Incompressible Newtonian Fluids

Cartesian Coordinates (x, y, z):

$$\Phi_v = 2 \left[\left(\frac{\partial u_x}{\partial x} \right)^2 + \left(\frac{\partial u_y}{\partial y} \right)^2 + \left(\frac{\partial u_z}{\partial z} \right)^2 \right] +$$

$$\left[\frac{\partial u_x}{\partial y} + \frac{\partial u_y}{\partial x} \right]^2 + \left[\frac{\partial u_z}{\partial y} + \frac{\partial u_y}{\partial z} \right]^2 + \left[\frac{\partial u_x}{\partial z} + \frac{\partial u_z}{\partial x} \right]^2$$

Cylindrical Coordinates (r, θ, z):

$$\Phi_v = 2 \left[\left(\frac{\partial u_r}{\partial r} \right)^2 + \left(\frac{1}{r} \frac{\partial u_\theta}{\partial \theta} + \frac{u_r}{r} \right)^2 + \left(\frac{\partial u_z}{\partial z} \right)^2 \right] +$$

$$\left[r \frac{\partial}{\partial r} \left(\frac{u_\theta}{r} \right) + \frac{1}{r} \frac{\partial u_r}{\partial \theta} \right]^2 + \left[\frac{1}{r} \frac{\partial u_z}{\partial \theta} + \frac{\partial u_\theta}{\partial z} \right]^2 + \left[\frac{\partial u_r}{\partial z} + \frac{\partial u_z}{\partial r} \right]^2$$

Spherical Coordinates (r, θ, ϕ):

$$\Phi_v = 2 \left[\left(\frac{\partial u_r}{\partial r} \right)^2 + \left(\frac{1}{r} \frac{\partial u_\theta}{\partial \theta} + \frac{u_r}{r} \right)^2 + \left(\frac{1}{r \sin\theta} \frac{\partial u_\phi}{\partial \phi} + \frac{u_r + u_\theta \cot\theta}{r} \right)^2 \right] +$$

$$\left[r \frac{\partial}{\partial r} \left(\frac{u_\theta}{r} \right) + \frac{1}{r} \frac{\partial u_r}{\partial \theta} \right]^2 + \left[\frac{\sin\theta}{r} \frac{\partial}{\partial \theta} \left(\frac{u_\phi}{\sin\theta} \right) + \frac{1}{r \sin\theta} \frac{\partial u_\theta}{\partial \phi} \right]^2 +$$

$$\left[\frac{1}{r \sin\theta} \frac{\partial u_r}{\partial \phi} + r \frac{\partial}{\partial r} \frac{\partial}{\partial r} \left(\frac{u_\phi}{r} \right) \right]^2$$

5.2 MODEL SIMPLIFICATION

In order to be able to obtain analytical solutions we must first simplify the balance equations. Although the balance equations are fundamental and rigorous, they are nonlinear, non-unique, complex and difficult to solve. In other words, they do not have a general solution and so far, only particular solutions for special problems have been found.

Therefore, the balance equations must be simplified sufficiently in oder to arrive at an analytical solution of the problem under consideration. The simplifications done on a system are typically based on the scale of the variables, an estimate of its maximum order of magnitude. As discussed in the previous chapter, scaling is the process of identifying the correct order of magnitude of the various unknowns. These magnitudes are often referred to as characteristic values, i.e., characteristic times, characteristic length, etc. When a variable is scaled with respect to its characteristic magnitude (scale) the new dimensionless variable will be of order 1, i.e. ($\sim O(1)$). For example, if we scale the x-velocity field, u_x, within a system, with respect to a characteristic velocity, U_0, we can generate a dimensionless velocity, or scaled velocity, given by

$$\hat{u}_x = \frac{u_x}{U_0} \tag{5.31}$$

Using the above relation, the original variable can be expressed in terms of the dimensionless variable and its characteristic value as,

$$u_x = U_0 \hat{u}_x \tag{5.32}$$

By substituting the new variables into the original equations we will acquire information that allows the simplification of a specific model. Length and time scales, for example, can lead to geometrical simplifications such as a reduction in dimensionality.

■ **EXAMPLE 5.1.**

Object submerged in a fluid. Consider an object with a characteristic length L and a thermal conductivity k that is submerged in a fluid of constant temperature T_∞ and convection coefficient h (see Fig.5.9). If a heat balance is made on the surface of the object, it must be equivalent to the heat by conduction, i.e.,

$$-k\frac{\partial T}{\partial n}\Big|_S = h\left(T_S - T_\infty\right) \tag{5.33}$$

The maximum value possible for the temperature gradient must be the difference between the central temperature, T_c, and the surface temperature, T_S,

$$\Delta T \sim T_c - T_S \tag{5.34}$$

giving us a characteristic temperature difference[3]. Here, the length variable is the normal distance Δn and has a characteristic length L. We can now approach the scaling of this problems in two ways. The first and quickest is to simply substitute the variables into the original equations, often referred to a *order of magnitude analysis*.

[3]Characteristic temperatures are always given in terms of temperature differences. For example, the characteristic temperature of the melt of an amorphous polymer in an extrusion operation is $\Delta T = T_h - T_g$, or the difference between the heater temperature and the glass transition temperature of the polymer.

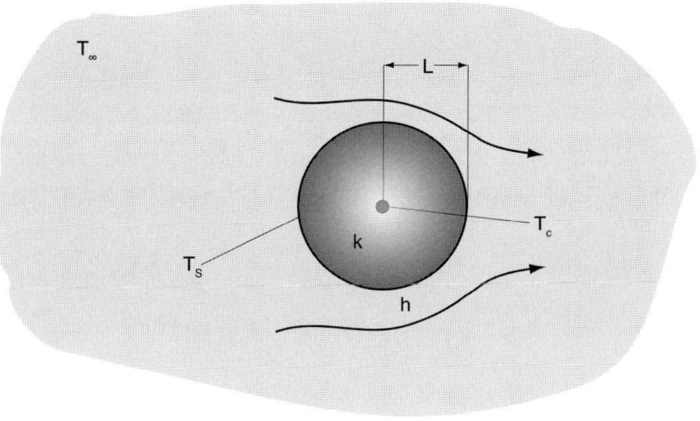

Figure 5.9: Schematic of a body submerged in a fluid.

The second is to express the original equations in terms of dimensionless variables. The order of magnitude analysis results in a scaled conduction given by,

$$k\frac{\partial T}{\partial n}\Big|_S \sim k\frac{T_c - T_S}{L} \tag{5.35}$$

reducing the problem to,

$$k\frac{T_c - T_S}{L} \sim h\left(T_S - T_\infty\right) \tag{5.36}$$

or in a more convenient way,

$$Bi = \frac{hL}{k} \sim \frac{T_c - T_S}{T_S - T_\infty} \tag{5.37}$$

where Bi is the *Biot number*.

When $Bi \ll 1$, the solid can be considered isothermal, which means that we *reduced the dimensionality* of the problem from (x, y, z), to a zero dimensional or *lumped model* [2, 11]. On the other hand, if $Bi \gg 1$, the fluid can be considered isothermal and $T_S = T_\infty$, which changes the convection boundary condition to a thermal equilibrium condition.

The same can be deduced if we scale the problem by expressing the governing equations in dimensionless form. Again, we choose the same characteristic values for normal distance and temperature, allowing us to generate dimensionless variables as

$$\hat{T} = \frac{T}{T_c - T_S}, \qquad \hat{n} = \frac{n}{L} \tag{5.38}$$

which can be solved to give,

$$T = (T_c - T_S)\hat{T}, \qquad n = L\hat{n} \tag{5.39}$$

Substituting these into the original equations results in,

$$-\frac{k}{Lh}\frac{\partial \hat{T}}{\partial \hat{n}} = \frac{T_S - T_\infty}{T_c - T_S} \tag{5.40}$$

or,

$$Bi\frac{\partial \hat{T}}{\partial \hat{n}} = \Theta \tag{5.41}$$

Again, since $\partial \hat{T}/\partial \hat{n}$ is of order one the same analysis done above applies here.

5.2.1 Reduction in Dimensionality

The number of special coordinates, or dimensionality of a problem, can be reduced using three basic strategies: symmetry, aspect ratio and series resistances.

Symmetry is the easiest to apply. It is based on the correct selection of the coordinate system for a given problem. For example, a temperature field with circular symmetry can be described using just the coordinates (r, z), instead of (x, y, z). In addition, symmetry can help to get rid of special variables that are not required by the conservation equations and interfacial conditions. For example, the velocity field in a tube, according to the Navier-Stokes and continuity equations, can have the functional form $u_z(r)$.

The ratio of two linear dimensions of an object is called the *aspect ratio*. There are a number of possible simplifications when the aspect ratio of an object or region is large (or small). For example, for the classical *fin approximation*, the thickness of the fin is small compared with the length, therefore the temperature will be assumed to change in the direction of the length only.

Finally, it is possible to reduce the dimensionality of a problem by determining which *rate processes in series* is the controlling step. As shown for $Bi \ll 1$, the convection controls the cooling process and conduction is so fast that the solid is considered isothermal, reducing the dimensionality from (x, y, z) to a zero dimensional problem or lumped mass method.

Characteristic times are a key factor in formulating conduction or diffusion models, because they determine how fast a system can respond to changes imposed at a boundary. In other words, if the temperature or concentration is perturbed at some location, it is important to estimate the finite time required for the temperature or concentration changes to be noticed at a given distance from the original perturbation. The time involved in a stagnant medium is the characteristic time for conduction or diffusion, therefore this is the most widely used characteristic time in transport models [3, 6].

■ **EXAMPLE 5.2.**

Temperature development in an extruder channel during melting. In this example, we illustrate reduction in dimensionality of the energy equation to find an equation that would reveal the change of the melt temperature through the gap between the solid bed and extruder barrel during melting, as schematically depicted in Fig. 5.10. To simplify the problem, we assume to have constant properties and a Newtonian viscosity.

The thickness between the solid and the barrel is small compared to the screw channel, which indicates that a reduction in dimensionality can be performed. Initially, it can be assumed that the velocity field is unidirectional, i.e. $u_x(y)$. The energy equation is then reduced to,

$$\rho C_p u_x \frac{\partial T}{\partial x} = k\left(\frac{\partial^2 T}{\partial x^2} + \frac{\partial^2 T}{\partial y^2}\right) + \mu\left(\frac{\partial u_x}{\partial y}\right)^2 \tag{5.42}$$

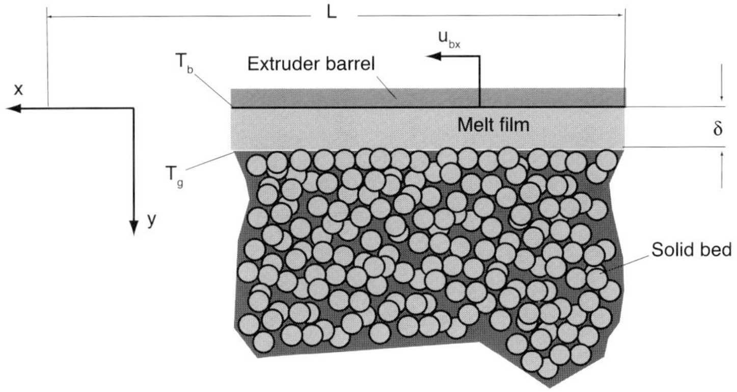

Figure 5.10: Schematic diagram of the melt film during melting in extruders.

By choosing characteristic variables for temperatures, velocities and lengths we can reduce the dimensionality even further. The temperature is scaled based on the maximum gradient, the length with the gap thickness and screw channel depth and the velocity with the barrel x-velocity,

$$\theta = \frac{T - T_g}{T_b - T_g}, \qquad \eta = \frac{y}{\delta},$$
$$\xi = \frac{x}{L}, \qquad \bar{u} = \frac{u_x}{u_{bx}}$$

and the energy equation will be,

$$\rho C_p u_{bx} \frac{\delta^2}{L} \bar{u} \frac{\partial \theta}{\partial \xi} = \frac{\delta^2}{L^2} \frac{\partial^2 \theta}{\partial \xi^2} + \frac{\partial^2 \theta}{\partial \eta^2} + \frac{\mu u_{bx}^2}{k(T_b - T_g)} \left(\frac{\partial \bar{u}}{\partial \eta} \right)^2 \tag{5.43}$$

which indicates that for the small aspect ratio, δ/L, two extra terms can be neglected, the conduction and convection in the x-direction,

$$\frac{\partial \theta^2}{\partial \eta^2} + \frac{\mu u_{bx}^2}{k(T_b - T_g)} \left(\frac{\partial \bar{u}}{\partial \eta} \right)^2 = 0 \tag{5.44}$$

The last step is to compare the two remaining terms: conduction and viscous dissipation. The two derivatives, according to the scaling parameter, are of order 1. The remaining term, $Br = \mu u_{bx}^2 / k(T_b - T_g)$, is the Brinkman number, which indictates whether the viscous dissipation is important or not. For $Br \ll 1$, the conduction is dominant, while for $Br > 1$, the viscous dissipation has to be included, which is the case in most polymer processing operations.

5.2.2 Lubrication Approximation

lubrication approximation Now, let's consider flows in which a second component and the inertial effects are nearly zero. Liquid flows in long, narrow channels or thin films often have these characteristics of being nearly unidirectional and dominated by viscous stresses.

Let's use the steady, two-dimensional flow in a thin channel or a narrow gap between solid objects as schematically represented in Fig. 5.11. The channel height or gap width

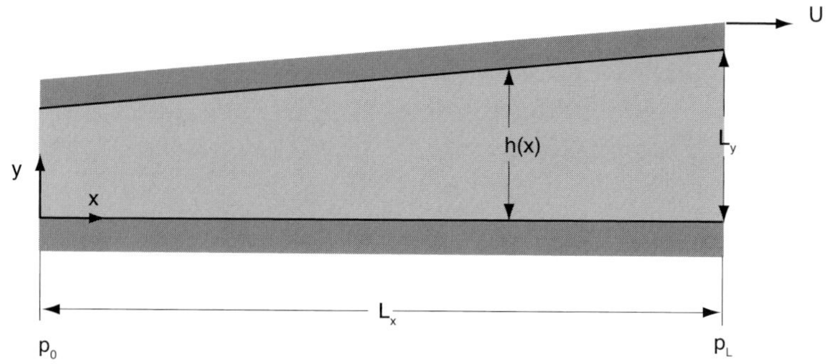

Figure 5.11: Schematic diagram of the lubrication problem.

varies with the position, and there may be a relative motion between the solid surfaces. This type of flow is very common for the oil between bearings. The original solution came from the field of tribology and is therefore often referred to as the *lubrication approximation.*

For this type of flow, the momentum equations (for a Newtonian fluid) are reduced to the steady Navier-Stokes equations, i.e.

$$\frac{\partial u_x}{\partial x} + \frac{\partial u_y}{\partial y} = 0 \tag{5.45}$$

$$\begin{aligned} \rho\left(u_x\frac{\partial u_x}{\partial x} + u_y\frac{\partial u_x}{\partial y}\right) &= -\frac{\partial p}{\partial x} + \mu\left(\frac{\partial^2 u_x}{\partial x^2} + \frac{\partial^2 u_x}{\partial y^2}\right) \\ \rho\left(u_x\frac{\partial u_y}{\partial x} + u_y\frac{\partial u_y}{\partial y}\right) &= -\frac{\partial p}{\partial y} + \mu\left(\frac{\partial^2 u_y}{\partial x^2} + \frac{\partial^2 u_y}{\partial y^2}\right) \end{aligned} \tag{5.46}$$

The lubrication approximation depends on two basic conditions, one *geometric* and one *dynamic*. The geometric requirement is revealed by the continuity equation. If L_x and L_y represents the length scales for the velocity variations in the x- and y-directions, respectively, and let U and V be the respective scales for u_z and u_y. From the continuity equation we obtain

$$\frac{V}{U} \sim \frac{L_y}{L_x} \tag{5.47}$$

In order to neglect pressure variation in the y-direction all the terms in the $y-$momentum equation must be small, in other words $V/U \ll 1$. From the continuity scale analysis we get that the geometric requirement is,

$$\frac{L_y}{L_x} \ll 1 \tag{5.48}$$

which holds for thin films and channels. The consequences of this geometric constrain in the Navier-Stokes equations are,

$$\frac{\partial p}{\partial y} \ll \frac{\partial p}{\partial x} \quad \text{and} \quad \frac{\partial^2 u_x}{\partial x^2} \ll \frac{\partial^2 u_x}{\partial y^2} \tag{5.49}$$

In addition, the continuity equation also tells us that the two inertia terms in the x-momentum equation are of similar magnitude, i.e.,

$$u_y \frac{\partial u_x}{\partial y} \sim \frac{VU}{L_y} \sim \frac{U^2}{L_x} \sim u_x \frac{\partial u_x}{\partial x} \tag{5.50}$$

These inertia effects can be neglected, i.e.,

$$\rho u_x \frac{\partial u_x}{\partial x} \ll \mu \frac{\partial^2 u_x}{\partial y^2} \quad \text{and} \quad \rho u_y \frac{\partial u_x}{\partial x} \ll \mu \frac{\partial^2 u_x}{\partial y^2} \tag{5.51}$$

only if $\rho U^2/L_x \ll \mu U/L_y^2$ or

$$\left(\frac{\rho U L_y}{\mu}\right)\left(\frac{L_y}{L_x}\right) = Re \left(\frac{L_y}{L_x}\right) \ll 1 \tag{5.52}$$

which is the dynamic requirement for the lubrication approximation. The x-momentum (Navier-Stokes) equation is then reduced to,

$$\frac{\partial^2 u_x}{\partial y^2} = \frac{1}{\mu}\frac{dp}{dx} \tag{5.53}$$

for $p = p(x)$ only.

5.3 SIMPLE MODELS IN POLYMER PROCESSING

There are only a few exact or analytical solutions of the momentum balance equations, and most of those are for situations in which the flow is unidirectional; that is, the flow has only one nonzero velocity component. Some of these are illustrated below. We end the section with a presentation of the , which today is widely accepted to model the flows that occur during mold filling processes.

5.3.1 Pressure Driven Flow of a Newtonian Fluid Through a Slit

One of the most common flows in polymer processing is the pressure driven flow between two parallel plates. When deriving the equations that govern *slit flow* we use the notation presented in Fig. 5.12 and consider a steady fully developed flow; a flow where the entrance effects are ignored.

This flow is unidirectional, that is, there is only one nonzero velocity component. The continuity for an incompressible flow is reduced to,

$$\frac{du_z}{dz} = 0 \tag{5.54}$$

The z-momentum equation for a Newtonian, incompressible flow (Navier-Stokes equations) is,

$$-\frac{\partial p}{\partial z} + \mu \frac{\partial^2 u_z}{\partial y^2} = 0 \tag{5.55}$$

and the x- and y-components of the equations of motion are reduced to,

$$-\frac{\partial p}{\partial x} = -\frac{\partial p}{\partial y} = 0 \tag{5.56}$$

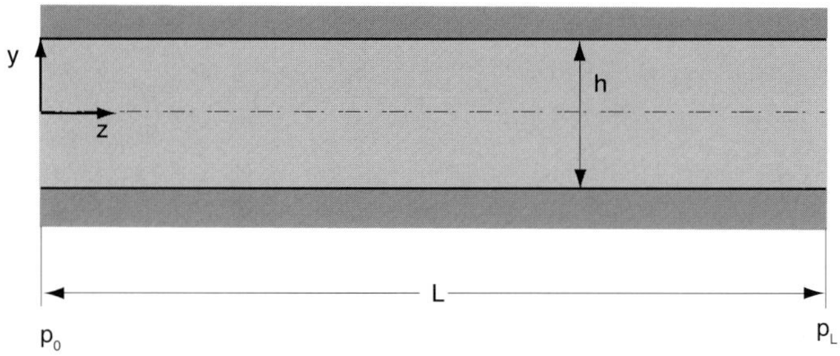

Figure 5.12: Schematic diagram of pressure flow through a slit.

This relation indicates that for this fully developed flow, the total pressure is a function of z alone. Additionally, since $\mathbf{u}$ does not vary with z, the pressure gradient, $\partial p/\partial z$, must be a constant. Therefore,

$$\frac{dp}{dz} = \frac{\Delta p}{L} \tag{5.57}$$

The momentum equation can now be written as,

$$\frac{1}{\mu}\frac{\Delta p}{L} = \frac{\partial^2 u_z}{\partial y^2} \tag{5.58}$$

As boundary conditions, two no-slip conditions given by $u_z(\pm h/2) = 0$ are used in this problem. Integrating twice and evaluating the two integration constants with the boundary conditions gives,

$$\begin{aligned} u_z(y) &= \frac{h^2}{8\mu}\frac{dp}{dz}\left[1 - \left(\frac{2y}{h}\right)^2\right] \\ &= \frac{h^2}{8\mu}\frac{\Delta p}{L}\left[1 - \left(\frac{2y}{h}\right)^2\right] \end{aligned} \tag{5.59}$$

Also note that the same profile will result if one of the non-slip boundary conditions is replaced by a symmetry condition at $y = 0$, namely $du_z/dy = 0$. The mean velocity in the channel is obtained integrating the above equation,

$$\bar{u}_z = \frac{1}{h}\int_0^h u_z(y)dy = \frac{h^2}{12\mu}\frac{dp}{dz} \tag{5.60}$$

and the volumetric flow rate,

$$Q = hW\bar{u}_z = \frac{Wh^3\Delta p}{12\mu L} \tag{5.61}$$

where W is the width of the channel.

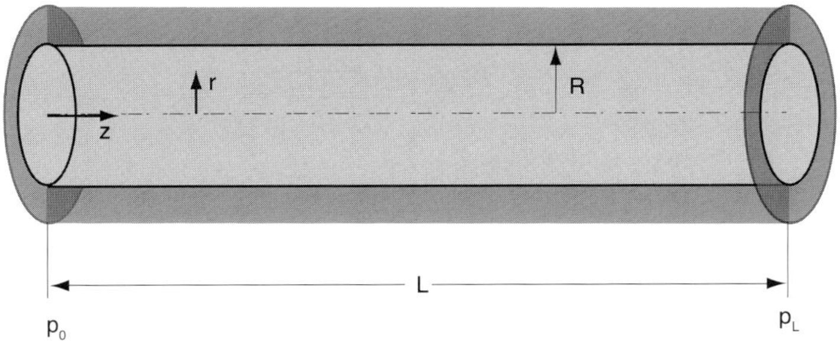

Figure 5.13: Schematic diagram of pressure flow through a tube.

5.3.2 Flow of a Power Law Fluid in a Straight Circular Tube (Hagen-Poiseuille Equation)

Tube flow is encountered in several polymer processes, such in extrusion dies and sprue and runner systems inside injection molds. When deriving the equations for pressure driven flow in tubes, also known as Hagen-Poiseuille flow, we assume that the flow is steady, fully developed, with no entrance effects and axis-symmetric (see Fig.5.13).

Thus, we have $u_z = u_z(r)$, $u_r = u_\theta = 0$ and $p = p(z)$. With this type of velocity field, the only non-vanishing component of the rate-of-deformation tensor is the zr-component. It follows that for the generalized Newtonian flow, τ_{zr} is the only nonzero component of the viscous stress, and that $\tau_{zr} = \tau_{zr}(r)$. The z-momentum equation is then reduced to,

$$\frac{1}{r}\frac{d}{dr}\left(r\tau_{zr}\right) = \frac{dp}{dz} \tag{5.62}$$

However, since $p = p(z)$ and $\tau_{zr} = \tau_{zr}(r)$, the above equation is satisfied only if both sides are constant and can be integrated to obtain,

$$r\tau_{zr} = \frac{dp}{dz}\frac{r^2}{2} + c_1 \tag{5.63}$$

At this point, a symmetry argument at $r = 0$ leads to the conclusion that $\tau_{zr} = 0$ because the stress must be finite. Hence, we must satisfy $c_1 = 0$. For a power law fluid it is found that,

$$\tau_{zr} = -m\left|\frac{du_z}{dr}\right|^n \tag{5.64}$$

The minus sign in this equation is required due to the fact that the pressure flow is in the direction of the flow ($dp/dz < 0$), indicating that $\tau_{zr} \leq 0$. Combining the above equations and solving for the velocity gradient gives

$$\frac{du_z}{dr} = -\left(-\frac{1}{2m}\frac{dp}{dz}\right)^{1/n} r^{1/n} \tag{5.65}$$

Integrating this equation and using the no-slip condition, at $r = R$, to evaluate the integration constant, the velocity as a function of r is obtained,

$$u_z(r) = \left(\frac{3n+1}{n+1}\right)\left[1 - \left(\frac{r}{R}\right)^{(n+1)/n}\right]\bar{u}_z \tag{5.66}$$

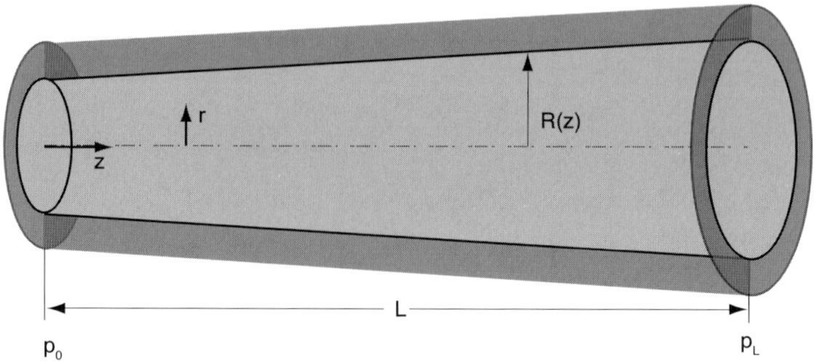

Figure 5.14: Schematic diagram of slightly tapered tube.

where the mean velocity, $\bar{u}_z$, is defined as,

$$\bar{u}_z = \frac{2}{R^2}\int_0^R u_z r dr = \left(\frac{n}{3n+1}\right)\left[-\frac{R^{n+1}}{2m}\frac{dp}{dz}\right]^{1/n} \tag{5.67}$$

Finally, the volumetric flow rate is given by

$$\begin{aligned} Q = \pi R^2 \bar{u}_z &= \left(\frac{n\pi}{3n+1}\right)\left[-\frac{R^{3n+1}}{2m}\frac{dp}{dz}\right]^{1/n} \\ &= \left(\frac{n\pi}{3n+1}\right)\left[-\frac{R^{3n+1}}{2m}\frac{\Delta p}{L}\right]^{1/n} \end{aligned} \tag{5.68}$$

5.3.3 Flow of a Power Law Fluid in a Slightly Tapered Tube

Based on the lubrication approximation, the momentum equations to solve the flow through a slightly tapered tube are the same equations that we use to solve for the equations that pertain to the straight circular tube, i.e.,

$$\frac{1}{r}\frac{d}{dr}(r\tau_{zr}) = \frac{dp}{dz} \tag{5.69}$$

This means that the solution for the velocity is the same and is applied at each distance z down the tube. Replacing R by $R(z)$ modifies the equations to,

$$u_z(r) = \frac{R(z)}{1+1/n}\left(\frac{R(z)\Delta p}{2mL}\right)^{1/n}\left[1-\left(\frac{r}{R}\right)^{1/n+1}\right] \tag{5.70}$$

$R(z)$ is obtained from the geometry,

$$R(z) = -\left(\frac{R_0 - R_L}{L}\right) z + R_0 \tag{5.71}$$

The volumetric flow rate will be,

$$Q = \left(\frac{\pi R^3(z)}{1/n+3}\right)\left(\frac{R(z)\Delta p}{2mL}\right) \tag{5.72}$$

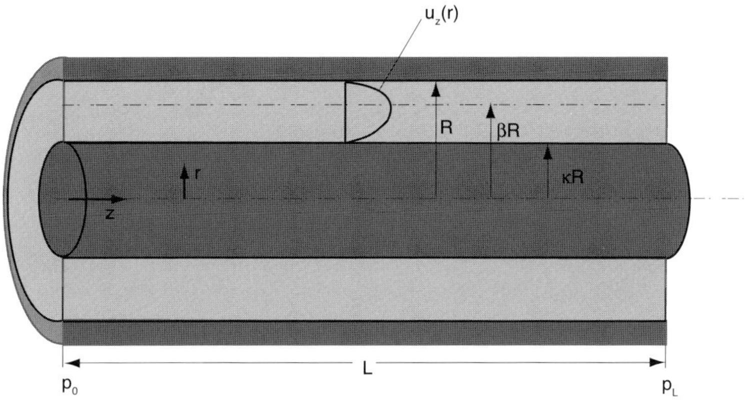

Figure 5.15: Schematic diagram of pressure flow through an annulus.

This equation gives a first order differential equation for the pressure,

$$\left(\frac{Q\,(1/n+1)}{\pi}\right)^n \frac{R^{-3n-1}}{2m} = -\frac{dp}{dz} \tag{5.73}$$

which can be now integrated between $p = p_0$ at $z = 0$ and $p = p_L$ at $z = L$, i.e.,

$$p_0 - p_L = \frac{2mL}{3n}\left[\frac{Q}{\pi}\left(\frac{1}{n+3}\right)\right]^n \left(\frac{R_L^{-3n} - R_0^{-3n}}{R_0 - R_L}\right) \tag{5.74}$$

5.3.4 Volumetric Flow Rate of a Power Law Fluid in Axial Annular Flow

Annular flow is encountered in pipe extrusion dies, wire coating dies and film blowing dies. In the problem under consideration, a Power law fluid is flowing through an annular gap between two coaxial cylinders of radii κR and R, with $\kappa < 1$ as schematically depicted in Fig. 5.15. The maximum in the velocity profile is located at $r = \beta R$, where β is a constant to be determined. Due to the geometrical characteristics and ignoring entrance effects, the flow is unidirectional, i.e., $\mathbf{u} = (u_r, u_\theta, u_z) = (0, 0, u_{z(r)})$.

The z-momentum equation is then reduced to,

$$\frac{1}{r}\frac{d}{dr}(r\tau_{zr}) = \frac{dp}{dz} \tag{5.75}$$

Integrating this equation we obtain,

$$r\tau_{zr} = \frac{dp}{dz}\frac{r^2}{2} + c_1 \tag{5.76}$$

The constant c_1 cannot be set equal to zero, because $\kappa R \le r \le R$. However, β can be used rather than c_1,

$$r\tau_{zr} = \frac{\Delta p R}{2L}\left(\frac{r}{R} - \beta^2\frac{R}{r}\right) \tag{5.77}$$

which makes β the new integration constant. The power-law expression for the shear stress is given by

$$\begin{aligned} \tau_{zr} &= -m\left(\frac{du_z}{dr}\right)^n \quad \text{if} \quad \kappa R \leq r \leq \beta R \\ \tau_{zr} &= m\left(-\frac{du_z}{dr}\right)^n \quad \text{if} \quad \beta R \leq r \leq R \end{aligned} \tag{5.78}$$

Substitution of these expressions into the momentum equation leads to differential equations for the velocity distribution in the two regions. Integrating these equations with boundary conditions, $u_z = 0$ at $r = \kappa R$ and at $r = R$, leads to

$$\begin{aligned} u_z &= R\left[\frac{\Delta p R}{2mL}\right]^{1/n} \int_\kappa^\xi \left(\frac{\beta^2}{\xi'} - \xi'\right)^{1/n} d\xi' \quad \text{if} \quad \kappa \leq \xi \leq \beta \\ u_z &= R\left[\frac{\Delta p R}{2mL}\right]^{1/n} \int_\xi^1 \left(\xi' - \frac{\beta^2}{\xi'}\right)^{1/n} d\xi' \quad \text{if} \quad \beta \leq \xi \leq 1 \end{aligned} \tag{5.79}$$

where $\xi = r/R$. In order to find the parameter β, the above equations must match at the location of the maximum velocity,

$$\int_\kappa^\xi \left(\frac{\beta^2}{\xi'} - \xi'\right)^{1/n} d\xi' = \int_\xi^1 \left(\xi' - \frac{\beta^2}{\xi'}\right)^{1/n} d\xi' \tag{5.80}$$

This equation is a relation between β, the geometrical parameter κ and the power-law exponent n. The volumetric flow rate in the annulus becomes,

$$\begin{aligned} Q &= 2\pi \int_{\kappa R}^R u_z r dr \\ &= \pi R^3 \left[\frac{\Delta p R}{2mL}\right]^{1/n} \int_\kappa^1 \left|\beta^2 - \xi'^2\right|^{1/n+1} \xi'^{-1/n} d\xi\prime \\ &= \frac{\pi R^{3+1/n}}{1/n+3}\left(\frac{\Delta p}{2mL}\right)^{1/n} \left[\left(1-\beta^2\right)^{1+1/n} - \kappa^{1-1/n}\left(\beta^2-\kappa^2\right)^{1+1/n}\right] \end{aligned} \tag{5.81}$$

5.3.5 Radial Flow Between two Parallel Discs – Newtonian Model

Radial flow between parallel discs is a very common flow type encountered in polymer processing, particularly during injection mold filling. In this section, we seek the velocity profile, flow rates and pressure for this type of flow using the notation presented in Fig. 5.16.

Let us consider a Newtonian fluid that is flowing due to a pressure gradient between two parallel disks that are separated by a distance $2h$. The velocity and pressure fields that we will solve for are $u_r = u_r(z, r)$ and $p = p(r)$. According to the Newtonian fluid model, the stress components are,

$$\begin{aligned} \tau_{rr} &= -2\mu\frac{\partial u_r}{\partial r} \\ \tau_{\theta\theta} &= -2\mu\frac{u_r}{r} \\ \tau_{rz} = \tau_{zr} &= -\mu\frac{\partial u_r}{\partial z} \end{aligned} \tag{5.82}$$

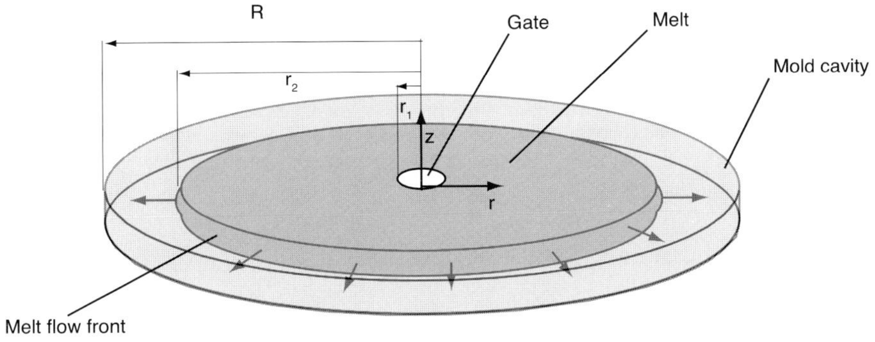

Figure 5.16: Schematic diagram of a center-gated disc-shaped mold during filling.

The continuity equation is reduced to,

$$\frac{1}{r}\frac{\partial}{\partial r}\left(r u_r\right) = 0 \tag{5.83}$$

which indicates that $r u_r$ must be a function of z only, $f(z)$. Therefore, from the continuity equation,

$$u_r = \frac{f(z)}{r} \tag{5.84}$$

the stresses are reduced to,

$$\begin{aligned} \tau_{rr} &= -\,2\mu\frac{f(z)}{r^2} \\ \tau_{\theta\theta} &= +\,2\mu\frac{f(z)}{r^2} \\ \tau_{rz} = \tau_{zr} &= -\,\frac{\mu}{r}\frac{df(z)}{dz} \end{aligned} \tag{5.85}$$

Neglecting the inertia effects, the momentum equation becomes,

$$-\frac{1}{r}\frac{\partial}{\partial r}\left(r\tau_{rr}\right) - \frac{\partial \tau_{zr}}{\partial z} + \frac{\tau_{\theta\theta}}{r} - \frac{\partial p}{\partial r} = 0 \tag{5.86}$$

which is reduced to

$$-\frac{dp}{dr} + \frac{\mu}{r}\frac{d^2 f(z)}{dz^2} = 0 \tag{5.87}$$

This equation can be integrated, because the pressure is only a function of r. The constants of integration can be solved for by using the following boundary conditions

$$f(\pm h) = 0 \tag{5.88}$$

For the specific case where the gate is at r_1 and the front at r_2, the velocity field is given by,

$$u_r\left(r, z\right) = \frac{h^2 \Delta p}{2\mu r \ln\left(r_2/r_1\right)}\left[1 - \left(\frac{z}{h}\right)^2\right] \tag{5.89}$$

The volumetric flow rate is found by integrating this equation over the cross sectional area,

$$Q = \int_0^{2\pi} \int_{-h}^{+h} u_z(r, z)\, r d\theta dz = \frac{4\pi h^3 \Delta p}{3\mu \ln(r_2/r_1)} \tag{5.90}$$

The above equation can also be used to solve for pressure drop from the gate to the flow front,

$$\Delta p = \frac{3Q\mu \ln(r_2/r_1)}{4\pi h^3} \tag{5.91}$$

Δp is the boundary condition when solving for the pressure distribution within the disc, by integrating eqn. (5.87),

$$p = \frac{\Delta p \ln(r/r_2)}{\ln(r_1/r_2)} \tag{5.92}$$

In order to predict the position of the flow front, r_2, as a function of time, we first perform a simple mass (or volume) balance,

$$2h\pi(r_2^2 - r_1^2) = Qt \tag{5.93}$$

which can be solved for r_2 as,

$$r_2 = \sqrt{\frac{Qt}{2h\pi} + r_1^2} \tag{5.94}$$

The above equations can now be used to plot the pressure requirement, or pressure at the gate, for a given flow rate as a function of time. They can also be used to plot for the pressure distribution within the disc at various points in time or flow front positions. In addition, the same equations can be used to solve for flow rates for given injection pressures.

■ **EXAMPLE 5.3.**

Predicting pressure profiles in a disc-shaped mold using a Newtonian model. To show how the above equations are used, let us consider a disc-shaped cavity of R =150 mm, a gate radius, r_1, of 5 mm, and a cavity thickness of 2 mm, i.e., h =1 mm. Assuming a Newtonian viscosity μ =6,400 Pa-s and constant volumetric flow rate Q =50 cm^3/s predict the position of the flow front, r_2, as a function of time, as well as the pressure distribution inside the disc mold.

Equations (5.92) and (5.94) can easily be solved using the given data. Figure 5.17 presents the computed flow front positions with the corresponding pressure profiles.

5.3.6 The Hele-Shaw model

Today, the most widely used model simplification in polymer processing simulation is the Hele-Shaw model [5]. It applies to flows in "narrow" gaps such as injection mold filling, compression molding, some extrusion dies, extruders, bearings, etc. The major assumptions for the lubrication approximation are that the gap is small, such that $h \ll L$, and that the gaps vary slowly such that

$$\frac{\partial h}{\partial x} \ll 1 \quad \text{and} \quad \frac{\partial h}{\partial y} \ll 1 \tag{5.95}$$

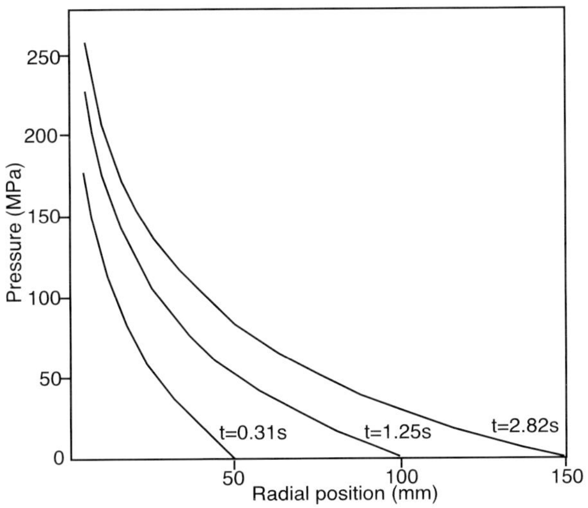

Figure 5.17: Radial pressure profile as a function of time in a disc-shaped mold computed using a Newtonian viscosity model.

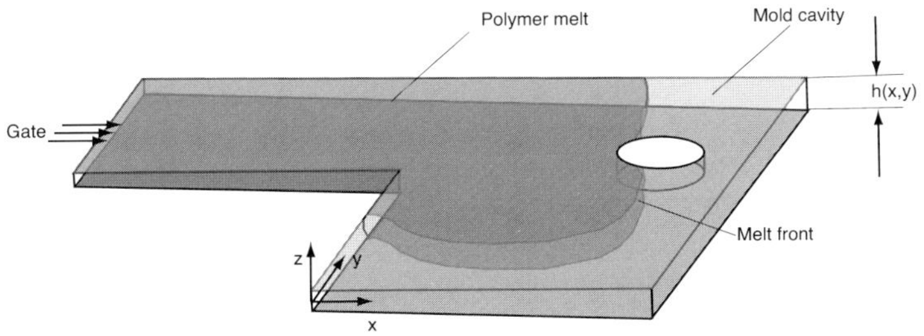

Figure 5.18: Schematic diagram of a mold filling process.

A schematic diagram of a typical flow described by the Hele-Shaw model is presented in Fig. 5.18.

We start the derivation with an order of magnitude analysis of the continuity equation

$$\frac{\partial u_x}{\partial x} + \frac{\partial u_y}{\partial y} + \frac{\partial u_z}{\partial z} = 0 \tag{5.96}$$

The characteristic values for the variables present in eqn. (5.96) are given by

$$u_x, u_y \sim U_c \qquad u_z \sim U_z$$
$$x, y \sim L \qquad z \sim h$$

Substituting these into the x and y terms of eqn. (5.96) results in

$$\frac{\partial u_x}{\partial x}, \frac{\partial u_y}{\partial y} \sim \frac{U_c}{L} \tag{5.97}$$

and into the z term in

$$\frac{\partial u_z}{\partial z} \sim \frac{U_z}{h} \tag{5.98}$$

With the continuity equation and the scales for the x- and y-velocities, we can solve for the z-velocity scale as

$$U_z = \left(\frac{h}{L}\right) U_c \tag{5.99}$$

Hence, $u_z \ll u_x, u_y$ and u_z can be ignored. We must point out that this velocity plays a significant role in heat transfer and orientation in the flow front region, because the free flow front is dominated by what is usually referred to as a fountain flow effect.

Next, an order of magnitude analysis is performed to simplify the momentum balance. This is illustrated using the x-component of the equation of motion in terms of stress

$$\rho\left(\frac{\partial u_x}{\partial t} + u_x\frac{\partial u_x}{\partial x} + u_y\frac{\partial u_x}{\partial y} + u_z\frac{\partial u_x}{\partial z}\right) = -\frac{\partial p}{\partial x} + \frac{\partial \tau_{xx}}{\partial x} + \frac{\partial \tau_{yx}}{\partial y} + \frac{\partial \tau_{zx}}{\partial z} \tag{5.100}$$

An order of magnitude of the inertia terms leads to

$$\rho u_x \frac{\partial u_x}{\partial x} \sim \rho\frac{U_c^2}{L}$$
$$\rho u_z \frac{\partial u_x}{\partial z} \sim \rho\frac{U_z U_c}{h} \sim \rho\frac{U_c^2}{L}$$

This order of the magnitude in the stress terms leads to

$$\frac{\partial \tau_{xx}}{\partial x} \sim \frac{\partial \tau_{yx}}{\partial y} \sim \frac{\eta U_c}{L^2}$$
$$\frac{\partial \tau_{zx}}{\partial z} \sim \frac{\eta U_c}{h^2}$$

For the flow of polymer melts, the Reynolds number, $Re = \rho U_c h/\eta$, is usually $\sim 10^{-5}$, with the exception of the reaction injection molding process, RIM, where $Re \rightarrow 1 - 100$ at the gate. The geometric and dynamic conditions of the lubrication approximation, applied to the Hele-Shaw model, will simplify the momentum equations by neglecting the inertia terms and the viscous terms containing τ_{xx} and τ_{yx}. Similarly, the y-momentum equation is simplified giving the following system of equations,

$$\begin{aligned}\frac{\partial p}{\partial x} &= \frac{\partial \tau_{zx}}{\partial z}\\ \frac{\partial p}{\partial y} &= \frac{\partial \tau_{zy}}{\partial z}\end{aligned} \tag{5.101}$$

In addition, since the velocity in the z-direction is small compared to the x- and y-directions (eqn. (5.99)), from the momentum equation in the z-direction we get that

$$\frac{\partial p}{\partial z} << \frac{\partial p}{\partial x}, \frac{\partial p}{\partial y} \tag{5.102}$$

which indicates that $p = p(x, y)$ only.

We are now left to deal with the constitutive equation. For a generalized Newtonian fluid, we can write

$$\boldsymbol{\tau} = \eta \dot{\boldsymbol{\gamma}} \tag{5.103}$$

where the viscosity is a function of temperature and magnitude of the rate of deformation tensor $\eta = \eta(\dot{\gamma}, T)$. Performing an order of magnitude analysis to the terms in the rate of deformation tensor, reduces its magnitude to

$$\dot{\gamma} = \sqrt{\left(\frac{\partial u_x}{\partial z}\right)^2 + \left(\frac{\partial u_y}{\partial z}\right)^2} \tag{5.104}$$

Integrating the stress with respect to the z-direction we get,

$$\tau_{zx} = \frac{\partial p}{\partial x} + c_1 \tag{5.105}$$

Symmetry for the velocity profile will set $c_1 = 0$. With the generalized Newtonian fluid constitutive equation, we get that

$$\frac{\partial u_x}{\partial z} = \frac{\partial p}{\partial x}\frac{z}{\eta} \tag{5.106}$$

Since this model is used to simulate the flow between thin gaps, such as in injection and compression molding, we are not interested in the detail of the flow field through the thickness of the part. Instead, we are more interested in the flow field and flow advancement in the planar (xy) directions. For this, it is convenient to integrate the velocity across the gap in order to compute the gap-wise average velocities,

$$\bar{u}_x = \frac{1}{h/2}\int_0^{h/2} u_x(z)dz \tag{5.107}$$

Substituting eqn. (5.106) into eqn. (5.107) results in

$$\bar{u}_x = \frac{1}{h/2}\left(-\frac{\partial p}{\partial x}\right)\int_0^{h/2}\int_{h/2}^{z}\frac{z'dz'}{\eta(z')}dz \tag{5.108}$$

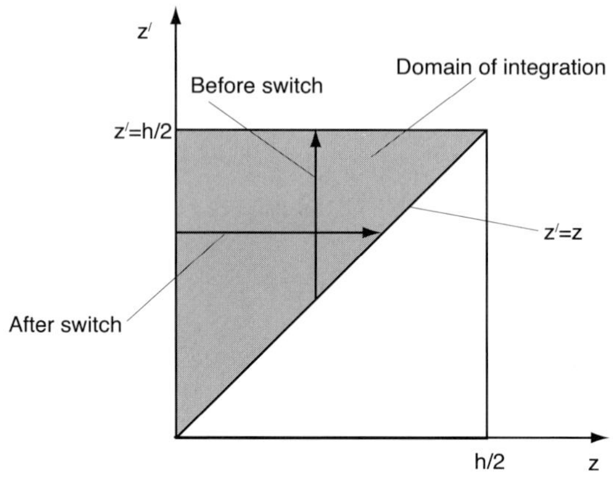

Figure 5.19: Diagram to illustrate order of integration change.

To simplify the integration we can switch the order of integration. The order of integration change, schematically depicted Fig. 5.19.

Changes the above integral from,

$$\int_0^{h/2}\int_z^{h/2} dz'dz \rightarrow \int_0^{h/2}\int_0^{z'} dzdz' \tag{5.109}$$

Hence, we can write

$$\bar{u}_x = \frac{-1}{h/2}\frac{\partial p}{\partial x}\int_0^{h/2}\int_0^{z'}\frac{z'}{\eta(z')}dzdz' \tag{5.110}$$

and integrate once to have

$$\bar{u}_x = -\frac{2}{h}\frac{\partial p}{\partial x}S \tag{5.111}$$

where S is usually referred to as the *flow conductance*, given by,

$$S = \int_0^{h/2}\frac{z'^2}{\eta(z')}dz' \tag{5.112}$$

Using this, we can define the gap-wise average velocities as

$$\begin{aligned}\bar{u}_x &= -\frac{2S}{h}\frac{\partial p}{\partial x}\\ \bar{u}_y &= -\frac{2S}{h}\frac{\partial p}{\partial y}\end{aligned} \tag{5.113}$$

where $S = S(x, y)$. It is important to point out that,

$$\frac{u_y(x,y,z)}{u_x(x,y,z)} = \frac{\dfrac{\partial p}{\partial y}(x,y)}{\dfrac{\partial p}{\partial x}(x,y)}$$

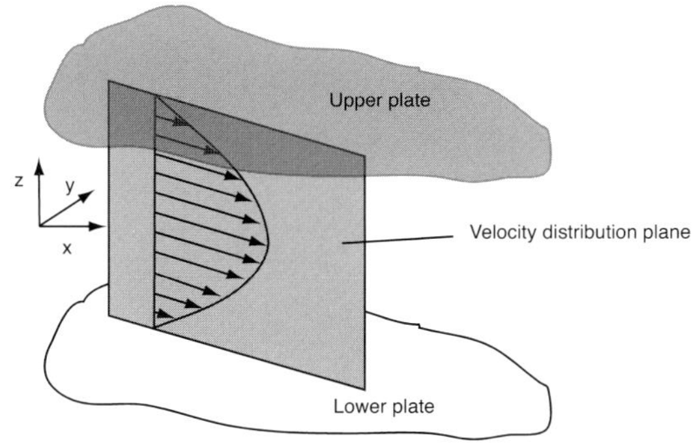

Figure 5.20: Typical velocity profile for a Hele-Shaw flow.

which means that although the velocities change in magnitude through the gap, they do not change in direction (see Fig. 5.20); a prime characteristic of Hele-Shaw type flows.

Now we take the continuity equation and integrate it over z

$$\frac{\partial}{\partial x}\int_0^{h/2} u_x dz + \frac{\partial}{\partial y}\int_0^{h/2} u_y dz = 0 \tag{5.114}$$

The *integrated continuity equation* is a weaker form of the full continuity equation. This is noticed in numerical solutions of mold filling problems, where continuity is never fully satisfied. However, this violation of continuity is insignificant and will not hinder the solution of practical mold filling problems. The integrated continuity equation reduces to

$$\frac{\partial}{\partial x}\left(\frac{h}{2}\bar{u}_x\right) + \frac{\partial}{\partial y}\left(\frac{h}{2}\bar{u}_y\right) = 0 \tag{5.115}$$

Substituting the gap-wise average velocities, we get

$$\frac{\partial}{\partial x}\left(S\frac{\partial p}{\partial x}\right) + \frac{\partial}{\partial y}\left(S\frac{\partial p}{\partial y}\right) = 0 \tag{5.116}$$

There are various special forms and simplifications of the above equation and they are given below. In subsequent chapters of this book we will illustrate how the various forms of the Hele-Shaw model are implemented to solve realistic mold filling problems.

Newtonian-isothermal Hele-Shaw model. A special form of the Hele-Shaw type flow governing equations is the isothermal Newtonian case where $\eta(z) = \mu$. This simplification leads to flow a conductance given by

$$S = \frac{h^3}{24\mu} \tag{5.117}$$

and and a governing equation for pressure expressed by

$$\frac{\partial^2 p}{\partial x^2} + \frac{\partial^2 p}{\partial y^2} = 0 \tag{5.118}$$

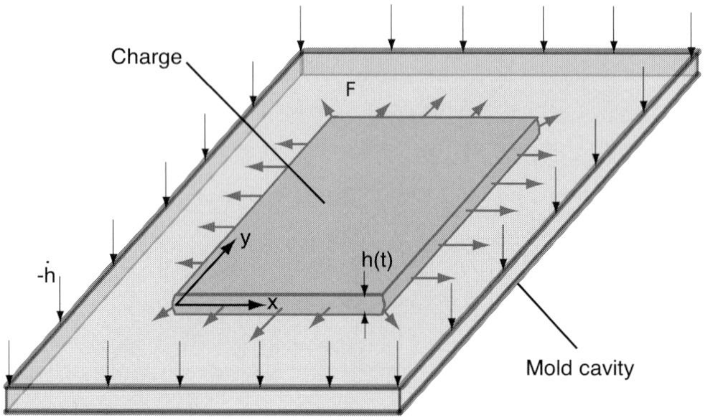

Figure 5.21: Schematic diagram of the compression molding process.

The velocity field can be calculated using

$$\begin{aligned}\bar{u}_x &= -\frac{h^2}{12\mu}\frac{\partial p}{\partial x}\\ \bar{u}_y &= -\frac{h^2}{12\mu}\frac{\partial p}{\partial y}\end{aligned} \tag{5.119}$$

Generalized Newtonian Hele-Shaw model using a power law viscosity. For a generalized Newtonian fluid with a power law fluid, $\eta = m\dot{\gamma}^{n-1}$, the viscosity is given by,

$$\eta = m^s z^{1-s}\left[\left(\frac{\partial p}{\partial x}\right)^2 + \left(\frac{\partial p}{\partial y}\right)^2\right]^{(s-1)/2} \tag{5.120}$$

where $s = 1/n$. This viscosity definition also leads to a closed form solution for the flow conductance, given by,

$$S = \frac{h^{s+2}}{(s+2)m^s 2^{s+1}}\left[\left(\frac{\partial p}{\partial x}\right)^2 + \left(\frac{\partial p}{\partial y}\right)^2\right]^{(s-1)/2} \tag{5.121}$$

Hele-Shaw model for compression molding of a Newtonian fluid. A special case, where the z-velocity component plays a significant role and must be included, is the compression molding process. The process is schematically depicted in Fig. 5.21.

Here, the extra term that represents the closing speed must be included in the continuity equation,

$$\frac{\partial \bar{u}_x}{\partial x} + \frac{\partial \bar{u}_y}{\partial y} + \frac{-\dot{h}}{h} = 0 \tag{5.122}$$

Using the velocity definition given by,

$$\begin{aligned}\bar{u}_x &= -\frac{h^2}{12\mu}\frac{\partial p}{\partial x}\\ \bar{u}_y &= -\frac{h^2}{12\mu}\frac{\partial p}{\partial y}\end{aligned} \tag{5.123}$$

the continuity equation results in,

$$\frac{\partial^2 p}{\partial x^2} + \frac{\partial^2 p}{\partial y^2} = 12\mu \frac{\dot{h}}{h^3} \tag{5.124}$$

5.3.7 Cooling or Heating in Polymer Processing

Cooling or heating is of great concern in polymer processing. Due to the low thermal conductivity of polymers, the cooling and heating steps of a process control the cycle time. Cooling or heating processes take place inside molds during injection and compression molding, respectively. This heating or cooling process is a contact process during which a polymer melt is pressed against a mold surface allowing for an effective heat transfer between mold and melt. During some extrusion processes, such as extrusion of fibers or films (during film blowing), cooling takes place in water or air, where we must rely on the heat transfer coefficient between the flowing media and the polymer surface. In other extrusion processes the polymer soon comes into contact with a metal surface, such as a sizing sleeve during extrusion of plastic pipes.

Heating and cooling often take place while the polymer melt flows, making viscous dissipation an influencing factor during the process. However, since most plastic parts are thin, the conduction often occurs only across the thickness and the viscous heating is a result of shear within the narrow gap of a die or mold cavity. For such cases, the equations reduce to,

$$\rho C_p \frac{\partial T}{\partial t} = k \frac{\partial^2 T}{\partial x^2} + \eta \left(\frac{\partial u_z}{\partial x} \right)^2 \tag{5.125}$$

There are various special cases for the above equation, some of which are described below.

Cooling or heation of a semi-infinite slab. Although very thick parts are not an issue in polymer processing, we can still solve this problem to illustrate time scales associated with various thicknesses. In a semi-infinite slab, we have a cooling or heating process that takes place in a space from $x = 0$ to $x = \infty$. At $t = 0$, the temperature throughout the slab is T_0 and the surface temperature is suddenly lowered or raised to T_S.

For this problem, the above equation reduces to

$$\frac{\partial T}{\partial t} = \alpha \frac{\partial^2 T}{\partial x^2} \tag{5.126}$$

Bird, Stewart and Lighfoot [3] present a solution to this problem given by,

$$\frac{T - T_S}{T_0 - T_S} = \text{erf}\left(\frac{x}{2\sqrt{\alpha t}} \right) \tag{5.127}$$

where $\text{erf}\left(\frac{x}{2\sqrt{\alpha t}} \right)$ is given by,

$$\text{erf}\left(\frac{x}{2\sqrt{\alpha t}} \right) = \frac{2}{\sqrt{\pi}} \int_0^{\frac{x}{2\sqrt{\alpha t}}} e^{-u^2} du \tag{5.128}$$

Figure 5.23 shows the dimensionless temperature as a function of dimensionless time $\frac{x}{2\sqrt{\alpha t}}$. From the above equations and figure, we can define a heat penetration thickness.

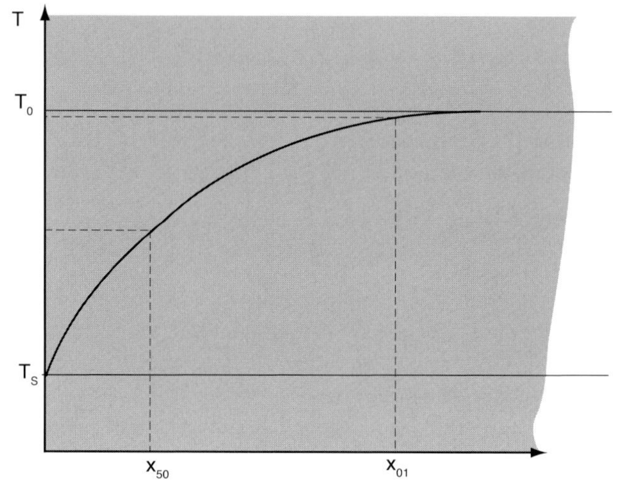

Figure 5.22: Schematic of a semi-infinite cooling body. Denoted are depths at which 50% and 1% of the temperature difference is felt.

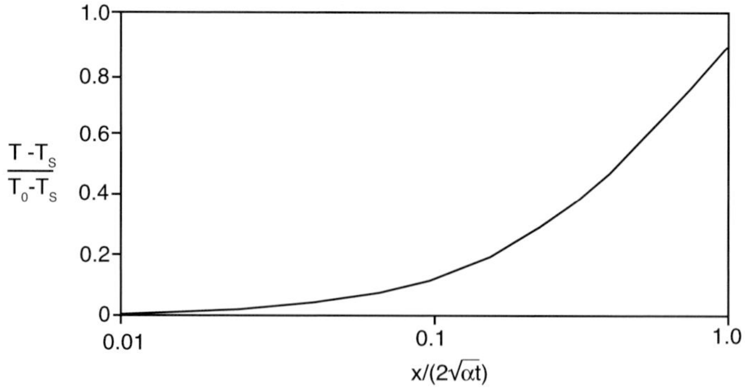

Figure 5.23: Dimensionless temperature as a function of dimensionless time and tickness.

Table 5.7: Penetration thickness and characteristic times in heating and cooling polymers

L	$t = L^2/\alpha$
100 μm	0.025 s
1 mm	2.5 s
2 mm	10 s
10 mm	250 s

For example, Fig. 5.22 presents two depths, one where 1% and another where 50% of the temperature differential is felt. The 1% temperature differential is defined by,

$$T = T_0 + 0.01(T_S - T_0) \tag{5.129}$$

or,

$$0.99 = \text{erf}\left(\frac{x_{01}}{2\sqrt{\alpha t}}\right) \tag{5.130}$$

which can be used to solve for the given time that leads to a 1% temperature change for a given depth x_{01},

$$t_{01} = \frac{x_{01}^2}{13.25\alpha} \tag{5.131}$$

The same analysis can be carried out for a 50% thermal penetration time,

$$t_{50} = \frac{x_{50}^2}{0.92\alpha} \tag{5.132}$$

Hence, the time when most of the temperature difference is felt by a part of a given thickness is of the order,

$$t = \frac{L^2}{\alpha} \tag{5.133}$$

which can be used as a characteristic time for a thermal event that takes place through diffusion. Since polymers have a thermal diffusivity of about 10^{-7}m^2/s, we can easily compute the characteristic times for heating or cooling as a function of part thickness, $2L$. Some characteristic times are presented as a function of thickness in Table 5.7.

With a characteristic time for heat conduction we can now define a dimensionless time using,

$$Fo = \frac{L^2}{\alpha t} \tag{5.134}$$

which is the well known Fourier number.

Cooling and heating of a finite thickness plate. A more accurate solution of the above problem is to determine the cooling process of the actual part, hence, one of finite thickness $2L$. For the heating process of a finite thickness plate we can solve eqn. (5.126) to give,

$$\frac{T - T_0}{T_S - T_0} = 1 - \frac{4}{\pi}\sum_{n=1}^{\infty}\frac{(-1)^{n-1}}{2n-1}\cos\left[\frac{(2n-1)\pi x}{L}\right] \times \exp\left(-\left[\frac{(2n-1)\pi}{2}^2 Fo\right]\right) \tag{5.135}$$

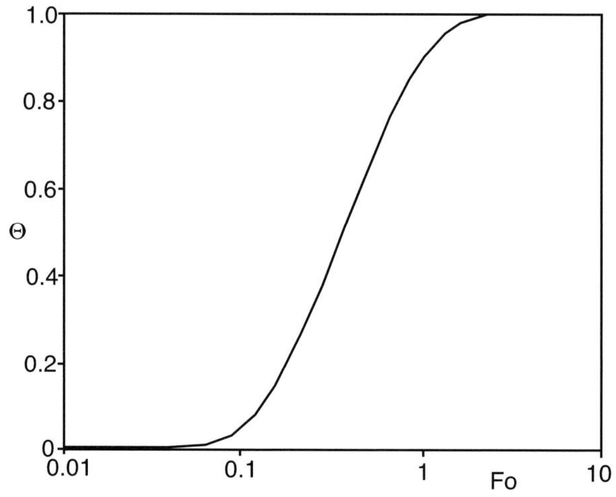

Figure 5.24: Center-line temperature history during heating of a finite thickness plate. Note that cooling is represented by the same curve using $1 - \Theta$ as the dimensionless temperature.

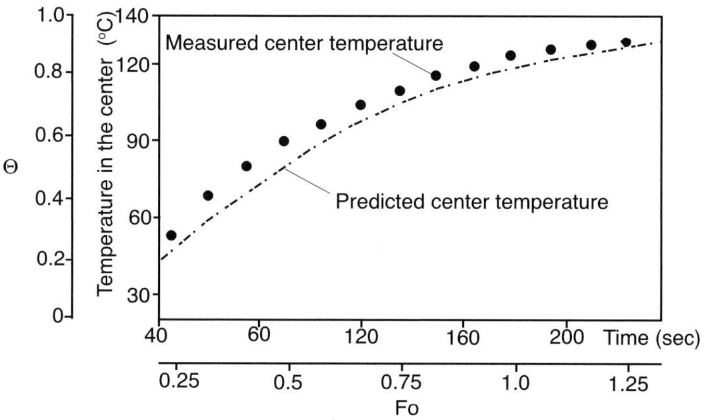

Figure 5.25: Experimental and computed center-line temperature history during heating of an 8 mm thick PMMA plate. The initial temperature T_0=20°C and the heater temperature T_S=140°C. [7]

Figure 5.24 presents the temperature history at the center of the plate and Fig. 5.25 shows a comparison between the prediction and a measured temperature development in an 8 mm thick PMMA plate. As can be seen, the model does a good job in approximating reality.

Cooling and heating of a finite thickness plate using convection. As mentioned earlier, cooling with air or water is very common in polymer processing. For example, the cooling of a film during film blowing is controlled by air blown from a ring located near the die exit. In addition, many extrusion operations extrude into a bath of running chilled water. Here, the controlling parameter is the heat transfer coefficient h, or in dimensionless

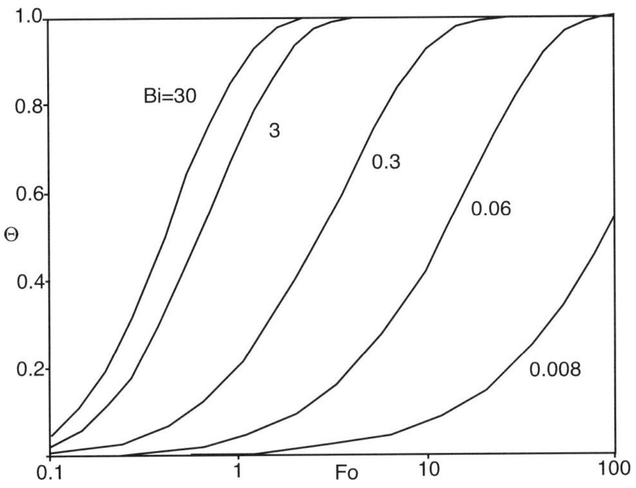

Figure 5.26: Center-line temperature histories of finite thickness plates during convective heating for various Biot numbers.

form the Biot number, Bi, given by,

$$Bi = \frac{hL}{k} \tag{5.136}$$

An approximate solution for the convective cooling of a plate of finite thickness is given by Agassant *et al.* [1],

$$\frac{T - T_f}{T_0 - T_f} \approx e^{(-BiFo)} \cos\left(\sqrt{Bi}\frac{x}{L}\right) \tag{5.137}$$

The center-line temperature for plates of finite thickness is given in Fig. 5.26 and a comparison between the prediction and experiments for an 8 mm thick PMMA plate cooled with a heat transfer coefficient, h, of 33 W/m^2/K is given in Fig. 5.27. As can be seen, theory and experiment are in relatively good agreement.

Problems

5.1 Derive the continuity equation in cylindrical coordinates.

5.2 Derive the x-direction momentum balance presented in eqn. (5.13).

5.3 Derive the equations for pressure driven flow through a slit using a shear thinning power law viscosity model.

- **a)** Derive the equation that describes the velocity distribution across the slit.
- **b)** Solve for the volumetric throughput through the slit.

5.4 Derive the equations for a combination of pressure flow and shear flow within a slit using a Newtonian viscosity model.

- **a)** Derive the equation that describes the velocity distribution across the slit.
- **b)** Solve for the volumetric throughput through the slit.

5.5 Solve the above problem using a power law viscosity model.

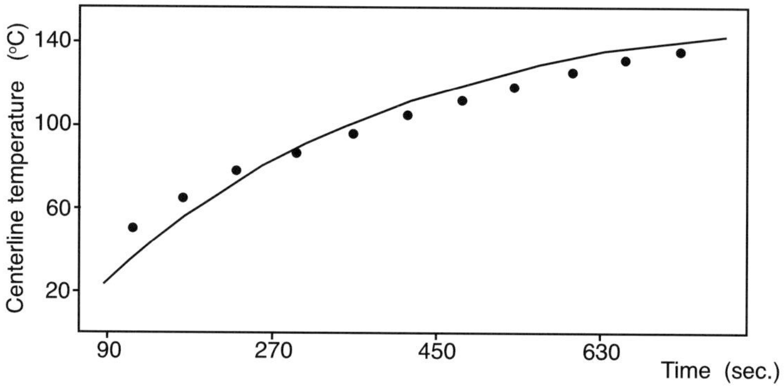

Figure 5.27: Center-line temperature history of an 8 mm thick PMMA plate during convective heating inside an oven set at 155°C. The initial temperature was 20°C. The predictions correspond to a Biot number, Bi=1.3 or a corresponding heat transfer coefficient, h=33 W/m^2/K.[7]

5.6 Solve for the Hagen-Poiseuille equation in a tube for a Newtonian model.

5.7 Using the Hele-Shaw model, analyze a compression molding problem where the melt is allowed to move only in the x-direction. Use a dimension in the x-direction of $2L$, and in the y-direction of W. The gapwise thickness is h. The two flow fronts are located at $x = \pm L$.

a) Sketch a clear diagram of the process.

b) State your assumptions.

c) Solve for the gapwise average velocity.

d) Determine the pressure distribution inside the mold and mold closing force.

5.8 Derive the equation for flow conductance for a power law fluid given in eqn.(5.121).

5.9 Derive the equation for the steady state temperature profile in a simple shear flow with viscous dissipation. Assume a Newtonian viscosity model.

a) Assume equal temperatures at the upper and lower surfaces.

b) Assume different temperatures at the upper and lower surfaces.

c) Plot the temperature distribution in dimensionless form for various Brinkman numbers.

5.10 Derive the equation for the steady state temperature profile in pressure driven slit flow with viscous dissipation. Assume a Newtonian viscosity model.

a) Assume equal temperatures at the upper and lower surfaces.

b) Assume different temperatures at the upper and lower surfaces.

c) Plot the temperature distribution in dimensionless form for various Brinkman numbers.

5.11 Derive the equation for the steady state temperature profile in a combined simple shear-pressure driven slit flow with viscous dissipation. Assume a Newtonian viscosity model.

a) Assume equal temperatures at the upper and lower surfaces.

b) Assume different temperatures at the upper and lower surfaces.

c) Plot the temperature distribution in dimensionless form for various Brinkman numbers.

5.12 Derive the analytical solution for the temperature distribution caused by viscous heating within a Couette flow between concentric cylinders assuming a constant viscosity μ.

REFERENCES

1. J.F. Agassant, P. Avenas, J.-Ph. Sergent, and P.J. Carreau. *Polymer Processing - Principles and Modeling*. Hanser Publishers, Munich, 1991.
2. A. Bejan. *Heat Transfer*. John Wiley & Sons, New York, 1993.
3. R.B. Bird, W.E. Steward, and E.N. Lightfoot. *Transport Phenomena*. John Wiley & Sons, New York, 2nd edition, 2002.
4. W.M. Deen. *Analysis of Transport Phenomena*. Oxford University Press, Oxford, 1998.
5. H.S. Hele-Shaw. *Proc. Roy. Inst.*, 16:49, 1899.
6. F.P. Incropera and D.P. DeWitt. *Fundamentals of Heat and Mass Transfer*. John Wiley & Sons, New York, 1996.
7. H. Potente. Kunststofftechnologie 2. Fachbereich 10 Maschinentechnik I - Technologie der Kunststoffe, Universitaet-Gesamthochschule Paderborn, 2000.

CHAPTER 6

ANALYSES BASED ON ANALYTICAL SOLUTIONS

Order and simplification are the first steps toward the mastery of a subject.

—Thomas Mann

Although all polymer processes involve complex phenomena that are non-isothermal, non-Newtonian and often viscoelastic, most of them can be simplified sufficiently to allow the construction of analytical models. These analytical models involve one or more of the simple flows derived in the previous chapter. These *back of the envelope* models allow us to predict pressures, velocity fields, temperature fields, melting and solidification times, cycle times, etc. The models that are derived will aid the student or engineer to better understand the process under consideration, allowing for optimization of processing conditions, and even geometries and part performance.

This chapter attempts to cover the most important polymer processes. First we derive solutions to non-isothermal approximations of various polymer processes. We begin with a Newtonian analysis of the metering or pumping section of the single screw extruder, followed by the flow in several common extrusion dies, including the analysis using non-Newtonian shear thinning polymer melts. Within this section, we also solve for flow and deformation in a fiber spinning operation using a viscoelastic flow model. The next sections present a detailed analysis of isothermal, Newtonian and non-Newtonian flow in two roll calendering systems. This is followed by the analysis of various problems and injection molding problems. From this point on, non-isothermal problems are introduced, which are exemplified using melting and solidification problems, ending with melting in a plasticating single screw extruder and the curing kinetics of elastomers and thermoseting polymers.

6.1 SINGLE SCREW EXTRUSION—ISOTHERMAL FLOW PROBLEMS

Most flows that take place during polymer processing can be simplified and modeled isothermally. When a system reaches steady state, a polymer melt can be considered isothermal, unless viscous dissipation plays a significant role. As was discussed in Chapter 4, the significance of viscous dissipation during processing is assessed using the Brinkman number given by

$$Br = \frac{\eta u_0^2}{k\Delta T} \tag{6.1}$$

where η is a characteristic viscosity, u_0 a characteristic screw speed, k the thermal conductivity of the melt and ΔT a characteristic temperature variation within the process. In the example below, we attempt to determine if a process can be considered isothermal or if viscous dissipation is significant.

■ **EXAMPLE 6.1.**

Determining the effect of viscous dissipation in the metering section of a single screw extruder. Consider a 60 mm diameter extruder with a 4 mm channel depth and a screw speed of 60 rpm. The melt used in this extrusion system is a polycarbonate with a viscosity of 100 Pa-s, a thermal conductivity of 0.2 W/m/K and a heater temperature of 300°C. To assess the effect of viscous heating we can choose a temperature difference, ΔT of 30K. This simply means that the heater temperature is 30K above the melting temperature of the polymer. For this system, the Brinkman number becomes

$$Br = \frac{\eta(\pi D n)^2}{k\Delta T} = \frac{100 \text{ Pa-s}(0.188 \text{ m/s})^2}{0.2 \text{ W/m/K}(30\text{K})} = 0.59 \tag{6.2}$$

which means that the heat is conducted out much faster than it is generated by viscous heating, making the isothermal assumption a plausible one.

This can be easily checked by assuming that the flow inside this section of the screw can be modeled using a simple shear flow, and that most of the conduction occurs through the channel thickness direction. For such a case, the energy equation in that direction, say the y-direction, reduces to

$$0 = k\frac{\partial^2 T}{\partial y^2} + \dot{Q} \tag{6.3}$$

where $\dot{Q}$ is the heat generation by viscous heating, which for simple shear flow reduces to

$$\dot{Q} = \mu\left(\frac{\partial u}{\partial y}\right)^2 = \mu\left(\frac{\pi D n}{h}\right)^2 \tag{6.4}$$

If we integrate eqn. (6.3) two times and use a boundary condition of T_0 =300°C on the screw and barrel surfaces, we get

$$T = T_0 + \frac{\mu}{2k}\frac{(\pi D n)^2}{h^2}\left(2hy - y^2\right) \tag{6.5}$$

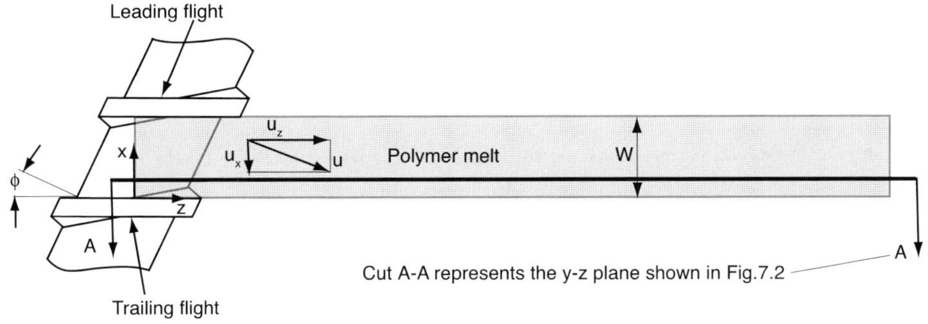

Figure 6.1: Unwrapped screw channel.

Setting y =0.002 m, which is located in the middle of the channel and using the above data, we get a temperature of 306.6^{o}C, a 6.6 K temperature rise. Although there is a measurable temperature rise in this system, it is not significant enough to warrant a non-isothermal analysis.

6.1.1 Newtonian Flow in the Metering Section of a Single Screw Extruder

Analyzing the flow in a single screw extruder using analytical solutions can only be done if we assume a Newtonian polymer melt. As can be seen in these sections, the flow inside the screw channel is a three dimensional flow made up of a combination of pressure and drag flows. Non-Newtonian flow can be solved for using numerical techniques and will be covered in Chapters 8 to 11 of this book.

The geometry of a single screw extruder can be simplified by unwinding or unwrapping the material from the channel as schematically depicted in Fig. 6.1. By unwinding the channel contents we are assuming that the effects caused by the curvature of the screw are negligible. This is true for most screw geometries where the channel is shallow. Furthermore, if we assume that the barrel rotates instead of the screw, we can model the flow inside the channel using a combination of shear and pressure flow between parallel plates.

Figure 6.2 shows a cross-section of the unwrapped channel in the yz-plane. The length L is the helical length of the channel, which for a square pitch screw can be computed using $L = (\text{number of turns})D/\sin(\phi)$. The lower surface of the channel is the screw root and is assigned a zero velocity and the upper surface is in contact with the barrel which is given a velocity $u = \pi Dn$. Because of the helical geometry of the screw, this velocity is broken down further into x and z components. The x-component is given by $u_x = u\sin(\phi)$ and is referred to as the cross-flow component. The z-component which is given by $u_z = u\cos(\phi)$, is the down-channel component and is the one that leads to pumping by dragging the polymer down the channel of the screw. Since the polymer is dragged against a die, the pressure will build-up moving in the down-channel direction. This results in a pressure flow that moves in the opposite direction of the drag flow component.

The volumetric flow is given by,

$$Q_T = Q_D + Q_P \tag{6.6}$$

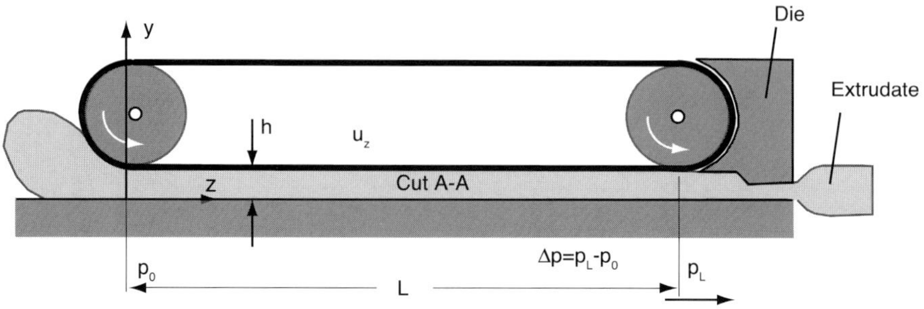

Figure 6.2: Model of a hypothetical viscosity pump.

where Q_D is the drag flow due to simple shear and Q_P is the pressure flow, given by the following equations,

$$\begin{aligned} Q_D &= \frac{1}{2} u_z h W \\ Q_P &= -\frac{W h^3 \Delta p}{12 \mu L} \end{aligned} \tag{6.7}$$

respectively. The total volumetric flow as a function of Δp is then given by

$$Q_T = \frac{1}{2} u_z h W - \frac{W h^3 \Delta p}{12 \mu L} \tag{6.8}$$

and finally

$$Q_T = \frac{\pi D n h W}{2} \cos\phi - \frac{W h^3 \Delta p}{12 \mu L} \tag{6.9}$$

As can be seen in the above equations, the resulting total flow is a combination of drag and pressure flows. Depending on the restriction of the die, various types of flows can develop inside the screw channel. Figure 6.3 schematically depicts the different situations that may arise. At closed discharge, which occurs when the die is plugged, the net flow in the down-channel direction is zero, at which point the maximum pressure build-up is achieved. At open discharge, when the die is absent and the extruder is pumping into the atmosphere, the flow in the down-channel direction reduces to a simple shear flow. It should be noted that the above equations neglect the effect of leakage flow over the flight, hence they over-predict the net material throughput as well as the maximum pressure build-up. Furthermore, since there is a no-slip condition between the polymer and the flight walls, the velocity profiles depicted in Fig. 6.3 are only valid away from the flights. This effect further contributes to over-prediction of the volumetric throughput of a single screw extruder.

To correct this effect, which significantly affects extruders with a deep channel screw, Tadmor and Gogos [25] present the following modification

$$Q_T = \frac{\pi D n h W}{2} \cos\phi F_D - \frac{W h^3 \Delta p}{12 \mu L} F_P \tag{6.10}$$

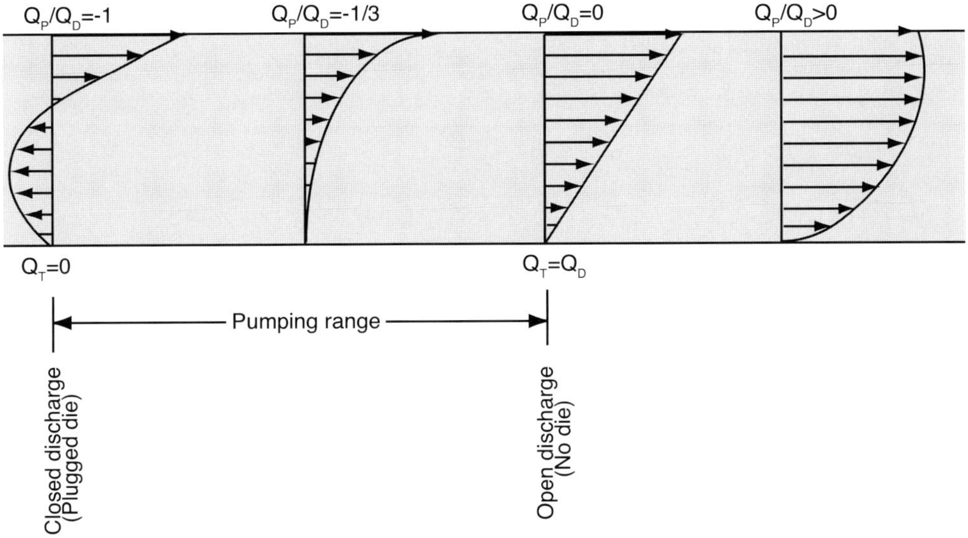

Figure 6.3: Down-channel velocity profiles for different pumping situations with a single screw extruder.

where F_D and F_P are correction factors that account for the flow reduction down the channel of the screw and can be computed using

$$F_D = \frac{16W}{\pi^3 h} \sum_{i=1,3,5}^{\infty} \frac{1}{i^3} \tanh\left(\frac{i\pi h}{2W}\right) \tag{6.11}$$

$$F_P = 1 - \frac{192h}{\pi^3 W} \sum_{i=1,3,5}^{\infty} \frac{1}{i^5} \tanh\left(\frac{i\pi W}{2h}\right) \tag{6.12}$$

It should be noted that the correction is less than 5% for channels that have an aspect ratio, W/h, larger than 10.

6.1.2 Cross Channel Flow in a Single Screw Extruder

The cross channel flow is derived in a similar fashion as the down channel flow. This flow is driven by the x-component of the velocity, which creates a shear flow in that direction. However, since the shear flow pumps the material against the trailing flight of the screw channel, it results in a pressure increase that creates a counteracting pressure flow which leads to a net flow of zero[1]. The flow rate per unit depth at any arbitrary position along the z-axis can be defined by

$$q_x = -\frac{u_x h}{2} - \frac{h^3}{12\mu}\frac{\partial p}{\partial x} = 0 \tag{6.13}$$

[1]This assumption is not completely true, since some of the material flows over the screw flight into the regions of lower pressure in the up-channel direction.

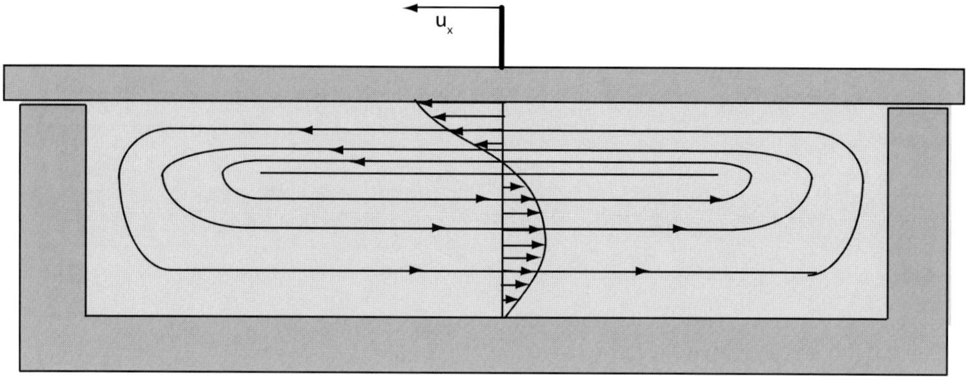

Figure 6.4: View in the down channel direction depicting the resulting cross flow.

Here, we can solve for the pressure gradient $\frac{\partial p}{\partial x}$ to be

$$\frac{\partial p}{\partial x} = -\frac{6\mu u_x}{h^2} \tag{6.14}$$

Once the pressure is known, we can compute the velocity profile across the thickness of the channel using

$$u_x(y) = -\frac{u_x y}{h} - \frac{1}{2\mu}\left(-\frac{6\mu u_x}{h^2}\right)\left(hy - y^2\right) \tag{6.15}$$

This velocity profile is schematically depicted in Fig. 6.4. As shown in the figure, the cross flow generates a recirculating flow, which performs a stirring and mixing action important in extruders for blending as well as melting.

If we combine the flow generated by the down channel and cross channel flows, a net flow is generated in axial or machine direction (u_l) of the extruder, schematically depicted in Fig. 6.5. As can be seen, at open discharge, the maximum axial flow is generated, whereas at closed discharge, the axial flow is zero. From the velocity profiles presented in Fig. 6.5 we can easily deduce, which path a particle flowing with the polymer melt will take.

Due to the combination of cross channel and down channel flows, peculiar particle paths develop for the various die restrictions. The paths that form for various situations are presented in Fig. 6.6. When the particle flows near the barrel surface of the channel, it moves at its fastest speed and in a direction nearly perpendicular to the axial direction of the screw. As the particle approaches the screw flight, it submerges and approaches the screw root, at which point it travels back at a slower speed, until it reaches the leading flight of the screw, which causes the particle to rise once more and travel in the down channel direction. Depending on the die restriction, the path changes. For example, for the closed discharge situation, the particle simply travels on a path perpendicular to the axial direction of the screw, recirculating between the barrel surface and screw root.

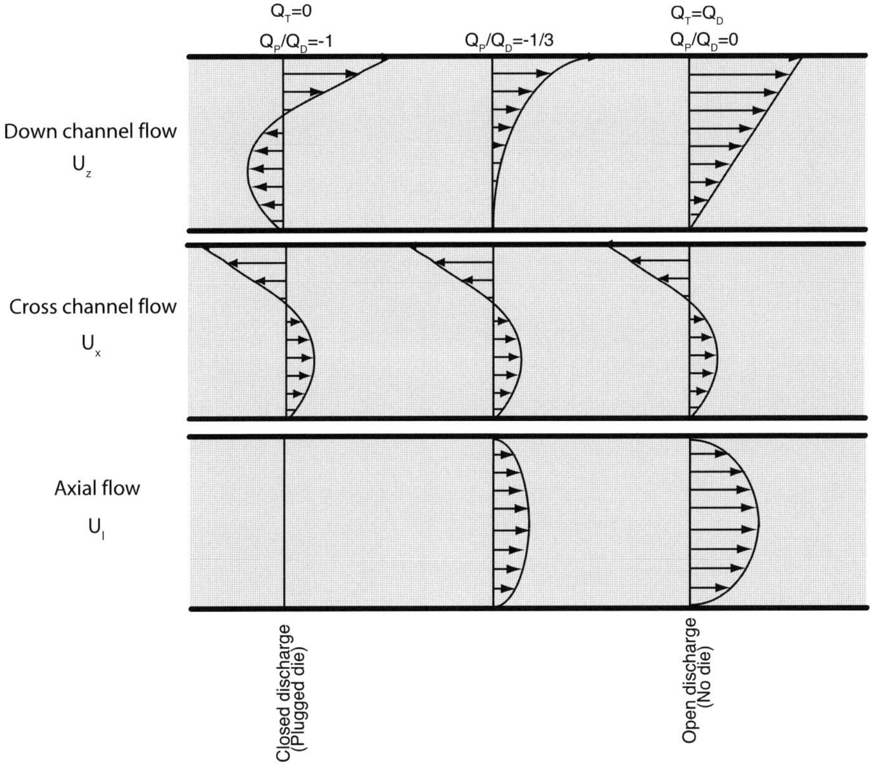

Figure 6.5: Down channel, cross channel and axial velocity profiles for various situations that arise in a single screw extruder.

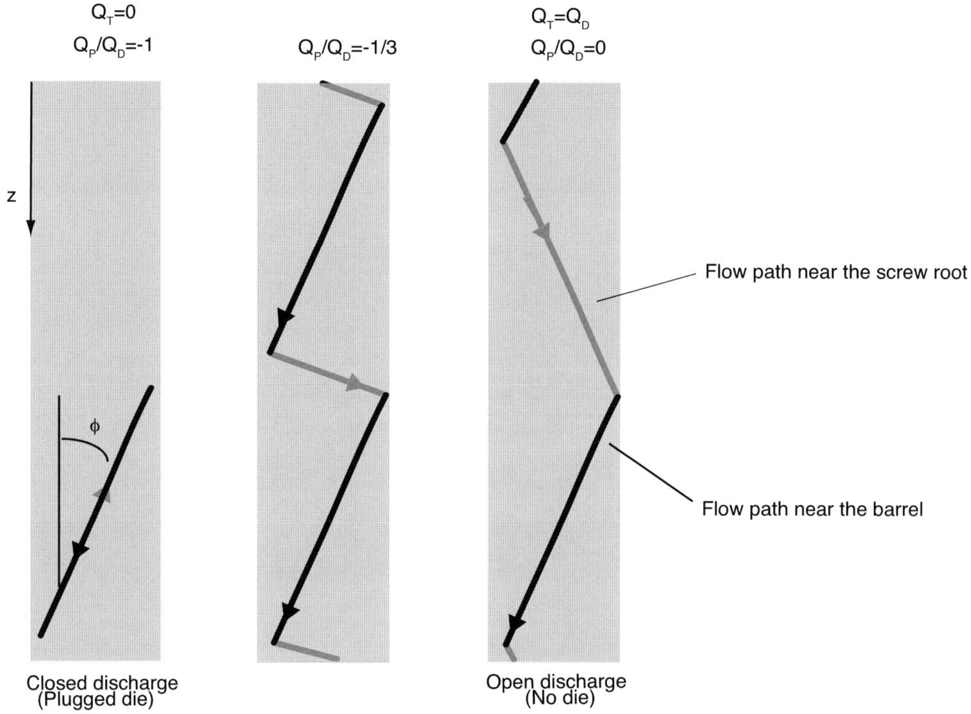

Figure 6.6: Fluid particle paths in a screw channel.

6.1.3 Newtonian Isothermal Screw and Die Characteristic Curves

When extruding a Newtonian fluid through a die, the throughput is directly proportional to the pressure build-up in the extruder and inversely proportional to the viscosity as stated in

$$Q_{die} = \frac{\aleph}{\mu}\Delta p \tag{6.16}$$

where $\aleph$ is a proportionality constant related to the geometry of the die, i.e., for a capillary die of length l and radius R, $\aleph = \pi R^4/8/l$. Equation (6.16) is commonly referred to as the *die characteristic curve*. If we equate the volumetric throughput of the extruder and die, using eqns. (6.10) and (6.16) we get

$$\frac{\pi DnhW}{2}\cos\phi F_D - \frac{Wh^3\Delta p}{12\mu L}F_P = \frac{\aleph}{\mu}\Delta p \tag{6.17}$$

which can be solved for the pressure build-up, $\Delta p = \Delta p_D$, corresponding to a specific die as

$$\Delta p_D = \frac{6\mu\pi Dn\cos\phi hWLF_D}{12\aleph L + Wh^3F_P} \tag{6.18}$$

Substituting eqn. (6.18) into eqn. (6.16), we arrive at the volumetric throughput for a single screw pump with a particular die, described by

$$Q = \left[\frac{\pi Dn\cos\phi hWF_D}{2 + \frac{Wh^3F_P}{6\aleph L}}\right] \tag{6.19}$$

which is commonly referred to as the *operating point*. This concept is more clearly depicted in Fig. 6.7.

As one can imagine, there are numerous types of die restrictions. A die that is used to manufacture a thick sheet of polystyrene is significantly less restrictive than a die that is used to manufacture a thin polyethylene teraphthalate film. To account for the variation of die restrictions the appropriate screw design for a specific application must be chosen. Figure 6.8 presents two types of dies, a restricted and a less restricted die, along with two screw characteristic curves for a deep channel screw and a shallow channel screw. As can be seen, the deep channel screw has a higher productivity when used with a less restricted die, and the shallow screw works best with a high restriction die. It is obvious that a deep screw carries more material and therefore has a higher productivity at open discharge, whereas a shallow screw carries a smaller overall amount of melt, resulting in lower productivity at open discharge. On the other hand, a shallow screw has higher rates of deformation at the same screw speed, which leads to higher shear stresses. This results in larger pressure build-up, which is needed for the high restriction dies. It is therefore necessary to asses each case on an individual basis and design the screw appropriately.

In order to maximize the throughput for a particular screw-die combination we set the variation of eqn. (6.19) with respect to channel depth, h, to zero as

$$\frac{\partial Q}{\partial h} = 0 \tag{6.20}$$

which results in an optimal channel depth, h^{optimum} for a specific die restriction $\aleph$ of

$$h^{\text{optimum}} = \left[\frac{6L\aleph}{W}\right]^{1/3} \tag{6.21}$$

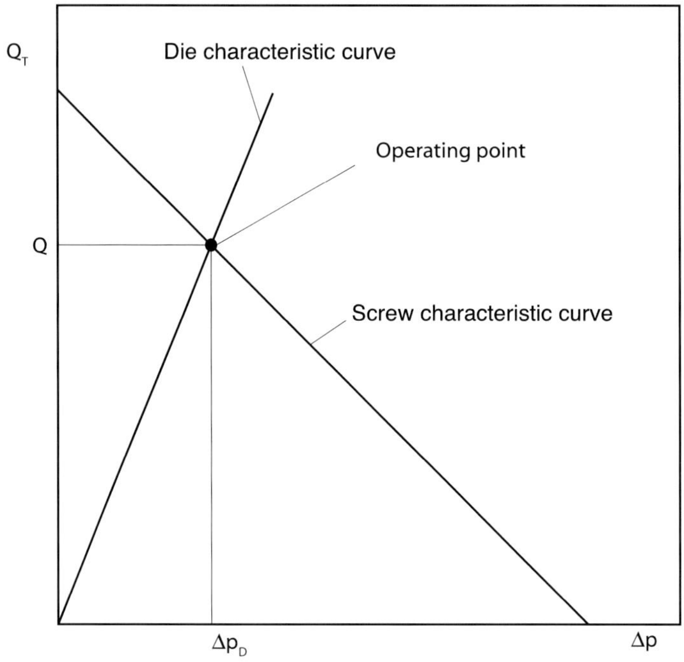

Figure 6.7: Screw and die characteristic curves.

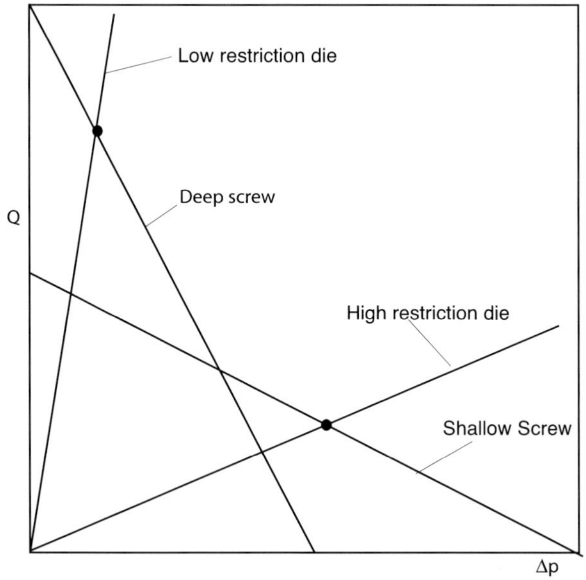

Figure 6.8: Screw and die characteristic curves for various screws and dies.

Similarly, we can solve for the optimum helix angle, ϕ, by setting the variation of eqn. (6.19) with respect to the helix angle to zero, $\partial Q/\partial\phi = 0$. The helix angle is embeded in the channel length, L, term

$$L = \frac{Z}{\sin\phi} \tag{6.22}$$

where Z is the axial length of the extruder's metering section. After differentiation we get

$$\sin^2\phi^{\text{optimum}} = \frac{1}{2 + \dfrac{\pi D h^3}{12 Z \aleph}} \tag{6.23}$$

■ **EXAMPLE 6.2.**

Optimum extruder geometry. You are given the task to find the optimum screw geometry of a 45 mm diameter extruder used for a 3 cm diameter pipe extrusion operation. The pipe's die land length is 100 mm and die opening gap is 2 mm. Determine the optimum channel depth in the metering section, and the optimum screw helix angle. Assume a Newtonian isothermal flow and an extrusion metering section that is 5 turns long. Since the die gap is much smaller than the pipe diameter and die length, for the solution of this problem we can assume pressure driven slit flow, which for a Newtonian fluid is governed by

$$Q_d = \frac{W_d h_d^3 \Delta p}{12\mu L_d} \tag{6.24}$$

where we deduce that the die restriction constant is $\aleph = \dfrac{W_d h_d^3}{12 L_d}$, which is substituted into eqn. (6.21) assuming a square pitch($\phi = 17.65^o$) and a channel width of 40 mm (for a 5 mm flight width) to give,

$$\begin{aligned} h^{\text{optimum}} &= h_d \left[\frac{6(5D/\sin 17.65^o)W_d}{12 L_d W}\right]^{1/3} \\ &= 2\text{ mm}\left[\frac{6(5\times 45\text{ mm}/\sin 17.65^o)\pi(30\text{ mm})}{12(100\text{ mm})(40\text{ mm})}\right] \\ &= 4.1\text{ mm} \end{aligned} \tag{6.25}$$

It is interesting to point out that for a die with a pressure flow through a slit, or sets of slits, the optimum channel depth is directly proportional to the die gap. Decreasing the die gap by a certain percentage will result in an optimum channel depth that is reduced by the same percentage. To determine the optimum helix angle we can re-write eqn. (6.23) for this specific application,

$$\begin{aligned} \sin^2\phi^{\text{optimum}} &= \frac{1}{2 + \dfrac{\pi D h^3 L_d}{Z W_d h_d^3}} \\ &= \frac{1}{2 + \dfrac{\pi(45\text{ mm})(4.1\text{ mm})^3(100\text{ mm})}{(5\times 45\text{ mm})(\pi 30\text{ mm})(2\text{ mm})^3}} \\ &= 0.129 \end{aligned} \tag{6.26}$$

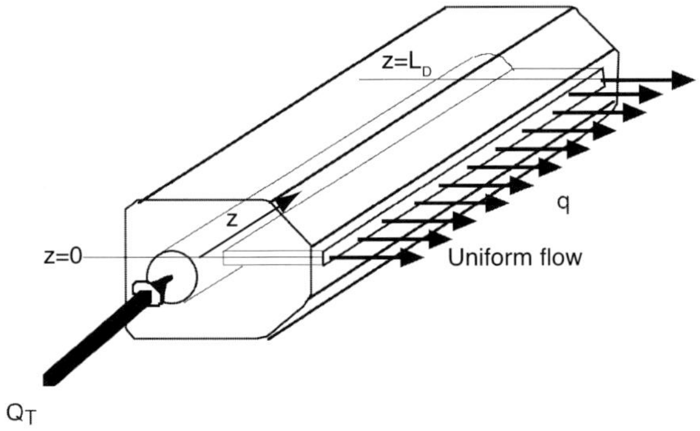

Figure 6.9: Schematic diagram of an end-fed sheeting die.

which results in $\phi^{\text{optimum}} = 21^o$, compared to 17.65^o for a square pitch screw. We note that here we used the optimum channel depth.

6.2 EXTRUSION DIES—ISOTHERMAL FLOW PROBLEMS

The flow in many extrusion dies can be approximated with one, or a combination, of simplified models such as slit flow, Hagen Poiseulle flow, annular flow, simple shear flow, etc. A few of these are presented in the following sections using non-Newtonian as well as Newtonian flow models.

6.2.1 End-Fed Sheeting Die

The end-fed-sheeting die, as presented in Fig. 6.9, is a simple geometry that can be used to extrude films and sheets. To illustrate the complexities of die design, we will modify the die, as shown in the figure, in order to extrude a sheet or film with a uniform thickness. In order to achieve this we must determine the length of the approach zone or die land as a function of the manifold direction, as depicted in the model shown in Fig. 6.10.

For this specific example, the manifold diameter will be kept constant and we will assume a Newtonian isothermal flow, with a constant viscosity μ. The flow of the manifold can be represented using the Hagen-Poiseuille equation as,

$$Q = \frac{\pi R^4}{8\mu}\left(-\frac{dp}{dz}\right) \tag{6.27}$$

and the flow in the die land (per unit width) can be modeled using the slit flow equation,

$$q = \frac{h^3}{12\mu}\left(-\frac{dp}{dl}\right) = \frac{h^3}{12\mu}\left(\frac{p(z)}{L_L(z)}\right) \tag{6.28}$$

A manifold that generates a uniform sheet must deliver a constant throughput along the die land. Performing a flow balance within the differential element, presented in Fig. 6.11,

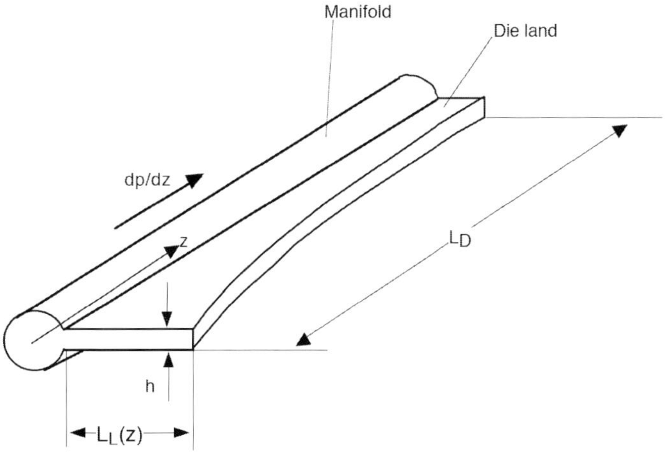

Figure 6.10: Schematic of the manifold and die land in an end-fed sheeting die.

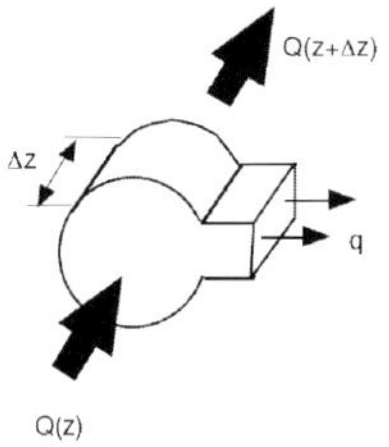

Figure 6.11: Differential element of the manifold in an end-fed sheeting die.

results in

$$\frac{dQ}{dz} = -q = \text{constant} \tag{6.29}$$

Integrating this equation and letting $Q = Q_T$ at $z = 0$ and $Q = 0$ at $z = L_D$ we get,

$$Q(z) = Q_T \left(1 - \frac{z}{L_D}\right) \tag{6.30}$$

Therefore,

$$\frac{dQ}{dz} = -\frac{Q_T}{L_D} = -\frac{h^3}{12\mu}\frac{p(z)}{L_L(z)} \tag{6.31}$$

which results in

$$L_L(z) = \frac{h^3}{12\mu}\frac{L_D}{Q_T}p(z) \tag{6.32}$$

where the pressure as a function of z must be solved for. The manifold equation can be written as,

$$\frac{dp}{dz} = -\frac{8\mu}{\pi R^4}Q_T\left(1 - \frac{z}{L_D}\right) \tag{6.33}$$

and integrated from $p = p_0$ at $z = 0$,

$$p(z) = p_0 - \frac{8\mu Q_T L_D}{\pi R^4}\left[\left(\frac{z}{L_D}\right) - \frac{1}{2}\left(\frac{z}{L_D}\right)^2\right] \tag{6.34}$$

which can be substituted into eqn. (6.32) to give

$$L_L(z) = \frac{h^3}{12\mu}\frac{L_D p_0}{Q_T} - \frac{2h^3 L_D^2}{3\pi R^4}\left[\left(\frac{z}{L_D}\right) - \frac{1}{2}\left(\frac{z}{L_D}\right)^2\right] \tag{6.35}$$

Note that the die design equation has pressure, volumetric flow rate and viscosity embedded inside and can therefore lead to unrealistic results. This is due to the fact that the flow, Q_T, was specified when formulating the equations. However, the die design will be balanced for any volumetric throughput. Hence, during die design it is appropriate to specify the land length at the beginning of the manifold, $L_L(0)$, and pick appropriate combinations of viscosity, flow rate and pressure,

$$L_L(0) = h^3 L_D \frac{p_0}{12\mu Q_T} \tag{6.36}$$

■ **EXAMPLE 6.3.**

End-fed sheeting die. Design a 1000 mm wide end-fed sheeting die with a 1 mm die land gap for a polycarbonate film. For the solution of the problem assume a manifold diameter of 15 mm and the longest portion of the length should be 50 mm. Using the above information we can write,

$$L_L(z) = 50\text{ mm} - \frac{2(1\text{ mm})^3(1000)^2}{3\pi(10\text{ mm})^4}\left[\left(\frac{z}{1000\text{ mm}}\right) - \frac{1}{2}\left(\frac{z}{1000\text{ mm}}\right)^2\right] \tag{6.37}$$

or

$$L_L(z) = 50\text{ mm} - 10.6\left[\left(\frac{z}{1000\text{ mm}}\right) - \frac{1}{2}\left(\frac{z}{1000\text{ mm}}\right)^2\right] \tag{6.38}$$

Hence, the land length starts at 50 mm length and reduces to 39.4 mm at the opposite end of the die.

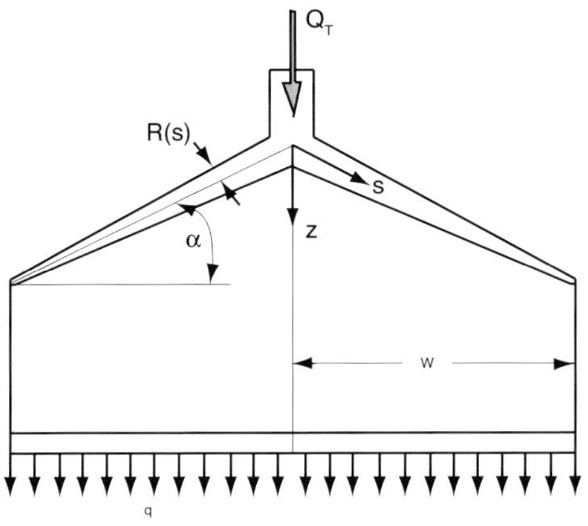

Figure 6.12: Schematic diagram of a coat hanger sheeting die.

6.2.2 Coat Hanger Die

Perhaps a more common sheeting die is the so-called coat hanger die, presented in detail in Chapter 3. For a given manifold angle α we must determine the manifold radius profile, $R(s)$, such that a uniform sheet or film is extruded through the die lips.

Using the nomenclature presented in Fig. 6.12 and assuming a land thickness of h we can assume the land length to be described by slit flow and the manifold by the Hagen-Poiseuille flow with a variable radius as

$$q = -\frac{h^3}{12\mu}\left(\frac{dp}{dz}\right) \tag{6.39}$$

and

$$Q(s) = \frac{\pi R(s)^4}{8\mu}\left(-\frac{dp}{ds}\right) \tag{6.40}$$

Equation (6.39) can be rewritten as

$$\frac{Q_T}{2W} = \frac{-h^3}{12\mu}\left(-\frac{dp}{dz}\right) \tag{6.41}$$

which can be solved for the pressure gradient in the die land

$$\frac{dp}{dz} = -\frac{6\mu Q_T}{Wh^3} \tag{6.42}$$

Here too, we can cut a small element out of the manifold area and can relate the pressure drop in the s-direction to the drop in the z-direction using

$$p(s) + \frac{dp}{ds}\Delta s = p(s) + \frac{dp}{dz}\Delta z \tag{6.43}$$

Combining the definition of pressure gradient in the die land, eqn. (6.42), with eqn. (6.43) and using geometry we get

$$\frac{dp}{ds} = -\frac{6\mu Q_T}{Wh^3}\sin\alpha \tag{6.44}$$

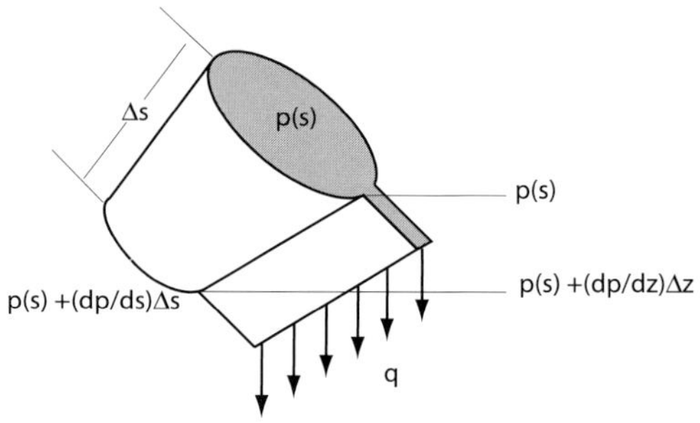

Figure 6.13: Differential element of the manifold in the coat hanger sheeting die.

which can be integrated to become

$$p(s) = p_0 - \frac{6\mu Q_T}{Wh^3} \sin \alpha s \tag{6.45}$$

and a mass balance results in

$$\frac{dQ}{ds} = -\frac{Q_T}{2W} \cos \alpha \tag{6.46}$$

Using the boundary condition that $Q = \frac{Q_T}{2}$ at $s = 0$ and $Q = 0$ at $s = W/\cos\alpha$, we can integrate eqn. (6.46) to be

$$Q(s) = Q_T \left(\frac{1}{2} - \frac{\cos \alpha}{2W} s \right) \tag{6.47}$$

We can now set the manifold equation, eqn. (6.40), with the pressure gradient defined in eqn. (6.46), equal to eqn. (6.47)

$$Q_T \left(\frac{1}{2} - \frac{\cos \alpha}{2W} s \right) = \frac{\pi R(s)^4}{8\mu} \left(\frac{6\mu Q_T}{Wh^3} \sin \alpha \right) \tag{6.48}$$

which, can be used to solve for the manifold radius profile

$$R(s) = \left[\frac{2\,(1 - s \cos \alpha / W)}{3\pi \sin \alpha} \right]^{1/4} \tag{6.49}$$

A cross-head tubing die is equivalent to the coat hanger die by wrapping it around a cylinder as can be recognized in the schematic presented in Fig. 3.17. If we follow the same derivation but for a shear thinning power law melt, we get

$$R(s) = \left[\frac{[(3 + 1/n/\pi)]^n h^{2n+1} (W - s \cos \alpha)^n}{2^n (2 + 1/n)^n (-\sin \alpha)} \right]^{1/(3n+1)} \tag{6.50}$$

which for a Newtonian fluid with $n = 1$ reduces to eqn. (6.49).

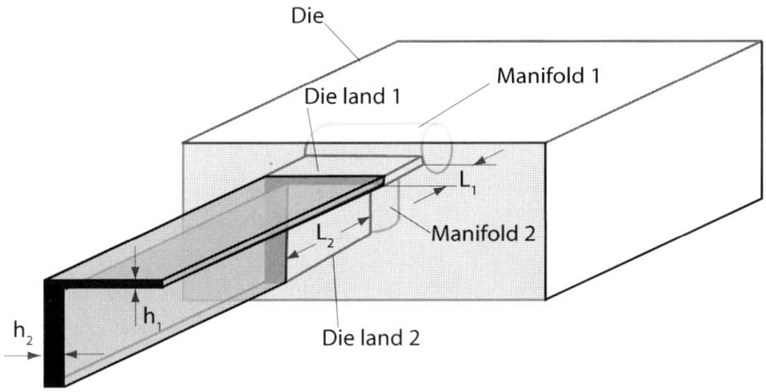

Figure 6.14: Schematic diagram of die with two different die land lengths and thicknesses.

6.2.3 Extrusion Die with Variable Die Land Thicknesses

When designing plastic parts it is often recommended that the part have uniform thickness. This is especially true for semi-crystalline polymers where thickness variations lead to variable cooling times, and those in turn to variations in the degree of crystallinity in the final part. Variations in crystallinity result in shrinkage variations, which lead to warpage. However, it is often necessary to design parts in which a thickness variation is inevitable, i.e., extrusion profiles with thickness variations as shown in Fig. 6.14.

The die land thickness differences can be compensated by using different land lengths such that the speed of the emerging melt is constant, resulting in a uniform product. If we assume a power-law viscosity model, a uniform pressure in the manifold and an isothermal die and melt, the average speed of the melt emerging from the die is

$$\bar{u_i} = \frac{h_i}{2(s+2)}\left(\frac{h_i \Delta p}{2mL_i}\right)^s \tag{6.51}$$

where $s = 1/n$. In order to achieve a uniform, product we must satisfy

$$\bar{u_1} = \bar{u_2} \tag{6.52}$$

or

$$\frac{h_1}{2(s+2)}\left(\frac{h_1 \Delta p}{2mL_1}\right)^s = \frac{h_2}{2(s+2)}\left(\frac{h_2 \Delta p}{2mL_2}\right)^s \tag{6.53}$$

which can be rearranged to become,

$$\left(\frac{L_1}{L_2}\right) = \left(\frac{h_1}{h_2}\right)^{1+n} \tag{6.54}$$

■ **EXAMPLE 6.4.**

Die design with two die land thicknesses. Determine the die land length ratios, L_2/L_1 for a die land thickness ratio, h_2/h_1 of 3, for various power-law indeces. Using eqn. (6.54), we can easily solve for the land length ratios for several power-law

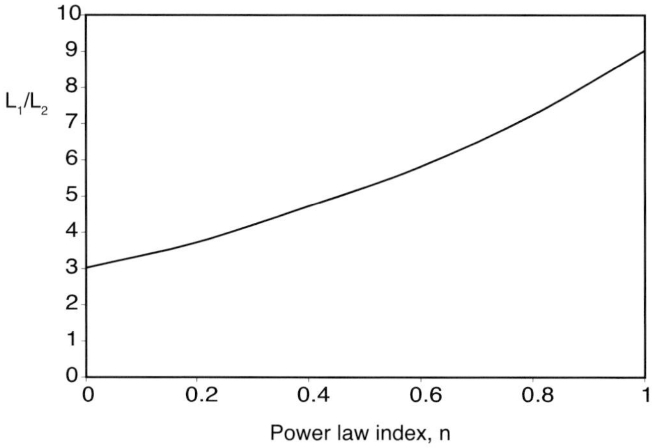

Figure 6.15: Land length ratios as a function of power law index for a die with a land height ratio of 3.

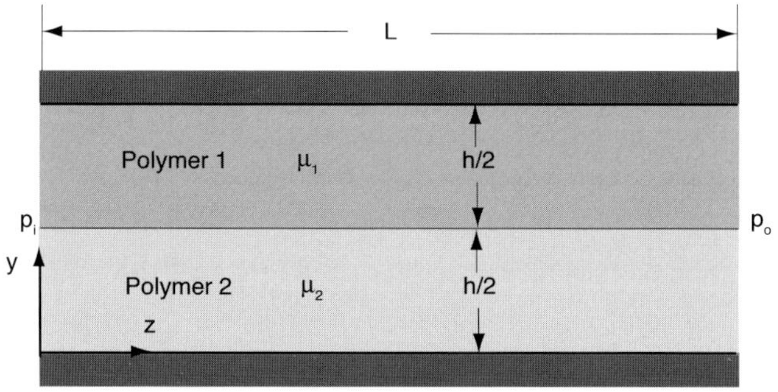

Figure 6.16: Schematic diagram of a two polymer layer system in a co-extrusion die.

indeces. This is presented graphically in Fig. 6.15. Note that while a Newtonian fluid requires a land length ratio of 9, a Bingham fluid, with a power law index of zero, requires a land length ratio of only 3. Hence, die design is very sensitive to the shear thinning behavior of the polymer melt, and must always be accounted for.

6.2.4 Pressure Flow of Two Immiscible Fluids with Different Viscosities

Pressure flow of two immiscible fluids with different viscosities that flow as separate layers between parallel plates are often encountered inside dies during co-extrusion when producing multi-layer films. Such a system is schematically depicted in Fig. 6.16, which presents two layers of thickness $h/2$ and viscosities μ_1 and μ_2, respectively.

When solving this problem, we first assume that the melts are both Newtonian fluids and that there is no velocity component in the y-direction. If we also assume that $\partial/\partial x = 0$, the continuity equation reduces to

$$\frac{\partial u_z}{\partial z} = 0 \tag{6.55}$$

The momentum balance for both layers can be written as

$$0 = -\frac{\partial p}{\partial z} + \mu_1 \frac{\partial^2 u_z^1}{\partial y^2} \tag{6.56}$$

$$0 = -\frac{\partial p}{\partial z} + \mu_2 \frac{\partial^2 u_z^2}{\partial y^2} \tag{6.57}$$

respectively, which can be integrated to become,

$$u_z^1 = \frac{1}{2\mu_1}\frac{\partial p}{\partial z} y^2 + c_1 y + c_2 \tag{6.58}$$

and

$$u_z^2 = \frac{1}{2\mu_2}\frac{\partial p}{\partial z} y^2 + c_3 y + c_4 \tag{6.59}$$

Using the boundary condition that $u_z^1 = 0$ at $y = 0$ results in $c_2 = 0$, and using $u_z^2 = 0$ at $y = h$ gives,

$$0 = \frac{1}{2\mu_1}\frac{\partial p}{\partial z} h^2 + c_3 h + c_4 \tag{6.60}$$

Furthermore, assuming negligible surface tension, we can assume that the stresses match at the melt-melt interface, $y = h/2$,

$$\tau_{zy}^1 = \tau_{zy}^2 \tag{6.61}$$

which gives,

$$\mu_1 \frac{\partial u_z^1}{\partial y} = \mu_2 \frac{\partial u_z^2}{\partial y} \tag{6.62}$$

which can also be written as,

$$\mu_1 c_1 = \mu_2 c_3 \tag{6.63}$$

The final condition is that the velocity in both melt layers match at the interface, $y = h/2$,

$$\frac{1}{2\mu_1}\frac{\partial p}{\partial z} h^2/4 + c_1 h/2 = \frac{1}{2\mu_2}\frac{\partial p}{\partial z} h^2/4 + c_3 h/2 + c_4 \tag{6.64}$$

Combining eqns. (6.60), (6.63) and (6.64) we get,

$$c_1 = \frac{1}{4\mu_1}\left[\frac{3\mu_1 + \mu_2}{\mu_1 + \mu_2}\right]\left(\frac{\partial p}{\partial z}\right) h \tag{6.65}$$

$$c_3 = \frac{1}{4\mu_2}\left[\frac{3\mu_1 + \mu_2}{\mu_1 + \mu_2}\right]\left(\frac{\partial p}{\partial z}\right) h \tag{6.66}$$

and

$$c_4 = \frac{1}{4\mu_2}\left[2 - \frac{3\mu_1 + \mu_2}{\mu_1 + \mu_2}\right]\left(\frac{\partial p}{\partial z}\right) h^2 \tag{6.67}$$

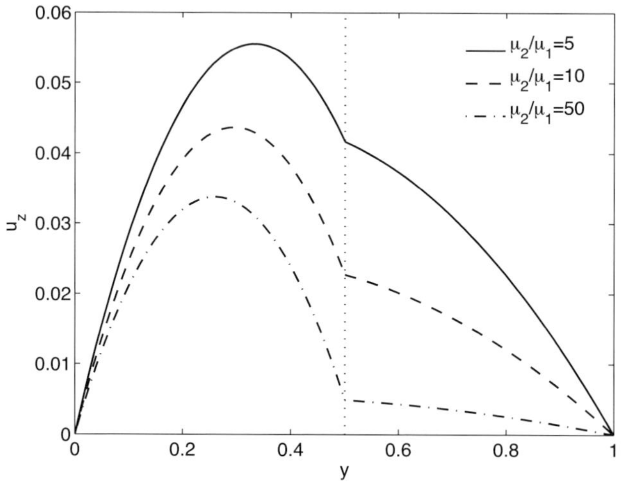

Figure 6.17: Velocity distribution in a two immiscible layer system with different viscosities and viscosity ratios μ_2/μ_1 of 5, 10 and 50.

which results in,

$$u_z^1 = \frac{1}{2\mu_1}\frac{\partial p}{\partial z}y^2 - \frac{1}{4\mu_1}\left[\frac{3\mu_1+\mu_2}{\mu_1+\mu_2}\right]\left(\frac{\partial p}{\partial z}\right)hy \tag{6.68}$$

and

$$u_z^2 = -\frac{1}{2\mu_2}\frac{\partial p}{\partial z}\left[h^2-y^2\right] + \frac{1}{4\mu_2}\left[\frac{3\mu_1+\mu_2}{\mu_1+\mu_2}\right]\left(\frac{\partial p}{\partial z}\right)\left[h^2-hy\right] \tag{6.69}$$

Figure 6.17 presents several velocity distributions within the two-layer system for different viscosity ratios, μ_2/μ_1. For this solution, a gap separation, and pressure gradient of unity was chosen.

6.2.5 Fiber Spinning

The process of fiber spinning, described in Chapter 3 and schematically represented in Fig. 6.18, will be modeled in this section using first a Newtonian model followed by a shear thinning model. To simplify the analysis, it is customary to set the origin of the coordinate system at the location of largest diameter of the extrudate. Since the distance from the spinnerette to the point of largest swell is very small, only a few die diameters, this simplification will not introduce large problems in the solution.

If we take the schematic of a differential fiber element presented in Fig. 6.19, we can define the fiber geometry by the function $R(x)$ and the unit normal vector $\mathbf{n}$. The continuity equation tells us that the volumetric flow rate through any cross-section along the x-direction must be Q

$$Q = \pi R(x)^2 u_x \tag{6.70}$$

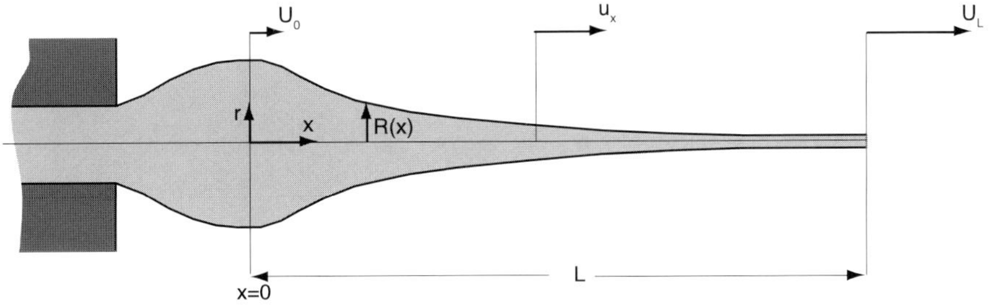

Figure 6.18: Schematic diagram of the fiber spinning process in the post-extrusion die region.

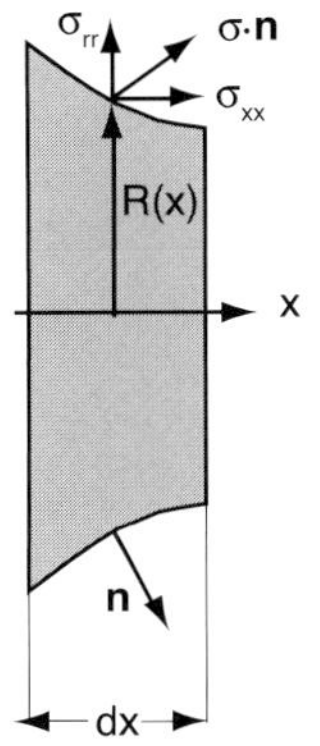

Figure 6.19: Differential element of a fiber during spinning.

We further assume that surface tension is negligible and that in steady state the surface will only move tangentially, which means that

$$\mathbf{u} \cdot \mathbf{n} = 0 \tag{6.71}$$

The components of the normal vector are described by the geometry of the fiber as

$$n_x = -\frac{dR}{dx}\left[1 + \left(\frac{dR}{dr}\right)^2\right]^{-1/2} \tag{6.72}$$

and

$$n_r = \left[1 + \left(\frac{dR}{dr}\right)^2\right]^{-1/2} \tag{6.73}$$

Due to the negligible effects of surface tension, we can assume that the stress boundary condition is

$$\boldsymbol{\sigma} \cdot \mathbf{n} = 0 \tag{6.74}$$

which for each direction can be written as

$$\sigma_{rx} n_r + \sigma_{xx} n_x = 0 \tag{6.75}$$

and

$$\sigma_{rr} n_r + \sigma_{rx} n_x = 0 \tag{6.76}$$

Due to the fact that the fiber is being pulled in the x-direction, we should expect a non-zero σ_{xx} at the free surface. Hence, we can write

$$\sigma_{rx} = -\sigma_{xx}\frac{n_x}{n_r} = \frac{dR}{dr}\sigma_{xx} \tag{6.77}$$

It is clear that only the x-component of the equation of motion plays a significant role in a fiber spinning problem

$$\rho\left(u_r\frac{\partial u_x}{\partial r} + u_x\frac{\partial u_x}{\partial x}\right) = \frac{1}{r}\frac{\partial}{\partial r}(r\sigma_{rx}) + \frac{\partial \sigma_{xx}}{\partial x} \tag{6.78}$$

The first term in the above equation drops out since u_x is not a function of r. Using eqn. (6.77) and rearranging somewhat, the equation of motion becomes

$$\rho u_x\frac{\partial u_x}{\partial x} = \frac{2}{R}\frac{dR}{dr}\sigma_{xx} + \frac{d\sigma_{xx}}{dx} \tag{6.79}$$

For a total stress, σ_{xx}, in a Newtonian approximation we write the constitutive relation

$$\sigma_{xx} = -p + 2\mu\frac{du_x}{dx} \tag{6.80}$$

We can show that the isotropic pressure p is given by

$$p = -(\sigma_{xx} + \sigma_{rr} + \sigma_{\theta\theta})/3 \tag{6.81}$$

However, we can assume that $\sigma_{rr} = 0$ and $\sigma_{\theta\theta} = 0$. Hence, we can write

$$\sigma_{xx} = 3\mu \frac{du_x}{dx} \tag{6.82}$$

Combining the above equation with the continuity equation, eqn. (6.70), the momentum balance presented in eqn. (6.79) becomes

$$\frac{d}{dx}(u_x)^2 = 12\frac{\mu}{\rho R}\frac{dR}{dr}\frac{du_x}{dx} + 6\frac{\mu}{\rho}\frac{d^2u_x}{dx^2} \tag{6.83}$$

If we neglect the effect inertia has on the stretching fiber and drop the inertial term in the above equation, it can be solved as

$$u_x = c_1 e^{c_2 x} \tag{6.84}$$

With boundary conditions $u_x = U_0$ at $x = 0$, and $u_x = U_L$ at $x = L$ we get

$$u_x = U_0 e^{x \ln D_R/L} = U_0 D_R^{x/L} \tag{6.85}$$

where D_R is the *draw down ratio* defined by

$$D_R = \frac{U_L}{U_0} \tag{6.86}$$

Using the continuity equation we can now write

$$R(x) = R_0 D_R^{-\frac{x}{2L}} \tag{6.87}$$

The derivation of the fiber spinning equations for a non-Newtonian shear thinning viscosity using a power law model are also derived. For a total stress, σ_{xx}, in a power law fluid, we write the constitutive relation

$$\sigma_{xx} = -p + 2m(3)^{(n-1)/2}\left(\frac{du_x}{dx}\right)^n \tag{6.88}$$

This leads to the velocity distribution

$$u_x = U_0\left[1 + \left(D_R^{(n-1)/n} - 1\right)\frac{x}{L}\right]^{n/(n-1)} \tag{6.89}$$

6.2.6 Viscoelastic Fiber Spinning Model

It is appropriate at this time to introduce viscoelastic flow analysis. Fiber spinning is one of the few processes that can be analyzed using analytical viscoelastic models. Here, we follow the approach developed by Denn and Fisher [4]. Neglecting inertia, we can start with the momentum balance by modifying eqn. (6.79) as,

$$\frac{2}{R}\frac{dR}{dr}\sigma_{xx} + \frac{d\sigma_{xx}}{dx} = \frac{1}{R^2}\frac{d}{dx}(R^2\sigma_{xx}) = 0 \tag{6.90}$$

Using the continuity balance and and setting the take-force to,

$$F = (\pi R_L^2 \sigma_{xx})|_L \tag{6.91}$$

we can integrate eqn. (6.90) to give,

$$\sigma_{xx} = \frac{\rho F u_x}{\dot{m}} \tag{6.92}$$

Since $\sigma_{rr} \approx 0$ we can write,

$$\sigma_{xx} - \sigma_{rr} = \tau_{xx} - \tau_{rr} = \frac{\rho F u_x}{\dot{m}} \tag{6.93}$$

which is the *first normal stress difference*. For the two stress components, Denn and Fisher [4] used the White-Metzner constitutive model,

$$\tau_{xx} + \lambda\left(u_x\frac{d\tau_{xx}}{dx} - 2\tau_{xx}\frac{du_x}{dx}\right) = -\eta\dot{\gamma}_{xx} = -2\eta\frac{du_x}{dx} \tag{6.94}$$

and

$$\tau_{rr} + \lambda\left(u_x\frac{d\tau_{rr}}{dx} - \tau_{rr}\frac{du_x}{dx}\right) = -\eta\dot{\gamma}_{rr} = -2\eta\frac{du_r}{dr} = \eta\frac{du_z}{dz} \tag{6.95}$$

respectively. Here, λ is the relaxation time defined by,

$$\lambda = \frac{\eta}{G} \tag{6.96}$$

where G is the elastic shear modulus. Using the power law model to define the viscosity η, we can combine the two constitutive equations, eqns. (6.94) and (6.95), to give the dimensionless equation,

$$\bar{U} + (\alpha\bar{U} - 3\epsilon)\left(\frac{d\bar{U}}{d\xi}\right)^n - 2\alpha^2\bar{U}\left(\frac{d\bar{U}}{d\xi}\right)^{2n} - n\alpha\bar{U}^2\left(\frac{d^2\bar{U}}{d\xi^2}\right)\left(\frac{d\bar{U}}{d\xi}\right)^{n-1} = 0 \tag{6.97}$$

where, $\bar{U} = u_x/U_0$, $\xi = x/L$ and, α and ϵ are dimensionless rheological and force parameters, defined by,

$$\alpha = \frac{m(3)^{(n-1)/2}}{G}\left(\frac{U_0}{L}\right)^n \tag{6.98}$$

and

$$\epsilon = \frac{m\dot{m}(3)^{(n-1)/2}}{\rho F L}\left(\frac{U_0}{L}\right)^{n-1} \tag{6.99}$$

respectively. To solve the problem, we need one additional boundary condition that $\tau_{xx} = \tau_0$ at $x = 0$, which is difficult to estimate. However, Denn and Fisher [4] solved eqn. (6.97) with the velocity boundary conditions of the Newtonian problem, given above. Figure 6.20 presents a plot of eqn (6.97) with various values of α. The graph also presents experimental results for the fiber spinning of polystyrene at 170°C. The fiber had a value of α between 0.2 and 0.3, but the theoretical prediction compares with the experiments for a value of α between 0.4 and 0.5. Phan-Thien had better agreement between the experiments done with polystyrene and low density polyethylene by using the Phan-Thien-Tanner model [22].

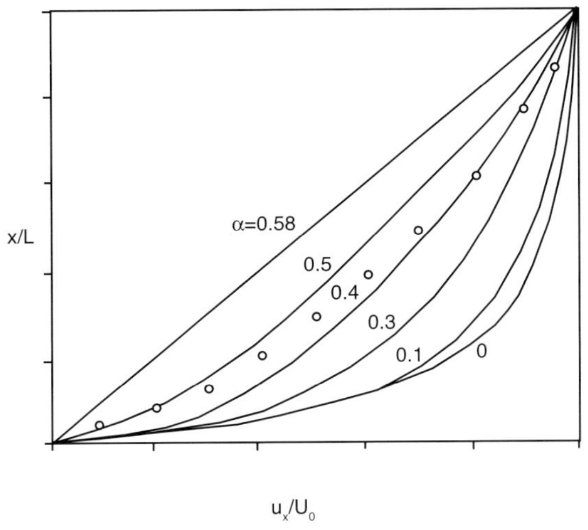

Figure 6.20: Comparison between experimental and computed velocity profiles during fiber spinning using Denn and Fisher's viscoelastic model [4].

6.3 PROCESSES THAT INVOLVE MEMBRANE STRETCHING

There are numerous processes that involve the stretching of a membrane such as film blowing, film casting, extrusion blow molding, injection blow molding, themoforming, etc. In this section we will address two very important processes: the film blowing process and the thermoforming process.

6.3.1 Film Blowing

Despite the non-isothermal nature of the film blowing process we will develop here an isothermal model to show general effects and interactions during the process. In the derivation we follow Pearson and Petrie's approach [20], [19] and [21]. Even this Newtonian isothermal model requires an iterative solution and numerical integration. Figure 6.21 presents the notation used when deriving the model.

A common form of analyzing film blowing is by setting-up a coordinate system, ξ, that moves with the moving melt on the inner surface of the bubble, and that is oriented with the film as shown in Fig. 6.21. Using the moving coordinates, we can define the three non-zero terms of the local rate of deformation tensor as

$$\begin{aligned}\dot{\gamma}_{11} &= 2\frac{\partial u_1}{\partial \xi_1}\\ \dot{\gamma}_{22} &= 2\frac{\partial u_2}{\partial \xi_2}\\ \dot{\gamma}_{33} &= 2\frac{\partial u_3}{\partial \xi_3}\end{aligned} \tag{6.100}$$

For an incompressible fluid, these three components must add-up to zero

$$\dot{\gamma}_{11} + \dot{\gamma}_{22} + \dot{\gamma}_{33} = 0 \tag{6.101}$$

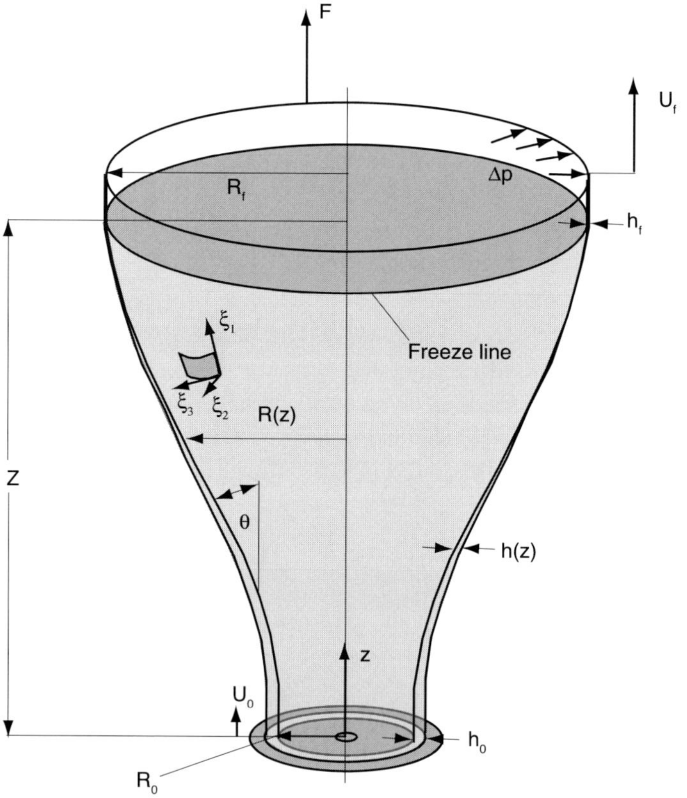

Figure 6.21: Schematic diagram of the film blowing process.

For a thin film, where $d\xi_1 = dz/\cos\theta$, the $\dot{\gamma}_{22}$ term can be defined by

$$\dot{\gamma}_{22} = 2\frac{u_2}{h} = \frac{2}{h}\frac{dh}{dt} = \frac{2}{h}\frac{dh}{d\xi_1}\frac{d\xi_1}{dt} = \frac{2u_1}{h}\frac{dh}{d\xi_1} = \frac{2u_1\cos\theta}{h}\frac{dh}{dz} \tag{6.102}$$

The rate of expansion in the circumferential direction is defined by the rate of growth of the circumference

$$u_3 = 2\pi\frac{dR}{dt} \tag{6.103}$$

divided by the local circumference, $2\pi R$, to become

$$\dot{\gamma}_{33} = \frac{1}{R}\frac{dR}{dt} = \frac{2u_1}{R}\frac{dR}{d\xi_1} = \frac{2u_1\cos\theta}{R}\frac{dR}{dz} \tag{6.104}$$

and finally

$$\dot{\gamma}_{11} = -\dot{\gamma}_{22} - \dot{\gamma}_{33} = -\frac{2u_1\cos\theta}{h}\frac{dh}{dz} - \frac{2\pi u_1\cos\theta}{R}\frac{dR}{dz} \tag{6.105}$$

It is possible to relate the total volumetric throughput, Q, to u_1 using

$$Q = 2\pi R h u_1 \tag{6.106}$$

We can now write

$$\begin{aligned}
\dot{\gamma}_{11} &= -\frac{Q\cos\theta}{\pi R h}\frac{1}{h}\frac{dh}{dz} - \frac{1}{R}\frac{dR}{dz} \\
\dot{\gamma}_{22} &= \frac{Q\cos\theta}{\pi R h}\frac{1}{h}\frac{dh}{dz} \\
\dot{\gamma}_{33} &= \frac{Q\cos\theta}{\pi R h}\frac{1}{R}\frac{dR}{dz}
\end{aligned} \tag{6.107}$$

The total stress in the ξ-coordinate system is written as

$$\sigma_{ii} = p - \mu\dot{\gamma}_{ii} \tag{6.108}$$

Since surface tension is neglected and no external forces act on the bubble

$$\sigma_{22} = 0 \tag{6.109}$$

Hence

$$p = \mu\dot{\gamma}_{22} = \frac{Q\mu\cos\theta}{\pi R h^2}\frac{dh}{dz} \tag{6.110}$$

The two stresses become

$$\sigma_{11} = -\frac{\mu Q\cos\theta}{\pi R h}\left(\frac{2}{h}\frac{dh}{dz} + \frac{1}{R}\frac{dR}{dz}\right) \tag{6.111}$$

and

$$\sigma_{33} = \frac{\mu Q\cos\theta}{\pi R h}\left(\frac{1}{R}\frac{dR}{dz} - \frac{1}{h}\frac{dh}{dz}\right) \tag{6.112}$$

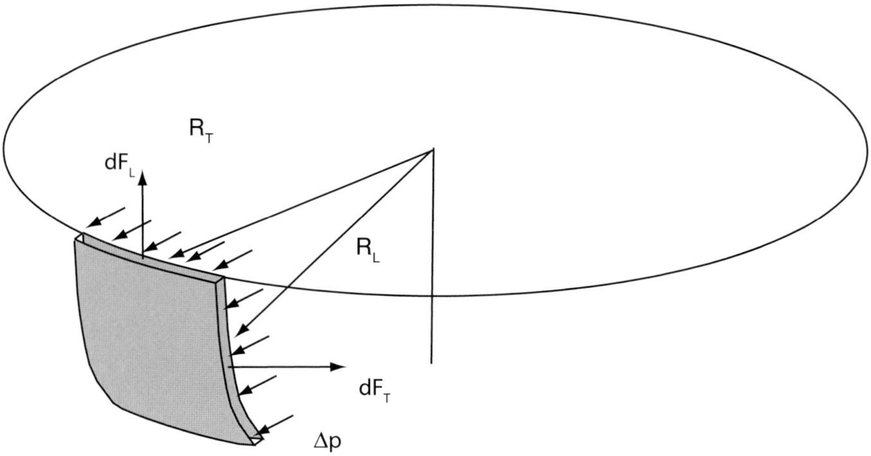

Figure 6.22: Forces acting on a film element.

It is necessary to perform a force balance for the bubble in order to determine the radius, $R(z)$, and the thickness, $h(z)$, of the bubble. The longitudinal force is computed using

$$F_L = 2\pi R h \sigma_{11} \tag{6.113}$$

and for the small fluid element defined in Fig. 6.22, the transverse force is defined by

$$dF_T = h\xi_1\sigma_{33} \tag{6.114}$$

A force balance about the differential element results in

$$\Delta p = h\left(\frac{\sigma_{11}}{R_L} + \frac{\sigma_{33}}{R_T}\right) \tag{6.115}$$

where R_L and R_T are defined in the ξ-coordinate system. In a cylindrical-polar coordinate system we can write

$$R_L = -\frac{\sec^3\theta}{d^2R/dz^2} \tag{6.116}$$

and

$$R_T = R\sec\theta \tag{6.117}$$

The bubble will grow to a maximum —or final— radius, R_f, when it freezes at a position $z = Z$, which is called the *freeze-line*. Since the bubble is pulled by a force F_Z, which is usually referred to as the *draw force*, we can perform a force balance between a position z and Z to give,

$$F_Z = 2\pi R\cos\theta h\sigma_{11} + \pi(R_f^2 - R^2)\Delta p \tag{6.118}$$

For convenience we define the following dimensionless parameters,

- Draw down ratio: $D_R = \dfrac{U_f}{U_0}$
- Dimensionless pressure: $B = \dfrac{\pi R_0^3 \Delta p}{\mu Q}$
- Blow-up ratio: $BUR = \dfrac{R_f}{R_0}$
- Dimensionless stress: $T = \dfrac{R_0 F_Z}{\mu Q} - B(BUR)^2$
- Dimensionless take-up force: $\hat{F} = \dfrac{R_0 F_Z}{\mu Q}$
- Dimensionless radius: $\hat{R} = R/R_0$
- Dimensionless axial direction: $\hat{z} = z/R_0$
- Dimensionless thickness: $\hat{h} = h/R_0$
- Thickness ratio: $h_0/h_f = D_r BUR$

Using these dimensionless parameters, defining $d\hat{R}/d\hat{z} = \tan\theta$, and combining the above equations yields two dimensionless differential equations

$$2\hat{R}^2(T + \hat{R}^2 B)\frac{d^2\hat{R}}{d\hat{z}^2} = 6\frac{d\hat{R}}{d\hat{z}} + \hat{R}\left(1 + \left[\frac{d\hat{R}}{d\hat{z}}\right]^2\right)\left(T - 3\hat{R}^2 B\right) \tag{6.119}$$

and

$$\frac{1}{\hat{h}}\frac{d\hat{h}}{d\hat{z}} = -\frac{1}{2\hat{R}}\frac{d\hat{R}}{d\hat{z}} - \frac{\left(1 + \left[\frac{d\hat{R}}{d\hat{z}}\right]^2\right)\left(T + \hat{R}^2 B\right)}{4} \tag{6.120}$$

As boundary conditions we specify that $\hat{R} = 1$ at $\hat{z} = 0$, $d\hat{R}/d\hat{z} = 0$ at $\hat{z} = Z/R_0$ and $\hat{h} = h_0/R_0$ at $\hat{z} = 0$.

Since T depends on BUR, we must first specify BUR and iterate until a solution of $\hat{R}(\hat{z})$ is found that agrees with the choice of BUR. Hence, we must integrate eqn. (6.119) numerically with each choice of BUR. After the correct value of BUR has been found, we numerically integrate eqn. (6.120). Figures 6.23 and 6.24 present solutions for a fixed value of $\hat{Z} = 20$ and a fixed value of $B = 0.1$, respectively.

■ **EXAMPLE 6.5.**

Film blowing. A tubular 50 μm thick low density polyethylene film is blown with a draw ratio of 5 at a flow rate of 50 g/s. The annular die has a diameter of 15 mm and a die gap of 1 mm. Calculate the required pressure inside the bubble and draw force to pull the bubble. Assume a Newtonian viscosity of 800 Pa-s, a density of 920 kg/m^3 and a freeze line at 300 mm.

Since we know the thickness reduction and the draw ratio of the film, we can compute the blow-up ratio,

$$BUR = (h_0/h_f)/D_R = (1000\ \mu\text{m}/50\ \mu\text{m})/5 = 4 \tag{6.121}$$

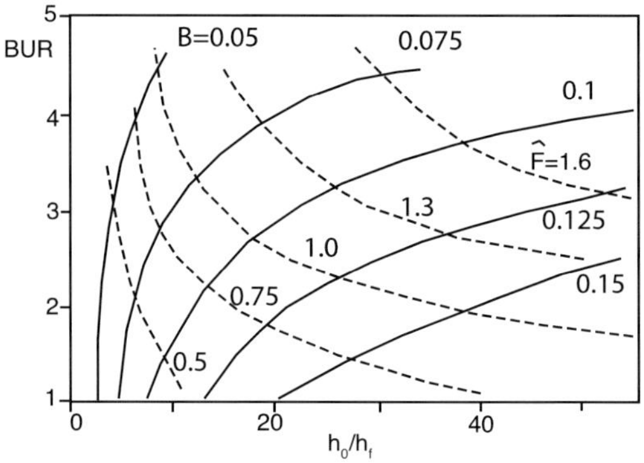

Figure 6.23: Predicted film blowing process using an isothermal Newtonian model for a dimensionless freezing line at $\hat{Z} = 20$ [21].

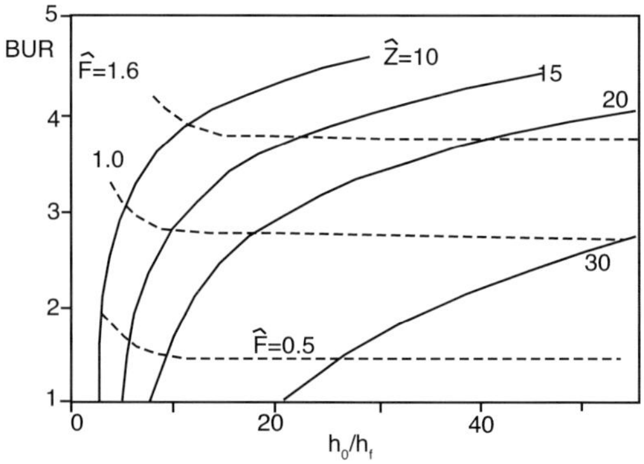

Figure 6.24: Predicted film blowing process using an isothermal Newtonian model for a dimensionless pressure $B = 0.1$ [21].

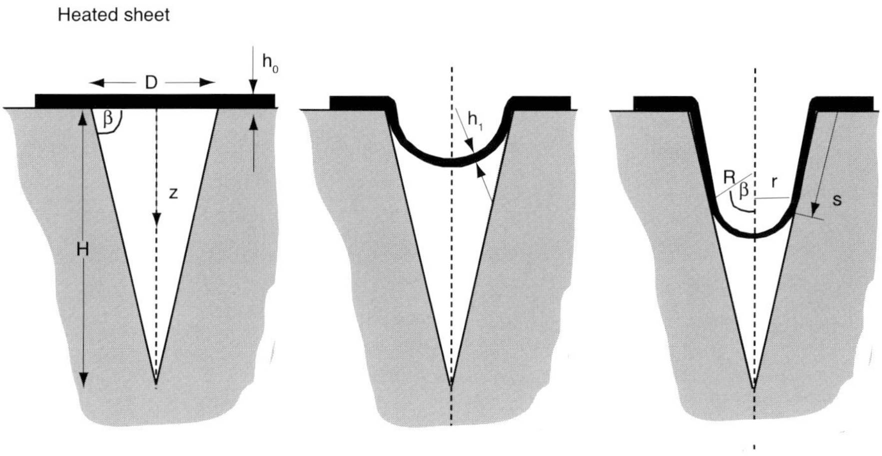

Figure 6.25: Schematic diagram of the thermoforming process of a conical geometry.

Next, we can compute the dimensionless freeze line using,

$$\hat{Z} = Z/R_0 = 300 \text{ mm}/15 \text{ mm} = 20 \tag{6.122}$$

which allows us to use Fig. 6.23 to get $B = 0.075$ and $\hat{F} = 1.3$, which results in $\Delta p = 307$ Pa and $F_z = 3.77$ N.

6.3.2 Thermoforming

A simple approximation of the thermoforming process is based on a mass balance principle. To illustrate this concept, let us consider the thermoforming process of a conical object, as schematically depicted in Fig. 6.25.

For the solution, we assume the notation presented in in Fig. 6.25. As shown in the figure, at an arbitrary point in time the bubble will contact the mold at a height z and will have a radius R, which is determined by the mold geometry

$$R = \frac{H - \sin\beta}{\sin\beta\tan\beta} \tag{6.123}$$

where H is the depth of the cone, s the contact point along the cone's wall and β the angle described in Fig. 6.25. The surface area of the cone at that point in time is given by

$$A = 2\pi R^2(1 + \cos\beta) \tag{6.124}$$

If we perform a mass balance as the bubble advances a distance Δs, we get

$$2\pi R(1 + \cos\beta)h|_s - 2\pi R^2(1 + \cos\beta)h|_{s+\Delta s} = 2\pi r h(s)\Delta s) \tag{6.125}$$

with $r = R\sin\beta$, the above equation results in

$$-\frac{d}{ds}(R^2 h(s)) = \frac{Rh(s)\sin\beta}{1 + \cos\beta} \tag{6.126}$$

differentiating eqn. (6.123), we get

$$\frac{dR}{ds} = \frac{1}{\tan\beta} \tag{6.127}$$

We can combine the above equations to get

$$\frac{dh}{h(s)} = \left(2 - \frac{\tan\beta\sin\beta}{1+\cos\beta}\right)\sin\beta\frac{ds}{H - s\sin\beta} \tag{6.128}$$

which can be integrated using $h(0) = h_1$ as a boundary condition,

$$\frac{h}{h_1} = \left(1 - \frac{s}{H}\sin\beta\right)^{\sec\beta - 1} \tag{6.129}$$

where h_1 is the thickeness of the bubble when it first makes contact with the cone wall. The initial thickness of the sheet, h_0, can be related to h_1 using

$$\frac{\pi D^2 h_0}{4} = \frac{\pi D^2(1+\cos\beta)}{2\sin^2\beta}h_1 \tag{6.130}$$

Finally, we can write the thickness distribution using

$$\frac{h}{h_0} = \frac{1+\cos\beta}{2}\left(1 - \frac{s}{H}\sin\beta\right)^{\sec\beta - 1} \tag{6.131}$$

This equation can be extended to simulate the thermoforming process of a truncated cone, which is a more realistic geometry encountered in the thermoforming industry.

6.4 CALENDERING – ISOTHERMAL FLOW PROBLEMS

As discussed in Chapter 3, the calendering process is used to squeeze a mass of polymeric material through a set of high-precision rolls to form a sheet or film. In this section, we will derive the well known model developed by Gaskell [11] and by McKelvey [15]. For the derivation, let us consider the notation and set-up presented in Fig. 6.26.

6.4.1 Newtonian Model of Calendering

In Gaskell's treatment, a Newtonian flow was assumed with a very small gap-to-radius ratio, $h \ll R$. This assumption allows us to assume the well known lubrication approximation with only velocity components $u_x(y)$. In addition, Gaskell's model assumes that a very large bank of melt exists in the feed side of the calender. The continuity equation and momentum balance reduce to

$$\frac{du_x}{dx} = 0 \tag{6.132}$$

and

$$\frac{\partial p}{\partial x} = \mu\frac{\partial^2 u_x}{\partial y^2} \tag{6.133}$$

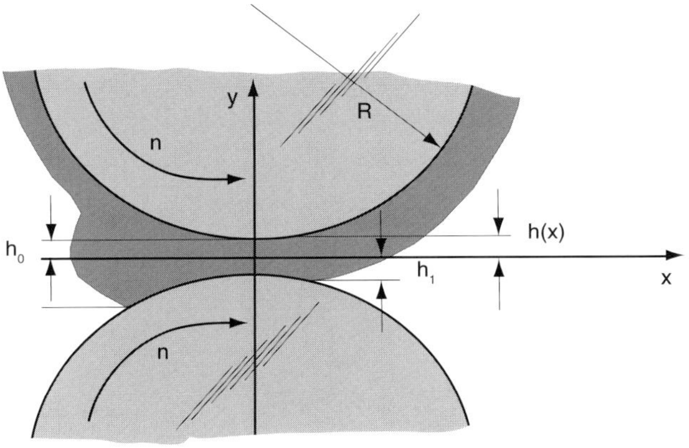

Figure 6.26: Schematic diagram of a two roll calendering system in the nip region.

respectively. Integrating eqn. (6.133) twice with the boundary conditions $u_x = U$ at $y = h(x)$ and $\frac{\partial u_x}{\partial y} = 0$ at $y = 0$, results in

$$u_x = U + \left[\frac{y^2 - h^2(x)}{2\mu}\right]\frac{dp}{dx} \tag{6.134}$$

where $U = 2\pi nR$ is the speed on the roll surface. Using the velocity profile we can compute the flow rate per unit width as

$$q = 2\int_0^h u_x dy = 2h\left[U - \frac{h^2}{3\mu}\frac{dp}{dx}\right] \tag{6.135}$$

which will not vary with x. The pressure distribution is unknown and will be solved for next. In order to do this, we require that the velocity at the outlet be uniform and equal to the roll surface speed, $u_x(y) = U$. A uniform velocity implies no shear stress, $\tau_{yx} = 0$, which means that the pressure gradient should also be zero at that point. Hence, at that position the flow rate can be expressed as

$$q = 2h_1 U \tag{6.136}$$

We can combine eqns. (6.135) and (6.136) to give

$$\frac{\partial p}{\partial x} = \frac{3\mu U}{h_1^2}\left(1 - \frac{h_1}{h}\right)\left(\frac{h_1}{h}\right)^2 \tag{6.137}$$

This equation implies that the pressure gradient vanishes at $x = x_1$ as well as at $x = -x_1$, at which point, as will be shown later, the pressure is at a maximum.

The half-gap between the rolls is defined by

$$h = h_0 + R - \sqrt{R^2 - x^2}, \tag{6.138}$$

but since we can assume that $x \ll R$, the term $\sqrt{R^2 - x^2}$ can be approximated using the first two terms of the binomial series. This results in

$$\frac{h}{h_0} = 1 + \xi^2 \tag{6.139}$$

where, $\xi^2 = \dfrac{x^2}{2Rh_0}$. Now we can integrate eqn. (6.137) to give

$$p = \frac{3\mu U}{4h_0}\sqrt{\frac{R}{2h_0}}\left(\left[\frac{\xi^2 - 1 - 5\lambda^2 - 3\lambda^2\xi^2}{(1+\xi^2)^2}\right]\xi + (1-3\lambda^2)\tan^{-1}\xi + C(\lambda)\right) \tag{6.140}$$

where, $\lambda = \dfrac{x_1^2}{2Rh_0}$ and $C(\lambda)$ is obtained by letting $p = 0$ at $\xi = \lambda$

$$C(\lambda) = \frac{(1+3\lambda^2)}{(1+\lambda^2)}\lambda - (1-3\lambda^2)\tan^{-1}\lambda \tag{6.141}$$

McKelvey [15] approximated $C(\lambda) \approx 5\lambda^3$. The maximum pressure occurs at $x = -x_1$ $(\xi = \lambda)$

$$p_{max} = \frac{3\mu U}{4h_0}\sqrt{\frac{R}{2h_0}}[2C(\lambda)] \approx \frac{15\mu U\lambda^3}{2h_0}\sqrt{\frac{R}{2h_0}} \tag{6.142}$$

Figure 6.27 presents a dimensionless pressure, p/p_{max}, as a function of dimensionless x-direction, ξ, for various values of λ, computed using the above equations. Figure 6.28 compares experimental pressure measurements to a curve computed using Gaskell's Newtonian model [11]. The two curves were matched by choosing the best value of λ to match the position of maximum pressure. The predicted pressure is very accurate at values of ξ larger than $-\lambda$ but is not very good before $-\lambda$. Although a shear thinning model would help achieve a better match between experiments and prediction [13], still within that region the accuracy of the models remain poor. Another aspect that should be pointed out at this point is that the maximum pressure, p_{max}, is very sensitive to λ, e.g., doubling λ increases p_{max} 8 times. It is reasonable to assume that $p \rightarrow 0$ when $\xi \rightarrow -\infty$, which means that λ must have a specific value, namely, $\lambda = 0.475$.

It should be noted that the pressure distribution goes to zero in the up-stream position ξ_2, where the material makes contact with both rolls. This position can be determined for any value of λ by letting the pressure in eqn. (6.140) go to zero. Figure 6.29 presents a graph of position of first contact, ξ_2, and the position where the sheet separates from the rolls, λ.

With the above equations, the velocity distribution between the rolls becomes

$$\hat{u}_x = 1 + \frac{3}{2}\frac{(1-\eta^2)(\lambda^2 - \xi^2)}{(1+\xi^2)} \tag{6.143}$$

where, $\hat{u}_x = u_x/U$ and $\eta = y/h$. Equation (6.143) can be used to determine that a stagnation point, $u_x(0) = 0$, exists at a position $\xi_s = -\sqrt{2+3\lambda^2}$. Figure 6.30 presents the flow pattern that develops in the nip region as predicted by eqn. (6.143). As can be seen, a recirculation pattern develops due to the backflow caused by the pressure build-up as the polymer is forced through the nip region. The calendering process was modeled in 2D using the RBFCM in Chapter 11 of this book for a Newtonian as well as power law viscosity models. Comparison with the analytical solutions reveal that the lubrication approximation does an excellent job when modeling the process.

We will calculate the power consumption as well as predict the temperature rise within the material due to viscous heating. In order to compute the power consumption, we need

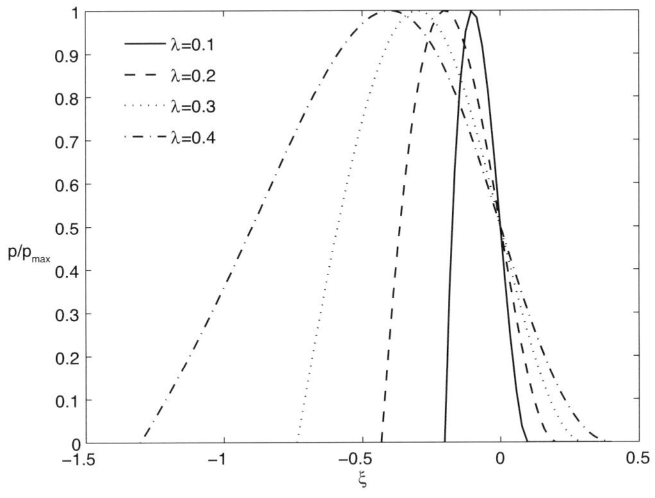

Figure 6.27: Computed pressure distribution between the rolls for various values of λ.

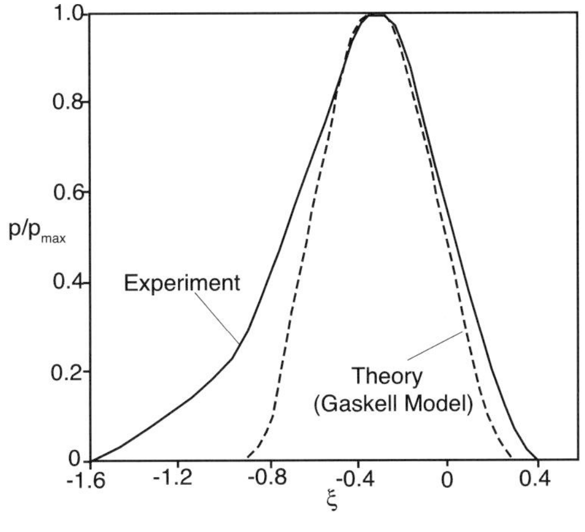

Figure 6.28: Comparison of theoretical and experimental pressure profiles [15]. The experiments were performed by Bergen and Scott [2] with roll diameters of 10 in, a gap in the nip region of 0.025 in and a speed U =5 in/s. The measured viscosity was 3.2×10^9 P (Poise, 1 P=0.1 Pa-s).

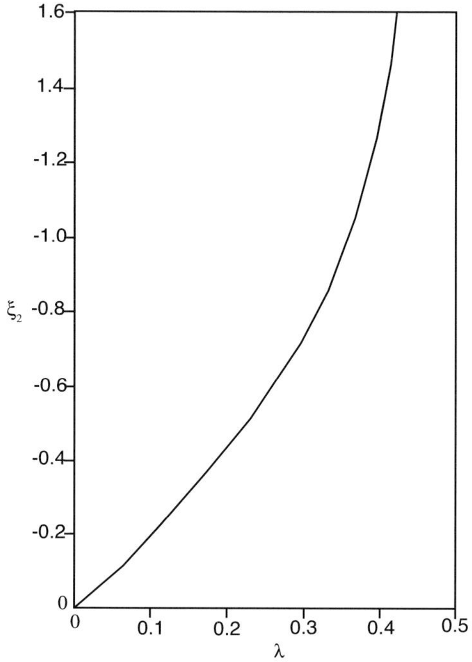

Figure 6.29: Relation between the position of first contact, ξ_2, and the position of sheet separation, λ.

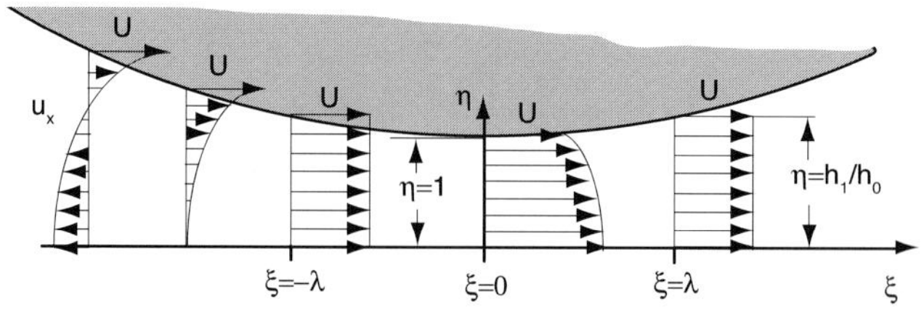

Figure 6.30: Flow pattern that develops in the nip region of a two-roll calendering system.

to integrate the product of the shear stress and the roll surface speed over the surface of the roll. The rate of deformation can be computed using eqn. (6.143) as

$$\dot{\gamma}_{yx}(\eta) = \frac{3U(\xi^2 - \lambda^2)}{h_0(1-\xi^2)^2}\eta \tag{6.144}$$

and the stress

$$\tau_{yx}(\eta) = \mu\frac{3U(\xi^2 - \lambda^2)}{h_0(1-\xi^2)^2}\eta \tag{6.145}$$

The maximum rate of deformation and shear stress occur at the roll surface, $\eta = 1$, at $\xi = 0$ where the gap is smallest

$$\dot{\gamma}_{max}(\eta) = \frac{3U\lambda^2}{h_0} \tag{6.146}$$

and

$$\tau_{max}(\eta) = \mu\frac{3U\lambda^2}{h_0}, \tag{6.147}$$

respectively. However, the overall maximum of stress and rate of deformation occurs at $\xi = \xi_2$ when $\xi_2 > -\sqrt{1+2\lambda^2}$, and at $\xi = \sqrt{1+2\lambda^2}$, if $\xi_2 < \sqrt{1+2\lambda^2}$. We can compute the overall power requirement by integrating $U\tau_{yx}$ along the surface of the roll, $\eta = 1$

$$P = 3\mu W U^2\sqrt{\frac{2R}{h_0}}\mathcal{F}(\lambda) \tag{6.148}$$

where W is the width of the rolls and

$$\mathcal{F}(\lambda) = (1-\lambda^2)[\tan^{-1}\lambda - \tan^{-1}\xi_{max}] - \left[\frac{(\lambda - \xi_{max})(1-\xi_{max}\lambda)}{(1-\xi_{max}^2)}\right] \tag{6.149}$$

Of importance to the mechanical design of the calendering system and to the prediction of the film thickness uniformity is the force separating the two rolls, F. This is computed by integrating the pressure over the area of interest on the surface of the roll

$$F = \frac{3\mu URW}{4h_0}\mathcal{G}(\lambda) \tag{6.150}$$

where $\mathcal{G}$ is given by

$$\mathcal{G}(\lambda) = \left(\frac{\lambda - \xi_2}{1-\xi_2^2}\right)[-\xi_2 - \lambda - 5\lambda^3(1+\xi_2^2)] + (1-2\lambda^2)(\lambda\tan^{-1}\lambda - \xi_2\tan^{-1}\xi_2) \tag{6.151}$$

Both functions $\mathcal{F}(\lambda)$ and $\mathcal{G}(\lambda)$ are shown in Fig. 6.31.

Finally, if from an adiabatic energy balance we assume that the power goes into heat generation, we can estimate the temperature rise within the material to be

$$\Delta T = \frac{P}{\rho Q W C_p} \tag{6.152}$$

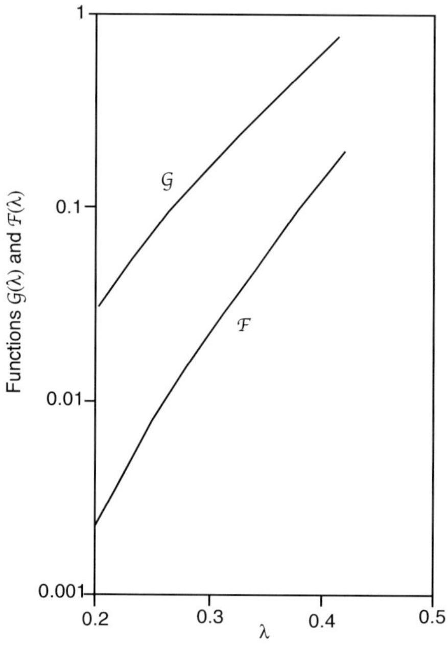

Figure 6.31: Power and force functions $\mathcal{F}(\lambda)$ and $\mathcal{G}(\lambda)$ used in eqns. (6.149) and (6.151).

■ **EXAMPLE 6.6.**

Calendering problem with a Newtonian viscosity polymer. A calender system with $R = 10$ cm, $w = 100$ cm, $h_0 = 0.1$ mm operates at a speed of $U = 40$ cm/s and produces a sheet thickeness $h_1 = 0.0218$ cm. The viscosity of the material is given as 1000 Pa-s. Estimate the maximum pressure developed in the material, the power required to operate the system, the roll separating force and the adiabatic temperature rise within the material.

Since the final sheet thickness is given we can compute λ using eqn. (6.139) as

$$\frac{h_1}{h_0} = 1 + \lambda^2 \tag{6.153}$$

resulting in λ =0.3. Equation (6.142) becomes

$$p_{max} \approx \frac{15(1000 \text{ Pa-s})(0.40 \text{ m/s})(0.3)^3}{0.0001 \text{ m}} \sqrt{\frac{0.1 \text{ m}}{0.0001 \text{ m}}} = 18.1 \text{ MPa} \tag{6.154}$$

The power is computed using eqn. (6.148) with $\mathcal{F}(0.3)$ =0.043

$$P = 3(1000 \text{ Pa-s})(0.4 \text{ m/s})^2(1 \text{ m})\sqrt{\frac{2(0.1 \text{ m})}{(0.0001 \text{ m})}}\mathcal{F}(0.3) = 923 \text{ W} \tag{6.155}$$

The separating force is computed using eqn. (6.150) with $\mathcal{G}(0.3)$=0.16

$$F = \frac{3(1000 \text{ Pa})(0.4 \text{ m/s})(0.1 \text{ m})(1 \text{ m})}{4(0.0001 \text{ m})}\mathcal{G}(0.3) = 48 \text{ kN} \tag{6.156}$$

Using a volumetric flow rate of $Q = 2Uh_1W$ =8.72×10^5 m^3/s, we can use eqn. (6.152) with a typical specific heat of 1000 J/kg/K and density of 1000 kg/m^3 to compute the adiabatic temperature rise,

$$\Delta T = \frac{923 \text{ W}}{(1000 \text{ kg/m}^3)(8.72 \times 10^5 \text{ m}^3/s)(1 \text{ m})(1500 \text{ J/kg/K})} = 7 \text{ K}. \tag{6.157}$$

EXAMPLE 6.7.

Calendering problem with *floating* roll. In a set of calendering rolls, weighing 500 kg each, the upper roll rests on top of the calendered polymer. The calender dimensions are R =0.15 m and W =2.0 m. For a material with a Newtonian viscosity of 1000 Pa-s and a speed of 0.1 m/s, what is the final sheet thickness?

To solve this problem, we begin with eqn. (6.150) and substitute $\mathcal{G}(\lambda)$ with a value of $\lambda = 0.475$

$$F = 1.23\frac{3\mu URW}{4h_0} \tag{6.158}$$

We can solve for h_0 with values of $F = 500 \times 9.81$ N, $U = 0.1$ m/s, R =0.15 m, W =2.0 m and μ =1000 Pa-s

$$h_0 = 1.23\frac{3\mu URW}{4F} = 0.0022 \text{ m (2.2 mm or 85.8 mils)} \tag{6.159}$$

6.4.2 Shear Thinning Model of Calendering

As with the Newtonian model, we assume a lubrication approximation, where the momentum balance reduces to

$$\frac{\partial p}{\partial x} = \frac{\partial \tau_{xy}}{\partial y} \tag{6.160}$$

If we assume a power law model, the shear stress τ_{xy} can be written as

$$\tau_{xy} = m\left|\frac{\partial u_x}{\partial y}\right|^{n-1}\frac{\partial u_x}{\partial y} \tag{6.161}$$

The absolute value in eqn. (6.161) is to avoid taking the root of a negative number. From the Newtonian solution we can see that there are two regions, one where the velocity gradient is positive, $\xi < \lambda$, and one where the velocity gradient is negative, $\xi > -\lambda$. In each region, the above equation must be integrated separately, resulting in two velocity distributions

$$u_x = U + \frac{1}{n/(1+n)}\left(\frac{1}{m}\frac{dp}{dx}\right)^{1/n}\left[y^{n/(1+n)} - h^{n/(1+n)}(x)\right] \tag{6.162}$$

for the region with the negative velocity gradient, and

$$u_x = U - \frac{1}{n/(1+n)}\left(-\frac{1}{m}\frac{dp}{dx}\right)^{1/n}\left[y^{n/(1+n)} - h^{n/(1+n)}(x)\right] \tag{6.163}$$

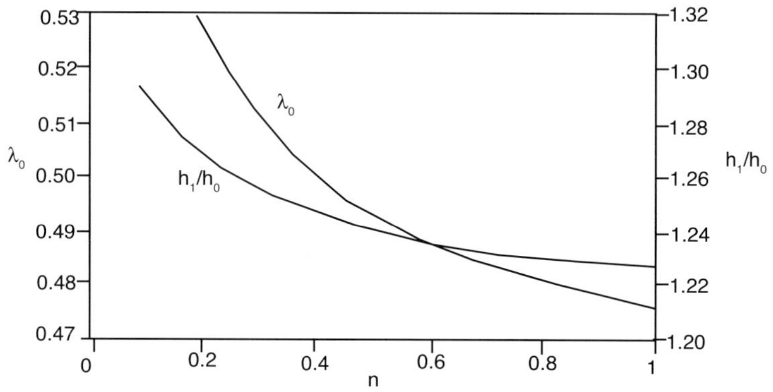

Figure 6.32: Function λ_0 and sheet thickness as a function of power law index, n.

for the region with the positive velocity gradient. Either equation can be used to solve for the pressure gradient

$$\frac{d\hat{p}}{d\xi} = -\left(\frac{2n+1}{n}\right)^n \sqrt{\frac{2R}{h_0}} \frac{(\lambda^2 - \xi^2)|\lambda^2 - \xi^2|^{n-1}}{(1+\xi^2)^{2n+1}} \tag{6.164}$$

where $\hat{p}$ is a power law dimensionless pressure defined by

$$\hat{p} = \frac{p}{m}\left(\frac{h_0}{U}\right)^n \tag{6.165}$$

Equation (6.164) can be integrated to become

$$\hat{p} = \sqrt{\frac{2R}{h_0}}\left(\frac{2n+1}{n}\right)^n \int_{-\lambda_0}^{\lambda_0} \frac{(\lambda_0^2 - \xi^2)^n}{(1+\xi^2)^{1+2n}} d\xi = \sqrt{\frac{2R}{h_0}} \mathcal{P}(n) \tag{6.166}$$

where λ_0 is the position where the integral vanishes

$$0 = \int_{-\infty}^{\lambda_0} \frac{(\lambda_0^2 - \xi^2)|\lambda_0^2 - \xi^2|^{n-1}}{(1+\xi^2)^{2n+1}} \tag{6.167}$$

Figure 6.32 presents λ_0 as a function of power law index, n. The figure also presents the ratio of final sheet thickness to the nip separation as a function of n.

We can also compute the roll separating force, F, and the power required to drive the system, P,

$$F = WRm\left(\frac{U}{h_0}\right)^n \mathcal{F}(n) \tag{6.168}$$

and

$$P = WU^2 m\sqrt{Rh_0}\left(\frac{U}{h_0}\right)^{n-1} \mathcal{E}(n) \tag{6.169}$$

Figure 6.33 [16] presents the functions $\mathcal{P}$, $\mathcal{F}$ and $\mathcal{E}$ as a function of the power law index.

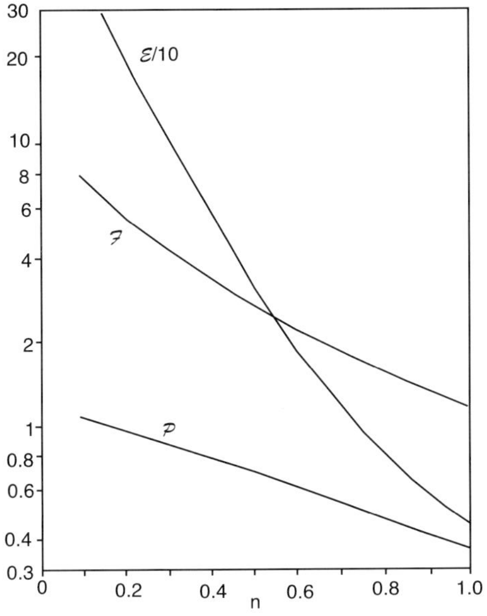

Figure 6.33: Pressure ($\mathcal{P}(n)$), force ($\mathcal{F}(n)$) and power ($\mathcal{E}(n)$) functions as a function of power law index, n.

Unkrüer [18] performed experimental studies on the flow development during the calendering process of unplasticized polyvinyl chloride to produce thin films. Among other things he compared the measured maximum pressure that develops between the rolls to the analytical value predicted by the shear thinning model presented above. Figure 6.34 shows this comparison between experiment and theory and their rather good agreement.

6.4.3 Calender Fed with a Finite Sheet Thickness

All the above problems relate to the calendering process where a large mass of polymer melt is fed into the calender. In some industrial applications, a finite polymer sheet of thickness h_f is fed to the calendering rolls, as depicted in Fig. 6.35.

To solve this problem, eqn. (6.166) is replaced by

$$\hat{p} = \sqrt{\frac{2R}{h_0}} \left(\frac{2n+1}{n}\right)^n \int_{-\xi_f}^{\lambda_0} \frac{(\lambda_0^2 - \xi^2)^n}{(1+\xi^2)^{1+2n}} d\xi = \sqrt{\frac{2R}{h_0}} \mathcal{P}(n) \qquad (6.170)$$

and eqn. (6.167) becomes

$$0 = \int_{-\xi_f}^{-\xi} \frac{(\lambda_0^2 - \xi^2)|\lambda_0^2 - \xi^2|^{n-1}}{(1+\xi^2)^{2n+1}} \qquad (6.171)$$

The position where the sheet being fed enters the system can be computed using

$$\xi_f = \sqrt{\frac{h_f}{h_0} - 1} \qquad (6.172)$$

Figure 6.36 presents a plot of final sheet thickness as a function of fed sheet thickness for a Newtonian polymer and a shear thinning polymer with a power law index of 0.25.

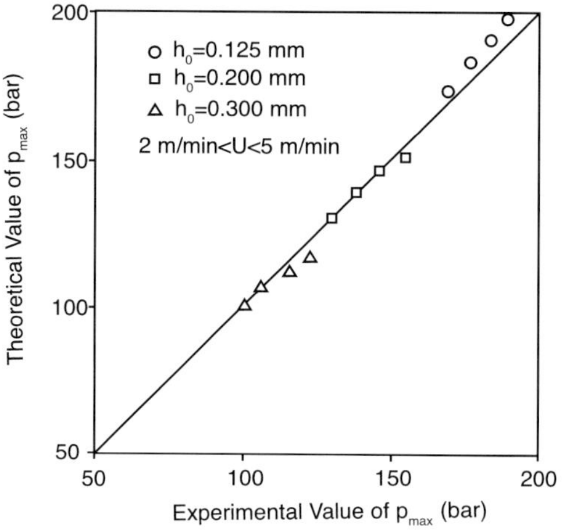

Figure 6.34: Comparison between experiments and theoretical predictions of maximum pressure between the rolls during the calendering process of an unplasticized PVC film. A power law index, n, of 0.1505 and a consistency index, m, of 155.2 kPa-s were used in the power law model of the viscosity.

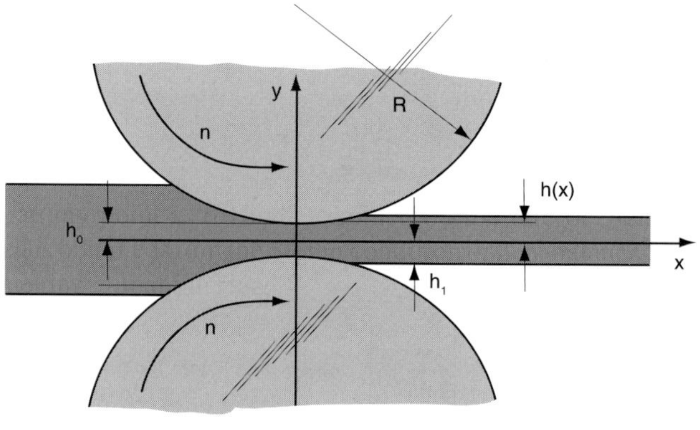

Figure 6.35: Schematic diagram of a two roll calendering system in the nip region fed with a finite sheet.

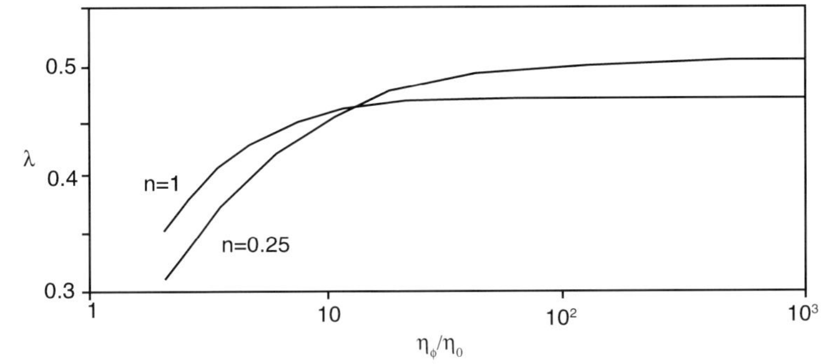

Figure 6.36: Calendered thickness as a function of fed sheet thickeness for a Newtonian fluid and a shear thinning fluid with $n = 0.25$.

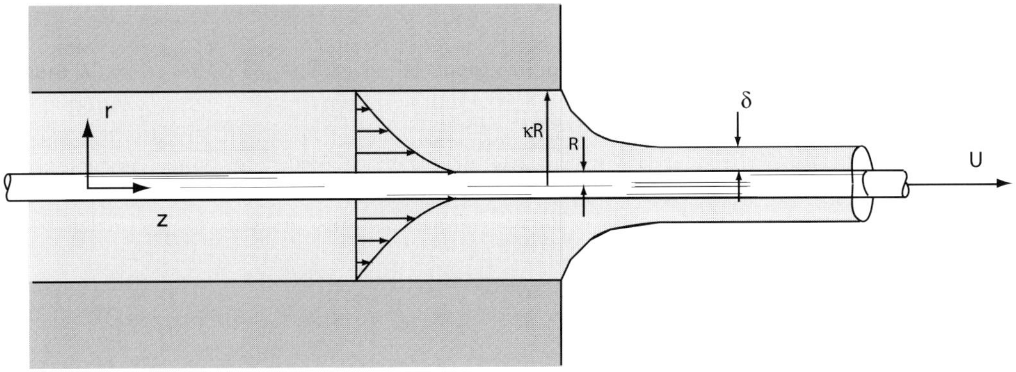

Figure 6.37: Schematic diagram of a wire coating die.

6.5 COATING PROCESSES

Various coating processes have been discussed in Chapter 3. During a coating operation a liquid is continuously applied to a surface or web, which can be a wire, paper, board, plastic films, metal films, etc. There are numerous types of coating processes and variations, for some of which analytical solutions are presented in this section.

6.5.1 Wire Coating Die

One important aspect of wire-coating is the thickness distribution of the polymer on the surface of the wire as well as the velocity distribution within the die. A simplified wire coating process is presented in the Fig. 6.37, where the wire radius is defined by R and the annulus radius by κR. This type of flow is often referred to as an *axial annular Couette flow*.

The most important assumptions when solving this problem are a steady, incompressible and isothermal flow. Let us now consider a power-law fluid, but neglect the elastic effects. Furthermore, for the solution of this specific problem, let us assume that the flow is primarily driven by drag and that there are no significant pressure drops across the die.

The unidirectional flow condition will simplify the momentum equation to

$$\frac{d}{dr}\left(r\tau_{rz}\right) = 0 \tag{6.173}$$

using the power-law model and two integrations, we obtain the velocity as

$$u_z = \left(\frac{c_1}{m}\right)^s \frac{r^{1-s}}{s-1} - c_2 \tag{6.174}$$

If we impose the boundary conditions

$$\begin{aligned} u_z\left(R\right) &= U \\ u_z\left(\kappa R\right) &= 0 \end{aligned} \tag{6.175}$$

the velocity profile will become

$$\frac{u_z\left(r\right)}{U} = \frac{\xi^{1-s} - 1}{\kappa^{1-s} - 1} \tag{6.176}$$

where $\xi = r/R$ and $s = 1/\mathrm{n}$.

In addition, from the definition of the stress tensor, using the power law model for the viscosity, τ_{rz} becomes

$$\tau_{rz} = \frac{mU^n(s-1)^n}{R^n(\kappa^{1-s}-1)^n}\left(\frac{R}{r}\right) \tag{6.177}$$

In order to relate the thickness of the coating to R and U, we must perform a mass balance on the region starting at the exit of the die and ending where the fluid has reached the same velocity as the wire. Here, we must assume that the polymer density is constant, although the melt undergoes density changes as it solidifies. The mass balance is written as

$$2\pi \int_{\kappa R}^{R} u_z(r) r dr = U\pi\left[(\kappa R + \delta)^2 - (\kappa R)^2\right] \tag{6.178}$$

where δ is the coating thickness. By substituting the velocity profile into the integral in the above equation, an equation for the coating thickness is obtained

$$\frac{2R^2}{\kappa^{1-s}-1}\left[\frac{1-\kappa^{3-s}}{3-s} - \frac{1-\kappa^2}{2}\right] = \left[(\kappa R + \delta)^2 - (\kappa R)^2\right] \tag{6.179}$$

which can be written as follows

$$\frac{2R^2\left(1-\kappa^{3-s}\right)}{\left(\kappa^{1-s}-1\right)\left(3-s\right)} - \frac{R^2\left(1-\kappa^2\right)}{\left(\kappa^{1-s}-1\right)} - 2\kappa R\delta - \delta^2 = 0 \tag{6.180}$$

This non-linear equation relates R and δ, which can be used to obtain the flow rate and the coating thickness. Finally, the force needed to pull the wire through the die obtained from integrating the stress at the polymer-die surface interface over the area of the die is defined as,

$$F = (2\pi\kappa RL)\left[\tau_{rz}\right]_{r=\kappa R} = (2\pi\kappa RL)\left[\frac{mU^n(s-1)^n}{R^n(\kappa^{1-s}-1)^n}\frac{1}{\kappa}\right] \tag{6.181}$$

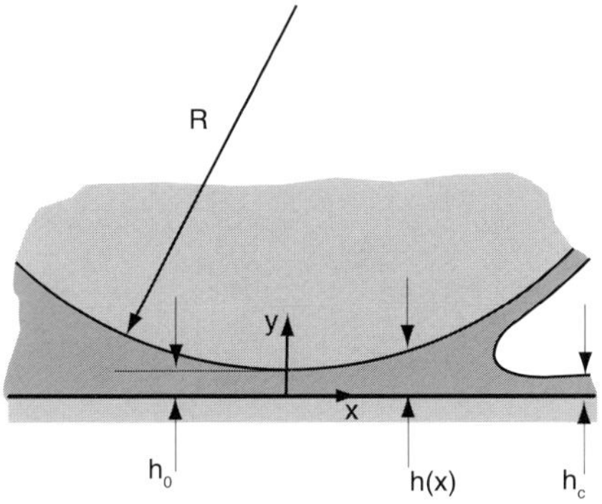

Figure 6.38: Schematic diagram of roll coating.

6.5.2 Roll Coating

In principle, roll coating is very similar to calendering, the difference being that the fluid does not separate from the roll. The fluid continues adhered to both surfaces, the substrate or web and the roll. In the analysis presented below, we follow the model presented by Middleman [16]. His derivation assumes that the coating fluid evenly wets both surfaces as schematically depicted in Fig. 6.38, as well as a Netwonian, isothermal and steady state flow.

Using these assumptions and the notation presented in Fig. 6.38 the equation of motion reduces to

$$-\frac{\partial p}{\partial x} + \mu \frac{\partial^2 u_x}{\partial y^2} = 0 \tag{6.182}$$

To generalize the solution, we introduce the dimensionless variables, $\xi = x/(Rh_0)^{1/2}$, $\eta = y/h_0$, $\hat{u}_x = u_x/u$, $\hat{p} = (ph_0^{3/2})/(\mu u R^{1/2})$. The momentum balance reduces to

$$\frac{\partial \hat{p}}{\partial \xi} = \frac{\partial^2 \hat{u}_x}{\partial \eta^2} \tag{6.183}$$

For small gaps, for which $h_0 \ll R$, the pressure is independen of η. After integrating eqn. (6.183) twice we get

$$\hat{u}_x = \frac{1}{2}\left(\frac{\partial \hat{p}}{\partial \xi}\right)\eta^2 + c_1\eta + c_2 \tag{6.184}$$

If we approximate the gap with a quadratic equation

$$\hat{h}(\xi) = h(x)/h_0 = 1 + \xi^2/2 \tag{6.185}$$

and apply the boundary conditions $\hat{u}_x = 1$ at $\eta = 0$ and at $\eta = \hat{h}(\xi)$, we get

$$\hat{u}_x = 1 + \frac{1}{2}\left(\frac{\partial \hat{p}}{\partial \xi}\right)(\eta^2 - \hat{h}(\xi)\eta) \tag{6.186}$$

At this point, the pressure and pressure gradient are unknown. However, we can perform a mass balance by using eqn. (6.186) to compute the volumetric flow rate

$$Q = W \int_0^{h(x)} u_x dy \tag{6.187}$$

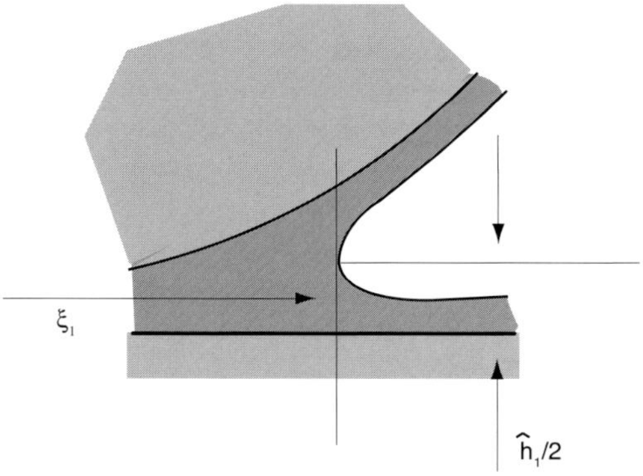

Figure 6.39: Schematic diagram of separation region.

and in dimensionless form

$$\hat{Q} = \frac{Q}{uWh_0} = \int_0^{\hat{h}(\xi)} \hat{u}_x dy \tag{6.188}$$

which can be integrated to give

$$\hat{Q} = \hat{h}(\xi) - \frac{1}{12}\left(\frac{d\hat{p}}{d\xi}\right)\hat{h}(\xi)^3 \tag{6.189}$$

The above equation can be written as

$$\frac{d\hat{p}}{d\xi} = 12\frac{\hat{h}(\xi) - \hat{Q}}{\hat{h}(\xi)^3} \tag{6.190}$$

$$\hat{p}(\xi) = 12\int_{\xi_0}^{\xi} \frac{\hat{h}(\xi) - \hat{Q}}{\hat{h}(\xi)^3} d\xi \tag{6.191}$$

where ξ_0 is the value of ξ at $\hat{p} = 0$. Since we are assuming that the liquid splits evenly to the web and the roll, we can write

$$Q = 2uh_c \tag{6.192}$$

and in dimensionless form

$$\hat{Q} = \frac{2h_c}{h_0} \tag{6.193}$$

At the separation point located at $\xi = \xi_1$ and $\eta = \hat{h}(\xi_1)/2$, depicted in Fig. 6.39, the velocity goes to zero.

Hence, after combining eqns. (6.186) and (6.190), we can write

$$\hat{u}_x = 1 - \frac{3}{2}\frac{\hat{h}(\xi) - \hat{Q}}{\hat{h}(\xi)} = 0 \tag{6.194}$$

which can be solved to give

$$\hat{h_1} = 3\hat{Q} \tag{6.195}$$

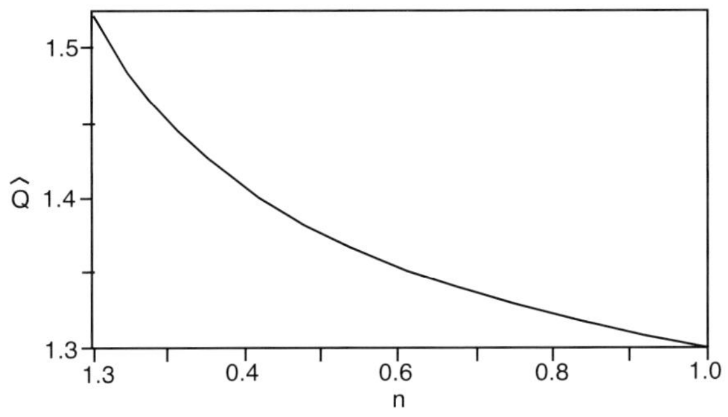

Figure 6.40: Dimensionless coating thickness as a function of power law index.

If we integrate eqn. (6.191), assuming that $\xi_0 = -\infty$, we get

$$\hat{p}(\xi) = \frac{(6 - (9/2)\hat{Q})\xi}{1 + \xi^2/2} - \frac{3\hat{Q}\xi}{(1 + \xi^2/2)^2} + \left(\frac{12}{\sqrt{2}} - \frac{9\hat{Q}}{\sqrt{2}}\right)\tan^{-1}\left(\frac{\xi}{\sqrt{2}}\right) + \frac{6\pi}{\sqrt{2}}\left(1 - \frac{3\hat{Q}}{4}\right) \tag{6.196}$$

Middleman suggests the simplest model for the separation region based on the assumption that the fluid splits at a point where $\hat{u}_x = 0$ and $\hat{p} = 0$. The above equation can be solved numerically to give

$$\hat{Q} = 1.30 \tag{6.197}$$

For a power law solution the equation, of motion is non-dimensionalized in a similar manner as with the Newtonian solution, except that dimensionless pressure is defined as

$$\hat{p} = \left(\frac{h_0}{u}\right)^n \left(\frac{h_0}{R}\right)^{1/2} \frac{p}{m} \tag{6.198}$$

When using the power law model, the pressure distribution cannot be evaluated explicitly, and is left in integral form as

$$\hat{p} = 2^{1+n}\left(\frac{1 + 2n}{n}\right)^n \int_{-\infty}^{\xi} \frac{(\hat{h}(\xi) - \hat{Q})^{n-1}(\hat{h}(\xi) - \hat{Q})}{\hat{h}(\xi)^{1+2n}} d\xi \tag{6.199}$$

Similar to the Newtonian solution, $\hat{Q}$ is obtained by assuming that at the separation point the velocity goes to zero. If we let $\hat{u}_x(\xi_1, \hat{h}(\xi_1)/2) = 0$

$$\xi_1 = \left[2\left(\frac{2n + 1}{n}\hat{Q} - 1\right)\right]^{1/2} \tag{6.200}$$

We can solve for $\hat{Q}$ by trial and error after setting the limit in eqn. (6.199) to ξ_1. Figure 6.40 presents the value for $\hat{Q}$ as a function of power law index. As can be seen, the shear thinning effect is not very large.

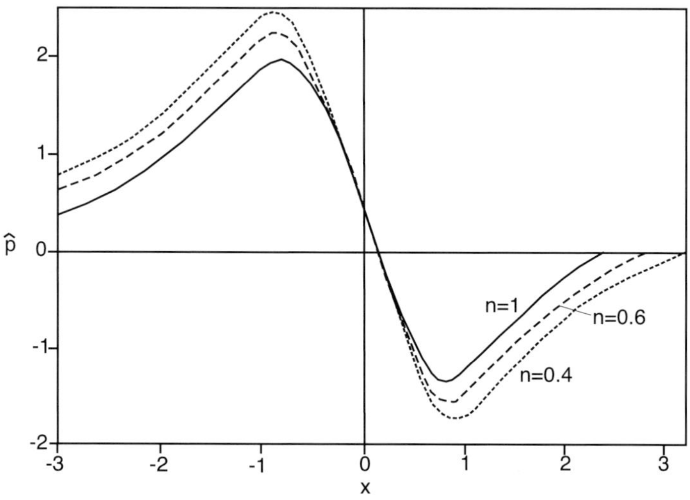

Figure 6.41: Dimensionless pressure distribution for various power law indices.

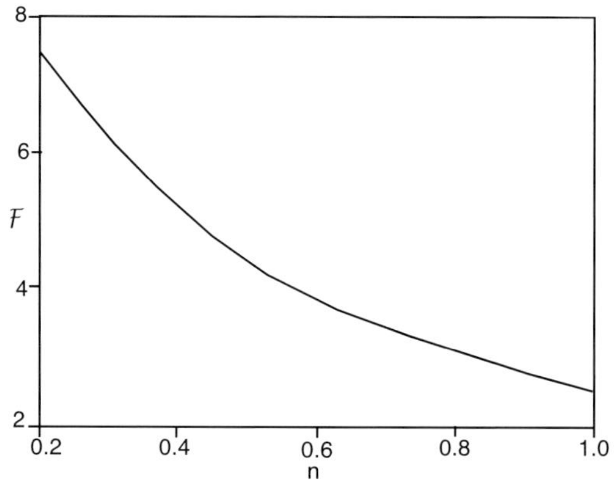

Figure 6.42: Dimensionless force as a function of power law index.

The pressure profile can now be solved for and is presented in dimensionless form for various power law indices in Fig. 6.41.

From the pressure distribution we can now compute the roll separating force using

$$F/W = mR\left(\frac{u}{h_0}\right)^n \mathcal{F}(n) \tag{6.201}$$

where $\mathcal{F}(n)$ is presented in Fig. 6.42.

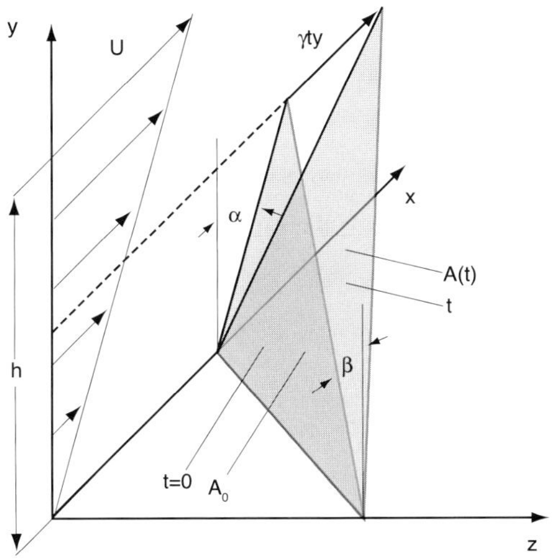

Figure 6.43: Change in interfacial area after applying a shear strain γ.

6.6 MIXING — ISOTHERMAL FLOW PROBLEMS

The basic concepts of mixing have already been discussed in Chapter 3 of this book. This section presents various examples and models that show how a mixing process can be analyzed and assessed.

6.6.1 Effect of Orientation on Distributive Mixing — Erwin's Ideal Mixer

Erwin [8] developed the theoretical background to assess the effect orientation has on the distributive mixing in single screw extruders. His starting point was the equation that relates the growth of an interface between fluids undergoing shear flow (Fig. 6.43), and described by

$$\frac{A}{A_0} = \sqrt{1 - 2\gamma \cos\alpha \cos\beta + \cos^2\alpha\gamma^2} \tag{6.202}$$

For a simple geometry ($\beta = 0$) undergoing simple shear flow, the above equation simplifies to

$$\frac{A}{A_0} = \gamma \cos\alpha \tag{6.203}$$

Let us consider the best case scenario, where the initial angle $\alpha = 0$. During shear, the surface is reoriented and will eventually approach 90°; hence, the mixing becomes very ineffective as the fluid interfaces are deformed. However, if during the flow, where the total shear, γ, is imposed upon the surfaces, the flow is interrupted in the middle so that the surface can be reoriented back to the initial position, we get

$$\frac{A_1}{A_0} = \frac{\gamma}{2} \tag{6.204}$$

of area increase for the first half and

$$\frac{A_2}{A_1} = \frac{\gamma}{2} \tag{6.205}$$

for the second half. Therefore, with one interruption used to reorient the surface we get

$$\frac{A_1}{A_0}\frac{A_2}{A_1} = \frac{\gamma}{2}\frac{\gamma}{2} = \left(\frac{\gamma}{2}\right)^2 \tag{6.206}$$

If the flow is interrupted twice, dividing the total shear into 3 sections we get

$$\frac{A_3}{A_0} = \left(\frac{\gamma}{3}\right)^3 \tag{6.207}$$

Hence, in general we can write

$$\frac{A}{A_0} = \left(\frac{\gamma}{N}\right)^N \tag{6.208}$$

where N is the number of interruptions plus 1. Each interruption could be a mixingmixing,interru section in a single screw extruder, such as pins or rhomboids, which attempt to randomnize surfaces that have been oriented by the shear imposed by the channel flow.

Erwin [9] and Ng [17] demonstrated this in an experimental study that involved placing black and white polyethylene blocks in a Couette device (Fig. 6.44a). Figure 6.44b shows that after applying a small amount of shear, the surfaces that were originally oriented in the radial direction have stretched a certain amount and have changed their orientation. It is clear from the photograph that the same surface tends to align with the planes of shear, reducing the mixing efficiency of the process. Hence, in order to increase the effectiveness of the mixer, Erwin [10] and Ng [17] took the Couette content and cut it into new blocks, that were placed inside the Couette device rotated by 90^o(Fig. 6.44c). This changed the orientation of the surfaces back to a position where they can more effectively feel the effects of deformation. They repeated this procedure several times, by performing 1,2, and 3 interruptions. Figure 6.45 compares the measured values of surface increase to eqn. (6.208).

Using this concept, Erwin [9] demonstrated that the upper bound for the ideal mixer is found in a mixer that applies a plane strain extensional flow or pure shear flow to the fluid and where the surfaces are maintained ideally oriented during the whole process; this occurs when $N = \infty$ and each time an infinitesimal amount of shear is applied. In such a system the growth of the interfacial areas follows the relation given by

$$\frac{A}{A_0} = e^{\gamma/2} \tag{6.209}$$

In Erwin's ideal mixer, the amount of mixing increases in an exponential fashion, compared to a linear increase, if the orientation of the fluids' interfaces remain undisturbed.

6.6.2 Predicting the Striation Thickness in a Couette Flow System – Shear Thinning Model

For this problem, we will consider a Couette device, schematically represented in Fig. 6.46 in which the inner cylinder rotates at a speed Ω and with a secondary component made up of a black tracer line with an initial thickness δ_0. As the tracer line is deformed, it spirals

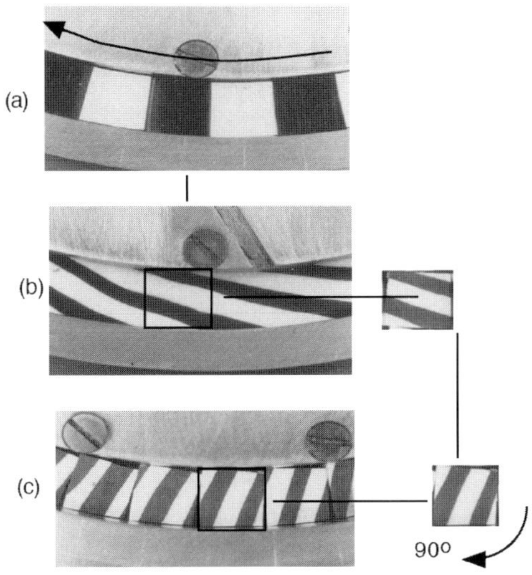

Figure 6.44: Couette flow and reorientation scheme, after Ng [17].

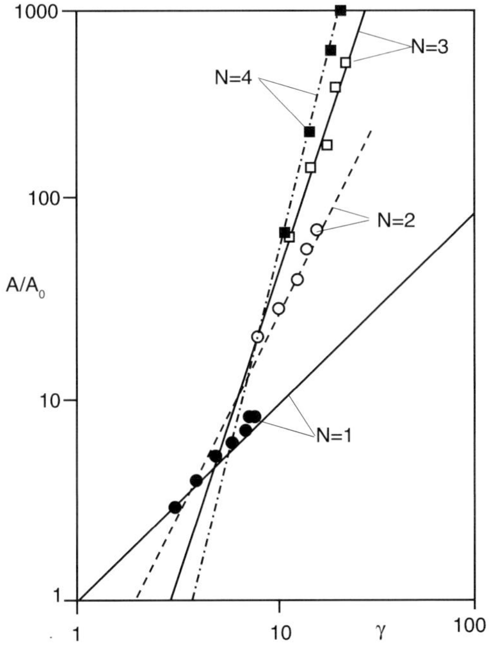

Figure 6.45: Increase in interfacial area as function of shear and number of interruptions.

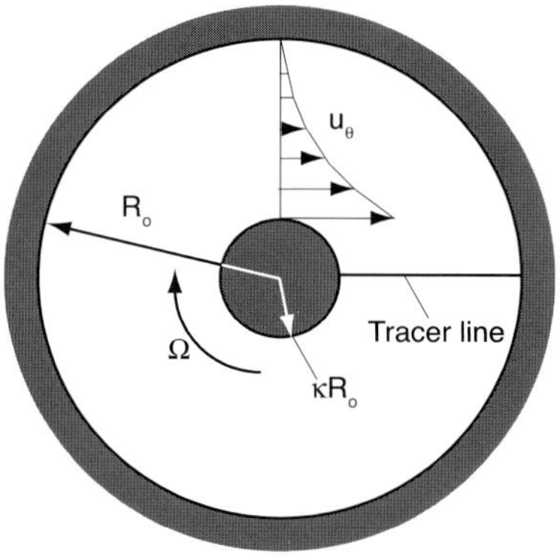

Figure 6.46: Schematic diagram of a Couette flow mixer with a tracer line.

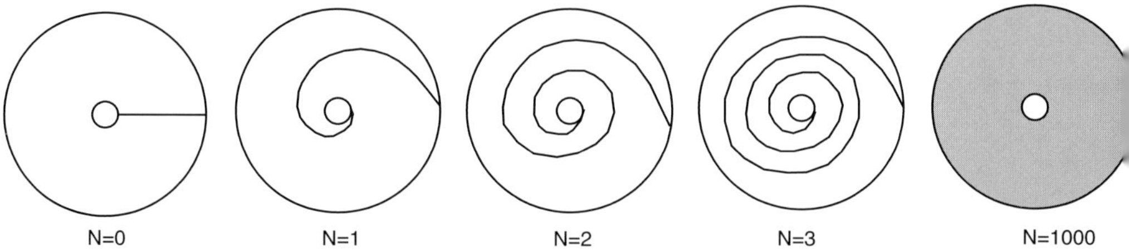

Figure 6.47: Non-diffusive mixing progress within a Couette mixing device (after McKelvey [15]).

around the inner cylinder as it gradually reduces its thickness ($\delta(t)$), as depicted for 1,2,3 and 1000 revolutions of the inner cylinder in Fig. 6.47 [15]. The thickness of the secondary component is referred to as the striation thickness. Eventually, the striation thickness is so small that the naked eye can no longer distinguish it. Approximately at this point we consider the mixing process to be complete. Here, we will derive the equations that will allow us to predict the striation thickness of the tracer line.

Using the notation described in Fig. 6.46 and the solution presented for Couette flow in Chapter 4, we write

$$u_\theta = r\Omega \frac{1-(R_o/r)^{2/n}}{1-\kappa^{-2/n}} \tag{6.210}$$

Since $u_\theta = r(d\theta/dt)$, we can integrate eqn. (6.210) to give

$$\theta = \Omega \frac{1-(R_o/r)^{2/n}}{1-\kappa^{-2/n}} t \tag{6.211}$$

Assuming $(r+dr)^{2/n} - r^{2/n} \approx (2/n)r^{(2-n)/n}dr$, after one revolution, $t = 2\pi/\Omega$, a distance dr on the tracer line has grown to

$$d\theta = \frac{4\pi}{n(\kappa^{-2/n}-1)} \frac{R_o^{2/n}}{r^{(n+2)/n}} dr \tag{6.212}$$

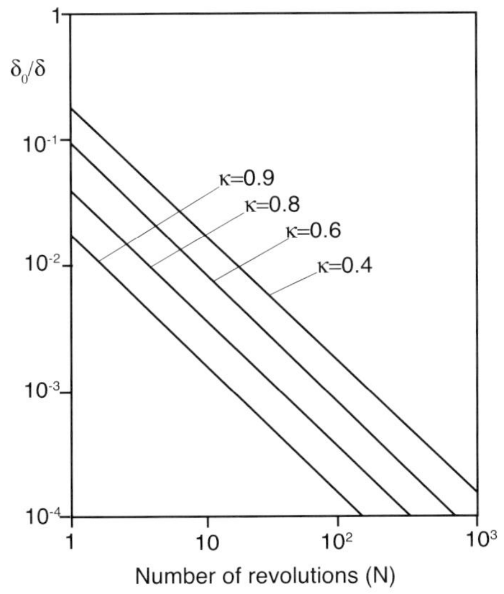

Figure 6.48: Striation thickness reduction as a function of number of revolutions in a Couette device.

After N revolutions, the length will be $Nd\theta$ and the total length of the tracer line will be

$$L_s = \int [dr^2 + r^2(Nd\theta)^2]^{1/2} \tag{6.213}$$

where L_s is the length of the striation or tracer line. Combining eqns. (6.212) and (6.213) we get

$$L_s = \int_{\kappa R_o}^{R_o} \left[1 + \frac{16\pi^2 N^2 R_o^{4/n}}{n^2(\kappa^{-2/n} - 1)^2} r^{-4/n}\right]^{1/2} dr \tag{6.214}$$

The striation thickness reduction can now be written as

$$\left(\frac{\delta}{\delta_0}\right) = \frac{R_0 - \kappa r}{L_s} \tag{6.215}$$

Since $dr^2 \gg (rNd\theta)^2$, we get

$$\left(\frac{\delta}{\delta_0}\right) = \left[\frac{2-n}{4\pi N}\right] \left[\frac{(1-\kappa)(\kappa^{-2/n} - 1)}{(\kappa^{(n-2)/n} - 1)}\right] \tag{6.216}$$

Figure 6.48 presents the reduction of the striation thickness as a function of number of revolutions, as well as ratio of inner to outer cylinder radius, κ for a Newtonian fluid. If we were to plot the striation thickness for a shear thinning fluid, say a power law index $n = 0.5$, the naked eye would not be able to distinguish between the Newtonian and the shear thinning results.

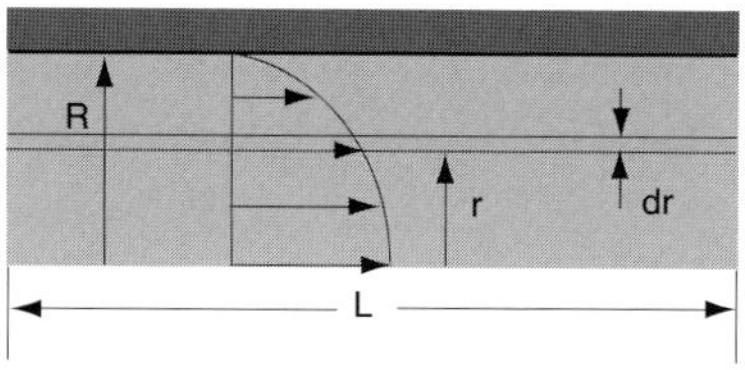

Figure 6.49: Schematic diagram of tube flow.

6.6.3 Residence Time Distribution of a Fluid Inside a Tube

The residence time distribution is a way to asses the mixing quality of a continuous flowing system such as an extruder. The residence time distribution is a measure of how much time the melt spends inside a continuous mixing device. The fluid particles entering the tube on one plane at the same time, will leave at different times according to their radial position. For example, a melt flowing through a tube has a large residence time near the wall, and the shortest residence time at the center. To illustrate this concept, let us consider the flow through a tube of radius R, length L and a volumetric throughput Q.

If we consider the tube geometry and notation presented in Fig. 6.49 we can show that the time, t, that every fluid particle spends inside the tube is defined by

$$t = \frac{L}{u_z(r)} \tag{6.217}$$

Residence time distribution functions were developed by Danckwerts [3] and are defined as *external* or *internal* RTD functions. The external RTD function $f(t)$ is defined such that $f(t)dt$ is the fraction of fluid exiting the system with a residence timebetween t and $t + dt$ and the internal RTD function $g(t)$ is defined such that $g(t)dt$ is the fraction of the fluid in the system with a residence time between t and $t + dt$.

For the external RTD, we can write

$$f(t)dt = \frac{dQ}{Q} = \frac{u_z(r)2\pi r dr}{\pi R^2 U} \tag{6.218}$$

where U is the average speed inside the tube, as is defined by

$$U = \frac{Q}{\pi R^2} \tag{6.219}$$

The average residence time distribution $\bar{t}$ can be computed using

$$\bar{t} = \frac{L}{U} \tag{6.220}$$

Using eqn. (6.219) and the analytical solution for Newtonian Hagen-Poiseuille flow given by

$$u_z(r) = 2U\left[1 - (\frac{r}{R})^2\right] = \frac{L}{t} \tag{6.221}$$

we can write

$$\frac{2rdr}{R^2} = \frac{\bar{t}}{2t^2}dt \tag{6.222}$$

and

$$\frac{u_z(r)}{U} = \frac{\bar{t}}{t} \tag{6.223}$$

If we solve for dt in eqn. (6.218), substitute it into eqn. (6.222), and using eqn. (6.223) we can solve for the residence time distribution as

$$f(t) = \frac{1}{2}\frac{\bar{t}^2}{t^3} \tag{6.224}$$

It is more common to assess mixing using the *cumulative residence time distribution*, defined by

$$F(t) = \int_{\bar{t}/2}^{t} f(t)dt \tag{6.225}$$

where the lower limit of the integral is the shortest residence time, i.e., the time it takes for the first fluid particles, at $r = 0$, to appear at the other end of the pipe. After integrating we get,

$$F(t) = 1 - \frac{1}{4}\left(\frac{t}{\bar{t}}\right)^{-2} \tag{6.226}$$

For a shear thinning fluid following a power law model, the residence time distribution becomes

$$f(t) = \frac{n(1+3n)}{2(1+n)^2}\frac{\bar{t}^2}{t^3}\left(1 - \frac{\bar{t}}{2t}\right)^{\frac{n-1}{1+n}} \tag{6.227}$$

and the cumulative residence time distribution

$$F(t) = \left(1 + \frac{n(1+2n)}{(1+2n)(1+n)}\frac{\bar{t}}{2t}\right)\left(1 - \frac{(1+n)(1+2n)}{(1+2n)(1+n)}\frac{\bar{t}}{2t}\right)^{\frac{n}{1+n}} \tag{6.228}$$

Figure 6.50 presents the cumulative residence time distribution for a tube with a Newtonian model and for a shear thinning fluid with power law indices of 0.5 and 0.1. Plug flow, which represents the worst mixing scenario, is also presented in the graph. A Bingham fluid, with a power law index of 0, would result in plug flow.

Similarly, we can derive the external cumulative residence time distribution $G(t)$ defined by,

$$G(t) = \int_{0}^{t} g(t)dt \tag{6.229}$$

6.6.4 Residence Time Distribution Inside the Ideal Mixer

The shape of the residence time or cumulative residence time distributions are used when optimizing the mixing ability of a system. Often, this shape is compared to the residence time in an ideal or *perfect mixer*. Such a mixer is a well stirred tank, as depicted in Fig. 6.51(a). Here, two components, a primary and secondary component, are fed to the tank at a total flow rate Q. The output can be regarded as a flow rate Q with a concentration $(1 - C_0)$ of

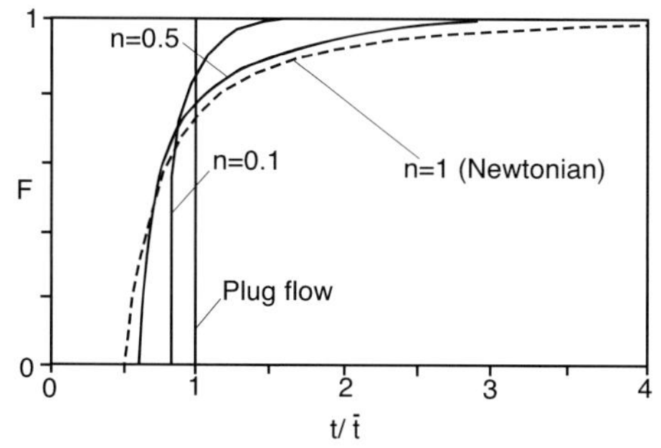

Figure 6.50: Cumulative residence time distributions for tube flow with a Newtonian fluid, and power law fluids with power law indeces of 0.5 and 0.1. Plug flow, which corresponds to a Bingham fluid with a power law index of 0 is also shown.

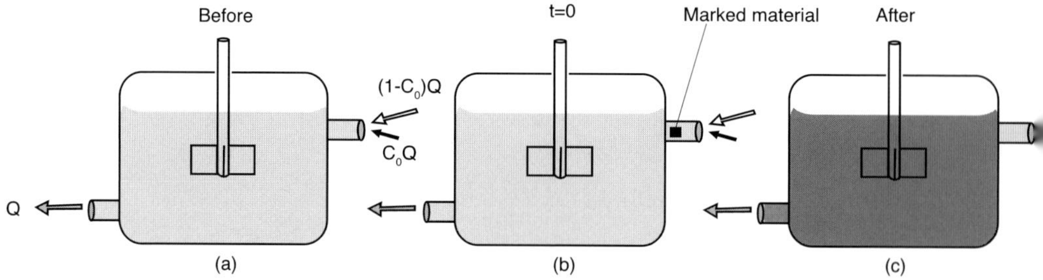

Figure 6.51: Flow through a stirred tank. (a) Before the marked material is added. (b) Instant the marked material is added, $t = 0$. (c) After the adding marked material, t.

the primary component and a concentration C_0 of the secondary component. We assume that the stirring tank is so rigorous that the contents are homogeneous at a concentration equal to the output.

At some point in time, which we will call $t = 0$, we mark a portion of the secondary component entering the tank (Fig. 6.51b), say with a pigment, so we can distinguish it from the material that was in the tank before $t = 0$ (Fig. 6.51a), as well as the material that enters the tank after (Fig. 6.51c) the marked material is inserted. We now call $F(t)$ the fraction of the material that was in the tank the instant the marked material is added, and has left the tank at time t. Hence, the volumetric output of the marked secondary component is $F(t)C_0Q$. The concentration of the marked secondary component in the output is $C(t)$. We can now write the balance

$$QC(t) = F(t)C_0Q \tag{6.230}$$

or

$$F(t) = \frac{C(t)}{C_0} \tag{6.231}$$

If we assume a perfectly homogeneous stirring tank, a balance of the marked material inside the tank will give

$$V\frac{dC}{dt} = QC_0 - QC \tag{6.232}$$

with the boundary condition $C(0) = 0$. With $\bar{t} = V/Q$, the above equation can be integrated to become

$$C(t) = C_0\left(1 - e^{-t/\bar{t}}\right) \tag{6.233}$$

or

$$F(t) = \left(1 - e^{-t/\bar{t}}\right) \tag{6.234}$$

Figure 6.52 compares the cumulative residence time distributions of a perfect mixer to Poiseuille flow. Disregarding the fact that the stirring tank's output starts at $t = 0$, we can see that the overall shape of the curve with the prefect mixer is much broader, pointing to a more homogeneous output, or simply, a better mixer.

6.7 INJECTION MOLDING – ISOTHERMAL FLOW PROBLEMS

Injection molding is a rather complex process during which non-Newtonian as well as non-isothermal effects play significant roles. Here, we present a couple of problems that are relatively simple to allow an analytical solution. Injection molding is discussed further in the next chapters.

6.7.1 Balancing the Runner System in Multi-Cavity Injection Molds

Inevitably, in multi-cavity injection molds, some of the mold cavity gates are located further than others from the sprue that delivers the melt from the plasticating unit. If the runner system that distributes the melt from the sprue to the individual cavities has a constant

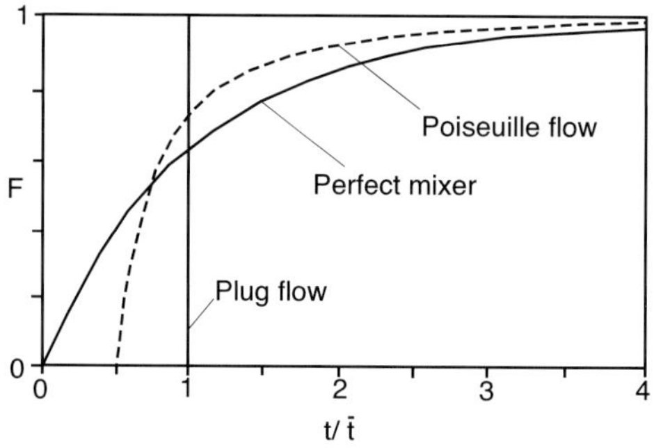

Figure 6.52: Cumulative residence time distribution for a stirring tank with perfect mixing.

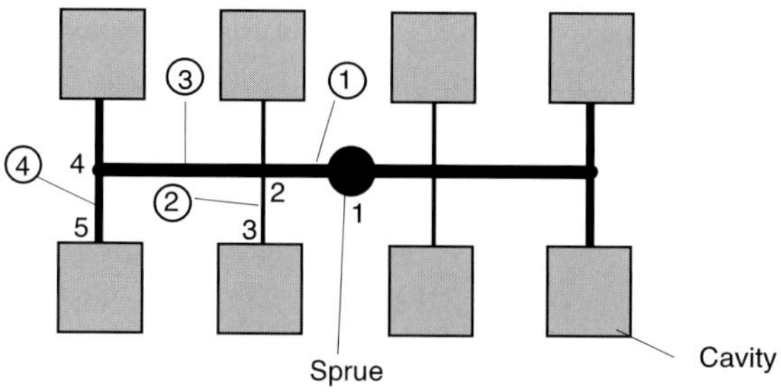

Figure 6.53: Schematic diagram of an eight cavity mold system with various runner lengths and radii.

diameter, each cavity would receive the melt at a different time and pressure. Differences in pressure will result in variable shrinkage, part weight and appearance. In order to avoid part inconsistencies the runner system has to be appropriately designed. For example, let us consider a multi-cavity system, schematically depicted in Fig. 6.53. In the figure, each portion of the runner is labeled with a number inside a circle, and the juncture between the runners are labeled with a number. In order to balance such a runner system we must assure that the flow-rate into each cavity is the same for the cavities near the gate and the cavities far away. Hence, for the case shown in Fig. 6.53 we must satisfy

$$Q_2 = Q_4 \tag{6.235}$$

The flow through runners 1 and 3 must be

$$Q_3 = 2Q_4 \tag{6.236}$$

and

$$Q_1 = 2Q_2 + 2Q_4, \tag{6.237}$$

respectively. Assuming an isothermal flow inside the runner system[2], the flow rate in each runner can be approximated using the Hagen-Poiseuille equation for a power law fluid

$$Q_i = \frac{\pi R^3}{s+3}\left(\frac{R\Delta p_i}{2mL_i}\right) \tag{6.238}$$

where Δp_i is the pressure drop between the junctures, specific to each runner within the system.

■ EXAMPLE 6.8.

Sample balancing problem. Let us consider the multi-cavity injection molding process shown in Fig. 6.54. To achieve equal part quality, the filling time for all cavities must be balanced. For the case in question, we need to balance the cavities by solving for the runner radius R_2. For a balanced runner system, the flow rates into all cavities must match. For a given flow rate Q, length L, and radius R_1, solve for the pressures at the runner system junctures. Assume an isothermal flow of a non-Newtonian shear thinning polymer. Compute the radius R_2 for a part molded of polystyrene with a consistency index (m) of 2.8×10^4 Pa-s^n and a power law index (n) of 0.28. Use values of $L = 10$ cm, $R_1 = 3$ mm, and $Q = 20$ cm^3/s.

Assuming a Hagen Poiseuille flow, we can write the following equations for the 4 runner system sections

- Section 1: $4Q = \frac{\pi(1.5R_1)^3}{s+3}\left[\frac{1.5R_1(p_1 - p_2)}{2mL}\right]^s$
- Section 2: $2Q = \frac{\pi(1.5R_1)^3}{s+3}\left[\frac{1.5R_1(p_2 - p_3)}{4mL}\right]^s$
- Section 3: $Q = \frac{\pi R_2^3}{s+3}\left[\frac{R_1(p_2 - 0)}{4mL}\right]^s$

[2]This is only true for a *hot runner system*. However, in many injection molding processes the runner system is directly inside the cooled mold, and the flow is not isothermal.

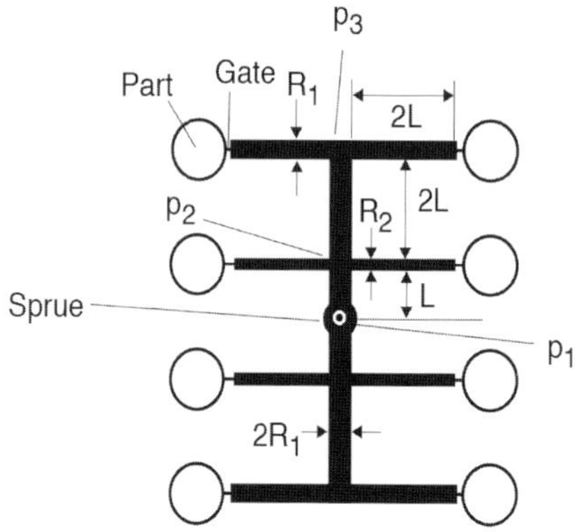

Figure 6.54: Runner system lay out.

- Section 4: $Q = \dfrac{\pi R_1^3}{s+3}\left[\dfrac{R_1(p_3-0)}{4mL}\right]^s$

The unknown parameters, p_1, p_2, p_3, and R_2 can be obtained using the listed equations. For the given values, a radius, R_2, of 2.34 mm would result in a balanced runner system, with pressures $p_1 = 562$ bar, $p_2 = 460$ bar, and $p_3 = 292$ bar. For comparison, had we assumed a Newtonian viscosity where $\mu = m$, a radius, R_2, of 2.76 mm would have resulted in a balanced system, with much higher pressures of $p_1 = 63030$ bar, $p_2 = 49120$ bar, and $p_3 = 35210$ bar.

6.7.2 Radial Flow Between Two Parallel discs

In Chapter 5 of this book we derived the equations that govern the pressure flow between two parallel discs for a Newtonian fluid. In a similar fashion, we can derive the equations that govern flow rate, gate pressure, and pressure distributions for disc-shaped cavities filling with a shear thinning fluid. For the equations presented in this section, we assumed a power law viscosity. For the velocity distribution we have

$$u_r(r,z) = -\frac{2n+1}{n+1}\frac{Q}{4\pi n h^2}\left(z^{1+1/n} - h^{1+1/n}\right) \tag{6.239}$$

where

$$Q = \frac{4\pi n h^{2+1/n}}{2n+1}\left(-\frac{r^n}{m}\frac{dp}{dr}\right)^{1/n} \tag{6.240}$$

with a pressure distribution of

$$p(r) = \frac{m}{(1-n)h}\left[\frac{(2n+1)Q}{4\pi n h^2}\right]^n \left(R^{1-n} - r^{1-n}\right) \tag{6.241}$$

where r is the radial position in the disc mold. By substituting $r = r_1$, we can compute the pressure at the gate.

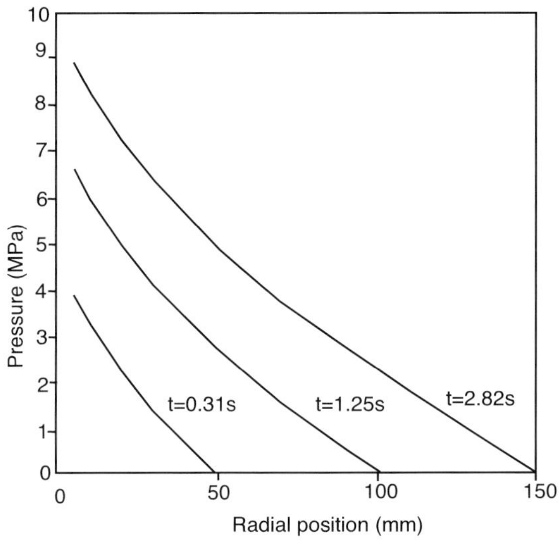

Figure 6.55: Radial pressure profile as a function of time in a disc-shaped mold computed using a shear thinning viscosity model (after Agassant [1]).

■ **EXAMPLE 6.9.**

Predicting pressure profiles in a disc-shaped mold using a shear thinning power law model [1]. We can solve the problem presented in example 5.3 for a shear thinning polymer with power law viscosity model. We will choose the same viscosity used in the previous example as the consistency index, $m = 6,400$ Pa-s^n, in the power law model, with a power law index $n = 0.39$. With a constant volumetric flow rate, Q, we get the same flow front location in time as in the previous problem, and we can use eqns. (6.239) to (6.241) to predict the required gate pressure and pressure profile throughout the disc.

Figure 6.55 presents the pressure profiles within the material for various melt flow front locations. First of all, we can see that the shear thinning behavior of the polymer has caused the pressure requirement to go down significantly (by a factor of 30). The curves presented in Fig. 6.55 also reveal that the shape of the curves was also affected when compared to the Newtonian profiles.

Figure 6.56 presents a comparison of the pressure at the gate for the Newtonian and shear thinning case. The figure also shows the effect of temperature [1]. We can see the effect that the cooling has on the pressure requirements. This is caused by a reduction of thickness due to the growth of a solidified layer on the mold surface, as well as an increase in viscosity due to a drop in overall temperature. For a better comparison, Fig. 6.57 presents the pressure requirements for the shear thinning and non-isothermal cases [1].

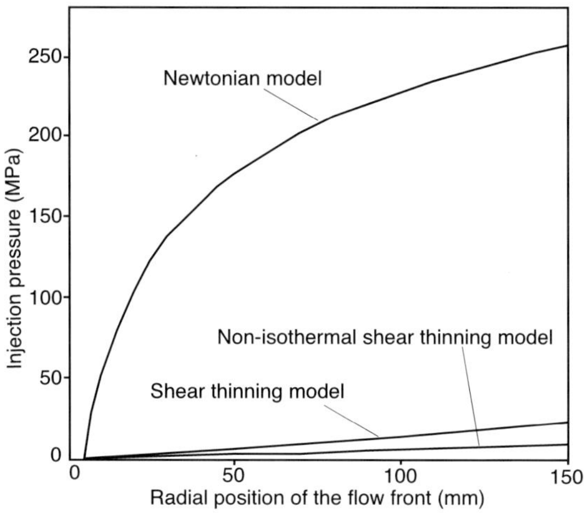

Figure 6.56: Pressure at the gate as a function of melt flow front position in a disc-shaped mold (After Agassant [1]).

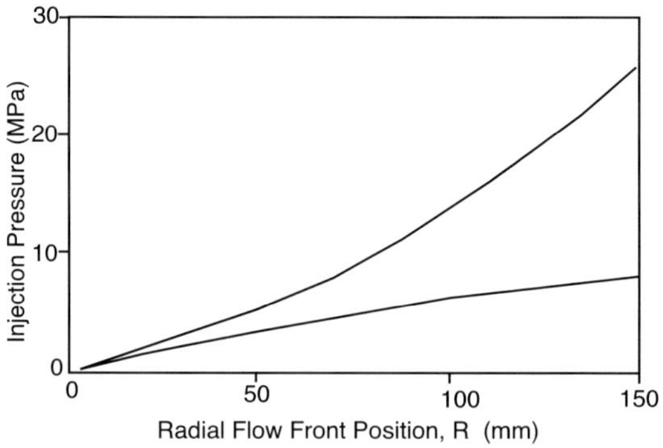

Figure 6.57: Comparison between the analytical shear thinning solution and a non-isothermal shear thinning solution of pressure at the gate as a function of melt flow front position in a disc-shaped mold (after Agassant [1]).

6.8 NON-ISOTHERMAL FLOWS

Although we analyze most polymer processes as isothermal problems, many are non-isothermal even at steady state conditions. The non-isothermal effects during flow are often difficult to analyze, and make analytical solutions cumbersome or, in many cases impossible. The non-isothermal behavior is complicated further when the energy equation and the momentum balance are fully coupled. This occurs when viscous dissipation is sufficiently high to raise the temperature enough to affect the viscosity of the melt.

Hence, when solving a non-isothermal problem the question arises −is this a problem where the equations of motion and energy are coupled? To address this question we can go back to Example 6.1, a simple shear flow system was analyzed to decide whether it can be addressed as an isothermal problem or not. In a simple shear flow, the maximum temperature will occur at the center of the melt. By substituting $y = h/2$ into eqn. (6.5), we get an equation that will help us estimate the temperature rise

$$\Delta T_{max} = \frac{3}{8}\frac{\eta u_o^2}{k} \tag{6.242}$$

When analyzing non-isothermal flow problems, we often assume that the viscosity decays exponentially with temperature following the relation

$$\eta = \eta_0 e^{-a(T-T_0)} \tag{6.243}$$

We can determine the change of viscosity with respect to temperature change by differentiating eqn. (6.243)

$$\frac{\partial \eta}{\partial T} = -\eta_0 a e^{-a\Delta T} \tag{6.244}$$

Here, we can see that an increase in temperature will reduce the viscosity by an amount controlled by the material constant a −the temperature dependence of the viscosity− and the actual temperature rise, ΔT. Hence, the effect is controlled by the product $a\Delta T$. Taking eqn. (6.242) and dropping the $3/8$ term we can say that

$$a\Delta T \propto a\frac{\eta u_o^2}{k} \tag{6.245}$$

Equation (6.245) is the well known Nahme-Griffith number

$$Na = a\frac{\eta u_o^2}{k} \tag{6.246}$$

which is a measure of the degree of coupling between the energy equation and the momentum balance. For the two problems that follow, we will assume that the energy equation and the equation of motion are not coupled, hence $Na \ll 1$.

6.8.1 Non-Isothermal Shear Flow

A common flow problem in polymer processing is a shear flow with a temperature gradient as depicted in Fig. 6.58. For example, this type of flow occurs within the melt film that develops during melting with drag flow removal, as will be discussed later in this chapter.

In the problem we are addressing in this section, the two plates are assigned two different temperatures, T_0 on the lower plate and T_1 on the upper moving plate. In addition, we are assuming a Newtonian flow with an exponential temperature dependence, $\mu = \mu_0 e^{-a(T-T_0)}$,

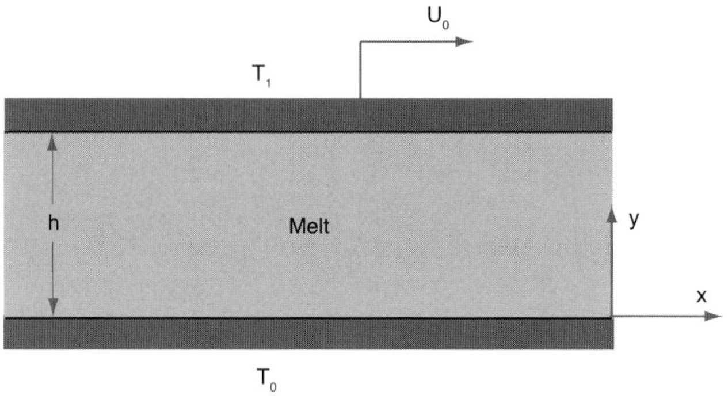

Figure 6.58: Schematic diagram of a shear flow problem with an imposed temperature gradient.

constant thermal properties and negligible viscous heating. In such a system, the energy balance reduces to

$$0 = k\frac{\partial^2 T}{\partial y^2} \tag{6.247}$$

which can be integrated using the boundary conditions given above to become

$$(T - T_0) = (T_1 - T_0)\left(\frac{y}{h}\right) \tag{6.248}$$

or

$$\Theta = \Theta_1\left(\frac{y}{h}\right) \tag{6.249}$$

The equation of motion reduces to

$$0 = \frac{\partial \tau_{yx}}{\partial y} \tag{6.250}$$

which after integration gives

$$\tau_{yx} = C_1 \tag{6.251}$$

We can now make use of the constitutive equation

$$\tau_{yx} = \mu_0 e^{-a\Theta}\frac{\partial u_x}{\partial y} = C_1 \tag{6.252}$$

which can be rewritten as

$$\frac{\partial u_x}{\partial y} = \frac{C_1}{\mu_0}e^{a\Theta} \tag{6.253}$$

and integrated using the boundary conditions, $u_x = 0$ and $T = T_0$ at $y = 0$, as well as $u_x = U_0$ and $T = T_1$ at $y = h$, to give

$$u_x = U_0\left(\frac{1 - e^{\Omega(y/h)}}{1 - e^{\Omega}}\right) \tag{6.254}$$

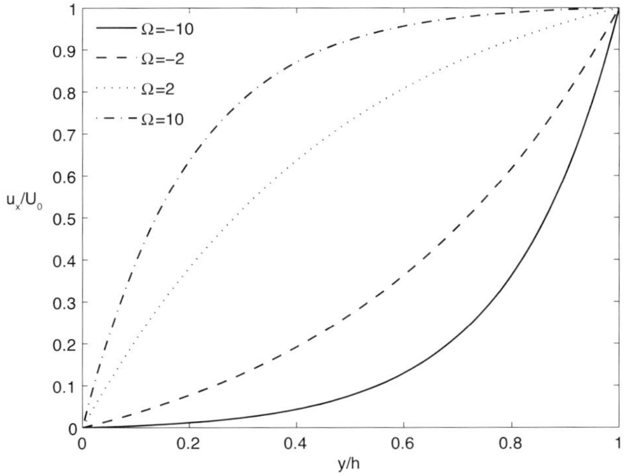

Figure 6.59: Dimensionless shear flow results with various dimensionless temperature imbalances.

where $\Omega = a\Theta_1$.

To illustrate the effect of thermal gradients and temperature dependent viscosity, we can plot a dimensionless velocity, u_x/U_0 as a function of dimensionless position, y/h, for various values of thermal imbalance between the surfaces, Θ_1. Note that Ω, the product between the temperature dependence of the viscosity and the temperature imbalance is also a dimensionless quantity. This gives a fully dimensionless graph that can be used to assess many case scenarios. Figure 6.59 presents dimensionless velocity distributions across the plates for various dimensionless temperature imbalances, Ω.

6.8.2 Non-Isothermal Pressure Flow Through a Slit

For this non-isothermal flow consider a Newtonian fluid between two parallel plates separated by a distance h. Again we consider the notation presented in Fig. 6.58, however, with both upper and lower plates being fixed. We choose the same exponential viscosity model used in the previous section. We are to solve for the velocity profile between the two plates with an imposed pressure gradient in the x-direction and a temperature gradient in the y-direction.

Again, neglecting viscous heating results in the linear temperature profile, eqn. (6.249), presented in the previous section. In this case, with the imposed pressure gradient, the equation of motion becomes

$$\frac{\partial p}{\partial x} = \frac{\partial \tau_{yx}}{\partial y} \tag{6.255}$$

which after integration leads to

$$\tau_{yx} = \left(\frac{\partial p}{\partial x}\right) y + C_1 \tag{6.256}$$

The stress can be defined using the constitutive equation $\tau_{yx} = \mu \left(\frac{\partial p}{\partial x} \right)$, and letting $\frac{\partial p}{\partial x} = \frac{\Delta p}{L}$ results in

$$\mu_0 e^{(\frac{-\Theta_1 a}{h} y)} \frac{du_x}{dy} = \frac{\Delta p}{L} + C_1 \tag{6.257}$$

Defining $\Omega = \Theta_1 a$ and $\hat{y} = y/h$, we can write

$$\frac{du_x}{dy} = \frac{1}{\mu_0} \frac{\Delta p}{L} y e^{\Omega \hat{y}} + \frac{C_1}{\mu_0} e^{\Omega \hat{y}} \tag{6.258}$$

which can be integrated to give

$$u_x = \frac{h^2 \Delta p}{\mu_0 L \Omega^2} e^{\Omega \hat{y}} (\Omega \hat{y} - 1) + \frac{C_1}{\mu_0 \Omega} e^{\Omega \hat{y}} + C_2 \tag{6.259}$$

Using the boundary conditions $u_x = 0$, at $\hat{y} = 0$ and $\hat{y} = 1$, we get the following velocity distribution

$$u_x = \frac{h^2 \Delta p}{\mu_0 L \Omega} \left[e^{\Omega \hat{y}} \left(\hat{y} + \frac{e^{\Omega}}{1 - e^{\Omega}} \right) - \frac{e^{\Omega}}{1 - e^{\Omega}} \right] \tag{6.260}$$

If we further assume a dimensionless velocity

$$\hat{u}_x = u_x \frac{\mu_0 L \Omega}{h^2 \Delta p} \tag{6.261}$$

Equation (6.262) reduces to

$$\hat{u}_x = \left[e^{\Omega \hat{y}} \left(\hat{y} + \frac{e^{\Omega}}{1 - e^{\Omega}} \right) - \frac{e^{\Omega}}{1 - e^{\Omega}} \right] \tag{6.262}$$

Figure 6.60 presents the dimensionless pressure flow velocity profile for various positive values of Ω. It should be noted that negative values of Ω lead to symmetric velocity profiles as the one generated by the positive values.

6.9 MELTING AND SOLIDIFICATION

Melting is an important step in any polymer process. Before the material can be shaped into its final form it must first be softened or molten. For example, during thermoforming the sheet is heated using radiative heaters to a temperature high enough that the sheet can be stretched and formed into, or over, the cavity that will give it its final shape. In extrusion and injection molding the pellets first move from the hopper into the plasticating region of the screw, where they are molten and subsequently pushed out of the die or into the mold cavity.

In the plastication step of the process, melting is critical in controlling cycle time. Also, during thermoforming, the heating of the sheet is the most time consuming step of the whole process. For example, if we consider the melting of an infinite slab, at an initial temperature of T_0, as presented in Fig. 6.61, the heat supplied by the hot wall, set at a heater temperature of T_h, will create a layer of molten polymer of thickness, $X(t)$.

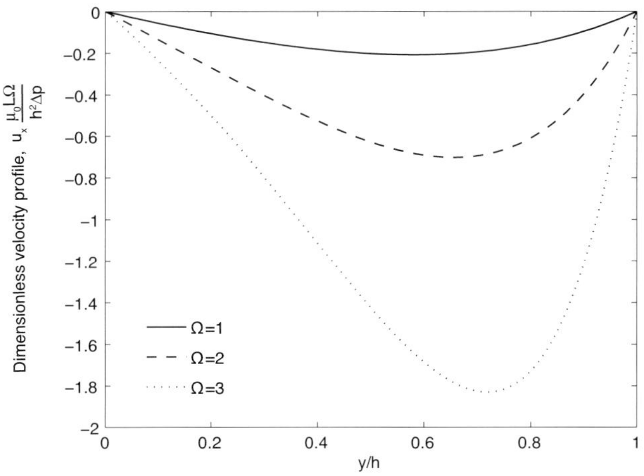

Figure 6.60: Dimensionless pressure flow velocity profile for various dimensionless temperature imbalances.

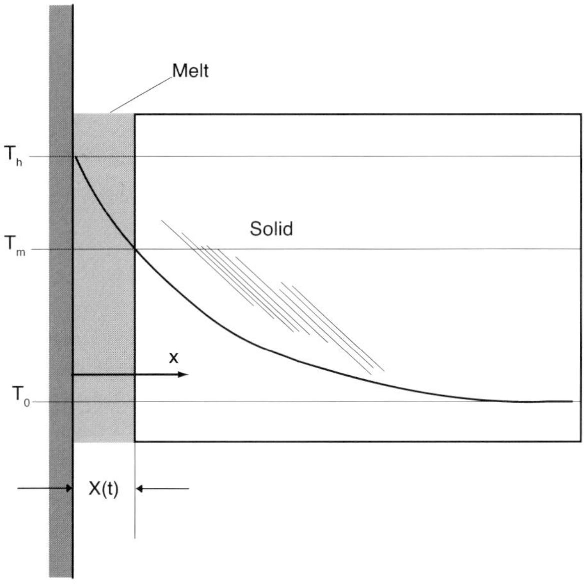

Figure 6.61: Schematic diagram of the melting process of an infinite slab.

For such a case, where the temperature can be computed as a function of time using

$$\frac{T - T_h}{T_0 - T_h} = \text{erf}\left(\frac{x}{\sqrt{4\alpha t}}\right) \tag{6.263}$$

where $\alpha = \frac{k}{\rho C_p}$ is the thermal diffusivity; the error function, erf, is defined by

$$\text{erf}\left(\frac{x}{\sqrt{4\alpha t}}\right) = \frac{2}{\sqrt{\pi}} \int_0^{\frac{x}{\sqrt{4\alpha t}}} e^{-s^2} ds \tag{6.264}$$

The equation clearly demonstrates that the molten layer, with temperatures above T_m for semi-crystalline polymers and above T_g for amorphous polymers, will follow the following relation

$$X(t) \propto \sqrt{\alpha t} \tag{6.265}$$

for amorphous polymers. The melting and solidification process for semi-crystalline materials is more complex because the heat of fusion or heat of crystallization, nucleation rate, etc. When measuring the specific heat as the material crystallizes, a peak representing the heat of fusion is detected, shown in Fig. 2.8. Although theoretical predictions of melting and solidification in semi-crystalline polymers predict a similar growth rate as with amorphous polymers

$$X(t) \propto f(\lambda, \alpha)\sqrt{t} \tag{6.266}$$

where $f(\lambda, \alpha)$ represents heat of fusion effects, experimental evidence [14] of solidification has demonstrated that the growth rate of the solidified crystallized layer in semi-crystalline polymers is finite at the beginning of the solidification process. This is mainly due to the fact that at the beginning the nucleation can only occur at a finite rate. Hence, the solution presented in eqn. (6.266) as well as the widely used Stefan condition, discussed later, can only be used for melting but do not hold for cooling and solidification of semi-crystalline polymers. This can be seen in Fig. 6.62, which shows the measured thickness of crystallized layers as a function of time for polypropylene plates quenched at three different temperatures. For further reading on this important topic the reader is encouraged to consult the literature [6, 12].

In either case, the growth rate of the molten layer during conduction melting follows the relation

$$\frac{dX(t)}{dt} \propto \frac{1}{\sqrt{t}} \tag{6.267}$$

which means that the growth rate of the molten layer, $X(t)$ during melting is infinite at $t = 0$, but rapidly decreases as the molten layer increases in thickness. For example, the growth rate at 10 seconds, is only 32% of the growth rate at 1 second, and the growth rate at 1 minute is only 13% of the growth rate at 1 second. Hence, even after only 10 seconds the melting rate is quite inefficient, and can only be increased by continuously maintaining the thickness of the molten layer small. This can be achieved by various *melt removal schemes*, namely, the *pressure flow melt removal* and the *drag flow melt removal* techniques, presented in Figs. 6.63 and 6.64, respectively.

During the pressure flow melt removal scheme a force is applied on the melting solid resulting in a pressure build-up within the melt film, causing the melt film to be squeezed

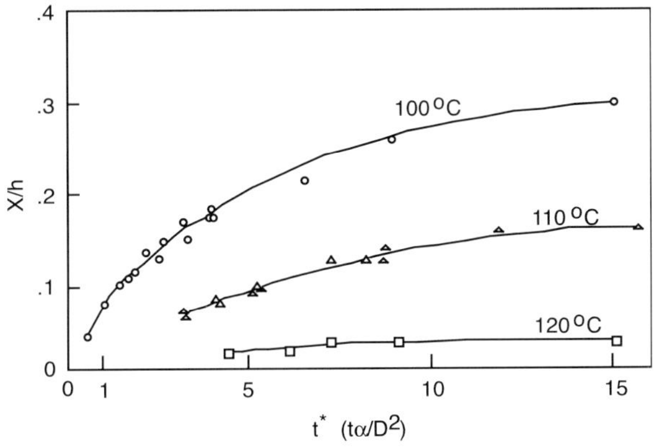

Figure 6.62: Dimensionless thickness of the crystallized layers as a function of dimensionless time for various temperatures of the quenching surface.

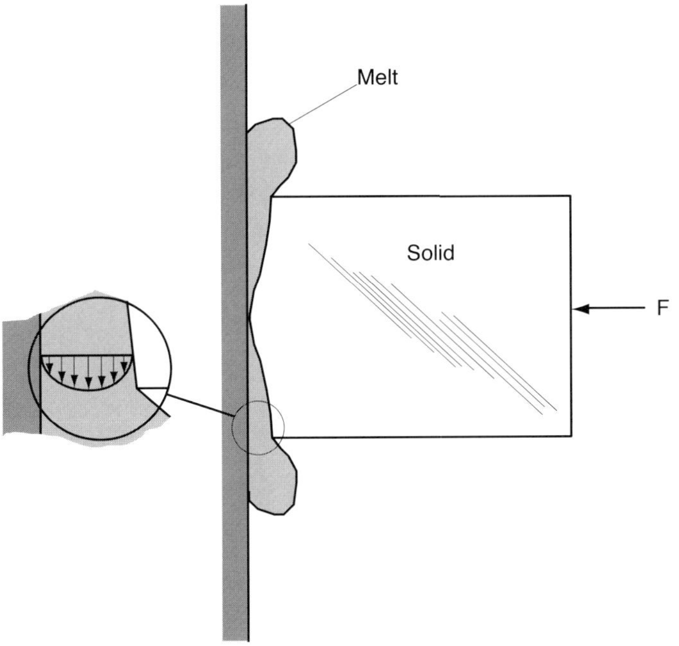

Figure 6.63: Schematic diagram of the melting process with pressure flow melt removal.

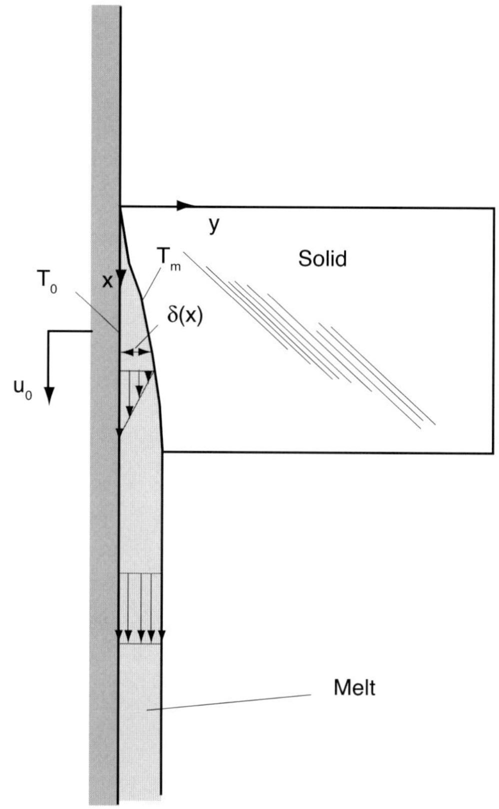

Figure 6.64: Schematic diagram of the melting process with drag flow melt removal.

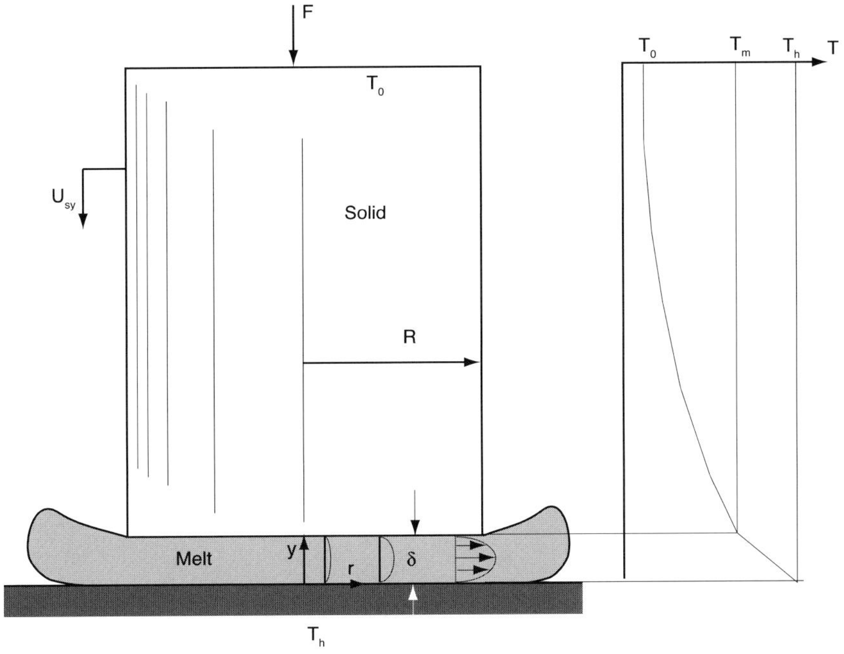

Figure 6.65: Schematic diagram of the melting process of a circular bar with pressure flow melt removal.

out from under the solid body. On the other hand, in the drag flow melt removal scheme, the melt is transported out from under the solid by the moving heated surface. This mechanism exists in the melting section of a single screw extruder. Both of these cases are addressed in more detail in the the following sections.

6.9.1 Melting with Pressure Flow Melt Removal

Melting with pressure flow melt removal is the less common of the two melt removal schemes. However, it can be seen during butt welding, where a plastic is pressed against a hot surface and the melt is squeezed from under the part. Another application of industrial relevance is a special form of polyester fiber manufacturing process, where a plastic bar is pressed against a heated screen [24]. Here, we will follow the derivation by Stammers and Beek [24]. First, let us consider a circular polymer bar of radius R pressed against a heated metal surface using a force F as depicted in Fig. 6.65

To simplify the problem, we can assume that the polymer bar moves at a constant speed U_{sy}, and that a film of constant thickness, δ, exists between the bar and the heated plate. In addition, we assume that the polymer melt is Newtonian and that the viscosity is independent of temperature. The Newtonian assumption is justified by low rates of deformation that develop in this relatively slow flow problem. Furthermore, due to these low rates of deformation we can assume that the convective and viscous dissipation effects are negligible.

We can perform a mass balance as

$$\rho_s \pi R^2(-U_{sy}) = 2\pi r \delta \hat{u_r} \rho_m \tag{6.268}$$

and we can compute the total melting rate as the material displaced by the moving bar

$$w_T = \pi(-U_{sy})\rho_s R^2 \tag{6.269}$$

In eqn. (6.268), $\hat{u_r}$ is the mean radial velocity at position r, and can be computed using

$$\hat{u_r} = \frac{(-U_{sy})r}{2\delta}\hat{\rho} = \frac{1}{\delta}\int_0^\delta u_r dy \tag{6.270}$$

where $\hat{\rho} = \rho_s/\rho_m$ is a dimensionless density.

For this problem, the momentum balance reduces to a simple form of the radial component of the equation of motion

$$\frac{dp}{dr} = \mu\frac{d^2 u_r}{dy^2} \tag{6.271}$$

We can now integrate eqn. (6.271) using a no-slip boundary condition on the heated metal surface, $u_r(0) = 0$, and on the polymer bar, $u_r(\delta) = 0$, to give

$$u_r(y) = \frac{1}{2\mu}\frac{dp}{dr}(y-\delta)y \tag{6.272}$$

Integrating eqn. (6.270) results in

$$\hat{u_r} = \frac{-\delta^2}{12\mu}\frac{dp}{dr} = \frac{(-U_{sy})r}{2\delta}\hat{\rho} \tag{6.273}$$

which can be solved for the pressure gradient

$$-\frac{dp}{dr} = \frac{6\mu\hat{\rho}(-U_{sy})r}{\delta^3} \tag{6.274}$$

Equation (6.274) can be integrated using the boundary condition $p(R) = 0$ to give the pressure profile

$$p(r) = \frac{3\mu\hat{\rho}(-U_{sy})}{\delta^3}(R^2 - r^2) \tag{6.275}$$

The pressure profile can be related to the force F used to press the polymer bar against the heated plate using,

$$F = \int_0^r 2\pi r p(r) dr = \frac{3\pi\mu\hat{\rho}(-U_{sy})R^4}{2\delta^3} \tag{6.276}$$

When examining eqn. (6.276) it is clear that the only two unknowns are the speed at which the solid moves, U_{sy}, and the melt film thickness, δ. To solve for melt film thickness we can perform a energy balance by setting the heat conduction through the thickness of the film equal to the heat of fusion of the melting material and the energy required to raise its temperature from T_0 to T_m

$$k_m\frac{(T_h - T_m)}{\delta} = \rho_s(-U_{sy})\left[\lambda + C_s(T_m - T_0)\right] \tag{6.277}$$

We can solve for δ and substitute it into eqn. (6.276) and solve for our last unknown, U_{sy},

$$(-U_{sy}) = \frac{0.67872}{R}\left[\frac{F}{\mu\hat{\rho}}\right]^{1/4}\left[\frac{k_m(T_h - T_m)}{\rho_s\left[\lambda + C_s(T_m - T_0)\right]}\right]^{3/4} \tag{6.278}$$

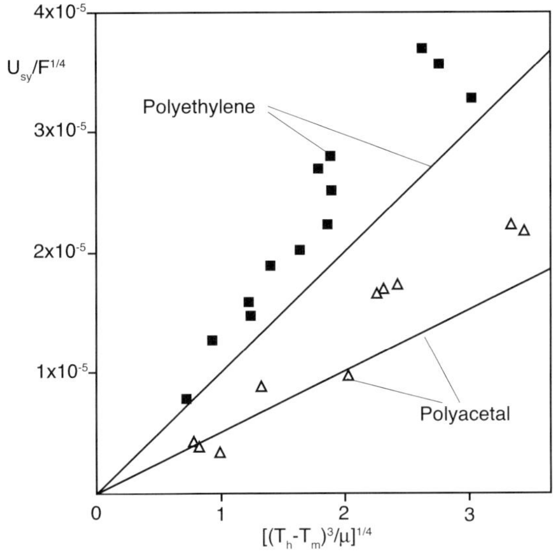

Figure 6.66: Comparison of experimental melting rates to the predicted (solid lines) rates using eqn. (6.278).

If we take the time to analyze the various terms in eqn. (6.278), we can gain significant insight. For example, we can see that the rate of melting is proportional to $F^{1/4}$, simply because an increase in force will reduce the thickness of the melt film, which in turn requires a larger pressure to cause it to flow from under the cylinder. On the other hand, the rate of melting is inversely proportional to R. Stammers and Beek [24] compared this equation to experimental results and found good agreement. Figure 6.66 presents a plot of $U_{sy}/F^{1/4}$ versus $[(T_h - T_m)^3/\mu]^{1/4}$ and shows the linear relation and the slope predicted by eqn. (6.278).

6.9.2 Melting with Drag Flow Melt Removal

Let us consider the case described in Fig. 6.64 of an isotropic homogeneous solid infinite slab of width W, pushed against a heated moving plate. Here, we want to solve for the temperatures, velocities and melting rates at steady-state conditions. The problem is essentially two-dimensional, which means that the velocity and temperature fields are functions of x and y. The melt film thickness, $\delta(x)$, is very small at $x = 0$ and it increases in the x-direction; however, its actual shape is unknown. Heat is conducted from the heated plate, T_0, to the solid-melt interface at $T = T_m$. Here, we are considering the more general case of a semi-crystalline polymer and will follow the assumptions and derivation from Tadmor [25, 27]. For simplicity, in his derivation Tadmor assumed constant thermal properties. Furthermore, as a geometric constraint, Tadmor assumed that the film thickness is much smaller than its width, $\delta/W \ll 1$. In addition, all body forces and inertial effects within the film, $Re \ll 1$, are also negigible . Together, these assumptions justify using the lubrication approximation.

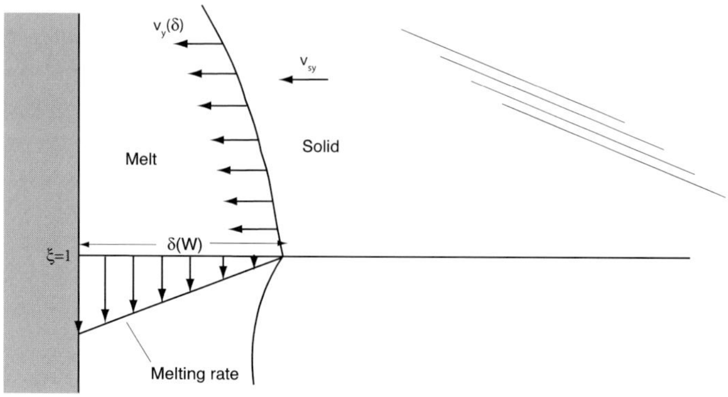

Figure 6.67: Schematic diagram of the melting at the interface.

The equations of continuity reduce to

$$\begin{aligned}\frac{\partial u_x}{\partial x} + \frac{\partial u_y}{\partial y} &= 0 \\ \frac{\partial \tau_{xy}}{\partial y} &= 0\end{aligned} \tag{6.279}$$

The energy equation within the melt will include the conductive and the viscous dissipation terms as follows

$$k_m \frac{\partial^2 T}{\partial y^2} - \tau_{xy} \frac{\partial u_x}{\partial y} = 0 \tag{6.280}$$

The boundary conditions for the momentum and energy balance equations are

$$\begin{aligned}&u_x(0) = U_0 \quad u_x(\delta) = 0 \\ &u_y(0) = 0 \\ &T(0) = T_0 \quad T(\delta) = T_m\end{aligned} \tag{6.281}$$

The velocity at any point $u_y(\delta)$ is determined by the rate of melting at the interface (Fig. 6.67), which is obtained from the Stefan condition or heat balance between conduction and the rate of melting at that interface,

$$k_m \left[-\frac{\partial T}{\partial y} \right]_{y=\delta} = \rho_m \left(-u_y(\delta) \right) \lambda + k_s \left[-\frac{\partial T}{\partial y} \right]_{y=\delta} \tag{6.282}$$

where λ is the heat of fusion and k_s, k_m are the thermal conductivity of the solid and melt, respectively.

In order to reduce eqn. (6.282) we must find the temperature distribution for the solid. The energy balance for the solid is given by

$$\rho_s C_s u_{sy} \frac{\partial T}{\partial y} = k_s \frac{\partial^2 T}{\partial y^2} \tag{6.283}$$

which can be solved by integration using the boundary conditions $T(\delta) = T_m$ and $T(\infty) = T_{s0}$, which results in the temperature profile

$$T = T_{s0} + (T_m - T_{s0})\, e^{\left(\dfrac{u_{sy}(y\delta)}{\alpha_s}\right)} \tag{6.284}$$

The rate of heat conduction out the solid-melt interface will be

$$-k_s \left[\frac{\partial T}{\partial y}\right]_{y=\delta} = -\left(T_m - T_{s0}\right) u_y(\delta)\rho_m C_s \tag{6.285}$$

where the mass balance in the interface, $u_{sy}\rho_s = u_y(\delta)\rho_m$, was used. Equation (6.282) can now be written as

$$k_m \left[\frac{\partial T}{\partial y}\right]_{y=\delta} = \rho_m u_y(\delta)\lambda' \tag{6.286}$$

where $\lambda' = \lambda + C_s\left(T_m - T_{s0}\right)$ is the energy required to bring the solid from an initial temperature T_{s0} to T_m and to melt it at that temperature.

The variables in the balance eqns. (6.279) to (6.283) can be scaled into dimensionless variables as

$$\Theta = \frac{T - T_m}{T_0 - T_m}$$

$$\xi = \frac{x}{W} \qquad \zeta = \frac{y}{\delta} \tag{6.287}$$

$$\hat{u}_x = \frac{u_x}{U_0} \qquad \hat{u}_y = \frac{u_y}{U_0(\delta_0/W)}$$

where the characteristic film thickness, δ_0, is determined from a scaling analysis of eqn. (6.286), using characteristic values, and reduces to

$$\delta_0 = \left(\frac{k_m\left(T_0 - T_m\right)W}{\lambda'\rho_m U_0}\right) \tag{6.288}$$

The dimensionless form of the boundary conditions eqn. (6.281) are

$$\begin{aligned} &\Theta(0) = 1 \quad \hat{u}_x(0) = 1 \quad \hat{u}_y(0) = 0 \\ &\Theta(1) = 0 \quad \hat{u}_x(1) = 0 \end{aligned} \tag{6.289}$$

Melting model using a Newtonian fluid with temperature independent viscosity μ. For a Newtonian fluid the equation of motion reduces to

$$\frac{\partial^2 \hat{u}_x}{\partial \zeta^2} = 0 \tag{6.290}$$

which with the boundary conditions, eqn. (6.289), has the solution

$$\hat{u}_x = 1 - \zeta \tag{6.291}$$

The energy equation, which can be solved independently from the equation of motion, reduces to

$$\frac{\partial^2 \Theta}{\partial \zeta^2} + Br \left(\frac{\partial \hat{u}_x}{\partial \zeta} \right)^2 = 0 \tag{6.292}$$

where the Brinkman number, Br, is defined by

$$Br = \frac{\mu U_0^2}{k_m (T_0 - T_m)} \tag{6.293}$$

Integrating eqn. (6.292) with the corresponding boundary conditions yields the temperature profile

$$\Theta = (1 - \zeta) + \frac{Br}{2} \zeta (1 - \zeta) \tag{6.294}$$

This temperature profile can be used to find the y-velocity component at the interface, i.e.,

$$\left(\frac{\partial \Theta}{\partial \zeta} \right)_{\zeta=1} = \frac{\delta}{\delta_0} \hat{u}_y(1) \tag{6.295}$$

to obtain

$$\hat{u}_y(1) = -\frac{\delta_0}{\delta} \left(1 + \frac{Br}{2} \right) \tag{6.296}$$

Integrating the continuity equation and using eqn. (6.291), it is found that

$$\hat{u}_y(1) = -\frac{1}{2} \frac{\dot{\delta}}{\delta_0} \tag{6.297}$$

Combining these two equations and integrating results in the film thickness profile

$$\delta(\xi) = \delta_0 \sqrt{(4 + 2Br)\,\xi} \tag{6.298}$$

The rate of melting, per unit width, is given by

$$w_L(x) = \rho_m U_0 \delta \int_0^1 \hat{u}_x d\zeta = \frac{U_0 \delta}{2} \rho_m \tag{6.299}$$

which results in

$$w_L = \left[\frac{U_0 \rho_m \left(k_m (T_0 - T_m) + \mu U_0^2 / 2 \right)}{\lambda'} W \right]^{1/2} \tag{6.300}$$

In this analysis, we neglected the convection in the film. Tadmor *et al.* [25, 27] accounted approximately for convection by including in λ' the heat needed to bring the melt from T_m to the mean temperature, i.e.,

$$\lambda'' = \lambda + C_s (T_m - T_{s0}) + C_s (T_0 - T_m) \bar{\Theta} \tag{6.301}$$

where the mean temperature is defined by

$$\bar{\Theta} = \frac{\int_0^1 \hat{u}_x \Theta d\zeta}{\int_0^1 \hat{u}_x d\zeta} = \frac{2}{3} + \frac{Br}{12} \tag{6.302}$$

In addition, w_L must be reduced by a factor of $\sqrt{2}$ because the newly melted material must be removed from the interface, allowing it to flow into the film at that point, keeping the film thickness constant, i.e.,

$$w_L = \left[\frac{U_0 \rho_m \left(k_m (T_0 - T_m) + \mu U_0^2/2\right)}{2\lambda''} W\right]^{1/2} \tag{6.303}$$

Power law model fluid with temperature dependent viscosity $m_0 = e^{(-a(T-T_m))}$**.** The rate of melting is strongly dependent on the shear thinning behavior and the temperature dependent viscosity of the polymer melt. However, we can simplify the problem significantly by assuming that the viscous dissipation is low enough that the temperature profile used to compute the viscosity is linear, i.e.,

$$\Theta = 1 - \zeta, \tag{6.304}$$

such that the equation of motion reduces to

$$\frac{\partial}{\partial \zeta}\left(e^{b(1-\zeta)}\left(-\frac{\partial \hat{u}_x}{\partial \zeta}\right)^n\right) = 0 \tag{6.305}$$

Solving this equation, the local velocity profile (derived in the previous sections) becomes

$$\hat{u}_x = \frac{e^{b'\zeta} - e^{b'}}{1 - e^{b'}} \tag{6.306}$$

where $b' = b/n = -a(T_0 - T_m)/n$. The energy equation without convection, will be

$$\frac{\partial^2 \Theta}{\partial \zeta^2} + Br\left(\frac{\delta_0}{\delta}\right)^{n-1} e^{b(1-\zeta)}\left(-\frac{\partial \hat{u}_x}{\partial \zeta}\right)^{n+1} = 0 \tag{6.307}$$

where the Brinkman number is defined by

$$Br = \frac{m_0 U_0^{(3n+1)/2} \rho_m^{(n-1)/2} \lambda'^{(n-1)/2}}{(T_0 - T_m)^{(n+1)/2} k_m^{(n+1)/2} W^{(n-1)/2}} \tag{6.308}$$

Integrating the equation will give the temperature profile

$$\Theta = (1-\zeta) + Br\left(\frac{\delta_0}{\delta}\right)^{n-1}\left(\frac{b'}{1-e^{-b'}}\right)^{n+1}\frac{e^{-b'}}{b'^2}\left[1 - e^{b'} - \zeta\left(1 - e^{b'}\right)\right] \tag{6.309}$$

with these equations, and similar to the Newtonian case, the film profile can be obtained

$$\delta(\xi) = \delta_0 \left[\frac{4\left[1 + Br\left(\frac{\delta_0}{\bar{\delta}}\right)^{n-1}\left(\frac{b'}{1-e^{-b'}}\right)^{n+1}\left(\frac{b' - 1 + e^{-b'}}{b'^2}\right)\right]\xi}{U_2}\right]^{1/2} \tag{6.310}$$

where $\bar{\delta}$ is an assumed value of the mean melt thickness and

$$U_2 = 2\frac{1 - b' - e^{-b'}}{b'\left(e^{-b'} - 1\right)} \tag{6.311}$$

By substitution of the parameters we find

$$\delta = \left[\frac{2\left(2k_m(T_0 - T_m) + U_1\right)\xi}{U_2 \rho_m U_0 \lambda'} \right] \tag{6.312}$$

where

$$U_1 = \frac{2m_0 U_0^{n+1}}{\bar{\delta}^{n-1}} \left(\frac{b'}{1 - e^{-b'}} \right)^{n+1} \frac{b' - 1 + e^{-b'}}{b'^2} \tag{6.313}$$

And the rate of melting, per unit width, is given by

$$w_L(\xi) = \left[\frac{\rho_m U_0 U_2 \left(k_m(T_0 - T_m) + U_1/2\right)\xi}{\lambda'} \right]^{1/2} \tag{6.314}$$

Convection can be accounted for, using the same approximation used for the Newtonian case [25, 27], then λ' is replaced by λ'' (eqn. (6.301)), and $w_L(\xi)$ is reduced by a factor of $\sqrt{2}$, i.e.,

$$w_L(\xi) = \left[\frac{\rho_m U_0 U_2 \left(k_m(T_0 - T_m) + U_1/2\right)\xi}{2\lambda''} \right]^{1/2} \tag{6.315}$$

where the mean temperature of the film is

$$\bar{\Theta} = \frac{b'/2 + e^{-b'}(1 + 1/b') - 1/b'}{b' + e^{-b'} - 1} \tag{6.316}$$

6.9.3 Melting Zone in a Plasticating Single Screw Extruder

One of the most important aspects when designing the screw geometry for a single screw extruder is to control the melting of the pellets within the screw. Here, we will present the model developed by Tadmor [27], which is based on observations of the state of the material along the screw channel. If we unwrap the channel contents several characteristics can be recognized, as schematically represented in Fig. 6.68. The first section of the screw, the solids conveying zone, compacts and pressurizes the polymer pellets and, as discussed in Chapter 3 of this book, is responsible for starting the motion in the down-channel direction. The first portion of the solids conveying zone is cooled with chilled water in order to avoid melt formation on the barrel surface. However, after the chilled region, due to friction of the pellets against each other and against the barrel surface, as well as the effect of the nearby heaters, a melt film forms between the bed of solid pellets and the barrel. One to three turns later, a melt pool forms against the trailing flight of the screw. These few turns, between the melt film formation and the start of the melt pool is often referred to as the *delay zone*. The delay zone has been described by various researchers in the past, and was quantitatively described by Tadmor and Klein [27]. Noriega *et. al* [18] were able to clearly discern its existence and attributed it to a *solid bed saturation process* where the melt pool does not form until the melt has completely filled the gaps between the pellets. From that point on, a solid bed profile develops as schematically depicted in Fig. 6.68.

Cutting a cross-section A-A of the channel contents in Fig. 6.68, leads to a cross-sectional view that is shown Fig. 6.69. Furthermore, to help visualize and to aid in the mass balances, we can break the channel contents in the melting section down to a differential element shown in Fig. 6.70.

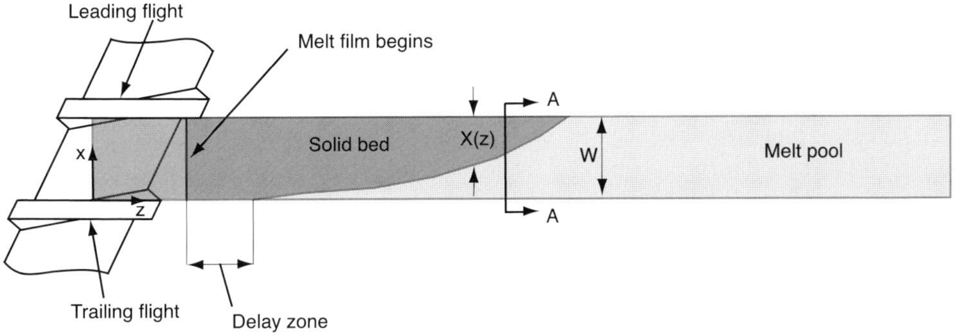

Figure 6.68: Schematic diagram of unwrapped channel contents of a single screw extruder.

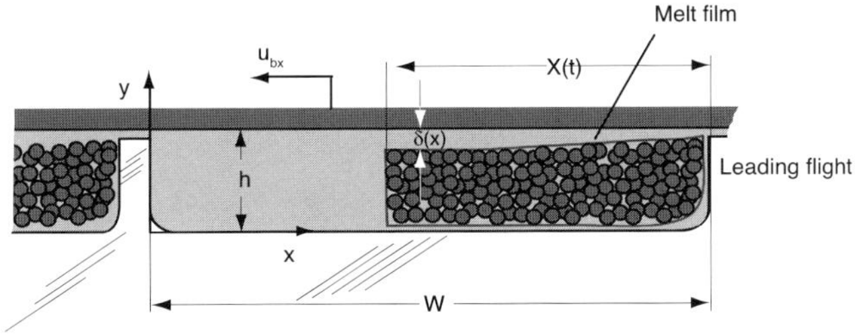

Figure 6.69: Schematic diagram of a channel cross-section in the melting zone of a single screw extruder.

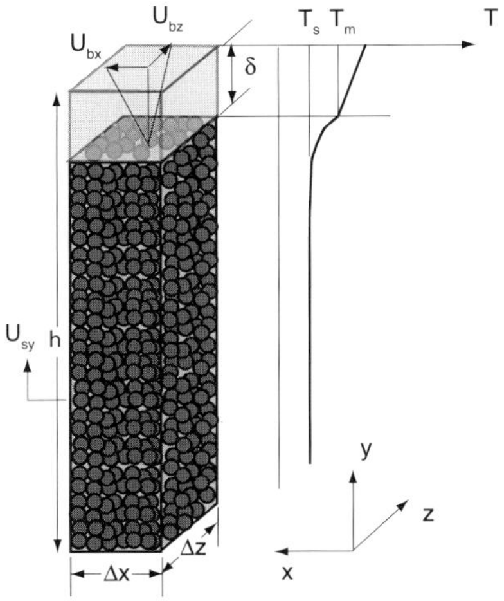

Figure 6.70: Differential element of the channel contents in the melting section of a single screw extruder. The element includes the temperature distribution across the channel as well as isothermal velocity distributions.

Once the melt film forms, the conveying mechanism changes at the barrel surface where viscous drag is now dominant, but frictional drag is still important at the root of the screw and the flights. The thickness of the melt film continues to increase as the plug proceeds down the channel until it attains a value of several times the flight clearance. At this point, the melt film thickness stays nearly constant and the melt is scraped off and accumulated on the side of the trailing or pushing flight. The axial distance from where the melt film appears until melt begins to accumulate at the pushing flight is referred to as the delay zone.

The Tadmor model assumes Newtonian fluids and shallow channels. The channel cross section and that of the solid bed are assumed to be rectangular. The width of the solid bed profile is denoted by $X(z)$, which is the the main objective that we are seeking with the model. The solid bed that develops at steady state conditions is the focal interest here. Furthermore, Tadmor assumed that melting only occurs at the barrel surface and the solid bed is homogeneous, continuous and deformable.

From observation, it is clear that the solid bed has a local down channel velocity u_{sz} and a local velocity component into the melt film of u_{sy}. As before, we resolve the barrel surface velocity U_b into a down channel, u_{bz}, and a cross channel component, u_{bx}. Using Tadmor's notation, we define the relative velocity between barrel surface and solid bed as

$$|U_j| = \sqrt{u_{bx}^2 + (u_{bz} - u_{sz})^2} \tag{6.317}$$

which controls the rate of viscous dissipation. The rate of melting per unit down channel distance is obtained directly from the Newtonian model of melting with drag removal

$$w_L = \left[\frac{U_0 \rho_m \left(k_m(T_0 - T_m) + \mu U_0^2/2\right)}{2\lambda''} W\right]^{1/2} \tag{6.318}$$

with X replacing W, T_b replacing T_0, u_{bx} replacing U_0 in the first term and U_j replacing U_0 in the second term, we have

$$w_L(z) = \left[\frac{u_{bx} \rho_m \left(k_m(T_b - T_m) + \mu U_j^2/2\right)}{2\left[\lambda + C_s(T_m - T_{s0}) + C_m \bar{\Theta}(T_b - T_m)\right]} X\right]^{1/2} \tag{6.319}$$

The change in the solid width is obtained by a differential mass balance as follows

$$\rho_s u_{sz}(h - \delta)X|_z - \rho_s u_{sz}(h - \delta)X|_{z+\Delta z} = w_L \Delta z \tag{6.320}$$

Neglecting the film thickness change in the down channel direction and taking the limit as $\Delta z \to 0$ we get

$$-\frac{d(HX)}{dz} = \frac{w_L(z)}{\rho_s u_{sz}} \tag{6.321}$$

By substituting the definition of $w_L(z)$, it reduces to

$$-\frac{d(HX)}{dz} = \frac{\Phi\sqrt{X}}{\rho_s u_{sz}} \tag{6.322}$$

where

$$\Phi = \left[\frac{u_{bx} \rho_m \left(k_m(T_b - T_m) + \mu U_j^2/2\right)}{2\left[\lambda + C_s(T_m - T_{s0}) + C_m \bar{\Theta}(T_b - T_m)\right]}\right]^{1/2} \tag{6.323}$$

For a constant channel depth, eqn. (6.322) can be integrated to give

$$\frac{X_2}{W} = \frac{X_1}{W}\left[1 - \frac{\psi(z_2 - z_1)}{2H}\right]^2 \tag{6.324}$$

where X_1 and X_2 are the widths of the solid bed at locations z_1 and z_2, respectively and

$$\psi = \frac{\Phi}{u_{sz}\rho_s\sqrt{X_1}} \tag{6.325}$$

Hence, for a constant channel depth we can determine the length of the channel required to melt the solid bed from eqn. (6.324). For a tapered channel of constant taper, eqn. (6.322) can be written as

$$-\frac{d(HX)}{dz} = \frac{\Phi\sqrt{X}}{A\rho_s u_{sz}} \tag{6.326}$$

where

$$A = -\frac{dH}{dz} \tag{6.327}$$

which integrated will give

$$\frac{X_2}{W} = \frac{X_1}{W}\left[\frac{\psi}{A} - \left(\frac{\psi}{A} - 1\right)\sqrt{\frac{H_1}{H_2}}\right]^2 \tag{6.328}$$

where X_1 and X_2 are the widths of the solid bed at down channel locations corresponding to heights H_1 and H_2. Equations (6.324) and (6.328) represent the basic equations for the melting model of a extruder. The total length of melting for a channel of constant depth is

$$z_T = \frac{2H}{\psi} \tag{6.329}$$

and for the tapered channel is

$$z_T = \frac{H}{\psi}\left(2 - \frac{A}{\psi}\right) \tag{6.330}$$

■ EXAMPLE 6.10.

Solid bed profile prediction for a plasticating single screw extruder. In this example we would like to use Tadmor's model to predict the solid bed profile of a low density polyethylene in a plasticating single screw extruder, based on experiments published by Tadmor and Klein[27]. In their experiments they used the following screw geometry:

- Square pitch screw, $D = 63.5$ mm, L/D =26.5 and W =54.16
- Feed zone - 12.5 turns and h_1 =9.4 mm
- Transition zone - 9.5 turns and constant taper from h_1 =9.4 mm to h_2 =3.23 mm

- Metering zone - 4.5 turns and h_2 =3.23 mm

The processing conditions used in the experiments were:

- Initial temperature of pellets: T_0 =24°C
- Barrel or heater temperature:T_b =149°C
- Pressure buildup: $\Delta p = 204$ bar
- Screw speed: $N = 60$ rpm
- Mass throughput: $\dot{m} = 61.8$ kg/hr

The material properties of the PE-LD to be used in the calculations are for a power law viscosity model and thermal properties independent of temperature and pressure, except for the melt density:

- Viscosity: $m_0 = 56,000$ Pa-s^n, $n = 0.345$, $a = 0.01$/K and T_m =110°C
- Thermal properties: $k_m = 0.1817$ W/m/K, $C_m = 2,596$ J/kg/K, $C_s = 2,763$ J/kg/K, $\rho_{bulk} = 595$ kg/m^3, $\rho_s = 915.1$ kg/m^3, $\rho_m = 852.7 + 5.018 \times 10^{-7}p - 0.4756T$, and $\lambda = 129.8$ kJ/kg.

From the above data, we can first compute the barrel velocity to be $U_b = \pi DN$=0.1995 m/s. Note that this speed is on the low end of realistic speeds used in industry for low density polyethylene. PE-LD usually can have screw speeds that lead to velocities up to of 1 m/s. Other polymers can take up to 0.5 m/s, and PVC about 0.2 m/s. The down channel and cross channel velocities become

$$u_{bz} = U_b \cos\phi \tag{6.331}$$

and

$$u_{bx} = U_b \sin\phi, \tag{6.332}$$

respectively. The solid bed velocity is computed using

$$u_{sz} = \frac{\dot{m}}{\rho_{bulk} W h} = 0.0567 \text{ m/s} \tag{6.333}$$

The relative speed between the solids bed and the barrel is calculated using

$$U_j = \sqrt{u_{bx}^2 + (u_{bz} - u_{sz})^2} = 0.1465 \text{ m/s} \tag{6.334}$$

We can now compute

$$b' = -a(T_0 - T_m)/n = -1.1304 \tag{6.335}$$

$$U_2 = 2\frac{1 - b' - e^{-b'}}{b'\left(e^{-b'} - 1\right)} = 2\left[\frac{1 + 1.1304 - e^{1.1304}}{-1.1304\left(e^{1.1304} - 1\right)}\right] = 0.8155 \tag{6.336}$$

and

$$\bar{\Theta} = \frac{b'/2 + e^{-b'}(1 + 1/b') - 1/b'}{b' + e^{-b'} - 1} = 0.7002 \tag{6.337}$$

Table 6.1: Thermal Properties for Selected Polymeric Materials

Turns	z	δ (m)	U_1(N/s)	ψ	X(m)	X/W
7.0	1.466	0.0002967	13.63	0.003169	0.05416	1.000
7.5	1.571	0.0002905	13.44	0.003158	0.05227	0.965
8.0	1.675	0.0002844	13.26	0.003147	0.05227	0.931
8.5	1.780	0.0002783	13.07	0.003137	0.05043	0.898
9.0	1.885	0.0002723	12.88	0.003126	0.04863	0.866
9.5	1.990	0.0002664	12.70	0.003115	0.04688	0.834
10.0	2.094	0.0002605	12.52	0.003105	0.04517	0.803
10.5	2.199	0.0002548	12.33	0.003094	0.04350	0.773
11.0	2.304	0.0002490	12.15	0.003083	0.04188	0.744
11.5	2.408	0.0002434	11.97	0.003073	0.04030	0.716
12.0	2.513	0.0002378	11.79	0.003062	0.03876	0.688
12.5	2.618	0.0002323	11.61	0.003052	0.03580	0.661

Neglecting the pressure effects, we can compute the melt density using an average temperature of 129.5°C as

$$\rho_m = 852.7 + -0.4756T = 852.7 - 0.4756(129.5) = 791.1 \text{ kg/m}^3 \tag{6.338}$$

The rest of the equations depend from each other and will reduce to:

$$U_1 = \frac{2m_0 U_0^{n+1}}{\bar{\delta}^{n-1}} \left(\frac{b'}{1-e^{-b'}}\right)^{n+1} \frac{b' - 1 + e^{-b'}}{b'^2} = 2786.8(\delta)^{0.655} \tag{6.339}$$

$$\begin{aligned} \delta &= \left[\frac{2\left(2k_m(T_0 - T_m) + U_1\right)\xi}{U_2 \rho_m U_0 \lambda'}\right] \\ &= 2.418 \times 10^{-4}[(14.7 + U_1)\xi]^{1/2} \end{aligned} \tag{6.340}$$

$$\begin{aligned} \Phi &= \left[\frac{U_{bx}\rho_m\left(k_m(T_b - T_m) + \mu U_j^2/2\right)}{2\left[\lambda + C_s(T_m - T_{s0}) + C_m\bar{\Theta}(T_b - T_m)\right]}\right]^{1/2} \\ &= 4.72 \times 10^{-3}[14.17 + U_1]^{1/2} \end{aligned} \tag{6.341}$$

$$\begin{aligned} \psi &= \frac{\Phi}{U_{sz}\rho_s\sqrt{X_1}} \\ &= 6.01 \times 10^{-4}[14.17 + U_1]^{1/2} \end{aligned} \tag{6.342}$$

For the constant channel depth region in the solids section, we compute with the initial position of the melt film of 1.26 m, we can write

$$\frac{X}{W} = [1 - 53.19\psi(z - 1.26)]^2 \tag{6.343}$$

Equations (6.339), (6.340), (6.342) and (6.343) must now be solved simultaneously starting at the end of the 6th turn. The results are presented in Table 6.1.

The remaining two sections are computed in a similar fashion. Figure 6.71 presents a comparison between the measured and the predicted solid bed profiles.

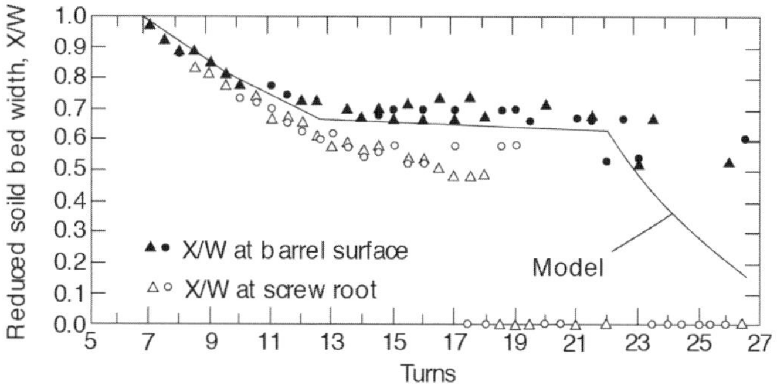

Figure 6.71: Predicted and experimental solids bed profile [27].

6.10 CURING REACTIONS DURING PROCESSING

While thermoplastics solidify and melt by lowering and raising the temperature below and above transition temperatures, thermosets and elastomers undergo an irreversible cross-linking or curing reaction during processing. When processing these materials it is important to predict the cycle time, in order to know when to demold a part. In addition, predicting the curing reaction allows us to optimize, control and understand the process. In this section, we follow the procedure used by Enns and Gillham [7].

As discussed in Chapters 2 and 4, a general cure kinetic equation that describes the reaction is given by

$$\frac{dc}{dt} = ae^{-E/RT} f(c) \tag{6.344}$$

where the function $f(c)$ describes the nature and order of the reaction. For a zero order reaction $f(c) = 1$, a first order reaction $f(c) = 1 - c$, and a second order reaction $f(c) = (1 - c)^2$. A first order autocatalytic reaction is described by $f(c) = (B + c)(1 - c)$ and a second order autocatalytic reaction by $f(c) = (B + c)(1 - c)^2$. For the sake of the analysis here, let us assume a first order reaction such that

$$\frac{dc}{dt} = ae^{-E/RT}(1 - c) \tag{6.345}$$

We can rearrange eqn. (6.345) and form the integral

$$t = \int_0^{c(t)} \frac{ae^{E/RT}}{1 - c} dc \tag{6.346}$$

which after integration results in

$$t = -ae^{E/RT} \ln(1 - c) \tag{6.347}$$

This equation can be used to solve for the time to gelation

$$t_{gel} = -ae^{E/RT} \ln(1 - c_g) \tag{6.348}$$

and using DiBenedetto's equation,[3] which relates the glass transition temperature of a curing resin as a function of the degree of cure [5]

$$\frac{T_g - T_{g0}}{T_{g0}} = \frac{(E_x/E_m - F_x/F_m)c}{1 - (1 - F_x/F_m)c} \qquad (6.349)$$

where E_x/E_m is the ratio of lattice for cross-linked and uncross-linked resins and F_x/F_m is the ratio of molecular mobilities. The ratio $\frac{E_x/E_m}{F_x/F_m}$ is determined from measurements of the minimum and maximum glass transition temperature as

$$\frac{T_{g1}}{T_{g0}} = \frac{E_x/E_m}{F_x/F_m} \qquad (6.350)$$

The individual ratios, E_x/E_m and F_x/F_m, are then determined by fitting eqn. (6.349) to measured data of glass transition temperatures as a function of cure. We can take eqn. (6.349) and solve for the degree of cure

$$c = \frac{\hat{T}_g}{E_x/E_m - 1 + (1 - F_x/F_m)\hat{T}_g} \qquad (6.351)$$

which in turn can be substituted into eqn. (6.347) to give

$$t_{vitr} = -ae^{E/RT}\ln\left(\frac{(F_x/F_m)\hat{T}_g}{E_x/E_m - 1 + (1 - F_x/F_m)\hat{T}_g}\right) \qquad (6.352)$$

We are now in the position to construct a time-temperature-transformation (TTT) diagram as presented in Chapter 2. Figure 6.72 presents a TTT diagram generated for an epoxy with $E_x/E_m = 0.34$, $F_x/F_m = 0.19$, $T_{g0} = 254$ K, $T_{g1} = 439$ K, $a = 4.5 \times 10^6 \mathrm{min}^{-1}$ and $R = 1.987$ cal/mol/K.

6.11 CONCLUDING REMARKS

This chapter gave an overview of how to simplify complex processes sufficiently to allow the use of analytical models for their analysis and optimization. These models are based on mass, momentum, energy and kinetic balance equations, with simplified constitutive models. At one point, as the complexity and the depth of these models increases by introducing more realistic geometries and conditions, the problems will no longer have an analytical solution, and in many cases become non-linear. This requires the use of numerical techniques which will be covered in the third part of this book, and for the student of polymer processing, perhaps in a more advanced course.

Problems

6.1 Plot the screw characteristic curve at 40, 80 and 120 rpm screw speed for a 45 mm square pitch extruder with a channel depth of 4 mm in the metering section and a flight width of 4 mm. The metering section of the extruder is 5 turns long. Assume a Newtonian viscosity μ =1000 Pa-s.

[3]Another form of this equation was presented in Chapter 2 as $T_g = T_{g0} + \frac{(T_{g1} - T_{g0})\lambda c}{1 - (a - \lambda)c}$.

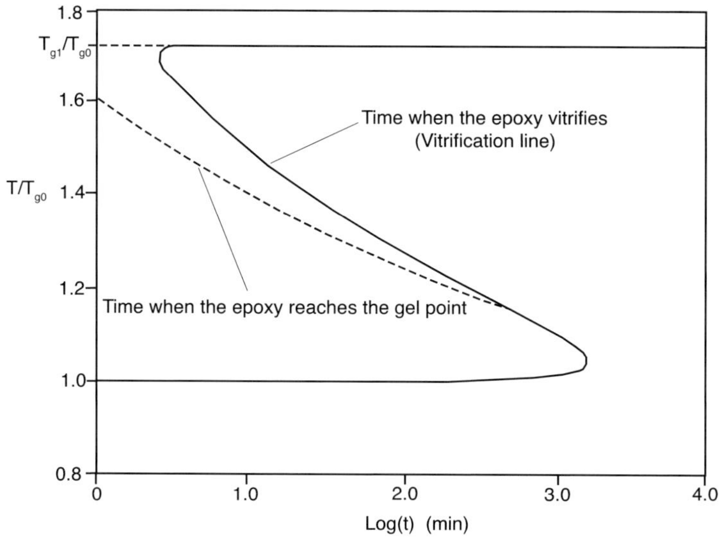

Figure 6.72: Predicted TTT behavior for an epoxy using a first order reaction kinetic model.

6.2 You are to use a square pitch 60 mm diameter extruder to extrude fibers through a spinnerette with 100 capillaries. Each capillary has a 0.2 mm diameter and is 1mm long.

a) What is the ideal channel depth in the metering section for this application? Assume a metering section that is 6 turn long and has a flight width of 4.5 mm.

b) What is the ideal helix angle for this application?

6.3 Derive the equations for a wire coating operation with a Newtonian fluid model and the notation presented in Fig. 6.37.

6.4 Formulate the die design equation for an end-fed sheeting die with a constant manifold diameter and a variable die land gap. Assume a Newtonian viscosity.

6.5 Formulate the die design equation for an end-fed sheeting die such as the one solved for in this chapter. Unlike that die, the new die should have a variable radius manifold as presented in Fig. 6.73. In this example, the axial distance from the manifold center to the die lips must be maintained constant. Assume a Newtonian viscosity, μ.

6.6 Using the geometry and notation given in Fig. 6.14, relate the die land length ratio, L_1/L_2 for a die land thickness $h_1/h_2 = 2$ to the power law index of a shear thinning polymer.

6.7 The flow in a wire coating die can be modeled using a combination of drag and pressure flow. Derive an expression for the velocity field inside the die.

a) Derive an expression for the volumetric flowrate through the die, Q.

b) Derive an expression for the wire coating thickness.

6.8 Derive an equation that will give the coating thickness in a blade coating operation schematically depicted in Fig. 6.74. Assume a Newtonian fluid and neglect inertia effects.

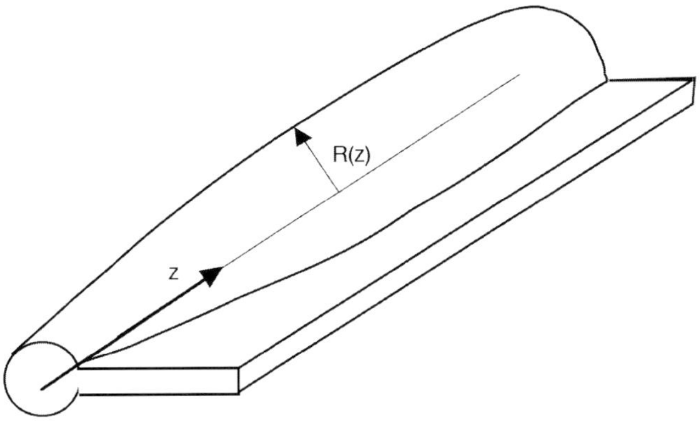

Figure 6.73: Schematic diagram of an end-fed sheeting die with a variable manifold diameter.

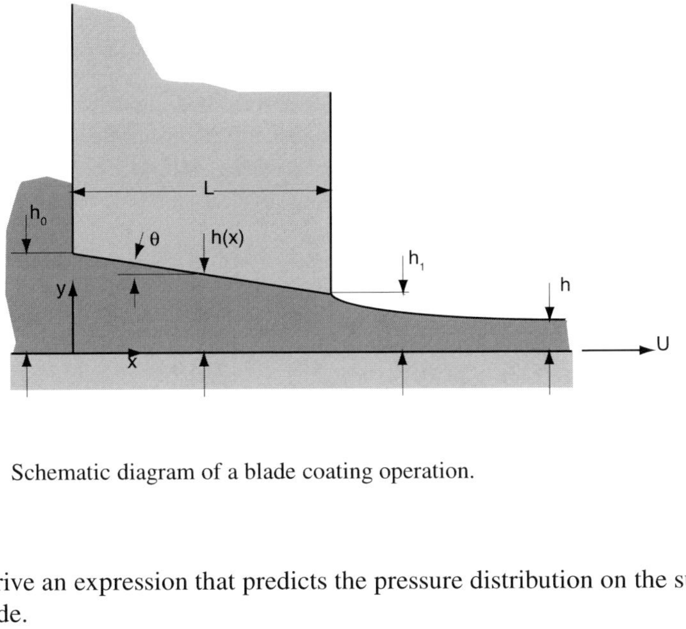

Figure 6.74: Schematic diagram of a blade coating operation.

a) Derive an expression that predicts the pressure distribution on the surface of the blade.

b) Derive an expression that predicts the velocity field under the blade.

6.9 Derive the die design equation (eqn. (6.50)) for a coat hanger die assuming a shear thinning power law viscosity model.

6.10 A polyamide 66 fiber is extruded into air and drawn to a draw down ratio, D_R, of 100 and a length of 3 m. The take-up velocity is 1km/min and the final radius of the fiber is 100 μm.

a) Using the data above as characteristic values for a fiber spinning operation, determine if inertia is significant in the momentum balance.

b) What force is required to draw the fiber?

c) Should we have included inertial effects in the analysis?

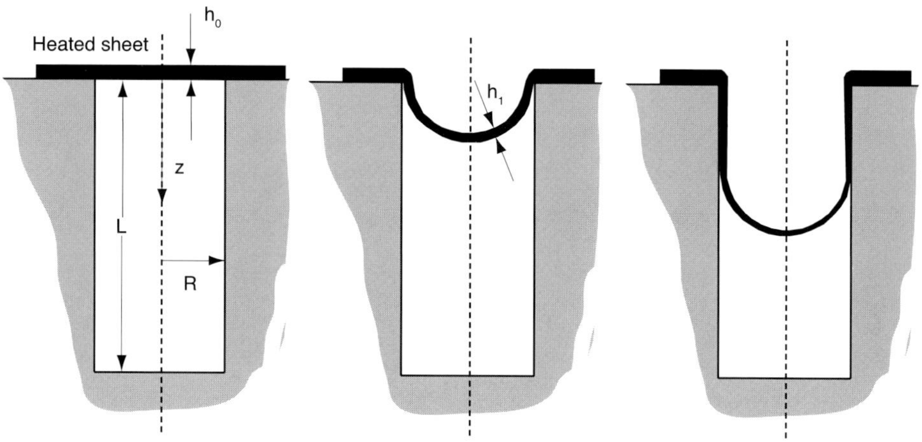

Figure 6.75: Schematic diagram of the thermoforming process of a cylindrical geometry.

6.11 Derive eqn. (6.89), which describes the velocity distribution along a spinning fiber using a shear thinning power law viscosity model.

6.12 Equation (6.89) can be used to predict the velocity distribution along a spinning fiber using a shear thinning power law viscosity model.

a) Plot the dimensionless velocity distribution along the fiber for power law indeces of 0.2, 0.5 and 1.0.

b) Spearot and Metzner [23] performed fiber spinning experiments on polyethylene melts. Find Spearot and Metzner's journal paper in your library and compare their data to eqn. (6.89). How does the equation compare to the experiments?

6.13 You are asked to extrude a 15 kg/h of a 100 μm thick polyethylene plastic bag with a lay-flat witdth of 20 cm. The annular extrusion die has diameter, R_0 of 5 cm and a gap, h_0, of 1 mm. Assume an isothermal Newtonian process up to the freezeline with a viscosity $\mu = 3 \times 10^4$ Pa-s and a density ρ =950 kg/m^3.

a) Determine the operating conditions.

b) How would the process change if you were to increase the bubble pressure by 20%?

6.14 Derive the equation that describes the thickness distribution after thermoforming the constant diameter tubular geometry described in Fig. 6.75. Use the notation presented in the figure. Assume that the sheet stops stretching once it touches the mold wall.

6.15 Derive the equation that describes the thickness distribution after thermoforming the long trough described in Fig. 6.76. The process is similar to the one schematically depicted in Fig. 6.75. Assume that the height of the trough, H, is much larger than the depth, L.

6.16 A 15 cm diameter, 2 m wide calendering system is used to produce a PVC sheet with a Newtonian viscosity of 10,000 Pa-s. The roll speed, U is 6 m/s and the distance between the nips is 5 mm.

a) What is the thickness of the PVC sheet?

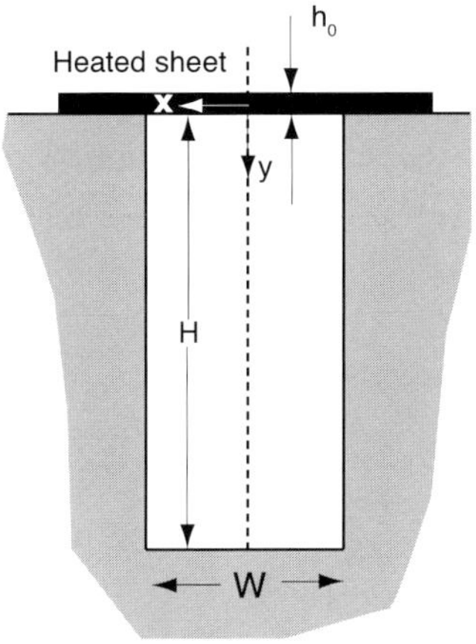

Figure 6.76: Schematic diagram of the thermoforming process of a deep trough.

b) What is the maximum pressure between the rolls?
c) What is the maximum roll separating force?
d) Calculate the maximum roll deflection for a steel roll with a thickness of 5mm.
e) How much power is required to drive the system?
f) Estimate the maximum temperature rise within the PVC.

6.17 A 15 cm diameter, 2 m wide calendering system is used to produce a PVC sheet with a shear thinning power law viscosity with a consistency index, m=17,000 Pa-s^n and a power law index, n=0.26. The roll speed, U=6 m/s and the distance between the nips is 5 mm.

a) What is the thickness of the PVC sheet?
b) What is the maximum pressure between the rolls?
c) What is the maximum roll separating force?
d) Calculate the maximum roll deflection for a steel roll with a thickness of 5mm.
e) How much power is required to drive the system?
f) Estimate the maximum temperature rise within the PVC.

6.18 Solve for the pressure distribution for the two previous problems and compare them to each other. On the same graph, plot the dimensionless pressure p/p^{max} for both cases.

6.19 Derive a model to predict flow and presssure distributions for the case of calendering a sheet of finite thickeness and Newtonian viscosity.

6.20 Internal batch mixers can sometimes be simplified to a system such as the one shown in Fig. 6.77. For $h \ll R$, the system can be further simplified by unwrapping the material from the cylinder and modeling it using rectangular coordinates.

a) What is the velocity distribution accross the gap far from the lip?
b) What is the pressure rise inside the system?

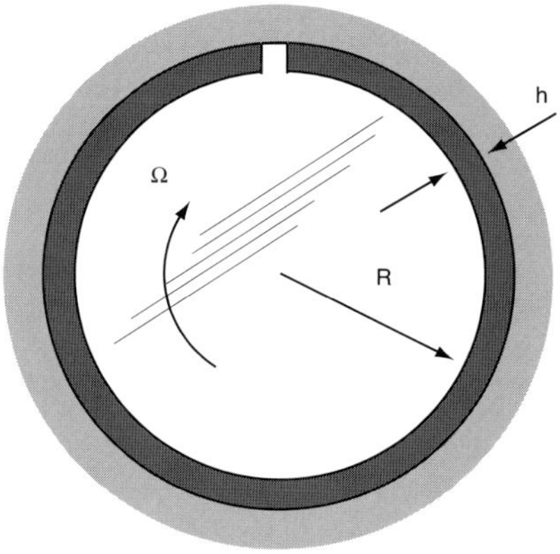

Figure 6.77: Schematic diagram of a simplified internal batch mixer geometry.

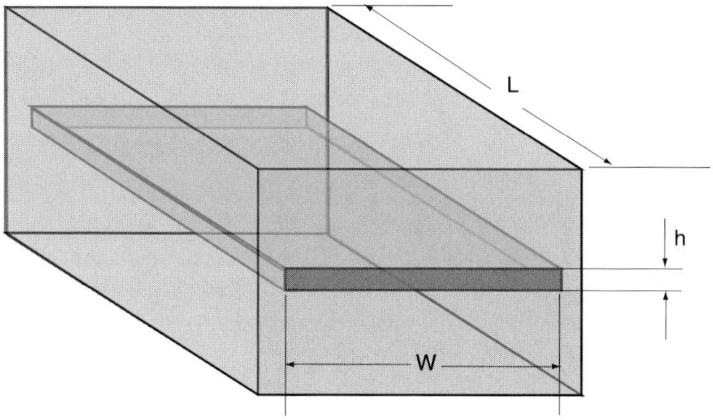

Figure 6.78: Schematic diagram of a pressure driven slit die.

6.21 Derive the equation for the cumulative residence time distribution, $F(t)$, for the fluid driven by pressure flow inside a slit. Assume a volumetric flow rate of Q and a Newtonian viscosity of μ. Use the notation used in the schematic of Fig. 6.78.

6.22 Derive the cumulative residence time distribution, $F(t)$, for pressure driven flow inside a slit of a power law fluid. Use the notation presented in Fig. 6.78.

6.23 For the slit-shaped cavity with a unidirectional flow, presented in Fig. 6.79, derive an expression for the pressure at the gate during mold filling for a constant injection speed. Assume a volumetric flow rate Q and an isothermal flow of a shear thinning polymer with a power law model.

a) Assuming that the polymer is a polystyrene with a consistence index of $m = 2.8 \times 10^4$ Pa-s^n and a power law index $n = 0.28$, with an injection speed of $Q = 30$ cm^3/s, mold geometry $h = 1$mm, $L = 15$ cm and $W = 20$ mm, determine the filling time, and plot the pressure at the gate as a function of mold filling. Use the geometry and notation presented in Fig. 6.79.

b) For the above numerical values, determine if viscous dissipation effects are significant.

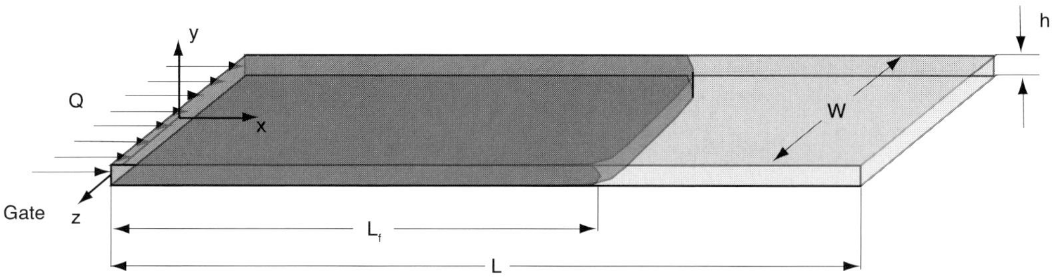

Figure 6.79: Schematic diagram of a slit-shaped injection mold cavity.

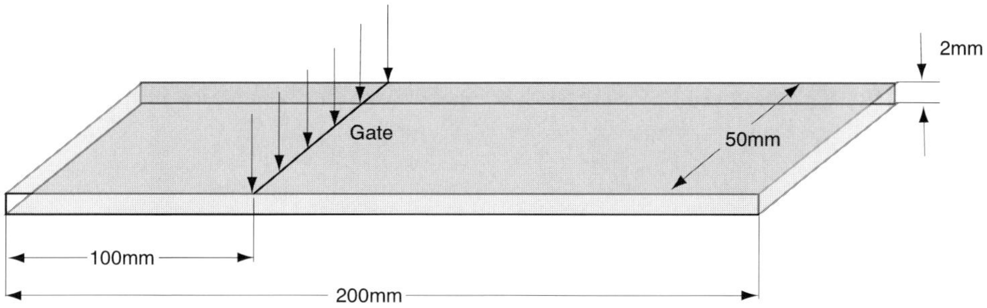

Figure 6.80: Schematic diagram of an off-center gated slit-shaped injection mold.

6.24 A long thin part is bein injection molded with the dimension and gate location shown in Fig. 6.80. The mold is filled at a constant flow rate. Estimate the maximum pressure at the gate and the clamp force at the instant just before the mold fills for a filling time of 2 seconds. Assume an isothermal flow of a shear thinning power law high-density polyethylene with a consistence index of 2×10^4Pa-s^n and a power law index of 0.41.

6.25 For compression molding inside the disc-shaped cavity shown in Fig. 6.81, derive an expression for the flow field and pressure distribution inside the mold during filling for an arbitrary flow front location R_f. Assume a Newtonian fluid and the notation found in the figure.

a) Derive an expression for the closing force as a function of flow front position.

b) Assume you are compression molding polypropylene with a viscosity $\mu = 1000$ Pa-s, inside a cavity with a 50 cm radius and an initial 25 mm thick, 20 cm radius initial charge. Assume a closing speed of 10 mm/s.

c) For the above numerical values, determine if viscous dissipation effects are significant.

6.26 Determine what the viscous dissipation effects are in a melting with pressure removal problem. Use the notation of Fig. 6.65.

a) Use material properties that pertain to polyethylene to assess the effects of viscous heating.

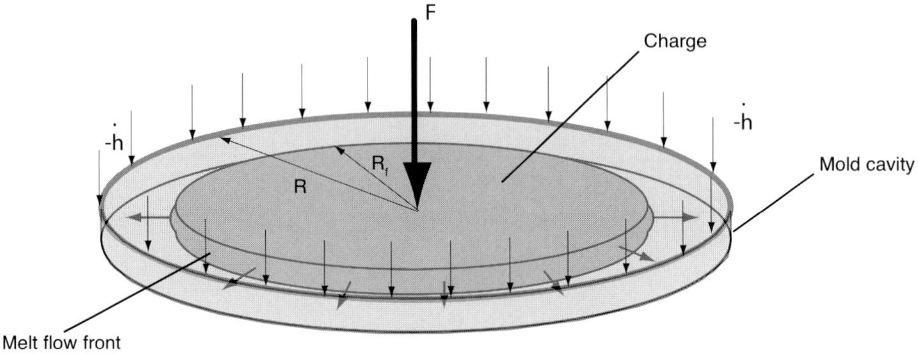

Figure 6.81: Schematic diagram of a disc-shaped compression mold cavity.

b) In your university library, find the paper on melting with pressure melt removal by Stammers and Beek [24]. Which of Stammers and Beek's assumptions, if any, could be changed to achieve a more accurate solution?

6.27 Using the data from Example 6.10, estimate the screw length required to melt the solid bed for a constant channel depth extruder. Estimate the screw length required if the entire screw were tapered.

6.28 A square pitch plasticating extruder has the following geometry:

- $D = 50$ mm
- $W = 42$ mm
- $L/D = 25$
- Feed section: $h = 12$ mm, 5 turns
- Transition: constant taper, 15 turns
- Metering section: $h = 4$ mm, 5 turns

The material being extruded is polypropylene with the following properties: $\rho =$ 900 kg/m^3, T_m =430 K, $k_m = 0.5$ W/m/K, C_p =2300 J/kg/K, μ =2000 Pa-s, and $\lambda = 23.4 \times 10^4$ J/kg. The operating conditions are: $n = 50$ rpm, T_0 =373 K (Pellet temperature inside the hopper), T_b =500 K and $\dot{m} = 2.5 \times 10^{-2}$ kg/s. Melting starts at the end of the third turn of the feed section. For a Newtonian, temperature independent viscosity, find the solid bed profile and calculate the percent of solids in the output.

6.29 Solve the above plasticating extruder problem with a mass throughput $\dot{m} = 1.8 \times 10^2$ kg/s.

6.30 Solve the above plasticating extruder problem with a mass throughput $\dot{m} = 3.6 \times 10^2$ kg/s.

6.31 Construct a time-temperature-transformation diagram for a thermoset that follows a second order reaction kinetic model described by

$$\frac{dc}{dt} = ae^{-E/RT}(1 - c^2)$$

Assume $E_x/E_m = 0.34$, $F_x/F_m = 0.19$, T_{g0} =254 K, T_{g1} =439 K, $a = 4.5 \times 10^6$ min^{-1}, and $R = 1.987$ cal/mol/K.

a) Plot the rate of curing as a function of degree of cure.

b) Plot the degree of cure and rate of curing as a function of time for a couple of sample processing temperatures.

c) When constructing the TTT diagram, please include lines for 20%, 50%, 90% and 95% cure.

6.32 Construct a time-temperature-transformation diagram for a thermoset that follows an autocatalytic second order reaction kinetic model described by

$$\frac{dc}{dt} = ae^{-E/RT}(B + c)(1 - c)^2$$

Assume the kinetic properties given in the previous problem and $B = 0.5$.

a) Plot the rate of curing as a function of degree of cure

b) Plot the degree of cure and rate of curing as a function of time for a couple of sample processing temperatures.

c) When constructing the TTT diagram, please include lines for 20%, 50%, 90% and 95% cure.

6.33 You are to pultrude a thin fiber reinforced epoxy plate at a rate of 1cm/s. Assume kinetic properties given in the last two problems and that the material is best represented with the autocatalytic second order reaction kinetic model of the previous problem. You are asked to design the die and the process to manufacture this product.

a) Draw a clear diagram of the process.

b) Fully state your assumptions, boundary conditions and simplifications.

c) What should the pultrusion die length be if you want the epoxy to be 90%, 95%, 99% and 99.9% cured when the plate emerges from the die.

REFERENCES

1. J.F. Agassant, P. Avenas, J.-Ph. Sergent, and P.J. Carreau. *Polymer Processing - Principles and Modeling*. Hanser Publishers, Munich, 1991.
2. J.T. Bergen and G.W. Scott. Pressure distribution in the calendering of plastic materials. *J. Appl. Mech.*, 18:101, 1951.
3. P.V. Danckwerts. *Chem. Eng. Sci.*, 2(1):1, 1953.
4. M.M. Denn and R.J. Fisher. The mechanics and stability of isothermal melt spinning. *AIChE J.*, 22:236, 1976.
5. A.T. DiBenedetto and L.E. Nielsen. *J. Macromol. Sci., Rev. Macromol. Chem.*, C3:69, 1969.
6. G. Eder and H. Janeschitz-Kriegl. *Polymer Bulletin*, 11:93, 1984.
7. J.B. Enns and J.K. Gillman. Time-temperature-transformation (ttt) cure diagram: Modeling the cure behavior of thermosets. *J. Appl. Polym. Sci.*, 28:2567–2591, 1983.
8. L. Erwin. *Polym. Eng. Sci.*, 18:1044, 1978.
9. L. Erwin. *Polym. Eng. Sci.*, 18:572, 1978.
10. L. Erwin. *Polym. Eng. Sci*, 18:738, 1978.
11. R.E. Gaskell. The calendering of plastic materials. *J. Appl. Mech.*, 17:334–336, 1950.

12. H. Janeschitz-Kriegl and G. Krobath. *Intern. Polym. Proc.*, 3:175, 1988.

13. C. Kiparissides and J. Vlachopoulis. Finite element analysis of calendering. *Polym.. Eng. Sci.*, 16:712–719, 1976.

14. G. Krobath, S. Liedauer, and H. Janeschitz-Kriegl. *Polymer Bulletin*, 14:1, 1985.

15. J.M. McKelvey. *Polymer Processing*. John Wiley and Sons, Inc., New York, 1962.

16. S. Middleman. *Fundamentals of Polymer Processing*. McGraw-Hill Book Co., New York, 1977.

17. K.Y. Ng. Master's thesis, University of Wisconsin-Madison, Madison, Wisconsin, 1979.

18. M.d.P. Noriega. PhD thesis, University of Wisconsin-Madison, 2003.

19. J.R. Pearson and C.J.S. Petrie. The flow of a tubular film. part i. formal mathematical representation. *J. Fluid Mech.*, 40(1), 1970.

20. J.R. Pearson and C.J.S. Petrie. A fluid-mechanical analysis of the film-blowing process. *Plast. Polym.*, 38:85, 1970.

21. J.R.A. Pearson and C.J.S. Petrie. The flow of a tubular film. part ii. interpretation of the model and discussion of solutions. *J. Fluid Mech.*, 42:609, 1970.

22. N. Phan-Thien. A nonlinear network viscoelastic model. *J. Rheol.*, 31(8):259–283, 1978.

23. J.A. Spearot and A.B. Metzner. Isothermal spinning of molten polyethylenes. *Trans. Soc. Rheol.*, 16:495, 1972.

24. E. Stammers and W.J. Beek. The melting of a polymer on a hot surface. *Polym. Eng. Sci.*, 9(1):49–55, 1969.

25. Z. Tadmor, I.J. Duvdevani, and I. Klein. Melting in plasticating extruders - theory and experiments. *Polym. Eng. Sci.*, 7:198–217, 1967.

26. Z. Tadmor and C.G. Gogos. *Principles of Polymer Processing*. John Wiley & Sons, New York, 1979.

27. Z. Tadmor and I. Klein. *Engineering Principles of Plasticating Extrusion*. Van Nostrand Reinhold Company, New York, 1970.

28. W. Unkrüer. *Beitrag zur Ermittlung des Druckverlaufes und der Fließvorgänge im Walzenspalt bei der Kalenderverarbeitung von PVC-hart zu Folien*. PhD thesis, RWTH-Aachen, 1970.

PART III

NUMERICAL TECHNIQUES

CHAPTER 7

INTRODUCTION TO NUMERICAL ANALYSIS

Part of the inhumanity of the computer is that, once it is competently programmed and working smoothly, it is completely honest.

—Isaac Asimov

In the previous chapters we understood how transport phenomena provides the physical and analytical tools needed to analyze all changes that a polymeric material undergoes during processing. Unfortunately, often the analysis does not result in a simple model or equation.

Often, a model that represents a polymer process is in the form of an algebraic equation, a set of non linear partial differential equations and/or an integral equation, which do not have analytical solutions. Numerical analysis will provide methods for obtaining useful solutions to those mathematical problems [10]. Such methods will give an approximate but satisfactory solution to the problem, which can be used to interprete and understand the problem. In the past decades, and thanks to the evolution of the high-speed digital computers, numerical simulation has evolved together with the sciences of fluid mechanics [10, 18, 26, 27], heat transfer [18, 26], transport phenomena and, of course, polymer rheology [16] and processing [3, 20].

The aim of this part of the book is to present the main and current numerical techniques that are used in polymer processesing. This chapter presents basic principles, such as error, interpolation and numerical integration, that serve as a foundation to numerical techniques, such as finite differences, finite elements, boundary elements, and radial basis functions collocation methods.

7.1 DISCRETIZATION AND ERROR

The analysis of polymer processing is reduced to the balance equations, mass or continuity, energy, momentum and species and to some constitutive equations such as viscosity models, thermal conductivity models, etc. Our main interest is to solve this coupled nonlinear system of equations as accurately as possible with the least amount of computational effort. In order to do this, we simplify the geometry, we apply boundary and initial conditions, we make some physical simplifications and finally we chose an appropriate constitutive equations for the problem. At the end, we will arrive at a mathematical formulation for the problem represented by a certain function, say $f(\mathbf{x}, T, p, \mathbf{u}, ...)$, valid for a domain V. Due to the fact that it is impossible to obtain an exact solution over the entire domain, we must introduce discretization, for example, a *grid*. The grid is just a domain partition, such as points for finite difference methods, or elements for finite elements. Independent of whether the domain is divided into elements or points, the solution of the problem is always reduced to a discreet solution of the problem variables at the points or nodal pointsinxxnodes. The choice of grid, i.e., type of element, number of points or nodes, directly affects the solution of the problem.

Thus, the first step in computing a numerical simulation is the construction of a grid. A well-constructed grid can result in a very accurate solution, but a poorly constructed grid is a major contributor to error in the solution. For example, the lack of convergence to a desired level normally is a result of a poor grid quality. The reader is encouraged to seek more information on grid types or grid generation in the literature [6, 21, 38, 37].

We must keep in mind that, when using numerical techniques, we are only finding an *approximate* solution to the problem for the given nodes inside the domain. The *error* of this approximate solution is defined as the difference between the solution and the true value, i.e.,

$$\text{True value} = \text{approximation} + \text{error} \tag{7.1}$$

Typically, the error in equation (7.1) is called *truncation error*. This is a hypothetical definition of the error, because it excludes *gross errors*, which are caused by unpredictable mistakes (for example, human or mechanical mistakes in an experiment), and *round off errors*, which are a result of having only a finite number of digits. The truncation error is the error that is implied in an approximation. More in-depth definitions on error can be found in the literature, as presented by Achieser [2], Hildebrand [10] or Davis [8].

7.2 INTERPOLATION

Every time we need a specific value of a mathematical function or a property in a table, such as specific heat in a cooling problem, we perform linear interpolation to obtain the desired value. Such a linear interpolation is the simplest way to compute data between known values. The given data can be a result of a physical measurement or numerical calculation that is difficult to process. Hence, in order to find the unknown function that correctly satisfies the data, measured or computed, we are going to set basis functions, such as a line or a first order polynomial, and solve for the coefficients of these functions.

When interpolating, we know a value of a function, $f(x)$, at discrete points x_i, where $i = 1, 2, ..., n$, and $x_i < ... < x_n$. We also define $\Delta x_i = x_i - x_{i-1}$ that can be equal for all points. We would like to find an analytical expression for $f(x)$ that allows us to calculate its value at any arbitrary point, x. If the arbitrary point x is between x_1 and x_n, the problem is called interpolation, however, if it is outside that range, it is called extrapolation. Use of extrapolation can be dangerous, because it assumes that a trend will continue beyond the available data [10, 19].

Linear interpolation, or first order interpolation, is the simplest form of solving the problem, since it assumes a linear behavior between two adjacent points; this is clearly depicted in Fig. 7.1. Here, we only use information given by the two points. Often, this is a good approximation, however, if a better interpolation is desired, we need to include data points beyond the two points used to interpolate. The inclusion of more points increases the order, and may include derivatives of the function at those points.

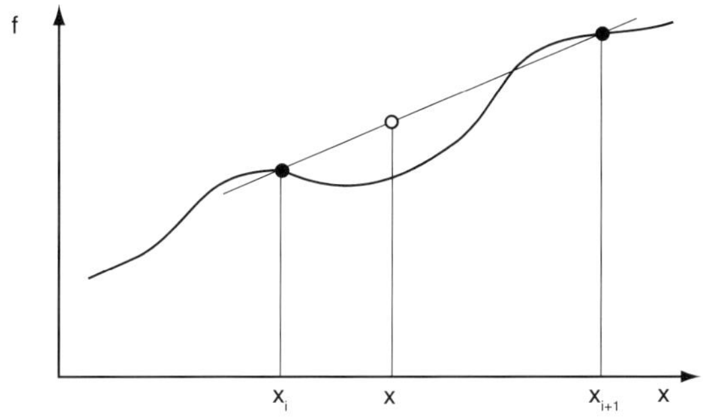

Figure 7.1: Linear interpolation.

Interpolation consists of finding the correlation between the known points according to the selected basis functions. Hence, we need to search for appropriate equations that fit the behavior of our function $f(x)$. For example, in linear interpolation, the chosen function is a straight line. The most commonly used functional forms are polynomials, rational functions, trigonometric functions and radial functions [10, 19, 21].

7.2.1 Polynomial and Lagrange Interpolation

A common form of interpolation is using polynomial basis functions, often referred to as Lagrange interpolation. The order of the polynomial increases with the number of points used; two points fit a unique line (first order polynomial), three points a unique quadratic polynomial, and in general, for any set of n points there is a unique polynomial of degree $n-1$. Hence, the objective is to look for a linear combination of coefficients, that according to the polynomial, will give an expression for the unknown function. Let us begin with a first and second order interpolation for a function $u(x)$, which could represent velocity measurements on a one dimensional grid shown in Fig. 7.2.

Initially, we want to find the velocity for an arbitrary point between i and $i+1$, using only the two adjacent points. Here, the highest polynomial is a first order polynomial or a

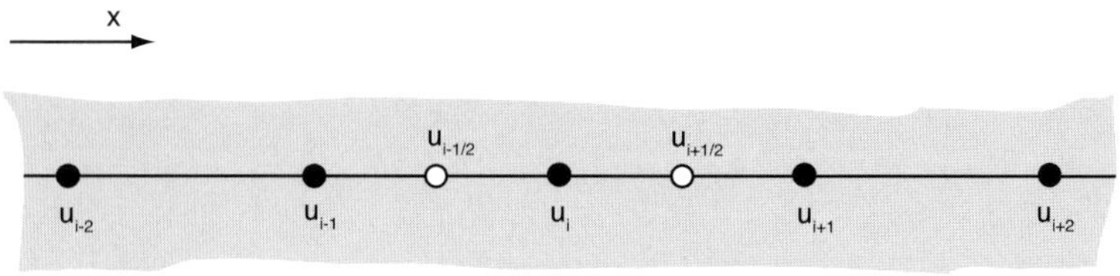

Figure 7.2: Schematic of a one-dimensional grid for $u(x)$.

straight line. In other words, $u(x)$ must be a function of the discreet values of u on points i and $i+1$, as

$$u = b_1 u_{i+1} + b_2 u_i \tag{7.2}$$

and it must fit a first order polynomial, as

$$u = a_1 x + a_2 \tag{7.3}$$

Substituting eqn. (7.3) into eqn. (7.2) results in

$$u = a_1 \left(b_1 x_{i+1} + b_2 x_i\right) + a_2 \left(b_1 + b_2\right) \tag{7.4}$$

which gives us a system of two equations with two unknowns, b_1 and b_2

$$\begin{bmatrix} x_{i+1} & x_i \\ 1 & 1 \end{bmatrix} \begin{pmatrix} b_1 \\ b_2 \end{pmatrix} = \begin{pmatrix} x \\ 1 \end{pmatrix} \tag{7.5}$$

Solving the linear algebraic system (eqn. (7.5)) results in a linear interpolation equation given by

$$u = u_i \frac{x_{i+1} - x}{x_{i+1} - x_i} + u_{i+1} \frac{x - x_i}{x_{i+1} - x_i} \tag{7.6}$$

Following the same procedure for a point x that is between $i-1$ and i, the desired relation will be

$$u = u_{i-1} \frac{x - x_i}{x_{i-1} - x_i} + u_i \frac{x_{i-1} - x}{x_{i-1} - x_i} \tag{7.7}$$

If the points between the intervals are located in the center of the interval (points $i - 1/2$ and $i + 1/2$ in Figure 7.2, both eqns. (7.6) and (7.7) will be reduced to the simple midpoint rule, i.e.,

$$u_{i+1/2} = \frac{1}{2}\left(u_i + u_{i+1}\right) \tag{7.8}$$

$$u_{i-1/2} = \frac{1}{2}\left(u_i + u_{i-1}\right) \tag{7.9}$$

Similarly, we fit a second order polynomial using three points

$$u = b_1 u_{i-1} + b_2 u_i + b_3 u_{i+1} \tag{7.10}$$

with a corresponding second order polynomial required for the interpolation

$$u = a_1 x^2 + a_2 x + a_3 \tag{7.11}$$

After solving for the constants we obtain the following second order interpolation function,

$$\begin{aligned} u(x) = {} & u_{i-1}\frac{(x - x_i)(x - x_{i+1})}{(x_{i-1} - x_i)(x_{i-1} - x_{i+1})} + u_i \frac{(x - x_{i-1})(x - x_{i+1})}{(x_i - x_{i-1})(x_i - x_{i+1})} \\ & + u_{i+1}\frac{(x - x_{i-1})(x - x_i)}{(x_{i+1} - x_i)(x_{i+1} - x_{i-1})} \end{aligned} \tag{7.12}$$

According to eqns. (7.6), (7.7) and (7.12), we can generalize for an arbitrary polynomial interpolation of order $n-1$ (n number of points), by saying that the function $u(x)$ is approximately equal to a linear combination of the known values u_i multiplied by interpolating functions, N_i,

$$u(x) = \sum_{i=1}^{n} N_i u_i \tag{7.13}$$

Depending on the nature of the interpolation functions, the different types of schemes are reached. For the example in Fig. 7.2, the interpolation functions were polynomials. In eqn. (7.13), n is the number of points used in the interpolation, and therefore, $n - 1$ is the order of the interpolation.

Every time that the number of points is increased to do a higher order interpolation, the order of the equation system that is reached will also increase (i.e., linear system for first order interpolation, quadratic system for second order interpolation). It is not recommended to use the same procedure that we used above, to find first and second order interpolation functions when performing a higher order interpolation. There are better methodologies that allow to find a $n - 1$ degree polynomial that satisfies the interpolation. The most common is called Newton's interpolation formula with divided differences [10], which give a general type of $(n - 1)$th degree polynomials called *Lagrange Polynomials*. When these polynomials are used in eqn. (7.13), Lagrange interpolation is obtained [19, 21], i.e.,

$$u(x) = \sum_{i=1}^{n} L_i u_i \tag{7.14}$$

where L_i are the Lagrange polynomials defined as,

$$L_i = \frac{(x - x_1)\dots(x - x_{i-1})(x - x_{i+1})\dots(x - x_n)}{(x_i - x_1)\dots(x_i - x_{i-1})(x_i - x_{i+1})\dots(x_i - x_n)} \tag{7.15}$$

In general, this method works well for interpolation; however, not for extrapolation, as long as not too many points are used [19]. If too many points are used, the resulting polynomial will inevitably lead to oscillations [10, 19, 21].

■ EXAMPLE 7.1.

First, second and 5th order polynomial interpolation for the specific heat capacity of a semi-crystalline thermoplastic (PA6). When performing a heat transfer simulation (heating or cooling) for a thermoplastic, the complete course of the specific heat capacity as a function of temperature is needed. A common way to do this

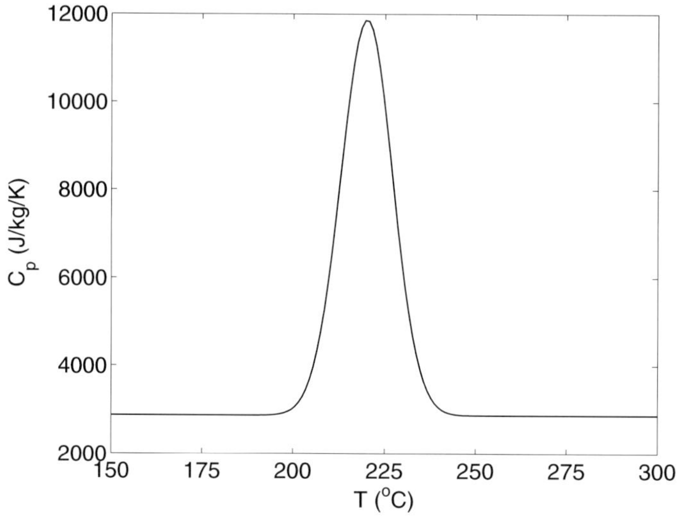

Figure 7.3: Specific heat for a polyamide 6 thermoplastic.

is to take a measured heat capacity, and use a polynomial interpolation between the experimental data to obtain the specific heat capacity for any temperature.

Figure 7.3 shows a typical specific heat measurement of a semi-crystalline thermoplastic (PA6) with a melting temperature around 220°C. Let us assume that we can obtain this continuous curve from a continuous equation. The increase in the C_p represents the heat of fusion of the transition between the semi-crystalline solid to a melt, and is represented with N measurements or discrete points (see Table 7.1). A first, second and 5th order polynomial interpolation will be performed in order to obtain the specific heat for three different temperatures.

The results for each interpolation are shown in Table 7.2 and Fig. 7.4 For 190°C and 250°C, the specific heat is locally a smooth function, therefore the three different schemes give very accurate results. The problem is for the 220°C, this temperature is right in the middle of the melting. The linear interpolation can use only two points, whereas the 2nd and 5th interpolation use more points following the behavior closer. For this particular case the 5th order interpolation is the one that mimics best the complete curve obtained from the equation. However, as already mentioned, we must be careful because a polynomial interpolation of a very high order can lead to oscillations [19].

It should be mentioned here that these concepts of interpolation are commonly applied in numerical analysis, where the values are functions defined at the nodes as a result of a domain discretization (grid). Hence, if the value of the function on a point between nodes $i+1$ and $i-1$ is needed, we can use a first (eqns. (7.6) or (7.7)) or a second order interpolation (eqn. (7.12)). For example, let us define the derivative of $u(x)$ at x_i in terms of the discreet values at locations $i+1/2$ and $i-1/2$, i.e.,

$$\left(\frac{du}{dx}\right)_i \approx f\left(u_{i+1/2}, u_{i-1/2}\right) \tag{7.16}$$

Table 7.1: Specific heat capacity of a polyamide 6 thermoplastic.

Temperature (^oC)	C_p (J/kg/K)
100.0000	2870.0000
110.5263	2870.0000
121.0526	2870.0000
131.5789	2870.0000
142.1052	2870.0000
152.6315	2870.0000
163.1578	2870.0000
173.6842	2870.0000
184.2105	2870.0246
194.7368	2885.2214
205.2631	3895.8068
215.7894	10407.8670
226.3157	8909.5687
236.8421	3397.6402
247.3684	2875.0262
257.8947	2870.0052
268.4210	2870.0000
278.9473	2870.0000
289.4736	2870.0000
300.0000	2870.0000

Table 7.2: Interpolated specific heat capacity (J/kg/K) for each temperature.

T (^oC)	1st	2nd	5th	Eqn.
190.00	2872.6992	2871.8667	2871.3654	2871.1767
220.00	10907.7835	11574.2922	11656.6442	12000.0000
250.00	2875.0205	2844.6609	2874.3131	2871.1445

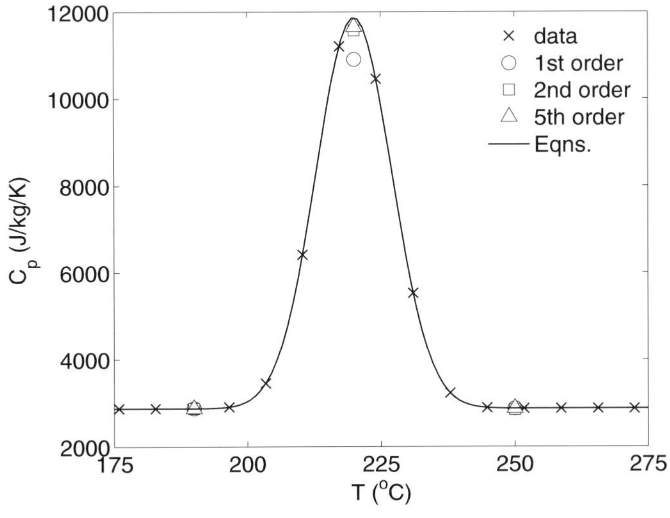

Figure 7.4: Interpolated specific heat.

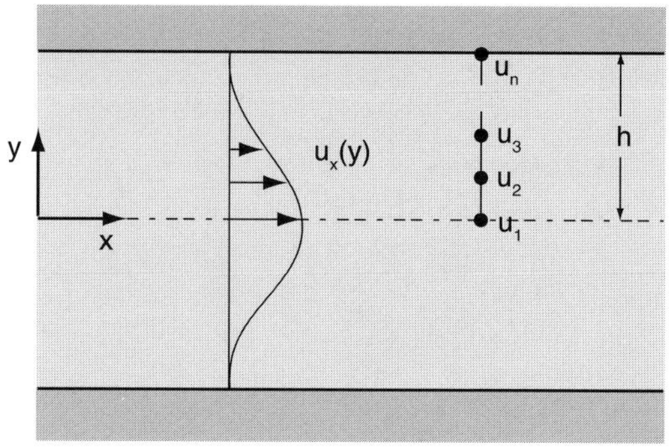

Figure 7.5: Schematic diagram of flow through a slit.

The order of the chosen interpolation to obtain $u_{i+1/2}$ and $u_{i-1/2}$ will affect the value of the derivative, the order of the approximation, and the error. Lagrange's interpolation formula is very useful to perform changes in the order of the interpolation and the points involved. For example, when using a second order interpolation we can use the set $\begin{bmatrix} x_{i-1} & x_i & x_{i+1} \end{bmatrix}$ for both interpolations ($u_{i+1/2}$ and $u_{i-1/2}$). Or we can choose different points for each interpolation, $\begin{bmatrix} x_{i-2} & x_{i-1} & x_i \end{bmatrix}$ for $u_{i-1/2}$ and $\begin{bmatrix} x_i & x_{i+1} & x_{i+2} \end{bmatrix}$ for $u_{i+1/2}$ (see Fig. 7.2), keeping the second order for the interpolation, but increasing the order in the derivative.

■ **EXAMPLE 7.2.**

Interpolation for the velocity gradient of a laminar flow of an incompressible power law fluid in a narrow slit. Let's assume that the velocity gradient in the slit shown in Fig. 7.5 can be calculated in a discrete way as follows

$$\frac{du}{dy} \approx \frac{u_{i+1/2} - u_{i-1/2}}{\Delta y} + O(\Delta y^n) \tag{7.17}$$

where the $i \pm 1/2$ nodes are imaginary located between nodes. From Chapter 5 we know that the velocity and velocity gradient profile of a power law fluid in a narrow slit are

$$u(y) = \left[\frac{(p_0 - p_L)h}{mL}\right]^{1/n} \frac{h}{1/n+1}\left[1 - \left(\frac{y}{h}\right)^{1/n+1}\right] \tag{7.18}$$

$$\frac{du}{dy} = -\left[\frac{(p_0 - p_L)y}{mL}\right]^{1/n} \tag{7.19}$$

A total of 10 equally spaced nodes will be distributed through the thickness and the velocity of each node will be calculated using eqn. (7.18). Then, eqn. (7.17) will be used to calculate the derivative using two different interpolation schemes for the $i \pm 1/2$ nodes: a first order interpolation using the two neighboring nodes and a second order interpolation using $\begin{bmatrix} i-2 & i-1 & i \end{bmatrix}$ for $i - 1/2$ and $\begin{bmatrix} i & i+1 & i+2 \end{bmatrix}$ for the $i + 1/2$.

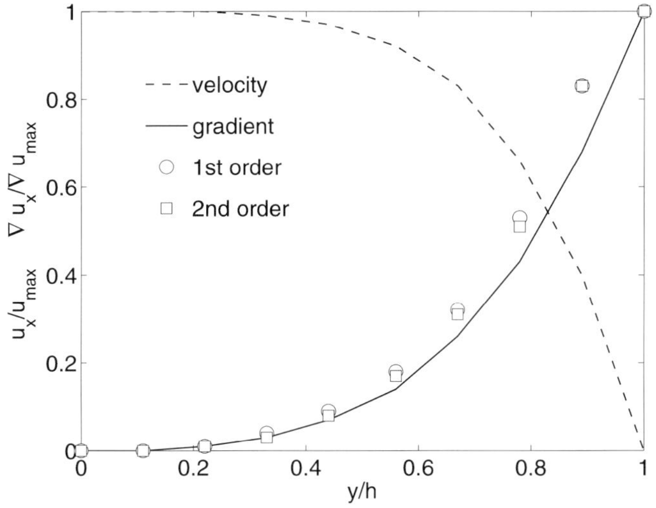

Figure 7.6: Normalized velocity and velocity gradient for the power law fluid on a narrow slit.

The first order interpolation with two neighboring nodes leads to the mid point rule for each imaginary node ($i \pm 1/2$). The derivative will be of the second order

$$\frac{du}{dy} \approx \frac{u_{i+1} - u_{i-1}}{2\Delta y} \tag{7.20}$$

On the other hand, using Langrange's interpolation formula for the second order interpolation we will obtain

$$u_{i+1/2} = \frac{1}{8}\left(3u_i + 6u_{i+1} - u_{i+2}\right) \tag{7.21}$$

$$u_{i-1/2} = \frac{1}{8}\left(3u_i + 6u_{i-1} - u_{i-2}\right) \tag{7.22}$$

and the third order derivative,

$$\frac{du}{dy} \approx \frac{-u_{i+2} + 6u_{i+1} - 6u_{i-1} + u_{i-2}}{8\Delta y} \tag{7.23}$$

Figure 7.6 illustrates the velocity profile from the analytical solution and the comparison between the analytical and the numerical velocity gradient. The velocity has been normalized with respect to the maximum value at the center of the slit ($y = 0$), while the gradients are normalized with the maximum value at the wall ($y = h$).

In this particular problem, in which the diffusion effects drive the momentum transport, by increasing the order in the interpolation, which finally represents an increased order in the derivative, the approximation given in eqn. (7.17) is improved. This is related with the idea of the form in which the information is traveling (explained later in Chapter 8). Both interpolations were chosen in a way that the function in the imaginary nodes, $i \pm 1/2$, was expressed in terms of nodes which are at both sides of the imaginary nodes. In other words, the information for the function at this imaginary node is comming from left and right. The way the information travels is analogous to diffusion, where the information travels in all directions.

Finally, Fig. 7.6 reveals the fact that even by increasing the order for the interpolation, the derivative approximation, when compared with the analytical, value is not improved. The assumed approximation of eqn. (7.17) has an extra term, which is a function of the node separation, Δy. What is important to recall here is that the approximation depends directly on the number of nodes, and these two approximations will be improved by increasing the number of nodes, thus decreasing distance between nodes.

7.2.2 Hermite Interpolations

In the previous discussion of interpolation we stated the problem for known values of a function at n points within a domain. In some cases, values of both, the function $f(x)$ and its derivative $f'(x)$ are available. Hence, it is required to find an interpolation formula which utilizes $2n$ data points. One way to solve this problem is to use polynomial interpolation since for $2n$ known values, a $2n-1$ order polynomial must exist. However, we must find this polynomial, $y(x)$, in a way that it has the same value and the same derivative as $f(x)$ at all n points. One function that fulfills this is [10]

$$y(x) = \sum_{i=1}^{n} h_i(x) f(x_i) + \sum_{i=1}^{n} \bar{h}_i(x) f'(x_i) \tag{7.24}$$

where $h_i(x)$ and $\bar{h}_i(x)$ are polynomials of maximum $(2n-1)$th degree. Two requirements must be satisfied for these two polynomials:

$$y(x_i) = f(x_i) \text{ if } h_i(x_j) = \delta_{ij} \text{ and } \bar{h}_i(x_j) = 0 \tag{7.25}$$

and

$$y'(x_i) = f'(x_i) \text{ if } h_i(x_j) = 0 \text{ and } \bar{h}_i(x_j) = \delta_{ij} \tag{7.26}$$

for $i \in [1, n]$ and $j \in [1, n]$, where δ_{ij} is the Kronecker delta. The Kronecker delta equals 1 when $i = j$, and 0 when $i \neq j$.

Recall that Lagrange polynomials $L_i(x)$ are polynomials of degree $n-1$, which are equal to the Kronecker delta at each of the points, and the function $[L_i(x)]^2$ is a polynomial of degree $2n-2$, which is also equal to the Kronecker delta at each of the points and whose derivative vanishes at x_j when $i \neq j$. Therefore, because $h_i(x)$ and $\bar{h}_i(x)$ are polynomials of degree $2n-1$, they can be written as

$$h_i(x) = r_i(x) [L_i(x)]^2 \tag{7.27}$$

$$\bar{h}_i(x) = s_i(x) [L_i(x)]^2 \tag{7.28}$$

where $r_i(x)$ and $s_i(x)$ are linear functions of x. In order to satisfy the conditions of eqns. (7.27) and (7.28), they must be defined as [10]

$$r_i(x) = 1 - 2L_i'(x_i)(x - x_i) \tag{7.29}$$

$$s_i(x) = x - x_i \tag{7.30}$$

These definitions result in the *Hermite's interpolation formula*,

$$y(x) = \sum_{i=1}^{n} h_i(x) f(x_i) + \sum_{i=1}^{n} \bar{h}_i(x) f'(x_i) \tag{7.31}$$

where

$$h_i(x) = [1 - 2L_i'(x_i)(x - x_i)] [L_i(x)]^2 \tag{7.32}$$

and

$$\bar{h}_i(x) = (x - x_i) [L_i(x)]^2 \tag{7.33}$$

■ **EXAMPLE 7.3.**

Hermite interpolation for the velocity and velocity gradient of a power law fluid in a narow slit. Consider the analytical expressions for the velocity and the velocity gradient for the flow of an incompressible power law fluid through a narrow slit due to a pressure gradient to represent velocity measurements. With eqns. (7.17) and (7.18), the velocity and its gradient are evaluated in 10 equally spaced points through the thickness. An expression to interpolate both, the velocity and its gradient, is required for a point $i + 1/2$ located between point i and $i + 1$.

In order to use Hermite interpolation, we must first chosse the order for the interpolation of $h_i(x)$ and $\bar{h}_i(x)$. For simplicity, let's use a first order interpolation, $n = 2$, for the Lagrange polynomials involved in these two terms. Using Hermite interpolation formula (eqn. (7.31)) and eqns. (7.32) and (7.33) we obtain

$$\begin{aligned} y_x = & \left[1 - 2\frac{x - x_i}{x_i - x_{i+1}}\right] \left[\frac{x - x_{i+1}}{x_i - x_{i+1}}\right]^2 f_i + \\ & \left[1 - 2\frac{x - x_{i+1}}{x_{i+1} - x_i}\right] \left[\frac{x - x_i}{x_{i+1} - x_i}\right]^2 f_{i+1} + \\ & (x - x_i) \left[\frac{x - x_{i+1}}{x_i - x_{i+1}}\right]^2 f_i' + (x - x_{i+1}) \left[\frac{x - x_i}{x_{i+1} - x_i}\right]^2 f_{i+1}' \end{aligned} \tag{7.34}$$

Deriving this equation with respect to x, an expression for the interpolated gradient is obtained, i.e.,

$$\begin{aligned} y'(x) = & \left[2\left(1 - 2\frac{x - x_i}{x_i - x_{i+1}}\right) \frac{x - x_{i+1}}{(x_i - x_{i+1})^2} - 2\frac{(x - x_{i+1})^2}{(x_i - x_{i+1})^3}\right] f_i + \\ & \left[2\left(1 - 2\frac{x - x_{i+1}}{x_{i+1} - x_i}\right) \frac{x - x_i}{(x_{i+1} - x_i)^2} - 2\frac{(x - x_i)^2}{(x_{i+1} - x_i)^3}\right] f_{i+1} + \\ & \left[2\frac{(x - x_i)(x - x_{i+1})}{(x_i - x_{i+1})^2} + \left(\frac{x - x_{i+1}}{x_i - x_{i+1}}\right)^2\right] f_i' + \\ & \left[2\frac{(x - x_i)(x - x_{i+1})}{(x_{i+1} - x_i)^2} + \left(\frac{x - x_i}{x_{i+1} - x_i}\right)^2\right] f_{i+1}' \end{aligned} \tag{7.35}$$

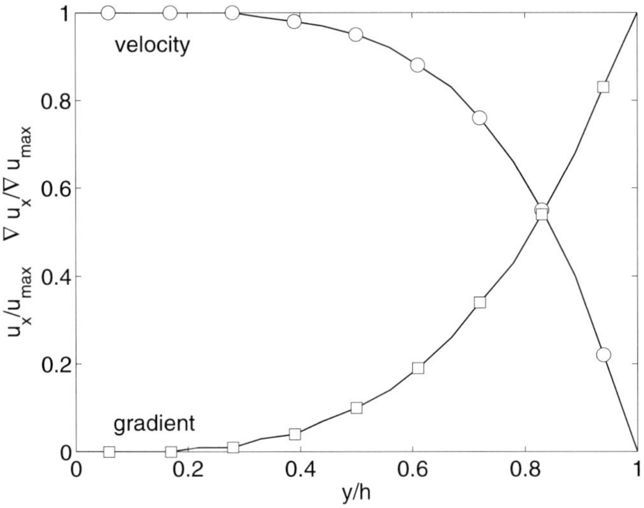

Figure 7.7: Analytical and interpolated velocity and its gradient for the power law fluid through a narrow slit.

These equations are reduced to

$$y_{i+1/2} = \frac{1}{2}\left(f_i + f_{i+1}\right) + \frac{\Delta x}{8}\left(f_i' + f_{i+1}'\right) \tag{7.36}$$

$$y_{i+1/2}' = \frac{3}{2\Delta x}\left(f_{i+1} - f_i\right) - \frac{1}{4}\left(f_i' + f_{i+1}'\right) \tag{7.37}$$

for the mid point $i + 1/2$, where Δx is the distance between nodes. Figure 7.7 ilustrates the normalized velocity and velocity gradient calculated using the analytical expressions (solid line) and the Hermite interpolation for the $i + 1/2$ points. It can be seen that the 1st order Lagrange polynomial renders a satisfactory interpolation.

7.2.3 Cubic Splines

Instead of using a specific polynomial $f(x, y)$ over an interval $[(x_0, y_0)(x_n, y_n)]$, we can divide the domain into $n - 1$ intervals defined by the points:

$$[(x_0, y_0), (x_1, y_1), (x_2, y_2), \ldots, (x_{n-1}, y_{n-1}), (x_n, y_n)]$$

as depicted in Fig. 7.8, and approximate the function using different polynomials on each interval. If we use a third degree polynomial in these intervals , so that the approximation agrees for every point n in the domain, the first and second derivative will be continuous on that domain. This type of approximation is called *cubic spline interpolation*.

Here, a cubic polynomial is fitted between adjacent points with the condition that second-order continuity is maintained between adjacent polynomials. Each spline is expressed in parametric form as

$$\mathcal{Y}(\xi) = a_i + b_i\xi + c_i\xi^2 + d_i\xi^3 \tag{7.38}$$

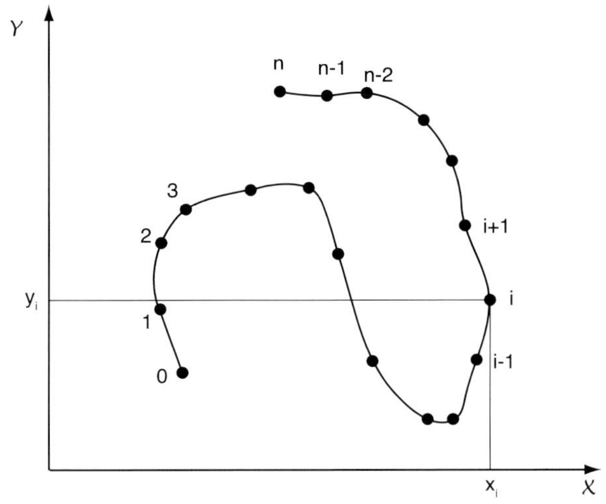

Figure 7.8: Representation of a curve in a 2D space.

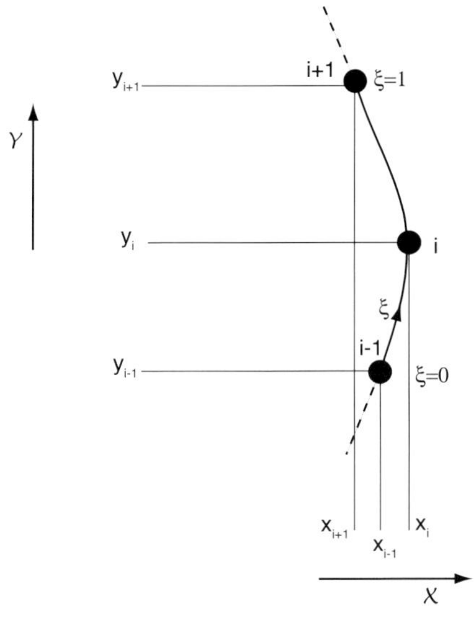

Figure 7.9: Geometric definition od a cubic spline in a 2D space.

for the y-coordinate description of interval i, and

$$\mathcal{X}(\xi) = e_i + f_i\xi + g_i\xi^2 + h_i\xi^3 \tag{7.39}$$

for the x-coordinate description of the interval, where the parameter ξ varies between 0 and 1 between the two control points of each segment, (y_i, x_i) and (y_{i+1}, x_{i+1}), as schematically depicted in Fig. 7.9.

If we concentrate on $\mathcal{Y}(\xi)$, we can say that

$$\begin{aligned}\mathcal{Y}_i(0) &= y_i = a_i \\ \mathcal{Y}_i(1) &= y_{i+1} = a_i + b_i + c_i + d_i\end{aligned} \tag{7.40}$$

The derivatives, which must be continuous between segments i and $i+1$, give

$$\begin{aligned}\mathcal{Y}'(0) &= D_i = b_i \\ \mathcal{Y}'(1) &= D_{i+1} = b_i + 2c_i + 3c_i\end{aligned} \tag{7.41}$$

Using these equations we can write,

$$\begin{aligned}a_i &= y_i \\ b_i &= D_i \\ c_i &= 3(y_{i+1} - y_i) - 2D_i - D_{i+1} \\ d_i &= 2(y_i - y_{i+1}) + D_i + D_{i+1}\end{aligned} \tag{7.42}$$

The derivatives and the second derivatives between segments $i-1$ and i must be continuous

$$\begin{aligned}\mathcal{Y}_{i-1}(1) &= y_i \\ \mathcal{Y}'_{i-1}(1) &= \mathcal{Y}'_i(0) \\ \mathcal{Y}''_{i-1}(1) &= \mathcal{Y}''_i(0)\end{aligned} \tag{7.43}$$

Finally, to match the number of equations and unknowns, we force the second derivatives to vanish at the endpoints,

$$\begin{aligned}\mathcal{Y}''_0(1) &= 0 \\ \mathcal{Y}''_n(1) &= 0\end{aligned} \tag{7.44}$$

For a series of segments where the first and last points do not match (open curve), the above equations can be re-arranged into a tridiagonal system of algebraic equations given by,

$$\begin{bmatrix} 2 & 1 & 0 & 0 & \cdots & 0 & 0 & 0 & 0 \\ 1 & 4 & 1 & 0 & \cdots & 0 & 0 & 0 & 0 \\ 0 & 1 & 4 & 1 & \cdots & 0 & 0 & 0 & 0 \\ \vdots & \vdots & \vdots & \vdots & \vdots & \vdots & \vdots & \vdots & \vdots \\ 0 & 0 & 0 & 0 & \cdots & 1 & 4 & 1 & 0 \\ 0 & 0 & 0 & 0 & \cdots & 0 & 1 & 4 & 1 \\ 0 & 0 & 0 & 0 & \cdots & 0 & 0 & 1 & 2 \end{bmatrix} \begin{pmatrix} D_0 \\ D_1 \\ D_2 \\ \vdots \\ D_{n-2} \\ D_{n-1} \\ D_n \end{pmatrix} = \begin{pmatrix} 3(y_1 - y_0) \\ 3(y_2 - y_0) \\ 3(y_3 - y_1) \\ \vdots \\ 3(y_n - y_{n-3}) \\ 3(y_n - y_{n-2}) \\ 3(y_n - y_{n-1}) \end{pmatrix} \tag{7.45}$$

For a closed curve the equation system takes the form,

$$\begin{bmatrix} 4 & 1 & 0 & 0 & \cdots & 0 & 0 & 0 & 0 \\ 1 & 4 & 1 & 0 & \cdots & 0 & 0 & 0 & 0 \\ 0 & 1 & 4 & 1 & \cdots & 0 & 0 & 0 & 0 \\ \vdots & \vdots & \vdots & \vdots & \vdots & \vdots & \vdots & \vdots & \vdots \\ 0 & 0 & 0 & 0 & \cdots & 1 & 4 & 1 & 0 \\ 0 & 0 & 0 & 0 & \cdots & 0 & 1 & 4 & 1 \\ 0 & 0 & 0 & 0 & \cdots & 0 & 0 & 1 & 4 \end{bmatrix} \begin{pmatrix} D_0 \\ D_1 \\ D_2 \\ \vdots \\ D_{n-2} \\ D_{n-1} \\ D_n \end{pmatrix} = \begin{pmatrix} 3(y_1 - y_n) \\ 3(y_2 - y_0) \\ 3(y_3 - y_1) \\ \vdots \\ 3(y_n - y_{n-3}) \\ 3(y_n - y_{n-2}) \\ 3(y_0 - y_{n-1}) \end{pmatrix} \tag{7.46}$$

The same procedure can be followed for the cubic splines that describe the x-coordinates of the segments $\mathcal{X}(\xi)$. With derivative values D_i known, we can now compute the cubic spline coefficients using eqn. (7.43). Cubic splines are ideal to represent free surfaces in polymer processing such as the bubble profile that develops during film blowing, the free surfaces during coating or the free flow fronts during mold filling [11].

7.2.4 Global and Radial Interpolation

For many applications, interpolations of functions of two or three variables defined in two- and three-dimensional domains must be considered. For example, *global interpolations* in two- and three-dimensional systems are analogous to polynomial interpolation in one-dimensional systems; however, global interpolants do not exist in 2- and 3D. This is a big drawback in numerical analysis because a basic tool available for one variable is not available for multivariable approximation [21]. The best developed aspect of this theory is that of piecewise polynomial approximation, associated with finite element and finite volume approximations for partial differential equations, which will be examined in detail in Chapters 9 and 10.

To interpolate a function $f_I(\mathbf{x})$, in a 2- and 3D domain, we can write

$$f_I(\mathbf{x}) = \sum_{i=1}^{N} a_i \psi_i(\mathbf{x}) \tag{7.47}$$

where $\mathbf{x}$ are the global coordinates. This interpolation must satisfy

$$f_I(\mathbf{x}_j) = \psi_j(\mathbf{x}_j) \quad \text{for} \quad 1 \leq j \leq N \tag{7.48}$$

which results in a system of equations for the unknown coefficients, i.e.,

$$\sum_{i=1}^{N} a_i \psi_i(\mathbf{x}_j) = f(\mathbf{x}_j) \quad \text{for} \quad 1 \leq j \leq N \tag{7.49}$$

The matrix $\boldsymbol{\Psi} = [\psi_i(\mathbf{x_j})]_1^N$ is called the *Gram matrix*, when the matrix is non-singular (has an inverse). If the matrix is non-singular, the equation $\boldsymbol{\Psi}\mathbf{a} = \mathbf{f}$ has a unique solution; however, according to Golberg [21], this fails to be true even in very simple cases [21].

EXAMPLE 7.4.

Linear interpolant for three points in a two dimensional domain. We want to obtain a linear interpolation for a function $f(x, y)$ at three points (x_j, y_j) for $j = 1, 2, 3$. The proposed interpolant will be,

$$f_I(x, y) = a_1 + a_2 x + a_3 y \tag{7.50}$$

and the Gram matrix will be defined by,

$$\boldsymbol{\Psi} = \begin{bmatrix} 1 & 1 & 1 \\ x_1 & x_2 & x_3 \\ y_1 & y_2 & y_3 \end{bmatrix} \tag{7.51}$$

In order for this matrix to be singular, the three points must be non-collinear. Otherwise an infinite number of planes exist that pass through the given line. It is difficult to find a general domain that satisfies the requirement that all the points be non-collinear. However, performing piece wise interpolation using interpolants in simple geometries with non-collinear points (triangles, rectangles) will avoid this problem. This is the basis of the finite element technique.

Global basis functions. Common global basis functions, where the interpolation functions for multi-dimensional domains can be obtained, come from expansions of *Pascal's triangle*. In 2D, Pascal's triangle is defined by,

$$\psi^g(\mathbf{x}) = \begin{array}{ccccccc} & & & 1 & & & \\ & & x & & y & & \\ & x^2 & & xy & & y^2 & \\ \vdots & \cdots & \cdots & \vdots & \cdots & \cdots & \vdots \end{array} \tag{7.52}$$

Pascal's triangle is often used to generate piecewise polynomial interpolations for various domains (triangles and rectangles in 2D and tetrahedrons, cubes and shells in 3D). In fact, most of the families of elements that are commonly used in finite elements, finite volumes and boundary elements, come from expansions of this triangle (more detail can be found in [67, 68]).

Radial basis functions. Radial interpolation uses *radial basis functions* in the linear combination that express the desired interpolated function, i.e.,

$$f_I(\mathbf{x}_j) = \sum_{i=1}^{N} \alpha_i \psi(r_{ij}) \quad \text{for} \quad 1 \leq j \leq N \tag{7.53}$$

where r_{ij} is the Euclidean distance between points defined by

$$r_{ij} = \sqrt{(x_j - x_i)^2 + (y_j - y_i)^2 + (z_j - z_i)^2} \tag{7.54}$$

A big advantage of this type of interpolation is that the matrices in the linear system of equations of coefficients, α_i, always have an inverse. A drawback is that in order to obtain the coefficients of the interpolation, we must solve very large systems of equations with full matrices. The most common radial functions are

$$\begin{aligned} \psi^r(r) &= 1 + r + r^2 + r^3 + \cdots + r^m && \text{power expansions} \\ &= r^{2m} \log r && \text{thin-plate splines} \\ &= \frac{1}{\sqrt{c^2 + r^2}} && \text{inverse multi-cuadratics} \end{aligned}$$

In Chapter 11 of this book we will use the thin spline radial function to develop the radial basis functions collocation method (RBFCM). A well known property of radial interpolation is that it renders a convenient way to calculate derivatives of the interpolated function. This is an advantage over other interpolation functions and it is used in other methods such us the dual reciprocity boundary elements [43], collocation techniques [24], RBFCM, etc.

For an interpolated function u,

$$u = [\psi]\,(\boldsymbol{\alpha}) \tag{7.55}$$

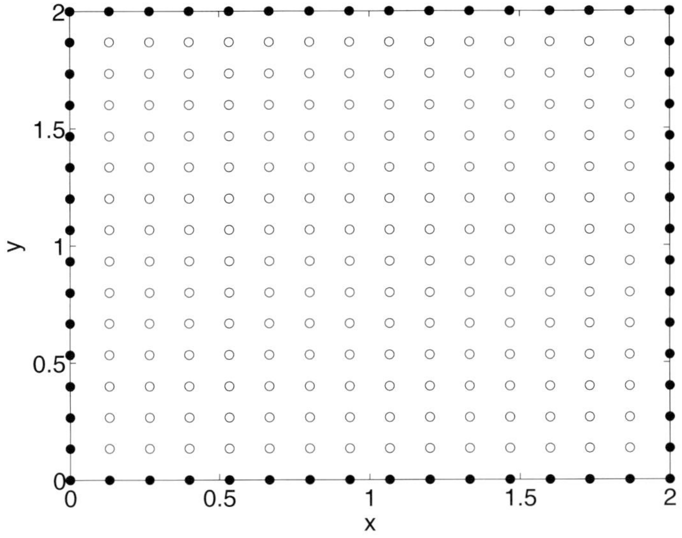

Figure 7.10: Geometry and mesh points where the values of u are known.

the derivatives of u can be approximated by

$$\frac{d^{(n)}u}{d\mathbf{x}^{(n)}} = \left[\frac{\mathbf{d}^{(\mathbf{n})}\boldsymbol{\psi}}{\mathbf{dx}^{(\mathbf{n})}}\right](\boldsymbol{\alpha}) \tag{7.56}$$

The elements of the matrix $\left[\frac{\mathbf{d}^{(\mathbf{n})}\boldsymbol{\psi}}{\mathbf{dx}^{(\mathbf{n})}}\right]$ are easily calculated depending of the radial functions that were selected.

■ **EXAMPLE 7.5.**

Two-dimensional radial interpolation. To illustrate the above concept, consider a 2×2 rectangle where a function $u(x, y)$ is defined by

$$u(x, y) = \sin x + \cos y \tag{7.57}$$

Two sets of discretizations were used to apply the method, a coarse (Fig. 7.10) with n nodes, and a finely discretized system with m nodes (Fig. 7.11). Let the values of u in the coarsely discretized system be known. These values will be used to interpolate between, in order to determine the values of u in the fine system. We calculate the coefficients for a radial interpolation as,

$$(\boldsymbol{\alpha}) = [\mathbf{F}^{\mathbf{R}}]^{-1} \cdot (\mathbf{u}) \tag{7.58}$$

where $(\mathbf{u})$ is the vector of n known values of u in the points depicted in Fig. 7.10 and $[\mathbf{F}^{\mathbf{R}}]$ is an $n \times n$ matrix of radial functions computed using $f^r = r_{ij}^2 \ln r_{ij}$.

Once the coefficient vector $(\boldsymbol{\alpha})$ has been solved for, we can use the radial interpolation to calculate the values of u for the finely discretized system shown in Fig. 7.11.

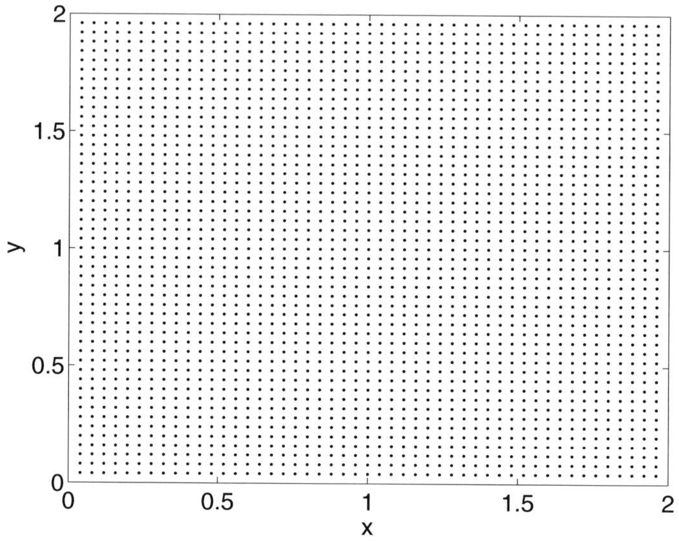

Figure 7.11: Mesh points where the values are interpolated.

Here, we use the same coefficients $\boldsymbol{\alpha}$ and the radial interpolating matrix is calculated from the desired or unknown point (fine discretization) to the points where the value is known (coarse discretization), i.e.,

$$(\mathbf{u}_I) = \left[\mathbf{F}_\mathbf{I}^\mathbf{R}\right] \cdot (\boldsymbol{\alpha}) \tag{7.59}$$

where $(\mathbf{u}_I)$ is a vector of m unknown values of u and $\left[\mathbf{F}_\mathbf{I}^\mathbf{R}\right]$ is an $m \times n$ matrix of radial functions, f^r. Figure 7.12 shows a comparison between the interpolated values and the values given by $u = \sin x + \cos y$. Here, we can observe the excellent performance of the radial interpolation.

One of the benefits of using radial interpolation is the way derivatives can be approximated. Once we solve for the coefficient vector $\boldsymbol{\alpha}$, the derivatives can be approximated using,

$$\frac{\partial \mathbf{u_I}}{\partial x} = \frac{\partial \mathbf{F}_\mathbf{I}^\mathbf{R}}{\partial x} \cdot \boldsymbol{\alpha} \tag{7.60}$$

$$\frac{\partial \mathbf{u_I}}{\partial y} = \frac{\partial \mathbf{F}_\mathbf{I}^\mathbf{R}}{\partial y} \cdot \boldsymbol{\alpha} \tag{7.61}$$

Figures 7.13 and 7.14 compare the interpolated and the analytical x- and y-derivatives of u. Again, the interpolation agrees with the analytical solution.

7.3 NUMERICAL INTEGRATION

Historically, numerical integration, sometimes also referred to as quadratures, has been a motivation for the development and advancement of numerical analysis. Integrals of

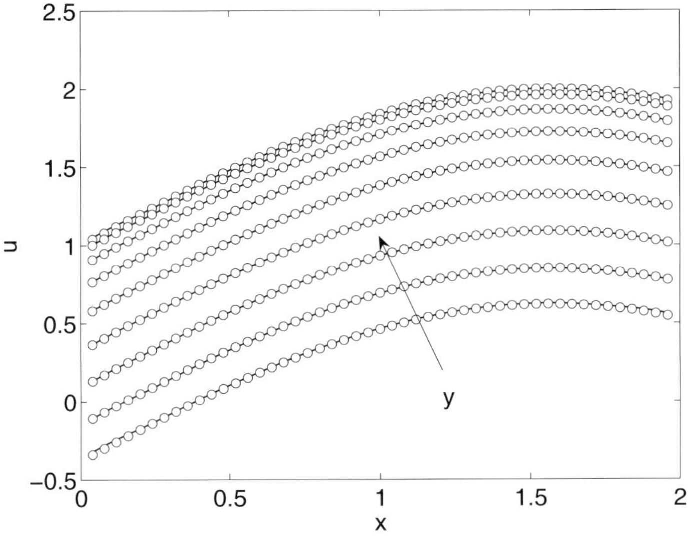

Figure 7.12: Interpolated values of u. Solid line represents the equation while the dots represent the interpolated values.

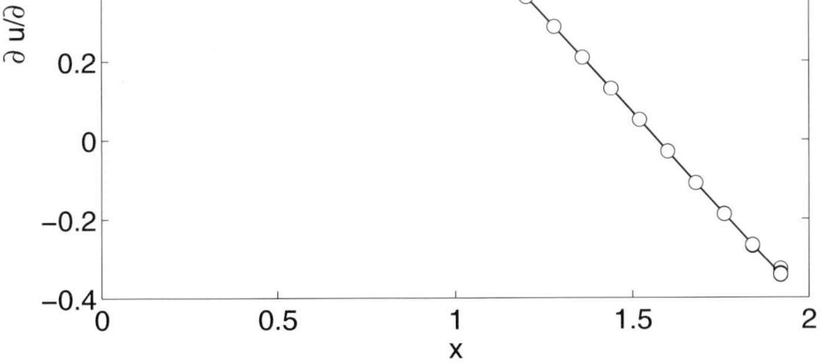

Figure 7.13: Interpolated values of $\partial u/\partial x$. Solid line represents the equation while the dots represent the interpolated values.

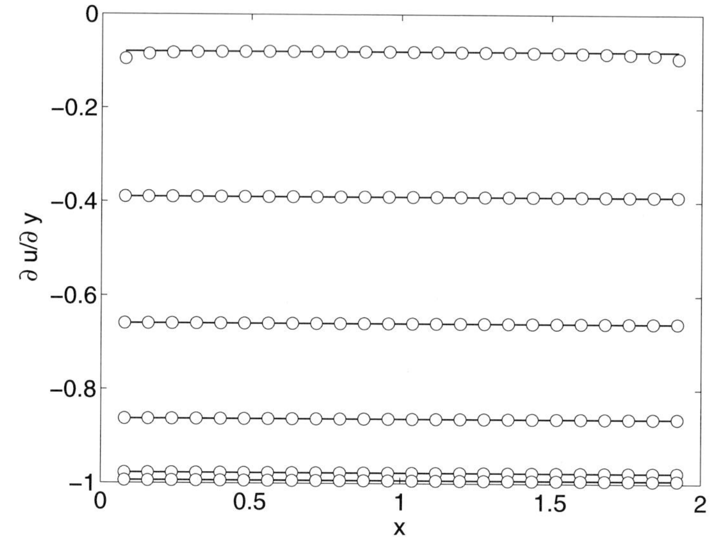

Figure 7.14: Interpolated values of $\partial u/\partial y$. Solid line represents the equation, while the dots represent the interpolated values.

elementary functions were not, in general, computed analytically, while derivatives were found easily. During the 18th and 19th centuries, a significant amount of work was put into finding methodologies that facilitate the solution for integral equations [10, 19]. In general, numerical integration uses the definition of the integral operator, i.e., for a continuous function $f(x)$ in the closed interval $[a, b]$, the integral of $f(x)$ between a and b is given by

$$\int_a^b f(x)dx = \lim_{\Delta x \to 0} \left(\sum_{i=1}^{n} f(x_i)\Delta x_i \right) \tag{7.62}$$

A good quadrature is one where the number of evaluations of the function is kept as small as possible in order to achieve an accurate solution. There are two different types of integrations methods, closed and open formula, as schematically depicted in Fig. 7.15. Those that use the value of the function at the lower and upper limits, $f(a)$ and $f(b)$, called closed formula and those that do not include these function values, called open formulas. The lattter are used when the function presents a singularity in one of the limits.

7.3.1 Classical Integration Methods

Classical integration methods are widely used techniques to evaluate integrals where a formula for $f(x)$ is not at hand such as evaluation of experimental data. The classical techniques often require that the spacing between the points is the same for all the points, as depicted in Fig. 7.16.

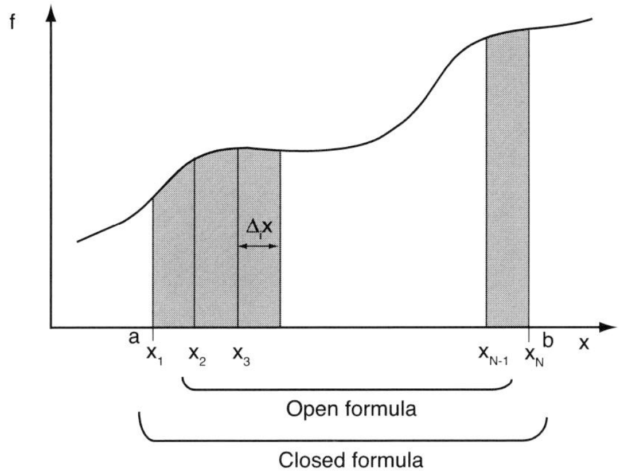

Figure 7.15: Closed and open numerical integration formulations.

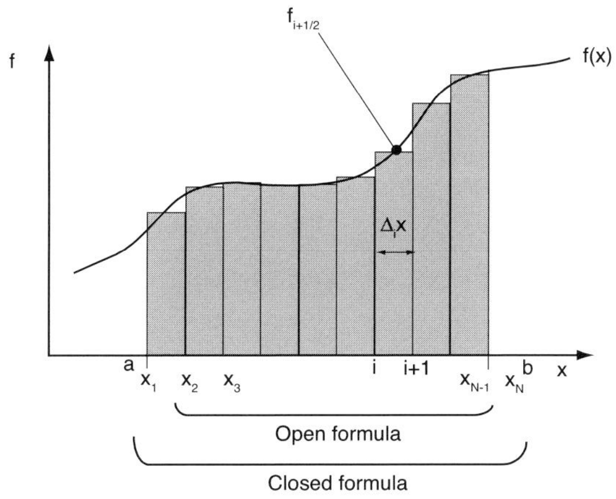

Figure 7.16: Linear numerical integration.

However, in some sets of data, the spacing between the points is not necessarily constant, for which we can use the most basic approximation of integration written as,

$$\int_a^b f(x)dx = \sum_{i=1}^{n-1} \int_{x_i}^{x_{i+1}} f(x)dx \approx \sum_{i=1}^{n-1} A_i \tag{7.63}$$

where the area of each rectangle i is defined as,

$$A_i = f_{i+1/2}\left(x_{i+1} - x_i\right) = f_{i+1/2}\Delta x_i \tag{7.64}$$

We simply need to find an expression for $f_{i+1/2}$. If we use linear interpolation, since the point is in the middle of the interval, we arrive to the mid point rule, and the integral between two limits x_i and x_{i+1} is given by

$$\int_{x_i}^{x_{i+1}} f(x)dx \approx \frac{\Delta x_i}{2}\left(f_i + f_{i+1}\right) + O(\Delta x_i^3 f'') \tag{7.65}$$

which tells us that the error is proportional to the cube of the interval size times the second derivative of the function at that point. This formula is the well known *trapezoidal rule*, which is a second order approximation made of a first order interpolation for the area, and a first order approximation for the interpolation of $f_{i+1/2}$. For constant intervals between the points we can include three points ($x_{i-1}, x_i \text{and} x_{i+1}$), and we can write

$$\int_{x_{i-1}}^{x_{i+1}} f(x)dx \approx \Delta x\left(\frac{1}{3}f_{i-1} + \frac{4}{3}f_i + \frac{1}{3}f_{i+1}\right) + O(\Delta x^5 f'''') \tag{7.66}$$

The above equation is the well known *Simpson's rule*, which is more accurate than the simpler trapezoidal rule.

■ **EXAMPLE 7.6.**

Total heat released of a cured thermoset polymer Q_T. As we saw in Chapter 2, the heat released by a thermoset or an elastomer during curing can be directly related to the degree of cure and the curing rate by

$$c = \frac{Q(t)}{Q_T} \qquad \frac{dc}{dt} = \frac{\dot{Q}}{Q_T} \tag{7.67}$$

where

$$Q(t) = \int_0^t \dot{Q}(t')dt' \tag{7.68}$$

and Q_T is the total heat that the polymer can release during conversion.

In order to obtain the degree of cure and rate of curing, we must first measure the reaction. This is typically done using a differential scanning calorimeter (DSC) as explained in Chapter 2. Typically, several dynamic tests are performed, where the temperature is increased at a constant rate and heat release rate ($\dot{Q}$) is measured until the conversion is finished. To obtain Q_T we must calculate the area under the curve $\dot{Q}$ versus t. Figure 7.17 shows four dynamic tests for a liquid silicone rubber at heating rates of 10, 5, 2.5 and 1 K/min. The trapezoidal rule was used to integrate the four curves. As expected, the total heat Q_T is the same (more or less) for all four tests. This is to be expected, since each curve was represented with approximately 400 data points.

7.3.2 Gaussian Quadratures

The problem with classical integration techniques is that they require a large number of integration points, which for two- and three-dimensional domains, where area and volume integrals are calculated, requires large amounts of storage and computation time.

To reduce the number of integration stations we must use an optimum number of points, selected along the *abscissas*, with their corresponding weighting coefficients, such as done with *Gaussian quadratures*. Using Gauss quadratures which can approximate the integral of any smooth function using

$$\int_a^b f(x)dx \approx \sum_{i=1}^{ngauss} f(x_{gi})w_i \tag{7.69}$$

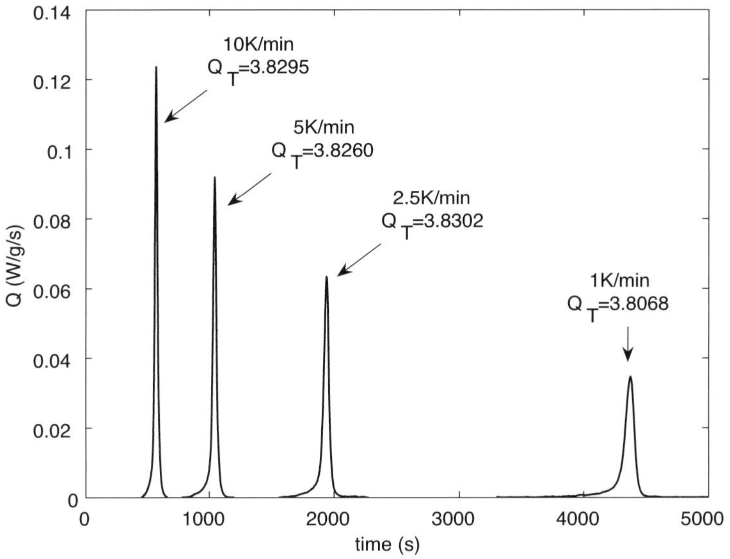

Figure 7.17: DSC measurements for a liquid silicone rubber at different rates.

where x_{gi} is the location of the Gauss points and w_i are the Gaussian weights. Depending on the number of integration stations and weighting function, the Gaussian quadratures will have different values for the abscissas and weights. Table 7.3 summarizes some abscissas and weights for the Gauss-Legendre quadrature technique, the most widely used quadrature used to evaluate integrals in the finite element and boundary element methods using

$$\int_{-1}^{+1} f(x)dx \approx \sum_{i=1}^{ngauss} f(x_i)w_i \tag{7.70}$$

for integration limits between -1 and +1. The abscissas and weights can be modified, if we integrate between the limits a and b, using

$$x_i^{ab} = \frac{1}{2}(b+a) + \frac{1}{2}(b-a)x_i \tag{7.71}$$

and

$$w_i^{ab} = \frac{1}{2}(b-a)w_i \tag{7.72}$$

■ **EXAMPLE 7.7.**

Numerical integration of a known function. To illustrate the Gauss quadrature integration technique and compare it to classical integration techniques, we will evaluate the following integral

$$I = \int_0^2 \left(3x^2 - 2x\right) dx = x^3 - x^2\Big|_0^2 = 4 \tag{7.73}$$

Table 7.3: Gauss-Legendre Abscissas and Weights.

$ngauss$	x_i	w_i
2	-1/$\sqrt{3}$	1
	1/$\sqrt{3}$	1
3	-0.77459667	0.55555556
	0	0.88888889
	0.77459667	0.55555556
4	-0.86113631	0.34785485
	-0.33998104	0.65214515
	0.33998104	0.65214515
	0.86113631	0.34785485
5	-0.90617985	0.23692689
	-0.53846931	0.47862867
	0	0.56888889
	0.53846931	0.47862867
	0.90617985	0.23692689
10	-0.97390653	0.06667134
	-0.86506337	0.14945135
	-0.67940957	0.21908636
	-0.43339539	0.26926672
	-0.14887434	0.29552422
	0.14887434	0.29552422
	0.43339539	0.26926672
	0.67940957	0.21908636
	0.86506337	0.14945135
	0.97390653	0.06667134

Table 7.4: Function $f(x) = 3x^2 - 2x$ evaluated at discrete x positions, x_i

Point number	x_i	$f(x_i)$
1	0	0
2	0.25	-0.3125
3	0.5	-0.25
4	0.75	0.1875
5	1.0	1
6	1.25	2.1875
7	1.5	3.75
8	1.75	5.6875
9	2.0	8

To use classical integration techniques, we will first generate a table of discrete points where we evaluate the function seperated by equal intervals of $\Delta x = 0.25$ (Table 7.4).

Let us first use the trapezoidal rule to calculate the above integral

$$\begin{aligned} I = 0.25 \Big[& \frac{1}{2}(0) + (-0.3175) + (-0.25) + 0.1875 + \\ & 1 + 2.1875 + 3.75 + 5.6875 + \frac{1}{2}(8) \Big] = 4.0625 \end{aligned} \tag{7.74}$$

which, over-predicts the integral by 1.56%. This is primarily due to the fact that the chosen points miss the lowest value of the function, $f(1/3) = -1/3$. Hence, data was used that did not accurately represent the function. For that same reason we do not expect Simpson's rule to perform much better. Simpson's rule gives

$$\begin{aligned} I = 0.25 \Big[& \frac{1}{3}(0) + \frac{4}{3}(-0.3175) + \frac{1}{3}(-0.25) + \frac{1}{3}(-0.25) + \\ & \frac{4}{3}(0.1875) + \frac{1}{3}(1) + \frac{1}{3}(1) + \frac{4}{3}(2.1875) + \\ & \frac{1}{3}(3.75) + \frac{1}{3}(3.75) + \frac{4}{3}(5.6875) + \frac{1}{3}(8) \Big] = 4.0416 \end{aligned} \tag{7.75}$$

which represents an error of 1.04%. Next we apply Gauss quadrature. Since the limits are not between -1 and +1, but 0 and 2, we must first modify the abscissa values. Using 8 digits of precision, these become $x_1^{ab} = 0.42264973$ and $x_2^{ab} = 1.57735027$ with their corresponding weights, $w_1^{ab} = 1$ and $w_2^{ab} = 1$. The integral can now be written in terms of Gauss quadrature as

$$\begin{aligned} I = & \left[3(0.42264973)^2 - 2(0.42264973)\right](1.0) + \\ & \left[3(1.57735027)^2 - 2(1.57735027)\right](1.0) = 4.0000000056 \end{aligned} \tag{7.76}$$

Had we used a higher precision for the Gauss points, the quadrature would have rendered a solution even closer to the exact integral.

7.4 DATA FITTING

Data fitting or modeling of data is often required to find a numerical representation for a set of data points. In polymer processing, we often want to fit complex models, such as viscosity, viscoelastic and cure kinetic models, to name a few, to data measured using rheometers, DSC or other material characterization instruments. This section will present some of these techniques with respective applications.

7.4.1 Least Squares Method

The least squares method is the best possible solution to an overdetermined system of equations, where we have more equations than unknowns. This is a very common situation in problems, in which, for example, we have a set of N measurements which we want to *fit* to an equation with M parameters ($M < N$). For example if we measure temperatures as a function of time

$$\begin{vmatrix} t & T \\ 0 & 20 \\ 10 & 28 \\ 20 & 32 \\ 30 & 40 \end{vmatrix} \tag{7.77}$$

and we want to fit these experimental data with the equation given by $T = B + Ct$. We will have an overdetermined linear system of equations given by

$$\begin{bmatrix} 1 & 0 \\ 1 & 10 \\ 1 & 20 \\ 1 & 30 \end{bmatrix} \begin{pmatrix} B \\ C \end{pmatrix} = \begin{pmatrix} 20 \\ 28 \\ 32 \\ 40 \end{pmatrix} \tag{7.78}$$

which can also be written as

$$\mathbf{Ax} = \mathbf{b} \tag{7.79}$$

An alternative is to find the best solution of B and C and minimize the error given by $\mathbf{e} = (\mathbf{Ax} - \mathbf{b})$. When minimizing the error, we usually write

$$\|\mathbf{e}\|_2 = \left(\mathbf{e}^T\mathbf{e}\right)^{1/2} \tag{7.80}$$

or

$$\begin{aligned} \|\mathbf{Ax} - \mathbf{b}\|^2 &= \left(\mathbf{e}^T\mathbf{e}\right)^{1/2} \\ &= (\mathbf{Ax} - \mathbf{b})^T (\mathbf{Ax} - \mathbf{b}) \\ &= (\mathbf{Ax})^T (\mathbf{Ax}) - 2\mathbf{x}^T\mathbf{A}^T\mathbf{b} + \mathbf{b}^T\mathbf{b} \\ \mathcal{X} &= \mathbf{x}^T\mathbf{A}^T\mathbf{Ax} - 2\mathbf{x}^T\mathbf{A}^T\mathbf{b} + \mathbf{b}^T\mathbf{b} \geq 0 \end{aligned} \tag{7.81}$$

The minimization is obtained from $d\mathcal{X}/d\mathbf{x} = 0$, which by convention is multiplied by $1/2$. This minimization reduces the problem to [9, 10, 19]

$$\frac{1}{2}\frac{d\mathcal{X}}{d\mathbf{x}} = \mathbf{A}^T\mathbf{Ax} - \mathbf{A}^T\mathbf{b} = 0 \tag{7.82}$$

Hence,

$$\mathbf{A}^T\mathbf{Ax} = \mathbf{A}^T\mathbf{b} \tag{7.83}$$

is used to find the vector $\mathbf{x}$ that best fits the data. We can also use this technique to fit the data to exponential or power equations given by

$$y = b\mathrm{e}^{Ct} \tag{7.84}$$

and

$$y = bx^{C} \tag{7.85}$$

by rewriting them as

$$\begin{aligned} \ln y &= \ln b + Ct \\ T &= B + Ct \end{aligned} \tag{7.86}$$

and

$$\begin{aligned} \ln y &= \ln b + C \ln x \\ T &= B + Ct \end{aligned} \tag{7.87}$$

respectively.

EXAMPLE 7.8.

Least squares method application. To illustrate the least squares method presented above, we will fit the data given in eqn. (7.77) by minimizing the error using eqn. (7.83)

$$\mathbf{A}^T\mathbf{A} = \begin{bmatrix} 4 & 60 \\ 60 & 1400 \end{bmatrix} \qquad \mathbf{A}^T\mathbf{b} = \begin{pmatrix} 120 \\ 2120 \end{pmatrix} \tag{7.88}$$

which can be solved for $B = 20.4$ and $C = 0.64$. The best fit is therefore given by

$$T = 20.4 + 0.64t \tag{7.89}$$

Figure 7.18 presents the data with the best fit linear equation given by the least squares method.

7.4.2 The Levenberg-Marquardt Method

The Levenberg-Marquardt method is used to model data with non-linear dependencies. Here, we may have m functions $f_1, f_2, f_3 ... f_m$ that depend on n parameters $p_1, p_2 ... p_n$ written in vector form as

$$\begin{aligned} \mathbf{f}^T &= (f_1 f_2 f_3 ... f_m) \\ \mathbf{p}^T &= (p_1 p_2 p_3 ... p_n) \end{aligned} \tag{7.90}$$

Since we are dealing with non-linear equations, this is an iterative procedure that begins by making an initial guess for $\mathbf{p}$, such as for example $(111....1)^T$. In each iteration step, the values of $\mathbf{p}$ are replaced by a new value of $\mathbf{p} + \mathbf{q}$. To determine, $\mathbf{p}$ we approximate the vector $\mathbf{f}(\mathbf{p} + \mathbf{q})$ by

$$\mathbf{f}(\mathbf{p} + \mathbf{q}) \approx \mathbf{f}(\mathbf{p}) + \mathbf{J}\mathbf{q} \tag{7.91}$$

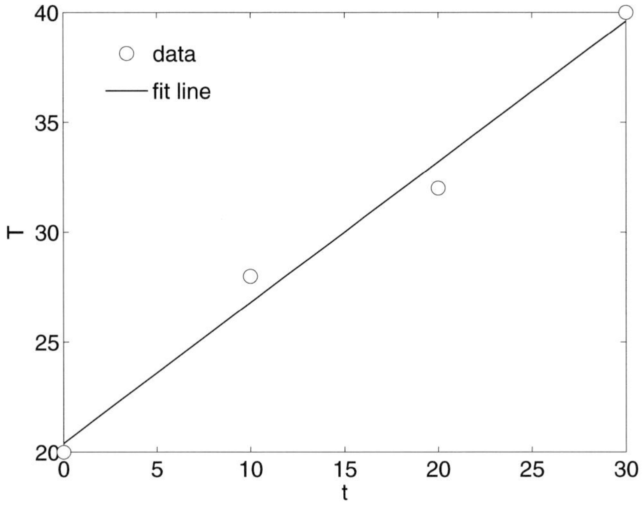

Figure 7.18: Fitting data in a line.

where, $\mathbf{J}$ is the Jacobian of $\mathbf{f}$

$$\mathbf{J} = \begin{bmatrix} \frac{\partial f_1}{\partial p_1} & \frac{\partial f_1}{\partial p_2} & \cdots & \frac{\partial f_1}{\partial p_n} \\ \frac{\partial f_2}{\partial p_1} & \frac{\partial f_2}{\partial p_2} & \cdots & \frac{\partial f_2}{\partial p_n} \\ \vdots & \vdots & \vdots & \vdots \\ \frac{\partial f_m}{\partial p_1} & \frac{\partial f_m}{\partial p_2} & \cdots & \frac{\partial f_m}{\partial p_n} \end{bmatrix} \tag{7.92}$$

As in the previous section, by minimizing the error we get

$$[\mathbf{J}^T\mathbf{J}]\mathbf{q} + \mathbf{J}^T\mathbf{f} \tag{7.93}$$

which can be solved for $\mathbf{q}$. The Levenberg-Marquardt method requires the introduction of a non-negative factor λ, which is used to modify the above equation to give

$$[\mathbf{J}^T\mathbf{J} + \lambda]\mathbf{q} + \mathbf{J}^T\mathbf{f} \tag{7.94}$$

Here too, we will minimize $\mathcal{X}^2$ to find the best fit parameters. We repeat the above procedure until $\mathcal{X}^2$ stops decreasing.

The following steps are followed:

- Guess $\mathbf{p}$
- Compute $\mathcal{X}^2(\mathbf{p})$
- Pick a value of λ (say (0.001))
- Determine $\mathbf{q}$ by solving the linear equations (7.94)

Table 7.5: Non-Newtonian Viscosity for Polystyrene at 453 K [5]

Shear rate $\dot{\gamma}$ (s^{-1})	Viscosity η (Pa-s)
0.3	14,800
0.4	14,600
0.5	14,300
1.0	12,000
3.0	8,000
6.0	5,500
10.0	4,000
20.0	2,800
30.0	2,300
60.0	1,500
100.0	1,100

- Check if $\mathcal{X}^2(\mathbf{p}+\mathbf{q}) \geq \mathcal{X}^2(\mathbf{p})$ and increase λ by a factor of 10. Go back and solve the linear equation 7.94
- Check if $\mathcal{X}^2(\mathbf{p}+\mathbf{q}) \leq \mathcal{X}^2(\mathbf{p})$ and decrease λ by a factor of 10
- Update $\mathbf{p} = \mathbf{p}+\mathbf{q}$. Go back and solve the linear equation 7.94

Details of this methodology can be found in the literature (see for instance [19]). The Levenberg-Marquardt technique, as many other numerical schemes, can be found as *freeware* on the world wide web [31].

EXAMPLE 7.9.

Fitting a Bird-Carreau model to viscosity data. Table 7.5 shows the measurements of Ballenger *et. al* [5] of the viscosity as a function of shear rate for polystyrene at 453 K.

For this particular data we can assume that $\eta_\infty = 0$, which reduces the Bird-Carreau model to,

$$\eta = \eta_0 \left[1 + (\lambda\dot{\gamma})^2\right]^{\frac{n-1}{2}} \tag{7.95}$$

To fit this data and obtain the three fitting parameters, we must use a non-linear fitting technique. In general, to perform this type of fitting requires a routine with the function that minimizes a certain quantity. In this case, we want to obtain a vector $\mathbf{x}$ with the three unknown parameters, i.e.,

$$\mathbf{x} = \begin{pmatrix} \eta_0 & \lambda & n \end{pmatrix} \tag{7.96}$$

and the function to minimize will be,

$$f_i = \eta_i - x(1)\left[1 + (x(2)\dot{\gamma}_i)^2\right]^{\frac{x(3)-1}{2}} \tag{7.97}$$

We used the MINPACK [31] routines to perform the fitting and the results are shown in Fig. 7.19. The fitted parameters for the PS data given above are $\eta_0 = 1.53 \times 10^4$ Pa-s, $\lambda = 1.11$s and $n = 0.45$.

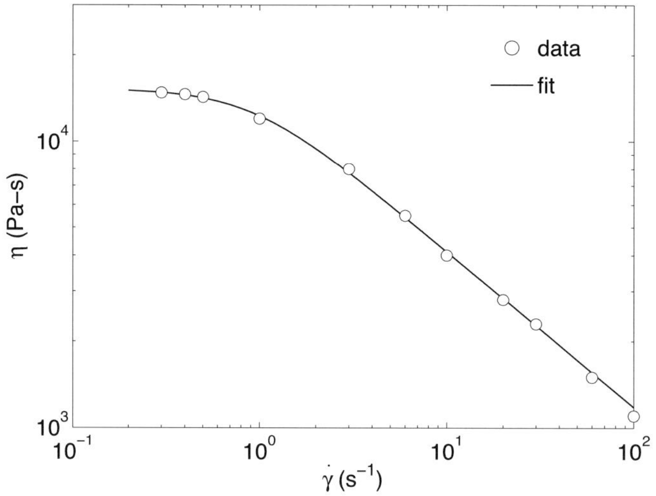

Figure 7.19: Non-Newtonian viscosity for PS at 453K.

EXAMPLE 7.10.

Modeling the vulcanization of liquid silicone rubber (LSR). Lopez *et al.* [28, 29] modeled the vulcanization of liquid silicone rubber using dynamic DSC tests. In order to do this, the Kissinger model [25, 26] and the Kamal-Sourour model were used to determine kinetic parameters that fit the experimental data. The idea is to generate a single set of parameters that accurately models the used scanning rates of 1, 2.5, 5, and 10 K/min.

The model used for fitting the data was the autocatalytic Kamal-Sourour model [13, 14],

$$\frac{dc}{dt} = (k_1 + k_2 c^m)(1 - c)^n \tag{7.98}$$

where k_i are the rate constants described by the Arrhenius equation

$$k_i = a_i e^{\left(-\frac{E_i}{RT}\right)} \tag{7.99}$$

where a_i are fittig parameter to the Kamal-Sourour model, E_i are the activation energy, T is the processing temperature in Kelvin, and R is the universal gas constant.

This model requires six parametes $(a_1, a_2, E_1, E_2, m, n)$ to be fitted to the experimental dynamic DSC data. Let $\mathbf{x}$ be the unknown vector parameter defined as,

$$\mathbf{x} = \begin{pmatrix} m & n & a_1 & a_2 & E_1 & E_2 \end{pmatrix} \tag{7.100}$$

In order to find a unique set of parameters for all rates, Lopez *et al.* proposed that the activation energy E_1 is determined through Kissinger's model, which is implemented to mathematically determine the other five parameters for the Kamal-Sourour model. The activation energy, E_2, is fitted as a constant, while the four remaining parameters,

m, n, a_1 and a_2, in the cure model are represented using a power series of the temperature [9] such as

$$x_i = b_{i1} + b_{i2}T + b_{i3}T^2 \tag{7.101}$$

where $i = 1, ..., 4$ and b_{ij} are the new fitting parameters. In other words, the technique looks for the 13 components of the $\mathbf{B}$ matrix defined in such a way that

$$\mathbf{x} = \begin{pmatrix} m & n & a_1 & a_2 & E_2 \end{pmatrix} = \begin{bmatrix} b_{11} & b_{12} & b_{13} \\ b_{21} & b_{22} & b_{23} \\ b_{31} & b_{32} & b_{33} \\ b_{41} & b_{42} & b_{43} \\ E_2 & 0 & 0 \end{bmatrix} \begin{pmatrix} 1 \\ T \\ T^2 \end{pmatrix} \tag{7.102}$$

In order to obtain a physical energy of activation, E_1, the Kissinger model [25, 26] was used. This model uses an nth-order expression to model the kinetic reaction defined as

$$\frac{dc}{dt} = k\left(1 - c\right)^n \tag{7.103}$$

Here, k is the rate constant defined by the Arrhenius equation, c is the extent of vulcanization or cure, and n is the order of the reaction.

Heat activated reactions show a variation of the position of the peak in the cure rate curves with varying heating rate. If the temperature rises during reaction, the reaction rate, dc/dt, will rise to a maximum value and then returns to zero as reaction completes. The temperature at which the reaction rate reaches its maximum is also the temperature of maximum deflection in differential thermal analysis. Dynamic DSC measurements at different heating rates are used to determine the activation energy of the material, because they show the effects both time and temperature have on the reaction.

The maximum rate of reaction occurs when the derivative of dc/dt is zero. If the temperature is raised at a constant rate, $\dot{T}$, then by differentiation of eqn. (7.103) and replacing k by an Arrhenius eqn. (7.99), the maximum rate of reaction can be defined as

$$\frac{d}{dt}\left(\frac{dc}{dt}\right) = \frac{dc}{dt}\left[\frac{E\dot{T}}{RT^2} - an(1-c)^{n-1}e^{(-E/RT)}\right] \tag{7.104}$$

The maximum rate of reaction occurs at the peak temperature of the heat flow from the thermal analysis, T_{peak}. The peak temperature can be defined by setting eqn. (7.104) to zero

$$\frac{E\dot{T}}{RT^2} = an(1-c)^{n-1}_{\text{peak}}e^{(-E/RT_{\text{peak}})} \tag{7.105}$$

The amount of unreacted material, $(1 - c)_{\text{peak}}$, is not determined directly through DSC analysis. Integration of eqn. (7.105), assuming a constant heating rate, gives an expression of the extent of reaction as a function of temperature. Approximations of this expression were obtained by Murray and White [32] through integration by parts. The expression for eqn. (7.105) after using Murray and White's approximations is written as

$$\frac{1}{n-1}\left[\frac{1}{(1-c)^{n-1}} - 1\right] = \frac{aRT^2}{E\dot{T}}\left[1 - \frac{2RT}{E}\right]e^{(-E/RT)} \tag{7.106}$$

Table 7.6: Vulcanization Data of LSR [28]

Heating Rate K/min	Sample mass mg	T_{peak} ^{o}C	Heat of Reaction J/g
1.0	19.2	94.9	3.887
1.0	13.3	94.7	3.824
2.5	23.0	102.0	3.856
2.5	18.6	101.3	3.799
5.0	13.9	107.9	3.847
5.0	11.9	108.3	3.826
10.0	15.8	115.8	3.861
10.0	23.4	115.8	3.740
Average			3.830
Standard deviation			0.040

Table 7.7: Fitting Parameters for the Vulcanization of LSR [28]

Parameter	$\times 1$	$\times T$	$\times T^2$
m	1.5533×10^2	-0.7588	9.3349×10^{-4}
n	1.7499×10^2	-0.8911	1.1426×10^{-3}
a_1	1.6554×10^{15}	2.8889×10^{12}	-1.8655×10^{10}
a_2	-5.8582×10^{-5}	2.4736×10^{-7}	-2.3529×10^{-10}
E_2	-36800.93	0	0
E_1	124850.53	0	0

The expression for the amount of unreacted material at T_{peak} for reaction orders other than zero or unity, can be simplified to

$$n(1-c)_{\text{peak}}^{n-1} = 1 + (1-n)\frac{2RT_{\text{peak}}}{E} \tag{7.107}$$

An expression for the activation energy can be found by substituting eqn. (7.107) into eqn. (7.104) and neglecting small quantities

$$-\frac{E}{R} = \frac{d}{d(1/T)}\left(\ln \frac{\dot{T}}{T_{\text{peak}}^2}\right) \tag{7.108}$$

The activation energy can be easily calculated by performing DSC tests at different scanning rates regardless of the order of the reaction.

Table 7.6 shows the repeatability for the vulcanization of a LSR formulation when different rates are used.

The data of heating rate and the temperature at which the maximum rate of reaction occurs, T_{peak}, was plotted, and fitted to a linear model. The activation energy, E_1, of a silicone rubber was calculated with the data from four dynamic scanning rates. The slope of the line corresponds to the negative ratio of the activation energy and the universal gas constant R (8.3145 J/gmol/K) as can be seen in Fig. 7.20.

Figure 7.21 shows the comparison between the experiments and the fitted model. The kinetic parameters are summarized in Table 7.7.

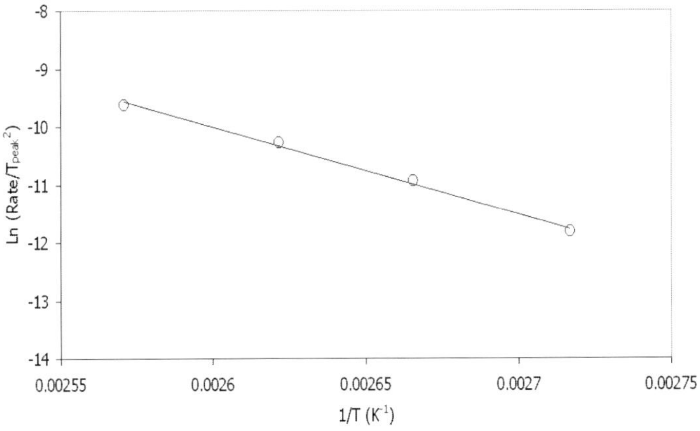

Figure 7.20: Kissinger model for a LSR formulation.

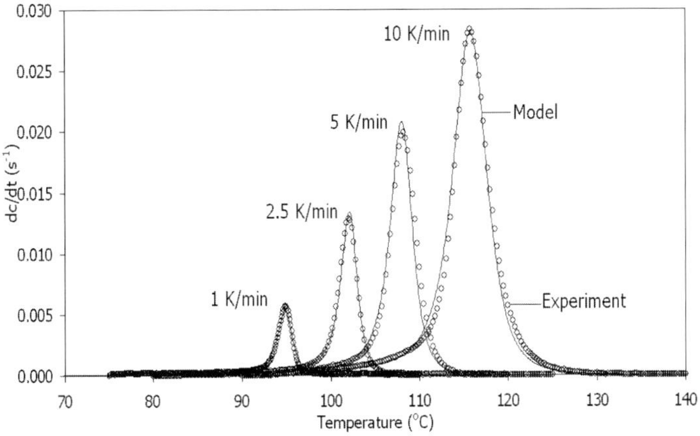

Figure 7.21: Fitted model and experimental data for a LSR formulation.

7.5 METHOD OF WEIGHTED RESIDUALS

Numerical methods such as the finite element method is based on the method of weighted residuals. To illustrate this method, we begin with boundary value partial differential equation (PDE) presented in the form

$$\frac{\partial}{\partial t}u(\mathbf{x},t) + \mathcal{L}u(\mathbf{x},t) = f(\mathbf{x}) \tag{7.109}$$

for $\mathbf{x} \in V$ and the corresponding boundary and initial conditions

$$\begin{aligned} u(\mathbf{x_S},t) &= u_S \\ u(\mathbf{x},0) &= u_0 \end{aligned} \tag{7.110}$$

for $\mathbf{x_S} \in S$. For the transient heat conduction problem, for example, the operator $\mathcal{L}$ is defined by

$$\mathcal{L} = -k\frac{\partial^2}{\partial x_j \partial x_j} \tag{7.111}$$

Equations such as eqn. (7.109) usually do not have an analytical solution; therefore, we will express an approximate solution, $\bar{u}$, as a linear combination of some known *basis* functions, ψ, similar to the procedure in the previous section

$$\bar{u}(\mathbf{x},t) = \sum_{i=1}^{N} c_i(t)\psi_i(\mathbf{x}) \tag{7.112}$$

Only a perfect analytical expression for $u(\mathbf{x},t)$ will satisfy eqn. (7.109). Hence, since our approximate solution will always have a residual, or error, we will try to make this residual as small as possible

$$\frac{\partial}{\partial t}\bar{u}(\mathbf{x},t) + \mathcal{L}\bar{u}(\mathbf{x},t) - f(\mathbf{x}) = R \neq 0 \tag{7.113}$$

Hence, we can require that the inner product of the residual and weighting functions, w_j go to zero over the domain [68]

$$\int_V R(\mathbf{x},t)w_j(\mathbf{x})d\mathbf{x} = 0 \tag{7.114}$$

This approximation is commonly called the *method of weighted residuals*; it was first described by Crandall [7] and is fully explained in many references [11, 12, 13, 19, 23, 68].

In principle, any set of independent functions, w_j, can be used for the purpose of weighting. According to the choice of the function, a different method is achieved. The most common choices are:

- *Point collocation.* The weighting functions are equal to delta functions located at n points, $w_j(\mathbf{x}) = \delta_j(\mathbf{x})$, which is equivalent to forcing the residual to zero at the n points [24].

- *The Galerkin Method.* The basis functions used in the approximation for $\bar{u}$ are also used for weighting, $w_j = \psi_j$. This method frequently leads to symmetric matrices and it is the most commonly used method in the finite element literature [14, 30].

In the mathematical literature, the Galerkin method is also known as *Galerkin-Bubnov*, while the case $w_j \neq \psi_j$ is commonly called *Petrov-Galerkin* [30, 68] and is used in special finite element formulations, such as those where the heat transfer is governed by convective effects. The application of Galerkin's method in the finite element method will be covered in detail in Chapter 9 of this textbook.

The choice of the basis functions used in the truncated series (eqn. 7.112) will lead to different types of methods. The most common set of methods are the finite element and the spectral methods. For the case of *finite element methods*, the domain is divided into small elements, and a basis function is specified in each element to interpolate the parameters throughout the element. The functions are locally defined within each element and can handle complex geometries.

In collocation methods, different sets of basis functions can also be used, for example *radial basis functions*, which are functions that depend only on the *euclidean* distance between collocation points i and j, r_{ij}.

EXAMPLE 7.11.

Fourier-Galerkin method. To illustrate the weighted residuals method we chose a Fourier-Galerkin method to solve a PDE of the form,

$$\frac{\partial u(x,t)}{\partial t} = \frac{\partial u(x,t)}{\partial x} \tag{7.115}$$

which is a hyperbolic PDE with an initial condition $u(x,0)$. For simplicity, we will assume a spatial domain $x \in [0, 2\pi]$, with a periodic boundary condition, where $u(0,t) = u(2\pi,t)$. To solve this with a Fourier-Galerkin weighted residuals method, we know that the basis and weighting functions are the same and in this case they are chosen as trigonometric polynomials

$$\bar{u} = \sum_k^N c_k(t)\psi_k(x) = \sum_k^N c_k(t)e^{ikx} \tag{7.116}$$

The weighting functions, $w_j(x) = e^{-ikx}/2\pi$, were chosen because they satisfy the orthogonality condition

$$\int_0^{2\pi} \psi_k(x)w_j(x)dx = \delta_{kj} \tag{7.117}$$

The weighted residual expression for this problem becomes

$$\begin{aligned} \frac{1}{2\pi}\int_0^{2\pi} \left[\left(\frac{\partial}{\partial t} - \frac{\partial}{\partial x}\right)\sum_k^N c_k(t)e^{ikx}\right] e^{-ijx}dx = 0 \\ \frac{1}{2\pi}\int_0^{2\pi} \left[\sum_k^N \left(\frac{dc_k}{dt} - ikc_k\right)e^{ikx}\right] e^{-ijx}dx0 \end{aligned} \tag{7.118}$$

with the orthogonality condition of the Fourier functions we get

$$\frac{dc_k}{dt} - ikc_k = 0 \text{ for } k = 1, ..., N \tag{7.119}$$

The N initial conditions for this system of equations will be found by applying the Galerkin method to the initial conditions for $u(x,0)$

$$c_k(0) = \int_0^{2\pi} u(x,0) w_k(x) dx = \frac{1}{2\pi}\int_0^{2\pi} u(x,0) e^{-ikx} dx \tag{7.120}$$

■ EXAMPLE 7.12.

The Chebyshev-collocation method. This example problem uses the Chebyshev-collocation method to approximate $u(x,t)$ in a domain $x \in [-1,1]$ as a solution to the PDE

$$\frac{\partial u(x,t)}{\partial t} = \frac{\partial^2 u(x,t)}{\partial x^2} \tag{7.121}$$

with homogeneous Dirichlet boundary conditions $u(1,t) = u(-1,t) = 0$ and some initial conditions $u(x,0)$. We want to proceed with a weighted residual method with the combination between collocation and Chebyshev polynomials as basis functions for the approximation. Note, that the Chebyshev polynomials [1, 4, 27]

$$\psi_j(x) = \cos\left(j \cos^{-1} x\right) \tag{7.122}$$

are a perfect choice because they satisfy the homogeneous boundary conditions at $x = 1$ and $x = -1$. The approximate $\bar{u}$ will be

$$\bar{u}(x,t) = \sum_k^N c_k(t)\psi_k(x) = \sum_k^N c_k(t)\cos(k \cos^{-1} x) \tag{7.123}$$

For collocation, we will replace the weighting functions by Dirac delta functions located at $N-1$ positions within the domain, i.e.,

$$w_j(x) = \delta(x - x_j) \tag{7.124}$$

The weighted residual expression for this Chebyshev-collocation will be

$$\int_{-1}^{1} \left[\frac{\partial \bar{u}}{\partial t} - \frac{\partial^2 \bar{u}}{\partial x^2}\right] \delta(x - x_j) dx = 0 \tag{7.125}$$

The definition and properties of the delta function will reduce this equation to the solution of the PDE for the approximate $\bar{u}$ at each collocation point x_j

$$\left[\frac{\partial \bar{u}}{\partial t} - \frac{\partial^2 \bar{u}}{\partial x^2}\right]_{x_j} = 0 \tag{7.126}$$

for $j = 1, ..., N-1$ collocation points, with boundary and initial conditions

$$\begin{aligned} \bar{u}(1,t) &= 0 \\ \bar{u}(-1,t) &= 0 \\ \bar{u}(x_k,0) &= u(x_k,0) \end{aligned} \tag{7.127}$$

■ **EXAMPLE 7.13.**

Solution of a steady-state one-dimensional diffusion with a source term using Garlerkin and collocation weighted residual techniques. In this example, we want to solve the PDE

$$\frac{d^2\phi}{d\xi^2} + Q(\xi) = 0 \tag{7.128}$$

where ϕ can, for example, represent the dimensionless temperature, $\xi \in [0,1]$ and $Q = 1$ for $\xi \in [0, 1/2]$ while $Q = 0$ for $\xi \in (1/2, 1]$. The boundary conditions will be $\phi(0) = \phi(1) = 0$, i.e., homogeneous Dirichlet boundary conditions. A good choice of the basis functions will be to use a *sin* Fourier series, because it will satisfy the boundary conditions. Such approximation will be

$$\phi \approx \bar{\phi} = \sum_{j=1}^{N} c_j \sin j\pi\xi \tag{7.129}$$

where ψ_j is represented by $\sin j\pi\xi$. The weighted residual formulation for this problem becomes

$$\int_0^1 w_k \left[\frac{d^2}{d\xi^2}\left(\sum_{j=1}^{N} c_j\psi_j\right) + Q\right] d\xi = 0 \tag{7.130}$$

For a Galerkin solution, the weighting functions will be the same basis functions and the weighted residual expression will reduce to

$$\sum_{j=1}^{N} \frac{-c_j(j\pi)^2}{2} \int_0^1 \sin(j\pi\xi')\sin(k\pi\xi')d\xi' + \int_0^{1/2} \sin k\pi\xi' d\xi' = 0$$

$$\sum_{j-1}^{N} \frac{-c_j(j\pi)^2}{2}\delta_{jk} - \frac{1}{k\pi}\left(\cos(k\pi/2) - 1\right) = 0 \tag{7.131}$$

which is an independent linear system of equations whose solution is

$$c_j = \frac{-2}{(j\pi)^3}\left(\cos(j\pi/2) - 1\right) \tag{7.132}$$

giving the solution for the approximate $\bar{\phi}$ as

$$\bar{\phi}(\xi) = \sum_{j=1}^{N} \frac{-2}{(j\pi)^3}\left(\cos(j\pi/2) - 1\right)\sin(j\pi\xi) \tag{7.133}$$

For a collocation solution, we need first to chose a set of points within the domain where we want to find the solution. At these locations, we will place delta functions for the weighted residual method. Here, we select N terms for the approximation and we need N points located at ξ_k ($k = 1, ..., N$). The collocation weighted residual

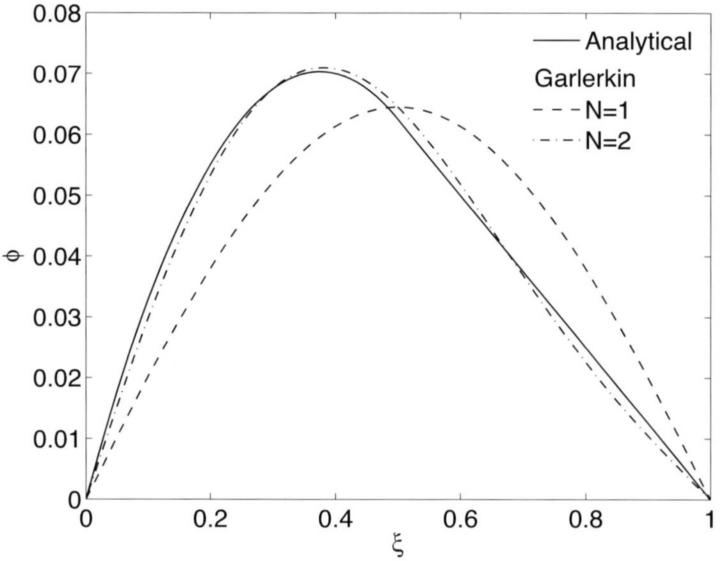

Figure 7.22: Comparison of Garlerkin's weighted residual and analytical solutions.

expression becomes,

$$\sum_{j=1}^{N} -c_j(j\pi)^2 \int_0^1 \sin(j\pi\xi')\delta(\xi'-\xi_k)d\xi' + \int_0^{1/2} \delta(\xi'-\xi_k) = 0$$

$$\sum_{j-1}^{N} -c_j(j\pi)^2 \sin(j\pi\xi_k) + \int_0^{1/2} \delta(\xi'-\xi_k) = 0 \tag{7.134}$$

Note that the second integral only plays a role if the point $\xi_k \in [0, 1/2]$. We now have a linear system of equations that will require a conventional matrix inversion method. If, for example $N = 2$, the system becomes

$$\begin{bmatrix} \pi^2 \sin\pi\xi_1 & (2\pi)^2 \sin 2\pi\xi_1 \\ \pi^2 \sin\pi\xi_2 & (2\pi)^2 \sin 2\pi\xi_2 \end{bmatrix} \begin{pmatrix} c_1 \\ c_2 \end{pmatrix} = \begin{pmatrix} 1 \\ 0 \end{pmatrix} \tag{7.135}$$

Galerkin employs the sinusoidal functions throughout the domain, using its periodicity and increasing the frequency as the number of terms are increased. The results using Galerkin's weighted residual are shown in Fig. 7.22, where only one and two terms were used. As can be seen, the results agree well with the analytical solution. In addition, eqn. (7.132) indicates that this coefficient decays rapidly as the number of terms increase. Figure 7.23 shows the predictions using collocation weighted residual methods using two and four points. Although, the solution is not as good as the Galerkin approximation, it is also interesting to note that few collocations points result in a fairly accurate solution. In fact, the solution is further improved by increasing the number of collocation points.

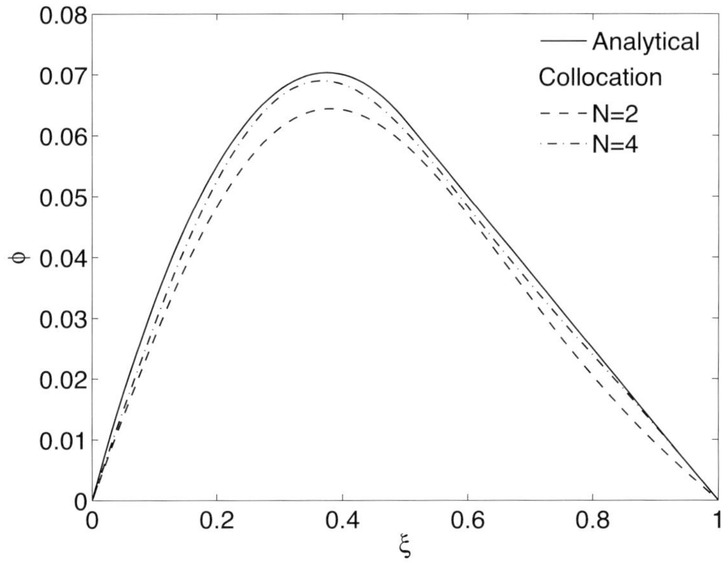

Figure 7.23: Comparison of the collocation weighted residual and analytical solutions.

Problems

7.1 Derive the equation for Simpson's integration rule.

7.2 Integrate the function $y = x^2$ between x =-1 and x =+1.

a) Use the trapezoidal rule using 4, 8 and 16 intervals. Plot the error as a function of number of intervals.

b) Use Simpson's rule using 4, 8 and 16 intervals. Plot the error as a function of number of intervals.

c) Use Gauss quadrature using 2 and 3 integration stations.

7.3 Integrate the function $y = x \ln x$ between x =0 and x =+1.

a) Use the trapezoidal rule using 4, 8 and 16 intervals. Plot the error as a function of number of intervals.

b) Use Simpson's rule using 4, 8 and 16 intervals. Plot the error as a function of number of intervals.

c) Use Gauss quadrature using 2, 4, and 10 integration stations. Plot the error as a function of number of integration stations.

7.4 Integrate the function $y = \ln x$ between x =0 and x =+1.

a) Use the trapezoidal rule. Plot the error as a function of number of intervals.

b) Use Gauss quadrature using 2, 4 and 10 integration stations. Plot the error as a function of number of integration stations.

7.5 Use numerical integration to determine the heat of fusion of the polyamide 6 represented in Fig. 7.3. Use the data given in Table 7.1.

a) Use the trapezoidal rule.

b) Use Simpson's rule.

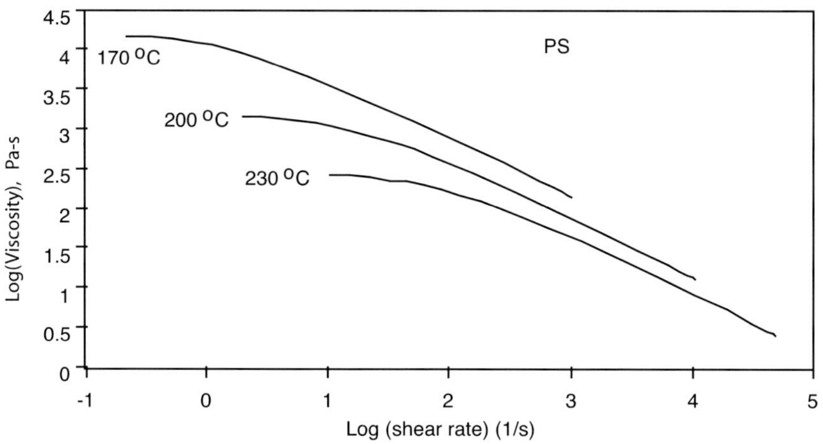

Figure 7.24: Viscosity curves for a polystyrene.

c) According to your results, give a range for the degree of crystallinity of your original sample.

7.6 Find a linear fit to the set of data points given below.

t	T
0	50
10	105
30	208
40	245
60	359

7.7 Derive the least squares matrix system to fit a quadratic equation of the form $y = a + bx + cx^2$.

7.8 Find a fit to the set of data points given below using the function $y = ax^b$.

x	y
0	0
5	2
10	10
15	20
20	30

7.9 Fit a power law model to the viscosity curve given in Fig. 7.24.

7.10 Fit a Bird-Carreau model to the viscosity curve given in Fig. 7.24. To solve this problem, download a non-linear fitting program from the world wide web.

7.11 Fit a set of cubic splines through the data set given below. Plot the curve with at least 5 points per segment.

i	x	y
1	1	1
2	2	1.2
3	3	1.5
4	4	2.5
5	3	4
6	1	3

REFERENCES

1. M. Abramowitz and I.A. Stegun. *Handbook of Mathematical Functions.* Dover Publications, New York, 1964.
2. N.I. Achieser. *Theory of Approximation.* Frederick Ungar Publising, New York, 1956.
3. J.F. Agassant, P. Avenas, J.Ph. Sergnet, and P.J. Carreau. *Polymer Processing, Principles and Modeling.* Hanser, New York, 1986.
4. G.B. Arfken and H.J. Weber. *Mathematical Methods for Physicists.* Academic Press, San Diego, 4th edition, 1995.
5. T.F. Ballenger, I.J. Chen, J.W. Crowder, G.E. Hagler, D.C. Bogue, and J.L. White. *J. Rheology,* 15:195, 1971.
6. R.V. Churchill. *Introduction to Complex Variables.* McGraw-Hill, New York, 1947.
7. S.H. Crandall. *Engineering Analysis.* McGraw-Hill, New York, 1956.
8. P.J. Davis. *Interpolation and Approximation.* Blaisdell Publishing Company, Waltham, 1963.
9. J. Escobar. *Algebra lineal para principiantes.* Serie Nabla-Delta, Medellín, Colombia, 1990.
10. J.H. Ferziger and M. Peric. *Computational Methods for Fluid Dynamics.* Springer, London, 1999.
11. B.A. Finlayson. *The Method of Weighted Residuals and Variational Principles.* Academic Press, New York, 1972.
12. B.A. Finlayson. *Nonlinear Analysis in Chemical Engineering.* McGraw-Hill, New York, 1980.
13. B.A. Finlayson and L.E. Scriven. The method of weighted residuals: a review. *Appl. Mech. Rev.,* 19:735, 1966.
14. B.G. Galerkin. Series solution of some problems of elastic equilibrium of rods and plates (russian). *Vest. Inzh. Tech.,* 19:897, 1915.
15. M.A. Golberg and C.S. Chen. *Discrete projection methods for integral equations.* Computational Mechanics Publications, Southampton, 1997.
16. G.V. Gordon and M.T. Shaw. *Computer Programs for Rheologists.* Hanser, New York, 1994.
17. J.P. Hernandez-Ortiz and T.A. Osswald. A novel cure reaction model fitting technique based on dsc scans. *J. Polym. Eng.,* 25(1):23, 2005.
18. F.B. Hildebrand. *Introduction to Numerical Analysis.* Dover Publishers, New York, 2nd edition, 1974.
19. T.J.R. Hughes. *The Finite Element Method.* Dover Publishers, New York, 2000.
20. C.L. Tucker III. *Computer Modeling for Polymer Processing, Fundamentals.* Hanser, Munich, 1989.

21. D.C. Ives. Conformal grid generation. In *Numerical Grid Generation*, page 107, New York, 1982. Elsevier.

22. M.R. Kamal. *Polym. Eng. Sci.*, 14:231, 1979.

23. M.R. Kamal and S. Sourour. *Polym. Eng. Sci.*, 13:59, 1973.

24. E.J. Kansa. Multiquadratics- a scattered data approximation scheme with applications to computational fluid mechanics. ii-solutions to parabolic, hyperbolic and elliptic partial differential equations. *Computers Math. Applic.*, 19:147, 1990.

25. H.E. Kissinger. *J. Research Natl. Bur. Standards*, 57:217, 1956.

26. H.E. Kissinger. *Analytical Chemistry*, 49(11):1702, 1957.

27. E. Kreyszig. *Advanced Engineering Mathematics*. John Wiley & Sons, New York, 8th edition, 1999.

28. L.M. Lopez. Modeling the vulcanization of liquid silicone rubber. Master's thesis, University of Wisconsin-Madison, Madison, 2004.

29. L.M. Lopez, A.B. Cosgrove, J.P. Hernandez-Ortiz, T.A. Osswald, and E. Camacho. In *SPE ANTEC*, Chicago, 2004.

30. S.C. Mikhlin. *Variational Methods in Mathematical Physics*. Macmillan, New York, 1964.

31. J.J. Moré, B.S. Garbow, and K.E. Hillstrom. User guide for minpack-1. Technical Report ANL-80-74, Argonne National Laboratory, 1980.

32. P. Murray and J. White. *J. Trans. Brit. Ceram. Soc.*, 54:151, 1955.

33. T.A. Osswald. *Numerical Methods for compression mold filling simulation*. PhD thesis, University of Illinois at Urbana-Champaign, Urbana-Champaign, 1986.

34. P.W. Partridge, C.A. Brebbia, and L.C. Wrobel. *The dual reciprocity boundary element method*. Computational Mechanics, Southampton, 1991.

35. S.V. Patankar. *Numerical Heat Transfer and Fluid Flow*. Hemisphere Publishers, New York, 1980.

36. W. H. Press, S. A. Teukolsky, W. T. Vetterling, and B. P. Flannery. *Numerical Recipes in Fortran 77*. Cambridge, 2nd edition, 1992.

37. R.E. Smith and L.J. Johnson. Automatic grid generation and flow solution for complex geometries. *AIAA J.*, 34:1120, 1996.

38. R.E. Smith and B.L. Weigel. Analytic and approximate boundary fitted coordinate systems for fluid flow simulation. *AIAA*, 80:192, 1980.

39. G. Strang. *Introduction to Applied Mathematics*. Wellesley-Cambridge Press, Wellesley, 1986.

40. J.C. Tannehill, D.A. Anderson, and R.H. Pletcher. *Computational Fluid Mechanics and Heat Transfer*. Taylor & Francis, Washington, 2nd edition, 1997.

41. H.K. Versteeg and W. Malalasekera. *An Introduction to Computational Fluid Dynamics: The Finite Volume Method*. Addison-Wesley Pub., New York, 1996.

42. O.C. Zienkiewicz. *Finite Element Methods in Stress Analysis*, chapter 13: Iso-parametric and associate elements families for two and three dimensional analysis. Tapir Press, Trondheim, 1969.

43. O.C. Zienkiewicz. *The Finite Element Method*. McGraw-Hill, London, 3rd edition, 1997.

CHAPTER 8

FINITE DIFFERENCE METHOD

I don't know, I don't care, and it doesn't make any difference!

—Albert Einstein

The finite difference method (FDM) is probably the easiest and oldest method to solve partial differential equations. For many simple applications it requires minimum theory, it is simple and it is fast. When a higher accuracy is desired, however, it requires more sophisticated methods, some of which will be presented in this chapter. The first step to be taken for a finite difference procedure is to replace the continuous domain by a finite difference mesh or grid. For example, if we want to solve partial differential equations (PDE) for two functions $\phi(x)$ and $u(x, y)$ in a 1D and 2D domain, respectively, we must generate a grid on the domain and replace the functions by functions evaluated at the discrete locations, $i\Delta x$ and $j\Delta y$, $\phi(i\Delta x)$ and $u(i\Delta x, j\Delta y)$, or ϕ_i and u_{ij}. Figure 8.1 illustrates typical 1D and 2D FDM grids.

A finite difference representation for a derivative can be introduced by recalling the definition of the derivative of a function $\phi(x)$

$$\frac{\partial \phi}{\partial x} = \lim_{\Delta x \to 0} \frac{\phi(x + \Delta x) - \phi(x)}{\Delta x} \tag{8.1}$$

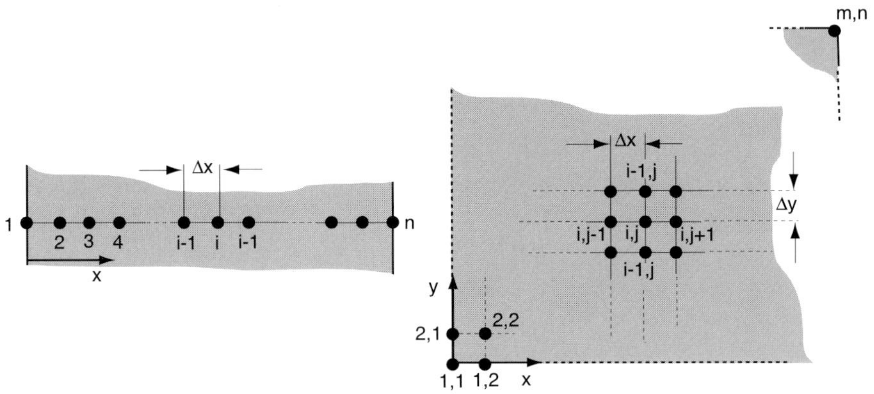

Figure 8.1: Typical 1D and 2D finite difference discretization.

which, for continuous functions, can be written as

$$\frac{\partial \phi}{\partial x} \approx \frac{\phi(x + \Delta x) - \phi(x)}{\Delta x} \tag{8.2}$$

This equation is a good approximation for a small but finite Δx. Using the i notation, this equation can be written as follows

$$\left.\frac{\partial \phi}{\partial x}\right|_i \approx \frac{\phi_{i+1} - \phi_i}{\Delta x} \tag{8.3}$$

which is generally referred to as a *forward difference* approximation. If we take a detailed look at this equation, we notice that we are approximating the derivative of the function at point x_i as a function of the values of the function at the discrete locations x_{i+1} and x_i. In the same way, we can obtain the value of the derivative by using x_i and x_{i-1}, i.e.,

$$\left.\frac{\partial \phi}{\partial x}\right|_i \approx \frac{\phi_i - \phi_{i-1}}{\Delta x} \tag{8.4}$$

which in turn is referred to as a *backward difference* approximation. Finally, eqns. (8.3) and (8.4) can be used to approximate the derivative at point i using the points located ahead (forward), $i + 1$, and behind (backward), $i - 1$, to get a *central difference* approximation

$$\left.\frac{\partial \phi}{\partial x}\right|_i \approx \frac{\phi_{i+1} - \phi_{i-1}}{2\Delta x} \tag{8.5}$$

In addition to the above *ad hoc* approximations there are many procedures to develop finite difference approximations for given PDE's with their respective finite difference grids. The most commonly used are

- Taylor-series expansions,
- Integral methods,

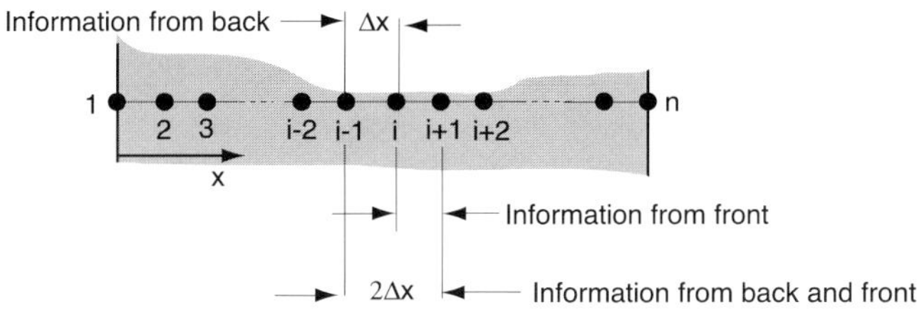

Figure 8.2: Schematic of the backward, forward and central difference schemes.

- Finite-volume approach,
- Polynomial fitting.

Typically, these methods arrive at the same finite difference representation for a given problem. However, we feel that Taylor-series expansions are easy to illustrate and we will therefore use them here in the derivation of finite difference equations. We encourage the student of polymer processing to look up the other techniques in the literature, for instance, integral methods and polynomial fitting from Tannehill, Anderson and Pletcher [26] or from Milne [16] and finite volume approach from Patankar [18], Versteeg and Malalasekera [27] or from Roache [20].

8.1 TAYLOR-SERIES EXPANSIONS

Taylor-series expansions allow the development of finite differences on a more formal basis. In addition, they provide tools to analyze the order of the approximation and the error of the final solution. In order to introduce the methodology, let's use a simple example by trying to obtain a finite difference expression for $\partial\phi/\partial x$ at a discrete point i, similar to those in eqns. (8.1) to (8.3). Initially, we are going to find an expression for this derivative using the values of ϕ at discrete locations $i-1$ and i (such as presented for a *backward difference* equation). Thus, we are looking for an expression such as

$$\left.\frac{\partial\phi}{\partial x}\right|_i = a\phi_{i-1} + b\phi_i \tag{8.6}$$

which is a simple linear interpolation (Chapter 7). Now, we proceed to expand ϕ_{i-1} in Taylor-series about the point i, i.e.,

$$\begin{aligned}\phi_i &= \phi_i \\ \phi_{i-1} &= \phi_i - \Delta x\left.\frac{\partial\phi}{\partial x}\right|_i + \frac{\Delta x^2}{2!}\left.\frac{\partial^2\phi}{\partial x^2}\right|_i - \frac{\Delta x^3}{3!}\left.\frac{\partial^3\phi}{\partial x^3}\right|_i + \ldots\end{aligned} \tag{8.7}$$

We can now solve for the value of the derivative by substituting these values in eqn. (8.6) as follows

$$\left.\frac{\partial\phi}{\partial x}\right|_i = a\left\{\phi_i - \Delta x\left.\frac{\partial\phi}{\partial x}\right|_i + \frac{\Delta x^2}{2!}\left.\frac{\partial^2\phi}{\partial x^2}\right|_i + \ldots\right\} + b\left\{\phi_i\right\} \tag{8.8}$$

which leads to a system of two equations with two unknowns

$$\left.\frac{\partial \phi}{\partial x}\right|_i = (a+b)\phi_i + a\left\{-\Delta x \left.\frac{\partial \phi}{\partial x}\right|_i + \frac{\Delta x^2}{2!}\left.\frac{\partial^2 \phi}{\partial x^2}\right|_i + \ldots\right\} \tag{8.9}$$

The fact that the sought formula is for the derivative leads to

$$a + b = 0 \tag{8.10}$$
$$-a\Delta x = 1 \tag{8.11}$$

and

$$\left.\frac{\partial \phi}{\partial x}\right|_i = \frac{\phi_i - \phi_{i-1}}{\Delta x} + \frac{\Delta x}{2!}\left.\frac{\partial^2 \phi}{\partial x^2}\right|_i + \ldots \tag{8.12}$$

If we collapse all the higher order terms we get

$$\left.\frac{\partial \phi}{\partial x}\right|_i = \frac{\phi_i - \phi_{i-1}}{\Delta x} + O(\Delta x) \tag{8.13}$$

Here, we obtained a backward finite difference expression for the derivative $\partial\phi/\partial x$ at a point i, and we get the order of the interpolation, i.e., all the terms that we collapsed in $O(\Delta x)$, which is a metric for the error built into the approximation. Similarly, we can obtain the forward difference approximation as

$$\left.\frac{\partial \phi}{\partial x}\right|_i = \frac{\phi_{i+1} - \phi_i}{\Delta x} + O(\Delta x) \tag{8.14}$$

and the central difference as

$$\left.\frac{\partial \phi}{\partial x}\right|_i = \frac{\phi_{i+1} - \phi_{i-1}}{2\Delta x} + O(\Delta x^2) \tag{8.15}$$

Note that due to the fact that the physical distance between the grid points, Δx, is small ($\Delta x \ll 1$), and that the central difference is of second order (Δx^2) nature, the central difference approximation leads to a much better solution than the first order backward and forward difference solutions.

A way to generalize the finite difference technique for any PDE is to use a Taylor table. For any PDE such as

$$\frac{\partial^{(n)}\phi}{\partial x^{(n)}} = \ldots + a_{i-2}\phi_{i-2} + a_{i-1}\phi_{i-1} + a_i\phi_i + a_{i+1}\phi_{i+1} + a_{i+2}\phi_{i+2} + \ldots \tag{8.16}$$

we construct a table where the columns correspond to the derivatives from the Taylor-series and the rows to the coefficients a_i used in the finite difference expression in eqn. (8.16). Table 8.1 shows the Taylor table for this equation.

The last row of the table will be used to generate equations needed to obtain the a_i coefficients. The sum of all the columns to the left of the sought derivative $\partial^{(n)}\phi/\partial x^{(n)}$

Table 8.1: Taylor-Series Table

	ϕ_i	$\frac{\partial \phi}{\partial x}$	$\cdots$	$\frac{\partial^{(n)} \phi}{\partial x^{(n)}}$	$\frac{\partial^{(n+1)} \phi}{\partial x^{(n+1)}}$	$\cdots$
$\vdots$	$\vdots$	$\vdots$	$\vdots$	$\vdots$	$\vdots$	$\vdots$
ϕ_{i-2}	a_{i-2}	$a_{i-2}(-2\Delta x)$	$\cdots$	$a_{i-2}\frac{(-2\Delta x^n)}{n!}$	$a_{i-2}\frac{(-2\Delta x^{n+1})}{(n+1)!}$	$\cdots$
ϕ_{i-1}	a_{i-1}	$a_{i-1}(-\Delta x)$	$\cdots$	$a_{i-1}\frac{(-\Delta x^n)}{n!}$	$a_{i-1}\frac{(-\Delta x^{n+1})}{(n+1)!}$	$\cdots$
ϕ_i	a_i	-	-	-	-	$\cdots$
ϕ_{i+1}	a_{i+1}	$a_{i+1}(\Delta x)$	$\cdots$	$a_{i+1}\frac{(\Delta x^n)}{n!}$	$a_{i+1}\frac{(\Delta x^{n+1})}{(n+1)!}$	$\cdots$
ϕ_{i+2}	a_{i+2}	$a_{i+2}(2\Delta x)$	$\cdots$	$a_{i+2}\frac{(2\Delta x^n)}{n!}$	$a_{i+2}\frac{(2\Delta x^{n+1})}{(n+1)!}$	$\cdots$
$\vdots$	$\vdots$	$\vdots$	$\vdots$	$\vdots$	$\vdots$	$\vdots$
	0	0	$\cdots$	1	?	$\cdots$

Table 8.2: Equation (8.18) Taylor-Series Table

	ϕ_i	$\frac{\partial \phi}{\partial x}$	$\frac{\partial^2 \phi}{\partial x^2}$	$\frac{\partial^3 \phi}{\partial x^3}$	$\frac{\partial^4 \phi}{\partial x^4}$
ϕ_{i-1}	a	$a(-\Delta x)$	$\frac{a(-\Delta x)^2}{2!}$	$\frac{a(-\Delta x)^3}{3!}$	$\frac{a(-\Delta x)^4}{4!}$
ϕ_i	b	-	-	-	-
ϕ_{i+1}	c	$c(\Delta x)$	$\frac{c(\Delta x)^2}{2!}$	$\frac{c(\Delta x)^3}{3!}$	$\frac{c(\Delta x)^4}{4!}$
	0	0	1	?	?

must be zero, while the column for the derivative of interest must add up to one, i.e.,

$$
\begin{aligned}
0 &= \cdots + a_{i-2} + a_{i-1} + a_i + a_{i+1} + a_{i+2} + \cdots \\
0 &= \cdots + a_{i-2}(-2\Delta x) + a_{i-1}(-\Delta x) + a_{i+1}(\Delta x) + a_{i+2}(2\Delta x) + \cdots \\
0 &= \cdots + a_{i-2}\frac{(-2\Delta x)^2}{2!} + a_{i-1}\frac{(-\Delta x)^2}{2!} + a_{i+1}\frac{(\Delta x)^2}{2!} + a_{i+2}\frac{(2\Delta x)^2}{2!} + \cdots \\
&\vdots \\
1 &= \cdots + a_{i-2}\frac{(-2\Delta x)^n}{n!} + a_{i-1}\frac{(-\Delta x)^n}{n!} + a_{i+1}\frac{(\Delta x)^n}{n!} + a_{i+2}\frac{(2\Delta x)^n}{n!} + \cdots
\end{aligned}
\tag{8.17}
$$

The columns to the right are used to complete the system of equations (sum equals zero) or to obtain the order of the approximation by calculating, after the coefficients a_i are obtained, the first column to the right whose sum is not zero.

■ **EXAMPLE 8.1.**

Second order finite difference for a second order derivative. Let's illustrate the Taylor-series by finding a finite difference for the second derivative

$$
\frac{\partial^2 \phi}{\partial x^2} = a\phi_{i-1} + b\phi_i + c\phi_{i+1} \tag{8.18}
$$

We have three unknowns (a, b, c) and the Taylor Table 8.2 can be generated.

In the Taylor table, the rows correspond to the interpolation variable coefficients for the PDE approximation and the columns to the terms of the Taylor-series of the corresponding coefficient. For this particular case, we need three equations that result from the sum over the columns. The sum over columns 2 to 4 will give the following equations

$$\begin{aligned} a + b + c &= 0 \\ a(-\Delta x) + c(\Delta x) &= 0 \\ \frac{(\Delta x)^2}{2!}(a + c) &= 1 \end{aligned} \tag{8.19}$$

which can be solved to give $a = c = 1/\Delta x^2$ and $b = -2/\Delta x^2$. The finite differences representation of eqn. (8.18) can now be written as

$$\frac{\partial^2 \phi}{\partial x^2} = \frac{\phi_{i-1} - 2\phi_i + \phi_{i+1}}{\Delta x^2} + O(\Delta x^n) \tag{8.20}$$

In order to obtain the order of approximation n, we first replace the values of (a, b, c) in the column corresponding to $\partial^3\phi/\partial x^3$ of Table 8.2

$$a\frac{-\Delta x^3}{3!} + c\frac{\Delta x^3}{3!} = 0 \tag{8.21}$$

Since no extra information of the approximation is given, we must proceed to the next column to the right of Table 8.2

$$a\frac{\Delta x^4}{4!} + c\frac{\Delta x^4}{4!} \neq 0 \tag{8.22}$$

which indicates that the above approximation of eqn. (8.20) is of the order $O(\Delta x^2)$.

■ EXAMPLE 8.2.

Using 4 grid points to represent $\partial\phi/\partial x$ Instead of using the first order (backward or forward) or the second order (central) finite difference approximation for the first derivative, let us calculate the derivative using four grid points (see Fig. 8.2)

$$\left.\frac{\partial \phi}{\partial x}\right|_i = a\phi_{i-2} + b\phi_{i-1} + c\phi_i + d\phi_{i+1} \tag{8.23}$$

Again, we use the Taylor table (Table 8.3).

Note that in this particular example, we need to include two additional columns to the right of the desired derivative column because we have four unknowns. The resulting linear set of equations is

$$\begin{aligned} a + b + c + d &= 0 \\ -2a\Delta x - b\Delta x + d\Delta x &= 1 \\ 4a\Delta x^2 + b\Delta x^2 + d\Delta x^2 &= 0 \\ -8a\Delta x^3 - b\Delta x^3 + d\Delta x^3 &= 0 \end{aligned} \tag{8.24}$$

which results in $a = 1/6\Delta x, b = -1/\Delta x, c = 1/2\Delta x$ and $1/3\Delta x$. Substituting this solution in the 5th column of Table 8.3 we obtain that its sum is not equal to zero, thus, we can get the order of the approximation from here. The final finite difference approximation for this case is

$$\frac{\partial \phi}{\partial x} = \frac{1}{6\Delta x}\left(\phi_{i-2} - \phi_{i-1} + 3\phi_i + 2\phi_{i+1}\right) + O(\Delta x^3) \tag{8.25}$$

with an order of approximation (or error) of $O(\Delta x^3)$.

Table 8.3: Taylor-Series Table for the 4 Points First Derivative FD

	ϕ_i	$\frac{\partial \phi}{\partial x}$	$\frac{\partial^2 \phi}{\partial x^2}$	$\frac{\partial^3 \phi}{\partial x^3}$	$\frac{\partial^4 \phi}{\partial x^4}$	$\frac{\partial^5 \phi}{\partial x^5}$
ϕ_{i-2}	a	$a(-2\Delta x)$	$a\frac{(-2\Delta x)}{2!}$	$a\frac{(-2\Delta x^3)}{3!}$	$a\frac{(-2\Delta x^4)}{4!}$	$a\frac{(-2\Delta x^5)}{5!}$
ϕ_{i-1}	b	$b(-\Delta x)$	$b\frac{(-\Delta x)}{2!}$	$b\frac{(-\Delta x^3)}{3!}$	$b\frac{(-\Delta x^4)}{4!}$	$b\frac{(-\Delta x^5)}{5!}$
ϕ_i	c	-	-	-	-	-
ϕ_{i+1}	d	$d(\Delta x)$	$d\frac{(\Delta x)}{2!}$	$d\frac{(\Delta x^3)}{3!}$	$d\frac{(\Delta x^4)}{4!}$	$d\frac{(\Delta x^5)}{5!}$
	0	1	0	0	?	?

8.2 NUMERICAL ISSUES

There are many numerical issues that we must discuss before proceeding to applications of FDM when solving polymer processing problems. The most important are,

- Truncation error
- Consistency
- Stability

As we can imagine, most of these issues are directly related to the order of the approximation used in the finite difference representation. In fact, the truncation error (as shown in Chapter 7) is the difference between the PDE and the FD representation, which is represented by the terms collapsed in $O(\Delta x^n)$. For problems represented by PDEs with more than one independent variable the truncation error will be the sum of the truncation error for each FD representation. For example, for a transient one dimensional PDE, where we use a first order approximation for the time derivative and a second order for the spatial derivative, we will have a truncation error that is $O(\Delta t) + O(\Delta x^2)$, which can also be written as $O(\Delta t, \Delta x^2)$.

The consistency of a finite difference approximation is the behavior of this representation when the mesh is refined. In a one dimensional case, for example, the mesh will indicate the value for Δx, which, as we discussed above, dictates the value of the truncation error. Thus, a finite difference representation of a PDE is said to be consistent if the truncation error goes to zero as the *grid size* (or Δx) goes to zero.

The stability of a FD representation deals with the behavior of the truncation error as the calculation proceeds in time or marches in space, typically, transient problems, and problems with convection-convection derivatives. A stable FD scheme will not allow the errors to grow as the solution proceeds in time or space. The issue of stability for transient problems will be analyzed in depth later in this chapter.

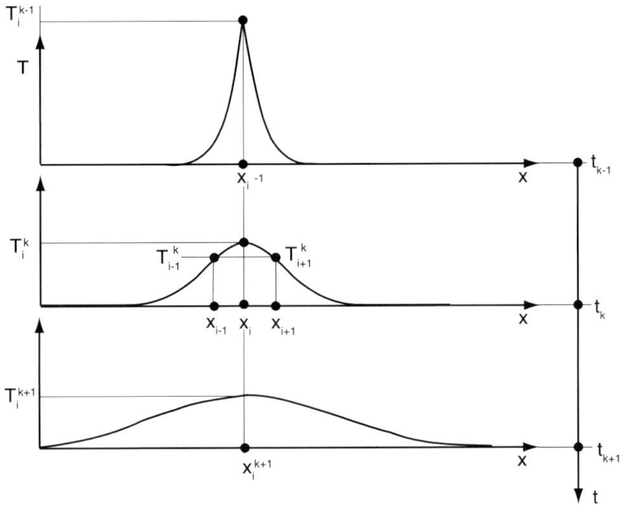

Figure 8.3: Info-travel concept in a 1D diffusion problem.

8.3 THE INFO-TRAVEL CONCEPT

Up to this point, we have discussed how to develop finite difference expressions for PDE's with any desired level of accuracy. We expect that the truncation error, the consistency and stability will be improved as the order of the approximation is increased. However, we have not mentioned the connection between the physics of the problem under consideration, and represented by the PDE, and the way we select the points for the finite difference expression, i.e., $i-2$, $i-1$, $i+1$, or $i+2$, as well as the order of the approximation, i.e., first order, second order, etc. In other words, the points, order and grid size are not arbitrarily chosen to approximate the derivatives at a point i, but should be chosen according to the physical principles of the problem. This concept is referred to as *info-travel* and will be discussed here.

The balance equations used to model polymer processes have, for the most part, first order derivatives in time, related with transient problems, and first and second order derivatives in space, related with convection and diffusive problems, respectively. Let us take the heat equation over an infinite domain as

$$\frac{\partial T}{\partial t} = \alpha \frac{\partial^2 T}{\partial x^2} \tag{8.26}$$

For this diffusive (conductive) problem, any perturbation in the temperature will move, or diffuse, symmetrically along the (x) axis. In other words the *information* will travel in all directions in space. Therefore, to approximate the second derivatives in space at a point i, we must use points that are symmetrically distributed in space, as shown in Fig. 8.3.

A good option will be to use a second order approximation for the space derivatives as follows

$$\frac{\partial^2 T}{\partial x^2} = \frac{T_{i+1,j} - 2T_{i,j} + T_{i-1,j}}{\Delta x^2} + O(\Delta x^2) \tag{8.27}$$

To increase the order of the approximation correctly, we must include the same number of points in any direction. This way we are accommodating, as best as possible, the way information travels in this problem. If the choice of points had been asymmetric, even by keeping the second order of the approximation (such as in eqn. (8.27)), the system is actually considering diffusion with a preferred direction, which is not realistic.

Following the info-travel concept for the time derivative, we can easily see that it would be unreasonable to use central or forward differences. Such a choice would mean that the derivative of temperature with respect to time at the kth time step depends on the value of temperature in the future, or the $(k+1)$th step. Hence, the correct way to express a time derivative should be by using the information in the past only, i.e.,

$$\frac{\partial T}{\partial t} = \frac{T_{i,j}^{k} - T_{i,j}^{k-1}}{\Delta t} + O(\Delta t) \tag{8.28}$$

The treatment of transient problems is covered in depth later in this chapter.

First order derivatives in space are not as straight forward as second derivatives, and must be dealt with according to the situation. We find these derivatives in many situations, for example, in the continuity equation

$$\frac{\partial u_i}{\partial x_i} = 0 \tag{8.29}$$

in convection terms

$$u_i \frac{\partial T}{\partial x_i} \text{ or } u_j \frac{\partial u_i}{\partial x_j} \tag{8.30}$$

or in boundary conditions,

$$-k \frac{\partial T}{\partial \mathbf{n}} = q_0 \text{ or } \frac{\partial u_i}{\partial \mathbf{n}} = 0 \tag{8.31}$$

The info-travel concept dictates the selection of the points and the order of the approximation for each individual case. The continuity equation comes from a mass balance on a differential element within the domain. Here, the mass can come and go through all sides of the differential element; hence, the information can travel in all directions, requiring a central difference formulation.

The convection terms, on the other hand, have a preferred direction, depending to the direction of the velocity that drives the convective term. Figure 8.4 presents the same temperature perturbation as in Fig. 8.3, but with an imposed velocity field, u_x.

For this problem, the energy balance with constant properties becomes

$$\left(\frac{\partial T}{\partial t} + u_j \frac{\partial T}{\partial x_j} \right) = \alpha \frac{\partial^2 T}{\partial x_j \partial x_j} \tag{8.32}$$

The convection term that controls the heat transfer is

$$u_x \frac{\partial T}{\partial x} \tag{8.33}$$

This velocity will indicate that the information travels from left to right, thus a correct FD expression will be one that uses the information from points to the left of any point i, where the derivative is wanted, for example

$$u_x \frac{\partial T}{\partial x} = u_{xi} \frac{T_i - T_{i-1}}{\Delta x} + O(\Delta x) \tag{8.34}$$

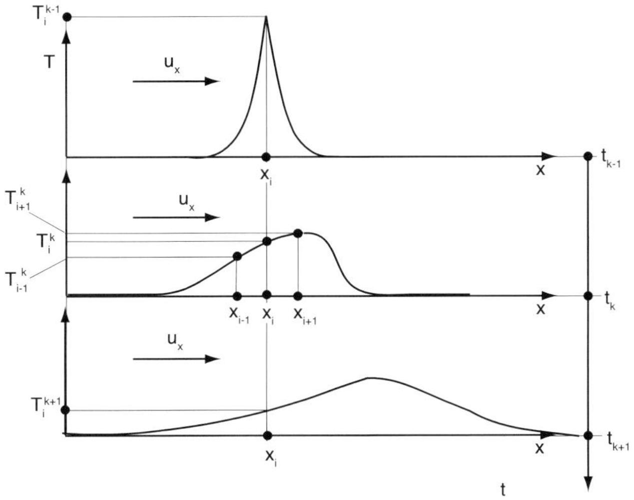

Figure 8.4: Info-travel concept with a combination of diffusion and convective terms.

Note that the above approximation is a first order approximation. If we were to use a central difference, we would increase the order, but contrary to what is expected, this choice will adversely affect the accuracy and stability of the solution due to the fact the information is forced to travel in a direction that is not supported by the physics of the problem. How convective problems are dealt will be discussed in more detail later in this chapter. The following sections will present steady state, transient and moving boundary problems with examples and applications.

8.4 STEADY-STATE PROBLEMS

This section will illustrate the tools taught in the above sections in the form of examples applied to steady state problems. Example 8.3 applies the finite difference method to a simple one-dimensional fin cooling problem and illustrates the nature of the system of equations that is normally achieved. Example 8.4 present a 2D compression molding problem where an iterative solution method is introduced.

■ **EXAMPLE 8.3.**

Steady-state temperature profile of a fin of uniform cross-sectional area. In a fin cooling process we want to obtain the temperature profile on the surface of the fin. The aspect ratio of the fin is large, $L/W \ll 1$, as shown in Fig. 8.5.

For a material with constant properties and heat transfer coefficient, the energy equation is reduced to

$$\frac{d^2\Theta}{dx^2} - m^2\Theta = 0 \tag{8.35}$$

where $\Theta = T(x) - T_\infty$, $m = hP/kA_c$, h is the convection heat transfer coefficient, $P = 2w+2t$ is the convection perimeter and $A_c = wt$ is the cross sectional conductive

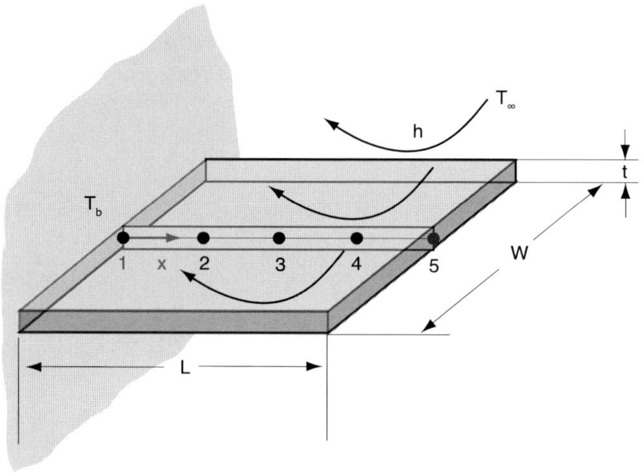

Figure 8.5: Schematic diagram and finite difference grid of a fin cooling problem.

area. The boundary conditions are defined by

$$\Theta(0) = T_0 - T_\infty = \Theta_0 \tag{8.36}$$

$$h\Theta(L) = -k\frac{d\Theta}{dx}\bigg|_{x=L} \tag{8.37}$$

The analytical temperature profile can be obtained as follows [2, 11]

$$\frac{\Theta}{\Theta_0} = \frac{\cosh m(L-x) + (h/mk)\sinh m(L-x)}{\cosh mL + (h/mk)\sinh mL} \tag{8.38}$$

Let's pretend that this solution is not known and proceed to arrive at a solution using finite differences. Here, the first step is to create the grid, which is illustrated in Fig. 8.5. For simplicity, we are going to use only $n = 5$ points (or nodes). The grid size is given by $\Delta x = L/(n-1)$. The FD expression of eqn. (8.35) is written as

$$\Theta_{i+1} - 2\Theta_i + \Theta_{i-1} - m^2\Delta x^2\Theta_i = 0 \tag{8.39}$$

Here, we are using a second order approximation for the second derivative using the correct info-travel concept for the conduction term. This equation comes from the energy balance within the domain, thus it will be used for the internal nodes $n = 2, 3$ and 4. The boundary condition for the first node is the temperature at the wall to which the fin is attached to

$$\Theta_1 = \Theta_0 \tag{8.40}$$

The boundary condition at $x = L$ or $n = 5$ has a first derivative that comes from the balance between the conduction and the convection at the end of the fin. We can use a backward finite difference to approximate this derivative, i.e.,

$$(\Theta_5 - \Theta_4) + \frac{h\Delta x}{k}\Theta_5 = 0 \tag{8.41}$$

Table 8.4: Example 8.3 Data

Parameter	Value
L	20.0 cm
w	20.0 cm
t	1.0 cm
h	1.0 W/K/m^2
k	10.0 W/m/K

which violates the info-travel concept for this specific boundary, but for very small Δx, the error introduced will be small. However, we can also use a central difference

$$(\Theta_6 - \Theta_4) + \frac{2h\Delta x}{k}\Theta_5 = 0 \tag{8.42}$$

which implies the generation of what is commonly called a *fictitious node*. This new node will introduce an additional unknown and therefore a new equation is required. This is done by considering node 5 as an internal node, for which eqn. (8.39) can be used.

The system of equations is as follows,

$$\begin{aligned}
\Theta_1 - \Theta_0 &= 0 \\
\Theta_1 - (2 + m^2\Delta x^2)\Theta_2 + \Theta_3 &= 0 \\
\Theta_2 - (2 + m^2\Delta x^2)\Theta_3 + \Theta_4 &= 0 \\
\Theta_3 - (2 + m^2\Delta x^2)\Theta_4 + \Theta_5 &= 0 \\
-\Theta_4 + (1 + h\Delta x/k)\Theta_5 &= 0
\end{aligned} \tag{8.43}$$

which in matrix form is

$$\begin{bmatrix} 1 & 0 & 0 & 0 & 0 \\ 1 & a & 1 & 0 & 0 \\ 0 & 1 & a & 1 & 0 \\ 0 & 0 & 1 & a & 1 \\ 0 & 0 & 0 & -1 & b \end{bmatrix} \begin{pmatrix} \Theta_1 \\ \Theta_2 \\ \Theta_3 \\ \Theta_4 \\ \Theta_5 \end{pmatrix} = \begin{pmatrix} \Theta_0 \\ 0 \\ 0 \\ 0 \\ 0 \end{pmatrix} \tag{8.44}$$

where $a = -(2 + m^2\Delta x^2)$ and $b = 1 + h\Delta x/k$. Note the tri-diagonal nature of the matrix for this linear system; in fact, this is a common characteristic of finite difference formulations for problems involving second derivatives. Such equation systems can be solved rapidly, using solution schemes such as the Thomas algorithm [19].

Figure 8.6 shows a comparison between the analytical solution and several discretizations for this problem using the data from Table 8.4.

As expected, the solution improves as the number of nodes increases (Δx decreases). Figure 8.7 shows how the truncation error decreases with Δx. The solid line in the figure is Δx^2 and the dots are the value of the truncation error from the simulation. The graph clearly indicates that the method is of second order accuracy in space.

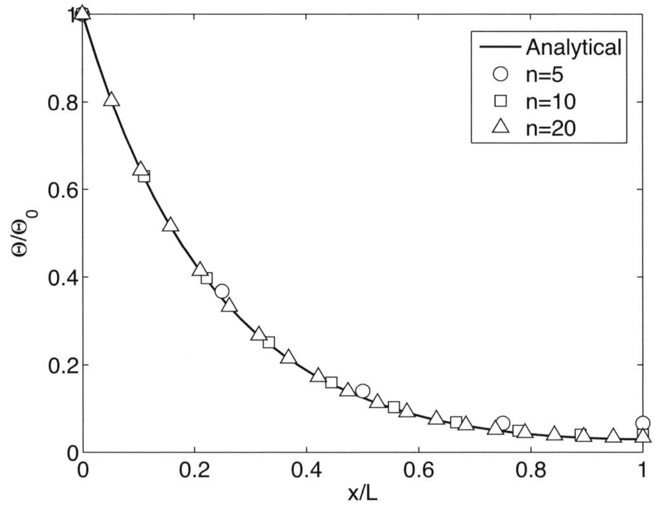

Figure 8.6: Steady state FDM and analytical temperature distribution along the cooling fin.

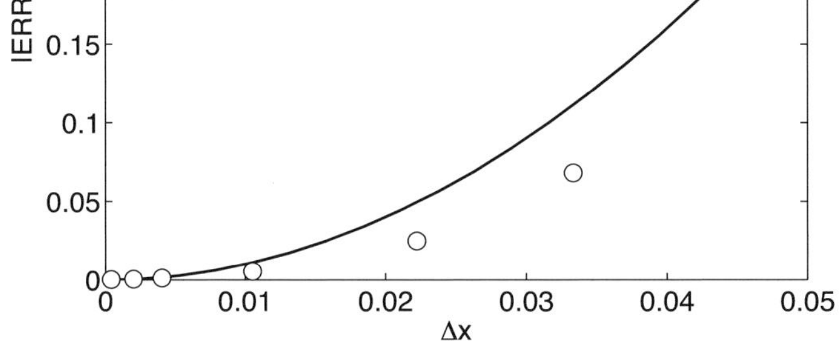

Figure 8.7: Error in the FD temperature distribution prediction along the cooling fin as a function of grid size Δx.

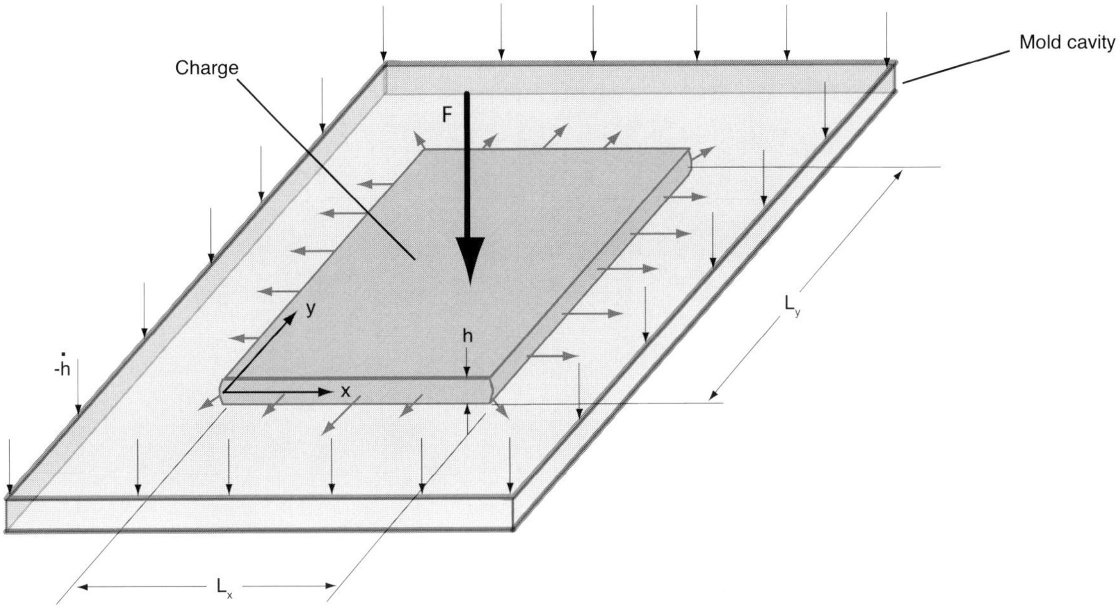

Figure 8.8: Schematic diagram of the compression molding of a rectangular charge.

Table 8.5: Example 8.4 Data

Parameter	Value
η	10000 Pa-s
L_x	0.2 m
L_y	0.4 m
h	0.005 m
$\dot{h}$	0.01 m/s

EXAMPLE 8.4.

Solution of the two-dimensional Poisson's equation: compression molding. In this example. we want to calculate the instantaneous pressure and velocity fields for the compression molding charge shown in the Fig. 8.8 with dimensions and parameters summarized in Table 8.5.

For the solution of this problem we will assume a Newtonian and isothermal flow. The *Hele-Shaw* approximation gives an expression for the mean velocity profile as a function of the pressure as follows (see Chapter 5)

$$\bar{u}_x = -\frac{h^2}{12\eta}\frac{\partial p}{\partial x} \tag{8.45}$$

$$\bar{u}_y = -\frac{h^2}{12\eta}\frac{\partial p}{\partial y} \tag{8.46}$$

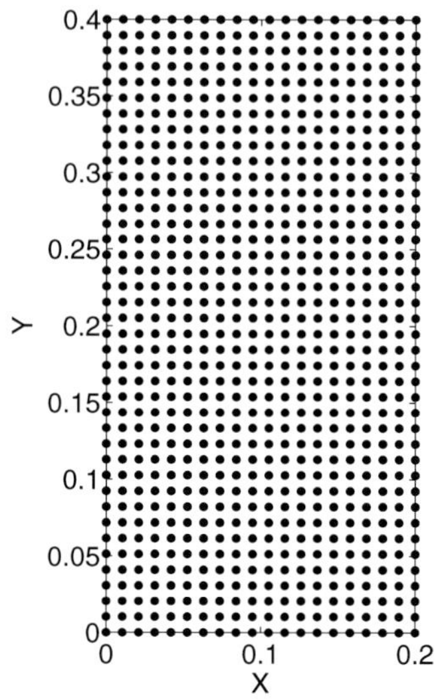

Figure 8.9: 2D FDM discretization.

Substituting these velocities into the integrated continuity equation results in Poisson's equation for the pressure as the governing equation

$$\frac{\partial^2 p}{\partial x^2} + \frac{\partial^2 p}{\partial y^2} = -12\eta \frac{\dot{h}}{h^3} \tag{8.47}$$

with the corresponding boundary conditions at the flow front

$$p(x, y) = 0 \text{ for } x, y \in \text{ front} \tag{8.48}$$

Using a second order approximation for the x- and y-directions, the FD expression of eqn. (8.47) is written as

$$\frac{p_{i-1,j} - 2p_{i,j} + p_{i+1,j}}{\Delta x^2} + \frac{p_{i,j-1} - 2p_{i,j} + p_{i,j+1}}{\Delta y^2} = -12\eta \frac{\dot{h}}{h^3} \tag{8.49}$$

which for the isometric mesh ($\Delta x = \Delta y$), shown in Fig. 8.9, can be rearranged as

$$p_{i-1,j} + p_{i+1,j} + p_{i,j-1} + p_{i,j+1} - 4p_{i,j} = -12\eta \frac{\dot{h}}{h^3} \Delta x^2 \tag{8.50}$$

One possible way to solve this problem is by obtaining a system of linear equations in a similar way as in Example 8.3. However, the equations must be organized carefully because the equations are coupled in the two directions. To avoid the complications given by the generation of the matrix for the linear system of equations, some iterative methodologies have been developed to solve for this type of problems. Such techniques solve the system of equations without the need of cumbersome matrix manipulation, such as LU-decomposition, matrix inversion, etc. [19].

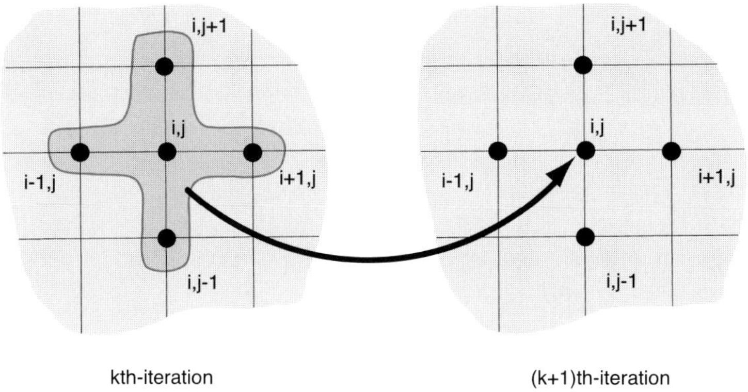

Figure 8.10: Schematic representation of the Jacobi iterative scheme.

An iterative method consists basically of first guessing a solution (a pressure field, $p^0_{i,j}$ in our example) and continuously replacing this guess until some convergence criterion is reached, i.e.,

$$\left|p^k_{i,j} - p^{k+1}_{i,j}\right| < \epsilon \tag{8.51}$$

There are three common schemes that can be used to improve the initial guess and the successive solutions. These are

- Jacobi iteration scheme
- Gaus-Seidel iteration scheme
- Successive over-relaxation (SOR).

Equation (8.50) can be rewritten to express $p_{i,j}$ as a function of the values of pressure at the neighboring nodes, resulting in the non-homogeneous function of Poisson's equation,

$$p_{i,j} = \frac{1}{4}\left(p_{i-1,j} + p_{i+1,j} + p_{i,j-1} + p_{i,j+1}\right) + f_{i,j} \tag{8.52}$$

where $f_{i,j} = 3\eta\dot{h}\Delta x^2/h^3$. The Jacobi iteration scheme is the slowest method because it uses old information for the improvement of the guess. It requires two arrays, one with the old iteration field, $p^k_{i,j}$, and one with the new improved guess, $p^{k+1}_{i,j}$, where k symbolizes the iteration number. The Jacobi iteration is as follows,

$$p^{k+1}_{i,j} = \frac{1}{4}\left(p^k_{i-1,j} + p^k_{i+1,j} + p^k_{i,j-1} + p^k_{i,j+1}\right) + f_{i,j} \tag{8.53}$$

Figure 8.10 illustrates the mechanism of iteration for the Jacobi iterative scheme (Algorithm 1).

Gauss-Seidel iteration is faster than Jacobi, because it uses new information from the already improved points, i.e., the points to the left and below i, j

$$p^{k+1}_{i,j} = \frac{1}{4}\left(p^{k+1}_{i-1,j} + p^k_{i+1,j} + p^{k+1}_{i,j-1} + p^k_{i,j+1}\right) + f_{i,j} \tag{8.54}$$

Algorithm 1 Jacobi iteration

```
program jacobi                              two arrays pold(nx,ny) and pnew(nx,ny)
  pold  = 0              initialization, initial guess and boundary conditions
  pnew = 0
  error = 1e6
  do while (error > tol)                                          Iteration loop
    error=0
    do j = 2,ny-1                                         only for internal nodes
      do i = 2,nx-1
        pnew(i,j) = 0.25*(pold(i-1,j)+pold(i+1,j)+
              pold(i,j-1)+pold(i,j+1))+f
        error = error + (pnew(i,j)-pold(i,j))**2
      enddo
    enddo
    pold=pnew
    error = sqrt(error)/nx/ny
  enddo
end program jacobi
```

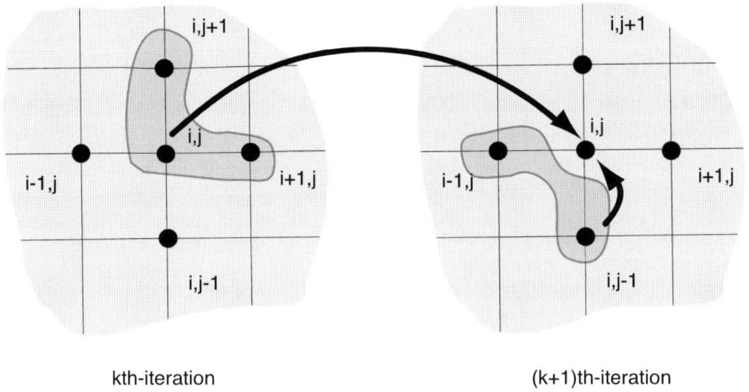

Figure 8.11: Schematic representation of the Gauss-Seidel iterative scheme.

Algorithm 2 Gauss-Seidel iteration

```
program gauss                                               one array p(nx,ny)
  p  = 0               initialization, initial guess and boundary conditions
  error = 1e6
  do while (error > tol)                                          Iteration loop
    error = 0
    do j = 2,ny-1                                         only for internal nodes
      do i = 2,nx-1
        pold = p(i,j)
        p(i,j) = 0.25*(p(i-1,j)+p(i+1,j)+p(i,j-1)+p(i,j+1))+f
        error = error + (p(i,j)-pold)**2
      enddo
    enddo
    error = sqrt(error)/nx/ny
  enddo
end program gauss
```

this is shown schematically in Fig. 8.11 (Algorithm 2).

SOR introduces a relaxation parameter, ω, into the iteration process. The correct selection of this parameter can improve the convergence up to 30 times when compared with Gauss-Seidel. SOR uses new information as well and starts as follows,

$$p_{i,j}^{k+1} = p_{i,j}^{k} + \omega \left[\frac{1}{4} \left(p_{i-1,j}^{k+1} + p_{i+1,j}^{k} + p_{i,j-1}^{k+1} + p_{i,j+1}^{k} \right) + f_{i,j} - p_{i,j}^{k} \right] \tag{8.55}$$

When $\omega < 1$, we have so-called *under-relaxation technique*, often used with non-linear problems. For example, when solving for the non-linear velocity distribution using a shear thinning power law model, the fastest solution is achieved when $\omega = n$ since $n < 1$. When $\omega > 1$, SOR becomes an *over-relaxation technique*.

After the pressure field is obtained, we can use a central FD expression to calculate the instantaneous velocity profile, i.e.,

$$\bar{u}_{xi,j} = -\frac{h^2}{12\eta}\frac{p_{i+1,j} - p_{i-1,j}}{2\Delta x} \tag{8.56}$$

$$\bar{u}_{yi,j} = -\frac{h^2}{12\eta}\frac{p_{i,j+1} - p_{i,j-1}}{2\Delta y} \tag{8.57}$$

Figure 8.12(b) shows the instantaneous pressure distribution and velocity profile for the top half of the geometry.

All three methods eventually arrived at the same result; however, each used different number of iterations to achieve an accurate solution. Figure 8.13 shows a comparison between the convergence process between Jacobi and Gauss-Seidel. The error plotted in in the Fig. 8.13 was computed using,

$$|\text{ERROR}| = \sqrt{\sum_{i,j}^{N} (p_{i,j}^{k} - p_{i,j}^{k-1})^2} \tag{8.58}$$

where N is the number of grid points in the finite difference discretization. The convergence of SOR for different values of the relaxation parameter, ω, is illustrated in Fig. 8.14 According to this analysis, the optimum value for ω was 1.8.

EXAMPLE 8.5.

One dimensional convection-diffusion problem. One problem illustrating issues that arise with combinations of conduction and convection is the one-dimensional problem in Fig. 8.15. Here, we have a heat transfer convection-diffusion problem, where the conduction which results from the temperature gradient and the flow velocity are both in the x-direction.

For the case where $D \gg L$ and assuming a material with constant properties, the energy balance reduces to

$$\rho C_p u_x \frac{\partial T}{\partial x} = k \frac{\partial^2 T}{\partial x^2} \tag{8.59}$$

with the hypothetical forced boundary conditions $T(0) = T_0$ and $T(L) = T_1$. The energy balance equation can be written in dimensionless form as

$$Pe \frac{\partial \Theta}{\partial \xi} = \frac{\partial^2 \Theta}{\partial \xi^2} \tag{8.60}$$

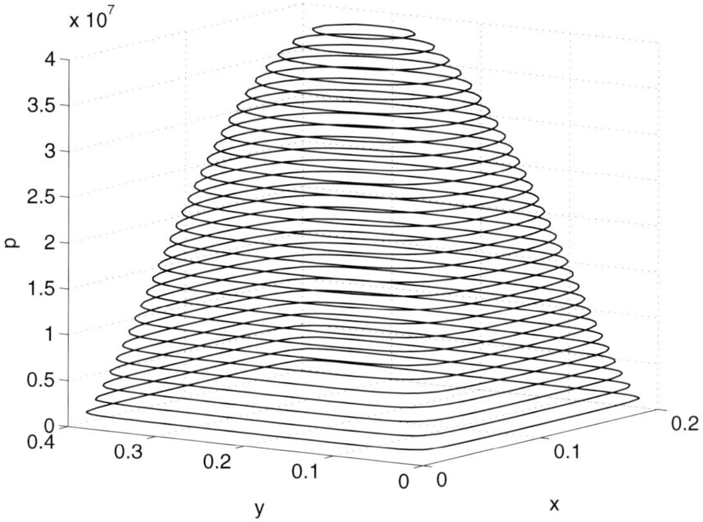

(a) Pressure field

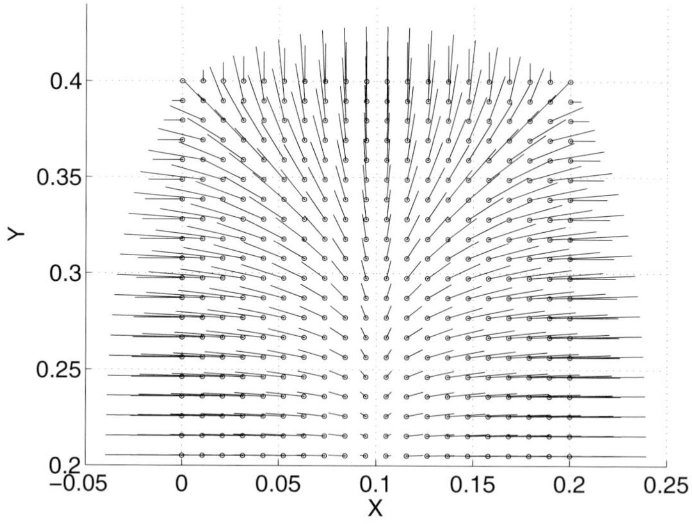

(b) Instantaneous velocity vectors

Figure 8.12: Predicted pressure and velocity fields for the compression molding FDM Example 8.4

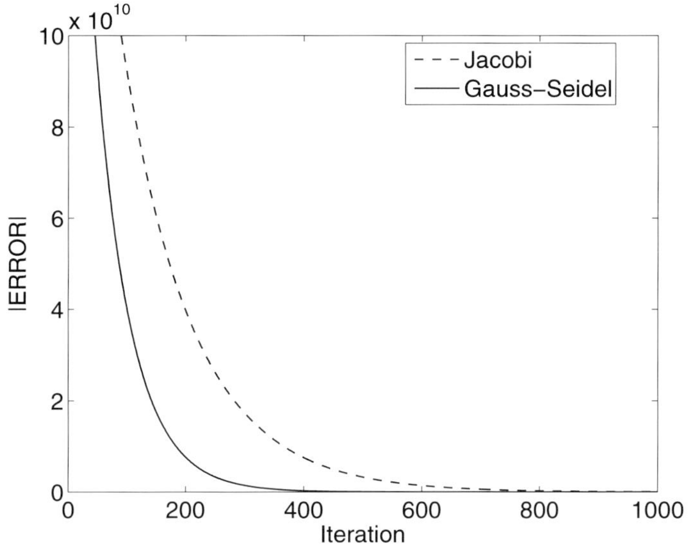

Figure 8.13: Convergence for the Jacobi and Gauss-Seidel iterative solution schemes for the FD compression molding problem.

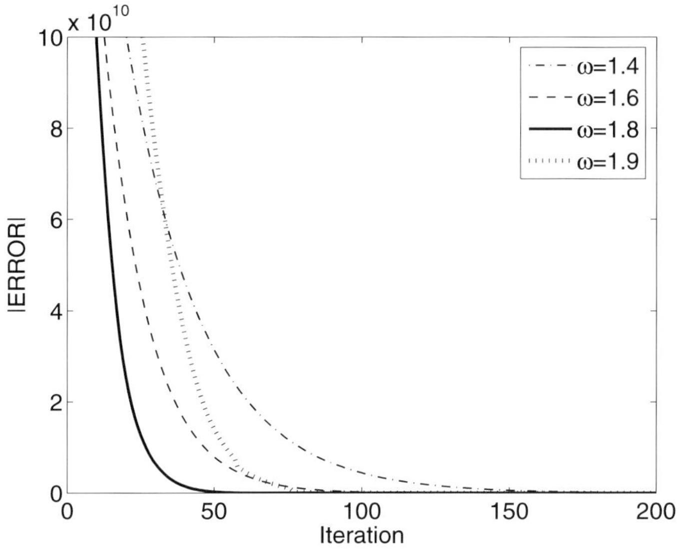

Figure 8.14: Convergence for the SOR iterative solution scheme with various over-relaxation parameters for the FD compression molding problem.

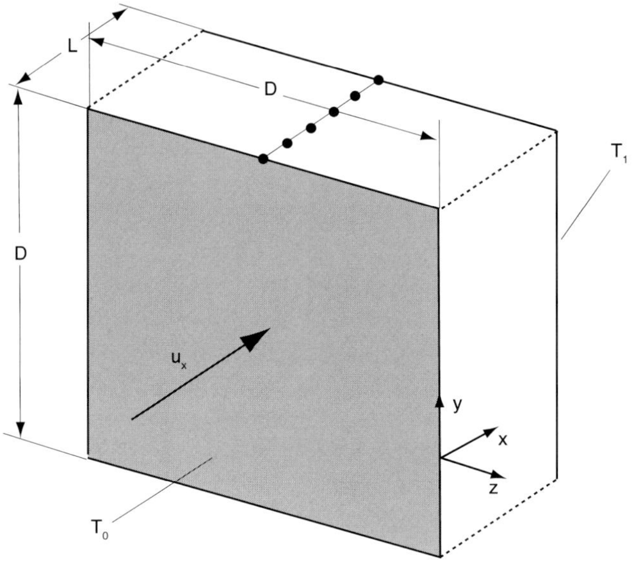

Figure 8.15: Schematic diagram of the convection-conduction problem.

where the *Peclet number* is defined by $Pe = \rho C_p u_x L/k$, the dimensionless temperature by $\Theta = (T - T_0)/(T_1 - T_0)$, and $\xi = x/L$. The boundary conditions in dimensionless form are

$$\Theta(0) = 0 \text{ and } \Theta(1) = 1 \tag{8.61}$$

There are general solutions for this problem, which are dictated by the value of the Peclet number. A problem dominated by diffusion (pure conduction), where $Pe \ll 1$, eqn. (8.60) reduces to

$$\frac{\partial^2 \Theta}{\partial \xi^2} \approx 0 \tag{8.62}$$

which can be integrated to become $\Theta = \xi$. The second type of problem is one where the convective term becomes significang, $Pe \gg 1$, where

$$\frac{\partial \Theta}{\partial \xi} = \frac{1}{Pe}\frac{\partial^2 \Theta}{\partial \xi^2} \approx 0 \tag{8.63}$$

Here, $\Theta \neq$ constant, because it cannot satisfy the boundary conditions. Thus, there is a boundary layer solution [3, 6, 21] given by

$$\Theta = \frac{e^{Pe\xi} - 1}{e^{Pe} - 1} \tag{8.64}$$

Figure 8.16 presents the temperature profile for different values of Pe as predicted by eqn. (8.64). The figure demonstrates how an increase in Peclet number leads to deviations in the temperature profile of the pure conductive problem. As a matter of fact, for $Pe \gg 1$ we will have an additional length scale in the problem, δ, the

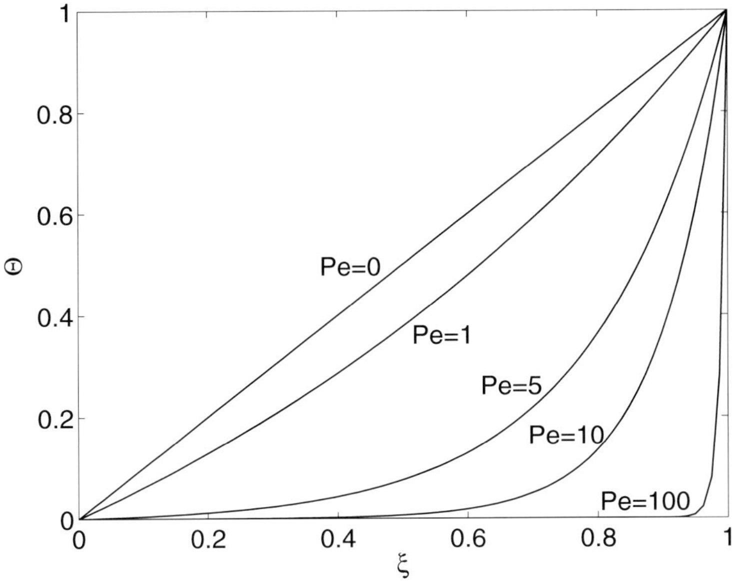

Figure 8.16: Temperature profile for the 1D convection-diffusion problem.

boundary layer thickness. We can use this length as the characteristic length for eqn. (8.59),

$$Pe_\delta \frac{\partial \Theta}{\partial \zeta} = \frac{\partial^2 \Theta}{\partial \zeta^2} \tag{8.65}$$

where $Pe_\delta = \rho C_p u_x \delta / k$ and $\zeta = x/\delta$. Analyzing this equation we know that $Pe_\delta \sim 1$ in the boundary layer, i.e., convection and diffusion are of the same order, hence, $\delta / L \sim 1/Pe$.

Let us proceed to generate a FD discretization of this problem using second order finite differences for the first and second derivative as follows

$$Pe \frac{\Theta_{i+1} - \Theta_{i-1}}{2\Delta\xi} = \frac{\Theta_{i+1} - 2\Theta_i + \Theta_{i-1}}{\Delta\xi^2} \tag{8.66}$$

which can be written as

$$\left(\frac{Pg}{2} - 1\right)\Theta_{i+1} + 2\Theta_i - \left(\frac{Pg}{2} + 1\right)\Theta_{i-1} = 0 \tag{8.67}$$

where $Pg = Pe\Delta\xi$ is the *grid Peclet number*. As boundary conditions we use $\Theta_1 = 0$ and $\Theta_n = 1$. The tri-diagonal matrix that is generated by applying the above equation to every internal grid point, can be solved for by back-substitution solution to give the value of Θ_i explicitly as

$$\Theta_i = \frac{1 - \left(\dfrac{1 + Pg/2}{1 - Pg/2}\right)^i}{1 - \left(\dfrac{1 + Pg/2}{1 - Pg/2}\right)^n} \tag{8.68}$$

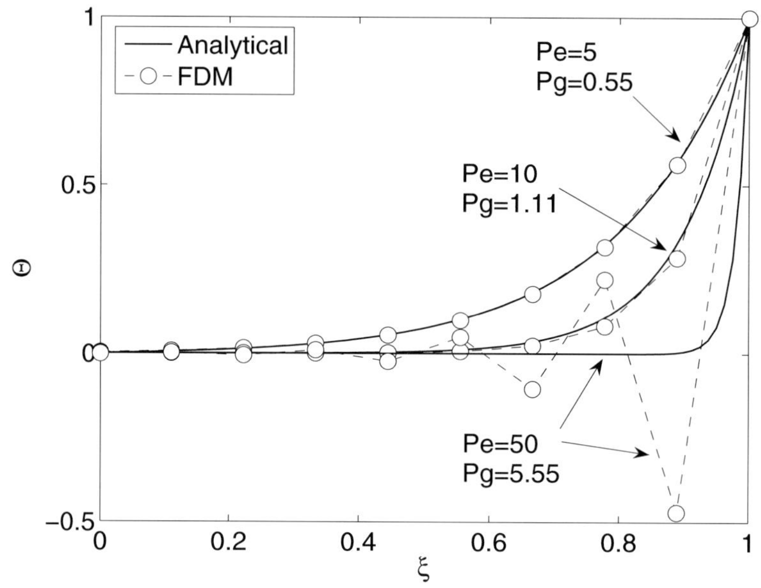

Figure 8.17: FD Temperature profile for the 1D convection-diffusion problem with a central difference convection term.

Note that $Pg > 2$ is critical, because the solution presents a sign change, which means the solution becomes unstable (see Figure 8.17). The root of the problem is explained by the info-travel concept. To generate the difference equation (eqn. (8.66)) we used a central finite difference for the convective derivative, which is incorrect, because the information of the convective term cannot travel in the upstream direction, but rather travels with the velocity u_x. This means that to generate the FD equation of a convective term, we only take points that are up-stream from the node under consideration. This concept is usually referred to as *up-winding technique*. For low Pe the solution is stable because diffusion controls and the information comes from all directions.

By using a backward finite differences, essentially *up-winding* the convective term, we get

$$Pe\frac{\Theta_i - \Theta_{i-1}}{\Delta\xi} = \frac{\Theta_{i+1} - 2\Theta_i + \Theta_{i-1}}{\Delta\xi^2} \tag{8.69}$$

with the explicit solution

$$\Theta_i = \frac{1 - (1 + Pg/2)^i}{1 - (1 + Pg/2)^n} \tag{8.70}$$

This equation does not present sign changes and the solution is free of spurious oscillations. However, the numerical and analytical results are noticeable different, due to an artificial diffusion that the FD expression introduces to the solution (see Fig. 8.18). This artificial diffusion is commonly known as *numerical diffusion*.

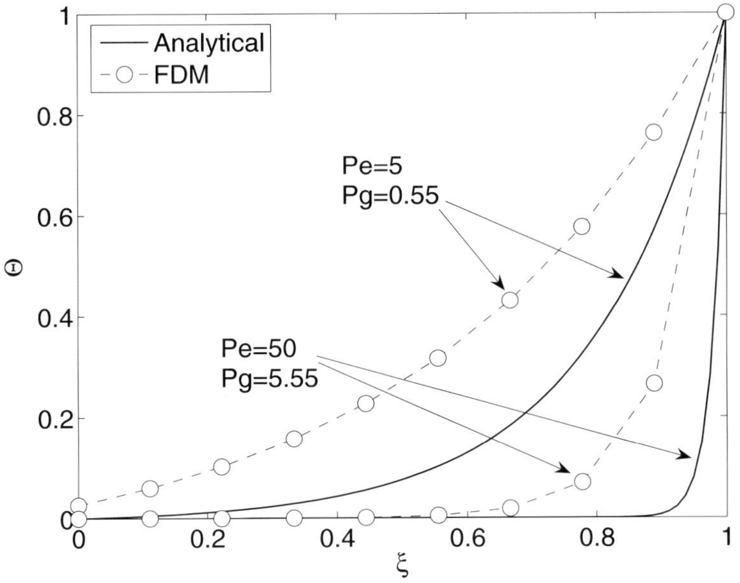

Figure 8.18: Temperature profile for the 1D convection-diffusion problem with an up-winded convection term.

8.5 TRANSIENT PROBLEMS

Transient problems begin with an initial condition and march forward in time in discrete time steps. We have discussed space derivatives, and now we will introduce the time derivative, or transient, term of the differential equation. Although the Taylor-series can also be used, it is more helpful to develop the FD with the integral method. The starting point is to take the general expression

$$\frac{d\phi}{dt} = f(\phi, t) \tag{8.71}$$

and integrate with respect to time in a small interval from t to $t + \Delta t$

$$\begin{aligned} \int_t^{t+\Delta t} \frac{d\phi}{dt'} dt' &= \int_t^{t+\Delta t} f(\phi, t')dt' \\ \phi(t + \Delta t) - \phi(t) &= \int_t^{t+\Delta t} f(\phi, t')dt' \end{aligned} \tag{8.72}$$

In order to approximate the remaining integral, we compute the product, as schematically depicted in Fig. 8.19, $f(\phi(t), t)\Delta t$ which simplifies eqn. (8.72) to

$$\begin{aligned} \phi(t + \Delta t) - \phi(t) &= f(\phi(t), t)\Delta t + O(\Delta t^2) \\ \frac{\phi(t + \Delta t) - \phi(t)}{\Delta t} &= f(\phi(t), t) + O(\Delta t) \end{aligned} \tag{8.73}$$

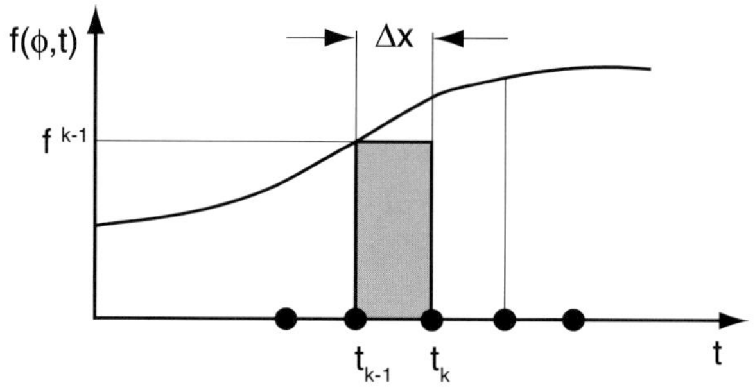

Figure 8.19: Explicit Euler time marching scheme.

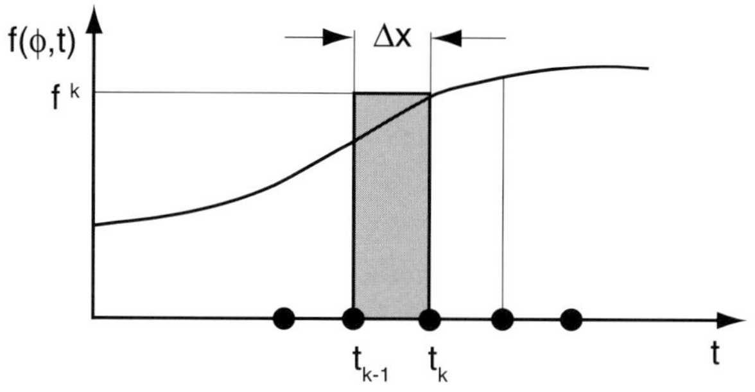

Figure 8.20: Implicit Euler time marching scheme.

or

$$\frac{\phi^{k+1} - \phi^k}{\Delta t} = f^k + O(\Delta t) \tag{8.74}$$

where the superscript k indicates time t and $k + 1$ time $t + \Delta t$. Equation (8.74) implies that the function at the right hand side is calculated in the past, making it a first order approximation method in Δt, known as *explicit Euler*.

Instead of using the function evaluated at the past, k, we can also construct a product based on the function evaluated at a future point of time, $k + 1$, as shown in Fig. 8.20. With this we obtain,

$$\frac{\phi^{k+1} - \phi^k}{\Delta t} = f^{k+1} + O(\Delta t) \tag{8.75}$$

which is also a first order method in time and due to the fact that the right hand side is evaluated in the future, it is called *implicit Euler*.

Both methods are first order in time but for practical purposes the explicit Euler is the easiest to apply due to the fact that the only unknown is the value ϕ^{k+1}; all other terms are evaluated in the kth time step, and due to prescribed initial conditions are always known. Hence, the value of ϕ^{k+1} can easily be solved for using eqn. (8.74), by marching forward in time. In the implicit Euler case, the whole right hand side of the equation is evaluated in the future, and must therefore be generated and solved for every time step. When marching

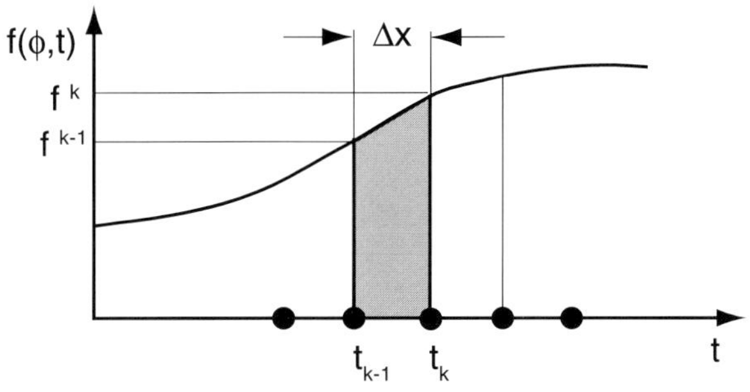

Figure 8.21: Crank-Nicholson time marching scheme.

forward in time, there is also the issue of stability that we must worry about. Since we are dealing with problems that evolve in time, we must assure that the truncation error does not grow in time. Explicit Euler has the problem of being conditionally stable, i.e., there are requirements that need to be fulfilled in order to make the scheme stable. On the other hand, implicit Euler is unconditionally stable. The conditions for stability in explicit Euler depends on the value of Δt and the nature of the function $f(\phi, t)$; details for this can be found in several references [10, 23, 26] and are discussed in the subsequent example problem.

If instead of implementing the values of the parameters either in the past or the future, we evaluate them at both t and $t + \Delta t$ and average the results as schematically depicted in Fig. 8.21, the results in an FD expression are given by,

$$\phi^{k+1} - \phi^k = \frac{1}{2}(f^{k+1} + f^k)\Delta t + O(\Delta t^3) \tag{8.76}$$

or

$$\frac{\phi^{k+1} - \phi^k}{\Delta t} = \frac{1}{2}(f^{k+1} + f^k) + O(\Delta t^2) \tag{8.77}$$

These equations are semi-implicit second order in time typically called *Adams-Moulton* (AM2) method or *Crank-Nicholson* (CN), when applied to diffusion problems, and due to the implicit nature of the procedure, the scheme is also unconditionally stable.

■ **EXAMPLE 8.6.**

Explicit Euler finite difference solution for a cooling semi-crystalline polymer plate. The cooling process is the dominating factor in many processes, which is also true for injection molding. In this example, we will illustrate how the explicit finite difference technique can be used to predict the cooling of a plate from temperatures above the melting point to the mold temperature. Such a solution is extremely useful, and with temperature dependent properties does not have an analytical solution. To simplify the problem somewhat, we assume that the polymer melt is injected fast enough into the cavity that it remains isothermal until the cavity is full. Although a fast injection speed would probably lead to viscous heating, we also neglect those effects. Hence, as an initial condition, we assume a constant temperature throughout the thickness of the plate, T_{inj}, and that $T_{inj} > T_m$, where T_m is the melting temperature.

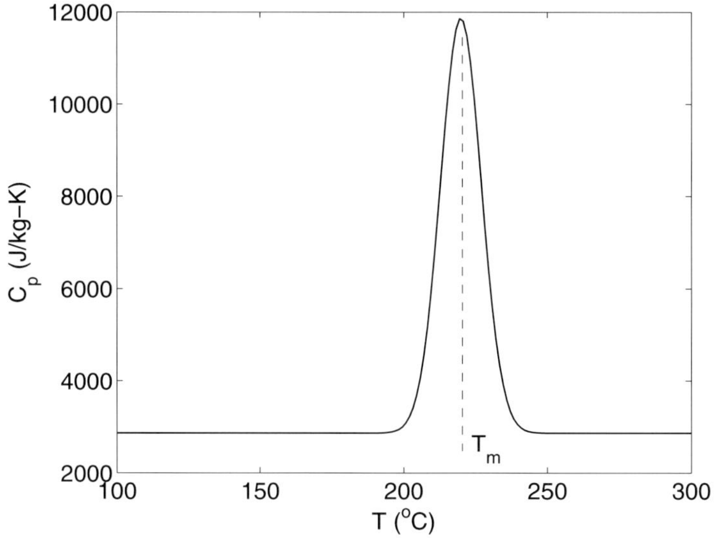

Figure 8.22: Specific heat as a function of temperature for a semi-crystalline thermoplastic (PA6).

The crystallization is modeled by an abrup increase in the specific heat, C_p around the melting temperature, as shown in Fig. 8.22. Assuming that the density, ρ, and thermal conductivity, k, remain constant[1], the one-dimensional energy equation becomes

$$\rho\frac{\partial}{\partial t}\left(C_p(T)T(t,x)\right) = k\frac{\partial^2 T(t,x)}{\partial x^2} \tag{8.78}$$

with $T(0,x) = T_{inj}$ and $T(t,0) = T(t,L) = T_{mold}$ as initial and boundary conditions, respectively.

For simplicity, we will assume that $\partial C_p/\partial t = 0$ and by using

$$\Theta = \frac{T - T_{mold}}{T_{inj} - T_{mold}} \text{ and } \xi = \frac{x}{L} \tag{8.79}$$

we can write eqn. (8.78) as

$$\frac{\partial \Theta}{\partial t} = \frac{\alpha(\Theta)}{L^2}\frac{\partial^2 \Theta}{\partial \xi^2} \tag{8.80}$$

where $\alpha(T) = k/\rho C_p(T)$ is the thermal diffusivity of the thermoplastic and $\alpha(\Theta)/L^2$ is the inverse of the Fourier number (Fo), a characteristic cooling time for the given material and geometry. The explicit Euler scheme for this equation reduces to

$$\Theta_i^{j+1} = \Theta_i^j + \frac{\alpha_i^j}{\alpha_g}\left(\Theta_{i+1}^j - 2\Theta_i^j + \Theta_{i-1}^j\right) \tag{8.81}$$

[1] In a numerical solution, we can include temperature dependent density and thermal conductivity. The temperature dependent density can be modeled interpolating throughout a pvT diagram. The temperature dependence of the thermal conductivity is not always available, as is the case for many properties used in modeling. Chapter 2 presents the Tait equation, which can be used to model the pvT behavior of a polymer.

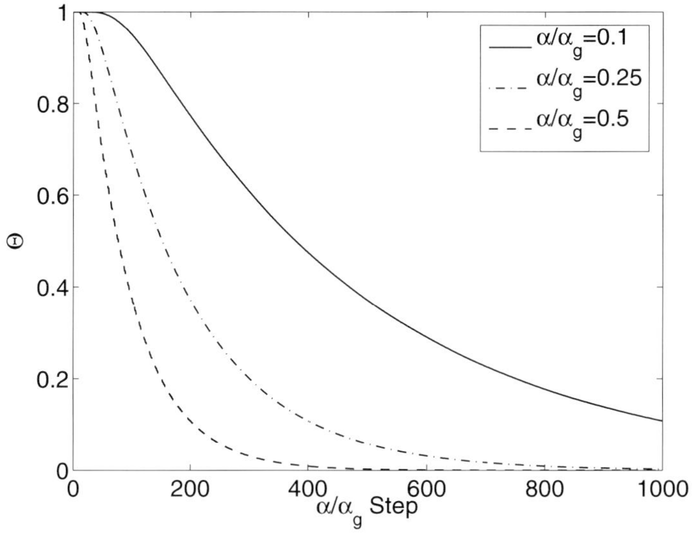

Figure 8.23: Stable solutions $\alpha/\alpha_g < 0.5$.

where $i = 2, ..., n-1$. Here, the subscript i represents the spatial grid node and the superscript j the time step, $\alpha_i^j = k/\rho C_p(T_i^j)$ and $\alpha_g = (\Delta\xi L)^2/\Delta t$ is the *grid diffusivity*. The initial condition will be $\Theta_i^0 = 1$ for $i = 1$ and n, and the boundary conditions are $\Theta_1^j = \Theta_n^j = 0$ for $j = 0, 1, 2, ...$

Before solving the complete problem, we need to explore the conditions that makes our explicit Euler expression eqn. (8.81) unstable. To do this, let's assume for the moment that the C_p is constant, and therefore $\alpha_i^j = \alpha$. Figure 8.23 shows how the center-line temperature evolves when many α/α_g steps are performed for different values of α/α_g. Figure 8.23 illustrates stable solutions, while Fig. 8.24 shows instabilities from the beginning of the simulation that grow with α/α_g steps. These instabilities are aggravated as $\alpha/\alpha_g > 0.5$. This is a well known limit for stability of explicit Euler schemes in transient diffusive problems [26]. Replacing the definitions of α and α_g, the stability condition is given by

$$\frac{\alpha}{\alpha_g} < 0.5 \Longrightarrow \Delta t < \frac{\Delta x^2}{2\alpha} \tag{8.82}$$

To solve correctly, the cooling of our the semi-crystalline material, i.e., including temperature effect such as $C_p(T)$ or $\alpha(T)$, every time step we must verify that eqn. (8.82) is satisfied. Figure 8.25 shows the evolution of the temperature with time for the properties and conditions given in Table 8.6 and Fig. 8.22 The delay in the temperature drop due to the crystallization can be clearly seen in Fig. 8.25, and Fig. 8.26, which illustrates the evolution of the center-line temperature. It should be pointed out here that the solution neglects the finite nucleation rate at the beginning of the cooling, and therefore overpredicts the speed of cooling.

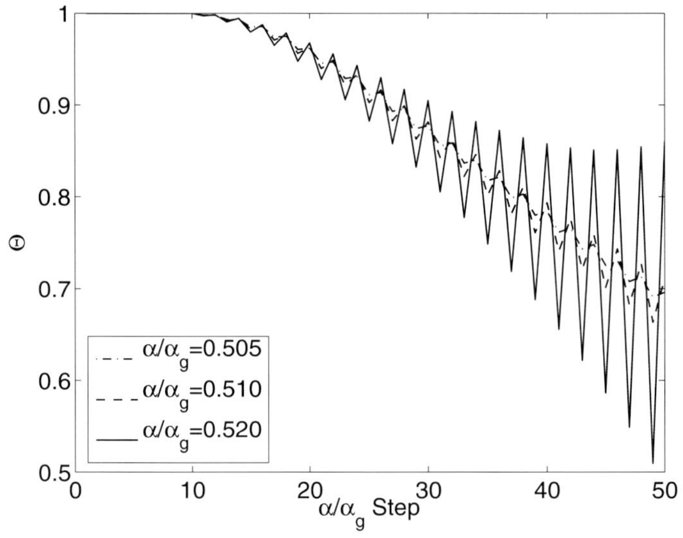

Figure 8.24: Evolution of the center line temperature for a constant C_p thermoplastic.

Table 8.6: Example 8.6 Data

Parameter	Value
ρ	1130 kg/m^3
k	0.135 W/m/K
T_m	220 ^{o}C
T_{inj}	260 ^{o}C
T_{mold}	30 ^{o}C
L	0.01 m

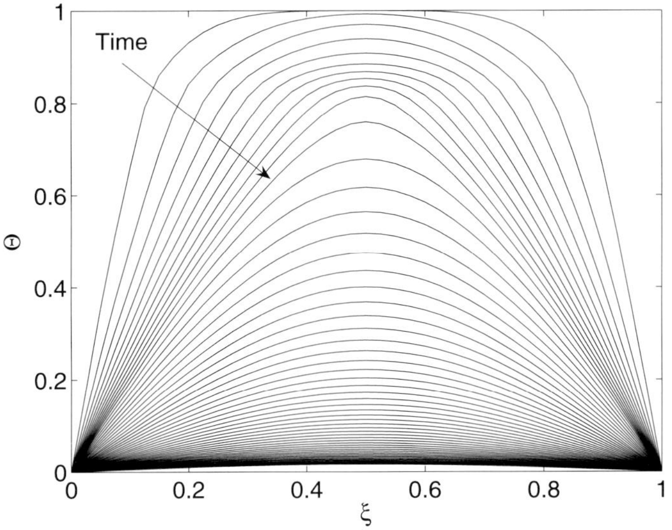

Figure 8.25: Evolution of the temperature profile for a cooling semi-crystalline plate.

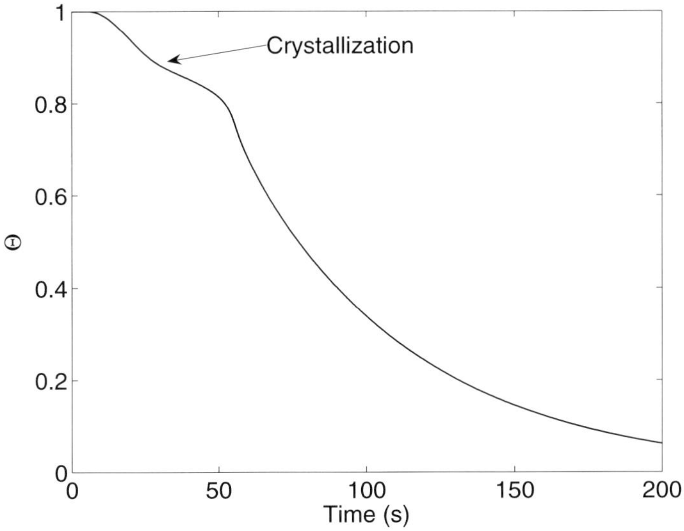

Figure 8.26: Evolution of the center-line temperature for a cooling semi-crystalline plate with variable $C_p(T)$.

■ EXAMPLE 8.7.

Implicit Euler finite difference solution for a cooling amorphous polymer plate. To illustrate the usage of implicit finite difference schemes, we will solve the cooling process of an amorphous polymer plate. Since an amorphous polymer does not go through a crystallization process we will assume a constant specific heat[2]. The implicit finite difference for this equation can be written as,

$$\begin{aligned}\Theta_i^{j+1} - \omega\frac{\alpha}{\alpha_g}\left(\Theta_{i+1}^{j+1} - 2\Theta_i^{j+1} + \Theta_{i-1}^{j+1}\right) = \\ \Theta_i^j + (1-\omega)\frac{\alpha}{\alpha_g}\left(\Theta_{i+1}^j - 2\Theta_i^j + \Theta_{i-1}^j\right)\end{aligned} \tag{8.83}$$

for $i = 2, ..., n-1$ and where $0 \leqslant \omega \leqslant 1$ is a factor that will determine the time scheme to be used. These are,

- $\omega = 1$ fully implicit Euler
- $\omega = 1/2$ Crank-Nicholson
- $\omega = 0$ fully explicit Euler

For implicit schemes, we will obtain a system of linear algebraic equations that must be solved. As mentioned in Example 8.1, one-dimensional diffusion problems generate tri-diagonal matrices, that can be solved for using the Thomas algorithm or other fast matrix routines. Equation (8.83) can be written as

$$a\Theta_{i-1}^{j+1} + b\Theta_i^{j+1} + a\Theta_{i-1}^{j+1} = f_i^j \tag{8.84}$$

for $i = 2, .., n-1$ and where

$$\begin{aligned}a &= -\omega\frac{\alpha}{\alpha_g} \\ b &= 1 + 2\omega\frac{\alpha}{\alpha_g} \\ f_i^j &= \Theta_i^j + (1-\omega)\frac{\alpha}{\alpha_g}\left(\Theta_{i+1}^j - 2\Theta_i^j + \Theta_{i-1}^j\right)\end{aligned} \tag{8.85}$$

and the initial and boundary conditions $\Theta_i^0 = 1$ for $i = 1$ and n, and $\Theta_1^j = \Theta_n^j = 0$ for $j = 0, 1, 2, ...$, respectively.

Implicit schemes are unconditionally stable, this is shown in Fig. 8.27 where the evolution of the temperature, in α/α_g steps, for values of α/α_g higher than 0.5 is shown. Higher values of α/α_g mean that we can use higher Δt, which at the end implies lower computational cost and faster solutions. The results in Fig 8.27 were obtained with the fully implicit Euler scheme, i.e., $\omega = 1$. The comparison between the implicit Euler and the Crank-Nicholson, $\omega = 0.5$ is illustrated in Fig. 8.28 for the center line temperature evolution. Although there is no apparent significance difference, we expect that the CN scheme is more accurate due to its second order nature.

[2]Typically, the specific heat of amorphous thermoplastics changes as it goes though the glass transition temperature, but for this specific application we will assume an average value. However, introducing a changing specific heat simply requires the use of an *if statement* that checks if the value of specific heat above or below T_g should be chosen.

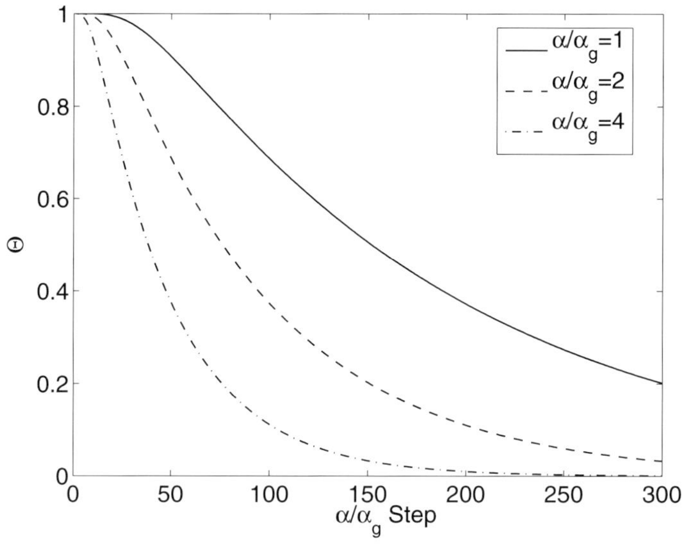

Figure 8.27: Implicit Euler solutions for a cooling amorphous thermoplastic plate.

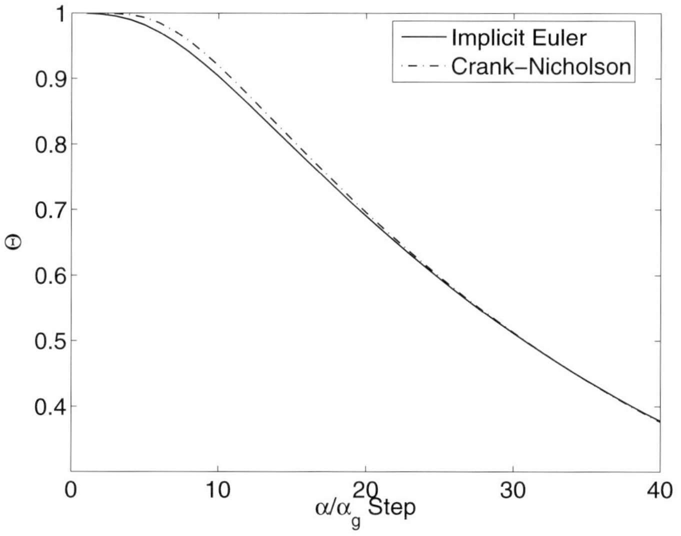

Figure 8.28: Comparison of implicit Euler and Crank-Nicholson solutions for a cooling amorphous thermoplastic plate.

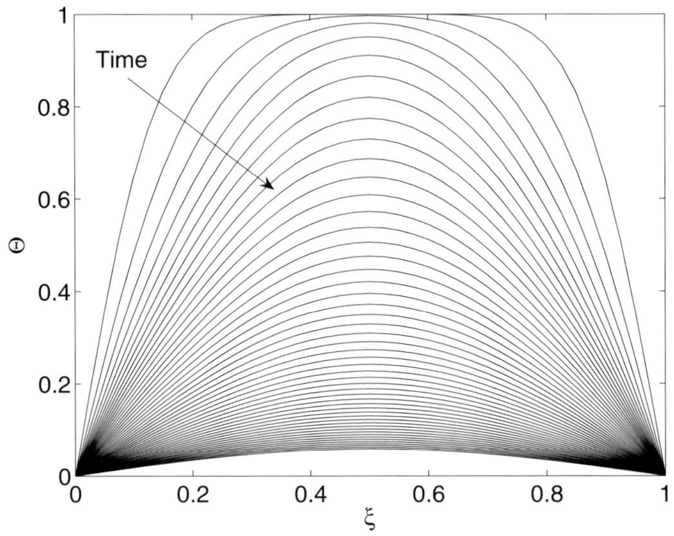

Figure 8.29: Evolution of the temperature profile for the cooling of an amorphous thermoplastic plate.

The evolution of the temperature profile as a function of time is illustrated in Fig. 8.29 for $\alpha/\alpha_g = 5$. As can be seen, due to the absence of crystallization, no delay is present in the cooling curves. If we are seeking the steady-state temperature profile of the thermoplastic after injection, and we want to achieve the steady-state profile in a few calculations, implicit methods are the correct choice because they allow the use of very large time steps, i.e., high values of α/α_g. Figure 8.30 shows the rapid evolution towards steady-state using implicit Euler with $\alpha/\alpha_g = 200$. This solution only requires 10 steps. When the same conditions are used with a CN scheme, although stable, the solution gives spurious oscillations, as shown in Fig. 8.31.

■ **EXAMPLE 8.8.**

Heat transfer during the curing process of a thermoset resin. In the previous examples we showed, how the finite difference technique can be used to predict cooling of thermoplastic materials. The technique can also be used to predict the curing reaction during solidification of thermosetting resins. Here, we present the work done by Barone and Caulk [1], who used an explicit finite difference technique to solve for the heat transfer and cure kinetics to predict the temperature and curing fields during processing of fiber reinforced unsaturated polyester during compression molding. When dealing with curing themosets, the energy equation has an exothermic heat generation, $\dot{Q}$. For curing thermosets, we can represent the exotherm with [1]

$$\dot{Q} = Q_T \frac{dc}{dt} \tag{8.86}$$

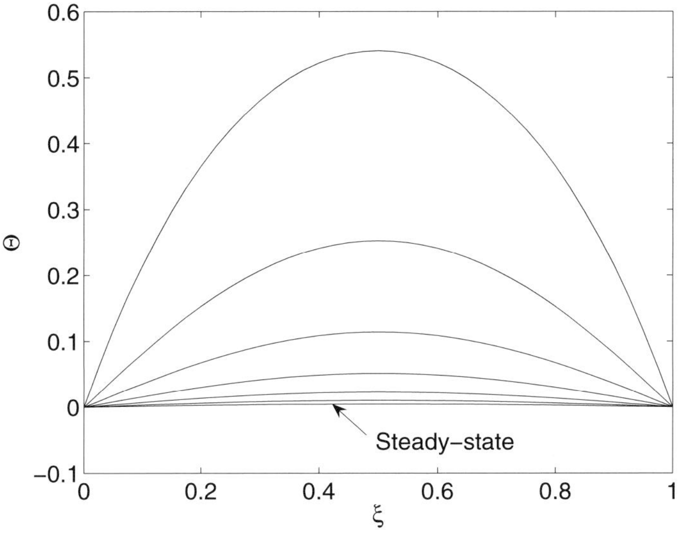

Figure 8.30: Steady-state temperature development using an implicit Euler scheme.

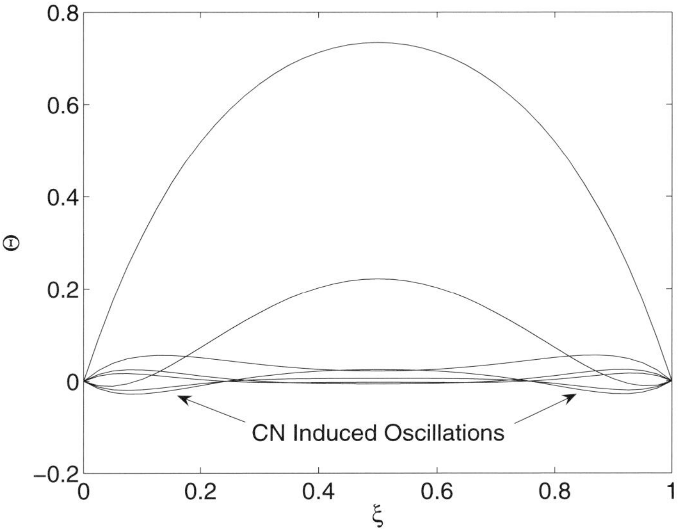

Figure 8.31: Steady-state temperature development using an implicit Crank-Nicholson scheme.

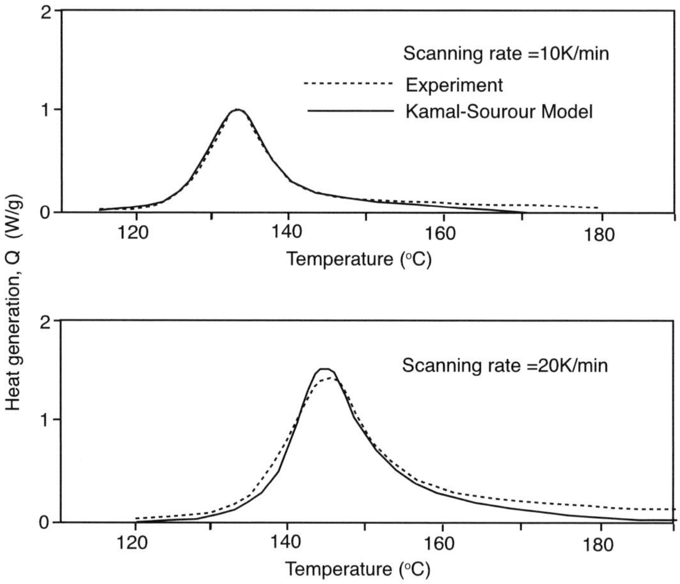

Figure 8.32: Comparison between measured (DSC) and computed (Kamal-Sourour) model of the heat generation during cure for two heating rates [1].

where dc/dt is the rate of curing, which can be represented with a reaction kinetic semi-empirical model, such as the one developed by Kamal and Sourour [13, 14]

$$\frac{dc}{dt} = (k_1 + k_2 c^m)(1 - c)^n \tag{8.87}$$

where m and n are the reaction order, k_1 and k_2 contain the temperature dependence of the curing reaction rate

$$\begin{aligned} k_1 &= a_1 e^{-b_1/RT} \\ k_2 &= a_2 e^{-b_2/RT} \end{aligned} \tag{8.88}$$

Here, R is the gas constant and b_1, b_2, a_1 and a_2 are constants that can be obtained by fitting the above equations to data measured with a differential scanning calorimeter (DSC) [9, 17].

Figure 8.32 presents the DSC scans for 10 and 20 K/min heating rates for an unsaturated polyester, with the theoretical prediction from the above equations [1]. Table 8.7 lists the properties and fitted parameters found by Barone and Caulk for a SMC material.

With these properties and values the finite difference technique was used to model the curing process in sheet molding compound (SMC) plates using a heat balance with an exothermic reaction as

$$\rho C_p \frac{\partial T}{\partial t} = k \frac{\partial^2 T}{\partial x^2} + \rho \dot{Q} \tag{8.89}$$

The above equation is solved similarly to the equations in the previous examples with the added exothermic reaction term. As mentioned earlier, Barone and Caulk used

Table 8.7: Sample Kinetic Parameters of Cure and Properties for SMC

Parameter	Value
a_1	4.9 10^{14}s^{-1}
a_2	6.2 10^5s^{-1}
b_1	140.0 kJ/mol
b_2	51.0 kJ/mol
m	1.3
n	2.7
Q_T	84.0 kJ/kg
ρ	1900.0 kg/m^3
C_p	1000.0 kJ/kg/K
k	0.53 W/m/K

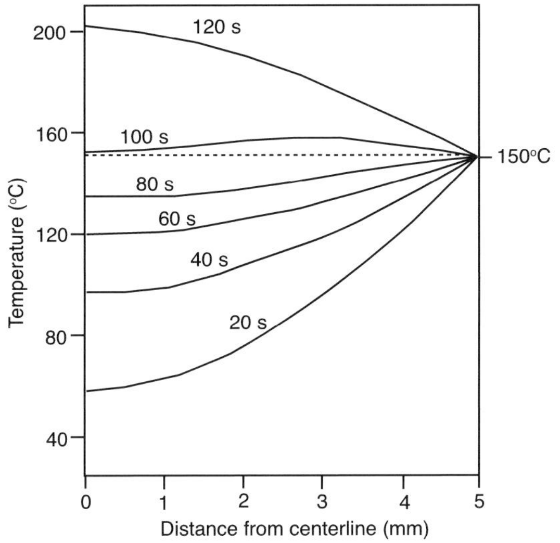

Figure 8.33: Temperature distribution for a curing 10 mm thick unsaturated polyester plate [1].

an explicit finite difference technique. Figs. 8.33 and 8.34 present the temperature and curing distributions across the thickness of a 10mm thick plate, respectively. The material constants presented in Table 8.7 were used for the calculations, with an initial temperature of 24°C and a mold temperature of 150°C. As can be seen, for this relatively thick part, the exotherm plays a significant role in the curing reaction, where the center of the plate has gone 40 K above the mold temperature, and curing variation through the thickness is significant. For a 2 mm plate thickness and the same mold temperature, Barone and Caulk demonstrated that the exotherm gives only a slight rise in part temperature and that the curing progresses evenly across the thickness.

Barone and Caulk performed the same calculation for several plate thicknesses and mold temperatures to predict the time for demolding. They defined demolding time as the time it takes for every point in the plate to have reached at least 80% cure. Figure 8.35 presents the processing window generated, where the time for demolding is plotted as a function of plate thickness and mold temperature. In addition, the

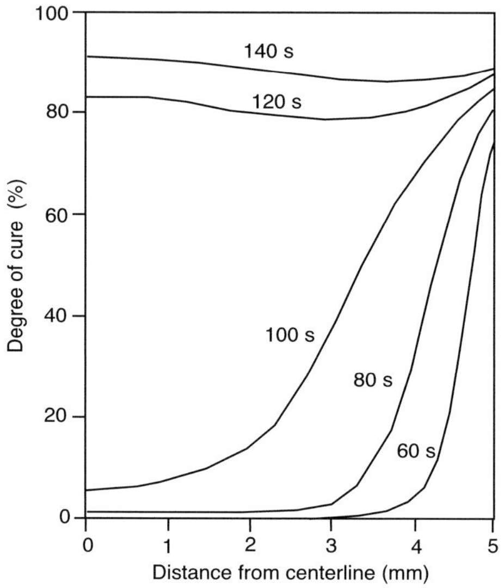

Figure 8.34: Degree of cure distribution for a curing 10 mm thick unsaturated polyester plate [1].

shaded area shows the region where the exotherm causes the center-temperature to exceed 200°C.

8.5.1 Higher Order Approximation Techniques

There are several methods that we can use to increase the order of approximation of the integral in eqn. (8.72). Two of the most common higher order explicit methods are the Adams-Bashforth (AB2) and the Runge-Kutta of second and fourth order. The Adams-Bashforth is a second order method that uses a combination of the past value of the function, as in the explicit method depicted in Fig. 8.19, and an average of the past two values, similar to the Crank-Nicholson method depicted in Fig. 8.21, and written as

$$\int_t^{t+\Delta t} f dt' = f^k \Delta t + \frac{1}{2}(f^k - f^{k-1})\Delta t + O(\Delta t^3) \tag{8.90}$$

which gives

$$\frac{\phi^{k+1} - \phi^k}{\Delta t} = \frac{3}{2} f^k - \frac{1}{2} f^{k-1} + O(\Delta t^2) \tag{8.91}$$

The method increases the order but the stability is compromized due to the extrapolation done by the the linear approximation between the previous times. This stability issue can be improved by adding an extra implicit step using an Adams-Moulton (AM2) as follows

$$\text{Step 1: AB2} \quad \phi^* = \phi^k + (3f^k - f^{k-1})\frac{\Delta t}{2} + O(\Delta t^2) \tag{8.92}$$

$$\text{Step 2: AM2} \quad \phi^{k+1} = \phi^k + (f^* + f^k)\frac{\Delta t}{2} + O(\Delta t^2) \tag{8.93}$$

where ϕ^* is a prediction of ϕ^{n+1}, and with $f^* = f(\phi^*, t + \Delta t)$ is a correction of this prediction. This method is called *Adams-Predictor-Corrector* (APC2), keeping the integral a second order approximation.

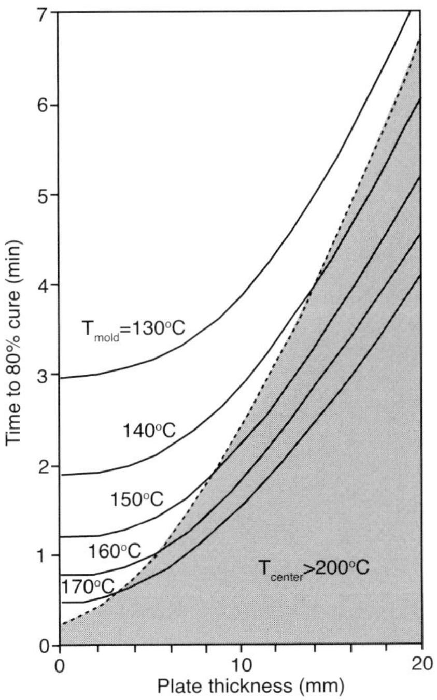

Figure 8.35: Cure time versus plate thickness for various mold temperatures. Shaded area indicates the conditions that lead to centerline temperatures above 200°C [1].

Runge-Kutta of the second (RK2) and fourth (RK4) order do not use an extrapolation between $k-1$ and k to find $k+1$, instead they use future points to do the extrapolation. The methodology is straight forward and can be summarized for the Runge-Kutta of second order (RK2) as [10]

$$\begin{aligned} K_1 &= f(\phi^k, t) \\ K_2 &= f(\phi^k + \Delta t K_1, t + \Delta t) \\ \phi^{k+1} &= \phi^k + (K_1 + K_2)\frac{\Delta t}{2} + O(\Delta t^2) \end{aligned} \tag{8.94}$$

and for the Runge-Kutta of fourth order (RK4) as

$$\begin{aligned} K_1 &= f(\phi^k, t) \\ K_2 &= \frac{1}{2} f(\phi^k + \Delta t K_1, t + 1/2\Delta t) \\ K_3 &= \frac{1}{2} f(\phi^k + \Delta t K_2, t + 1/2\Delta t) \\ K_4 &= \frac{1}{2} f(\phi^k + 2\Delta t K_3, t + \Delta t) \\ \phi^{k+1} &= \phi^k + (K_1 + 2K_2 + 2K_3 + K_4)\frac{\Delta t}{3} + O(\Delta t^4) \end{aligned} \tag{8.95}$$

These Runge-Kutta methods do not require information from the past, and are very versatile if the time steps need to be adjusted as the solution evolves. The stability of the RK2 is similar to the APC2, while the RK4 has less strong conditions for stability [10]. Both are ideal for initial value problems in time or in space,

■ **EXAMPLE 8.9.**

Fully developed flow in screw extruders. To illustrate the type of problem that can be solved using higher order approximation techniques we will present the work done by Griffith [9] in 1962. Grittith developed the governing equations for the fully developed flow of non-Newtonian fluids in the metering section of a screw. As discussed in Chapter 6 of this book, the flow in the metering section is a complex three-dimensional flow, that, when modeled with a non-Newtonian, shear thinning viscosity, does not have an analytical solution. Even if we simplify the flow into two components, a cross-channel and a down-channel component, the coupling of these two components through the rate of deformation dependent viscosity requires numerical techniques to arrive at a solution. Griffith used the usual unwrapped screw geometry schematically depicted in Fig. 8.36. Note that here we are using the coordinate system used by Griffith, not the one used in Chapter 6.

For the flow components that are significant in the present geometry, the important components of the stress tensor $\sigma_{ij} = -\delta_{ij}p + \tau_{ij}$, can be expressed as,

$$\begin{aligned} \sigma_{xy} &= \eta(\dot{\gamma})\frac{\partial u_y}{\partial x} \\ \sigma_{xz} &= \eta(\dot{\gamma})\frac{\partial u_z}{\partial x} \end{aligned} \tag{8.96}$$

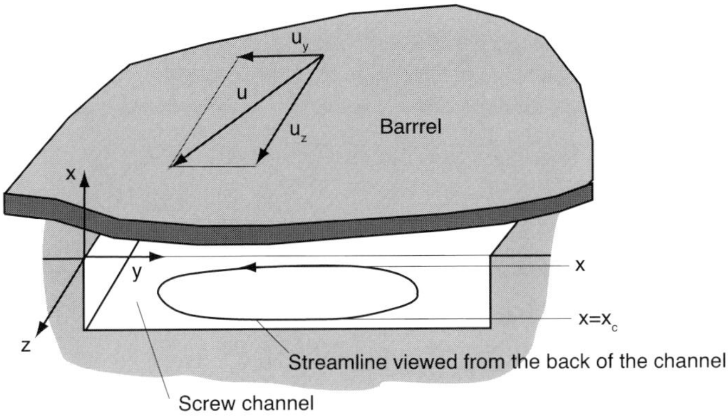

Figure 8.36: Unwrapped screw channel with conditions and dimensions [9].

where $\dot{\gamma}$ is the magnitude of the shear rate tensor defined by

$$\dot{\gamma} = \sqrt{\frac{1}{2}\left(\left(\frac{\partial u_y}{\partial x}\right)^2 + \left(\frac{\partial u_z}{\partial x}\right)^2\right)} \tag{8.97}$$

Neglecting the inertial terms and using eqn. (8.96), the momentum equations are reduced to

$$\begin{aligned} -\frac{\partial p}{\partial y} + \frac{\partial \sigma_{xy}}{\partial x} &= 0 \\ -\frac{\partial p}{\partial z} + \frac{\partial \sigma_{xz}}{\partial x} &= 0 \end{aligned} \tag{8.98}$$

$$\begin{aligned} (x - x_3)\frac{\partial p}{\partial y} - \eta\frac{\partial^2 u_y}{\partial x^2} &= 0 \\ (x - x_4)\frac{\partial p}{\partial z} - \eta\frac{\partial^2 u_z}{\partial x^2} &= 0 \end{aligned} \tag{8.99}$$

with the velocity boundary conditions

$$\begin{aligned} u_y = u_z = 0 \quad &\text{for} \quad x = 0 \\ u_y = u_2 \sin\phi \quad \text{and} \quad u_z = u_2 \cos\phi \quad &\text{for} \quad x = h \end{aligned} \tag{8.100}$$

Two other conditions, which relate to the places ($x = x_3$ and $x = x_4$) where the two velocity gradients equal zero, are

$$\begin{aligned} \frac{\partial u_y}{\partial x} &= 0 \quad \text{for} \quad x = x_3 \\ \frac{\partial u_z}{\partial x} &= 0 \quad \text{for} \quad x = x_4 \end{aligned} \tag{8.101}$$

Since the analysis of extrusion usually deals with steady state conditions, the energy equation reduces to a balance between heat conduction and viscous dissipation

$$-k\frac{\partial^2 T}{\partial x^2} = \sigma_{xi}\frac{\partial u_i}{\partial x} = \sigma_{xx}\frac{\partial u_x}{\partial x} + \sigma_{xy}\frac{\partial u_y}{\partial x} + \sigma_{xz}\frac{\partial u_z}{\partial x} \tag{8.102}$$

or,

$$-k\left.\frac{\partial^2 T}{\partial x^2}\right|_x - k\left.\frac{\partial^2 T}{\partial x^2}\right|_{x_c} = \sigma_{xi}\left.\frac{\partial u_i}{\partial x}\right|_x + \sigma_{xi}\left.\frac{\partial u_i}{\partial x}\right|_{x_c} \tag{8.103}$$

Here, the position x is the location of the streamline located near the barrel, and x_c the position of the same streamline located near the root of the screw. If the flow is fast enough, we can assume that the temperature along the streamline is constant as it travels near the barrel and returns to near the root of the screw. Hence, for the barrel surface we must take the boundary condition

$$\begin{aligned} T = \frac{T_1 + T_2}{2} \quad &\text{for} \quad x = h \\ \frac{\partial T}{\partial x} = 0 \quad &\text{for} \quad x = 0 \end{aligned} \tag{8.104}$$

where T_1 and T_2 are the temperatures at the root of the screw and the inside diameter of the barrel, respectively. Rearranging somewhat we can write

$$-k\left.\frac{\partial^2 T}{\partial x^2}\right|_x - k\left.\frac{\partial^2 T}{\partial x^2}\right|_{x_c} = 2\eta\left.\dot{\gamma}^2\right|_x + 2\eta\left.\dot{\gamma}^2\right|_{x_c} \tag{8.105}$$

We can relate x and x_c by performing a mass balance in the y-direction, for which the net flow must be zero[3]

$$\int_0^x u_y dx = -\int_{x_c}^h u_y dx \tag{8.106}$$

Summarizing, the model of the screw channel flow is governed by eqns. (8.99), (8.105) and (8.106) with boundary conditions eqns. (8.100), (8.101) and (8.104). The constitutive equation that was used by Griffith is a temperature dependent shear thinning fluid described by

$$\eta(T, \dot{\gamma}) = m e^{[-bn(2T - T_1 - T_2)/2]} \dot{\gamma}^{(n-1)} \tag{8.107}$$

The momentum balance now becomes

$$\begin{aligned} \frac{\partial u_y}{\partial x} &= e^{\Theta} G_y(x - x_3)\left[G_y^2(x - x_3)^2 + G_z^2(x - x_4)^2\right]^{(1-n)/2} \\ \frac{\partial u_z}{\partial x} &= e^{\Theta} G_z(x - x_4)\left[G_y^2(x - x_3)^2 + G_z^2(x - x_4)^2\right]^{(1-n)/2} \end{aligned} \tag{8.108}$$

[3]This assumption is not fully valid since, due to the flight clearance, there is leakage flow over the flight of the screw.

where the new dimensionless parameters, Θ for temperature and G_i for pressure gradient, are defined as

$$\begin{aligned}\Theta =& b\left(2T - T_1 - T_2\right)/2 \\ G_y =& \left(\frac{\Delta p}{\lambda}\right) h^{n+1}\left(m u_2^n\right)^{-1} \\ G_z =& \Delta p h^{n+1} \sin\phi_2 \left(L m f_\gamma f_d\right)^{-1}\left(\pi D_2 N\right)^{-n}\end{aligned} \tag{8.109}$$

In the above equations, $h = (D_1 - D_2)/2$ is the channel height, L is the axial length of the screw, f_γ is a curvature correction[4] and f_d is a correction for the wall effect given in Chapter 6.

To complete the dimensionless form of the eqn. (8.108), the lengths and coordinates are normalized with the channel depth, $\hat{x} = x/h$ and the velocities with $\pi N D_2$, $\hat{u_y} = u_y/\pi N D_2$ and $\hat{u_z} = u_z/\pi N D_2$. The boundary conditions for eqn. (8.108) are

$$\begin{aligned}\hat{u_y} = \hat{u_z} = 0 \quad &\text{for} \quad \hat{x} = 0 \\ \hat{u_y} = \sin\phi \quad \text{and} \quad \hat{u_z} = \cos\phi \quad &\text{for} \quad \hat{x} = 1\end{aligned} \tag{8.110}$$

for the velocities and

$$\begin{aligned}\frac{\partial \hat{u_y}}{\partial \hat{x}} &= 0 \quad \text{for} \quad \hat{x} = \hat{x_3} \\ \frac{\partial \hat{u_z}}{\partial \hat{x}} &= 0 \quad \text{for} \quad \hat{x} = \hat{x_4}\end{aligned} \tag{8.111}$$

for the velocity gradients. The energy equation is

$$\begin{aligned}\frac{\partial^2\Theta}{\partial x^2}\left[1 - \frac{\partial x}{\partial x_c}\right] - \frac{\partial\Theta}{\partial x}\frac{\partial x_c}{\partial x}\frac{\partial^2 x}{\partial x_c^2} =& \\ &- Bre^{\Theta}\{\left[G_y^2(x - x_3)^2 + G_z^2(x - x_4)^2\right]^{(n+1)/2n} \\ &- \frac{\partial x_c}{\partial x}\left[G_y^2(x - x_3)^2 + G_z^2(x - x_4)^2\right]^{(n+1)/2n}\}\end{aligned} \tag{8.112}$$

where Griffith defined the Brinkman number as

$$Br = \frac{m u_2^{n+1} h^{1-n}}{k b^{-1}} \tag{8.113}$$

The dimensionless boundary conditions for the energy balance are given by

$$\begin{aligned}\Theta = 0 \quad &\text{for} \quad x = 1 \\ \frac{\partial\Theta}{\partial x} = 0 \quad &\text{for} \quad x = x_c\end{aligned} \tag{8.114}$$

and the material balance becomes

$$\int_0^{\hat{x}} \hat{u_y} d\hat{x} = -\int_{\hat{x_c}}^1 \hat{u_y} d\hat{x} \tag{8.115}$$

[4]For shallow channels this correction factor is 1.

Assuming that a streamline has a constant temperature and that all viscous dissipation goes into local heating instead of being conducted to the screw and barrel, the energy equation becomes

$$\hat{u_y} Pe \frac{\partial \Theta}{\partial \hat{y}} = Br e^{\Theta} \left[G_y^2 (x - x_3)^2 + G_z^2 (\hat{x} - \hat{x_4})^2 \right]^{(n+1)/2n} \tag{8.116}$$

where the Peclet number $Pe = hu_2/\alpha$ and α is the thermal diffusivity.

Using the above assumptions, the adiabatic temperature rise for the material is given by

$$\frac{\Delta\Theta}{Br} = \left(\frac{W}{Pe} \right) \left[\frac{(6x-2)^2}{|x(3x-2)|} + \frac{(6x_c - 2)^2}{|x_c(3x_c - 2)|} \right] \tag{8.117}$$

where W is the channel width normalized width of the extruder channel, $W \approx \pi D/h$.

Griffith solved the equations for chosen values of Br, G_z, ϕ and n. The values of G_y, x_3, x_4 and Θ were estimated and an initial approximate velocity field was solved using Runge-Kutta integrations of the momentum balance. The temperatures were then obtained by solving the energy equation using Runge-Kutta integration, using the first solution of the velocity field. The new approximation of Θ was used to compute new viscosities and a new velocity field was solved for by solving the momentum balance equation once more. When integrating the energy equation, the previous approximation of Θ was used in the exponential term. After the computations converged, the model gave fluid throughput as a function of pressure gradients, temperature and velocity fields across the thickness, values of x_3 and x_4, as well as velocity derivatives and temperature gradients at the barrel surface, $\hat{x} = 1$.

Figures 8.37 and 8.38 [9] present velocity and temperature fields across the thickness, respectively, for various values of Br, and for $n = 1$ and $n = 0.6$. Griffith calculated the screw characteristic curves for Newtonian and non-Newtonian shear thinning fluids using various power law indices. Figure 8.39 presents these results and compares them to experiments performed with a carboxyl vinyl polymer ($n = 0.2$) and corn starch ($n = 1$).

8.6 THE RADIAL FLOW METHOD

To aid the polymer processing student and engineer in finding required injection pressures and clamping forces, Stevenson [22] derived the non-isothermal non-Newtonian equations for the flow in a disc and solved them using finite difference techniques. The outcome was a set of dimensionless groups and graphs that can be applied to any geometry after a lay-flat approximation. In his analysis, Stevenson represented the flow inside the cavity with a radial flow between two parallel plates. In order to use this representation, we must first *lay flat* the part to find the longest flow path as schematically depicted in Fig. 8.40.

Since the longest flow path may exceed the radius of the projected area that causes mold separating pressures, we must also find the radius of equivalent projected area, R_p, to compute a more accurate mold clamping force. However, to perform the calculations to predict velocities and pressure fields, we assume a disc geometry of radius R and thickeness h, schematically depicted in Fig. 8.41.

As a constitutive model for the momentum balance, Stevenson chose a temperature dependent power law model represented by

$$\eta(\dot{\gamma}, T) = m e^{-a(T-T_1)} \dot{\gamma}^{n-1} \tag{8.118}$$

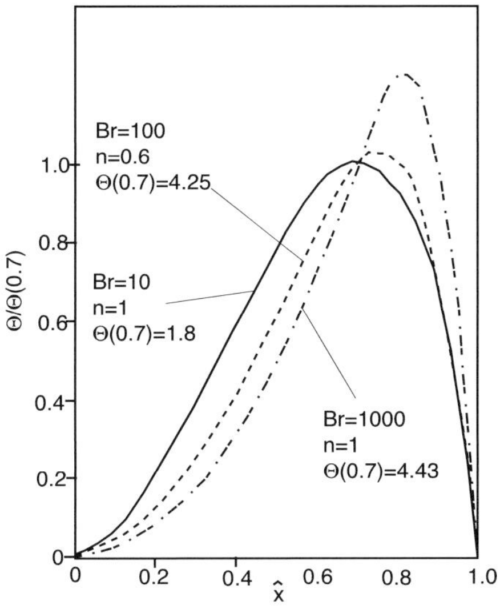

Figure 8.37: Temperature distributions across the thickness of the channel for various Brinckman numbers [9].

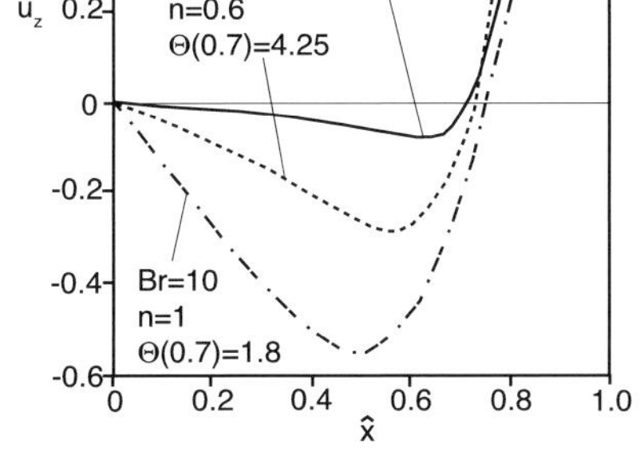

Figure 8.38: Velocity distributions across the thickness of the channel for various Brinckman numbers [9].

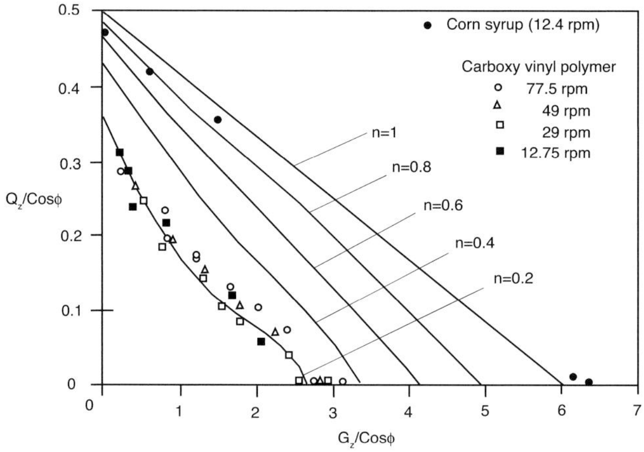

Figure 8.39: Screw characteristic curves for various power law indeces. Experimental values are shown for a carboxyl vinyl polymer with $n = 0.5$ and corn starch (Newtonian) with $n = 1$ [9].

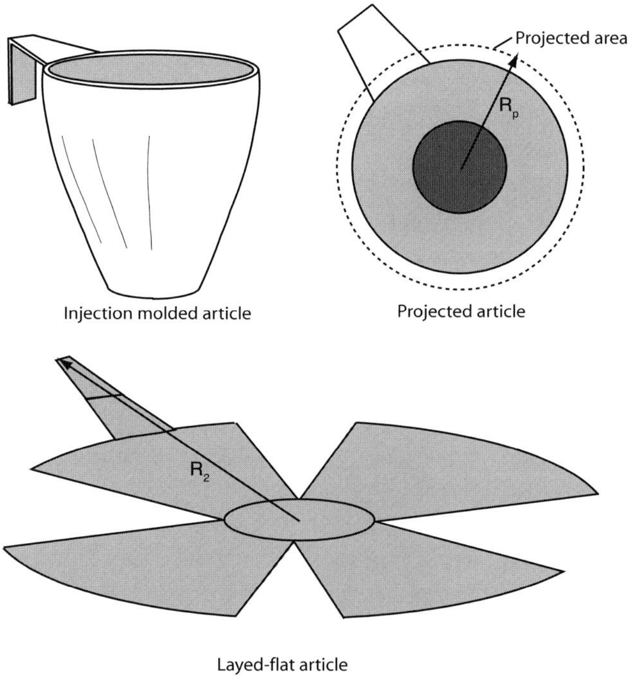

Figure 8.40: Schematic diagram of an injection molding item with its projected area and its lay-flat representation.

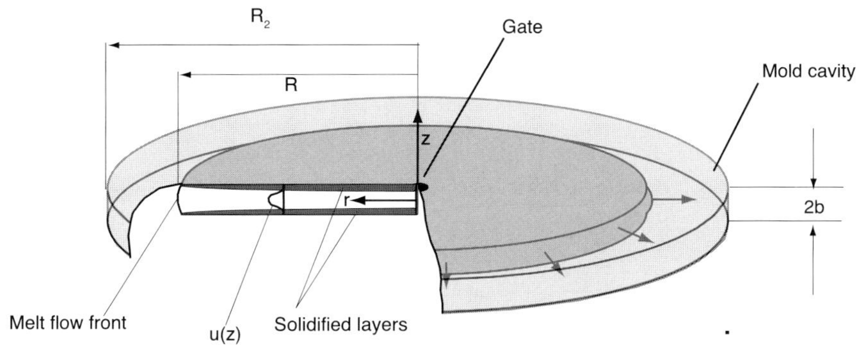

Figure 8.41: Schematic of the disc representation and nomenclature for an injection molded item.

The momentum balance for the disc geometry simplifies to

$$\frac{\partial p}{\partial r} = -\frac{\partial \tau_{rz}}{\partial z} \tag{8.119}$$

where the deviatoric stress tensor is defined by

$$\tau_{rz} = -\eta \frac{\partial u}{\partial z} \tag{8.120}$$

The energy balance for the geometry represented in Fig. 8.41 results in transient and convective terms, conduction through the thickness and viscous dissipation caused by the through-the-thickness shear components

$$\rho C_p \left(\frac{\partial T}{\partial t} + u \frac{\partial T}{\partial r} \right) = k \frac{\partial^2 T}{\partial z^2} + \eta \left(\frac{\partial u}{\partial z} \right)^2 \tag{8.121}$$

Assuming a characteristic viscosity of $\bar{\eta} = m e^{-aT_1} \bar{\dot{\gamma}}^{n-1}$, where the characteristic rate of deformation is taken as $\bar{\dot{\gamma}} = \bar{u}/b$, where t_f is the fill time and $\bar{u} = R_2/t_f$ the characteristic velocity, we can write the viscosity in dimensionless form as

$$\hat{\eta}(\hat{\dot{\gamma}}, \Theta) = \frac{\eta}{\bar{\eta}} = e^{\beta(1-\Theta)} \hat{\dot{\gamma}}^{n-1} \tag{8.122}$$

where

$$\beta = a(T_1 - T_w) \tag{8.123}$$

The dimensionless number β determines the intensity of the coupling between the energy equation and the momentum balance. With the dimensionless viscosity, and assuming a characteristic pressure of $\bar{p} = \bar{\eta}\bar{u}R_2/b^2$ (R_2 was chosen as the characteristic r-dimension and b as the characteristic z-dimension), the momentum balance can also be written in dimensionless form as

$$\frac{\partial \hat{p}}{\partial \hat{r}} = \frac{\partial}{\partial \hat{z}} \left[\left(e^{\beta(1-\Theta)} \right) \left(\frac{\partial \hat{u}}{\partial \hat{z}} \right)^n \right] \tag{8.124}$$

which can be used in the region $0 > \hat{z} > -1$ where the velocity gradient is positive.

Similarly, the energy balance can also be written in dimensionless form as

$$\frac{\partial \Theta}{\partial \tau} + \hat{u}\frac{\partial \Theta}{\partial \hat{r}} = \tau \left[\frac{\partial^2 \Theta}{\partial \hat{z}^2} + Bre^{\beta(1-\Theta)} \left(\frac{\partial \hat{u}}{\partial \hat{z}} \right)^{n+1} \right] \tag{8.125}$$

where the dimensionless time is defined by

$$\tau = \frac{t_f \alpha}{b^2} \tag{8.126}$$

the dimensionless temperature is

$$\Theta = \frac{T - T_w}{T_1 - T_w} \tag{8.127}$$

and the Brinkman number is

$$Br = \frac{me^{-aT_1}b^2}{k(T_1 - T_\omega)} \left(\frac{r_2}{t_f b} \right)^{n+1} \tag{8.128}$$

The boundary conditions for the dimensionless governing equations are given by

$$\begin{aligned}
&-1 < \hat{z} < 1 \quad && \hat{r} = 0 \quad && \hat{u}\hat{r} = (\hat{u}\hat{r})_I \quad && \Theta = 1 \\
&-1 < \hat{z} < 1 \quad && \hat{r} = \hat{r}_{int} \quad && \hat{p} = 0 \quad && \frac{\partial \Theta}{\partial \hat{r}} = 0 \\
&0 < \hat{r} < \hat{r}_{int} \quad && \hat{z} = 0 \quad && \frac{\partial \hat{u}}{\partial \hat{z}} = 0 \quad && \frac{\partial \Theta}{\partial \hat{z}} = 0 \\
&0 < \hat{r} < \hat{r}_{int} \quad && \hat{z} = \pm 1 \quad && \hat{u} = 0 \quad && \Theta = 0
\end{aligned} \tag{8.129}$$

where r_{int} is the interfacial radius or free flow front location during filling. The pressure and the clamping forces can be non-dimensionalized with the isothermal prediction of injection pressure and force as

$$\frac{\Delta p}{\Delta p_I} = f(\tau, \beta, Br, n) \tag{8.130}$$

and

$$\frac{F(R_2)}{F_I(R_2)} = g(\tau, \beta, Br, n) \tag{8.131}$$

respectively. The analytical equations for the injection pressure and clamping force for the isothermal case are given by

$$\Delta p_I = \frac{me^{-aT_1}}{1-n} \left[\frac{1+2n}{2n} \frac{R_2}{t_f b} \right]^n \frac{R_2}{b} \tag{8.132}$$

and

$$F_I(R_2) = \pi R_2^2 \left(\frac{1-n}{3-n} \right) \Delta p_I \tag{8.133}$$

Table 8.8: Material Properties of ABS Used in the Experiments

Parameter	Value
ρ	$1020\ \mathrm{kg/m^3}$
k	0.184 W/m/K
C_p	2,343 J/kg/K
α	$7.7\times10^{-7\mathrm{m}^2}$/s
m	29 MPas^n
n	0.29
a	0.01369/K

respectively. The functions $f(\tau, \beta, Br, n)$ and $g(\tau, \beta, Br, n)$ are the solutions of the dimensionless momentum and energy balance equations. Stevenson solved these equations using the finite difference formulation with a uniform grid in the r- and z-directions [28]. The time steps were chosen such that r_{int} concides with the r-direction grid points r_k. Hence, for a constant flow rate Q

$$\Delta t_{k-1} = 2\pi b \frac{\left[r_{1+(k-1)\Delta r}\right]^2 - \left[r_1 + (k-2)\Delta r\right]^2}{Q} \tag{8.134}$$

When solving for the energy equation an implicit FDM was used with a backward (upwinded) difference representation of the convective term. The viscous dissipation term was evaluated with velocity components from the previous time step. The equation of motion was integrated using a trapezoidal quadrature. Stevenson tested his model by comparing it to actual mold filling experiments of a disc with an ABS polymer. Table 8.8 presents data used for the calculations.

Figure 8.42 presents a comparison of predicted and experimental pressures as a function of volumetric flow rate, Q, for two injection temperatures and two mold thicknesses. The figure shows that the model over-predicts the injection pressure requirements by 20-30%. Stevenson stated that one possible source of error is that his model used the plasticating unit heater temperatures as the inlet temperature into the cavity, neglecting the effects of viscous heating in the sprue and runner system. Nevertheless, these predictions are good enough for most estimates of pressure and clamping force requirements. Stevenson also performed various analyses to see what effect viscous dissipation, convection and non-isothermal assumptions had on the solution of the problem. Figure 8.43 presents results from these analyses. It is clear that all the effects included in the model are important and should be incorporated when modeling injection molding.

The clamping force is solved for by integrating the pressure distribution as,

$$F(r) = 2\pi \int_0^r p(r')dr' \tag{8.135}$$

and in dimensionless form using

$$\hat{F}(\hat{r}) = 2 \int_0^{\hat{r}} \hat{p}(\hat{r}')d\hat{r}' \tag{8.136}$$

Figures 8.44, 8.45, 8.46 and 8.47 present the dimensionless injection pressures and clamping forces. Values from these graphs can be used to estimate injection pressure and

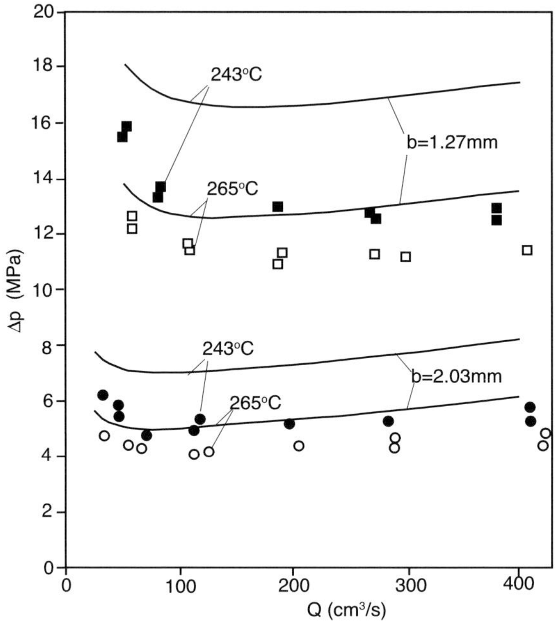

Figure 8.42: Comparison between theoretical predictions with experimental measurements of mold filling pressure requirements as a function of injection speed [28].

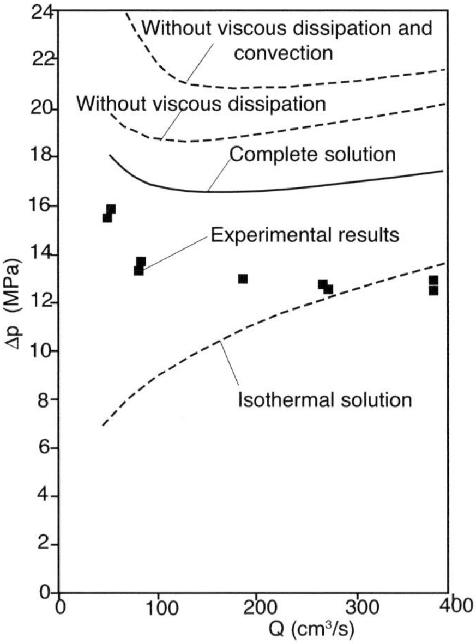

Figure 8.43: Comparison between different theoretical predictions with experimental measurements of mold filling pressure requirements as a function of injection speed [28].

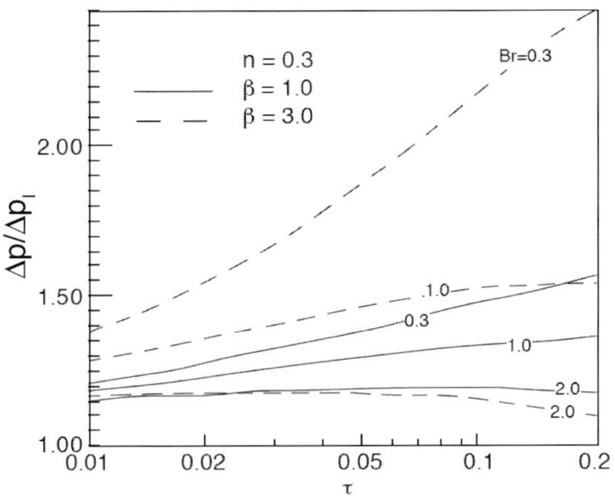

Figure 8.44: Dimensionless injection pressure as a function of β, Br and τ for $n = 0.3$.

clamping force using,

$$\Delta p = \left[\frac{\Delta p}{\Delta p_I}\right] \Delta p_I \tag{8.137}$$

and

$$F(Rp) = \left[\frac{F(R_2)}{F_I(R_2)}\right] \left[\frac{F(R_p)}{F(R_2)}\right] F_I(R_2) \tag{8.138}$$

where the ratio $F(R_p)/F(R_2)$ is found in Fig. 8.48.

■ EXAMPLE 8.10.

Sample application of the radial flow method. In this sample application, we are to determine the maximum clamping force and injection pressure required to mold an ABS suitcase shell with a filling time, t_f=2.5 s. For the calculation we will use the dimensions and geometry schematically depicted in Fig. 8.49, an injection temperature of 227°C (500 K), a mold temperature of 27°C (300 K) and the material properties given in Table 8.8.

We start this problem by first laying the suitcase flat and determining the required geometric factors (Fig. 8.50). From the suitcase geometry, the longest flow path, R_2, is 0.6 m and the radius of the projected area, R_p, is 0.32 m. Using the dimensions of the part and the conditions and properties given above, we can compute the four dimensionless groups that govern this problem

- $\beta = 0.01369/\text{ K}(500\text{ K} - 300\text{ K}) = 2.74$
- $\tau = \dfrac{2.5\text{ s}(0.184\text{ W/m/K})}{(0.001\text{ m})^2(1020\text{ kg/m}^3)(2,343\text{ J/kg/K})}$=0.192

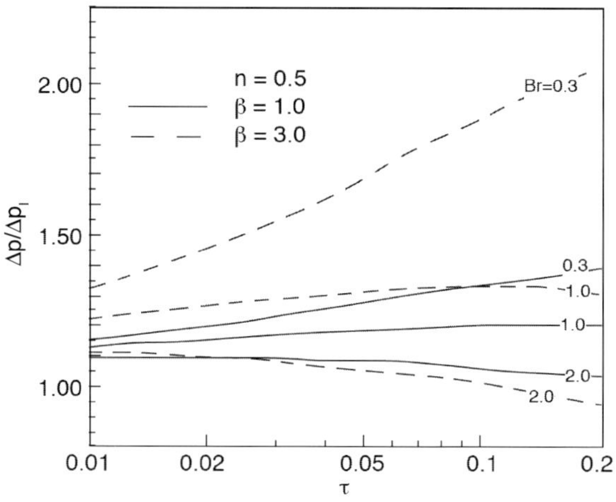

Figure 8.45: Dimensionless injection pressure as a function of β, Br and τ for $n = 0.5$.

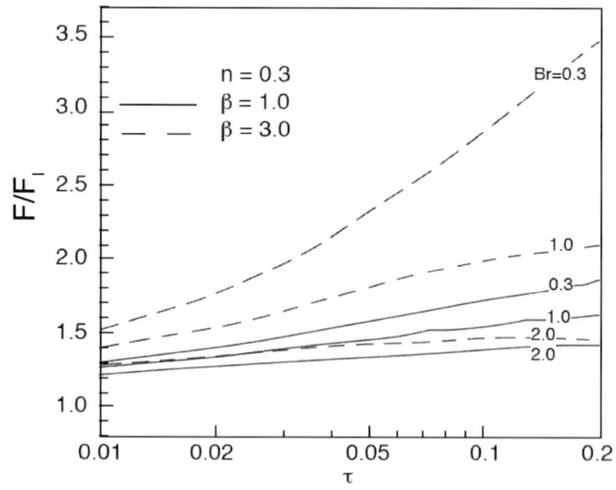

Figure 8.46: Dimensionless clamping force as a function of β, Br and τ for $n = 0.3$.

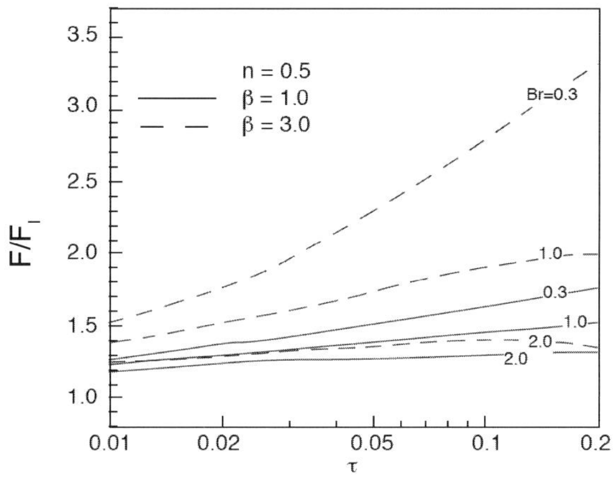

Figure 8.47: Dimensionless clamping force as a function of β, Br and τ for $n = 0.5$.

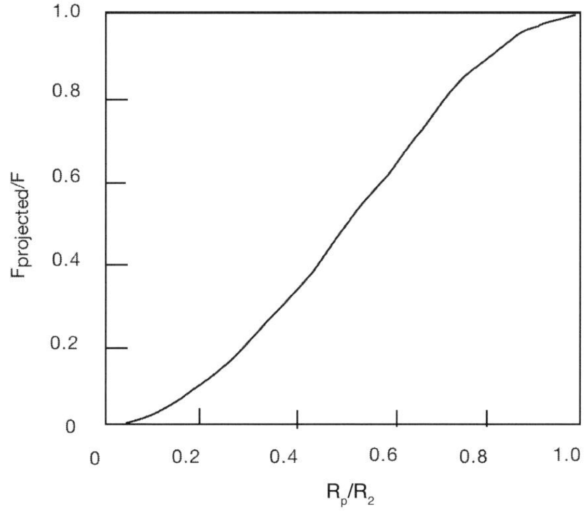

Figure 8.48: Clamping force correction for the projected area.

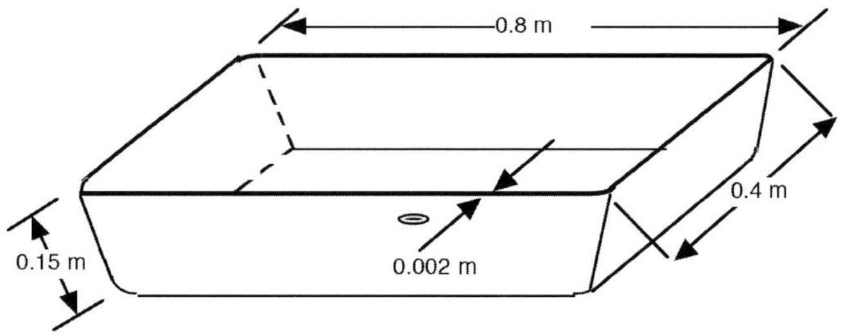

Figure 8.49: Suitcase geometry.

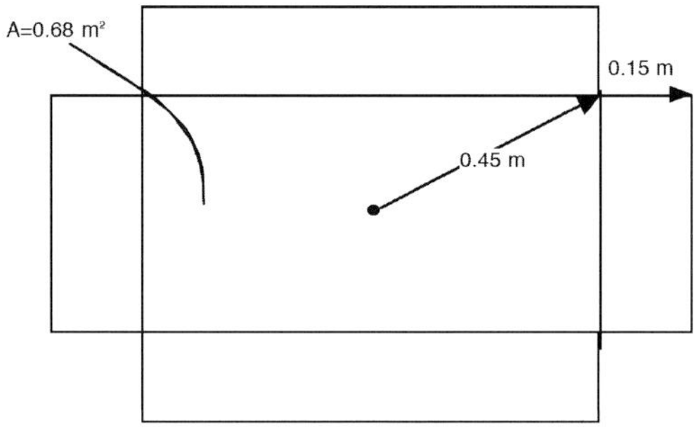

Figure 8.50: Layed-flat suitcase geometry.

- $Br = \dfrac{29 \times 10^6 \text{ Pa-s}^n e^{-0.01369/\text{ K}(500\text{ K})}(0.001\text{ m}^2)}{0.184\text{ W/m/K}(500\text{ K} - 300\text{ K})}\left[\dfrac{0.6\text{ m}}{2.5\text{s}(0.001\text{ m})}\right]^{0.29+1} = 0.987$

The isothermal injection pressure and clamping force are computed next,

- $\Delta p_I = \dfrac{29 \times 10^6 e^{-0.01369/\text{K}(500\text{ K})}}{1 - 0.29}\left[\dfrac{(1 + 2 \times 0.29)0.6\text{ m}}{2(0.29)(2.5\text{ s}(0.001\text{ m}))}\right]\left[\dfrac{0.6\text{ m}}{0.001\text{ m}}\right] = 171\text{ MPa}$

- $\pi(0.6\text{ m}^2)\dfrac{1 - 0.29}{3 - 0.29}(171 \times 10^6\text{ Pa}) = 50.7 \times 10^6\text{ N}$

We now look up $\Delta p/\Delta p_I$ and F/F_I in Figs. 8.44, 8.45, 8.46 and 8.47 Since the change between n=0.3 and n=0.5 is very small, we choose n=0.3. However, for other values of n we can interpolate or extrapolate. For β=2.74, we interpolate between β=1 and β=3. For β=1 we get $\Delta p/\Delta p_I$=1.36 and F/F_I=1.65, and for β=3 we get $\Delta p/\Delta p_I$=1.55 and F/F_I=2.1. Hence, for β=2.74 we get $\Delta p/\Delta p_I$=1.53 and F/F_I=2.04. Taking the product with the isothermal predictions, we get Δp=262 MPa and F=10.3×10^7 N or 10,300 metric tons.

Since the part exceeds the projected area, Fig. 8.48 can be used to correct the computed clamping force. The clamping force can be corrected for R_p= 0.32 m

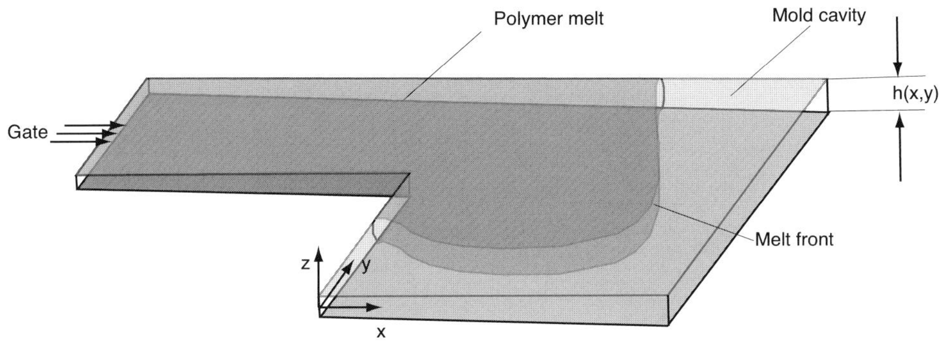

Figure 8.51: Schematic of a filling mold cavity with variable thickness.

using R_p/R_2=0.53 in the figure. Hence we get F_p=(0.52)10,300 metric tons = 5,356 metric tons.

For our suitcase cover, where the total volume is 1,360 cm^3 and total part area is 0.68 m^2, the above numbers are too high. A useful rule-of-thumb is a maximum allowable clamping force of 2 tons/in^2. Here, we have greatly exceeded that number. Normally, around 3,000 metric tons/m^2 are allowed in commercial injection molding machines. For example, a typical injection molding machine with a shot size of 2,000 cm^3 has a maximun clamping force of 630 metric tons with a maximun injection pressure of 1,400 bar. A machine with much larger clamping forces and injection pressures is suitable for much larger parts. For example, a machine with a shot size of 19,000 cm^3 allows a maximum clamping force of 6,000 metric tons with a maximum injection pressure of 1,700 bar. For this example, we must reduce the pressure and clamping force requirements. This can be acomplished by increasing the injection and mold temperatures or by reducing the filling time. Recommended injection temperatures for ABS are between 210 and 240°C and recommended mold temperatures are between 40 and 90°C. As can be seen, there is room for improvement in the processing conditions, so one must repeat the above procedure using new conditions.

8.7 FLOW ANALYSIS NETWORK

The Flow Analysis Network (FAN) developed by Broyer, Gutfinger and Tadmor in the 1970's [4, 24] is the finite difference precursor of today's mold filling simulations. The technique works well to predict mold filling patterns, injection pressures and clamping forces of two-dimensional or non-planar layed-flat part geometries. The method takes a geometry as the one depicted in Fig. 8.51 and represents it with a finite difference grid as shown in Fig. 8.52. Each finite difference grid point or node has its own control volume and each one of these control volumes, i, is assigned a fill factor, f_i. The fill factor is the fraction of the control volume that is filled with polymer.

As referenced in Fig. 8.52, at any given point during the mold filling process there are five different types of nodes. These are,

- Gate nodes (a). These nodes are full at the beginning of the mold filling simulation at which point they are assigned a fill factor of 1 ($f_a = 1$). They are either assigned a

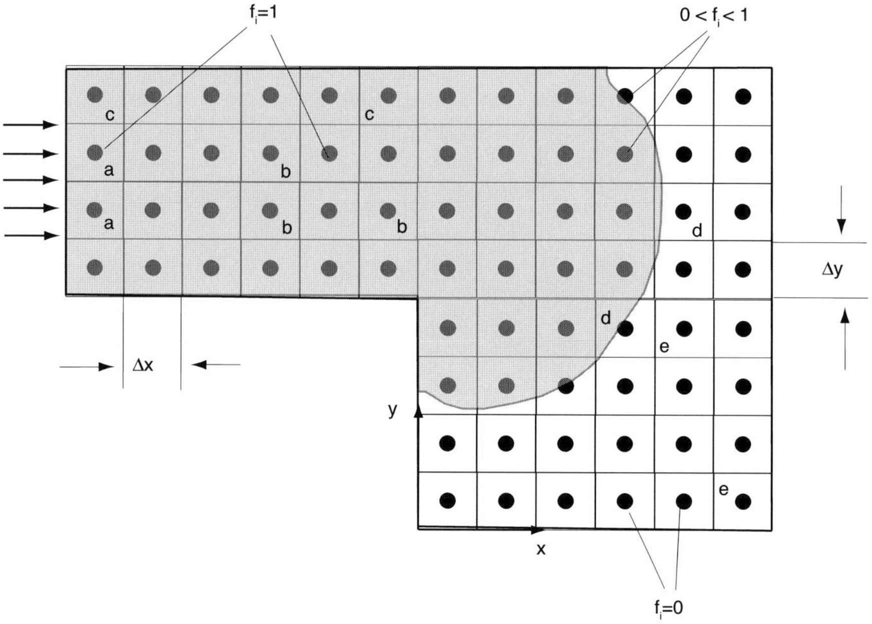

Figure 8.52: FAN discretization of a filling mold cavity with variable thickness.

pressure(controlled pressure) boundary condition or a flow rate (controlled injection speed).

- Full nodes (b). These nodes are the ones that have already filled during the mold filling process ($f_b = 1$). When solving the governing equations at any given time step, the pressure is unknown inside these control volumes.

- Mold edge nodes (c). These nodes have less than 4 neighbors. A neighbor that is missing on one side implies no flow across that edge, taking care of the $\partial p/\partial \mathbf{n} = \mathbf{0}$ boundary condition (natural or Neumann boundary condition).

- Melt front node (d). These are the nodes that are temporarily on the free flow front during mold filling, and are therefore partially filled ($0 < f_d < 1$). During that specific time step this node is assigned a zero pressure boundary condition, $p_d = 0$ (essential or Dirichlet boundary condition).

- Empty nodes (e). All nodes, except for the gate nodes, begin as empty nodes ($f_e = 0$). As the mold fills, these nodes change to type (d) and eventually to type (b). Empty nodes are assigned a zero pressure boundary condition, $p_e = 0$.

The Hele-Shaw model was used to describe the flow in the FAN formulation. For a Newtonian case, the flow is described by

$$\bar{u_x} = \frac{h(x,y)^2}{12\mu}\frac{\partial p}{\partial x} \tag{8.139}$$

$$\bar{u_y} = \frac{h(x,y)^2}{12\mu}\frac{\partial p}{\partial y} \tag{8.140}$$

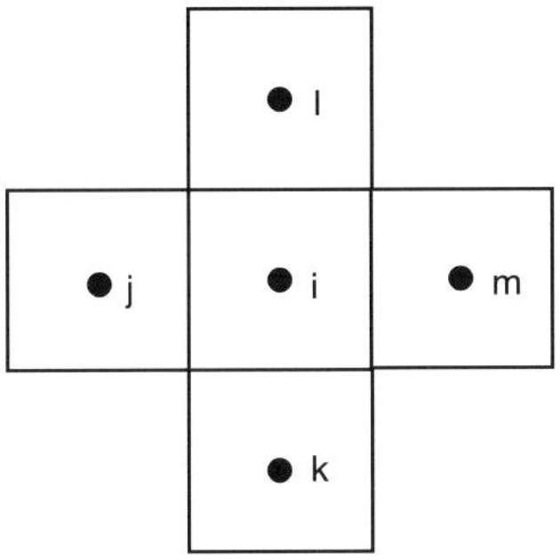

Figure 8.53: Local system used for mass balance.

The continuity equation is satisfied by performing a mass (volume) balance around each control volume, i. Using the notation found in Fig. 8.53 the mass balance is written as

$$q_{ij} + q_{il} + q_{ik} + q_{im} = 0 \tag{8.141}$$

where the volumetric flow rate, q_{ij} is given by

$$q_{ij} = \bar{u_{ij}}(\Delta x)h_{ij} = \frac{h_{ij}^3 \Delta x}{12\mu}\frac{p_i - p_j}{\Delta x} = \frac{h_{ij}^3}{12\mu}(p_i - p_j) \tag{8.142}$$

Equation (8.141) now becomes

$$\begin{aligned} &\frac{1}{12\mu}(h_{ij}^3 + h_{il}^3 + h_{ik}^3 + h_{im}^3)p_i + \\ &(-\frac{h_{ij}^3}{12\mu})p_j + (-\frac{h_{ik}^3}{12\mu})p_k + (-\frac{h_{il}^3}{12\mu})p_l + (-\frac{h_{im}^3}{12\mu})p_m = 0 \end{aligned} \tag{8.143}$$

If we perform a mass balance on all the control volumes in the system, we can write

$$[a]\{p\} = 0 \tag{8.144}$$

where

$$a_{ii} = \frac{1}{12\mu}(h_{ij}^3 + h_{il}^3 + h_{ik}^3 + h_{im}^3) \tag{8.145}$$

and

$$a_{ij} = -\frac{h_{ij}^3}{12\mu} \tag{8.146}$$

Note that all other entries in martrix $[a]$ are zero, $a_{in} = 0$. This matrix is symmetric and banded. The size of the band, if the cells are properly numbered, is very small compared to the size of a problem. We will discuss matrix storage, manipulation and solution in more detail in Chapter 9 of this book. Once the matrix system has been assembled, we can store and re-use it every time step. Every time step we apply the boundary conditions by setting all the pressures of the empty and partially filled nodes to zero. If the pressure on node i

is set to zero, $p_i = 0$, we eliminate the row and the column values that pertain to node i, ($a_{ij} = a_{ji} = 0$) and set the diagonal that pertains to that control volume to one ($a_{ii} = 1$). The natural boundary conditions along the mold edges are automatically taken into account.

The unknown pressures are acquired by solving the system of algebraic equations, eqn. (8.144), after the boundary conditions are applied. With a known pressure field we can solve for the flow rates between the nodes and a the fill factor in the partially filled nodes can be updated using the smallest time step required to fill the next node. How to determine the appropriate time step is shown in Algorithm 3. The flow fronts are advanced in Algorithm 4. At that point, new boundary conditions are applied to the set of algebraic equations to solve for the new pressure and flow fields. This is repeated until all control volumes are full.

Algorithm 3 FAN-Δt calculation

```
subroutine timestep
  Dt  = 1E10                      initialize time step size to a large number
  do while (f(i)<1))
    Qtotal(i)=q(i,j)+q(i,k)+q(i,l)+q(i,m)
     Dti=(V(i)-f(i)*V(i))/Qtotal(i)
     if(Dti.lt.Dt)Dt=Dti                    return the smallest time step
                                          required to fill a control volume
  enddo
end subroutine timestep
```

Algorithm 4 FAN- Flow front advancement

```
subroutine newflowfront
  do i=1,N
    f(i)=(f(i)*V(i)+Qtotal(i)*Dt)/V(i)
  enddo
end subroutine newflowfront
```

Tadmor [25] compared the flow analysis network to injection molding experimental results obtained by Krueger and Tadmor [15]. Tadmor and Krueger's experiments were performed injecting polystyrene into thin flat molds with inserts and variable thickness as shown in Fig. 8.54 The mold size was 3.8$\times$15.2 cm with thicknesses of 3.35 mm in the thick sections to 1.68 mm in the thin sections. As can be seen, the experiments and the simulation agree quite well, within experimental error and control volume size. Although the above procedure was presented for the injection molding process, it can also be used to model compression molding by modifying eqn. (8.141) to

$$q_{ij} + q_{il} + q_{ik} + q_{im} = -\dot{h}_i(\Delta x)^2 \tag{8.147}$$

The whole equation system can also be modified to model the flow of non-Newtonian shear thinning fluids. This results in a set of non-linear algebraic equations that must be solved by using iterative techniques. Non-planar parts can also be modeled using this procedure, by simply orienting the grid in space according to mid-plane surface of the injection molded part. However, note that we have used mold sides that align with the x- and y-coordinates. Although curved and slanted sides are also possible to simulate using FAN, the coefficients a_{ij} need to be adjusted accordingly, making the technique cumbersome and more difficult to implement than the more versatile finite element control volume approach discussed in the next chapter.

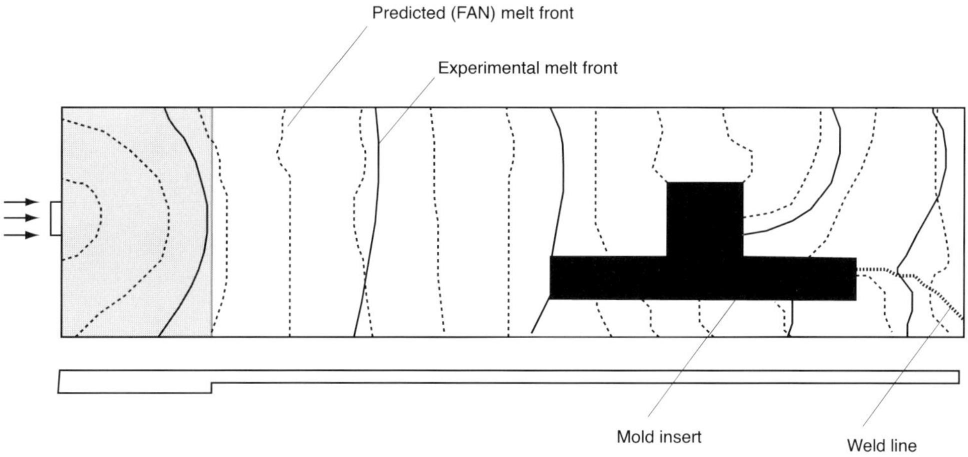

Figure 8.54: Experimental and simulated (FAN) melt front advancement in a shallow rectangular mold.

8.8 PREDICTING FIBER ORIENTATION – THE FOLGAR-TUCKER MODEL

The state of particle orientation at a point can be fully described by an orientation distribution function, $\psi(\phi, x, y)$. The distribution is defined such that the probability of a particle, located at x and y at time t, being oriented between angles ϕ_1 and ϕ_2, is given by

$$P(\phi_1 < \phi < \phi_2) = \int_{\phi_1}^{\phi_2} \psi(\phi, x, y) d\phi \tag{8.148}$$

This is graphically depicted in Fig. 8.55. For simplicity, the x, y, and t from the orientation distribution function can be dropped.

Since one end of a particle is indistinguishable from the other, the orientation distribution function must be periodic:

$$\psi(\phi) = \psi(\phi + \pi) \tag{8.149}$$

Since all particles are located between $-\pi/2$ and $\pi/2$, the orientation distribution function must be normalized such that

$$\int_{-\pi/2}^{\pi/2} \psi(\phi) d\phi = 1 \tag{8.150}$$

The orientation distribution function changes constantly as the particles travel within a deforming fluid element. Assuming the fiber density is homogeneous throughout the fluid and remains that way during processing, a balance around a differential element in the distribution function can be performed. This is graphically represented in Fig. 8.56. Here, the rate of change of the fiber density of the differential element, shown in the figure, should be the difference between the number of particles that move into and out of the control volume in a short time period Δt. This can be written as

$$\frac{\psi(\phi)\Delta\phi}{\Delta t} = \psi(\phi)\dot{\phi}(\phi) - \psi(\phi + \Delta\phi)\dot{\phi}(\phi + \Delta\phi) \tag{8.151}$$

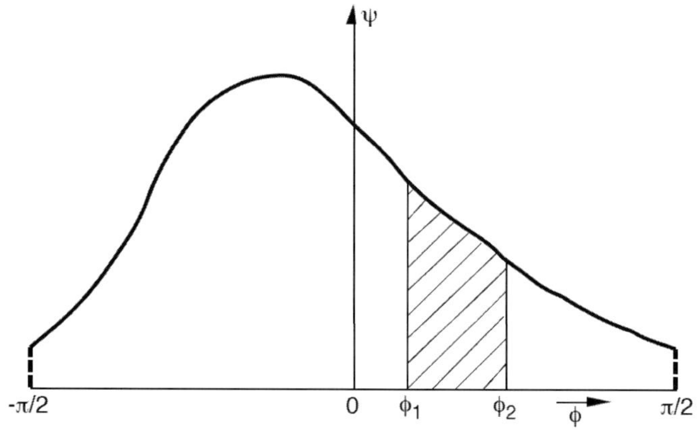

Figure 8.55: Orientation distribution functions.

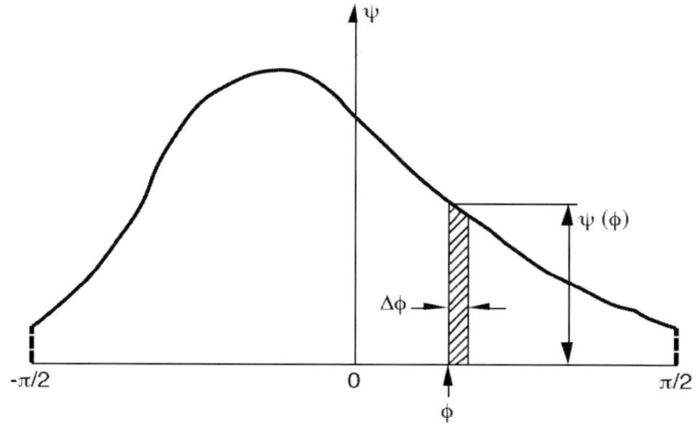

Figure 8.56: Differential element in an orientation distribution function.

Letting Δt and $\Delta\phi \rightarrow 0$ reduces eqn. (8.151) to

$$\frac{\partial \psi}{\partial t} = \frac{\partial}{\partial \phi}(\psi \dot{\phi}) \tag{8.152}$$

This expression is known as the *fiber density continuity equation*. It states that a fiber, which moves out of one angular position must move into a neighboring one, conserving the total number of fibers. If the initial distribution function, ψ_0, is known, an expression for the angular velocity of the particle, $\dot{\phi}$, must be found to solve for eqn. (8.152) and determine how the distribution function varies in time. The motion of the fibers can often be described by the motion of a rigid single rod in a planar flow.

A classic model that describes the motion of a rigid rod of finite aspect ratio and dilute suspensions is Jeffey's model [12]. In polymer processing, the dilute assumption is usually not a valid assumption. In compression molding, for example, in a charge with 20-50% fiber content by volume, the fibers are so closely packed that one cannot see through a resinless bed of fibers, even for very thin parts. This means that as a fiber rotates during flow, it bumps into its neighbors making the fiber-fiber interaction a major inhibitor of fiber rotation. Folgar and Tucker [7] derived a model for the orientation behavior of fibers in concentrated suspensions. For the case of planar flow Folgar and Tucker's model that describes the fiber rotation is written as

$$\dot{\phi} = \frac{C_I \dot{\gamma}}{\psi}\frac{\partial \psi}{\partial \phi} - \cos\phi \sin\phi \frac{\partial u_x}{\partial x} - \sin^2\phi \frac{\partial u_x}{\partial y} + \cos^2\phi \frac{\partial u_y}{\partial x} + \sin\phi\cos\phi \frac{\partial u_y}{\partial y} \quad (8.153)$$

Here, $\dot{\gamma}$ is the magnitude of the strain rate tensor and C_I is a phenomenological coefficient which models the interactions between the fibers, usually referred to as the *Folgar-Tucker interaction coefficient*. The coefficient varies between 0, for a fiber without interaction with its neighbors, and 1, for a closely packed bed of fibers. For a fiber reinforced polyester resin mat with 20-50% volume fiber content, C_I is usually between 0.03 and 0.06. When eqn. (8.153) is substituted into eqn. (8.152), the transient governing equation for fiber orientation distribution with fiber interaction built-in, becomes

$$\begin{aligned} \frac{\partial \psi}{\partial t} = &- C_I \dot{\gamma} \frac{\partial^2 \psi}{\partial \phi^2} - \frac{\partial \psi}{\partial \phi}\left(-\mathrm{sc}\frac{\partial u_x}{\partial x} - \mathrm{s}^2 \frac{\partial u_x}{\partial y} + \mathrm{c}^2 \frac{\partial u_y}{\partial x} + \mathrm{sc}\frac{\partial u_y}{\partial y}\right) \\ &- \psi \frac{\partial}{\partial \phi}\left(-\mathrm{sc}\frac{\partial u_x}{\partial x} - \mathrm{s}^2 \frac{\partial u_x}{\partial y} + \mathrm{c}^2 \frac{\partial u_y}{\partial x} + \mathrm{sc}\frac{\partial u_y}{\partial y}\right) \end{aligned} \quad (8.154)$$

where s and c represent $\sin\phi$ and $\cos\phi$, respectively.

The Folgar-Tucker model can easily be solved using finite difference methods. The above equation was solved using a Crank-Nicholson method for a uni-axial flow during compression molding process of sheet molding compound (SMC), where the only non-zero velocity gradient term is $\partial u_x / \partial x$. Assuming random initial fiber orientation distribution, $\psi_0 = 1/\pi$, the fiber orientation distribution functions were computed for different initial mold coverage and compared to experimental results. The model worked well when compared to the experiments. Figures 8.57, 8.58 and 8.59 compare the measured fiber orientation distributions to the calculated distributions using the Folgar-Tucker model for cases with 67%, 50%, and 33% initial charge mold coverage.

To illustrate the effect of fiber orientation on material properties of the final part, Fig. 8.60 [5] shows how the fiber orientation distributions that correspond to 67 50 and 33% initial mold coverage affect the stiffness of the finished plates. The Folgar-Tucker model has been implemented into various, commercially available compression mold filling simulation programs and successfully tested with several realistic compression molding applications.

8.9 CONCLUDING REMARKS

Although the finite difference method is a straight forward, easy method to numerically solve partial differential equations in transport phenomena, there are many problems when applying FDM to irregular geometries, moving boundaries, and complex three-dimensional systems. FDM is ideal when analyzing problems with simple geometries and it is often used in conjunction with other techniques such as finite elements. For example, mold filling

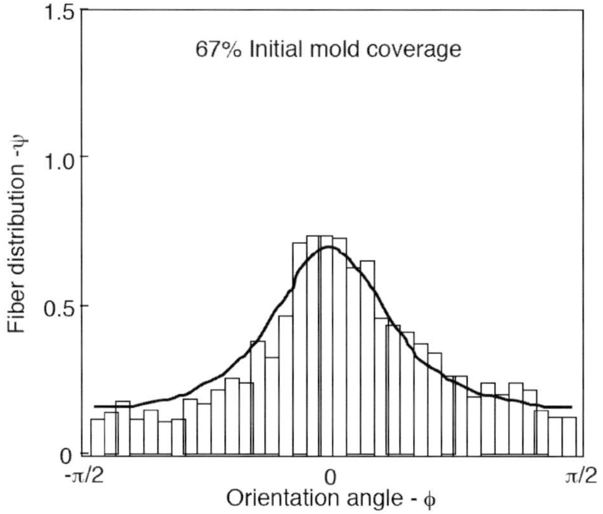

Figure 8.57: Predicted and experimental fiber orientation distribution function for a plate molded with an initial mold coverage of 67%.

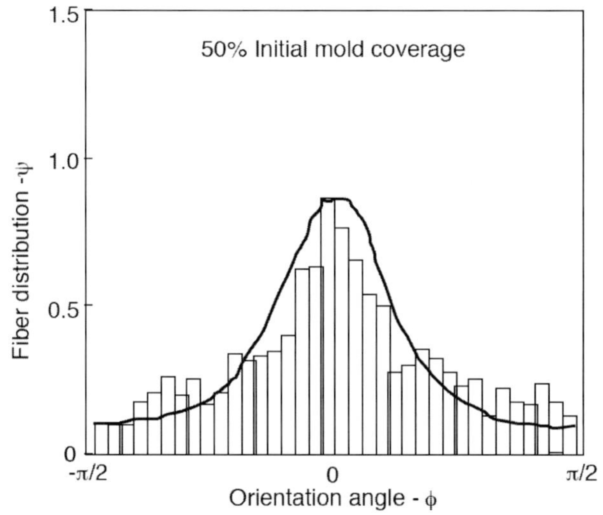

Figure 8.58: Predicted and experimental fiber orientation distribution function for a plate molded with an initial mold coverage of 50%.

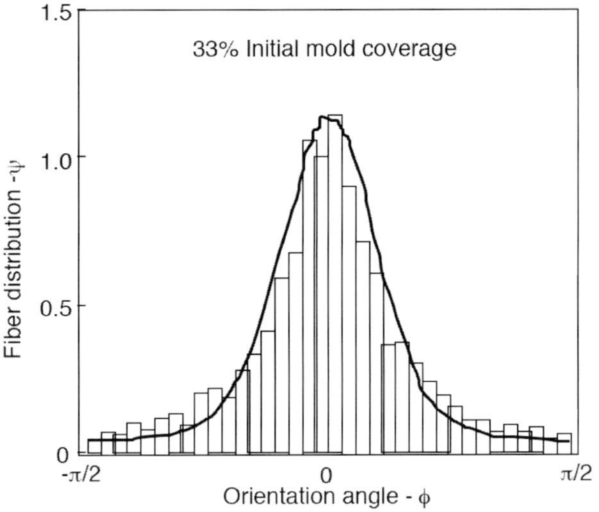

Figure 8.59: Predicted and experimental fiber orientation distribution function for a plate molded with an initial mold coverage of 33%.

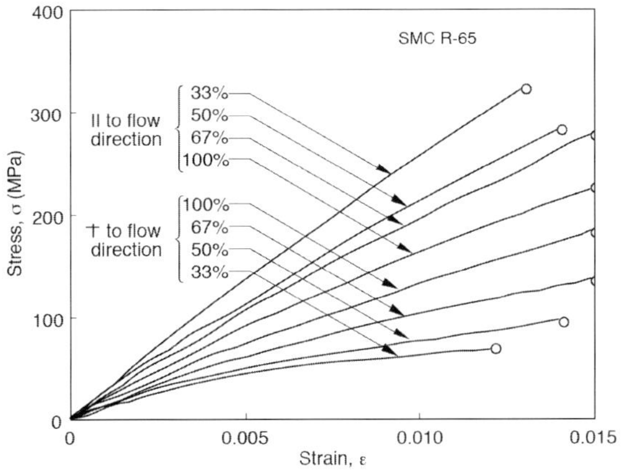

Figure 8.60: Stress-strain curves of 65% glass by volume SMC for various degrees of compression mold coverage[5].

simulations use the finite difference method to solve for the temperature profile needed in the Hele-Shaw approximation essentially solving a one-dimensional problem. On the other hand, the more complex 2D non-planar mold filling problem is solved for using the finite element method.

Problems

8.1 Use the Taylor tables to generate finite difference expressions for $\dfrac{du}{dx}$ and $\dfrac{d^2u}{dx^2}$.

a) Generate a third order approximation.

b) Generate a fourth order approximation.

8.2 Using the central difference approximation given by $\dfrac{du}{dx} \approx \dfrac{u_{i+1/2} - u_{i-1/2}}{\Delta x}$ do the following:

a) Use a second order polynomial interpolation to find $u_{i\pm1/2}$.

b) Use the Taylor table and determine the order of the approximation.

8.3 Using a power law model, determine the velocity profile in a pressure driven slit flow. Use Δp =1000 Pa, m =1000 Pa-s^n, $h = 1$ mm and a distance from entrance to exit of 1000 mm. Solve the problem using a 1D steady state finite difference solution for $n = 1$, 0.8, 0.6, 0.4, and 0.2.

8.4 Solve a coupled heat transfer and flow system of a temperature dependent viscosity melt that is driven by a pressure between two parallel plates. The viscosity is given by $\eta = \eta_0 e^{-a(T-T_0)}$. Use the dimensions given in the previous problem, where $\eta_0 = 1000$ Pa-s, $a = 0.04$ K^{-1}, the bottom plate temperature, T_0, of 140°C and the upper plate temperature of 160°C.

a) Find the temperature and velocity distributions between the plates.

b) Is viscous dissipation important in this problem?

8.5 Write a finite difference program that will compute the pressure and velocity fields for the 1 cm thick L-shaped charge depicted in Fig. 10.42. Assume a Newtonian viscosity of 500 Pa-s.

a) Plot the pressure distribution. You may use a software package or you may sketch it by hand.

b) Draw the nodal velocity vectors.

c) Will a knitline form in the re-entrant corner?

8.6 Write a small finite difference program to solve the one dimensional convection-diffusion example (8.59).

a) Use 10 (as in example), 20, 50 and 100 grid points and compare the error in the various solutions.

b) For the convection term, use grid points i, $i-1$ and $i-2$. How does the error compare to the simple backward difference solution?

8.7 Write an explicit finite difference program to compute the cooling of a 2 mm thick polystyrene ruler. Assume a thermal conductivity, k, of 0.15 W/m/K, a density, ρ, of 1050 kg/m^3 and a specific heat, C_p, before T_g of 800 J/kg/K and above T_g of 1,200 J/kg/K. Assume T_g =100°C, and initial temperature, T_0, of 150°C and a mold temperature, T_c, of 40°C.

a) Solve the problem with an average C_p=1000 J/kg/K.

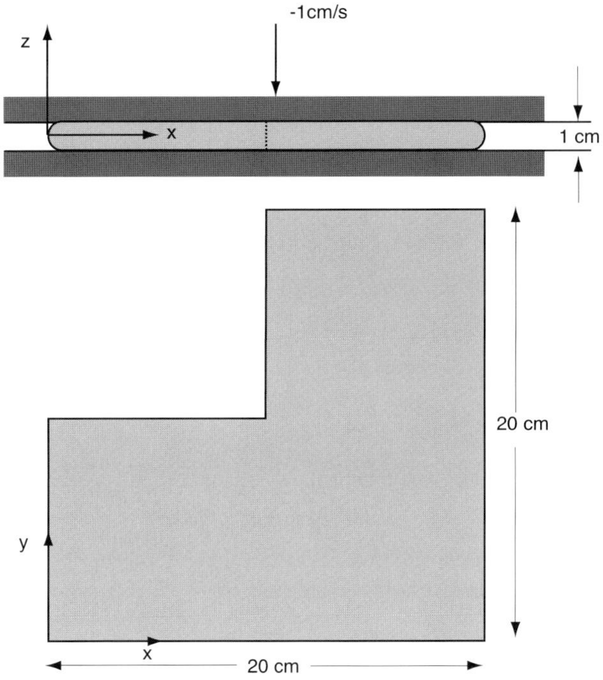

Figure 8.61: Schematic diagram of the compression molding of an L-shaped charge.

b) Solve the problem with a variable specific heat. How important is it to include the change in specific heat as the material goes through the glass transition temperature?

8.8 Write an explicit finite difference program to compute the cooling of a 2 mm thick polypropylene part. Assume a thermal conductivity, k, of 0.24 W/m/K, a density, ρ, of 905 kg/m^3 and a specific heat, C_p, of 1,930 J/kg/K. The melting temperature, T_m, is 160°C and the heat of fusion, λ, is 200 kJ/kg. Use the model that includes the heat of fusion, presented in Fig. 8.62, to compute the specific heat during the computation. Assume an initial temperature, T_0, of 180°C and a mold temperature, T_c, of 40°C.

a) Solve the problem with the model presented in the figure.

b) Solve the problem with a model for specific where the jump starts at 155°C and ends at 165°C. Compare the results.

8.9 Write a short explicit finite difference program to compute the fiber orientation function in a compression molding process with stretching only in the $x-$direction. Test the program with an example, where a 3 mm thick 50$\times$50 cm plate is compression molded with an initial mold coverage of 50%. Assume an interaction coefficient $C_I = 0.05$. Assume initial fiber orientation distribution that is random.

a) Determine the time step that will render an accurate solution.

b) Solve the problem with interaction coefficients of 0.01 and 0.1. How important is this factor?

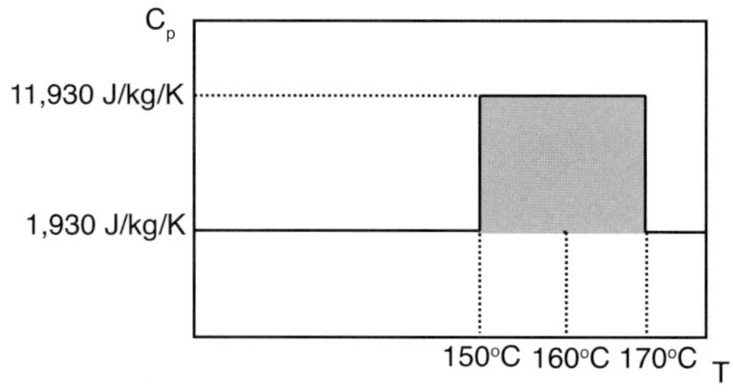

Figure 8.62: Model for the specific heat that includes the heat of fusion. Note that the gray area is equivalent to λ.

c) Solve for the interaction coefficient with initial mold coverages of 33%, 10% and a very thick charge with an initial mold coverage of 5%. Comment on your results.

8.10 Rework Example 8.10 to find clamping forces that are more realistic. Use an injection temperature of 240°C and a mold temprature of 90°C. Are your results reasonable?

REFERENCES

1. M.R. Barone and D.A. Caulk. *Int. J. Heat Mass Transfer*, 22:1021, 1979.
2. A. Bejan. *Heat Transfer*. John Wiley & Sons, New York, 1993.
3. A. Bejan. *Convection Heat Transfer*. John Wiley & Sons, New York, 1995.
4. E. Broyer, C. Gutfinger, and Z. Tadmor. A theoretical model for the cavity filling process in injection molding. *J. of Rheology*, 9(3):423, 1975.
5. C.Y. Chen and C.L. Tucker III. *Reinf. Plast. Comp*, 3:120, 1984.
6. W.M. Deen. *Analysis of Transport Phenomena*. Oxford University Press, Oxford, 1998.
7. F.P. Folgar and C.L. Tucker III. *Reinf. Plast. Comp.*, 3:98, 1984.
8. R.M. Griffith. Fully developed flow in screw extruders. *I & EC Fundamentals*, 1(3):180–187, 1962.
9. J.P. Hernandez-Ortiz and T.A. Osswald. A novel cure reaction model fitting technique based on dsc scans. *J. Polym. Eng.*, 25(1):23, 2005.
10. F.B. Hildebrand. *Introduction to Numerical Analysis*. Dover Publishers, New York, 2nd edition, 1974.
11. F.P. Incropera and D.P. DeWitt. *Fundamentals of Heat and Mass Transfer*. John Wiley & Sons, New York, 1996.
12. G.B. Jeffery. *Proc. Roy. Soc.*, A102:161, 1922.
13. M.R. Kamal. *Polym. Eng. Sci.*, 14:231, 1979.
14. M.R. Kamal and S. Sourour. *Polym. Eng. Sci.*, 13:59, 1973.
15. W.L. Krueger and Z. Tadmor. Injection molding into recangular cavity containing inserts. In *SPE ANTEC*, page 87, 1978.
16. W.E. Milne. *Numerical Solution of Differential Equations*. Dover Publications, New York, 1970.

17. T. A. Osswald and G. Menges. *Material Science of Polymers for Engineers.* Hanser Publishers, Munich, 2nd edition, 2003.

18. S.V. Patankar. *Numerical Heat Transfer and Fluid Flow.* Hemisphere Publishers, New York, 1980.

19. W. H. Press, S. A. Teukolsky, W. T. Vetterling, and B. P. Flannery. *Numerical Recipes in Fortran 77.* Cambridge, 2nd edition, 1992.

20. P.J. Roache. *Computational Fluid Mechanics.* Hermosa Publishers, New Mexico, 1982.

21. H. Schlichting and K. Gersten. *Boundary Layer Theory.* Springer, Berlin, 8th edition, 2000.

22. J.F. Stevenson. *Polym. Eng. Sci*, 18:577, 1978.

23. G. Strang. *Introduction to Applied Mathematics.* Wellesley-Cambridge Press, Wellesley, 1986.

24. Z. Tadmor, E. Broyer, and C. Gutfinger. Flow analysis network (fan) - a method for solving flow problems in polymer processing. *Poly. Eng. Sci.*, 14(9):660, 1974.

25. Z. Tadmor and C.G. Gogos. *Principles of Polymer Processing.* John Wiley & Sons, New York, 1979.

26. J.C. Tannehill, D.A. Anderson, and R.H. Pletcher. *Computational Fluid Mechanics and Heat Transfer.* Taylor & Francis, Washington, 2nd edition, 1997.

27. H.K. Versteeg and W. Malalasekera. *An Introduction to Computational Fluid Dynamics: The Finite Volume Method.* Addison-Wesley Pub., New York, 1996.

28. K.K. Wang, S.F. Shen, J.F. Stevenson, C.A. Hieber, and S. Chung. Computer-aided injection molding. NSF report, College on Engineering, New York, 1975.

CHAPTER 9

FINITE ELEMENT METHOD

A round man cannot be expected to fit in a square hole right away. He must have time to modify his shape.

—Mark Twain

The finite element method (FEM) was first developed in 1956 to numerically analyze stress problems [16] for the design of aircraft structures. Since then it has been modified to solve more general problems in solid mechanics, fluid flow, heat transfer, among others. In fact, due to its versatility, the method is being used to study coupled problems for applications with complex geometries where the solutions are highly non-linear.

9.1 ONE-DIMENSIONAL PROBLEMS

Instead of starting with a rigorous and mathematical development of the finite element technique, we proceed to present the finite element method through a solution of one-dimensional applications. To illustrate the technique, we will first find a numerical solution to a heat conduction problem with a volumetric heat source

$$k\frac{\partial^2 T}{\partial x^2} = \dot{Q} \tag{9.1}$$

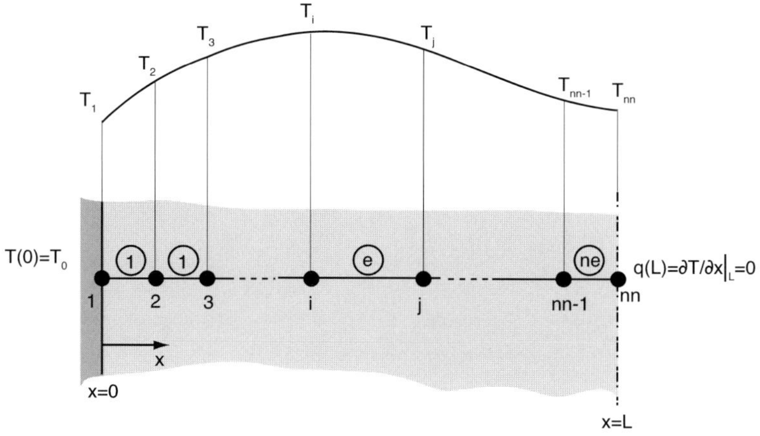

Figure 9.1: One dimensional finite element discretization.

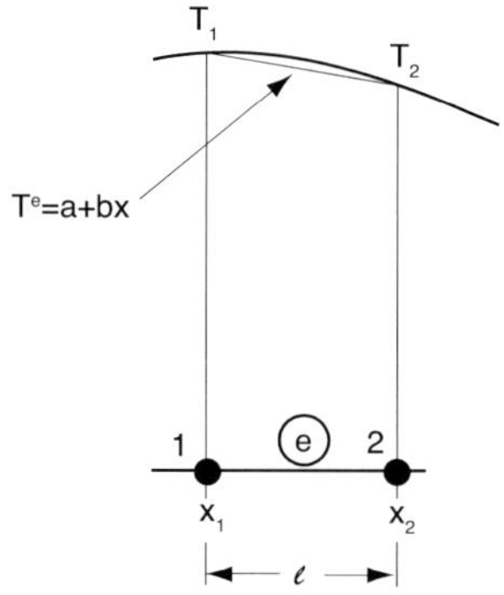

Figure 9.2: Schematic of a two-noded 1D linear element.

with a Dirichlet boundary condition, $T(0) = T_0$, and a Neumann boundary condition $q(L) = \partial T/\partial x = 0$. The internal heat source can be viscous dissipation during shear flow in a single screw extruder, as will be illustrated with an example at the end of this section.

9.1.1 One-Dimensional Finite Element Formulation

Our first step in the development of the technique is to discretize the domain by creating a mesh. Here, we will divide the domain into elements as illustrated in Fig. 9.1. The figure presents nodes and elements (element numbers are shown inside circles).

Since we chose a discretization with two-noded elements, we are assuming that within each element the unknown temperature function $T(x)$ is a simple linear polynomial that can be written as

$$T^e = a + bx \tag{9.2}$$

If we assume that each individual element has its own local numbering, as depicted in Fig. 9.2, we can evaluate eqn. (9.2) for each node using their respective nodal coordinates x_1 and x_2, with their corresponding temperatures T_1 and T_2

$$\begin{aligned} T_1 &= a + bx_1 \\ T_2 &= a + bx_2 \end{aligned} \tag{9.3}$$

Equation (9.3) can be written in matrix form as

$$\begin{pmatrix} T_1 \\ T_2 \end{pmatrix} = \begin{bmatrix} 1 & x_1 \\ 1 & x_2 \end{bmatrix} \begin{pmatrix} a \\ b \end{pmatrix} \tag{9.4}$$

which can be inverted to become

$$\begin{pmatrix} a \\ b \end{pmatrix} = \begin{bmatrix} 1 & x_1 \\ 1 & x_2 \end{bmatrix}^{-1} \begin{pmatrix} T_1 \\ T_2 \end{pmatrix} = \frac{1}{x_2 - x_1} \begin{bmatrix} x_2 & -x_1 \\ -1 & 1 \end{bmatrix} \begin{pmatrix} T_1 \\ T_2 \end{pmatrix} \tag{9.5}$$

If we write eqn. (9.2) in matrix form

$$T^e = \begin{bmatrix} 1 & x \end{bmatrix} \begin{pmatrix} a \\ b \end{pmatrix} \tag{9.6}$$

we can substitute eqn. (9.5) into (9.6), and letting $\ell = x_2 - x_1$, we get

$$T^e = \begin{bmatrix} 1 & x \end{bmatrix} \begin{bmatrix} \dfrac{x_2}{\ell} & -\dfrac{x_1}{\ell} \\ -\dfrac{1}{\ell} & \dfrac{1}{\ell} \end{bmatrix} \begin{pmatrix} T_1 \\ T_2 \end{pmatrix} \tag{9.7}$$

The above equation can be written in terms of Lagrange interpolation formula (Chapter 7)

$$T^e = N_1 T_1 + N_2 T_2 \tag{9.8}$$

where the interpolation functions N_1 and N_2 are defined by

$$\begin{aligned} N_1 &= \frac{x_2 - x}{\ell} \\ N_2 &= \frac{x - x_1}{\ell} \end{aligned} \tag{9.9}$$

These functions, which are commonly referred to as *shape functions*, have the unique properties that they equal 1 when evaluating them at their corresponding node, and zero when evaluating them at the opposite node, as shown in Fig. 9.3.

Now, that we have an expression of the approximated element temperature in terms of geometric parameters and the nodal values for temperature, we proceed to insert this approximation into the original governing equation, i.e.,

$$k \frac{\partial^2 T^e}{\partial x^2} \approx \dot{Q(x)} \tag{9.10}$$

or

$$k \frac{\partial^2}{\partial x^2} \left(\begin{bmatrix} N_1 & N_2 \end{bmatrix} \begin{pmatrix} T_1 \\ T_2 \end{pmatrix} \right) \approx \dot{Q(x)} \tag{9.11}$$

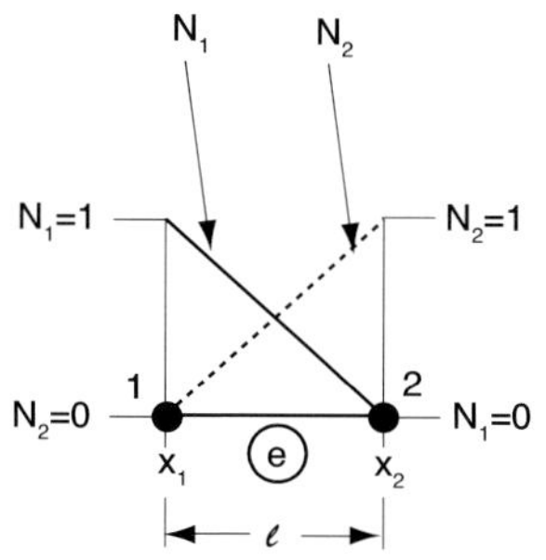

Figure 9.3: Shape functions in a two-noded 1D linear element.

> **Theorem 9.1.1.** ***Galerkin's Theorem:*** *Galerkin's theorem of weighted residuals. To generate n equations for the n unknowns in the function $\mathcal{F}(\phi_1, \phi_2, \phi_3 ... \phi_n)$ we must,*
>
> $$\int_V \mathcal{F}(\phi_1, \phi_2, \phi_3 ... \phi_n) W_i dV$$
>
> *where $W_1, W_2, W_3, ... W_i$ are independent weighting functions.*

When we inspect this equation we realize that there are two functional problems. First, there is one equation per element with two unknowns. Second, we are using a linear approximation for the temperature; however, we have second spatial derivatives of temperature. The first problem is solved by using *Garlekin's method of weighted residuals* (Theorem (9.1.1)).

Hence, we integrate eqn. (9.11) over the domain, V, and use two weighting functions W_1 and W_2, which results in

$$k \int \frac{\partial^2}{\partial x^2} \left(\begin{bmatrix} N_1 & N_2 \end{bmatrix} \begin{pmatrix} T_1 \\ T_2 \end{pmatrix} \right) W_i dx \approx \int \dot{Q}(x) W_i dx \tag{9.12}$$

where $i = 1, 2$. A set of ideal weighting functions that satisfy Garlerkin's theorem of weighted residuals are the shape functions, N_i, described above. The above equation can now be written as

$$k \int \frac{\partial^2 T^e}{\partial x^2} N_i dx \approx \int \dot{Q}(x) N_i dx \tag{9.13}$$

which for each element represents two equations with two unknowns. The second problem is solved by applying the *Green-Gauss transformation* (Theorem (9.1.2)).

Using the Green-Gauss transformation, eqn (9.13) can be transformed to

$$k \int \frac{\partial^2 T^e}{\partial x^2} N_i dx = -k \int \frac{\partial T^e}{\partial x} \frac{\partial N_i}{\partial x} dx + \oint_S N_i \frac{\partial T}{\partial x} n_x dS \tag{9.14}$$

Examining the last integral of this equation we deduce that it is equal to zero since at $x = 0$ temperature is known and no FEM equation is required, and at $x = L$, $\partial T / \partial x = 0$. The

Theorem 9.1.2. ***Green's Theorem:*** *Integration by parts*

$$\int_V \phi \frac{\partial \psi}{\partial x} dV = - \int_V \frac{\partial \phi}{\partial x} \psi dV + \oint_S \phi \psi n_x dS$$
$$\int_V \phi \frac{\partial \psi}{\partial y} dV = - \int_V \frac{\partial \phi}{\partial y} \psi dV + \oint_S \phi \psi n_y dS$$
$$\int_V \phi \frac{\partial \psi}{\partial z} dV = - \int_V \frac{\partial \phi}{\partial z} \psi dV + \oint_S \phi \psi n_z dS$$

The above can be written in generalized form as,

$$\int_V \phi \frac{\partial \psi}{\partial \mathbf{n}} dV = - \int_V \frac{\partial \phi}{\partial \mathbf{n}} \psi dV + \oint_S \phi \psi \cdot \mathbf{n} dS$$

finite element equation for this problem becomes

$$k \int \frac{\partial T^e}{\partial x} \frac{\partial N_i}{\partial x} dx = - \int \dot{Q} N_i dx \tag{9.15}$$

Since we dropped the last term in the equation, we are satisfying the adiabatic boundary condition (Neumann), $q(L) = 0$. On the other hand, we still must consider the Dirichlet boundary condition, $T(0) = T_0$. Since the Neumann boundary conditions is automatically satisfied, while the Dirichlet must be enforced, in the finite element language they are usually referred to as *natural* and *essential* boundary conditions.

Equation (9.15) is the final finite element expression of the one-dimensional heat equation problem. The temperature gradient in that equation can be expressed as

$$\frac{\partial T^e}{\partial x} = \begin{bmatrix} \frac{\partial N_1}{\partial x} & \frac{\partial N_2}{\partial x} \end{bmatrix} \begin{pmatrix} T_1 \\ T_2 \end{pmatrix} \tag{9.16}$$

which can be written as,

$$\frac{\partial T^e}{\partial x} = \mathbf{B} \begin{pmatrix} T_1 \\ T_2 \end{pmatrix} \tag{9.17}$$

where $\mathbf{B}$ is simply referred to as the *B-matrix*, and by differentiating the shape functions with respect to x, for the linear 1D element used here, is defined by

$$\mathbf{B} = \begin{bmatrix} -\frac{1}{\ell} & \frac{1}{\ell} \end{bmatrix} \tag{9.18}$$

Equation (9.15) can now be written as

$$k \left[\int \mathbf{B}^T \mathbf{B} dx \right] \begin{pmatrix} T_1 \\ T_2 \end{pmatrix} = - \int \dot{Q} \begin{pmatrix} N_1 \\ N_2 \end{pmatrix} dx \tag{9.19}$$

At this point, we have finished formulating the finite element to solve the problem of interest. Table 9.1 summarizes the steps that were followed to develop this finite element model. The scheme applies for all finite element derivation.

Table 9.1: Procedure to develop a finite element model

Governing equations:	$\mathcal{L}\phi = b(\mathbf{x}, t)$
Initial and boundary conditions:	$\phi(\mathbf{x}, 0) = \phi_0$ $\phi(\mathbf{x}, t) = \bar{\phi}$ for $\mathbf{x} \in S_1$ *(Essential)* $\partial\phi(\mathbf{x}, t)/\partial\mathbf{n} = \bar{q}$ for $\mathbf{x} \in S_2$ *(Natural)*
1. Choose an element to divide the domain into	*Shape functions* N_i
2. Approximate the element variables:	$\phi^e = \sum \phi_i N_i$
3. Galerkin-weighted residual expression:	$\int_V (\mathcal{L}\phi^e - b) N_i dV = 0$ Theorem (9.1.1)
4. Green-Gauss transformation	Theorem (9.1.2)

9.1.2 Numerical Implementation of a One-Dimenional Finite Element Formulation

Once a finite element formulation has been implemented in conjunction with a specific element type – either 1D, 2D or 3D – the task left is to numerically implement the technique and develop the computer program to solve for the unknown primary variables – in this case temperature. Equation (9.19) is a form that becomes very familiar to the person developing finite element models. In fact, for most problems that are governed by Poisson's equation, problems solving displacement fields in stress-strain problems and flow problems such as those encountered in polymer processing, the finite element equation system takes the form presented in eqn. (9.19). This equation is always re-written in the form

$$[\mathbf{K^e}] \cdot (\mathbf{T^e}) = (\mathbf{f^e}) \tag{9.20}$$

where $\mathbf{K^e}$ is called the *element stiffness matrix*, $\mathbf{T^e}$ the *displacement vector* and $\mathbf{f^e}$ the *force vector*, according to the original applications of FEM in the field of elasticity. In relation to eqn. (9.19), the terms are defined as

$$\begin{aligned} [\mathbf{K^e}] &= k \int [\mathbf{B}]^T [\mathbf{B}] dx \\ (\mathbf{f^e}) &= \int \dot{Q} \mathbf{N} dx \end{aligned} \tag{9.21}$$

Fron the definition of $\mathbf{B}$ in eqn. (9.18) we can see that for our linear element all the terms in the stiffness matrix are constants, and can be easily integrated

$$[\mathbf{K^e}] = k \int_{x_{\text{node-1}}}^{x_{\text{node-2}}} \begin{bmatrix} 1/l_e^2 & -1/l_e^2 \\ -1/l_e^2 & 1/l_e^2 \end{bmatrix} dx = k \begin{bmatrix} 1/l_e & -1/l_e \\ -1/l_e & 1/l_e \end{bmatrix} \tag{9.22}$$

The next step in the numerical implementation is to find a relation between each individual element stiffness matrix, element displacement vector and element force vector and the *global system*. Hence, we must generate a global stiffness matrix, global displacement vector and global force vector. This step in the implementation is actually only *data management* or *book keeping*.

To illustrate this concept, we make use of the small system with 4 elements and 5 nodal points, schematically presented in Fig. 9.4. To better demonstrate the data management techniques when *assembling the global system of equations* we have not enumerated the

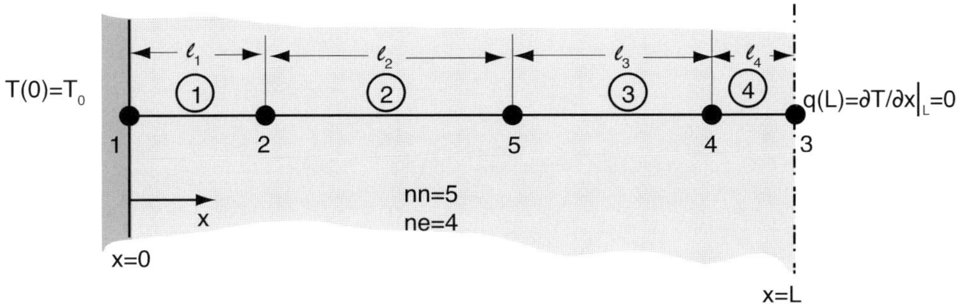

Figure 9.4: Schematic of a 1D domain finite element discretization with 4 linear elements and 5 nodal points.

nodes sequentially, as needed to reduce storage requirements and consequently computation time.

The next step is to generate the element stiffness matrices and force vectors. Symbolically, the 4 element stiffness matrices are written as

$$[\mathbf{K^1}] = k\begin{bmatrix} 1/\ell_1 & -1/\ell_1 \\ -1/\ell_1 & 1/\ell_1 \end{bmatrix}, [\mathbf{K^2}] = k\begin{bmatrix} 1/\ell_2 & -1/\ell_2 \\ -1/\ell_2 & 1/\ell_2 \end{bmatrix}$$

$$[\mathbf{K^3}] = k\begin{bmatrix} 1/\ell_3 & -1/\ell_3 \\ -1/\ell_3 & 1/\ell_3 \end{bmatrix}, [\mathbf{K^4}] = k\begin{bmatrix} 1/\ell_4 & -1/\ell_4 \\ -1/\ell_4 & 1/\ell_4 \end{bmatrix}$$

and the element force vectors as

$$(\mathbf{F^1}) = \begin{pmatrix} \frac{c\ell_1}{2} \\ \frac{c\ell_1}{2} \end{pmatrix} \qquad (\mathbf{F^2}) = \begin{pmatrix} \frac{c\ell_2}{2} \\ \frac{c\ell_2}{2} \end{pmatrix}$$

$$(\mathbf{F^3}) = \begin{pmatrix} \frac{c\ell_3}{2} \\ \frac{c\ell_3}{2} \end{pmatrix} \qquad (\mathbf{F^4}) = \begin{pmatrix} \frac{c\ell_4}{2} \\ \frac{c\ell_4}{2} \end{pmatrix}$$

To use computer storage more efficiently, the vector of unknown temperatures will eventually be stored in the *global force vector*, **f**. The next steps in the finite element procedure (Table 9.1) will be to form the global stiffness matrix and force vector, and to solve the resulting linear system of algebraic equations, as presented in Algorithm 5.

The characteristics of our 1D problem serve to generalize the finite element methodology, because, as we can imagine, for a complex two- or three-dimensional geometry it is impossible to generate isotropic meshes with nodes in sequential order. The heart of the data management schemes used by finite element programs is the so-called *connectivity matrix*, nee(i,j). This matrix describes, which nodes form a specific element – essentially giving the element connectivity. For the mesh presented in Fig. 9.4 the connectivity matrix

Algorithm 5 FE program

```
program FEM
  call input(xi,nee,bc)                 coordinates, connectivity and
                                                boundary conditions
  call assem-global(kglobal,fglobal)   Assemble global stiffness matrix
                                                and force vector
  call boundary-conditions
  call solve(kglobal,fglobal,solution)
  call out-file
end program FEM
```

is written as

$$\text{nee}(i,j) = \begin{bmatrix} 1 & 2 \\ 2 & 5 \\ 5 & 4 \\ 4 & 3 \end{bmatrix} \tag{9.23}$$

where the row number denotes the element number and the two numbers on that row represent node 1 and node 2 for that specific element. For example, element 3 is formed by nodes 5 and 4. The number of columns will vary according to the number of nodes in the element. A three-noded triangle will have 2 columns, a 20-noded brick, 20 columns, and so forth. Thus, subindex i varies according to the element number, ($i = 1, ..., 4$), and j according to the number of nodes per element, i.e., ($j = 1, 2$). Hence, this matrix is general for any type of element and any dimensional order.

The global stiffness matrix and force vector, which represent our equation system are formed by direct addition of the element stiffness matrices and force vectors. The corresponding position of an element component in the global system is given by the connectivity matrix. In two- or three-dimensional problems the positions are related to the connectivity matrix as well as the direction under consideration. The *global stiffness matrix* and *force vector assembly* technique is presented in Algorithm 6.

Although this algorithm is clear and simple, it presents the most ineffective way of storing the global stiffness matrix since it results in a full sparse matrix. Later in this section we will discuss how the storage space and computation time is minimized by using alternative storing schemes such as *banded matrices*.

For the 4 element system presented in Fig. 9.4, the global stiffness is given by

$$[\mathbf{K}] = k \begin{bmatrix} \ell_1/2 & -\ell_1/2 & 0 & 0 & 0 \\ -\ell_1/2 & \ell_1/2+\ell_2/2 & 0 & 0 & -\ell_2/2 \\ 0 & 0 & \ell_4/2 & -\ell_4/2 & 0 \\ 0 & 0 & -\ell_4/2 & \ell_3/2+\ell_4/2 & -\ell_3/2 \\ 0 & -\ell_2/2 & 0 & -\ell_3/2 & \ell_2/2+\ell_3/2 \end{bmatrix} \tag{9.24}$$

Algorithm 6 Global stiffness matrix and force vector assembly

```
subroutine assem-global(kglobal,fglobal)
  kglobal = 0; fglobal = 0
  do i = 1,ne                                                    Loop all elements
    call elem-k-f(kelem,felem)                     element stiffness matrix and
                                                                     force vector

                                              The following do loop adds the element
                                        stiffness matrices into the global stiffness
                                            The addition algorithm presented results
                                                                   in a full matrix
    do j = 1, nelem                                      nelem = nodes per element
      irow = nee(i,j)
      fglobal(irow) = fglobal(irow) + felem(j)               force vector addition
      do k = 1,nelem
        icol = nee(i,k)
        kglobal(irow,icol) = kglobal(irow,icol) + kelem(j,k)     stiffness matrix
                                                                         addition
      enddo
    enddo
  enddo
end subroutine assem-global
```

and the global force vector is

$$(\mathbf{F}) = \begin{pmatrix} \frac{c\ell_1}{2} \\ \frac{c\ell_1}{2} + \frac{c\ell_2}{2} \\ \frac{c\ell_3}{2} \\ \frac{c\ell_3}{2} + \frac{c\ell_4}{2} \\ \frac{c\ell_2}{2} + \frac{c\ell_3}{2} \end{pmatrix} \tag{9.25}$$

At this point, it is important to point out that the equations are coupled only in those locations where two nodes share the same element. Nodes that do not share the same element are uncoupled and their corresponding positions in the global stiffness matrix is zero. Systems discretized into a large number of elements result in a large number of nodes that are uncoupled, which leads to very *sparce matrices*, i.e., matrices filled with many non-diagonal zeros. This is a well known attribute of FE equation systems and contributes to reduced computation times during the solution process. Consequently, the number of non-zero non-diagonal terms can be reduced as much as possible by maintaining the element nodal point numbers as close as possible, i.e., optimizing the connectivity matrix, eqn. (9.23), in such a way that the difference between the numbers in a specific row is minimized. This way the matrix will have most of the non-zero terms collapsed within a band near the diagonal. This type of matrix is very common when dealing with finite elements and is generally referred to a *banded matrix*. As will be discussed below, there are very efficient banded matrix storage schemes and solvers, minimizing computation times.

Before we proceed to our discussion of global stiffness matrix storage schemes, we will discuss the last aspect of the finite element implementation, namely, the application of the boundary conditions. As discussed earlier, the *natural boundary conditions* are imbedded in the finite element equation system – it is implied that every boundary node without an

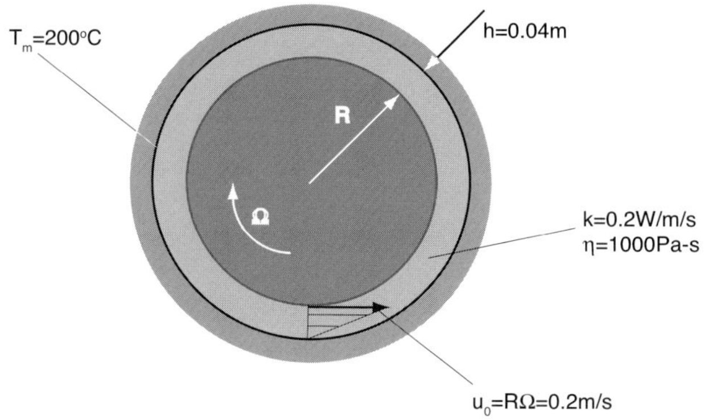

Figure 9.5: Schematic diagram of a Couette device.

essential boundary condition has a natural boundary condition. In our case, the essential boundary condition is the prescribed temperature on node 1. However, in two- and three-dimensional systems there may be many nodes (at least one), at which essential boundary conditions must be applied. For such nodes, the equations must be modified by adding the product of the known temperatures and their corresponding coefficients to the global force vector. Finally, the rows and columns of the corresponding nodes are eliminated by setting them to zero (since they already exist in the force vector), the diagonal is set to one and the component of the force vector is set equal to the applied boundary condition. The scheme that is used to apply boundary conditions is presented in Algorithm 7.

Algorithm 7 Boundary Conditions

```
subroutine boundary-conditions
  do k = 1,nbc
    i = ibc(k)
    fglobal = fglobal - kglobal(:,i)*bc(k)              modify the force vector
    kglobal(i,:)  = 0.0; kglobal(:,i) = 0.0      eliminate the columns and rows
                                                of the known boundary condition
    kglobal(i,i) = 1.0; fglobal(i) = bc(k)      boundary condition is T =bc(k)
  enddo
end subroutine boundary-conditions
```

■ **EXAMPLE 9.1.**

Temperature distribution in a polymer confined in a narrow-gap Couette device. To illustrate the techniques presented in the last sections, in this example we will model the heat transfer within a Couette device shown in Fig. 9.5. In the analysis we will assume that viscous dissipation plays a significant role and we are seeking the temperature profile across the gap with the effects of viscous heating.

If we assume a small gap, $h \ll R$, the energy equation is reduced to a one-dimensional balance between the conduction and the viscous dissipation, schematically depicted in Fig. 9.6. The governing equation becomes

$$k\frac{d^2T}{dx^2} + \dot{Q}_{\text{viscous heating}} = 0 \tag{9.26}$$

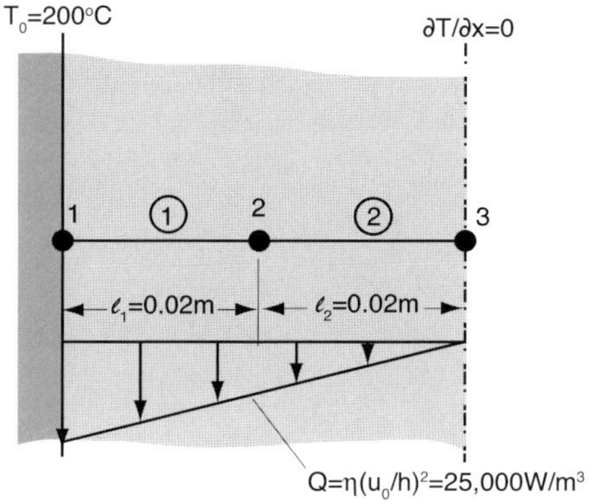

Figure 9.6: Simplified model of the Couette device of Fig. 9.5 with its finite element discretization.

where the viscous heating is given by

$$\dot{Q}_{\text{viscous heating}} = \eta \left(\frac{du_y}{dx} \right)^2 \tag{9.27}$$

This problem has an analytical solution (see Chapter 5) defined by

$$T(x) = T_m + \frac{\eta}{2k} \frac{u_0^2}{h^2} \left(2hx - x^2\right) \tag{9.28}$$

where $u_0 = R\Omega$.

The dimensions, properties and processing variables are $h = 0.04$ m, $\eta = 1000$ Pa-s, $k = 0.2$ J/m/K, $u_0 = 0.2$ m/s and $T_m = 200^o$C, respectively. Viscous heating can be estimated from the above values as

$$\dot{Q} \approx \eta \left(\frac{\Delta u_y}{\Delta x} \right)^2 = 25.000 \text{ W/m}^3 \tag{9.29}$$

The element stiffness matrix, $\mathbf{K^e}$, for the two elements in this problem are given by

$$[\mathbf{K^e}] = k \begin{bmatrix} 1/\ell & -1/\ell \\ -1/\ell & 1/\ell \end{bmatrix} = \begin{bmatrix} 10 & -10 \\ -10 & 10 \end{bmatrix} \tag{9.30}$$

and the element force vector is calculated as

$$(\mathbf{f^e}) = \begin{pmatrix} \dfrac{\dot{Q}\ell_e}{2} \\ \dfrac{\dot{Q}\ell_e}{2} \end{pmatrix} = \begin{pmatrix} 250 \\ 250 \end{pmatrix} \tag{9.31}$$

Using the connectivity matrix for this system given by

$$\text{nee}(i,j) = \begin{bmatrix} 1 & 2 \\ 2 & 3 \end{bmatrix} \tag{9.32}$$

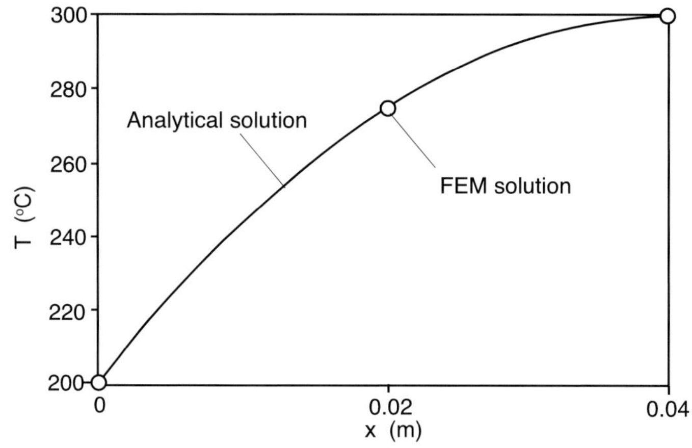

Figure 9.7: Comparison between analytical and finite element temperature distributions across the gap in a Couette devise with viscous heating .

and the algorithms presented above, we can assemble the global stiffness matrix and the force vector to become

$$\begin{bmatrix} 10 & -10 & 0 \\ -10 & 20 & -10 \\ 0 & -10 & 10 \end{bmatrix} \begin{pmatrix} T_1 \\ T_2 \\ T_3 \end{pmatrix} = \begin{pmatrix} 250 \\ 500 \\ 250 \end{pmatrix} \tag{9.33}$$

Since the Dirichlet (essential) boundary condition is that $T_1 = 200$, and we must complete the system above by implementing this condition applying the boundary condition algorithm

$$\begin{bmatrix} 1 & 0 & 0 \\ 0 & 20 & -10 \\ 0 & -10 & 10 \end{bmatrix} \begin{pmatrix} T_1 \\ T_2 \\ T_3 \end{pmatrix} = \begin{pmatrix} 200 \\ 500 - (200)(-10) \\ 250 - (200)(0) \end{pmatrix}$$

$$\begin{bmatrix} 1 & 0 & 0 \\ 0 & 20 & -10 \\ 0 & -10 & 10 \end{bmatrix} \begin{pmatrix} T_1 \\ T_2 \\ T_3 \end{pmatrix} = \begin{pmatrix} 200 \\ 2500 \\ 250 \end{pmatrix} \tag{9.34}$$

The Neumann (natural) boundary condition $q_x = 0$ is automatically satisfied. The above system of algebraic equations can easily be solved to give $T_1 = 200$, $T_2 = 275$ and $T_3 = 300$. A comparison between the analytical finite element solutions is shown in Fig. 9.7. As can be seen, the agreement is excellent.

9.1.3 Matrix Storage Schemes

There are many matrix storage schemes that can be used to minimize computer storage usage and computation time. One that is widely accepted is the *banded matrix storage scheme*. Here, we will restrict our discussion to this scheme. However, some problems in polymer processing, especially when dealing with three-dimensional flows, require very large amounts of storage, and consequently computation time. The latter is especially true when *core memory* has been exceeded and the computer is forced to physically swap

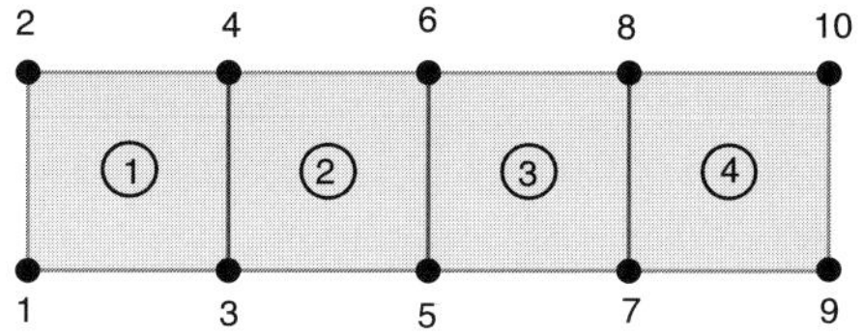

Figure 9.8: Finite element mesh with 4 elements and 10 nodes.

memory back and forth between core and hard disc. Once the computer is resorting to memory swapping, the computational efficiency of a system is compromised.

Finite element stiffness matrices are always symmetric and banded. For example, the mesh presented in Fig. 9.8 has a stiffness matrix given by

$$[\mathbf{K}] = \begin{bmatrix} K_{11} & K_{12} & K_{13} & K_{14} & 0 & 0 & 0 & 0 & 0 & 0 \\ K_{21} & K_{22} & K_{23} & K_{24} & 0 & 0 & 0 & 0 & 0 & 0 \\ K_{31} & K_{32} & K_{33} & K_{34} & K_{35} & K_{36} & 0 & 0 & 0 & 0 \\ K_{41} & K_{42} & K_{43} & K_{44} & K_{45} & K_{46} & 0 & 0 & 0 & 0 \\ 0 & 0 & K_{53} & K_{54} & K_{55} & K_{56} & K_{57} & K_{58} & 0 & 0 \\ 0 & 0 & K_{63} & K_{64} & K_{65} & K_{66} & K_{67} & K_{68} & 0 & 0 \\ 0 & 0 & 0 & 0 & K_{75} & K_{76} & K_{77} & K_{78} & K_{79} & K_{710} \\ 0 & 0 & 0 & 0 & K_{85} & K_{86} & K_{87} & K_{88} & K_{89} & K_{810} \\ 0 & 0 & 0 & 0 & 0 & 0 & K_{97} & K_{98} & K_{99} & K_{910} \\ 0 & 0 & 0 & 0 & 0 & 0 & K_{107} & K_{108} & K_{109} & K_{1010} \end{bmatrix} \tag{9.35}$$

The matrix presented in the above equation is clearly a banded matrix, with a *bandwidth* of 4. Note that the bandwidth is the maximum difference between node numbers of the elements of a given mesh times the number of degrees of freedom per node. Algorithm 8 computes the bandwidth of any mesh with $nelem$ nodes per element.

Algorithm 8 Compute the bandwidth of a banded Global Stiffness matrix

```
subroutine band(nee)
  nband = 0
  do i = 1, ne                                          Loop all elements
    do j = 1, nelem                               nelem = nodes per element
      do k = 1, nelem
        idiff = abs(nee(i,j) - nee(i,k))
        if ( idiff .gt.  nband ) nband = idiff
      enddo
    enddo
  enddo
  nband = nband + 1
end subroutine band
```

By examining the matrix it becomes obvious that within the upper part of the matrix, above or to the right of the bandwidth, all terms are zero and therefore do not need to be stored, or operated on during the solution process. In addition, since the stiffness matrix is symmetric, we do not need to store the stiffness matrix components that are below the diagonal. Hence, a more effective way of storing the above matrix is as

$$[\mathbf{K}] = \begin{bmatrix} K_{11} & K_{12} & K_{13} & K_{14} \\ K_{22} & K_{23} & K_{24} & 0 \\ K_{33} & K_{34} & K_{35} & K_{36} \\ K_{44} & K_{45} & K_{46} & 0 \\ K_{55} & K_{56} & K_{57} & K_{58} \\ K_{66} & K_{67} & K_{68} & 0 \\ K_{77} & K_{78} & K_{79} & K_{710} \\ K_{88} & K_{89} & K_{810} & 0 \\ K_{99} & K_{910} & 0 & 0 \\ K_{1010} & 0 & 0 & 0 \end{bmatrix} \tag{9.36}$$

With a known bandwidth, it is quite straight forward to add the element stiffness matrices into the global system using the scheme presented in Algorithm 9.

Algorithm 9 Global stiffness matrix and force vector assembly for a banded matrix

```
subroutine assem-global(kglobal,fglobal)
  kglobal = 0; fglobal = 0
  do i = 1,ne                                               Loop all elements
    call elem-k-f(kelem,felem)                    element stiffness matrix and
                                                                 force vector
    do j = 1, nelem                                   nelem = nodes per element
      irow = nee(i,j)
      fglobal(irow) = fglobal(irow) + felem(j)             force vector addition
      do k = 1,nelem
        icol = nee(i,k)-irow+1                   Add only upper part of the matrix
        if ( icol .gt.  0) kglobal(irow,icol) = kglobal(irow,icol) + kelem(j,k)
stiffness matrix
      enddo
    enddo
  enddo
end subroutine assem-global
```

9.1.4 Transient Problems

The most common methodology when solving transient problems using the finite element method, is to perform the usual Garlerkin weighted residual formulation on the spatial derivatives, body forces and time derivative terms, and then using a finite difference scheme to approximate the time derivative. The development, techniques and limitations that we introduced in Chapter 8 will apply here. The time discretization, explicit and implicit methods, stability, numerical diffusion etc., have all been discussed in detail in that chapter.

For a general partial differential equation, we can write

$$\frac{\partial}{\partial t}u(\mathbf{x},t) + \mathcal{L}u(\mathbf{x},t) = f(\mathbf{x},t) \tag{9.37}$$

where $\frac{\partial}{\partial t}u(\mathbf{x},t)$ represents the time derivative of the primary variables $u(\mathbf{x},t)$, $\mathcal{L}$ operates on $u(\mathbf{x},t)$, such as spatial derivatives and $f(\mathbf{x},t)$ is a forcing function such as an internal heat generation or body forces. With a transient problem, we must also divide the domain into finite elements and approximate the variables within the element using interpolation functions, or shape functions

$$u^e \approx \sum_i N_i(\mathbf{x}) u_i(\mathbf{x}, t) \tag{9.38}$$

and substitute the approximated representation of the variables into the governing equations, and apply Garlerkin's weighted residual to eqn. (9.37) as

$$\int_V \left(\frac{\partial u^e}{\partial t} + \mathcal{L} u^e - f \right) N_i dV = 0 \tag{9.39}$$

At this point, the time derivative is implemented using a finite difference scheme as

$$\int_V \frac{1}{\Delta t} \left(u^{e,j+1} - u^{e,j} \right) N_i dV + \left[\int_V \left(\mathcal{L} u^e - f \right) N_i dV \right]^k = 0 \tag{9.40}$$

In this equation, the superscripts j and k represent the time steps. If $k = j$ it will result in an explicit Euler, and if $k = j + 1$ we have an implicit Euler time transient finite difference technique.

We can illustrate this technique with a transient one-dimensional cooling (or heating) problem. Let's assume that the initial condition is a constant temperature across the thickness of T_0. In addition, we assume the physical properties such as density, ρ, specific heat, C_p, thermal conductivity, k, remain constant during the thermal process. This results in the following governing equation

$$\rho C_p \frac{\partial T}{\partial t} = k \frac{\partial^2 T}{\partial x^2} + \dot{Q} \tag{9.41}$$

with $T(0, x) = T_0$ an initial condition, and for example $T(t, 0) = T(t, L) = T_{mold}$ as boundary conditions at both boundaries. Our first step is to discretize the domain. To illustrate this, we assume the same two-noded linear element used in the previous sections, hence approximating the temperature distribution along the element using

$$T^e = N_1 T_1 + N_2 T_2 \tag{9.42}$$

where the interpolation functions N_1 and N_2 are given in eqn. (9.9). At this point we apply Garlerkin's method to each finite element (eqn.9.39), which results in

$$\rho C_p \int_{x_1}^{x_2} \frac{\partial T^e}{\partial t} N_i dx - k \int_{x_1}^{x_2} \frac{\partial^2 T^e}{\partial x^2} N_i dx = \int_{x_1}^{x_2} \dot{Q} N_i dx \tag{9.43}$$

Again, we must apply a Green-Gauss transformation described in Theorem (9.1.2) to the second spatial derivative terms to get

$$\rho C_p \int_{x_1}^{x_2} \frac{\partial T^e}{\partial t} N_i dx + k \int_{x_1}^{x_2} \frac{\partial T^e}{\partial x} \frac{\partial N_i}{\partial x} dx = 0 \tag{9.44}$$

We apply the finite difference scheme to the first integral of the above equation and treat the second integral in the same way as we did in the steady-state problem of the previous section. This results in

$$\begin{aligned} &\frac{\rho C_p}{\Delta t} \int_{x_1}^{x_2} \left(T^{j+1} - T^j \right) \begin{bmatrix} N_1 \\ N_2 \end{bmatrix} dx + \\ &\qquad \omega k \int_{x_1}^{x_2} [\mathbf{B}]^T [\mathbf{B}] dx \begin{bmatrix} T_1 \\ T_2 \end{bmatrix}^j + (1 - \omega) k \int_{x_1}^{x_2} [\mathbf{B}]^T [\mathbf{B}] dx \begin{bmatrix} T_1 \\ T_2 \end{bmatrix}^{j+1} = 0 \end{aligned} \tag{9.45}$$

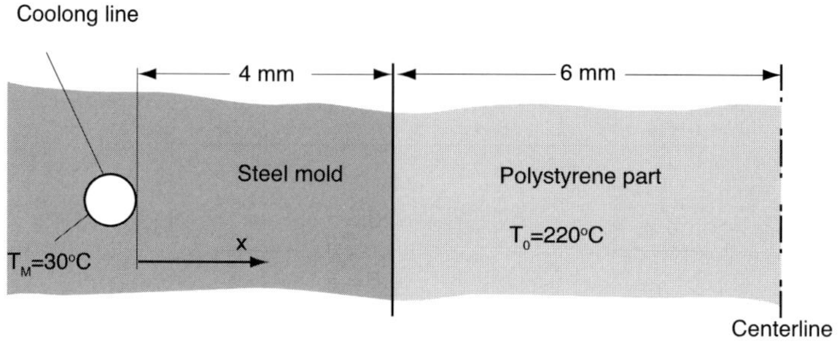

Figure 9.9: Schematic diagram of an injection molding cooling problem.

where ω is a constant that determines whether we are using an explicit Euler ($\omega = 1$), implicit Euler ($\omega = 0$) or a Crank-Nicholson ($\omega = 0.5$) approach. Note that the terms T^{j+1} and T^j are temperature fields represented by

$$T^{j+1} = N_1 T_1^{j+1} + N_2 T_2^{j+1} \tag{9.46}$$

and

$$T^j = N_1 T_1^j + N_2 T_2^j \tag{9.47}$$

respectively. This results in a set of linear algebraic equations represented by

$$[\mathbf{M^e}] \cdot (\mathbf{T^e_{j+1}}) - [\mathbf{M^e}] \cdot (\mathbf{T^e_j}) + \omega[\mathbf{K^e}] \cdot (\mathbf{T^e_j}) + (1-\omega)[\mathbf{K^e}] \cdot (\mathbf{T^e_{j+1}}) = (\mathbf{f^e}) \tag{9.48}$$

where $[\mathbf{M^e}]$ is the *element mass matrix* defined by

$$[\mathbf{M^e}] = \frac{\rho C_p}{\Delta t} \int_{x_1}^{x_2} \begin{bmatrix} N_1 N_1 & N_1 N_2 \\ N_2 N_1 & N_2 N_2 \end{bmatrix} dx \tag{9.49}$$

After integrating eqn. (9.49) we get

$$[\mathbf{M^e}] = \frac{\rho C_p}{\Delta t} \begin{bmatrix} \ell/3 & \ell/6 \\ \ell/6 & \ell/3 \end{bmatrix} \tag{9.50}$$

■ EXAMPLE 9.2.

Transient one dimensional heat transfer problem. To illustrate the transient one-dimensional heat transfer problem, the above formulation was used to solve the cooling of a thick injection molded part depicted in Fig. 9.9. In this problem we are assuming that the cooling line is maintained at 30°C and it is located 4 mm away from the mold surface. The polystyrene is injected at 220°C and we neglect cooling and viscous heating effects during mold filling.

Using the properties given in Table 9.2 and a 40 linear element (41 nodes) mesh, the temperature field as a function of time was computed and is presented in Figs. 9.10 and 9.11 Figure 9.10 shows the temperature profile the first second of cooling at time intervals of 0.05 seconds. At the beginning, the steel rapidly approaches the temperature of the polymer and forces the surface to cool down to nearly the cooling line temperature. Figure 9.11 presents the results after the first second, for a time span of 8 minutes, with time intervals of 10 seconds.

Table 9.2: Steel and Polystyrene Material Properties

Property	Value
ρ_{PS}	1100 kg/m^3
k_{PS}	0.16 W/m/K
$C_{p\text{PS}}$	1500 J/kg/K
ρ_{Steel}	7800 kg/m^3
k_{Steel}	70 W/m/K
$C_{p\text{Steel}}$	1000 J/kg/K

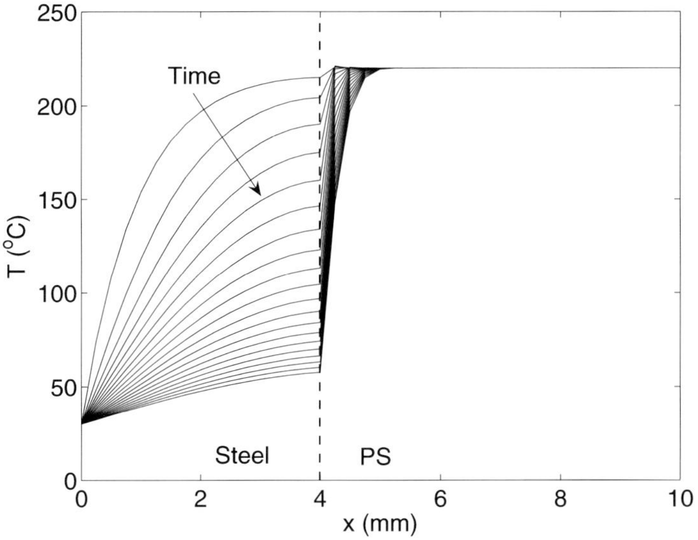

Figure 9.10: First second of cooling of a thick polystyrene part inside a steel mold. Lines represent the first second of cooling with 0.05 second time intervals.

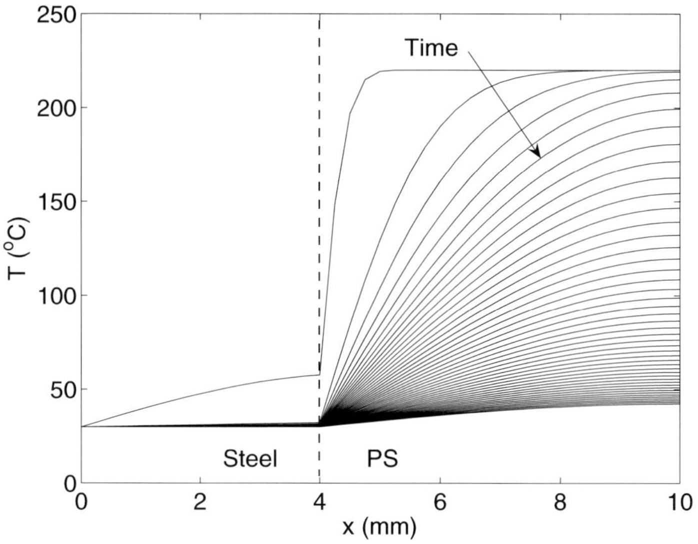

Figure 9.11: Cooling of a thick polystyrene part inside a steel mold. Lines represent 8 minutes of cooling with 10 second time intervals.

9.2 TWO-DIMENSIONAL PROBLEMS

In the last section we presented the finite element method using one-dimensional geometries as an example. Although in principle the technique is the same in two- and three-dimensions, this section will present two-dimensional problems using two types of finite elements. First, we will present the *constant strain triangle*, which is the simplest two-dimensional element and one that, as the name suggests, originated in the field of elasticity, but that has been widely used to solve governing equations in various field problems. The second element that will be used in the finite element implementation is the *isoparametric element*, which is much more flexible because it is used in conjunction with numerical integration, easily allowing the introduction of physical properties that vary in space.

9.2.1 Solution of Posisson's equation Using a Constant Strain Triangle

In this section, we will proceed to develop a finite element formulation for the two-dimensional Poisson's equation using a linear displacement, constant strain triangle. Poisson's equation has many applications in polymer processing, such as injection and compression mold filling, die flow, potential problems, heat transfer, etc. The general form of Poisson's equation in two-dimensions is

$$\frac{\partial^2 \phi}{\partial x^2} + \frac{\partial^2 \phi}{\partial y^2} = f(x, y, \phi) \tag{9.51}$$

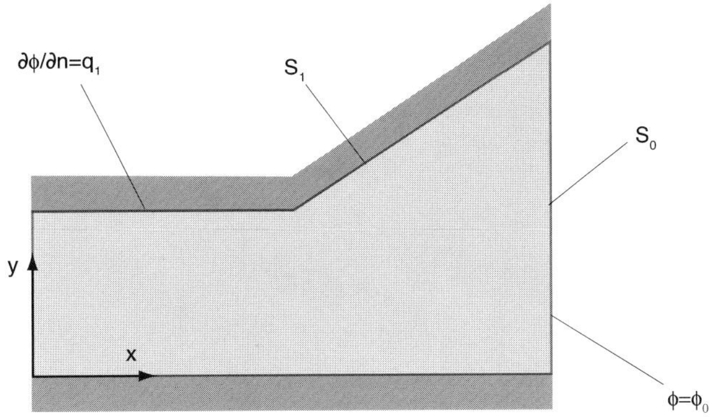

Figure 9.12: Schematic diagram of a two-dimensional domain and corresponding boundaries.

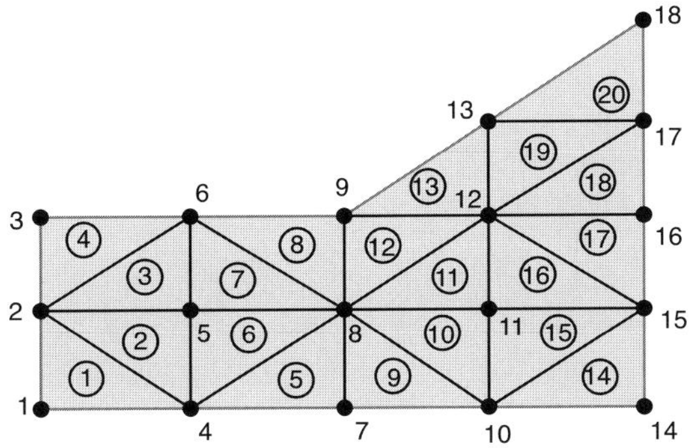

Figure 9.13: Finite element mesh using constant strain, or gradient, triangles for the domain presented in Fig. 9.12

throughout the domain, with essential and natural boundary conditions

$$\phi = \phi_0 \qquad \text{on } S_0 \tag{9.52}$$

$$\frac{\partial \phi}{\partial n} = q_1 \qquad \text{on } S_1 \tag{9.53}$$

respectively, as schematically depicted in Fig. 9.12.

The first step when formulating the finite element solution to the above equations, is to discretize the domain of interest into triangular elements, as schematically depicted in Fig. 9.13. In the constant strain triangles, represented in Fig. 9.14, the field variable within the element is approximated by,

$$\phi^e = a + bx + cy \tag{9.54}$$

which is a linear approximation for the field variable within the element and the gradient of this field variable is constant throughout the element.

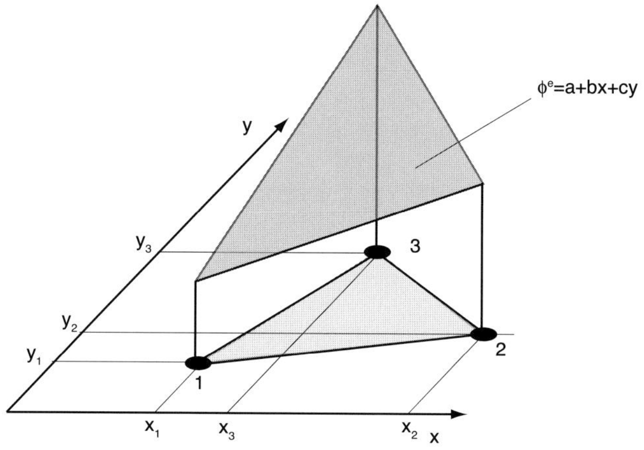

Figure 9.14: Schematic diagram of a two-dimensional triangular element with corresponding nodal coordinates and linear field variable approximation.

We can use eqn. (9.54) to evaluate the field variable, ϕ at every node location to get,

$$\begin{pmatrix} \phi_1 \\ \phi_2 \\ \phi_3 \end{pmatrix} = \begin{bmatrix} 1 & x_1 & y_1 \\ 1 & x_2 & y_2 \\ 1 & x_3 & y_3 \end{bmatrix} \begin{pmatrix} a \\ b \\ c \end{pmatrix} \tag{9.55}$$

The above equation can be used to solve for the coefficient vector

$$\begin{pmatrix} a \\ b \\ c \end{pmatrix} = [\mathbf{X}]^{-1} \begin{pmatrix} \phi_1 \\ \phi_2 \\ \phi_3 \end{pmatrix} \tag{9.56}$$

where

$$[\mathbf{X}]^{-1} = \frac{1}{|\mathbf{X}|} \begin{bmatrix} 2(x_2y_3 - x_3y_2) & 2(x_3y_1 - x_1y_3) & 2(x_1y_2 - x_2y_1) \\ y_2 - y_3 & y_3 - y_1 & y_1 - y_2 \\ x_3 - x_2 & x_1 - x_3 & x_2 - x_1 \end{bmatrix} \tag{9.57}$$

In the above equation $|\mathbf{X}|$ is the determinant of the $\mathbf{X}$ matrix and it is defined as

$$|\mathbf{X}| = (x_2y_3 + x_1y_2 + x_3y_1) - (x_2y_1 + x_3y_2 + x_1y_3) = 2\text{Area}^e \tag{9.58}$$

where, Area^e is the area of the element under consideration.

We can now rewrite eqn. (9.54) in terms of the nodal values of ϕ

$$\phi^e = \begin{bmatrix} 1 & x & y \end{bmatrix} [\mathbf{X}]^{-1} \begin{pmatrix} \phi_1 \\ \phi_2 \\ \phi_3 \end{pmatrix} \tag{9.59}$$

or

$$\phi^e = \begin{bmatrix} N_1 & N_2 & N_3 \end{bmatrix} \begin{pmatrix} \phi_1 \\ \phi_2 \\ \phi_3 \end{pmatrix} \tag{9.60}$$

The gradients can now be obtained by differentiating eqns. (9.59) or (9.60) with respect to x and y. The x-gradient is given by

$$\frac{\partial \phi^e}{\partial x} = \begin{bmatrix} 0 & 1 & 0 \end{bmatrix} [\mathbf{X}]^{-1} \begin{pmatrix} \phi_1 \\ \phi_2 \\ \phi_3 \end{pmatrix} = \begin{bmatrix} \frac{\partial N_1}{\partial x} & \frac{\partial N_2}{\partial x} & \frac{\partial N_3}{\partial x} \end{bmatrix} \begin{pmatrix} \phi_1 \\ \phi_2 \\ \phi_3 \end{pmatrix} \tag{9.61}$$

where[1]

$$\frac{\partial N_i}{\partial x} = \frac{1}{|\mathbf{X}|}(y_j - y_k) \tag{9.62}$$

Similarly, the gradient with respect to y is given by

$$\frac{\partial \phi^e}{\partial y} = \begin{bmatrix} 0 & 0 & 1 \end{bmatrix} [\mathbf{X}]^{-1} \begin{pmatrix} \phi_1 \\ \phi_2 \\ \phi_3 \end{pmatrix} = \begin{bmatrix} \frac{\partial N_1}{\partial y} & \frac{\partial N_2}{\partial y} & \frac{\partial N_3}{\partial y} \end{bmatrix} \begin{pmatrix} \phi_1 \\ \phi_2 \\ \phi_3 \end{pmatrix} \tag{9.63}$$

where

$$\frac{\partial N_i}{\partial y} = \frac{1}{|\mathbf{X}|}(x_j - x_k) \tag{9.64}$$

We can now substitute the approximated linear function of the field variable into the governing eqn. (9.51), and form the weighted residual Galerkin finite element expression as

$$\int_V \left(\frac{\partial^2 \phi^e}{\partial x^2} + \frac{\partial^2 \phi^e}{\partial y^2} - f \right) N_i dV = 0 \tag{9.65}$$

The first two terms of the above equation are transformed using the Green-Gauss Theorem (9.1.2) which results i

$$\int_V \left(\frac{\partial \phi^e}{\partial x}\frac{\partial N_i}{\partial x} + \frac{\partial \phi^e}{\partial y}\frac{\partial N_i}{\partial y} \right) dV = \int_V f N_i dV + \oint_S \frac{\partial \phi^e}{\partial n} N_i dS \tag{9.66}$$

Equation (9.66) can also be written in terms of the *B matrix*

$$\int_V [\mathbf{B}]^T [\mathbf{B}] dV \begin{pmatrix} \phi_1 \\ \phi_2 \\ \phi_3 \end{pmatrix} = -\int_V f N_i dV + \oint_S \frac{\partial \phi^e}{\partial n} N_i dS \tag{9.67}$$

where

$$[\mathbf{B}] = \frac{1}{\text{Area}^e} \begin{bmatrix} y_2 - y_3 & y_3 - y_1 & y_1 - y_2 \\ x_3 - x_2 & x_1 - x_3 & x_2 - x_1 \end{bmatrix} \tag{9.68}$$

The second integral on the right hand side of eqn. (9.67) can be evaluated for problems with a prescribed Neumann boundary condition, such as heat flow when solving conduction problems. For the Hele-Shaw approximation used to model some die flow and mold filling problems, where $\partial p / \partial n = 0$, this term is dropped from the equation.

[1]The usual notation is used ijk, jki, kij, or 123, 231, 312.

9.2.2 Transient Heat Conduction Problem Using Constant Strain Triangle

Following the procedure used with the one-dimensional FEM model and using the constant strain triangle element developed in the previous section, we can now formulate the finite element equations for a transient conduction problem with internal heat generation rate per unit volume of $\dot{Q}$. The governing equation is given by

$$\frac{\partial T}{\partial t} = \alpha \frac{\partial^2 T}{\partial x^2} + \frac{\partial^2 T}{\partial y^2} + \dot{Q} \tag{9.69}$$

If we use an implicit finite difference time stepping procedure, we must evaluate the spatial derivatives in the next time step $j+1$. In order to arrive at a linear set of algebraic equations that can be solved using standard matrix routines, we would the like the final set of equation to be of the form

$$[\bar{\mathbf{K}}^{\mathbf{e}}] \cdot (\mathbf{T}^{\mathbf{e}}_{\mathbf{j+1}}) = (\bar{\mathbf{f}}^{\mathbf{e}}) \tag{9.70}$$

where $[\bar{\mathbf{K}}^{\mathbf{e}}]$ is a combined stiffness and mass matrix, given by

$$[\bar{\mathbf{K}}^{\mathbf{e}}] = \alpha[\mathbf{K}^{\mathbf{e}}] - \frac{1}{\Delta t}[\mathbf{M}^{\mathbf{e}}] \tag{9.71}$$

where the element mass matrix $[\mathbf{M}^{\mathbf{e}}]$ is defined by

$$[\mathbf{M}^{\mathbf{e}}] = \begin{bmatrix} \text{Area}^e/6 & \text{Area}^e/12 & \text{Area}^e/12 \\ \text{Area}^e/12 & \text{Area}^e/6 & \text{Area}^e/12 \\ \text{Area}^e/12 & \text{Area}^e/12 & \text{Area}^e/6 \end{bmatrix} \tag{9.72}$$

The element force vector $(\bar{\mathbf{f}}^{\mathbf{e}})$ is formed by the internal heat generation term and the temperatures in the previous time steps as

$$(\bar{\mathbf{f}}^{\mathbf{e}}) = -\frac{1}{\Delta t}[\mathbf{M}^{\mathbf{e}}] \cdot (\mathbf{T}^{\mathbf{e}}_{\mathbf{j}}) + (\dot{\mathbf{Q}}^{\mathbf{e}}) \tag{9.73}$$

where the internal heat generation term is given by

$$(\dot{\mathbf{Q}}^{\mathbf{e}}) = \begin{pmatrix} \dfrac{\dot{Q}\text{Area}^e}{3} \\ \dfrac{\dot{Q}\text{Area}^e}{3} \\ \dfrac{\dot{Q}\text{Area}^e}{3} \end{pmatrix} \tag{9.74}$$

9.2.3 Solution of Field Problems Using Isoparametric Quadrilateral Elements.

The previous section used the constant strain three-noded element to solve Poisson's equation with steady-state as well as transient terms. The same problems, as well as any field problems such as stress-strain and the flow momentum balance, can be formulated using *isoparametric elements*. With this type of element, the same (as the name suggests) shape functions used to represent the field variables are used to interpolate between the nodal coordinates and to transform from the xy coordinate system to a local element coordinate system. The first step is to discretize the domain presented in Fig. 9.12 using the isoparametric quadrilateral elements as shown in Fig. 9.15.

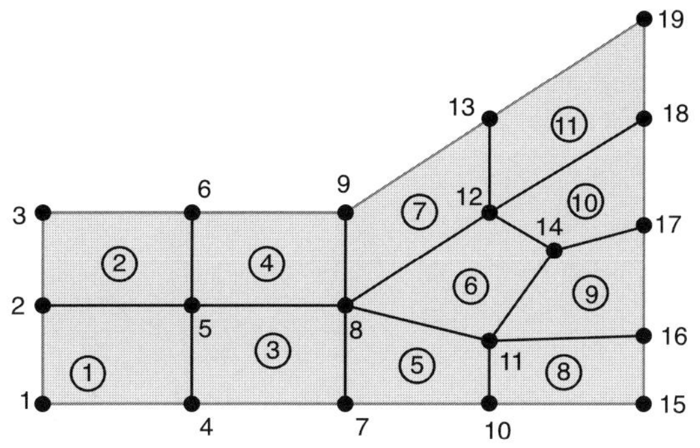

Figure 9.15: Finite element mesh using isoparametric quadrilateral elements for the domain presented in Fig. 9.12

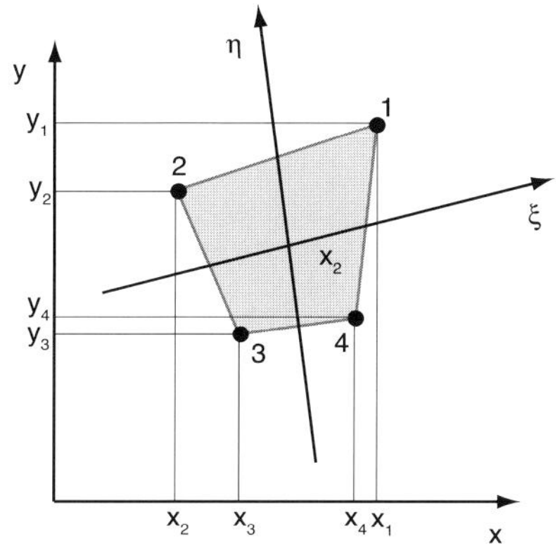

Figure 9.16: Isoparametric quadrilateral element in the xy-coordinate system.

If we pick an arbitrary element we can see that it is represented by the xy-coordinates of the four nodal points, as depicted in Fig. 9.16. The figure also shows a $\xi\eta$-coordinate system embedded within the element. In the $\xi\eta$, or local, coordinate system, we have a perfectly square element of area 2×2, where the element spreads between $-1 \geq \xi \leq 1$ and $-1 \geq \eta \leq 1$. This attribute allows us to easily allows us to use *Gauss quadrature* as a numerical integration scheme, where the limits vary between -1 and 1. The isoparametric element described in the $\xi\eta$-coordinate system is presented in Fig. 9.17.

If we assume nodal shape functions N_i that are equal 1 at (ξ_i, η_i) and zero at the other three nodes (ξ_j, η_j), we can define the field variable and the nodal coordinates in term of those shape functions and the nodal values of the parameters, i.e.,

$$\begin{aligned} \phi^e &= N_1\phi_1 + N_2\phi_2 + N_3\phi_3 + N_4\phi_4 \\ x^e &= N_1x_1 + N_2x_2 + N_3x_3 + N_4x_4 \\ y^e &= N_1y_1 + N_2y_2 + N_3y_3 + N_4y_4 \end{aligned} \tag{9.75}$$

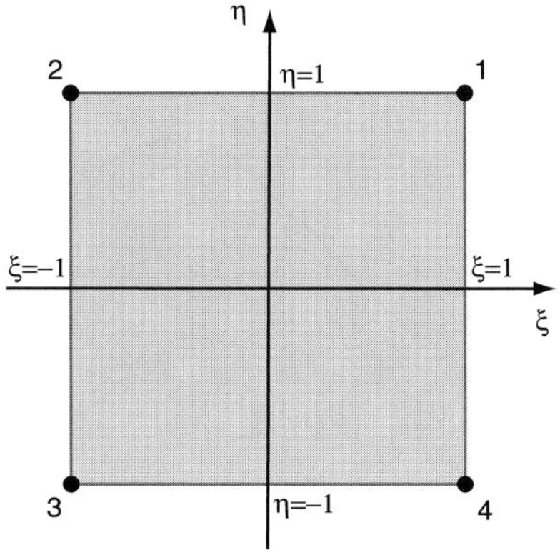

Figure 9.17: Isoparametric quadrilateral element in the $\xi\eta$-coordinate system.

where the shape functions, N_i, are defined as

$$\begin{aligned} N_1 &= \frac{1}{4}(1+\xi)(1+\eta) \\ N_2 &= \frac{1}{4}(1-\xi)(1+\eta) \\ N_3 &= \frac{1}{4}(1-\xi)(1-\eta) \\ N_4 &= \frac{1}{4}(1+\xi)(1-\eta) \end{aligned} \tag{9.76}$$

The finite element expression for a problem using isoparametric finite elements will be similar to the ones developed in previous examples. For example, a problem governed by Poisson's equation will result in the following finite element equation,

$$\int\int [\mathbf{B}]^T[\mathbf{B}]dxdy \begin{pmatrix} \phi_1 \\ \phi_2 \\ \phi_3 \\ \phi_4 \end{pmatrix} = -\int\int fN_i dxdy \tag{9.77}$$

where f represents a forcing function, such as internal heat generation, transient terms, etc. If we examine the above equation carefully, we note that the shape functions are in terms of ξ and η, but the derivatives of these functions as well as the integrations are in terms of x and y. Hence, the terms in the *B Matrix*, $\mathbf{B}$, which contains the derivatives of the shape functions with respect to x and y, can be transformed into derivatives of the shape functions with respect to ξ and η with the use of the *Jacobian transformation matrix* given by

$$\frac{\partial N_i}{\partial \mathbf{x}} = [\mathbf{J}]^{-1}\frac{\partial N_i}{\partial \boldsymbol{\xi}} \tag{9.78}$$

where the Jacobian transformation matrix in two dimensions is defined by

$$[\mathbf{J}] = \begin{bmatrix} \dfrac{\partial x}{\partial \xi} & \dfrac{\partial y}{\partial \xi} \\ \dfrac{\partial x}{\partial \eta} & \dfrac{\partial y}{\partial \eta} \end{bmatrix} \tag{9.79}$$

The second problem encountered is that we are integrating over x and y functions in the $\xi\eta$ system, hence, the terms $dxdy$ must be changed to $d\xi d\eta$. To do this, we use a special form of the *Jacobian transformation* given by

$$|\mathbf{J}|d\xi d\eta = dxdy \tag{9.80}$$

where $|\mathbf{J}|d$ is the determinant of the Jacobian matrix and is computed as

$$|\mathbf{J}| = \frac{\partial x}{\partial \xi}\frac{\partial y}{\partial \eta} - \frac{\partial y}{\partial \xi}\frac{\partial x}{\partial \eta} \tag{9.81}$$

Using these transformations, eqn. (9.77) becomes

$$\int_{-1}^{1}\int_{-1}^{1}[\mathbf{B}]^T[\mathbf{B}]|\mathbf{J}|d\xi d\eta \begin{pmatrix} p_1 \\ p_2 \\ p_3 \\ p_4 \end{pmatrix} = -\int_{-1}^{1}\int_{-1}^{1} fN_i|\mathbf{J}|d\xi d\eta \tag{9.82}$$

As explained earlier, the limits of integration of the above equation make it ideal for applying a Gauss integration scheme, discussed in Chapter 7. Equation (9.82) can therefore be approximated as

$$\left[\sum_{i=1}^{NGauss}\sum_{j=1}^{NGauss}[\mathbf{B}]_{ij}^T[\mathbf{B}]_{ij}|\mathbf{J}|_{ij}w_iw_j\right]\begin{pmatrix} \phi_1 \\ \phi_2 \\ \phi_3 \\ \phi_4 \end{pmatrix} = -\sum_{i=1}^{NGauss}\sum_{j=1}^{NGauss} fN_k|\mathbf{J}|_{i,j}w_iw_j \tag{9.83}$$

where $NGauss$ are the number of Gauss points for each direction, and w_i and w_j are the weights for the corresponding Gauss points. For example, if we choose $NGauss = 2$ we would have a 2×2 Gauss points integration schemes with 4 integration stations. The following example presents an algorithm that illustrates how the above integrals are evaluated.

■ EXAMPLE 9.3.

Solution of the two-dimensional Poisson's equation: compression molding. To illustrate the use of the four-noded isoparametric element, we can solve for the pressure distribution and velocity field during compression molding of an L-shaped polymer charge, shown in Fig. 9.18, with the physical and numerical data presented in Table 9.3.

As discussed in Chapter 8 of this book, the momentum balance and the continuity equation lead to the Hele-Shaw approximation given by

$$\frac{\partial^2 p}{\partial x^2} + \frac{\partial^2 p}{\partial y^2} = -12\mu\frac{\dot{h}}{h^3} \tag{9.84}$$

with homogeneous Dirichlet boundary conditions at the flow front

$$p(x, y) = 0 \text{ for } x, y \in \text{front} \tag{9.85}$$

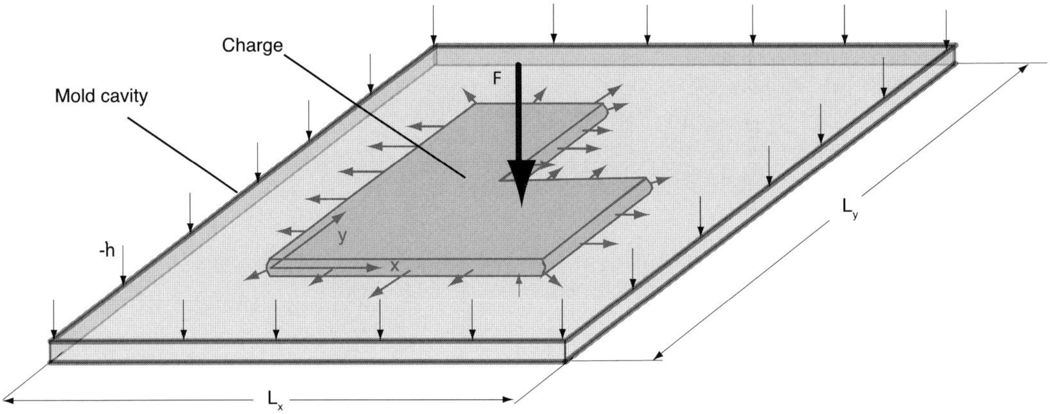

Figure 9.18: Schematic diagram of the compression molding of an L-shaped charge.

Table 9.3: Example 9.3 Data

Parameter	Value
μ	10,000 Pa-s
L_x	0.4 m
L_y	0.4 m
h	0.005 m
$\dot{h}$	0.01 m/s
ne	300
nn	341
Ngauss	2

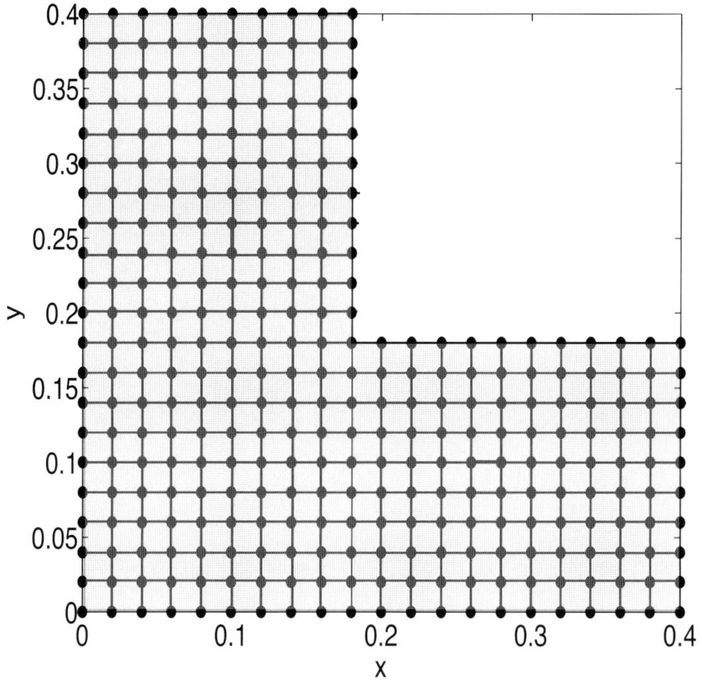

Figure 9.19: Four-noded isoparametric finite element mesh for the L-shape charge.

The average velocities are obtained from the pressure gradients as follows (see Chapter 5),

$$\bar{u}_x = -\frac{h^2}{12\mu}\frac{\partial p}{\partial x}$$
$$\bar{u}_y = -\frac{h^2}{12\mu}\frac{\partial p}{\partial y} \qquad (9.86)$$

A finite element program with four-noded isoparametric elements was used to solve the above governing equations and boundary conditions. Algorithm 10 presents the scheme used to evaluate the element stiffness matrices and force vectors using numerical integration.

Figure 9.19 presents the mesh used for the solution and Fig. 9.20 illustrates the pressure field for the charge, the instant when both mold halves make contact with the charge and squeezing begins. Figure 9.21 presents the instantaneous velocity field the moment the charge starts flowing at the beginning of squeezing flow.

9.2.4 Two Dimensional Penalty Formulation for Creeping Flow Problems

The isoparametric element works quite well to formulate the finite element equations for flow problems, such as flows with non-Newtonian shear thinning viscosity. Due to the flexibility that exists to integrate variables throughout the elements, the method lends itself

Algorithm 10 Computing the element stiffness matrix and force vector for a four-noded isoparametric element by numerical integration

```
subroutine elem-k-f
  do i = 1,NGauss                                            integration loops
    do j = 1,NGauss
      xi = xg(i); eta = xg(j)     xg(i)  and xg(j) store the Gauss points, i.e.,
                                     xg(1)=-0.5773502691 and xg(2)=0.5773502691
                                                                   when NGauss=2
      n(1) = 0.25(1+xi)(1+eta)                                   shape functions
      n(2) = 0.25(1-xi)(1+eta)
      n(3) = 0.25(1-xi)(1-eta)
      n(4) = 0.25(1+xi)(1-eta)

      dn(1,1) = 0.25(1+eta)        shape function derivatives with respect to ξ
      dn(1,2) = -0.25(1+eta)
      dn(1,3) = -0.25(1-eta)
      dn(1,4) = 0.25(1-eta)

      dn(2,1) = 0.25(1+xi)         shape function derivatives with respect to η
      dn(2,2) = 0.25(1-xi)
      dn(2,3) = -0.25(1-xi)
      dn(2,4) = -0.25(1+xi)

      do k =1,4                                              Gauss point position
        x = x+n(k)pe(1,k)                         pe:  element nodal coordinates
        y = y+n(k)pe(2,k)
      enddo

      do k1 = 1,2                                                       Jacobian
        do k2 = 1,4
          jac(1,k2) = jac(1,k2)+dn(1,k2)pe(k1,k2)
          jac(2,k2) = jac(2,k2)+dn(2,k2)pe(k1,k2)
        enddo
      enddo
      det = jac(1,1)jac(2,2)-jac(1,2)jac(2,1)

      invjac(1,1)=jac(2,2)/det                    inverse of the Jacobian matrix
      invjac(1,2)=-jac(1,2)/det
      invjac(2,1)=-jac(2,1)/det
      invjac(2,2)=jac(1,1)/det

      do k=1,4                                                          B matrix
        b(1,k)=invjac(1,1)dn(1,k)+invjac(1,2)dn(2,k)
        b(2,k)=invjac(2,1)dn(1,k)+invjac(2,2)dn(2,k)
      enddo
      bt = TRANSPOSE(b)

      kelem = kelem + MATMUL(bt,b)*det*w(i)*w(j)
      felem(:)  = felem(:)  - 12*mu*hdot*n(:)*det*w(i)*w(j)/h**3
                                                w(i) and w(j) are the weights,
                                        i.e., w(1)=1 and w(2)=1 when NGauss=2
    enddo
  enddo
end subroutine elem-k-f
```

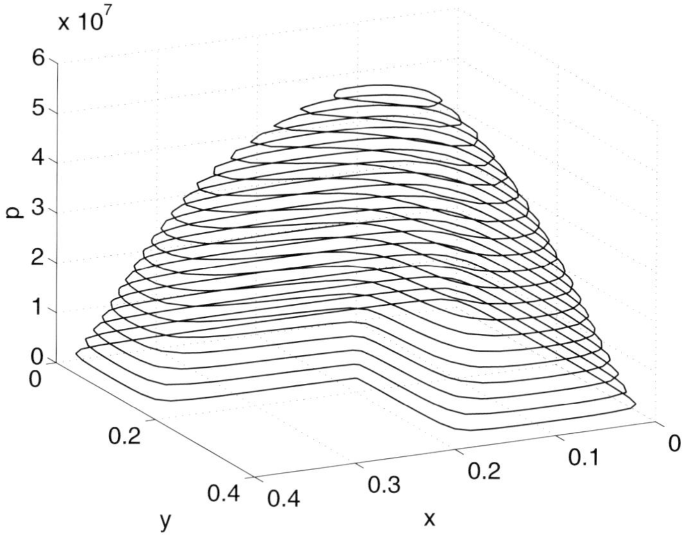

Figure 9.20: Pressure distribution within the L-shape charge the instant when both mold halves make contact with the charge.

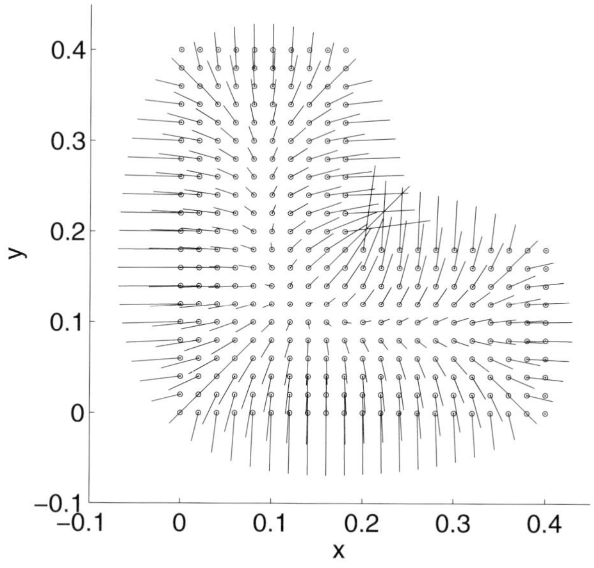

Figure 9.21: Instantaneous velocity field for the L-shape charge at the beginning of flow.

to simulate systems with variable and non-linear physical properties, as well as coupled flow and heat transfer problems. Here, we will restrict our discussion to non-Newtonian Stokes flow with a rate of deformation dependent viscosity. For such a general case, the continuity and momentum equations are given by

$$\frac{\partial u_x}{\partial x} + \frac{\partial u_y}{\partial y} = 0 \tag{9.87}$$

$$-\frac{\partial p}{\partial x} + \left[\frac{\partial \tau_{xx}}{\partial x} + \frac{\partial \tau_{yx}}{\partial y}\right] = f_x \tag{9.88}$$

$$-\frac{\partial p}{\partial y} + \left[\frac{\partial \tau_{xy}}{\partial x} + \frac{\partial \tau_{yy}}{\partial y}\right] = f_y \tag{9.89}$$

Using the *generalized Newtonian* constitutive equation, the deviatoric stress tensor is defined as

$$\tau_{ij} = \eta \dot{\gamma}_{ij} \tag{9.90}$$

where $\dot{\gamma}_{ij}$ is the shear rate defined as

$$\dot{\gamma}_{ij} = \frac{\partial u_i}{\partial x_j} + \frac{\partial u_j}{\partial x_i} \tag{9.91}$$

The momentum equations can therefore be written in terms of velocities

$$-\frac{\partial p}{\partial x} + \left[\frac{\partial}{\partial x}\left(2\eta\frac{\partial u_x}{\partial x}\right) + \frac{\partial}{\partial y}\left(\eta\left(\frac{\partial u_x}{\partial y} + \frac{\partial u_y}{\partial x}\right)\right)\right] = f_x \tag{9.92}$$

$$-\frac{\partial p}{\partial y} + \left[\frac{\partial}{\partial y}\left(2\eta\frac{\partial u_y}{\partial y}\right) + \frac{\partial}{\partial x}\left(\eta\left(\frac{\partial u_x}{\partial y} + \frac{\partial u_y}{\partial x}\right)\right)\right] = f_y \tag{9.93}$$

This results in a system of three equations with three variables: u_x, u_y and p. It is very common to eliminate the pressure from the set of equations by introducing a slight compressibility to the fluid. This approach is often referred to as the *penalty formulation*. The compressibility factor γ is introduced in the continuity equation as

$$\frac{\partial u_x}{\partial x} + \frac{\partial u_y}{\partial y} = \frac{p}{\gamma} \tag{9.94}$$

When $\gamma \to \infty$, the fluid is considered incompressible and eqns. (9.92) and (9.93) are fully satisfied. Replacing the continuity equation (9.94) in the momentum equations we get

$$-\gamma\frac{\partial}{\partial x}\left(\frac{\partial u_x}{\partial x} + \frac{\partial u_y}{\partial y}\right) + \left[\frac{\partial}{\partial x}\left(2\eta\frac{\partial u_x}{\partial x}\right) + \frac{\partial}{\partial y}\left(\eta\left(\frac{\partial u_x}{\partial y} + \frac{\partial u_y}{\partial x}\right)\right)\right] = f_x \tag{9.95}$$

$$-\gamma\frac{\partial}{\partial y}\left(\frac{\partial u_x}{\partial x} + \frac{\partial u_y}{\partial y}\right) + \left[\frac{\partial}{\partial y}\left(2\eta\frac{\partial u_y}{\partial y}\right) + \frac{\partial}{\partial x}\left(\eta\left(\frac{\partial u_x}{\partial y} + \frac{\partial u_y}{\partial x}\right)\right)\right] = f_y \tag{9.96}$$

At this point, we can proceed to the finite element formulation of the above governing equations. For this, we will use the isoparametric element presented in the last sections.

Using the same shape function definitions we can approximate the velocity field across the element using

$$
\begin{aligned}
u_x^e &= N_1 u_{x1} + N_2 u_{x2} + N_3 u_{x3} + N_4 u_{x4} \\
u_y^e &= N_1 u_{y1} + N_2 u_{y2} + N_3 u_{y3} + N_4 u_{y4}
\end{aligned} \tag{9.97}
$$

Next, we apply Galerkin's weighted residual method and reduce the order of integration of the various terms in the above equations using the Green-Gauss Theorem (9.1.2) for each element. For a simpler presentation we will deal with each term in the above equations separately. The terms of the x-component (eqn. (9.95)) of the penalty formulation momentum balance become

Term 1:

$$\int_V \frac{\partial}{\partial x}\left(2\eta \frac{\partial u_x^e}{\partial x}\right) N_i dV = -2\int_V \eta \frac{\partial u_x^e}{\partial x}\frac{\partial N_i}{\partial x} dV + 2\oint_S \eta \frac{\partial u_x^e}{\partial x} n_x N_i dS$$

Term 2:

$$
\begin{aligned}
\int_V \frac{\partial}{\partial y}\left[\eta(\frac{\partial u_x^e}{\partial y} + \frac{\partial u_y^e}{\partial x})\right] N_i dV &= -\int_V \eta(\frac{\partial u_x^e}{\partial y} + \frac{\partial u_y^e}{\partial x})\frac{\partial N_i}{\partial y} dV \\
&+ \oint_S \eta\left(\frac{\partial u_x^e}{\partial y} + \frac{\partial u_y^e}{\partial x}\right) N_i n_y dS
\end{aligned}
$$

Term 3:

$$
\begin{aligned}
-\gamma \int_V \frac{\partial}{\partial x}\left(\frac{\partial u_x^e}{\partial x} + \frac{\partial u_y^e}{\partial y}\right) N_i dV &= \gamma \int_V (\frac{\partial u_x^e}{\partial x} + \frac{\partial u_y^e}{\partial y})\frac{\partial N_i}{\partial x} dV \\
&- \gamma \oint_S \left(\frac{\partial u_x^e}{\partial x} + \frac{\partial u_y^e}{\partial y}\right) N_i n_x dS
\end{aligned}
$$

which results in

$$
\begin{aligned}
&-2\int_V \eta \frac{\partial u_x^e}{\partial x}\frac{\partial N_i}{\partial x} dV - \int_V \eta(\frac{\partial u_x^e}{\partial y} + \frac{\partial u_y^e}{\partial x})\frac{\partial N_i}{\partial y} dV \\
&+\gamma \int_V (\frac{\partial u_x^e}{\partial x} + \frac{\partial u_y^e}{\partial y})\frac{\partial N_i}{\partial x} dV = \\
&-2\oint_S \eta \frac{\partial u_x^e}{\partial x} n_x N_i dS - \oint_S \eta\left(\frac{\partial u_x^e}{\partial y} + \frac{\partial u_y^e}{\partial x}\right) N_i n_y dS \\
&+\oint_S t_x^e N_i dS + \int_V f_x N_i dV
\end{aligned} \tag{9.98}
$$

where $\mathbf{t} = p\mathbf{n} = (pn_x, pn_y)$. An expression for the y-momentum can be obtained in a similar way.

The integrals of the shape function derivatives lead to the following matrices

$$
\begin{aligned}
S_{ij}^{11} &= \int\int \frac{\partial N_i}{\partial x}\frac{\partial N_j}{\partial x} dxdy \\
S_{ij}^{12} &= \int\int \frac{\partial N_i}{\partial x}\frac{\partial N_j}{\partial y} dxdy \\
S_{ij}^{22} &= \int\int \frac{\partial N_i}{\partial y}\frac{\partial N_j}{\partial y} dxdy
\end{aligned} \tag{9.99}
$$

which can be used to define components of the stiffness matrix as

$$[\bar{\mathbf{K}}^{11}] = 2\eta[\mathbf{S}^{11}] + \eta[\mathbf{S}^{22}] \tag{9.100}$$

$$[\bar{\mathbf{K}}^{12}] = \eta[\mathbf{S}^{12}] \tag{9.101}$$

$$[\bar{\mathbf{K}}^{22}] = 2\eta[\mathbf{S}^{22}] + \eta[\mathbf{S}^{11}] \tag{9.102}$$

This results in an element stiffness matrix of

$$\begin{bmatrix} [\mathbf{K}^{11}] & [\mathbf{K}^{12}] \\ [\mathbf{K}^{12}]^T & [\mathbf{K}^{22}] \end{bmatrix} \begin{pmatrix} u_{x1} \\ u_{x2} \\ u_{x3} \\ u_{x4} \\ u_{y1} \\ u_{y2} \\ u_{y3} \\ u_{y4} \end{pmatrix} = \begin{pmatrix} F_1^x \\ F_2^x \\ F_3^x \\ F_4^x \\ F_1^y \\ F_2^y \\ F_3^y \\ F_4^y \end{pmatrix} \tag{9.103}$$

where

$$[\mathbf{K}^{11}] = [\bar{\mathbf{K}}^{11}] + \gamma[\mathbf{S}^{11}] \tag{9.104}$$

$$[\mathbf{K}^{12}] = [\bar{\mathbf{K}}^{12}] + \gamma[\mathbf{S}^{12}] \tag{9.105}$$

$$[\mathbf{K}^{22}] = [\bar{\mathbf{K}}^{22}] + \gamma[\mathbf{S}^{22}] \tag{9.106}$$

and

$$F_i^x = \int\int f_x N_i dxdy + \oint t_x N_i dS \tag{9.107}$$

$$F_i^y = \int\int f_y N_i dxdy + \oint t_y N_i dS \tag{9.108}$$

The above equations can be broken down into terms that contain the viscosity, η, and which pertain to the momentum balance, and those terms that contain the compressibility factor, γ, which pertain to the continuity equation

$$\left[\eta[\mathbf{K}^1] + \gamma[\mathbf{K}^2]\right](\mathbf{u}) = (\mathbf{F}) \tag{9.109}$$

In order to solve a flow problem, we must weaken the continuity equation portion of the final finite element system. This is done by under-integrating the terms that pertain to the continuity equation. By using only one Gauss integration station when integrating the continuity equation portion of the equation, and four Gauss integration stations (2×2) when integrating the momentum balance portion of the equation, we are able to soften the equations and solve them with a slight penalty on the pressure solution. If we use the same integration order on all the terms of the equation, we would arrive at the trivial solution that satisfies the equations, of zero velocity throughout the system.

The above equations can be used to simulate the flow of non-Newtonian fluids with a shear thinning viscosity. This requires an iterative under-relaxation scheme where a Newtonian solution is found first. The initial velocity field is used to compute rates of deformation and viscosity. Next, a corrected velocity field is computed with the updated velocities, at which point an under-relaxation is performed using

$$(\mathbf{u})_{k+1}^{\text{relaxed}} = \omega(\mathbf{u})_k^{\text{relaxed}} + (1-\omega)(\mathbf{u})_{k+1} \tag{9.110}$$

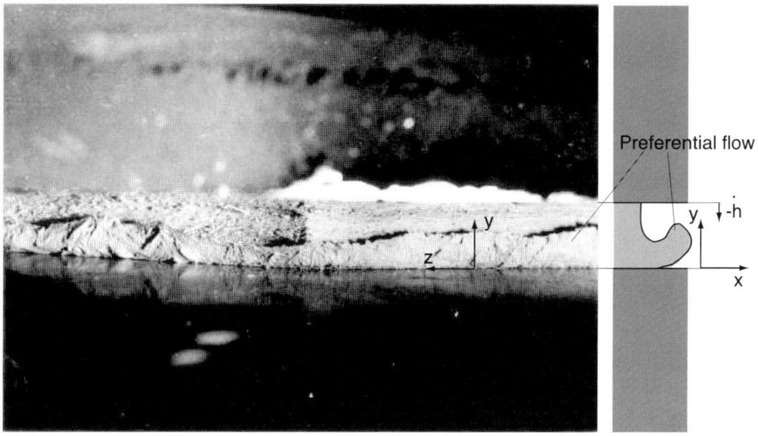

Figure 9.22: Preferential flow during flow SMC after 20 seconds of one-sided heating [12].

where k and $k+1$ denote the previous and current iteration steps, respectively. It has been found that when using a power law model, the power law index, n, works best as the relaxation parameter, $\omega = n$. The above finite element model can also be implemented with thermal effects by solving the momentum balance and energy equation simultaneously. If diffusion dominates when solving the energy equation (small Peclet number), the implementation is straight forward. However, when convection dominates diffusion, the solved temperature fields will present spurious oscillations, as discussed in Chapter 8, where a 1D convection-diffusion was solved with and without up-winding techniques. Later in this chapter we will present a 3D formulation with convective and diffusive terms. In the example that follows, the conduction occurs primarily through the thickness, and the convection in the planar direction of the part. Hence, no special treatment is necessary.

■ **EXAMPLE 9.4.**

Non-isothermal non-Newtonian flow during compression molding To study the non-isothermal effect during squeezing of sheet molding compound (SMC), Osswald *et.al* [12] solved the momentum and energy equations simultaneously using four noded isoparametric elements and the formulation presented above. Primarily, they wanted to study the effects that one-sided heating had on the flow front development of SMC charges during mold filling. One-sided heating occurs from the time when the SMC charge is laid on the mold surface and the time when the upper mold half makes contact with the top surface of the charge. During this period, a layer of hot material forms in the lower part of the SMC charge. This hot, and consequently low viscosity fluid, is squeezed out from under the charge, causing what is often referred to as preferential flow. Preferential flow leads to air entrapment, fiber-matrix separation, and, as a result, to SMC parts with weak edges. Figure 9.22 shows a photograph of an SMC flow front moving towards a mold opening. The picture clearly shows how the preferential flow is creating a swirl in the free flow front. This type of flow can be problematic since the edges, especially corners, are subjected to higher stresses and often give rise to cracks in SMC parts such as automotive body panels.

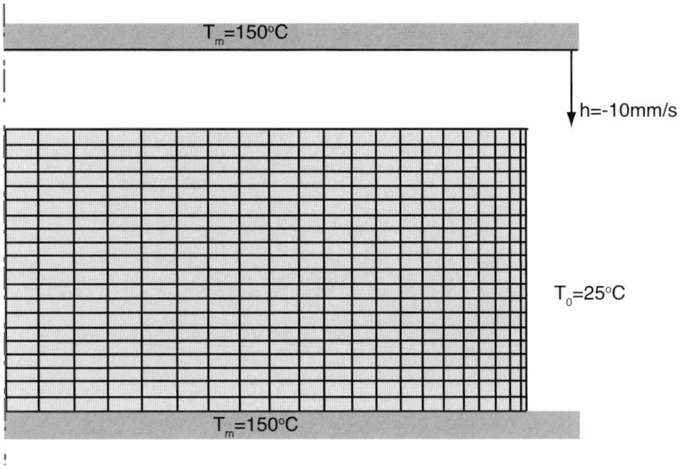

Figure 9.23: Finite element mesh of the cross-section of an SMC charge [12].

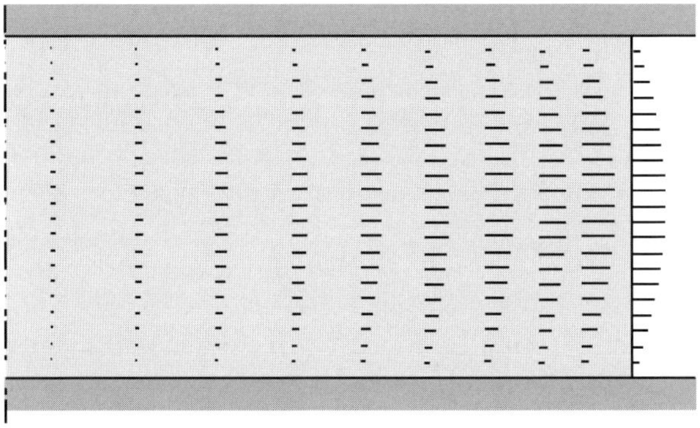

Figure 9.24: Velocity field during isothermal squeezing of an SMC charge [12].

The work by Osswald *et.al* [12] was done to understand this phenomenon. To simulate the effect shown in the photograph, they used the two-dimensional mesh and processing conditions presented in Fig. 9.23. Note that in order to better see the set-up and results, the mesh is shown distorted in the thickness direction of the charge. Since the thickness-to-length (L/D) ratio is very small, the heat transfer in the non-isothermal solution reduces to a 1D problem, and was solved using the finite difference technique.

Figure 9.24 presents the predicted velocity field for an isothermal flow. Here, the charge and the mold surfaces have the same temperature of 150°C. Using the velocity field, Osswald *et.al* [12] moved the nodes by a small time Δt. A new velocity field was computed using the updated mesh. These steps were repeated until the mesh became too distorted, leading to unreasonable values in the velocity field. Excessive element distortion is often detected when the *determinant of the Jacobian matrix*, $|J|$, is less or equal to zero.

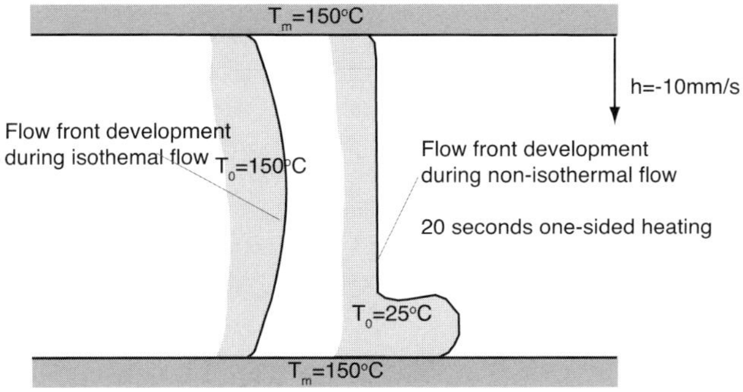

Figure 9.25: Flow front shape after isothermal (left) and non-isotheramal (right) flows. Non-isothermal flows are a result of a 20 second one-sided heating [12].

Figure 9.25 presents the shape of the flow front for the isothermal flow case, as well as for a non-isothermal case where a 20 second one-sided heating was used. Other cases showed that one-sided heating times larger than 10 seconds resulted in a preferential flow that formed a swirl in front of the charge. One-sided heating times less than 10 seconds led to preferential flow that eventually disappeared as the rest of flow front caught up with the faster moving melt.

9.3 THREE-DIMENSIONAL PROBLEMS

When modeling a system, we try to reduce the problem to a two-, and if possible, to a one-dimensional model. However, often it is not possible to reduce the dimensionality of a problem, forcing us to solve a full three-dimensional model. In principle, solving a problem in 3D using a finite element formulation work the same way as in 1D or 2D. However, set-up effort, and therefore engineering time, as well as computational costs go up drastically when a problem is solved using a full three dimensional model. Most developments as described for 1D and 2D finite element formulations are also valid for 3D. In this section, we will present several finite elements and formulations.

9.3.1 Three-dimensional Elements

Figure 9.26 illustrates two types of three-dimensional elements, the *prisms* and the *tetrahedrals*. The prisms shown are of the *serendipity* family and the coordinates for these elements are normalized the same way as the rectangular two-dimensional elements and the shape functions are just simple extensions of the 2D elements. The tetrahedral elements have *volume* normalized coordinates, which again are an extension of the 2D *area* normalized coordinates.

The shape functions for the prism linear elements (8 nodes) are

$$N_i = \frac{1}{8}(1+\xi_0)(1+\eta_0)(1+\zeta_0) \tag{9.111}$$

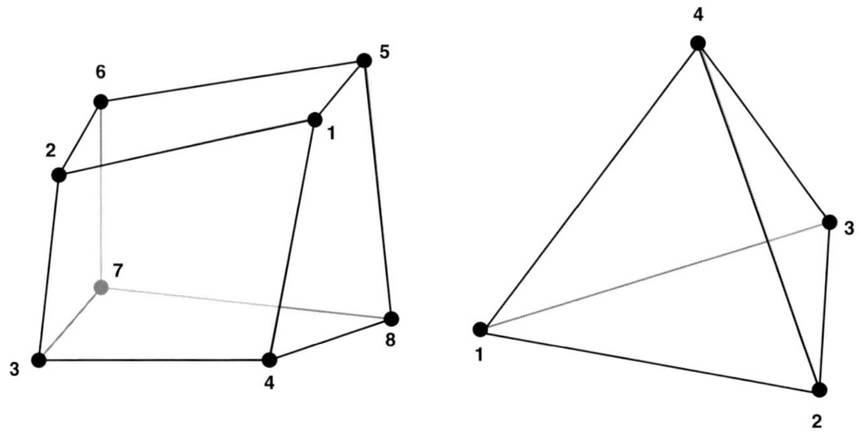

Figure 9.26: Three dimensional prism (left) and tetrahedral (right) elements.

and for the quadratic (20 nodes)

$$\begin{aligned} N_i &= \frac{1}{8}(1+\xi_0)(1+\eta_0)(1+\zeta_0)(\xi_0+\eta_0+\zeta_0-2) \text{ Corner nodes} \\ N_i &= \frac{1}{4}(1-\xi^2)(1+\eta_0)(1+\zeta_0) \text{ typical mid-side node: } \xi_i=0, \eta_i=\pm 1 \text{ and } \zeta_i=\pm 1 \end{aligned} \tag{9.112}$$

The volume normalized coordinates for the tetrahedral elements are defined as

$$\begin{aligned} x &= L_1x_1 + L_2x_2 + L_3x_3 + L_4x_4 \\ y &= L_1y_1 + L_2y_2 + L_3y_3 + L_4y_4 \\ z &= L_1z_1 + L_2z_2 + L_3z_3 + L_4z_4 \\ 1 &= L_1 + L_2 + L_3 + L_4 \end{aligned} \tag{9.113}$$

where the L_i area coordinates are defined below. The shape functions for the linear and quadratic elements are defined in the same way as eqns. (9.111) and (9.112).

Similar to the two-dimensional isoparametric element, for three dimensional elements we use a mapping of the normalized coordinates, ξ, η, ζ, (L_i volume coordinates for a tetrahedral element), in such a way that the cartesian coordinates will appear as a curvilinear set.

As in 2D, this mapping is achieved providing a one-to-one relationship between the Cartesian and the curvilinear coordinates, i.e.,

$$\begin{pmatrix} x \\ y \\ z \end{pmatrix} = f \begin{pmatrix} \xi \\ \eta \\ \zeta \end{pmatrix} \text{ or } g \begin{pmatrix} L_1 \\ L_2 \\ L_3 \\ L_4 \end{pmatrix} \tag{9.114}$$

As with 2D, the coordinates can be transformed using

$$x^e = N_1x_1 + N_2x_2 + ... + N_nx_n = \sum_{i=1}^{n} N_ix_i \tag{9.115}$$

where $i = 1, .., n$ are the element nodes and N_i are the shape functions.

In 3D we also need the two transformations used with the 2D isoparametric element. In the first place, the global derivatives of the formulation, $\partial N_i/\partial x$, must be expresses in terms of local derivatives, $\partial N_i/\partial \xi$. Second, the integration of volume (or surface) needs to be performed in the appropriate coordinate system with the correct limits of integration. The global and local derivatives are related through a *Jacobian transformation matrix* as follows

$$\begin{pmatrix} \dfrac{\partial N_i}{\partial \xi} \\ \dfrac{\partial N_i}{\partial \eta} \\ \dfrac{\partial N_i}{\partial \zeta} \end{pmatrix} = \mathbf{J} \begin{pmatrix} \dfrac{\partial N_i}{\partial x} \\ \dfrac{\partial N_i}{\partial y} \\ \dfrac{\partial N_i}{\partial z} \end{pmatrix} \tag{9.116}$$

where the Jacobian matrix is defined as,

$$\mathbf{J} = \begin{pmatrix} \dfrac{\partial x}{\partial \xi} & \dfrac{\partial y}{\partial \xi} & \dfrac{\partial z}{\partial \xi} \\ \dfrac{\partial x}{\partial \eta} & \dfrac{\partial y}{\partial \eta} & \dfrac{\partial z}{\partial \eta} \\ \dfrac{\partial x}{\partial \zeta} & \dfrac{\partial y}{\partial \zeta} & \dfrac{\partial z}{\partial \zeta} \end{pmatrix} = \begin{pmatrix} \sum \dfrac{\partial N_i}{\partial \xi} x_i & \sum \dfrac{\partial N_i}{\partial \xi} y_i & \sum \dfrac{\partial N_i}{\partial \xi} z_i \\ \sum \dfrac{\partial N_i}{\partial \eta} x_i & \sum \dfrac{\partial N_i}{\partial \eta} y_i & \sum \dfrac{\partial N_i}{\partial \eta} z_i \\ \sum \dfrac{\partial N_i}{\partial \zeta} x_i & \sum \dfrac{\partial N_i}{\partial \zeta} y_i & \sum \dfrac{\partial N_i}{\partial \zeta} z_i \end{pmatrix} \tag{9.117}$$

The transformation of the variables and the region with respect to which the integration is made involves the determinant of the Jacobian, det$\mathbf{J} = |\mathbf{J}|$. For instance, a volume element becomes

$$dV = dxdydz = |\mathbf{J}|d\xi d\eta d\zeta \tag{9.118}$$

where $|\mathbf{J}|$is the determinant of Jacobian matrix.The limits of integration for three-dimensional brick elements will be from -1 to $+1$

$$\int_{-1}^{1}\int_{-1}^{1}\int_{-1}^{1} f|\mathbf{J}|d\xi d\eta d\zeta \tag{9.119}$$

and for the tetrahedral elements

$$\int_{0}^{1}\int_{0}^{1-\eta}\int_{0}^{1-\eta-\xi} f|\mathbf{J}|d\xi d\eta d\zeta \tag{9.120}$$

The isoparametric coordinates are related to L_i by the following relations

$$\begin{aligned} \xi &= L_1 \\ \eta &= L_2 \\ \zeta &= L_3 \\ 1-\xi-\eta-\zeta &= L_4 \end{aligned} \tag{9.121}$$

9.3.2 Three-Dimensional Transient Heat Conduction Problem With Convection

There are many ways of solving the energy equation with convection effects. One that will be presented here is the widely accepted *streamline upwind Petrov-Galerkin method*

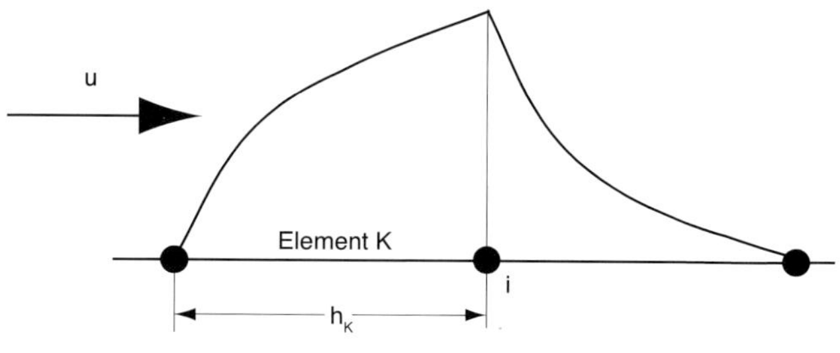

Figure 9.27: Schematic of the shape functions when using upwinding methods.

(SUPG) developed by Brooks and Hughes [3]. Essentially, the finite element equations remain the same; however, as shown here, modified shape functions are introduced on the upwind side of a nodal point. Hence, we have two interpolation, or shape, functions that define the temperature, or convected variable, distribution. One definition uses the conventional shape functions given by

$$T^e = (\mathbf{N}) \cdot (\mathbf{T^e}) \tag{9.122}$$

and the other uses a modified shape function that puts a higher weight on the upwind side of a nodal point and is given by

$$T^e = (\bar{\mathbf{W}}) \cdot (\mathbf{T^e}) \tag{9.123}$$

The modified weighting functions, $\bar{W}_i$, are given by

$$\bar{W}_i = N_i + \frac{\bar{K}}{\| \mathbf{u} \|^2} ([\mathbf{B}] \cdot (\hat{u})) \tag{9.124}$$

where

$$\bar{K} = \frac{\xi \| \hat{\mathbf{u}} \| h_K}{\sqrt{15}} \tag{9.125}$$

In the above equation, ξ is a function of the Kth element size, h_K and the Peclet number given by

$$\xi = \coth(Pe) - 1/Pe \tag{9.126}$$

and where the Peclet number is computed using

$$Pe = \| \hat{\mathbf{u}} \| h_K/(2\alpha) \tag{9.127}$$

Figure 9.27 shows a schematic of the weighting functions around a nodal point. It is clear that the side that lies in the upwind direction of the flow adds more weight that the side that lies in the down-wind direction.

An implicit finite difference time stepping procedure results in the linear set of algebraic equations of the form

$$[\bar{\mathbf{K}}^{\mathbf{e}}] \cdot (\mathbf{T}^{\mathbf{e}}_{\mathbf{j+1}}) = (\bar{\mathbf{f}}^{\mathbf{e}}) \tag{9.128}$$

where $[\bar{\mathbf{K}}^{\mathbf{e}}]$ is a combined stiffness, mass and convective matrix, given by

$$[\bar{\mathbf{K}}^{\mathbf{e}}] = \alpha[\mathbf{K}^{\mathbf{e}}] - \frac{1}{\Delta t}[\bar{\mathbf{M}}^{\mathbf{e}}] + [\bar{\mathbf{C}}^{\mathbf{e}}] \tag{9.129}$$

The element force vector $(\bar{\mathbf{f}}^{\mathbf{e}})$ is formed by the internal heat generation term and the temperatures in the previous time steps as

$$(\mathbf{f}^{\mathbf{e}}) = -\frac{1}{\Delta t}[\bar{\mathbf{M}}^{\mathbf{e}}] \cdot (\mathbf{T}^{\mathbf{e}}_{\mathbf{j}}) + (\dot{\mathbf{Q}}^{\mathbf{e}}) \tag{9.130}$$

The terms in the above matrices are defined by

$$[\mathbf{K}^{\mathbf{e}}] = \int\int\int [\mathbf{B}]^T[\mathbf{B}]dxdydz \tag{9.131}$$

$$[\bar{\mathbf{M}}^{\mathbf{e}}] = \int\int\int [\bar{\mathbf{W}}]^T[\mathbf{N}]dxdydz \tag{9.132}$$

$$[\bar{\mathbf{C}}^{\mathbf{e}}] = \int\int\int [\bar{\mathbf{W}}]([\mathbf{N}]^T \cdot (\mathbf{u}))[\mathbf{B}]dxdydz \tag{9.133}$$

and the internal heat generation term is defined by

$$(\dot{\mathbf{Q}}^{\mathbf{e}}) = \int\int\int [\bar{\mathbf{W}}](\dot{Q})dxdydz \tag{9.134}$$

The three-dimensional mold filling example given below uses this procedure to account for the energy balance.

9.3.3 Three-Dimensional Mixed Formulation for Creeping Flow Problems

In this chapter, we have derived the two-dimensional finite element penalty formulation for creeping flows where the pressure was eliminated by assuming a compressible flow. Here, we will use a mixed formulation, where the pressure is included among the unknown variables. In the mixed formulation, we use different order of approximation for the pressure as we will for the velocity. For instance, if tetrahedral elements are used, we can use a quadratic representation for the velocity (10 nodes) and a linear representation for the pressure (4 nodes). Hence, we must use different shape functions for the velocity and pressure. For such a formulation we can write

$$u_x = \sum_j u_{xj} N^1_j \tag{9.135}$$

$$u^e_y = \sum_j u_{yj} N^1_j \tag{9.136}$$

$$u^e_z = \sum_{j=1} u_{zj} N^1_j \tag{9.137}$$

$$p^e = \sum_{k=1} p_k N^2_k \tag{9.138}$$

Substituting the approximated velocities and pressures into the governing equations, applying Galerkin's method and Green-Gauss transformation where necessary, we arrive at the set of linear algebraic equations given by

$$\begin{bmatrix} [\mathbf{S_c}^{11}] & [\mathbf{S_c}^{22}] & [\mathbf{S_c}^{33}] & 0 \\ [\mathbf{S}^{11}] & [\mathbf{S}^{12}] & [\mathbf{S}^{13}] & [\mathbf{S_p}^{14}] \\ [\mathbf{S}^{12}]^T & [\mathbf{S}^{22}] & [\mathbf{S}^{23}] & [\mathbf{S_p}^{24}] \\ [\mathbf{S}^{13}]^T & [\mathbf{S}^{23}]^T & [\mathbf{S}^{33}] & [\mathbf{S_p}^{34}] \end{bmatrix} \begin{pmatrix} \mathbf{u_x} \\ \mathbf{u_y} \\ \mathbf{u_z} \\ \mathbf{p} \end{pmatrix} = \begin{pmatrix} \mathbf{F^x} \\ \mathbf{F^y} \\ \mathbf{F^z} \\ 0 \end{pmatrix} \tag{9.139}$$

Similar to the two-dimensional formulation for creeping flow, the terms in the above equation are defined by

$$\begin{aligned}
S_{cij}^{11} &= \int\int\int N_i^2 \frac{\partial N_j^1}{\partial x} dxdydz \\
S_{cij}^{12} &= \int\int\int N_i^2 \frac{\partial N_j^1}{\partial y} dxdydz \\
S_{cij}^{13} &= \int\int\int N_i^2 \frac{\partial N_j^1}{\partial z} dxdydz \\
S_{ij}^{11} &= \int\int\int \frac{\partial N_i^1}{\partial x} \frac{\partial N_j^1}{\partial x} dxdydz \\
S_{ij}^{12} &= \int\int\int \frac{\partial N_i^1}{\partial x} \frac{\partial N_j^1}{\partial y} dxdydz \\
S_{ij}^{13} &= \int\int\int \frac{\partial N_i^1}{\partial x} \frac{\partial N_j^1}{\partial z} dxdydz \\
S_{ij}^{22} &= \int\int\int \frac{\partial N_i^1}{\partial y} \frac{\partial N_j^1}{\partial y} dxdydz \\
S_{ij}^{23} &= \int\int\int \frac{\partial N_i^1}{\partial y} \frac{\partial N_j^1}{\partial z} dxdydz \\
S_{ij}^{33} &= \int\int\int \frac{\partial N_i^1}{\partial z} \frac{\partial N_j^1}{\partial z} dxdydz \\
S_{pij}^{11} &= \int\int\int \frac{\partial N_j^1}{\partial x} N_i^2 dxdydz \\
S_{pij}^{12} &= \int\int\int \frac{\partial N_j^1}{\partial y} N_i^2 dxdydz \\
S_{pij}^{13} &= \int\int\int \frac{\partial N_j^1}{\partial z} N_i^2 dxdydz
\end{aligned} \tag{9.140}$$

and the force vector is computed using

$$F_i^x = \int\int\int f_x N_i dxdydz + \oint t_x N_i^1 dS \tag{9.141}$$

$$F_i^y = \int\int\int f_y N_i dxdydz + \oint t_y N_i^1 dS \tag{9.142}$$

$$F_i^y = \int\int\int f_z N_i dxdydz + \oint t_z N_i^1 dS \tag{9.143}$$

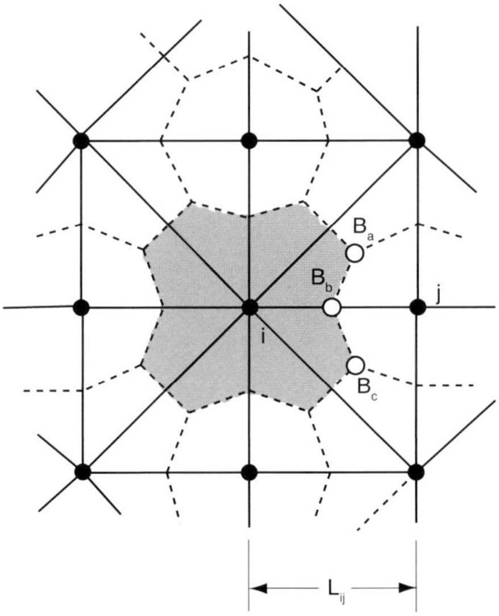

Figure 9.28: Schematic diagram of a control volume in a three-noded element mesh.

9.4 MOLD FILLING SIMULATIONS USING THE CONTROL VOLUME APPROACH

As discussed in Chapter 8, where we presented the *flow analysis network* (FAN), mold filling is an important aspect of injection and compression molding processes. During mold filling, the flow fronts will form weldlines (knitlines in compression molding), air pockets (which can lead to *dieseling*), molecular or fiber orientation, among others. The orientation in turn leads to anisotropic thermo-mechanical properties, which control the properties of the final part, as well as shrinkage and warpage. Consequently, mold filling simulation presents the information required to understand and optimize a process and the performance of a final product.

9.4.1 Two-Dimensional Mold Filling Simulation of Non-Planar Parts (2.5D Model)

Based on the flow analysis network, Wang *et al.*, [18] and Osswald [11] developed the finite element/control volume appproach (FEM-CVA) for injection and compression molding, respectively. Similar to FAN, FEM-CVA assigns a fill factor to every nodal point or nodal control volume. The nodal control volumes are constructed by connecting element centroids to element midsides, as shown in Fig. 9.28.

The control volumes are defined as follows:

- Gate nodes - Node through which the melt will enter the cavity. At the beginning of mold filling, entrance nodes start as full ($f_i = 1$)
- Filled nodes - Nodes which lie behind the flow front are considered filled ($f_i = 1$)
- Melt front nodes - Nodal control volumes containing the free flow front ($0 < f_i < 1$)

- Empty nodes - Node whose control volume lies ahead of the flow front and are still empty ($f_i = 0$)

The boundary conditions are defined in the same way as with the flow analysis network. The nodes whose control volumes are empty or partially filled are assigned a zero pressure, and the gate nodes are either assigned an injection pressure or an injection volume flow rate. Just as is the case with flow analysis network, a mass balance about each nodal control volume will lead to a linear set of algebraic equations, identical to the set finite element formulation of Poisson's or Laplace's equation. The mass balance (volume balance for incompressible fluids) is given by

$$\sum_{j=1}^{N} q_{ij} = 0 \tag{9.144}$$

for injection molding, or

$$\sum_{j=1}^{N} q_{ij} = -A_i \dot{h}_i \tag{9.145}$$

for compression molding.

The intra-nodal flow rates, q_{ij} are given by

$$q_{ij} = \frac{S_{ij}}{h_{ij}} \frac{p_j - p_i}{L_{ij}} (h_{ij} X_{ij}) \tag{9.146}$$

where S_{ij} is the flow conductance for the Hele-Shaw model defined in Chapter 5, p_i and p_j are nodal pressures, h_{ij} is the average gap height between nodes i and j, L_{ij} is the distance between nodes i and j and X_{ij} is the width of the effective window between the two nodes under consideration given by

$$X_{ij} = \overline{B_a B_b} \hat{U}_{ab} + \overline{B_b B_c} \hat{U}_{bc} \tag{9.147}$$

where points B_a, B_b and B_c are defined in Fig. 9.28 along with the other geometric parameters used in the above equations and where $\hat{U}_{ab}$ and $\hat{U}_{bc}$ are defined by

$$\hat{U}_{ab} = \mathbf{d}_{ij} \cdot \mathbf{n}_{ab}$$
$$\hat{U}_{bc} = \mathbf{d}_{ij} \cdot \mathbf{n}_{bc}$$

where $\mathbf{d}_{ij}$ represents a unit vector between nodes i and j and, $\mathbf{n}_{ab}$ and $\mathbf{n}_{bc}$, represent unit normal vectors of lines $B_a B_b$ and $B_b B_c$, respectively.

Once the boundary conditions are applied, the pressure field can be solved using the appropriate matrix solving routines. Note that for mold filling problems, there is a natural boundary condition that satisfies no flow across mold boundaries or shear edges, $\partial p / \partial n = 0$. Once the pressure field has been solved, it is used to perform a mass balance using eqn. (9.144) or (9.145). Once the flowrates across nodal control volume boundaries are known, a simulation program updates the nodal control volume fill factors using

$$f_i^{k+1} V_i^{k+1} = f_i^k V_i^k + \sum_j q_{ij} \Delta t \tag{9.148}$$

where Δt is the time step and the subscripts k and $k+1$ denote the time level. Note that the volume has been assigned a time step because, when modeling compression molding, the

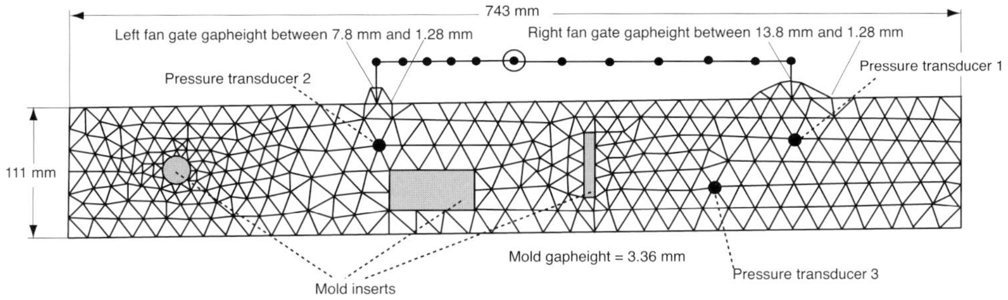

Figure 9.29: Triangular and tubular finite element representation of rectangular mold cavity with 3 inserts [18].

old and the new volumes, V_i^k and V_i^{k+1} are different since the element gap heights change with time. This accounts for the squeezing motion of the mold.

The optimum time step in a FEM-CVA simulation is the one that fills exactly one new control volume. Once the fill factors are updated, the simulation proceeds to solve for a new pressure and flow field, which is repeated until all fill factors are 1. While the FEM-CVA scheme does not know exactly, where the flow front lies, one can recover flow front information in post-processing quite accurately. One very common technique is for the simulation program to record the time when a node is half full, $f_i = 0.5$. This operation is performed when the nodal fill factors are updated; if the node has $f_i^k < 0.5$ and $f_i^{k+1} > 0.5$ then the time at which the fill factor was 0.5 is found by interpolating between t^k and t^{k+1}. These *half-times* are then treated as nodal data and the flow front or filling pattern at any time is drawn as a contour of the corresponding half-times, or isochronous curves.

■ EXAMPLE 9.5.

Injection molding filling of a two-gated rectangular mold. Wang and co-workers [18] implemented this technique into a simulation program to predict the non-Newtonian, non-isothermal injection mold filling process. They tested their technique with a two-gated rectangular mold with three inserts and variable runner diameters. They chose an unbalanced runner system on purpose to better illustrate the simulation program. In addition to the above formulation, they used two-noded elements to represent the runner system. Figure 9.29 presents the finite element mesh employed by Wang *et al.* [18] with the dimensions and location of the pressure transducers used to record pressure during mold filling. The fan gates were of variable thickness as pointed out in the figure, and the mold cavity was of constant thickness.

The material used in the experiments was an ABS polymer, whose viscosity was approximated with a shear thinning temperature dependent power law model. Figure 9.30 presents the experimental as well predicted filling pattern. Both filling patterns show relatively good qualitative agreement. Similar agreement was found in the predicted and measured pressure traces. Transducers 1 and 3 present very good agreement, whereas the predictions for transducer 2 seem to be consistently lower by about 20%. This discrepancy is explained by the fact that at the beginning, the flow rate out of the left runner is under-predicted. Such a discrepancy was attributed by Wang and co-workers to the fact that the *juncture losses* in the bends of the runner

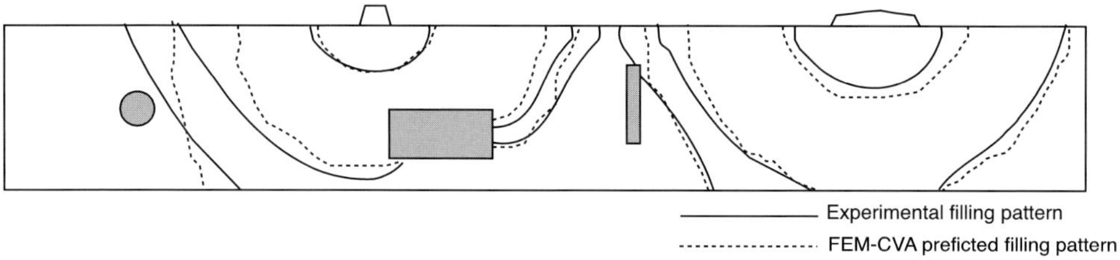

Figure 9.30: Comparison between experimental and FEM-CVA predicted filling patterns. [18].

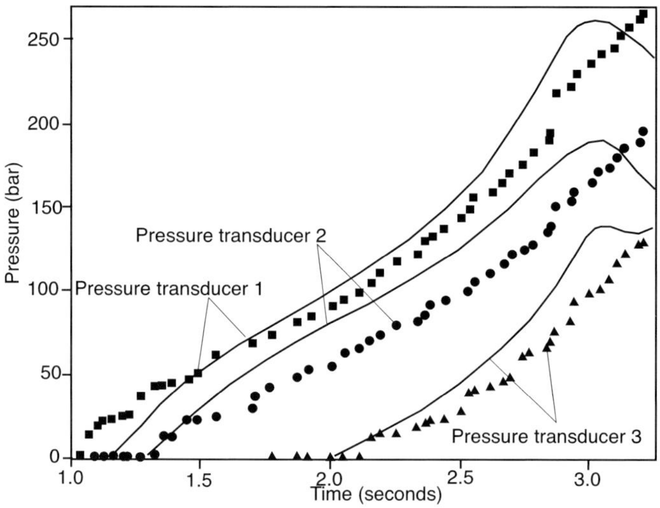

Figure 9.31: Comparison between experimental and FEM-CVA predicted pressure history at three pressure transducer locations. [18].

system, as well as the runner-gate and gate-cavity, interfaces were not taken into account by the simulation program.

■ **EXAMPLE 9.6.**

Compression Mold Filling of an Automotive Hood. The FEM-CVA was also implemented by Osswald and Tucker [11] and [12] to solve for the mold filling during compression molding of sheet molding compound. They used the Barone and Caulk model for compression molding of thin SMC parts. The Barone and Caulk model for thin charges results from a momentum balance where the dominant forces are driving pressures and a hydrodynamic friction coefficient, K_H, between the charge and the upper and lower mold surface. The momentum balance is written as

$$u_x = -\frac{h(t)}{2K_H}\frac{\partial p}{\partial x} \tag{9.149}$$

and,

$$u_y = -\frac{h(t)}{2K_H}\frac{\partial p}{\partial y} \tag{9.150}$$

for the x- and y-directions, respectively. Substituting these equations in an integrated continuity equation (as done in the Hele-Shaw model) we get

$$\frac{\partial^2 p}{\partial x^2} + \frac{\partial^2 p}{\partial x^2} = -\frac{2K_H(-\dot{h})}{h(t)^2} \tag{9.151}$$

Osswald and Tucker [12] compared their simulation with the mold filling process of a compression molded automotive hood. The final thickness of the hood was approximately 2.8 mm, but only 1.27 mm in the headlight area. Since the part is symmetric only half was simulated. In the experiments, short shots were produced by placing sheet metal shims on the mold stops. The shims prevented the mold from completely closing, causing the flow to stop at intermediate stages of mold filling. Figure 9.32 presents a comparison between the experimental and the predicted filling pattern. It should be noted that although the hood in the figure appears to be flat, a three-dimensional mesh was used to represent the curved hood geometry. However, each element within the mesh represented a local two-dimensional flow as presented in eqns. (9.149) to (9.151), oriented in 3D space. This type of approximation to model the flow in non-planar parts (2D flow in 3D space) is referred to as 2.5D flow. The figure shows excellent agreement between experimental and predicted filling patterns.

9.4.2 Full Three-Dimensional Mold Filling Simulation

Based on the control volume approach and using the three-dimensional finite element formulations for heat conduction with convection and momentum balance for non-Newtonian fluids presented earlier, Turng and Kim [10] and [17] developed a three-dimensional mold filling simulation using 4-noded tetrahedral elements. The nodal control volumes are defined by surfaces that connect element centroids and sides as schematically depicted in Fig. 9.33.

The element side surfaces are formed by lines that connect the centroid of the triangular side and the midpoint of the edge. Kim's definition of the control volume fill factors are the same as described in the previous section. Once the velocity field within a partially filled mold has been solved for, the melt front is advanced by updating the nodal fill factors. To test their simulation, Turng and Kim compared it to mold filling experiments done with the optical lenses shown in Fig. 9.34. The outside diameter of each lens was 96.19 mm and the height of the lens at the center was 19.87 mm. The thickest part of the lens was 10.50 mm at the outer rim of the lens. The thickness of the lens at the center was 6 mm. The lens was molded of a PMMA and the weight of each lens was 69.8 g.

The finite element mesh and boundary conditions used to represent the lens mold are presented in Fig. 9.35 [10]. The finite element mesh was generated using 90,352 four-noded tetrahedral elements with 17,355 nodal points, with 4 to 5 element rows across the thickness of the mold. Since the ram speed was constant during the mold filling process, the flow rate was assumed constant during the filling of the cavity. The filling time was approximately 4.9 seconds. Based on the surface area of the gate, a uniform inlet velocity of 120 mm/s was used as a boundary condition in order to have a 4.9 second fill time. The inlet temperature

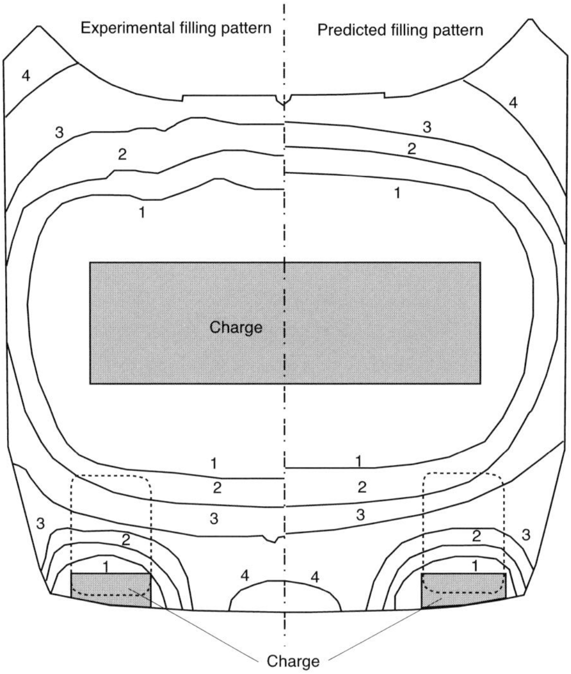

Figure 9.32: Comparison between experimental and FEM-CVA predicted filling patterns during compression molding of an automotive hood [12].

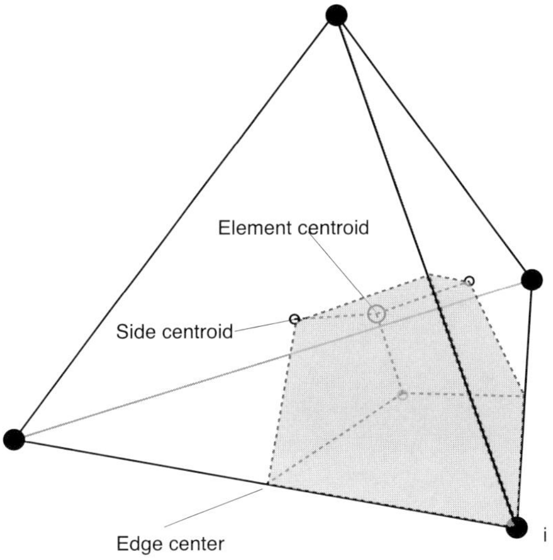

Figure 9.33: Contribution of a tetrahedral element to nodal control volume i [10].

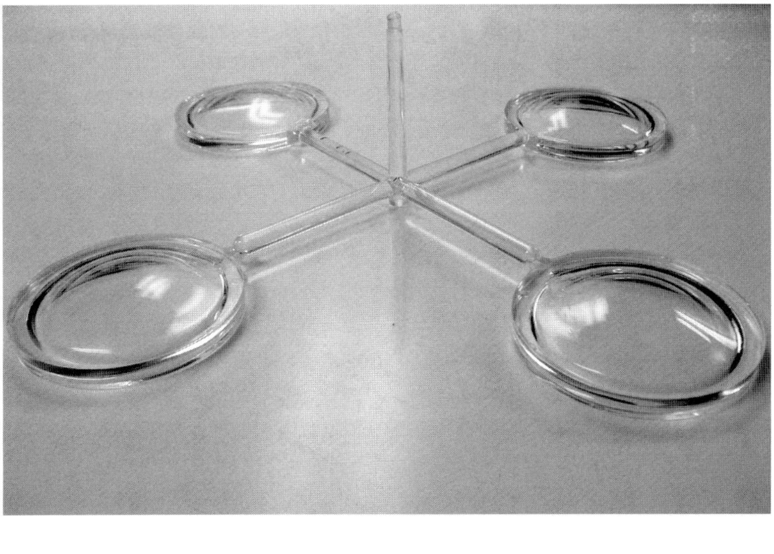

Figure 9.34: Photograph of optical lens parts molded by a four-cavity mold [10] (courtesy of 3M Precision Lens).

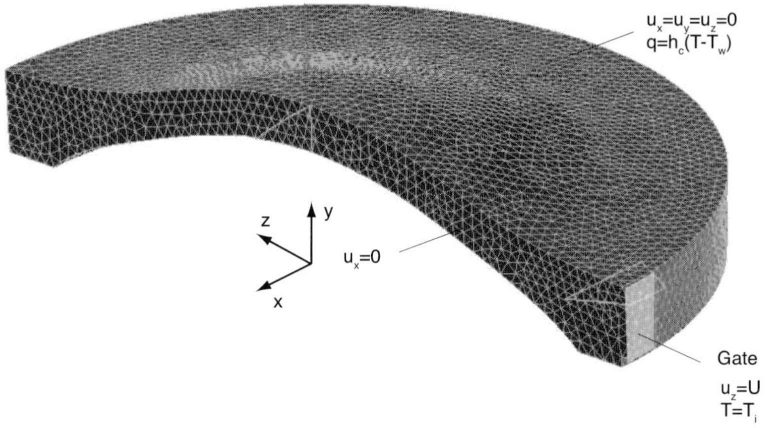

Figure 9.35: Finite element mesh of one half of the lens geometry and associated boundary conditions [10].

was assumed to be 505.37 K and a heat transfer coefficient of 4,000 W/m^2/K was applied at the mold surfaces.

The polymer melt viscosity was simulated using a cross-model given by

$$\eta(\dot{\gamma}, T, p) = \frac{\eta_0(T, p)}{1 + \left(\dfrac{\eta_0 \dot{\gamma}}{\tau^*}\right)^{1-n}} \tag{9.152}$$

where n is the power law index and τ^* is the stress level at which the viscosity is during the transition between zero shear-rate viscosity and the shear thinning region. For amorphous thermoplastics, it is common to use a WLF equation for the temperature dependence as

$$\eta_0(T, p) = D_1 e^{-\dfrac{A_1(T - T^*)}{A_2 + (T - T^*)}} \tag{9.153}$$

when $T > T^*$ and $\eta_0 = \infty$ when $T \leq T^*$, where T^* is the glass transition temperature. In the above equation

$$T^* = D_2 + D_3 p \tag{9.154}$$

and

$$A_2 = \hat{A}_2 + D_3 p \tag{9.155}$$

Turng and Kim used the material properties listed in Table 9.4.

Figure 9.36 presents a comparison of the experimental and numerical filling pattern. As can be seen, the agreement is excellent. It must be pointed out that the solution was dependent of the heat transfer coefficient between the mold wall and the flowing polymer melt.

Figure 9.37 presents the position of the flow front before the mold fills. It is evident from the shape of the flow front at that stage of filling that an intricate weldline is forming

Table 9.4: Material Properties for PMMA

Property	Value
ρ	1185kg/m^3
k	0.21 W/m/K
C_p	2300 J/kg/K
n	0.264
τ^*	9.9338×10^4
A_1	42.565
$\hat{A}_2$	51.6 K
D_1	5.86×10^{16}
D_2	377.15 K
D_3	0 K/Pa

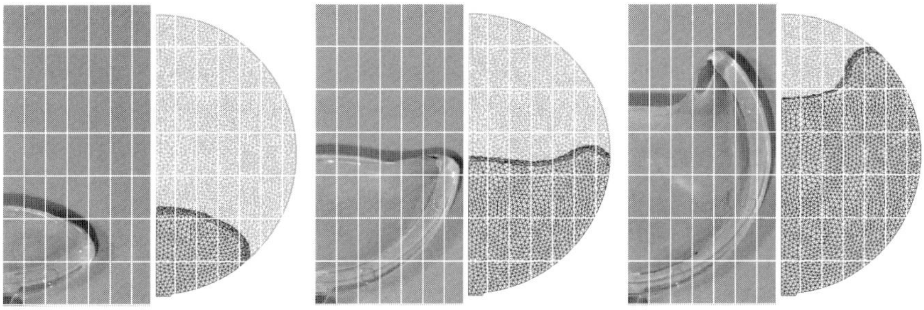

Figure 9.36: Experimental and predicted melt front advancement at fill times of 1.2, 2.4 and 3.6 seconds [10].

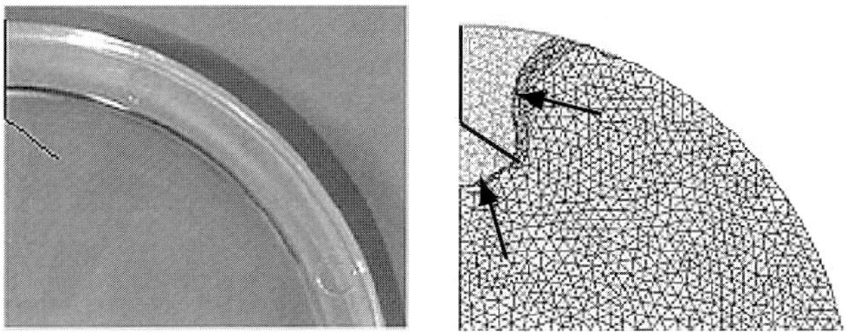

Figure 9.37: Melt front prediction at 4.6 seconds filling time, and weld line location in the final part [10].

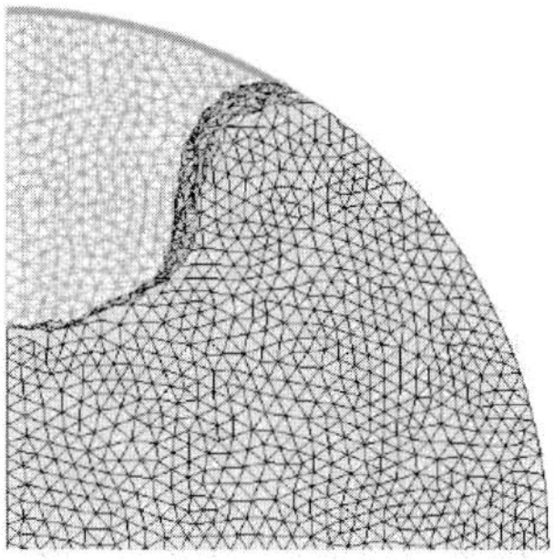

Figure 9.38: Melt front prediction at the end of filling after raising the mold temperature by 50 K [10].

in that region. The predicted temperature fields reflected up to 20 K of temperature rise due to viscous dissipation.

The same simulation was used to find the optimal conditions that would eliminate the weldline that formed during the end of filling. It was found that the flow was significantly affected when increasing the mold temperature by 50 K. This change not only created a flow that eliminated the weldline location, but also further reduced birefringence effects, a desirable factor when manufacturing optical lenses. Figure 9.38 presents the simulated flow front location at the end of filling after raising the mold temperature by 50 K. Here, we can clearly see the reduction in the race-tracking effect, that was pronounced with the original process conditions.

9.5 VISCOELASTIC FLUID FLOW

In general, as discussed in Chapter 5, the total stress tensor is defined by

$$\begin{aligned}\boldsymbol{\sigma} &= -p\boldsymbol{\delta} + \boldsymbol{\tau} \\ \sigma_{ij} &= -p\delta_{ij} + \tau_{ij}\end{aligned} \tag{9.156}$$

where $\boldsymbol{\delta}$ is the unit tensor and δ_{ij} is the Kronecker delta. With this total tensor, the conservation laws for an incompressible isothermal fluid flow yields

$$\frac{\partial u_i}{\partial x_i} = 0 \tag{9.157}$$

$$\frac{\partial^2 \sigma_{ij}}{\partial x_j} + \rho f_i = \rho \frac{Du_i}{Dt} \tag{9.158}$$

where $\mathbf{f}$ is the body force per unit mass. Here, we need to close the system of equations with a constitutive equation, which will relate the stress, $\boldsymbol{\tau}$, to the deformation experienced by the fluid. In some applications it is better to split these stress into a purely viscous component, $\boldsymbol{\tau}_N$, which is usually interpreted as the solvent contribution to the stress in the polymeric solution, or as the stress response associated with fast relaxation modes [2, 6, 13, 15], and an extra-stress which contains all the elastic components of the stress tensor. This split stress has a lot of impact on the mathematical nature of the full set of governing equations, making the equations numerically more stable [14]. The stress tensor can now be written as

$$\boldsymbol{\tau} = \boldsymbol{\tau_N} + \boldsymbol{\tau_V}$$
$$\tau_{ij} = \tau_{Nij} + \tau_{Vij} \tag{9.159}$$

where $\boldsymbol{\tau_V}$ denotes the *viscoelastic stress*, while $\boldsymbol{\tau_N}$ is the Newtonian component defined by

$$\boldsymbol{\tau_N} = \eta_N \dot{\boldsymbol{\gamma}} \tag{9.160}$$

For the viscoelastic stress, we can use differential or integral constitutive models (see Chapter 2). For differential models we have the general form

$$Y\boldsymbol{\tau_V} + \lambda_1 \boldsymbol{\tau_{V(1)}} + \lambda_2 \left(\dot{\boldsymbol{\gamma}} \cdot \boldsymbol{\tau_V} + \boldsymbol{\tau_V} \cdot \dot{\boldsymbol{\gamma}}\right) + \lambda_3 \left(\boldsymbol{\tau_V} \cdot \boldsymbol{\tau_V}\right) = \eta_V \dot{\boldsymbol{\gamma}} \tag{9.161}$$

where the upper-convective derivative $\boldsymbol{\tau_{V(1)}}$ is defined as

$$\boldsymbol{\tau_{V(1)}} = \frac{D\boldsymbol{\tau_V}}{Dt} - \left(\boldsymbol{\tau_V} \cdot \nabla \boldsymbol{u} + \nabla \boldsymbol{u}^T \cdot \boldsymbol{\tau_V}\right) \tag{9.162}$$

In order to solve viscoelastic problems, we must select the most convenient model for the stress and then proceed to develop the finite element formulation. Doue to the excess in non-linearity and coupling of the viscoelastic momentum equations, three distinct Galerkin formulations are used for the governing equations, i.e., we use different shape functions for the viscoelastic stress, the velocity and the pressure

$$\begin{aligned}
\boldsymbol{\tau_V^e} &= \sum_{k1=1}^{n_\tau} \boldsymbol{\tau_{V i}} N_i^1 \\
\boldsymbol{u}^e &= \sum_{k2=1}^{n_u} \boldsymbol{u}_i N_i^2 \\
p^e &= \sum_{k3=1}^{n_p} p_i N_i^3
\end{aligned} \tag{9.163}$$

where N_i^1, N_i^2 and N_i^3 represent the three different finite element shape functions and n_τ, n_u and n_p will be the order of the element for each variable.

The next step will be to formulate the Galerkin-weighted residual for each of the governing equations

$$\int_V \left[\frac{\partial u_i^e}{\partial x_i}\right] N_{k3}^3 dV = 0 \tag{9.164}$$

$$\int_V \left[\frac{\partial}{\partial x_i}\left(-p\delta_{ij} + \eta_N \dot{\gamma}_{ij}^e + \tau_{Vij}^e\right) + \rho\left(f_i^e - \frac{Du_i^e}{Dt}\right)\right] N_{k2}^2 dV = 0 \tag{9.165}$$

$$\begin{aligned}
&\int_V \left[\mathbf{Y}\tau_{Vij} + \lambda_1 (\tau_{Vij})_{(1)} + \lambda_2 \left(\dot{\gamma}_{il}\tau_{Vlj} + \tau_{Vil}\dot{\gamma}_{lj}\right) + \right. \\
&\qquad \left. \lambda_3 \left(\tau_{Vil}\tau_{Vlj}\right) - \eta_V \dot{\gamma}_{ij}\right] N_{k1}^1 = 0
\end{aligned} \tag{9.166}$$

Applying the Green-Gauss Theorem (9.1.2) to the residual of the momentum eqn. (9.165) we get

$$\int_V \left[N_j^2 \rho \left(\frac{Du_i^e}{Dt} - f_i^e \right) + \frac{\partial N_j^2}{\partial x_k} \left(-p^e \delta_{ik} + \eta_N \dot{\gamma}_{ik}^e + \tau_{Vik}^e \right) \right] dV = \oint_S N_j^2 \sigma_{ik}^e n_k dS \quad (9.167)$$

where n_k is the outward unit normal, S the arc length measured along the boundary and $\sigma_{ij} n_j$ is the traction at the boundary, which can be specified with this type of formulations. Formulations like this have been used widely in the literature to analyze viscoelastic flow problems [1, 4, 9, 14].

Using the upper-convective Maxwell fluid eqn. (9.166) it reduces to

$$\int_V \left[\tau_{Vij} + \lambda_1 (\tau_{Vij})_{(1)} - \eta_V \dot{\gamma}_{ij} \right] N_{k1}^1 = 0 \quad (9.168)$$

For a two-dimensional problem, we can define the following set of matrices and tensors that will help arrange the FE system

$$A_{ij} = \int_V N_i^1 N_j^1 dxdy,$$

$$B_{ij}^x = \int_V N_i^1 \frac{\partial N_j^2}{\partial x} dxdy, \qquad B_{ij}^y = \int_V N_i^1 \frac{\partial N_j^2}{\partial y} dxdy,$$

$$C_{ij}^x = \int_V N_i^3 \frac{\partial N_j^2}{\partial x} dxdy, \qquad C_{ij}^y = \int_V N_i^3 \frac{\partial N_j^2}{\partial y} dxdy,$$

$$D_{ijk}^x = \int_V N_i^1 N_j^3 \frac{\partial N_k^1}{\partial x} dxdy, \qquad D_{ijk}^y = \int_V N_i^1 N_j^2 \frac{\partial N_j^1}{\partial y} dxdy,$$

$$E_{ijk}^x = \int_V N_i^1 N_k^1 \frac{\partial N_k^2}{\partial x} dxdy, \qquad E_{ijk}^y = \int_V N_i^1 N_j^1 \frac{\partial N_j^2}{\partial y} dxdy,$$

$$F_{ij} = \int_V \left[2 \frac{\partial N_i^2}{\partial x} \frac{\partial N_j^2}{\partial x} + \frac{\partial N_i^2}{\partial y} \frac{\partial N_j^2}{\partial y} \right] dxdy$$

$$G_{ij} = \int_V \left[2 \frac{\partial N_i^2}{\partial y} \frac{\partial N_j^2}{\partial y} + \frac{\partial N_i^2}{\partial x} \frac{\partial N_j^2}{\partial x} \right] dxdy$$

$$H_{ij} = \int_V \frac{\partial N_i^2}{\partial x} \frac{\partial N_j^2}{\partial y} dxdy$$

$$I_{ijk}^x = \int_V N_i^2 N_j^2 \frac{\partial N_k^2}{\partial x} dxdy, \qquad I_{ijk}^y = \int_V N_i^2 N_j^2 \frac{\partial N_k^2}{\partial y} dxdy$$

Using these definitions, the Galerkin finite element formulation for the stress equations will be

$$A_{ij}\tau_{V11}^j + \lambda\left[\left(D_{ijk}^x u_1^j - 2E_{ijk}^x u_1^j + D_{ijk}^y u_2^j\right)\tau_{V11}^k - 2E_{ijk}^y u_1^j \tau_{V12}^k\right] = 2\eta_V B_{ij}^x u_1^j \tag{9.169}$$

$$A_{ij}\tau_{V22}^j + \lambda\left[\left(D_{ijk}^x u_1^j - 2E_{ijk}^y u_2^j + D_{ijk}^y u_2^j\right)\tau_{V22}^k - 2E_{ijk}^x u_2^j \tau_{V12}^k\right] = 2\eta_V B_{ij}^y u_2^j \tag{9.170}$$

$$A_{ij}\tau_{V12}^j + \lambda\left[\left[\left(D_{ijk}^x - E_{ijk}^x\right)u_1^j + \left(D_{ijk}^y - E_{ijk}^y\right)u_2^j\right]\tau_{V12}^k - E_{ijk}^y u_1^j \tau_{V22}^k - E_{ijk}^x u_2^j \tau_{V11}^k\right] = \eta_V\left(B_{ij}^y u_1^j + B_{ij}^x u_2^j\right) \tag{9.171}$$

and for the momentum equation

$$B_{ji}^x \tau_{V11}^j + B_{ji}^y \tau_{V12}^j + \eta_N F_{ij} u_1^j + \eta_N H_{ij} u_2^j - C_{ij}^x p^j + \rho\left(I_{ijk}^x u_1^j + I_{ijk}^y u_2^j\right)u_1^k = f_i^x \tag{9.172}$$

$$B_{ji}^x \tau_{V12}^j + B_{ji}^y \tau_{V22}^j + \eta_N H_{ij} u_1^j + \eta_N F_{ij} u_2^j - C_{ij}^y p^j + \rho\left(I_{ijk}^x u_1^j + I_{ijk}^y u_2^j\right)u_2^k = f_i^y \tag{9.173}$$

The continuity equation is given by

$$-C_{ij}^x u_1^j - C_{ij}^y u_2^j = 0 \tag{9.174}$$

Here, τ_{V11}^j, τ_{V12}^j, τ_{V22}^j, u_1^j, u_2^j and p^j represent the nodal values of the stress, velocity and pressure. Finally, the right hand side of the momentum equations contain the contribution of the body forces and the tractions imposed at the boundary

$$f_i^x = \int_V N_i^2 \rho f_x dxdy + \oint_S N_i^2\left(\sigma_{1j}n_j\right)dS \tag{9.175}$$

$$f_i^y = \int_V N_i^2 \rho f_y dxdy + \oint_S N_i^2\left(\sigma_{2j}n_j\right)dS \tag{9.176}$$

The convective terms in a viscoelastic flow become dominant at heigh Weissenberg numbers and must be dealt with in a similar manner as with diffusion problems that have large advective effects. Hence, the streamline upwind Petrov-Garlerkin (SUPG) method must also be used. In such problems, the conventional shape functions used as weighting functions are replaced by those that put larger weight on the elements that lie in the upwind side of a node.

■ **EXAMPLE 9.7.**

Viscoelastic flow effects in polymer coextrusion. In this example we will present work done by Dooley [7, 8] on the viscoelastic flow in multilayer polymer extrusion. Dooley performed extensive experimental work where he coextruded multilayer systems through various non-circular dies such as the teardrop channel presented in Fig. 9.39. For the specific example shown, 165 layers were coextruded through a feedblock to form a single multiple-layer structure inside the channel.

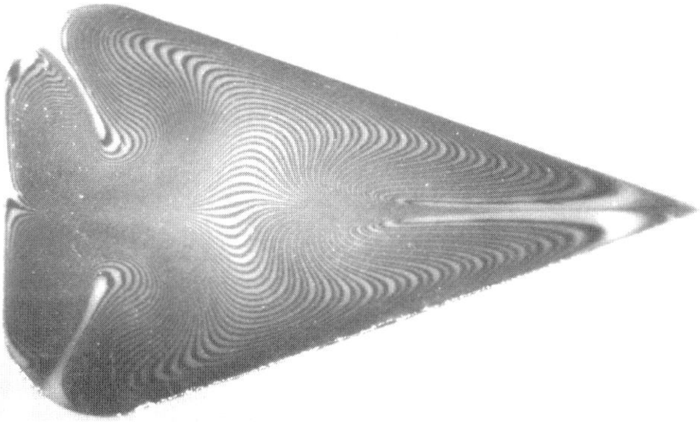

Figure 9.39: 165-layer polystyrene structure near the end of a tear drop channel geometry [8, 7].

For the solution of this problem, the momentum and continuity equations for the steady-state flow of an incompressible viscoelastic fluid are given by

$$-\frac{\partial p}{\partial x_i} + \frac{\partial \tau_{ij}}{\partial x_j} = 0 \tag{9.177}$$

$$\frac{\partial u_i}{\partial x_i} = 0 \tag{9.178}$$

where $\boldsymbol{\tau}$ is the viscoelastic stress tensor which can be decomposed as a discrete spectrum of N relaxation times as follows

$$\boldsymbol{\tau} = \sum_{i=1}^{N} \boldsymbol{\tau}_i \tag{9.179}$$

For each $\boldsymbol{\tau}_i$, a constitutive equation must be selected. Dooley and Dietsche [5] evaluated the White-Metzner, the Phan-Thien Tanner-1, and the Giesekus models, given by,

$$\boldsymbol{\tau}_i + \lambda_i \boldsymbol{\tau}_{i(1)} = \eta_i \dot{\boldsymbol{\gamma}} \tag{9.180}$$

$$\exp\left[\frac{\epsilon_i \lambda_i}{\eta_i} \mathrm{tr}(\boldsymbol{\tau}_i)\right] \boldsymbol{\tau}_i + \lambda_i \left[(1 - \frac{\xi_i}{2}) \boldsymbol{\tau}_{i(1)} + \frac{\xi_i}{2} \boldsymbol{\tau}_i^{(1)}\right] = \eta_i \dot{\boldsymbol{\gamma}} \tag{9.181}$$

and

$$\boldsymbol{\tau}_i + \frac{\alpha_i \lambda_i}{\eta_i} \boldsymbol{\tau}_i \cdot \boldsymbol{\tau}_i + \lambda_i \boldsymbol{\tau}_{i(1)} = \eta_i \dot{\boldsymbol{\gamma}} \tag{9.182}$$

respectively. The finite element technique was used to solve the above equations using quadratic elements for the velocities, while the stress and the pressure were approximated using linear interpolation functions. Figure 9.40 presents the mesh

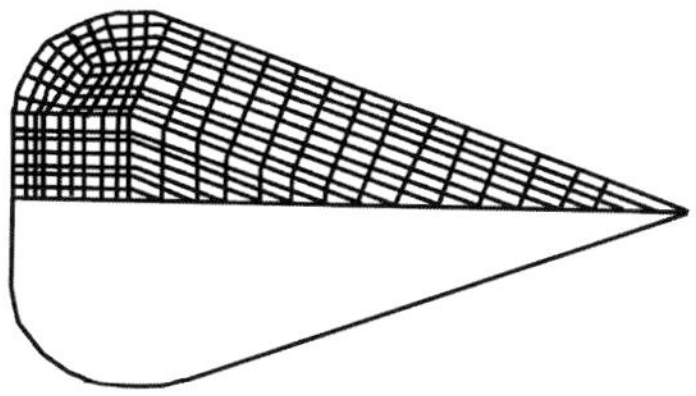

Figure 9.40: Finite element mesh used to simulate the teardrop channel geometry [7].

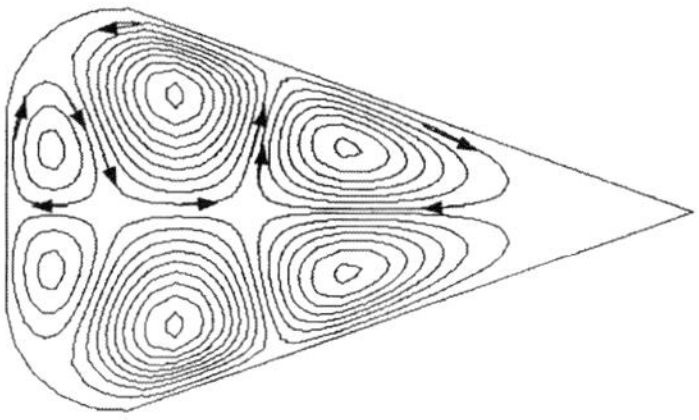

Figure 9.41: Predicted flow patterns for an elastic material in a teardrop channel geometry [7].

used to represent the teardrop channel geometry. When more than one relaxation time was used the split between the viscous and viscoelastic stress was performed.

Figure 9.41 presents the predicted secondary flow patterns that result from the vicoelastic flow effects. The Giesekus model with one relaxation time was used for the solution presented in the figure. For the simulation, a relaxation time, λ, of 0.06 seconds was used along with a viscosity, η, of 8,000 Pa-s and a constant α of 0.80. Similar results were achieved using the Phan-Thien Tanner-1 model. As expected, when the White-Metzner model was used, a flow without secondary patterns was predicted. This is due to the fact that the White-Metzner model has a second normal stress difference, N_2 of zero.

Problems

9.1 Integrate eqn. (9.49) to derive the terms in the one-dimensional element mass matrix given in eqn. (9.50).

9.2 Derive the equations that result in the element mass matrix for the constant strain triangle given in eqn. (9.72).

9.3 Write a 1D FEM program using 2-noded tube elements to balance complex runner systems in injection molding. Compare the simulation to the runner system presented in Chapter 6.

9.4 Derive the equations that result in the constant strain triangle element force vector that represents the internal heat generation term $\dot{Q}$ given in eqn. (9.72).

9.5 What would the constant strain finite element equations look like for the transient heat conduction problem with internal heat generation if you were to use a Crank-Nicholson time stepping scheme?

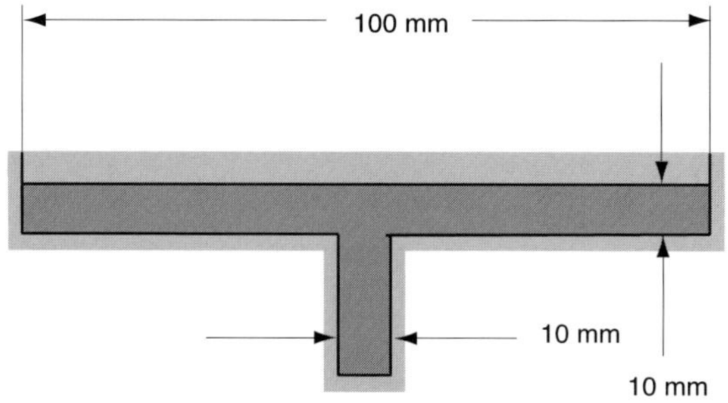

Figure 9.42: Ribbed part molded with unsaturated polyester.

9.6 Write a two-dimensional finite element program, using constant strain triangles and 1D tube elements, to predict the flow and pressure distribution in a variable thickness die. Use the Hele-Shaw model. Compare the FEM results with the analytical solution for an end-fed sheeting die.

9.7 Write a two dimensional FEM program using 2D 4-noded isoparametric elements to model compression molding of a non-Newtonian (power law) fluid. Use the geometry and parameters given for the L-shaped charge given in Fig. 9.18.

9.8 In your university library, find the paper Barone, M.R. and T.A. Osswald, *J. of Non-Newt. Fluid Mech.*, 26, 185-206, (1987), and write a 2D FEM program to simulate the compression molding process using the Barone-Caulk model presented in the paper. Compare your results to the BEM results presented in the paper.

9.9 Write a two-dimensional finite element program, using constant strain triangles, to predict the curing reaction of unsaturated polyester parts. Use the Kamal-Sourour model and the kinetic constants given in Chapter 8. Predict the degree of cure as a function of time of the ribbed cross-section given in Fig. 9.42 Assume an initial temperature of 25°C and a mold temperature of 150°C.

REFERENCES

1. G. Astarita and G. Marrucci. *Principles of Non-Newtonian Fluid Mechanics*. McGraw-Hill, London, 1974.
2. R. Byron Bird, Charles Curtiss, F. Robert C. Armstrong, and Ole Hassager. *Dynamics of Polymer Liquids: Kinetic Theory*, volume 2. John Wiley & Sons, 2nd edition, 1987.
3. A.N. Brooks and T.J.R. Hughes. Streamline upwind-Petrov-Garlerkin formulation for convection dominated flows with particular emphasis on the incompressible Navier-Stokes equations. *Comp. Meth. Appl. Mech. Eng.*, 32:199, 1982.
4. M.J. Crochet, A.R. Davis, and K. Walters. *Numerical Simulation of Non-Newtonian Flow*. Elsevier, Amsterdam, 1984.
5. L. Dietsche and J. Dooley. In *SPE ANTEC*, volume 53, page 188, 1995.
6. M. Doi and S.F. Edwards. *The Theory of Polymer Dynamics*. Oxford University Press, 1986.
7. J. Dooley. *Viscoelastic Flow Effects in Multilayer Polymer Coextrusion*. PhD thesis, TU Eindhoven, 2002.

8. J. Dooley and K. Hughes. In *SPE ANTEC*, volume 53, page 69, 1995.

9. M. Kawahara and N. Takeuchi. Mixed finite element method for analysis of viscoelastic fluid flow. *Comput. Fluids.*, 5:33, 1977.

10. S.-W. Kim. *Three-Dimensional Simulation for the Filling Stage of the Polymer Injection Molding Process Using the Finite Element Method*. PhD thesis, University of Wisconsin-Madison, 2005.

11. T.A. Osswald. *Numerical Methods for compression mold filling simulation*. PhD thesis, University of Illinois at Urbana-Champaign, Urbana-Champaign, 1986.

12. T.A. Osswald and C.L. Tucker III. Compression mold filling simulation for non-planar parts. *Intern. Polym. Proc.*, 5(2):79–87, 1990.

13. H.-C. Öttinger. *Stochastic Processes in Polymeric Fluids*. Springer, 1996.

14. J.F.T. Pittman and C.L. Tucker III. *Fundamentals of Computer Modeling for Polymer Processing*. Hanser Publishers, Munich, 1989.

15. M. Rubinstein and R.H. Colby. *Polymer Physics*. Oxford University Press, 2003.

16. M.J. Turner, R.W. Clough, H.C. Martin, and L.J. Topp. Stiffness and deflection analysis of complex structures. *Journal of the Aeronautical Sciences*, 23(9):805, 1956.

17. L.-S. Turng and S.-W. Kim. Three-dimensional simulation for the filling stage of the polymer injection molding process using the finite element method. *To be published*, 2005.

18. K.K. Wang, S.F. Shen, C. Cohen, C.A. Hieber, T.H. Kwon, and R.C. Ricketson. Computer aided design and fabrication of molds and computer control of injection molding - progress report no. 11. Technical report, Cornell University, 1985.

CHAPTER 10

BOUNDARY ELEMENT METHOD

> Do not worry about your difficulties in Mathematics. I can assure you mine are still greater.
>
> —Albert Einstein

The two main advantages of the boundary element method (BEM) over FDM and FEM are, first, that the integral equations generated when applying the method fully satisfy the partial differential equations that govern the problem, and second, that the integral representation generated for linear problems is a boundary-only integral formula. The limitations that result from the volume discretization of the other techniques, especially when dealing with complex geometries, which may include free surfaces, moving boundaries and solid inclusions, are not present when using linear BEM to solve a problem.

The BEM has been limited to linear problems because the fundamental solution or Green's function is required to obtain a boundary integral formula equivalent to the original partial differential equation of the problem. The non-homogeneous terms accounting for nonlinear effects and body forces are included in the formulation by means of domain integrals, making the method lose its boundary-only character. Techniques have been developed to approximate these domain integrals directly such as cell integration [12, 43], Monte Carlo integration [54], or indirectly by approximation of domain integrals to the boundary and then solving these new boundary-only integrals such as dual reciprocity [3, 17, 18, 43], and particular integral technique [3].

10.1 SCALAR FIELDS

bem,scalar field An effective method of formulating the boundary-value problems of potential theory is to represent the harmonic function by a single-layer or a double-layer potential generated by continuous source distributions, of initially unknown density, over the boundary S, and forcing these potentials to satisfy the prescribed boundary conditions of the problem. This procedure leads to the formulation of integral equations which define the source densities concerned. This method is usually called indirect method, and can be formulated in terms of a single-layer potential (equation of the first kind) or a double-layer potential (equation of a second kind) [21, 50, 51].

In engineering applications it is often convenient to obtain integral representations which directly involve the field and its fluxes, rather than equations for single- or double-layer densities. This methodology is commonly called the direct method. For Poisson's equation this can be done using the Green's identities for scalar fields. As we already know, Poisson's equation is widely used in transport phenomena and polymer processing, and it is defined as,

$$\nabla^2 u(\mathbf{x}, t) = \frac{\partial^2 u(\mathbf{x}, t)}{\partial x_j \partial x_j} = b(\mathbf{x}, t) \tag{10.1}$$

For example, for a constant thermal conductivity, k, the energy conservation equation can be written in this form for the temperature, i.e.,

$$\nabla^2 T = \frac{\partial^2 T}{\partial x_j \partial x_j} = b(\mathbf{x}, t) \tag{10.2}$$

where the non-homogeneous term $b(\mathbf{x}, t)$ is defined by

$$b(T, \mathbf{x}, t) = \frac{\rho}{k} \frac{D(C_p T)}{Dt} - \frac{1}{2k} \eta(\dot{\gamma}) \left(\dot{\gamma} : \dot{\gamma}\right) + \frac{\dot{Q}}{k} \tag{10.3}$$

where ρ is the density, C_p is the specific heat, η the viscosity and $\dot{Q}$ the heat generation per unit volume.

10.1.1 Green's Identities

Let V be a region in space bounded by a closed surface S (of Lyapunov-type [24, 50]), and $\mathbf{f}(\mathbf{x})$ be a vector field acting on this region. A Lyapunov-type surface is one that is smooth. The divergence (Gauss) theorem establishes that the total flux of the vector field across the closed surface must be equal to the volume integral of the divergence of the vector (see Theorem 10.1.1).

Defining $\mathbf{f} = \phi \nabla \psi$ into Gauss theorem and using the chain rule for the divergence of the vector, the so-called Green's first identity is obtained (Theorem (10.1.2)). This identity is also valid when we use it for a vector $\mathbf{g} = \psi \nabla \phi$, when we substract the fist identity for $\mathbf{g}$ to the first identity of $\mathbf{f}$ we obtain the Green's second identity (Theorem (10.1.3)).

Theorem 10.1.1. Divergence (Gauss) Theorem

$$\int_V \nabla \cdot \mathbf{f} dV = \int_S \mathbf{f} \cdot \mathbf{n} dS$$
$$\int_V \frac{\partial f_i}{\partial x_i} dV = \int_S f_i n_i dS$$

Theorem 10.1.2. Green's First Identity for Scalar Fields.

$$\int_S \phi \frac{\partial \psi}{\partial x_i} n_i dS = \int_V \frac{\partial \psi}{\partial x_i} \frac{\partial \phi}{\partial x_i} dV + \int_V \phi \frac{\partial^2 \psi}{\partial x_i \partial x_i} dV$$
$$\int_S \phi \nabla \psi \cdot \mathbf{n} dS = \int_V \nabla \psi \cdot \nabla \phi dV + \int_V \phi \nabla^2 \psi dV$$

Theorem 10.1.3. Green's Second Identity for Scalar Fields

$$\int_S \left(\phi \frac{\partial \psi}{\partial x_i} - \psi \frac{\partial \phi}{\partial x_i} \right) n_i dS = \int_V \left(\phi \frac{\partial^2 \psi}{\partial x_i \partial x_i} - \psi \frac{\partial^2 \phi}{\partial x_i \partial x_i} \right) dV$$
$$\int_S (\phi \nabla \psi - \psi \nabla \phi) \cdot \mathbf{n} dS = \int_V \left(\phi \nabla^2 \psi - \psi \nabla^2 \phi \right) dV$$

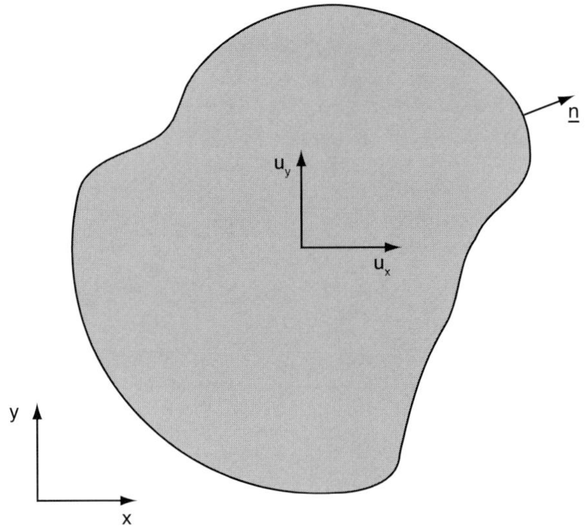

Figure 10.1: Schematic of the domain of interest divergence theorem nomenclature.

■ **EXAMPLE 10.1.**

Green's identities for a 2D Laplace's equation (heat conduction) Here, we will demonstrate how to develop Green's identities for a two-dimensional heat conduction problem, which for a material with constant properties is described by the Laplace equation for the temperature, i.e.,

$$\nabla^2 T = 0 \tag{10.4}$$

Using the domain definition and nomenclature presented in Fig. 10.1, the divergence theorem for a 2D vector **f** can be written as

$$\int_V \left(\frac{\partial f_x}{\partial x} + \frac{\partial f_y}{\partial y} \right) dV = \int_S (f_x n_x + f_y n_y)\, dS \tag{10.5}$$

To obtain the Green's identities, or integral representations, of the Laplace equation, we define the vectors $\mathbf{f} = \phi \nabla T$ and $\mathbf{g} = T \nabla \phi$. Here, ϕ is an additional function that we will define later. For now, the only requirement for this function is to be two times differentiable in space. Substituting our new definition of the vector **f**, we get

$$\int_V \left[\frac{\partial}{\partial x} \left(\phi \frac{\partial T}{\partial x} \right) + \frac{\partial}{\partial y} \left(\phi \frac{\partial T}{\partial y} \right) \right] dV = \int_S \left(\phi \frac{\partial T}{\partial x} n_x + \phi \frac{\partial T}{\partial y} n_y \right) dS \tag{10.6}$$

The chain rule of differentiation will give us the Green's first identity for vector $\mathbf{f} = \phi \nabla T$,

$$\begin{aligned} &\int_V \phi \left(\frac{\partial^2 T}{\partial x^2} + \frac{\partial^2 T}{\partial y^2} \right) dV + \int_V \left(\frac{\partial \phi}{\partial x} \frac{\partial T}{\partial x} + \frac{\partial \phi}{\partial y} \frac{\partial T}{\partial y} \right) dV \\ &\quad = \int_S \left(\phi \frac{\partial T}{\partial x} n_x + \phi \frac{\partial T}{\partial y} n_y \right) dS \end{aligned} \tag{10.7}$$

The first integral is zero, because the function T satisfies Laplace's equation, therefore

$$\int_V \left(\frac{\partial \phi}{\partial x}\frac{\partial T}{\partial x} + \frac{\partial \phi}{\partial y}\frac{\partial T}{\partial y} \right) dV = \int_S \left(\phi \frac{\partial T}{\partial x} n_x + \phi \frac{\partial T}{\partial y} n_y \right) dS \tag{10.8}$$

This is Green's first identity for the vector $\mathbf{f} = \phi \nabla T$. If we follow the same methodology, we will get the Green's first identity for $\mathbf{g} = T \nabla \phi$

$$\begin{aligned} \int_V T \left(\frac{\partial^2 \phi}{\partial x^2} + \frac{\partial^2 \phi}{\partial y^2} \right) dV + \int_V \left(\frac{\partial \phi}{\partial x}\frac{\partial T}{\partial x} + \frac{\partial \phi}{\partial y}\frac{\partial T}{\partial y} \right) dV \\ = \int_S \left(T \frac{\partial \phi}{\partial x} n_x + T \frac{\partial \phi}{\partial y} n_y \right) dS \end{aligned} \tag{10.9}$$

Here, we will not make the function ϕ satisfy Laplace's equation, therefore, we will conserve the first integral.

Subtracting the first identities for the two vectors will result in

$$-\int_V T \nabla^2 \phi dV = \int_S (\phi \nabla T - T \nabla \phi) \cdot \mathbf{n} dS \tag{10.10}$$

Notice that this is an integral representation of Laplace's equation for temperature. We will need to specify the extra function ϕ so it can become a complete representation. It is important to point out here that we have not made any approximation when deriving this formulation, making it an exact solution of the differential equation, $\nabla^2 T = 0$.

10.1.2 Green's Function or Fundamental Solution

The final result of Example 10.1 is Green's second identity for two vectors defined by $\mathbf{f} = \phi \nabla T$ and $\mathbf{g} = T \nabla \phi$. The only aspect that remains to be resolved is a correct selection of the extra function ϕ. The best selection is a function that satisfies a special form of Poisson's equation given by

$$\nabla^2 \phi = -\delta(\mathbf{x} - \mathbf{x_0}) \tag{10.11}$$

where $\delta(\mathbf{x} - \mathbf{x_0})$ is the Dirac delta function with its peak at the point $\mathbf{x_0}$ [14]. This choice has the added advantage that it will reduce even further our integral representation of Laplace's equation

$$\begin{aligned} -\int_V T(-\delta(\mathbf{x} - \mathbf{x_0}))\, dV &= \int_S (\phi \nabla T - T \nabla \phi) \cdot \mathbf{n} dS \\ T(\mathbf{x_0}) &= \int_S (\phi \nabla T - T \nabla \phi) \cdot \mathbf{n} dS \end{aligned} \tag{10.12}$$

for a point $\mathbf{x_0}$ located *inside* the domain V. We have now reduced the integral formulation for Laplace's equation to a boundary-only expression.

The function ϕ that satisfies eqn. (10.11) is called the *fundamental solution* or *Green's function*. Mathematically, the fundamental solution of a problem is the solution of the governing differential equation when the Dirac delta is acting as a forcing term [39, 52, 53]. Due to the infinite nature of the problem, no boundary conditions are needed and, providing that the Dirac delta or delta function is of a singular nature, Green's function or fundamental solution is also singular. In general, it is defined by

$$\mathcal{L}\phi^* = -\delta(\mathbf{x}, \boldsymbol{\xi}) \tag{10.13}$$

where $\mathcal{L}$ is a scalar differential operator and $\boldsymbol{\xi}$ is the source point. The physical interpretation of a Green's function in the case of Laplace's equation is that ϕ^* is the temperature distribution, for example, corresponding to an infinite heat source at $\boldsymbol{\xi}$.

Additionally, we can now make use of two useful properties of the delta function,

$$\lim_{V \to 0} \int_V \delta(\mathbf{x}, \boldsymbol{\xi}) dV = 1 \quad \text{and}$$
$$\int_V f(\mathbf{x}) \delta(\mathbf{x}, \boldsymbol{\xi}) dV = f(\boldsymbol{\xi}) \quad \text{if } \boldsymbol{\xi} \in V \tag{10.14}$$

Basically, when integrating the product of a function and the Dirac delta function, the Dirac delta function acts like a filter, resulting in the value of that function evaluated at the point where the Dirac delta function is applied.

The most common technique for the derivation of fundamental solutions is to use integral transforms, such as, Fourier, Laplace or Hankel transforms [29, 39]. For simple operators, such as the Laplacian, direct integration and the use of the properties of the Dirac delta are typically used to construct the fundamental solution. For the case of a two-dimensional Laplace equation we can use a two-dimensional Fourier transform, $F[]$, to get the fundamental solution as follows,

$$F\left[\nabla^2 \phi^*\right] = F\left[-\delta\right]$$
$$s^2 F[\phi^*] = -1$$
$$F[\phi^*] = -\frac{1}{s^2}$$

and the inverse Fourier

$$\phi^* = F^{-1}\left[-\frac{1}{s^2}\right] = -\frac{1}{2\pi} \ln r$$
$$\frac{\partial \phi^*}{\partial n} = \frac{1}{2\pi r} \frac{\partial r}{\partial n}$$

where r is the Euclidean distance between the location of the source and any point in the domain. Table 10.1 presents the most common Green's functions which can be used as basis for many problems in transport phenomena.

10.1.3 Integral Formulation of Poisson's Equation

We can now generate an equivalent integral formulation for Poisson's equation

$$\nabla^2 u(\mathbf{x}, t) = \frac{\partial^2 u}{\partial x_j \partial x_j} = b(u, \mathbf{x}, t) \tag{10.15}$$

in a domain V with a closed surface S (of the Lyapunov type [29, 40]), (see Fig. 10.2), which can have Neumann (i.e., constant temperature), Dirichlet (i.e., constant heat flux), Robin (i.e., convection heat condition), or any other type of boundary conditions on S.

The integral formulation for Poisson's equation is found the same way as for Laplace's equation (using Green's second identity, Theorem (10.1.3)), except that now the second volume integral is kept in Green's second identity. For a point $\mathbf{x_0} \in V$ the integral formulation

Table 10.1: Green's Function for Commonly Used Operators.

Equation	**2D**[a]	**3D**
Laplace $\nabla^2\phi^* = -\delta(\mathbf{x}, \boldsymbol{\xi})$	$\phi^* = \dfrac{-1}{2\pi r}\ln r$	$\phi^* = \dfrac{-1}{4\pi r}$
Helmholtz $(\nabla^2 + \lambda^2)\phi^* = -\delta(\mathbf{x}, \boldsymbol{\xi})$	$\phi^* = \dfrac{-1}{4\mathrm{i}}\mathrm{H}^{(1)}(\lambda r)$	$\phi^* = \dfrac{-1}{4\pi r}e^{-\mathrm{i}\lambda r}$
Modified Helmholtz $(\nabla^2 - \lambda^2)\phi^* = -\delta(\mathbf{x}, \boldsymbol{\xi})$	$\phi^* = \dfrac{-1}{4\pi}\mathrm{K}_0(\lambda r)$	$\phi^* = \dfrac{-1}{4\pi r}e^{\lambda r}$
Bi-harmonic $\nabla^4\phi^* = -\delta(\mathbf{x}, \boldsymbol{\xi})$	$\phi^* = \dfrac{-1}{8\pi}r^2\ln r$	$\phi^* = \dfrac{-r}{8\pi r}$

[a] Here, $\mathrm{H}^{(1)}$ and K_0 are the Hankel and Bessel functions respectively [1].

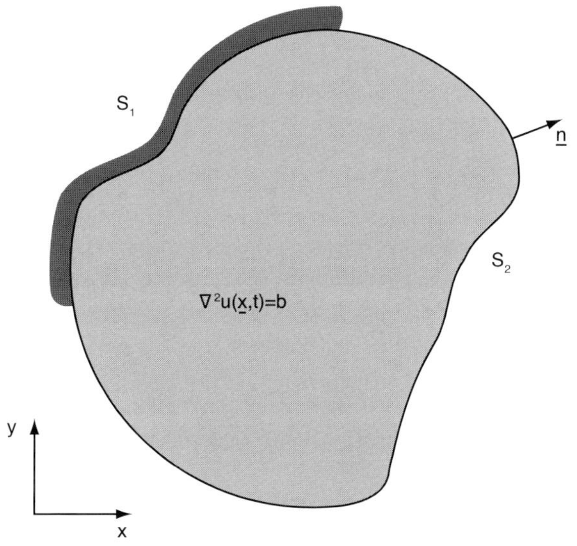

Figure 10.2: Schematic of the domain of interest.

will be

$$u(\mathbf{x_0}) = \int_S \phi^*(\mathbf{x}, \mathbf{x_0}) \frac{\partial u(\mathbf{x})}{\partial \mathbf{n}} dS - \int_S u(\mathbf{x}) \frac{\partial \phi^*(\mathbf{x}, \mathbf{x_0})}{\partial \mathbf{n}} dS + \int_V \phi^*(\mathbf{x}, \mathbf{x_0}) b(u, \mathbf{x}, t) dV \tag{10.16}$$

In the mathematical literature this equation is interpreted in the following way:

- The function $u(\mathbf{x_0})$ is a superposition of a single-layer potential of density $\partial u/\partial \mathbf{n}$ (first boundary integral),
- A double-layer potential of density $u(\mathbf{x})$ (second boundary integral) and
- A volume potential of density $b(u, \mathbf{x}, t)$ [29, 50, 51].

The use of this terminology reflects the fact that, as the point $\mathbf{x_0}$ approaches the *surface* of the domain, the double-layer potential has a discontinuity and it must be taken it into account by multiplying the field $u(\mathbf{x_0})$ by a coefficient.

In terms of heat transfer this can be physically interpreted that, if the infinite heat source is at the boundary (the infinite character is given by the delta function), then for a smooth surface only half of the delta function must be included. The general integral representation of Poisson's equation becomes,

$$c(\mathbf{x_0})u(\mathbf{x_0}) = \int_S \phi^*(\mathbf{x}, \mathbf{x_0}) \frac{\partial u(\mathbf{x})}{\partial \mathbf{n}} dS - \int_S u(\mathbf{x}) \frac{\partial \phi^*(\mathbf{x}, \mathbf{x_0})}{\partial \mathbf{n}} dS + \int_V \phi^*(\mathbf{x}, \mathbf{x_0}) b(u, \mathbf{x}, t) dV \tag{10.17}$$

where $c = 1$ for points inside the domain ($\mathbf{x_0} \in \mathbf{V}$), $c = 1/2$ for points in a smooth surface ($\mathbf{x_0} \in S$), and $c = 0$ for points outside the domain and surface.

For two-dimensional non-smooth surfaces, like the one shown in Fig. 10.3, the coefficient can be calculated by,

$$c(\mathbf{x_0}) = \frac{\beta}{2\pi} \tag{10.18}$$

This becomes cumbersome when we want to apply the method to any geometry. Fortunately there is a way to overcome the direct calculation of this coefficients as we will demonstrate in the numerical implementation.

10.1.4 BEM Numerical Implementation of the 2D Laplace Equation

Consider the two-dimensional Laplace equation for temperature in the domain presented in Fig. 10.4

$$\nabla^2 T = 0 \tag{10.19}$$

where we have boundaries with Dirichlet, $T = \bar{T}$, in S_1 and Neumann boundary conditions, $q = \partial T/\partial \mathbf{n} = \bar{q}$, in S_2, where we define $S = S_1 + S_2$ as the boundary of the domain V. Since with BEM we are required to apply both boundary conditions, the Dirichlet and Neumann boundary conditions, in the BEM literature they are not referred to as "essential" and "natural."

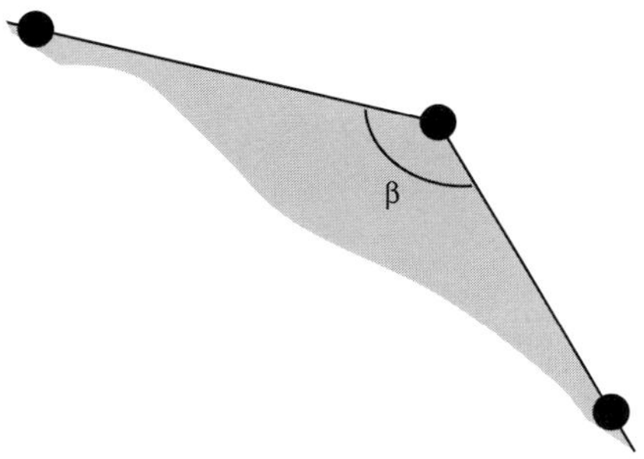

Figure 10.3: Corner and angle definition in a 2D domain.

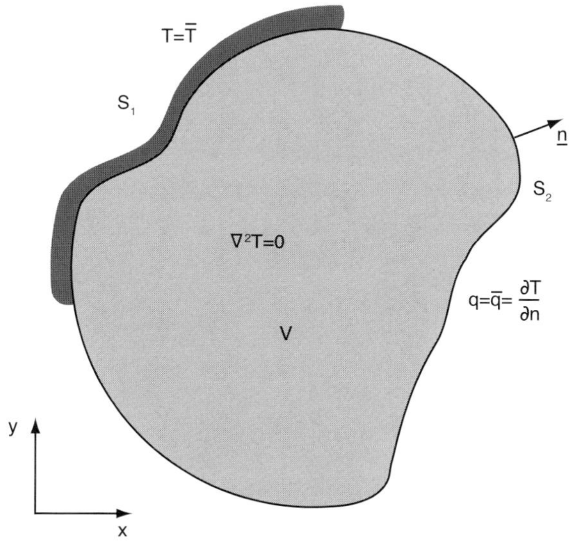

Figure 10.4: Schematic of the domain of interest with governing equation and boundary type definition.

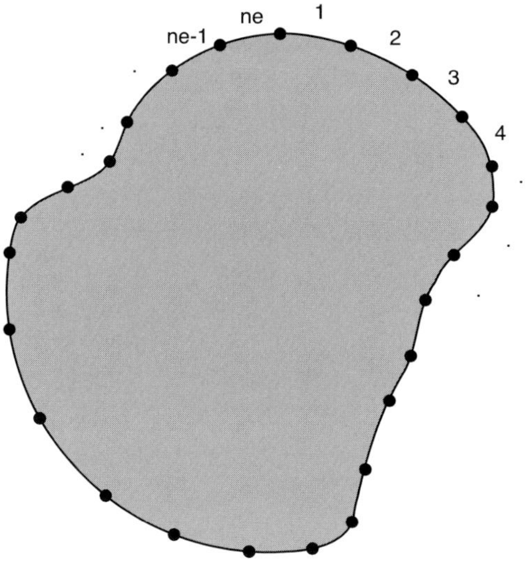

Figure 10.5: Schematic of the discretized domain of interest.

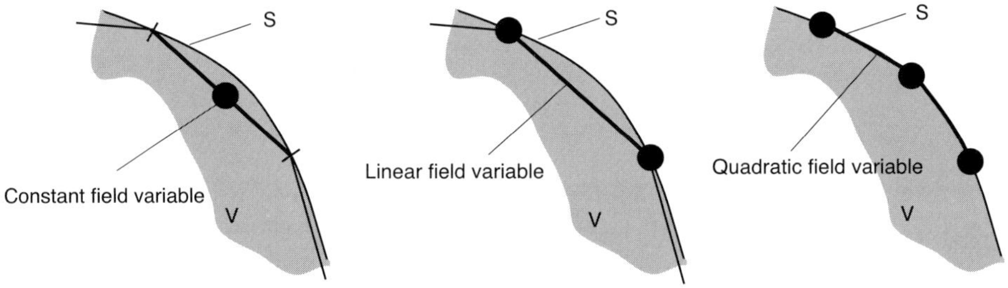

Figure 10.6: Schematic of 1D element types to represent 2D domains.

By using Green's identities and Green's functions, an equivalent integral equation is obtained as

$$c(\mathbf{x_o})T(\mathbf{x_o}) = \int_S \phi^*(\mathbf{x}, \mathbf{x_o})q(\mathbf{x})dS - \int_S q^*(\mathbf{x}, \mathbf{x_o})T(\mathbf{x})dS \tag{10.20}$$

where $q^* = \partial\phi^*/\partial\mathbf{n}$. It is impossible to find an exact general solution for the above integral equation, requiring us to approximate the integrals. The first step, as usual, is to divide the surface into smaller surface elements, or boundary elements. Similar to FEM, the selection of the element is related to the order of the approximation. Figure 10.5 shows a typical mesh for the domain in Fig. 10.4, while Fig. 10.6 illustrates three common two-dimensional elements that can be used for the discretization. If the surface is divided into NE elements, eqn. (10.20) can be written as

$$\frac{1}{2}T_i = \sum_{j=1}^{NE} \int_{S_j} \phi^* q^j dS - \sum_{j=1}^{NE} \int_{S_j} q^* T^j dS \tag{10.21}$$

for any node, i, in the surface, where a smooth surface was assumed (we will generalize later), and where the superscript $j = 1, ..., NE$ indicate the elements on the boundary.

If we use constant elements (one node), the value of the temperature and heat are considered constant and equal to the value at the mid point of the element. Therefore, we can place the values for temperature and heat flow outside of the integrals in eqn. (10.21) to get

$$\frac{1}{2}T_i + \sum_{j=1}^{NE} T_j \int_{S_j} q^* dS = \sum_{j=1}^{NE} q_j \int_{S_j} \phi^* dS \tag{10.22}$$

Here, we change the element superscript j by a subscript j because, for the constant elements, each node actually represents an element. This equation can be written in compact form as

$$\frac{1}{2}T_i + \bar{h}_{ij}T_j = g_{ij}q_j \tag{10.23}$$

where the matrices $\mathbf{h}$ and $\mathbf{g}$ are the integrals of the Green's functions from each node i to all the other points in the boundary j defined by

$$\bar{h}_{ij} = \int_{S_j} q^*(\mathbf{x_i}, \mathbf{x_j})dS \qquad g_{ij} = \int_{S_j} \phi^*(\mathbf{x_i}, \mathbf{x_j})dS \tag{10.24}$$

we can also collapse the coefficients in the small matrix $\mathbf{h}$ as follows

$$h_{ij} = \bar{h}_{ij} + \frac{1}{2}\delta_{ij} \tag{10.25}$$

which, once we apply eqn. (10.23) to every node $i = 1, ..., N$ on the surface, will allow us to write

$$\mathbf{HT} = \mathbf{Gq} \tag{10.26}$$

Each of these matrices are of dimension $N \times N$, and the vectors $\mathbf{u}$ and $\mathbf{q}$ are of dimension N, because with constant elements, the number of nodes and elements is the same. In eqn. (10.26), we have $2N$ unknowns with only N equations. In order to complete the system, we must use N boundary conditions, which can be all temperatures or heat fluxes, or combinations of heat fluxes and temperatures. The methodology of including the boundary conditions is simple and consists in an exchange of columns between the known values and the unknowns. After the exchange, eqn. (10.26) reduces to

$$\mathbf{Ax} = \mathbf{b} \tag{10.27}$$

where the vector $\mathbf{x}$ has all the remaining unknown boundary values of temperature and heat fluxes.

Equation (10.27) can be solved by using the usual solution schemes for linear algebraic equations. Once these values are obtained, the integral representation of Laplace's equation, eqn. (10.20), can be used to find the value of the temperature at any point within the domain, i.e.,

$$T_i = \sum_{j=1}^{N} G_{ij}q_j - \sum_{j=1}^{N} H_{ij}u_j \tag{10.28}$$

Here, the coefficients of the matrices **H** and **G** are calculated from the internal point or node i to the boundary nodes $j = 1, ..., N$. If the values of the fluxes are needed in some internal points, we just calculate them by using the derivatives of the temperature from the integral representation, i.e.,

$$q_i^x = \left[\frac{\partial T}{\partial x}\right]_i = \int_S \frac{\partial \phi^*}{\partial x} q dS - \int_S \frac{\partial q^*}{\partial x} u dS \tag{10.29}$$

$$q_i^y = \left[\frac{\partial T}{\partial y}\right]_i = \int_S \frac{\partial \phi^*}{\partial y} q dS - \int_S \frac{\partial q^*}{\partial y} u dS \tag{10.30}$$

Algorithm 11 shows the steps for a general boundary element calculation.

Algorithm 11 Boundary Element Method (BEM).

```
program BEM
  call input(x,y,nee,bc)     reading nodes' coordinates, element connectivity and
                                                              boundary conditions
  call BEM-assemble                                          BEM H and G matrices
  call boundary-condition                                         column exchange
  call solve                                                          x = A^-1 b
end program BEM
```

10.1.5 2D Linear Elements.

In order to increase the accuracy of the element and therefore reduce the number of boundary elements we must increase the order of the interpolation functions across the element. If we choose a linear element, similar to FEM, we can approximate any variable within the element with the use of a linear isoparametric interpolation as follows

$$x^e(\xi) = N_1(\xi)x_1 + N_2(\xi)x_2 \tag{10.31}$$

where the interpolation or shape functions are defined as

$$\begin{aligned} N_1(\xi) &= \frac{1}{2}(1-\xi) \\ N_2(\xi) &= \frac{1}{2}(1+\xi) \end{aligned} \tag{10.32}$$

for $\xi \in [-1, 1]$. Figure 10.7 shows a schematic of the linear 2D isoparametric element.

The values of the temperature, T, and the normal heat flux, q, are then interpolated using eqn. (10.31) as,

$$\begin{aligned} T^e(\xi) &= N_1(\xi)T_1 + N_2(\xi)T_2 && = \begin{bmatrix} N_1 & N_2 \end{bmatrix} \begin{pmatrix} T_1 \\ T_2 \end{pmatrix} \\ q^e(\xi) &= N_1(\xi)q_1 + N_2(\xi)q_2 && = \begin{bmatrix} N_1 & N_2 \end{bmatrix} \begin{pmatrix} q_1 \\ q_2 \end{pmatrix} \end{aligned} \tag{10.33}$$

The integral formulation is the same as for the constant element, i.e.,

$$\frac{1}{2}T_i = \sum_{j=1}^{NE} \int_{S_j} \phi^* q^j dS - \sum_{j=1}^{NE} \int_{S_j} q^* T^j dS \tag{10.34}$$

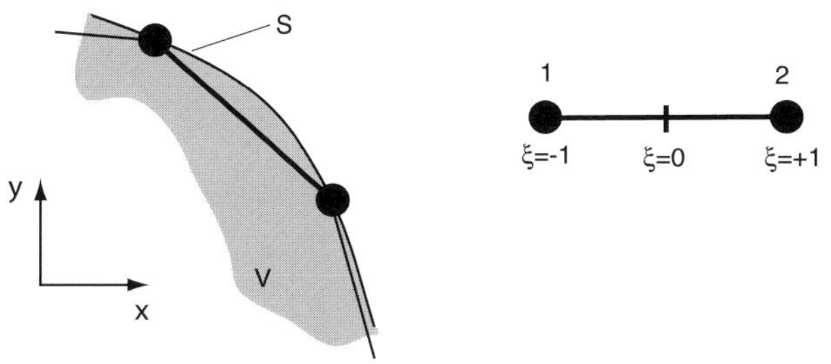

Figure 10.7: Isoparametric linear element difinition.

Using the interpolation eqn. (10.33) we get

$$\frac{1}{2}T_i = \sum_{j=1}^{NE} \int_{S_j} \phi^* \begin{bmatrix} N_1 & N_2 \end{bmatrix} dS \begin{pmatrix} q_1^j \\ q_2^j \end{pmatrix} - \sum_{j=1}^{NE} \int_{S_j} q^* \begin{bmatrix} N_1 & N_2 \end{bmatrix} dS \begin{pmatrix} T_1^j \\ T_2^j \end{pmatrix} \quad (10.35)$$

which again can be compacted to

$$\frac{1}{2}T_i = \sum_{j=1}^{NE} \begin{bmatrix} g_{ij}^1 & g_{ij}^2 \end{bmatrix} \begin{pmatrix} q_1^j \\ q_2^j \end{pmatrix} - \sum_{j=1}^{NE} \begin{bmatrix} \bar{h}_{ij}^1 & \bar{h}_{ij}^2 \end{bmatrix} \begin{pmatrix} T_1^j \\ T_2^j \end{pmatrix} \quad (10.36)$$

where

$$g_{ij}^k = \int_{S_j} \phi^*(\mathbf{x_i}, \mathbf{x^j}) N_k dS \qquad \bar{h}_{ij}^k = \int_{S_j} q^*(\mathbf{x_i}, \mathbf{x^j}) N_k dS \quad (10.37)$$

Figure 10.8 shows a schematic of the juncture between two elements $j-1$ and j, where we can see that node 1 of element $j-1$ is the same as node 2 of element j. According to the definition of the normal heat flux, $q = \partial T / \partial \mathbf{n}$, we can have two different values of the normal heat flux because the normal vector can be different for the two elements (as shown in the Fig. 10.8). However, we must assure the continuity of the temperature from one element to another, which implies that the value of the temperature is the same for node 1 of element $j-1$ and node 2 of element j, i.e., $T_1^{j-1} = T_2^j$. It is important to note that we have N values of temperatures and $2N$ values of normal heat fluxes. The integral equation for a specific node i will look as

$$\frac{1}{2}T_i + \begin{bmatrix} \bar{h}_{i1} & \bar{h}_{i2} & \dots & \bar{h}_{iN} \end{bmatrix} \begin{pmatrix} u_1 \\ u_2 \\ \dots \\ u_N \end{pmatrix} =$$

$$\begin{bmatrix} g_{i1}^1 & g_{i1}^2 & g_{i2}^1 & g_{i2}^2 & \dots & g_{iNE}^1 & g_{iNE}^2 \end{bmatrix} \begin{pmatrix} q_1^1 \\ q_2^1 \\ q_1^2 \\ q_2^2 \\ \dots \\ q_1^{NE} \\ q_2^{NE} \end{pmatrix} \quad (10.38)$$

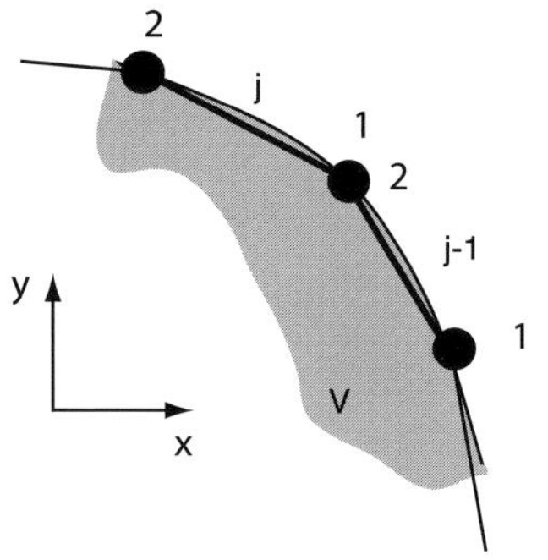

Figure 10.8: Linear element connectivity.

where $\bar{h}_{ij} = \bar{h}_{i1}^{j-1} + \bar{h}_{i2}^{j}$, and the system can be written as

$$\mathbf{HT} = \mathbf{Gq} \tag{10.39}$$

where $\mathbf{H} = \bar{\mathbf{H}} + 1/2\delta$. Now, matrix $\mathbf{G}$ is rectangular, of dimension $N \times 2NE$ and vector $\mathbf{q}$ of dimension $2NE$.

At this point, we can proceed with applying the boundary conditions. For the case of a smooth surface with a continuous normal vector, the values of the normal heat flux will be continuous and matrix $\mathbf{G}$ will be calculated in the same way that matrix $\mathbf{H}$ was found, which will reduce the system to N equations with $2N$ unknowns. This requires N boundary conditions. As with the constant element, the columns will be exchanged, forming a linear system of algebraic equations. For the case of surfaces with a discontinuous normal vector, we will have the following situations:

- the value of the heat flux is known for the two elements and the temperature is unknown,
- the temperature and one element flux are known and the remainder heat is unknown,
- the temperature is known and both heat element fluxes are unknown.

The two first cases, although cumbersome, will close the system of equations. However, the third case will imply the use of an extra equation or the use of a discontinuous element. When equivalent integral equations to partial differential equations are developed, it is required that the surface is of a *Lyapunov type* [29, 40]. For the purpose of this book, we will assume that this type of surfaces have the condition of having a continuous normal vector. The integral formulation also can be generated for *Kellog type* surfaces, which allow the existence of corners that are not too sharp. To avoid complications, we can assume that even for very sharp corners the normal vector is continuous, as depicted in Fig. 10.9.

This assumption, which is not very significant from the physical point of view, will reduce the system to two $N \times N$ matrices $\mathbf{H}$ and $\mathbf{G}$. With N boundary conditions, the system of equations will be closed.

Numerical calculation of the c coefficients. Even if we assume that the normal vector is locally continuous, we cannot assume that always the angle between elements is always π, resulting in $c = 1/2$ for every boundary nodal point. As we discussed before,

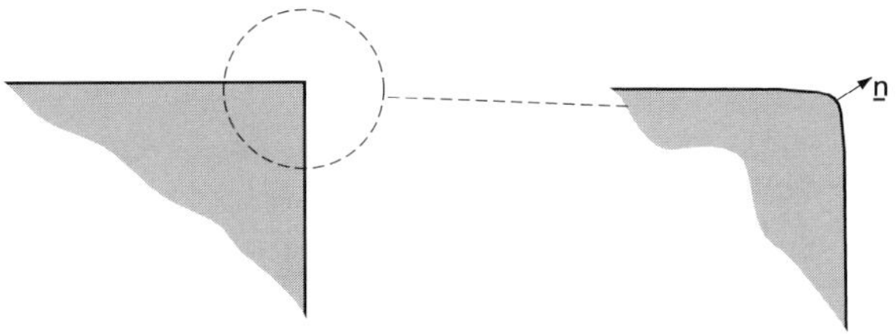

Figure 10.9: Continuous corner definition.

in two-dimensional problems we can use the angle β to directly calculate the coefficients (eqn. (10.18)). However, this will complicate the formulation since for every surface nodal point we must compute the angle, making it even worse for 3D.

A different way to obtain these coefficients is using the fact that the integral formulation developed from Green's identities does not have any restriction to have a uniform potential on the surface (such as a constant temperature). A constant potential will imply that the normal derivatives, q, must be zero, and the integral formulation reduces to

$$\mathbf{HT} = 0 \tag{10.40}$$

where $\mathbf{T}$ is the vector of constant potential (temperature). This equation implies that the sum of all the column values of matrix $\mathbf{H}$ must be zero. Therefore, the diagonal terms of the $\mathbf{H}$ matrix can be calculated using,

$$H_{ii} = \sum_{j=1}^{N} H_{ij} \qquad \text{for } i \neq j \tag{10.41}$$

which saves us the problem of calculating the coefficients c directly.

10.1.6 2D Quadratic Elements

To reduce the number of surface elements, to better represent the curvature, for more complicated geometries it is better to use quadratic elements to approximate the variables within the elements. The integral formulation will be the same with an additional term in the smaller matrices $\bar{\mathbf{h}}$ and $\mathbf{g}$. The potential and fluxes become

$$T^j(\xi) = \begin{bmatrix} N_1 & N_2 & N_3 \end{bmatrix} \begin{pmatrix} T_1^j \\ T_2^j \\ T_3^j \end{pmatrix}$$

$$q^j(\xi) = \begin{bmatrix} N_1 & N_2 & N_3 \end{bmatrix} \begin{pmatrix} q_1^j \\ q_2^j \\ q_3^j \end{pmatrix} \tag{10.42}$$

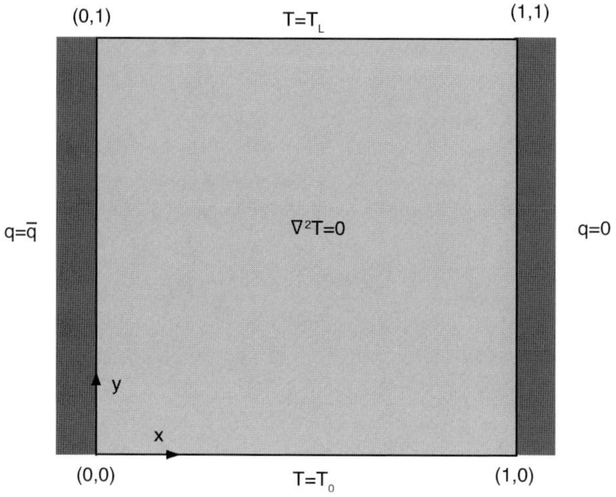

Figure 10.10: Schematic diagram of a square domain with dimensions, governing equations and boundary conditions.

where

$$\begin{aligned} N_1 &= \frac{1}{2}\xi(\xi - 1) \\ N_2 &= (1 - \xi)(1 + \xi) \\ N_3 &= \frac{1}{2}\xi(\xi + 1) \end{aligned} \tag{10.43}$$

With this interpolation, the small matrices will be defined by

$$g_{ij}^k = \int_{S_j} \phi^*(\mathbf{x_i}, \mathbf{x_j}) N_k dS \qquad \bar{h}_{ij}^k = \int_{S_j} q^*(\mathbf{x_i}, \mathbf{x_j}) N_k dS \tag{10.44}$$

for $k = 1, 2, 3$.

However, in eqn. (10.44), $dS = dxdy$, which requires us to change coordinates, similar to the isoparametric formulation in FEM

$$dS = \sqrt{\left(\frac{\partial x}{\partial \xi}\right)^2 + \left(\frac{\partial y}{\partial \xi}\right)^2} d\xi = |\mathbf{J}| d\xi \tag{10.45}$$

Here, $|\mathbf{J}|$ is the Jacobian of the transformation from $x - y$ to ξ.

■ **EXAMPLE 10.2.**

Heat equation in a square geometry with linear elements. Using the square section depicted in Fig. 10.10, we want to solve the steady-state conduction equation for a material of constant conductivity.

For such a case, the energy equation is reduced to Laplace's equation for temperature

$$\nabla^2 T = 0 \tag{10.46}$$

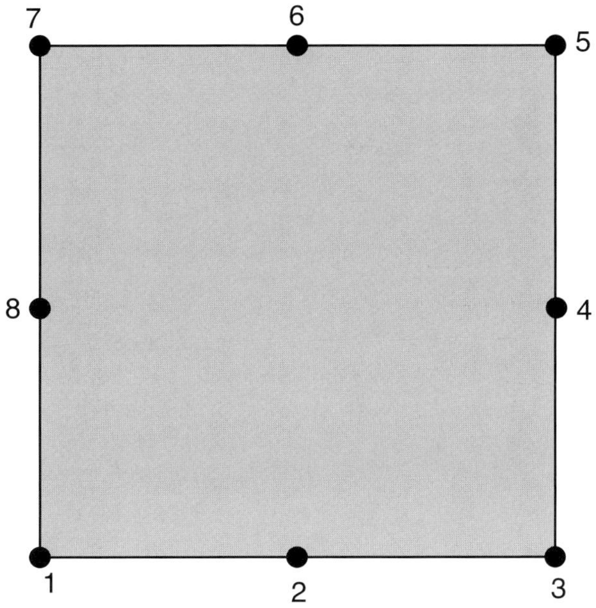

Figure 10.11: Schematic diagram of the square domain discretization.

with Dirichlet boundary conditions $T = T_0$ at $y = 0$ and $T = T_L$ at $y = L$, and adiabatic boundary conditions $q = \bar{q}$ at $x = 0$ and $q = 0$ at $x = L$. Using L as the characteristic length and $\Delta T = T_L - T_0$ as the characteristic temperature gradient, eqn. (10.46) is written as

$$\frac{\partial^2 \Theta}{\partial \xi^2} + \frac{\partial^2 \Theta}{\partial \eta^2} = 0 \tag{10.47}$$

where

$$\xi = \frac{x}{L} \qquad \eta = \frac{y}{L} \qquad \Theta = \frac{T - T_0}{T_L - T_0} \tag{10.48}$$

with the corresponding boundary conditions

$$\begin{aligned} \Theta(\xi, 0) &= 0 & \Theta(\xi, 1) &= 1 \\ q(0, \eta) &= \frac{\bar{q}}{q_c} & q(1, \eta) &= 0 \end{aligned} \tag{10.49}$$

where $q_c = k(T_L - T_0)/L$ is the characteristic heat flux at the left wall.

We now generate a mesh using linear elements. A typical mesh is shown in Fig. 10.11, according to this mesh, the elements have a connectivity matrix given by

$$\text{nee}(i,j) = \begin{bmatrix} 1 & 2 \\ 2 & 3 \\ 3 & 4 \\ 4 & 5 \\ 5 & 6 \\ 6 & 7 \\ 7 & 8 \\ 8 & 1 \end{bmatrix} \tag{10.50}$$

Similar to FEM, we must take care of how the nodes are numbered to avoid having normal vectors pointing in the wrong direction – the inside of the domain in this case. Here, we selected a positive normal vector pointing out of the domain, a notation that must be consistent. The terms in the matrices $\bar{\mathbf{h}}$ and $\mathbf{g}$ are defined as

$$g_{ij}^k = \int_{-1}^{+1} \phi^*(\mathbf{x_i}, \mathbf{x^j}) N_k |\mathbf{J}| d\xi \qquad \bar{h}_{ij}^k = \int_{-1}^{+1} q^*(\mathbf{x_i}, \mathbf{x^j}) N_k |\mathbf{J}| d\xi \tag{10.51}$$

here, $i = 1, ..., N$ represents the equation node, $j = 1, ..., NE$ the element and $k = 1, 2$ the shape function. Gaussian quadratures can be used and the integrals are approximated by

$$\begin{aligned} g_{ij}^k &= \sum_{ix=1}^{ngauss} \phi^*(\mathbf{x_i}, \mathbf{x^j}(ix)) N_k(ix) |\mathbf{J}|_{ix} w(ix) \\ \bar{h}_{ij}^k &= \sum_{ix=1}^{ngauss} q^*(\mathbf{x_i}, \mathbf{x^j}(ix)) N_k(ix) |\mathbf{J}|_{ix} w(ix) \end{aligned} \tag{10.52}$$

where $ix = 1, ..., ngauss$ is the location of the Gaussian point and $\mathbf{w}$ is the Gaussian weight vector. Algorithm 12 illustrates the methodology of assembling the matrices $\mathbf{H}$ and $\mathbf{G}$, whereas Algorithm 13 performs the calculation of the matrices in eqn. (10.52).

Figure 10.12 illustrates the dimensionless temperature distribution for a 1×1 square geometry with $T_0 = 100$ K, $T_L = 500$ K, $q(0, y) = 40$ W/m^2 and a thermal conductivity $k = 0.1$W/m/K. The results in this figure are calculated with 100 nodes and 20 Gauss points. Figures 10.13 and 10.14 show a comparison between the BEM solution, with 5 and 20 Gauss points, and the analytical solution for the mid-plane temperatures, $x = L/2$ and $y = L/2$.

10.1.7 Three-Dimensional Problems

For three-dimensional problems the integral formulations previously obtained are also valid and are implemented into two-dimensional elements that cover the domain surface as shown in Fig. 10.15. Here, we use triangular and rectangular elements as used with FEM. Again, depending on the number of nodes per element, we can have constant, linear and quadratic elements. To be able to represent any geometry it is best to use curvilinear isoparametric elements as schematically illustrated in Fig. 10.16.

The curvilinear elements will require, as in FEM, a transformation of coordinates from the cartesian (x, y, z) to the isoparametric (ξ_1, ξ_2, η), where ξ_1 and ξ_2 are the isoparametric

Algorithm 12 Assembling Boundary Element Matrices.

```
subroutine BEM-assemble
  Hbig = 0; Gbig = 0
  do i = 1,N+NIP                                               equation node i,
                                        for boundary (N) and internal nodes (NIP)
    xo = x(i); yo = y(i)      x(NN+NIP), y(NN+NIP) contain the nodes coordinates
    do j = 1,NE                                              loop over elements
      do k = 1,2
        xe(k) = x(nee(j,k))                           element nodes' coordinates
        ye(k) = y(nee(j,k))
      enddo
      call small-bem(xo,yo,xe,ye,g,h)                           h(1,2) and g(1,2)
      do k = 1,2
        jj = nee(j,k)                       element node position in the matrices
        Hbig(i,jj) = Hbig(i,jj) + h(1,k)
        Gbig(i,jj) = Gbig(i,jj) + g(1,k)              using the continuous normal
      enddo
    enddo
  enddo
  do i = 1, N+NIP                                          make H diagonal zero
    Hbig(i,i) = 0
  enddo
  do i = 1,N                                  calculation of the H diagonal terms
    do j = 1, N
      if (i /= j) Hbig(i,i) = Hbig(i,i) - Hbig(i,j)
    enddo
  enddo
  do i = N+1, N+NIP                                    internal nodes coefficient
    Hbig(i,i) = 1
  enddo
end subroutine BEM-assemble
```

Algorithm 13 Assembling Small Boundary Matrices with Linear Elements.

```
subroutine small-bem(xo,yo,xe,ye,g,h)
  h = 0; g = 0
  do ix = 1,ngauss                                             integration in ξ
    xi = gp(ix)
    n(1) = 0.5*(1-xi)                                   interpolation functions
    n(2) = 0.5*(1+xi)
    xgp = n(1)*xe(1) + n(2)*xe(2)                       gauss point coordinates
    ygp = n(1)*ye(1) + n(2)*xe(2)
    dn(1) = -0.5                            interpolation functions derivatives
    dn(2) =  0.5
    dxde = dn(1)*xe(1) + dn(2)*xe(2)          gauss point coordinate derivatives
    dyde = dn(1)*ye(1) + dn(2)*xe(2)
    jacobian = sqrt( dxde**2 + dyde**2 )                               Jacobian
    norm(1) = dxde/jacobian                                       normal vector
    norm(2) = dyde/jacobian
    dx = xgp - xo
    dy = ygp - yo
    r  = sqrt( dx**2 + dy**2 )
    drdn = dx*norm(1) + dy*norm(2)
    phi = -log(r)/2/pi                                    fundamental solutions
    qq  = -drdn/2/pi/r/r
    g = g + n*phi*jacobian*w(ix)                                       Integral
    h = h + n*qq*jacobian*w(ix)
  enddo
end subroutine small-bem
```

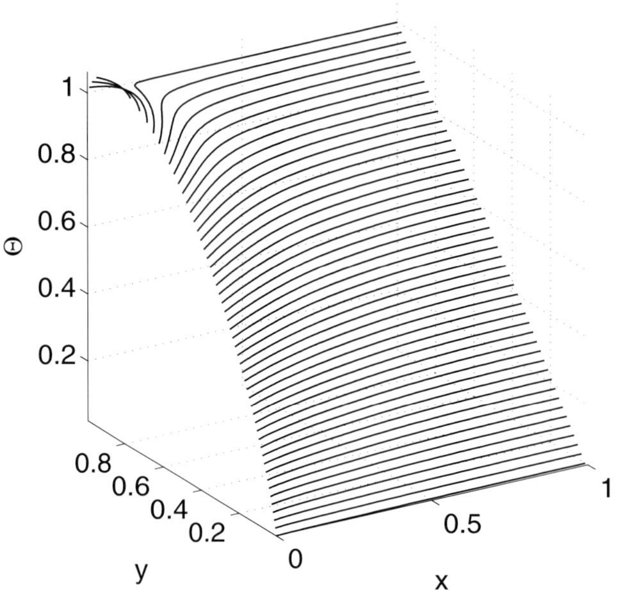

Figure 10.12: BEM Dimensionless temperature profile.

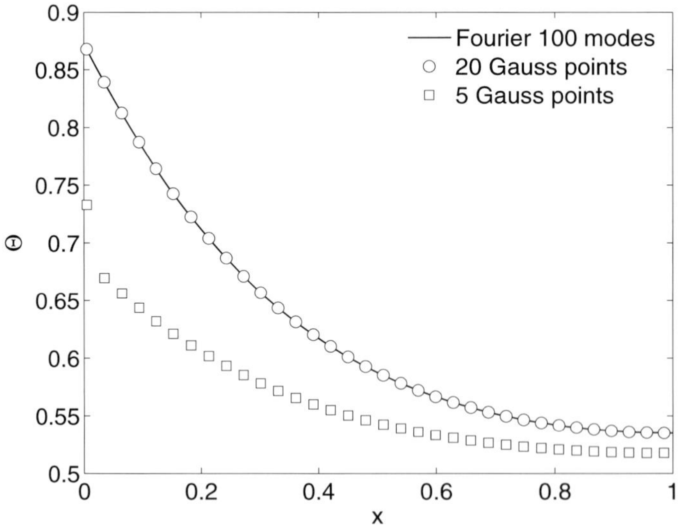

Figure 10.13: Comparision between the BEM solution and the exact solution for the temperature at $\eta = 0.5$.

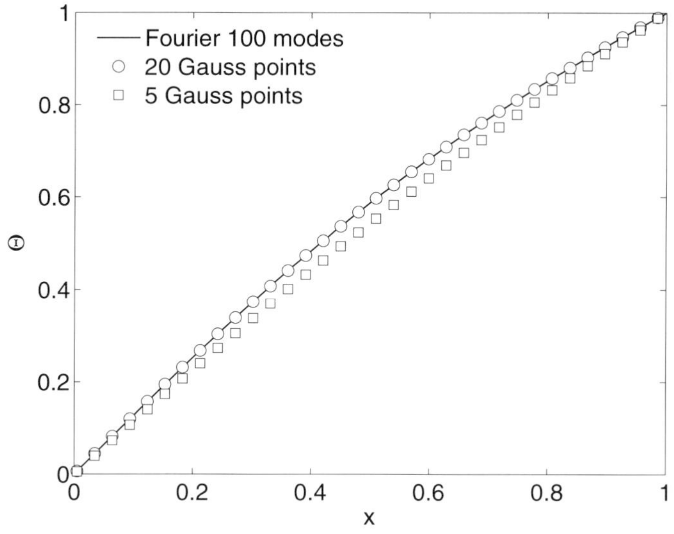

Figure 10.14: Comparision between the BEM solution and the exact solution for the temperature at $\xi = 0.5$.

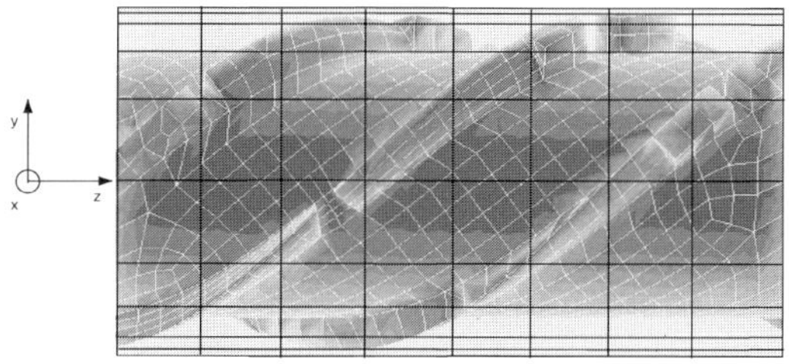

Figure 10.15: 3D single screw extruder barrel and mixing head discretization. White element delineations define the mixing head surface, and the black lines define the barrel surface representation. The mixing head surface is represented with triangular as well as quadrilateral elements.

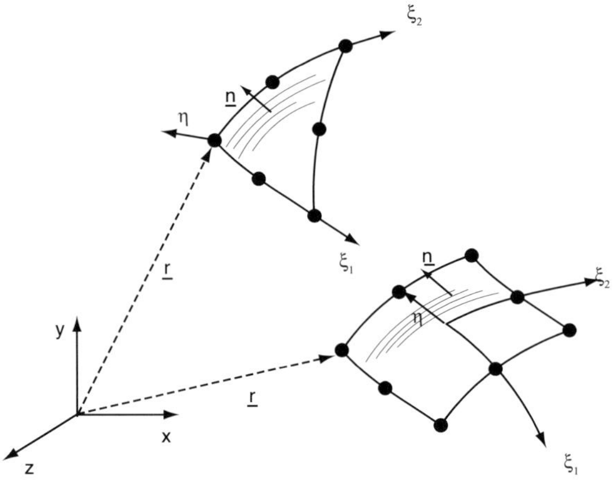

Figure 10.16: Schematic of 2D elements used to represent 3D geometries.

coordinates and η is the direction of the normal vector (Fig. 10.16). For a variable u, the transformation is given by

$$\begin{pmatrix} \dfrac{\partial u}{\partial \xi_1} \\ \dfrac{\partial u}{\partial \xi_2} \\ \dfrac{\partial u}{\partial \eta} \end{pmatrix} = \begin{bmatrix} \dfrac{\partial x}{\partial \xi_1} & \dfrac{\partial y}{\partial \xi_1} & \dfrac{\partial z}{\partial \xi_1} \\ \dfrac{\partial x}{\partial \xi_2} & \dfrac{\partial y}{\partial \xi_2} & \dfrac{\partial z}{\partial \xi_2} \\ \dfrac{\partial x}{\partial \eta} & \dfrac{\partial y}{\partial \eta} & \dfrac{\partial z}{\partial \eta} \end{bmatrix} \begin{pmatrix} \dfrac{\partial u}{\partial x} \\ \dfrac{\partial u}{\partial y} \\ \dfrac{\partial u}{\partial z} \end{pmatrix} \tag{10.53}$$

or

$$\begin{pmatrix} \dfrac{\partial u}{\partial \xi_1} \\ \dfrac{\partial u}{\partial \xi_2} \\ \dfrac{\partial u}{\partial \eta} \end{pmatrix} = [\mathbf{J}] \begin{pmatrix} \dfrac{\partial u}{\partial x} \\ \dfrac{\partial u}{\partial y} \\ \dfrac{\partial u}{\partial z} \end{pmatrix} \tag{10.54}$$

Extra transformations are needed for the differential elements in the integrals. For the differential volume we have

$$dV = |\mathbf{J}| d\xi_1 d\xi_2 d\eta \tag{10.55}$$

where $|\mathbf{J}|$ is the determinant of the Jacobian matrix defined by

$$|\mathbf{J}| = \left| \frac{\partial \mathbf{r}}{\partial \xi_1} \times \frac{\partial \mathbf{r}}{\partial \xi_2} \cdot \frac{\partial \mathbf{r}}{\partial \eta} \right| \tag{10.56}$$

The differential area element will be

$$dS = |\mathbf{G}| d\xi_1 d\xi_2 \tag{10.57}$$

where $|\mathbf{G}|$ is the magnitude of the reduced Jacobian vector defined by

$$|\mathbf{G}| = \left| \frac{\partial \mathbf{r}}{\partial \xi_1} \times \frac{\partial \mathbf{r}}{\partial \xi_2} \right| \tag{10.58}$$

This magnitude vector is calculated from the surface parametrization as follows [9, 29]

$$|\mathbf{G}| = \sqrt{G_1^2 + G_2^2 + G_3^2} \tag{10.59}$$

where

$$G_1 = \left(\frac{\partial y}{\partial \xi_1} \frac{\partial z}{\partial \xi_2} - \frac{\partial y}{\partial \xi_2} \frac{\partial z}{\partial \xi_1} \right) \tag{10.60}$$

$$G_2 = \left(\frac{\partial z}{\partial \xi_1} \frac{\partial x}{\partial \xi_2} - \frac{\partial z}{\partial \xi_2} \frac{\partial x}{\partial \xi_1} \right) \tag{10.61}$$

$$G_3 = \left(\frac{\partial x}{\partial \xi_1} \frac{\partial y}{\partial \xi_2} - \frac{\partial x}{\partial \xi_2} \frac{\partial y}{\partial \xi_1} \right) \tag{10.62}$$

10.2 MOMENTUM EQUATIONS

The momentum balance equations can be written in a form that is valid for the Navier-Stokes equations as well as low Reynolds number non-Newtonian flow equations:

$$\frac{\partial u_i}{\partial x_i} = 0 \tag{10.63}$$

$$-\frac{\partial p}{\partial x_i} + \mu \frac{\partial^2 u_i}{\partial x_j \partial x_j} = g_i \tag{10.64}$$

where $\mathbf{u}$ is the velocity field, p the pressure or the modified pressure, depending if gravity is included in the analysis [30]. Equation (10.63) is the continuity equation for incompressible fluids [5, 41]. For the Navier-Stokes equations, the pseudo-body force term g in eqn. (10.64) is defined as

$$g_i = \rho \left(\frac{\partial u_i}{\partial t} + u_j \frac{\partial u_i}{\partial x_j} \right) \tag{10.65}$$

while for the low Reynolds number flow of non-Newtonian fluids it is

$$g_i = -\frac{\partial \tau_{ij}^{(e)}}{\partial x_j} \tag{10.66}$$

where $\tau^{(e)}$ is the extra stress tensor that represents the non-Newtonian effects in the stress tensor. For inelastic generalized Newtonian fluids, this stress tensor is defined as

$$\tau_{ij}^{(e)} = (\eta(\dot{\gamma}) - \mu)\, \dot{\gamma}_{ij} \tag{10.67}$$

In this case, μ is an arbitrary constant, chosen as the zero shear rate viscosity. The expression for the non-Newtonian viscosity is a constitutive equation for a generalized Newtonian fluid, like the power law or Ostwald-de-Waele model [6]

$$\eta(\dot{\gamma}) = m\dot{\gamma}^{n-1} \tag{10.68}$$

where m is the consistency index and $n \in [0, 1]$ the power law index.

Theorem 10.2.1. Green's First Identity for a Flow Field $(\mathbf{u}, p)$

$$\frac{\mu}{2}\int_V \left(\frac{\partial u_i}{\partial x_j}+\frac{\partial u_j}{\partial x_i}\right)\left(\frac{\partial v_i}{\partial x_j}+\frac{\partial v_j}{\partial x_i}\right) dV +$$
$$\int_V \left(\frac{\partial^2 u_i}{\partial x_j \partial x_j}-\frac{\partial p}{\partial x_i}\right) v_i dV = \int_S \sigma_{ij} v_j n_i dS$$

where $\mathbf{n}$ *is an outward unit vector with respect to the surface* S *and* $\mathbf{v}$ *is an additional divergence-free velocity field.*

10.2.1 Green's Identities for the Momentum Equations

In order to obtain Green's identities for the flow field $(\mathbf{u}, p)$, a vector $\mathbf{z}$ is defined as the dot product of the stress tensor $\sigma(\mathbf{u}, p)$ and a second solenoidal vector field $\mathbf{v}$ (divergence-free). The divergence or Gauss' Theorem (10.1.1) is applied to the vector $\mathbf{z}$

$$\int_V \frac{\partial \sigma_{ij} v_j}{\partial x_i} dV = \int_S \sigma_{ij} v_j n_i dS \tag{10.69}$$

where, for an incompressible Newtonian fluid the stress tensor $\sigma(\mathbf{u}, p)$ is defined by

$$\sigma_{ij} = -p\delta_{ij} + \mu\left(\frac{\partial u_i}{\partial x_j}+\frac{\partial u_j}{\partial x_i}\right) \tag{10.70}$$

The chain rule for differentiating the volume integral of eqn. (10.69) and the following identities [50]

$$\frac{\partial \sigma_{ij}}{\partial x_j} v_i = \left(\frac{\partial^2 u_i}{\partial x_j \partial x_j}-\frac{\partial p}{\partial x_i}\right) v_i \tag{10.71}$$

$$\sigma_{ij}\frac{\partial v_i}{\partial x_j} = \frac{\mu}{2}\left(\frac{\partial u_i}{\partial x_j}+\frac{\partial u_j}{\partial x_i}\right)\left(\frac{\partial v_i}{\partial x_j}+\frac{\partial v_j}{\partial x_i}\right) \tag{10.72}$$

give Green's first identity for the flow field $(\mathbf{u}, p)$ shown in Theorem (10.2.1). Similar to Green's second identity for scalar fields, the second identity for the momentum equations is obtained by applying Green's first identity to a flow field $(\mathbf{v}, q)$ and subtract it from the first identity of the flow field $(\mathbf{u}, p)$ (see Theorem (10.2.2)).

10.2.2 Integral Formulation for the Momentum Equations

Similar to scalar field problems, in order to obtain an integral representation for the momentum eqns. (10.63) and (10.64) for the flow field $(\mathbf{u}, \mathbf{p})$, Green's formulae for the momentum equations (Theorems (10.2.1) and (10.2.2)) are used together with the fundamental singular solution of Stokes' equations, i.e.,

$$\begin{aligned} &\frac{\partial u_i^k}{\partial x_i} = 0 \\ &-\frac{\partial q^k}{\partial x_i} + \mu\frac{\partial^2 u_i^k}{\partial x_j \partial x_j} = -\delta(\mathbf{x}-\mathbf{x_o})\delta_{ik} \end{aligned} \tag{10.73}$$

Theorem 10.2.2. Green's Second Identity for a Flow Field $(\mathbf{u}, p)$ and $(\mathbf{v}, q)$

$$\int_V \left[\left(\frac{\partial^2 u_i}{\partial x_j \partial x_j} - \frac{\partial p}{\partial x_i} \right) v_i - \left(\frac{\partial^2 v_i}{\partial x_j \partial x_j} - \frac{\partial q}{\partial x_i} \right) u_i \right] dV = \int_S \left[\sigma_{ij} v_j n_i - \sigma^*_{ij} u_j n_i \right] dS$$

where

$$\sigma^*_{ij} = -q\delta_{ij} + \mu \left(\frac{\partial v_i}{\partial x_j} + \frac{\partial v_j}{\partial x_i} \right)$$

Here $u_i^k(\mathbf{x} - \mathbf{x_0})$ represents the velocity field generated by a point force in the k-direction at a point $\mathbf{x_0}$ and q^k is the corresponding pressure.

With this we obtain an integral representation for the momentum eqns. (10.63) and (10.64), which was first developed by Ladyzhenskaya in 1963 [40]. For a point $\mathbf{x_0} \in V$ we obtain

$$u_k(\mathbf{x_0}) = \int_S \sigma^*_{ij}(\mathbf{x}, \mathbf{x_0}) u_i(\mathbf{x}) n_j(\mathbf{x}) dS - \int_S u_i^k(\mathbf{x}, \mathbf{x_0}) \sigma_{ij}(\mathbf{x}) n_j(\mathbf{x}) dS + \int_V u_i^k(\mathbf{x}, \mathbf{x_0}) g_i(\mathbf{x}) dV \tag{10.74}$$

where

$$u_i^k(\mathbf{x}, \mathbf{x_0}) = -\frac{1}{8\pi\mu r} \left(\delta_{ik} + \frac{(\mathbf{x} - \mathbf{x_0})_i (\mathbf{x} - \mathbf{x_0})_k}{r^2} \right) \tag{10.75}$$

is the fundamental singular solution of the Stokes system of equations or Green's fundamental solution, known as the *Stokeslet*, located at the point $\mathbf{x_0}$ and oriented in the k-th direction, with a corresponding pressure

$$q^k(\mathbf{x}, \mathbf{x_0}) = \frac{1}{4\pi} \frac{\partial}{\partial n_k} \left(\frac{1}{r} \right) = -\frac{1}{4\pi} \frac{(\mathbf{x} - \mathbf{x_0})_k}{r^3} \tag{10.76}$$

and

$$\sigma^*(\mathbf{x}, \mathbf{x_0}) = -\, q^k \delta_{ij} + \mu \left(\frac{\partial u_i^k}{\partial x_j} + \frac{\partial u_j^k}{\partial x_i} \right) = -\, \frac{3}{4\pi} \frac{(\mathbf{x} - \mathbf{x_0})_i (\mathbf{x} - \mathbf{x_0})_j (\mathbf{x} - \mathbf{x_0})_k}{r^5} \tag{10.77}$$

is the symmetric component of a *Stokes doublet*, which is a fundamental singularity called *Stresslet* [29, 50]. The inner product between the Stresslet and the normal vector gives the *traction fundamental solution*,

$$K_{ij}(\mathbf{x}, \mathbf{x_0}) = -\frac{3}{4\pi} \frac{(\mathbf{x} - \mathbf{x_0})_i (\mathbf{x} - \mathbf{x_0})_j (\mathbf{x} - \mathbf{x_0})_k}{r^5} n_k \tag{10.78}$$

which simplifies eqn. (10.74) to

$$u_i(\mathbf{x_0}) = \int_S K_{ij}(\mathbf{x}, \mathbf{x_0}) u_j(\mathbf{x}) dS - \int_S u_i^j(\mathbf{x}, \mathbf{x_0}) t_j(\mathbf{x}) dS + \int_V u_i^j(\mathbf{x}, \mathbf{x_0}) g_j(\mathbf{x}) dV \tag{10.79}$$

where we define $\mathbf{t} = \sigma \cdot \mathbf{n}$ as the traction vector at the surface. Again, we define the surface integrals as hydrodynamic single- and double-layer potentials, while the domain integral is called a hydrodynamic volume potential. The use of this terminology is again due to the fact that the hydrodynamic double-layer potential is discontinuous as the point $\mathbf{x_0}$ crosses the surface S, thus we must complete the integral formulation as

$$c_{ij}(\mathbf{x_0}) u_j(\mathbf{x_0}) = \int_S K_{ij}(\mathbf{x}, \mathbf{x_0}) u_j(\mathbf{x}) dS - \int_S u_i^j(\mathbf{x}, \mathbf{x_0}) t_j(\mathbf{x}) dS + \int_V u_i^j(\mathbf{x}, \mathbf{x_0}) g_j(\mathbf{x}) dV \tag{10.80}$$

where $c_{ij}(\mathbf{x_0})$ is a second order tensor defined as

$$c_{ij} = \begin{pmatrix} c_i & 0 & 0 \\ 0 & c_i & 0 \\ 0 & 0 & c_i \end{pmatrix} \tag{10.81}$$

10.2.3 BEM Numerical Implementation of the Momentum Balance Equations

Similar to scalar problems, the first step of the BEM is to discretize the boundary into a series of elements over which the velocity and traction are assumed to vary according to some interpolation functions.

Once the boundary is divided into NE elements, eqn. (10.80) will be equivalent to

$$c_{ij} u_j = \sum_k \int_{S_k} K_{ij} u_j dS - \sum_k \int_S u_i^j t_j dS \tag{10.82}$$

where $k = 1, ..., NE$. Each element is defined by a number of points or nodes where the unknown values of the velocity or traction are sought. For our implementation here, we will use eight-noded isoparametric quadratic elements (Fig. 10.17).

The value of any variable at any point within the element is defined in terms of the node's values according to the isoparametric interpolation. As with finite elements, the coordinates and the velocity field for each element can be written as follows

$$\mathbf{x} = \sum_i N_i \mathbf{x}_i \tag{10.83}$$

$$\mathbf{u} = \sum_i N_i \mathbf{u}_i \tag{10.84}$$

where $i = 1, ..., 8$ and N_i are interpolation or shape functions given in terms of the local coordinates. For the 8-noded quadratic element the interpolation or shape functions for the

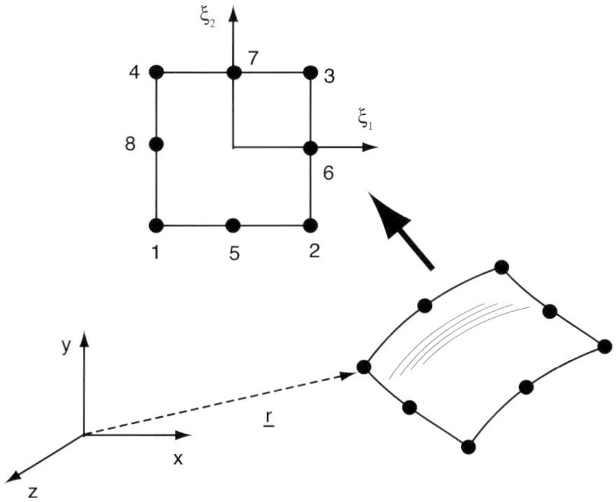

Figure 10.17: 2D isoparametric element definition.

corner nodes are defined by [34, 68, 67],

$$
\begin{aligned}
N_i &= \frac{1}{4}(1+\xi_{10})(1+\xi_{20})(\xi_{10}+\xi_{20}-1) \text{ corner nodes} \\
N_i &= \frac{1}{2}(1-\xi_1^2)(1+\xi_{20}) \text{ mid-side nodes with } \xi_{1,i}=0 \\
N_i &= \frac{1}{2}(1+\xi_{10})(1-\xi_2^2) \text{ mid-side nodes with } \xi_{2,i}=0
\end{aligned}
\tag{10.85}
$$

where $\xi_{10} = \xi_1 \xi_{1,i}$ and $\xi_{20} = \xi_2 \xi_{2,i}$. Equation (10.83) can be written in matrix form as

$$
\begin{pmatrix} x \\ y \\ z \end{pmatrix} = \begin{bmatrix} N_1 & 0 & 0 & N_2 & 0 & 0 & \dots & N_8 & 0 & 0 \\ 0 & N_1 & 0 & 0 & N_2 & 0 & \dots & 0 & N_8 & 0 \\ 0 & 0 & N_1 & 0 & 0 & N_2 & \dots & 0 & 0 & N_8 \end{bmatrix} \begin{pmatrix} x^1 \\ y^1 \\ z^1 \\ x^2 \\ y^2 \\ z^2 \\ \dots \\ x^8 \\ y^8 \\ z^8 \end{pmatrix}
\tag{10.86}
$$

or in more compact form,

$$
\mathbf{x} = \mathbf{N}\mathbf{x}^j \tag{10.87}
$$

where $\mathbf{N}$ is the matrix of isoparametric shape functions and $\mathbf{x}^j = (x^j, y^j, z^j)$ is the vector of nodal coordinates of the eight element nodes, where j denotes the node. Similarly, the velocity and traction fields can be expressed as

$$
\mathbf{u} = \mathbf{N}\mathbf{u}^j \tag{10.88}
$$

$$
\mathbf{t} = \mathbf{N}\mathbf{t}^j \tag{10.89}
$$

For any point in the domain and boundary, the fundamental solutions in the boundary integrals of eqn. (10.82) can be written in matrix form as

$$K_{ij} = \mathbf{h} = \begin{bmatrix} K_{11} & K_{12} & K_{13} \\ K_{21} & K_{22} & K_{23} \\ K_{31} & K_{32} & K_{33} \end{bmatrix} \tag{10.90}$$

and

$$u_i^j = \mathbf{g} = \begin{bmatrix} u_1^1 & u_1^2 & u_1^3 \\ u_2^1 & u_2^2 & u_2^3 \\ u_3^1 & u_3^2 & u_3^3 \end{bmatrix} \tag{10.91}$$

By substitution of eqns. (10.88) to (10.91) into eqn. (10.82), the boundary integral formula can be written as follows

$$\mathbf{c}\mathbf{u}_i = \bar{\mathbf{H}}_{ij}\mathbf{u}^j - \mathbf{G}_{ij}\mathbf{t}^j \tag{10.92}$$

where

$$\bar{\mathbf{H}}_{ij} = \int_{S_j} \mathbf{h}\mathbf{N} dS_j \tag{10.93}$$

$$\mathbf{G}_{ij} = \int_{S_j} \mathbf{g}\mathbf{N} dS_j \tag{10.94}$$

Here, similar to the scalar fields, the velocity is a continuous function; therefore there is a unique value of $\mathbf{u}$ in every node. Generally, this is not true for the traction vector. However, for a Lyapunov surface, where the normal is continuous, the tractions are also continuous. Equation (10.92) becomes

$$\mathbf{H}\mathbf{u} = \mathbf{G}\mathbf{t} \tag{10.95}$$

where $\mathbf{H} = \bar{\mathbf{H}} - \mathbf{c}$.

For a problem with N boundary nodes and NIP internal points

$$\mathbf{H} \in \mathbf{M}\,(3(N + NIP), 3(N + NIP))$$

and

$$\mathbf{G} \in \mathbf{M}\,(3(N + NIP), 3N)$$

Consequently, there are $3(N + NIP)$ velocity unknowns and $3N$ traction unknowns. This makes eqn. (10.95) a system of $3(N+NIP)$ equations with $3(N+NIP)+3N$ unknowns. Each boundary nodal point has either traction or velocity specified for each direction as a boundary condition, thus the system in eqn. (10.95) can ultimately be arranged into a solvable system of linear algebraic equations as

$$\mathbf{A}\mathbf{x} = \mathbf{b} \tag{10.96}$$

where the coefficient matrix $\mathbf{A}$ contains columns of matrices $\mathbf{H}$ or $\mathbf{G}$; it is fully populated and non-symmetric. The vector $\mathbf{x}$ has unknown traction or velocity and $\mathbf{b}$ is the vector obtained from the multiplication of the boundary conditions with the corresponding coefficients in $\mathbf{H}$ or $\mathbf{G}$.

10.2.4 Numerical Treatment of the Weakly Singular Integrals

It is easily seen that the Green functions or fundamental solutions in the matrices $\mathbf{H}$ and $\mathbf{G}$ go to infinity as the distance between the source and field point decreases, i.e., the Euclidean distance $r \to 0$. We already saw how the singular coefficients in the $\mathbf{H}$ matrix can be calculated from a constant potential over the surface or a rigid body motion. In other words, these terms are included in the calculation of the diagonal that includes the coefficient matrix $\mathbf{c}$. However, the weak singularity of the fundamental solutions in matrix $\mathbf{G}$ needs a special treatment. In particular, the weak singularity of the Stokeslet is of the order $O(r \log r)$ which can be dealt with a self-adaptive coordinate transformation called the Telles' transformation [64]. For example, consider the evaluation of an integral

$$\int_{-1}^{+1} f(x)dx \tag{10.97}$$

where the function $f(x)$ is weakly singular at x_0. The singularity can be cancelled off by forcing its Jacobian to be zero at the singular point in a new Telles space defined as [64]

$$x = a\gamma^3 + b\gamma^2 + c\gamma + d \tag{10.98}$$

where the constants in this third order polynomial are given by

$$\begin{aligned} a &= \frac{1}{Q} \qquad b = -\frac{3\bar{\gamma}}{Q} \\ c &= \frac{3\bar{\gamma}^2}{Q} \qquad d = \frac{3\bar{\gamma}}{Q} \end{aligned} \tag{10.99}$$

with

$$\begin{aligned} Q &= 1 + 3\bar{\gamma}^2 \\ \bar{\gamma} &= \left(x_0 x^* + |x^*|\right)^{1/3} + \left(x_0 x^* - |x^*|\right)^{1/3} + x_0 \\ x^* &= x_0^2 - 1 \end{aligned} \tag{10.100}$$

The integral is now calculated in terms of γ as follows

$$\int_{-1}^{+1} f\left[\frac{(\gamma - \bar{\gamma})^3 + \bar{\gamma}(\bar{\gamma}^2 + 3)}{Q}\right] \frac{3(\gamma - \bar{\gamma})^2}{Q} d\gamma \tag{10.101}$$

which can be evaluated using the standard Gauss quadrature. After the transformation, all standard Gauss points of the numerical quadrature are biased towards the singularity where the Jacobian is zero.

■ **EXAMPLE 10.3.**

Poiseuille flow of a Newtonian fluid in a circular tube. For a pressure driven flow of a Newtonian fluid in a circular tube, we can obtain an analytical solution as we already did in Chapter 5. Ignoring the entrance effects, the solution for the velocity field as a function of the radial direction (see Fig. 10.18) is as follows

$$u(r) = \frac{\tau_R R}{2\mu}\left[1 - \left(\frac{r}{R}\right)^2\right] \tag{10.102}$$

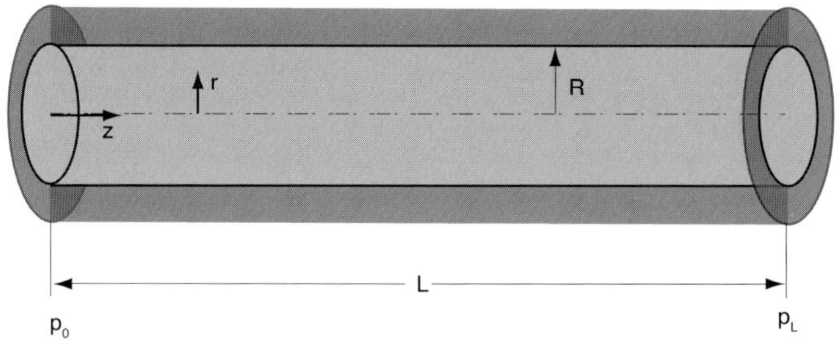

Figure 10.18: Schematic diagram of pressure flow through a tube.

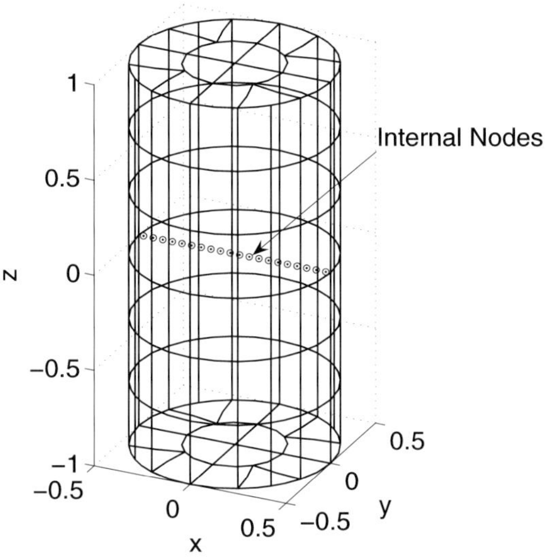

Figure 10.19: Typical BEM mesh and internal nodes location.

where μ is the Newtonian viscosity, R the tube radius and τ_R the shear stress at the tube walls defined by

$$\tau_R = \frac{p_0 - p_L}{2L} R \tag{10.103}$$

where p_0 is the pressure at the entrance, p_L the pressure at the end and L is the tube length.

To find a solution to this problem using BEM, we must solve the Stokes system of equations with their corresponding equivalent integral formulation eqn. (10.82) with traction boundary conditions at the entrance and end of the tube and with no-slip boundary conditions at the tube walls. We start by creating the surface mesh and by selecting the position of the internal points where we are seeking the solution. Figure 10.19 shows a typical BEM mesh with 8-noded quadratic elements.

The BEM methodology listed in Algorithm 11 will be applied again with the exception of the matrices dimensions. The BEM matrices assembly will be performed in a similar way as in the scalar case. Algorithms 14 and 15 show the new assembly methodology for the matrices and the integral calculation of the components, including Telles' transformation for matrix **G**.

Algorithm 14 Assembling Boundary Element Matrices for the Stokes Momentum Equations.

```
subroutine BEM-assemble
  Hbig = 0; Gbig = 0
  do i = 1,N+NIP                                              equation node i,
                                            for boundary (N) and internal nodes (NIP)
    ii = 3*i
    xo = xi(ii - 2:ii)                              xo(3) and xi(3*(NN+NIP)) contains
                                                                the nodes coordinates
    do j = 1,NE                                                   loop over elements
      telles = .false.              logical variable to check if Telles is needed
      do k = 1,8                                                  8 nodes per element
        kk = 3*nee(j,k); telles = ii.eq.kk.or.telles
        xe(3*k - 2:3*k) = xi(kk - 2:kk)       element nodes' coordinates xe(8*3)
      enddo
      call small-bem(telles,xo,xe,g,h)                        h(3,8*3) and g(3,8*3)
      do k = 1,8
        jj = 3*nee(j,k)                       element node position in the matrices
        Hbig(ii - 2:ii,jj - 2:jj) = Hbig(ii - 2:ii,jj - 2:jj) + h(:,3*k - 2:3*k)
        Gbig(ii - 2:ii,jj - 2:jj) = Gbig(ii - 2:ii,jj - 2:jj) + g(:,3*k - 2:3*k)
                                                          using the continuous normal
      enddo
    enddo
  enddo
  do i = 1, N+NIP                                             make H diagonal zero
    ii = 3*i
    Hbig(ii - 2:ii,ii - 2:ii) = 0
  enddo
  do i = 1,N                                      calculation of the H diagonal terms
    do j = 1, N
      ii = 3*i; jj = 3*j
      if (i /= j) Hbig(ii - 2:ii,ii - 2:ii) = Hbig(ii - 2:ii,ii - 2:ii) -  &
          Hbig(ii - 2:ii,jj - 2:jj)
    enddo
  enddo
  do i = 1, NIP                                           internal nodes coefficient
    ii = 3*NN + 3*i
    Hbig(ii - 2,ii - 2) = 1
    Hbig(ii - 1,ii - 1) = 1
    Hbig(ii,ii) = 1
  enddo
end subroutine BEM-assemble
```

Figure 10.20 shows a comparison between the BEM and the analytical solution for the velocity profile. As we can see, the results are satisfactory; as a matter of fact, Fig. 10.21 shows the error for two different discretizations. The error is always less than 1% for the relatively coarse mesh, while is less than 0.2% for the finer mesh.

Algorithm 15 Small Boundary Element Matrices (Integrals).

```
subroutine small-bem(telles,xo,xe,g,h)
  if (telles) call telles-position(xo,xe,epsis,nus)   it return the value of the
                                                isoparametric coordinates depending
                                                     on the location of the element
  do ix = 1,ngauss                                                  integral in xi1
    do iy = 1,ngauss                                                integral in xi2
      if (telles) then                                       Telles' transformation
        ep1s = epsis*epsis-1.;    ep2s = nus*nus-1.
        gam1 = (epsis*ep1s+abs(ep1s))**(1/3)+(epsis*ep1s-abs(ep1s))**(1/3)+epsis
        gam2 = (nus*ep2s+abs(ep2s))**(1/3)+(nus*ep2s-abs(ep2s))**(1/3)+nus
        xjac = 3.*(gp(i)-gam1)**2/(1.+3.*gam1**2)
        yjac = 3.*(gp(j)-gam2)**2/(1.+3.*gam2**2)
        zjac = xjac*yjac
        xi1 = ((gp(i)-gam1)**3+gam1*(gam1**2+3.))/(1.+3.*gam1**2)
        xi2 = ((gp(j)-gam2)**3+gam2*(gam2**2+3.))/(1.+3.*gam2**2)
      else
        zjac = 1.; xi1 = gp(ix); xi2 = gp(iy)                    regular Gauss point
      endif
      dn(5) = 0.5*(-2.*xi1)*(1.-xi2);   dn(6) = 0.5*(1.-xi2**2)
      dn(7) = 0.5*(-2.*xi1)*(1.+xi2);   dn(8) = 0.5*(-1.)*(1.-xi2**2)
      dn(1) = 0.25*(-1.)*(1.-xi2)-0.5*(dn(5)+dn(8))
      dn(2) = 0.25*(1.-xi2)-0.5*(dn(5)+dn(6))
      dn(3) = 0.25*(1.+xi2)-0.5*(dn(6)+dn(7))
      dn(4) = 0.25*(-1.)*(1.+xi2)-0.5*(dn(7)+dn(8))
      phi = 0.; phi(1,1::3) = dn; phi(2,2::3) = dn; phi(3,3::3) = dn
      dxde1 = matmul(phi,xe)                                      derivatives dxde1
      dn(5) = 0.5*(1.-xi1**2)*(-1.);   dn(6) = 0.5*(1.+xi1)*(-2.*xi2)
      dn(7) = 0.5*(1.-xi1**2);   dn(8) = 0.5*(1.-xi1)*(-2.*xi2)
      dn(1) = 0.25*(1.-xi1)*(-1.)-0.5*(dn(5)+dn(8))
      dn(2) = 0.25*(1.+xi1)*(-1.)-0.5*(dn(5)+dn(6))
      dn(3) = 0.25*(1.+xi1)-0.5*(dn(6)+dn(7))
      dn(4) = 0.25*(1.-xi1)-0.5*(dn(7)+dn(8))
      phi = 0.; phi(1,1::3) = dn; phi(2,2::3) = dn; phi(3,3::3) = dn
      dxde2 = matmul(phi,xe)                                      derivatives dxde2
      n(5) = 0.5d0*(1.d0-epsi**2)*(1.d0-nu);  n(6) = 0.5d0*(1.d0+epsi)*(1.d0-nu**2)
      n(7) = 0.5d0*(1.d0-epsi**2)*(1.d0+nu); n(8) = 0.5d0*(1.d0-epsi)*(1.d0-nu**2)
      n(1) = 0.25d0*(1.d0-epsi)*(1.d0-nu)-0.5d0*(n(5)+n(8))
      n(2) = 0.25d0*(1.d0+epsi)*(1.d0-nu)-0.5d0*(n(5)+n(6))
      n(3) = 0.25d0*(1.d0+epsi)*(1.d0+nu)-0.5d0*(n(6)+n(7))
      n(4) = 0.25d0*(1.d0-epsi)*(1.d0+nu)-0.5d0*(n(7)+n(8))
      phi = 0.d0; phi(1,1::3) = n; phi(2,2::3) = n; phi(3,3::3) = n
      x = matmul(phi,xe)                                     Gauss point coordinates
                                                   Jacobian, normal vector and "r"
      ee = dot-product(dxde1,dxde1);   gg = dot-product(dxde2,dxde2)
      ff = dot-product(dxde1,dxde2);   jac = sqrt(ee*gg-ff*ff)
      norm(1) = (dxde1(2)*dxde2(3)-dxde2(2)*dxde1(3))/jac
      norm(2) = (dxde2(1)*dxde1(3)-dxde1(1)*dxde2(3))/jac
      norm(3) = (dxde1(1)*dxde2(2)-dxde1(2)*dxde2(1))/jac
      dx = x - xo; r = sqrt(dot-product(dx,dx)); drdn = dot-product(dx,norm)
      do i = 1,3                                                   Green's function
        do k = 1,3
          uf(i,k) = ( (DELTA(i,k)/r) + dx(i)*dx(k)/r**3 )/(8.d0*pi*visc)
          tf(i,k) = -3.d0*( dx(i)*dx(k)*drdn )/(4.d0*pi*r**5)
        enddo
      enddo
      g = g + matmul(uf,phi)*jac*zjac*gw(ix)*gw(iy)              Integral summation
      h = h + matmul(tf,phi)*jac*w(ix)*w(iy)
    enddo
  enddo
end subroutine small-bem
```

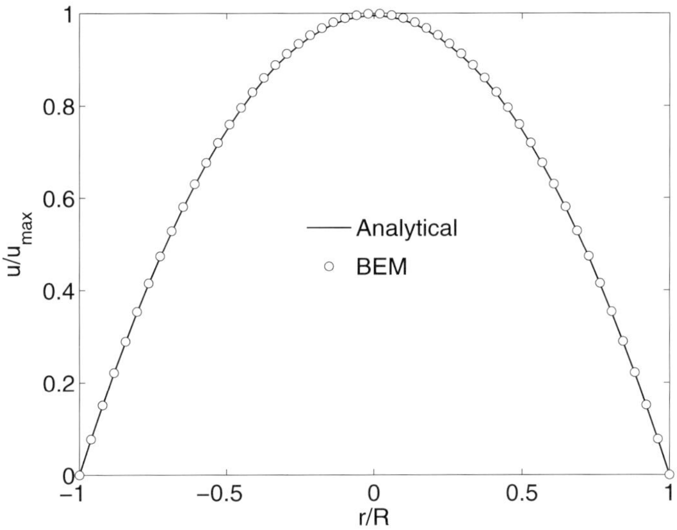

Figure 10.20: Comparision between the BEM and the analytical solution for the velocity.

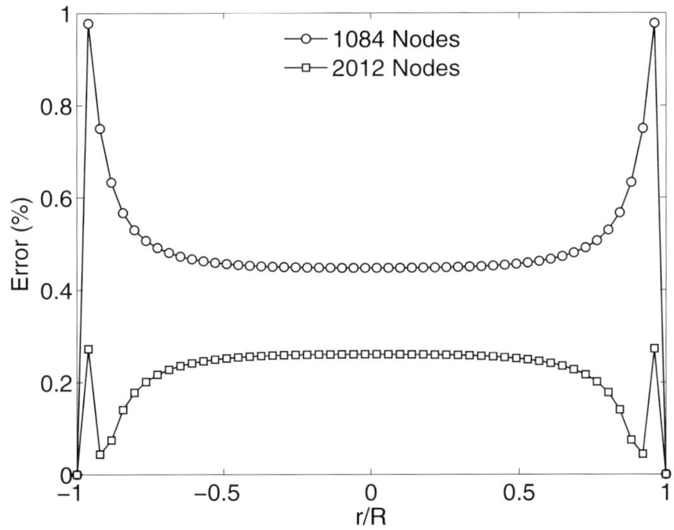

Figure 10.21: BEM error for two different discretizations.

10.2.5 Solids in Suspension

Low Reynolds number flows with boundary integral representation have been used to describe rheological and transport properties of suspensions of solid spherical particles, as well as for numerical solution of different problems, including particle-particle interaction, the motion of a particle near a fluid interface or a rigid wall, the motion of particles in a container, and others.

Boundary element methods can be used for particulate flows where direct[1] formulations can be used. The surface tractions on the solids are integrated to compute the hydrodynamic force and torque on those particles, which for suspended particles must be zero.

■ **EXAMPLE 10.4.**

Fiber motion – Jeffery orbits. The motion of ellipsoids in uniform, viscous shear flow of a Newtonian fluid was analyzed by Jeffery [32, 33] in 1922. For a prolate spheroid of aspect ratio a (defined as the ratio between the major axis and the minor axis) in simple shear flow, $\mathbf{u}^{\infty} = (z\dot{\gamma})$, the angular motion of the spheroid is described by

$$\tan\theta = \frac{Ka}{\sqrt{a^2\cos^2\phi + \sin^2\phi}} \tag{10.104}$$

and

$$\tan\phi = a\tan\left(2\pi\frac{t}{T}\right) \tag{10.105}$$

where θ is the angle between the fiber's major axis and the vorticity axis, i.e. y-axis, ϕ is the angle between the z-axis and the xz-projection of the fiber axis (see Fig. 10.22), T is the orbit period

$$T = \frac{2\pi}{\dot{\gamma}}\left(a + \frac{1}{a}\right) \tag{10.106}$$

and K is the orbit constant, determined by the initial orientation using

$$K = \tan\theta_0\sqrt{\cos^2\phi_0 + \frac{\sin^2\phi_0}{a}} \tag{10.107}$$

These equations predict that the spheroid will repeatedly rotate through the same orbit, the particle will not migrate across the streamline, and that the orbit period is independent of the initial orientation.

The BEM was implemented for the motion of a single rigid cylindrical fiber in simple shear. To avoid discontinuities in the normal vector, semi-spheres of the same cylinder radius were used to cap the cylinder, as schematically depicted in Fig. 10.23. The aspect ratio for the fiber is redefined as,

$$a = \frac{L+D}{D} \tag{10.108}$$

[1] Direct means that we relate in the integral equation velocities and tractions directly. There are some indirect integral formulations, because the velocity and the tractions are related indirectly by means of hydrodynamic potentials [29].

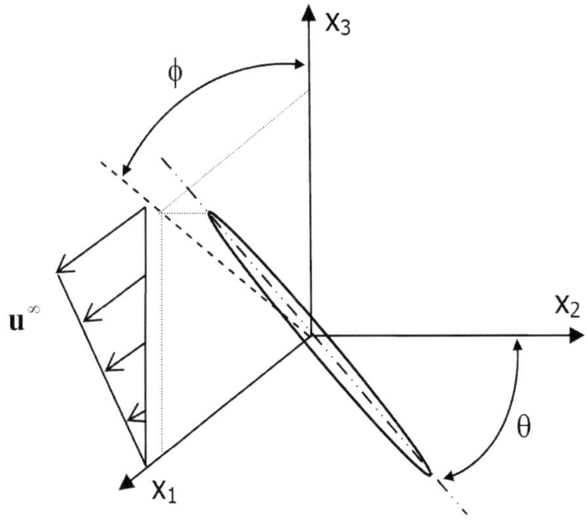

Figure 10.22: Prolate spheroid in shear flow.

The fiber is suspended in the liquid, which means that due to small time scales given by the pure viscous nature of the flow, the hydrodynamic force and torque on the particle are approximately zero [26, 51]. Numerically, this means that the velocity and traction fields on the particle are unknown, which differs from the previous examples where the velocity field was fixed and the integral equations were reduced to a system of linear equations in which velocities or tractions were unknown, depending on the boundary conditions of the problem. Although computationally expensive, direct integral formulations are an effective way to find the velocity and traction fields for suspended particles using a simple iterative procedure. Here, the initial tractions are assumed and then corrected, until the hydrodynamic force and torque are zero.

For the BEM simulations, the fiber length was set to 2 length units, the diameter to 0.2 length units and the shear rate to 2.0 reciprocal time units. These data imply an aspect ratio $a = 11$ and an orbit period $T = 843.34$ time units. Figure 10.24 shows the evolution of θ and ϕ in time for a fiber initially perpendicular to the vorticity axis, i.e., $\theta_0 = \pi/2$. This simulation requires a large number of elements on the fiber surface (500 elements with 1200 nodes) and a small time step (0.01 time units); it is computationally expensive (10 minutes per time step); however, the results agree with Jeffery's prediction. The path of the fiber during the simulation is illustrated in Fig. 10.25. Figures 10.26 and 10.27 show the evolution of the orientation angles as a function of time and the fiber path $\theta_0 = \pi/6(30^o)$.

■ **EXAMPLE 10.5.**

Viscosity of a sphere's suspension. The basic problem of suspension mechanics is to predict the macroscopic transport properties of a suspension, i.e., thermal conduc-

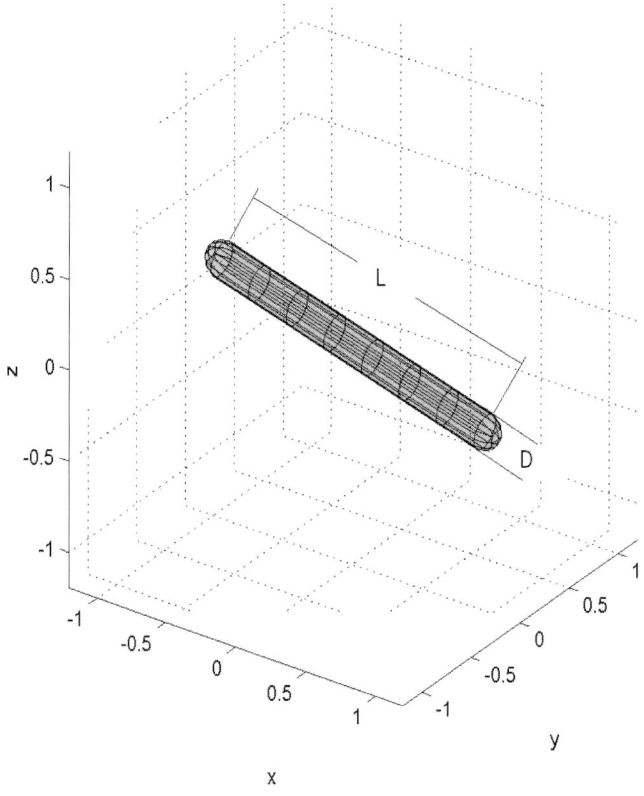

Figure 10.23: Fiber representation for the BEM simulation.

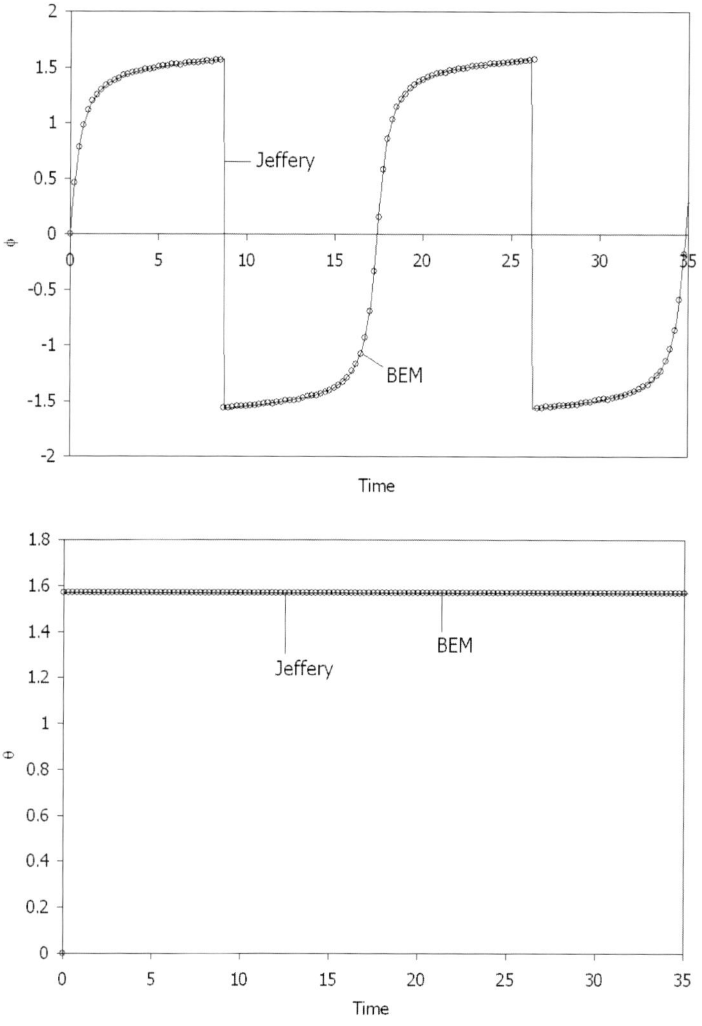

Figure 10.24: BEM and Jeffery orientation angles for $\theta_0 = \pi/2$.

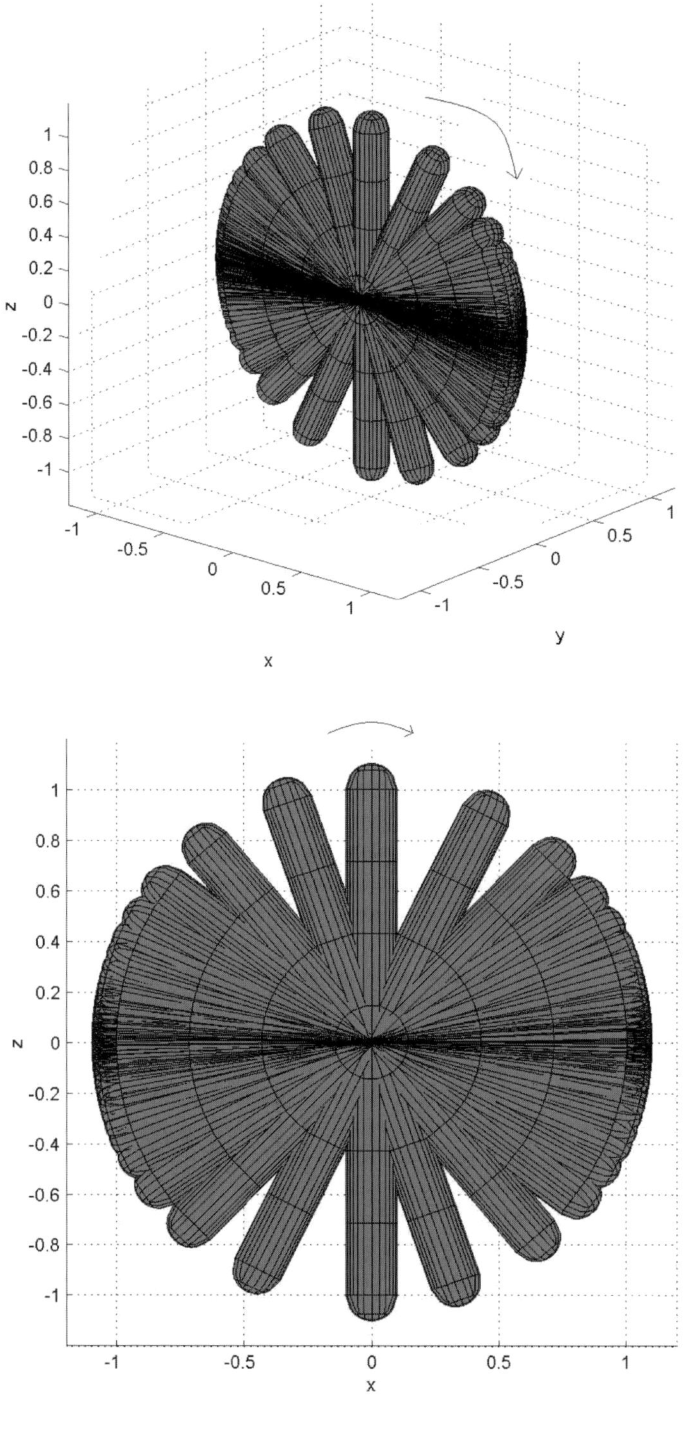

Figure 10.25: BEM predicted path for $\theta_0 = \pi/2$.

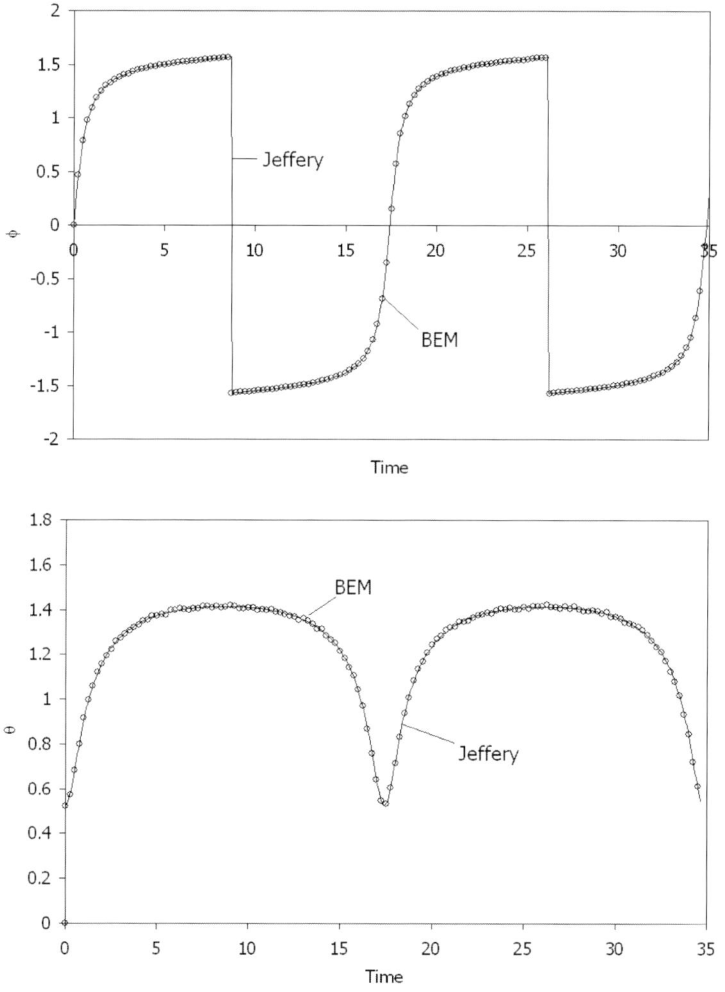

Figure 10.26: BEM and Jeffery orientation angles for $\theta_0 = \pi/6$.

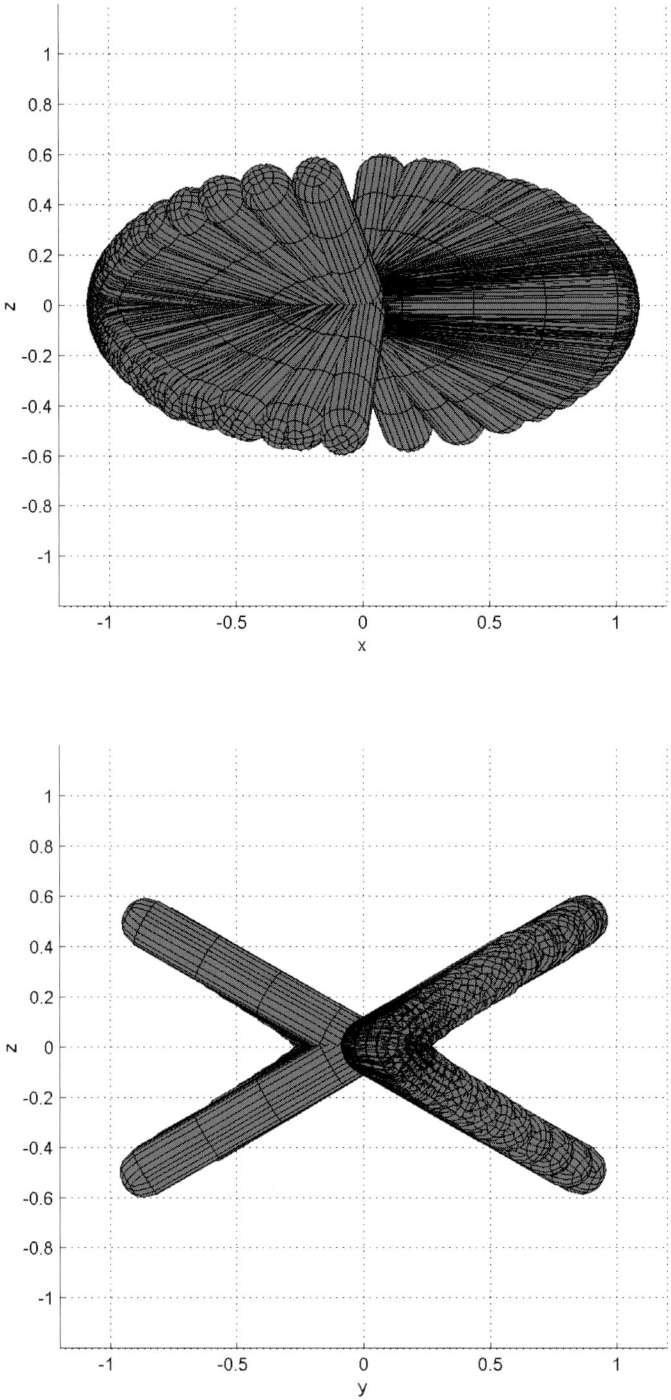

Figure 10.27: BEM predicted path for $\theta_0 = \pi/6$.

tivity, viscosity, sedimentation rate, etc., from the micro-structural mechanics. These flows are governed by at least three length scales: the size of the suspended particles, the average spacing between the particles, and the characteristic dimension of the container in which the flow occurs. A number of excellent reviews of the general subject of suspension rheology are available [16, 26]. Of special interest is the hydrodynamic treatment of the problem by Frisch and Simha [16] and Hermans [27]. Numerous models have been proposed to estimate the suspension viscosity. Most of them are a power series of the form

$$\frac{\mu}{\mu_0} = 1 + a_1\phi + a_2\phi^2 + ... \tag{10.109}$$

where ϕ is the volume concentration of the suspended solids. For dilute systems of spheres of equal size, where interaction effects are neglected, Einstein [15] arrives at the following formula

$$\frac{\mu}{\mu_0} = 1 + 2.5\phi \tag{10.110}$$

Einstein's formula holds for any type of linear viscometers, and can be derived by different methods [10, 26, 31]. For dilute systems, considering the first-order effect of the spheres interacting with one another, Guth and Simha [25] gave

$$\frac{\mu}{\mu_0} = 1 + 2.5\phi + 14.1\phi^2 \tag{10.111}$$

The direct boundary integral formulation was used to simulate suspended spheres in simple shear flow. The viscosity was then calculated by integration of the surface tractions on the moving wall. Figure 10.28 shows a typical mesh for the domain and spheres for these simulations; in this mesh, the box has dimensions of $1 \times 1 \times 1$ (Length units)^3 and 40 spheres of radius of 0.05 length units.

Initially, the spheres are positioned randomly in the box, periodic boundary conditions are used in the x- and y-direction and no-slip on the z-direction. The spheres move according to the flow field and the viscosity is calculated for several time steps, and for each configuration an average suspension viscosity is obtained. The box is divided into 216 elements with 650 nodes, and each sphere into 96 elements with 290 nodes. The computational time depends, as for any particulate simulation, on the number of spheres. Two different sphere radii were used in the simulations: 0.05 length units and 0.07 length units. In the same way, the box dimensions were set to $1 \times 1 \times 1$ (length units)^3 and $0.8 \times 0.8 \times 0.8$ (length units)^3. Each case was simulated with 10, 20, 30 and 40 spheres.

After 1000 strain units, the recorded viscosity was fitted to give

$$\frac{\mu}{\mu_0} = 1 + 2.5463\phi + 11.193\phi^2 \tag{10.112}$$

The numerical correlations given by the direct BEM simulations are similar to the expressions given earlier. In fact, the first coefficient in the power expansion is close to the one predicted by Einstein [15]. The second coefficient in the power expansion is between the value suggested by Guth and Simha [25] and one suggested by Vand [65, 66]. In Figure 10.29, the calculated BEM relative viscosity is collapsed for all cases.

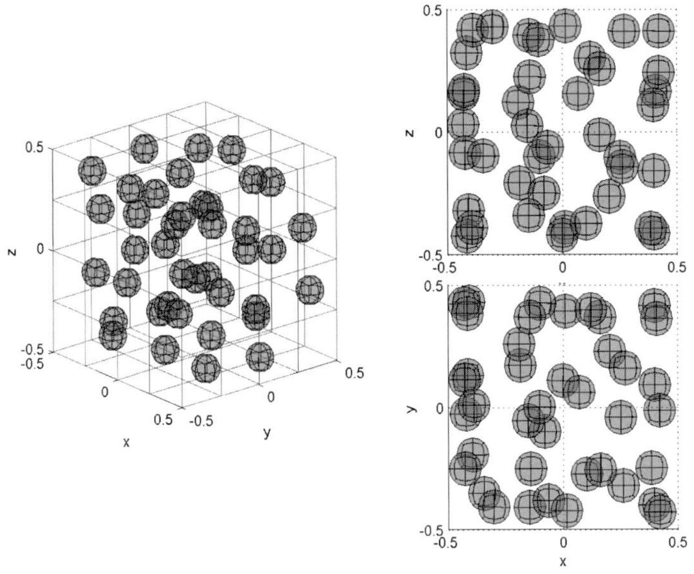

Figure 10.28: Spheres suspended in simple shear flow.

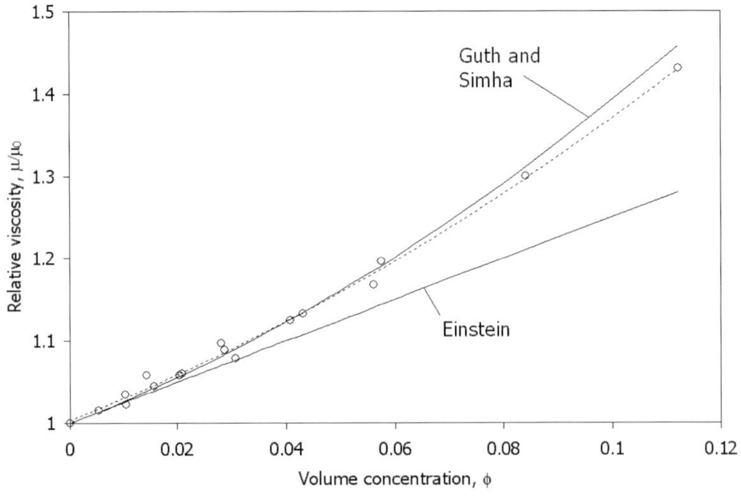

Figure 10.29: Calculated BEM relative viscosity.

10.3 COMMENTS OF NON-LINEAR PROBLEMS

When non-linearities are included in the analysis, we must also solve the domain integral in the integral formulations. Several methods have been developed to approximate this integral. As a matter of fact, at the international conferences on boundary elements, organized every year since 1978 [43], numerous papers on different and novel techniques to approximate the domain integral have been presented in order to make the BEM applicable to complex non-linear and time dependent problems. Many of these papers were pointing out the difficulties of extending the BEM to such applications. The main drawback in most of the techniques was the need to discretize the domain into a series of internal cells to deal with the terms not taken to the boundary by application of the fundamental solution, such as non-linear terms.

Some of these methods approximate the domain integrals to the boundary in order to eliminate the need for internal cells, i.e., boundary-only formulations. The dual reciprocity method (DRM) introduced by Nardini and Brebbia [42] is one of the most popular techniques. The method is closely related to the method of particular integrals technique (PIT), introduced by Ahmad and Banerjee [2], which also transforms domain integrals to boundary integrals. In the PIT method, a particular solution satisfying the non-homogeneous PDE is first found and then the remainder of the solution, satisfying the homogeneous PDE, is obtained by solving the corresponding integral equations. The boundary conditions for the homogeneous PDE must be adjusted to ensure that the total solution satisfies the boundary conditions of the original problem [2, 4, 44, 45]. The DRM also uses the concept of particular solutions, but instead of obtaining the particular solution and the homogeneous solution separately, it applies the divergence theorem to the domain integral terms and converts the domain integral into equivalent boundary integrals [43].

Two major disadvantages were encountered when applying the DRM and PIT to non-linear flow problems. First, the lack of convergence as the non linear terms in the problem become dominant. For the Navier-Stokes equations, Cheng *et al.*[11] and Power and Partridge [48, 49] reported problems when the Reynolds number was higher than 200. For non-Newtonian fluid flow, Davis [12] and Hernández [28] faced problems when the shear-thinning exponent was lower than 0.8. Finally, for thermal problems with natural convection problems (non-isothermal) when the Rayleigh number was higher than 103 [46, 47, 57, 58]. Second, in the PIT and DRM the resulting algebraic system consists of a series of matrix multiplications of fully populated matrices, which generates expensive computing times for complex problems.

When dealing with the BEM solution of large problems, it is common to use the method of domain decomposition, in which the original domain is divided into subregions, and finding the full integral representation formula to each region. At the interfaces between adjacent subregions, continuity conditions are enforced. Some authors refer to the subregion BEM formulation as the Green element method (GEM; see Taigbenu [62] and Taigbenu and Onyejekwe [63]). Popov and Power [44, 45] found that the DRM approximation of the volume potential of a highly nonlinear problem can be substantially improved by using the domain decomposition scheme. This decomposition technique solved the problems that were previously encountered in the DRM and PIT, i.e., high Reynolds number [17, 18, 46, 47], low shear-thinning exponents [19] and high Rayleigh numbers [20]. Although the method keeps the boundary-only character, it is necessary to construct internal divisions in the domain, which tends to become similar to a finite element mesh. The corresponding matching conditions, that are necessary to keep the system closed, i.e., continuity of the velocity and equilibrium of tractions between adjacent sub-domains, will lead to cumbersome over-determined systems or complicated discontinuous elements, which will require an internal mesh [44, 45, 46, 47]. In simple two- dimensional problems, the discontinuous elements will not be an impediment, while in full three-dimensional domains, the domain decomposition will be a difficult task.

As a consequence, the application of these types of methods for complex non-linear problems is limited. Both methods intend to keep the boundary-only character, which is perfect for small order nonlinearities, but require domain decomposition (complicated internal meshes) when dealing with high order nonlinearities. Techniques that directly approximate the domain integral have been developed over the years: Fourier expansions, the Galerkin vector technique, the multi reciprocity method, Monte Carlo integration and cell integration. In the early boundary element analysis, the evaluation of the domain integrals was mostly done by cell integration. The technique is effective and general, but causes the method to lose its boundary-only nature. It is the simplest way of computing the domain term by subdividing the domain into a series of internal cells, on each of which a numerical integration scheme, such as Gauss quadratures, can be applied. Several authors applied the technique for Newtonian, non-Newtonian and non-isothermal problems with very accurate results and without the restrictions finded by the techniques that approximate the domain integrals into boundary integrals [12, 35, 36, 38, 60, 61]. The domain discretization for the Cell-BEM technique is done by dividing the boundary into a specific type of elements, while the domain will have a different type of mesh. The internal cells are not required to be discretized all the way to the boundary in order to avoid the discontinuity of the kernels in the boundary and to avoid the necessity of recording which nodes are in the boundary and domain at the same time. As reported by several authors [7, 8, 12, 3], this does not affect the accuracy of the technique, in fact it is considered an advantage.

The difference between boundary and domain discretization increases the time needed for pre-processing of a specific problem. In addition, in moving boundary problems there is the necessity of re-meshing the internal cells, which implies the record of internal solutions and interpolations for transient problems. In the cell-BEM, the integral formulation is applied for both, the boundary and internal nodes, and for every node, the internal cells are used to approximate the domain integral (volume potential). A set of nonlinear equations is formed for the boundary and internal unknowns, and the equations are solved by successive iterations or by Newton's method [12, 17, 18, 28, 54]. In conclusion, the big inconvenience of the cell-BEM technique is the cumbersome pre-processing, two different meshes, and the re-meshing in moving boundary problems.

10.4 OTHER BOUNDARY ELEMENT APPLICATIONS

Numerous problems in polymer processing have been solved in the past years with the use of the boundary element method. In all these solutions, the complexity of the geometry was the primary reason why the technique was used. Some of these problems are illustrated in this section.

Gramann, Osswald and Rios [22, 23, 54] used BEM to simulate various mixing processes in two and three dimensions, an example of which is presented in Figs. 10.30 and 10.31. In both these systems the velocity and velocity gradients were computed as particles were tracked while traveling through the system.

The velocity gradients were used to compute the rate of deformation tensor, the magnitudes of the rate of deformation and vorticity tensors. The magnitudes of the rate of

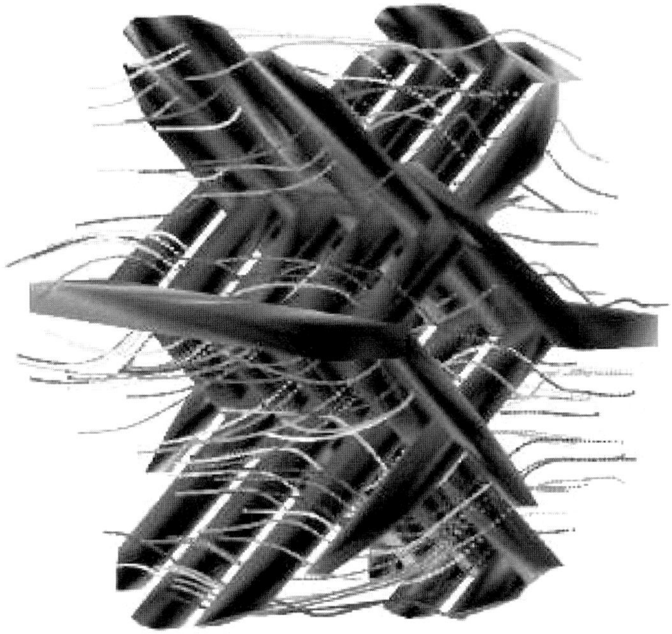

Figure 10.30: Deforming drop inside a rhomboidal mixing section.

Figure 10.31: Flow patterns inside a section of a static mixer.

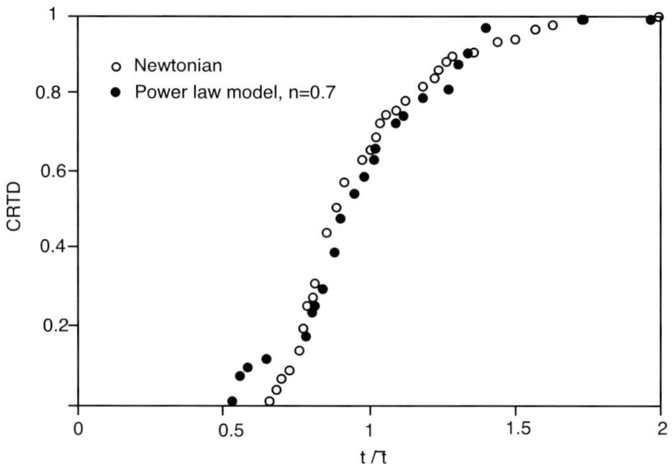

Figure 10.32: Flow patterns inside a section of a static mixer.

deformation and vorticity tensors were used to compute the flow number given by

$$\lambda = \frac{\dot{\gamma}}{\dot{\gamma} + \omega} \tag{10.113}$$

where λ denotes the flow number and equals 1 when the flow generated is elongational (ideal for dispersive mixing), 0.5 when the flow is dominated by shear, or 0 for rigid motion or pure rotational (a sign of poor mixing). From their studies it was found that although these mixers are excellent distributive mixers, they are primarily dominated by shear ($\lambda = 0.5$). In addition, when tracking the particles through the system, Gramann and Osswald recorded the time when the particles left the system at the outlet of the mixers. The ratio of number of points that have emerged from the mixer at an arbitrary point in time to the total number of points is the *cumulative residence time distribution function.* Figure 10.32 presents the cumulative residence time distribution function (CRTD) of the Kenics static mixer presented in Fig. 10.31 for a Newtonian polymer and a shear thinning polymer with a power law viscosity of $n = 0.7$. To simulate the power law behavior of the melt using BEM, Rios [54] developed a Monte Carlo technique where random points were sprinkled throughout the domain to account for the non-Newtonian non-linearities.

Rios and Osswald [56] used the boundary element to perform a comparative study of rhomboidal mixing sections. Rhomboidal mixing sections are defined by the pitch of the two cuts performed when machining the elements. For example, a 1D3D rhomboidal mixer has a cut with a pitch of one diameter per turn, and a second cut with a pitch of 3 diameters per turn as shown in Fig. 10.33. Rios and Osswald studied 9 different configurations. For each configurations they computed the CRTD and the *weighted average total strain* (WATS) defined by

$$\text{WATS} = \frac{\sum_i^N \gamma_i}{N} \tag{10.114}$$

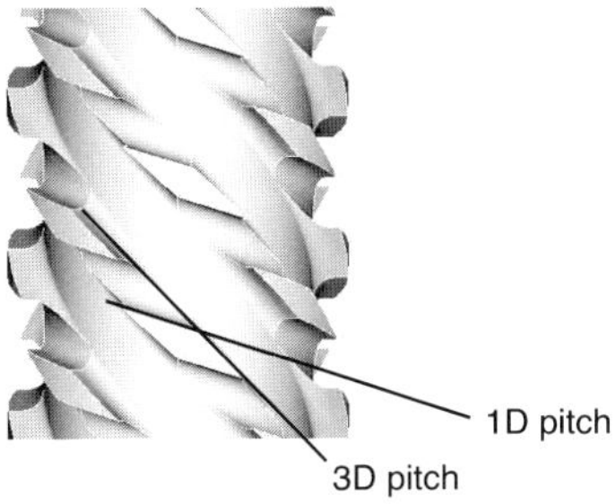

Figure 10.33: Geometric definition of a rhomboidal mixing head.

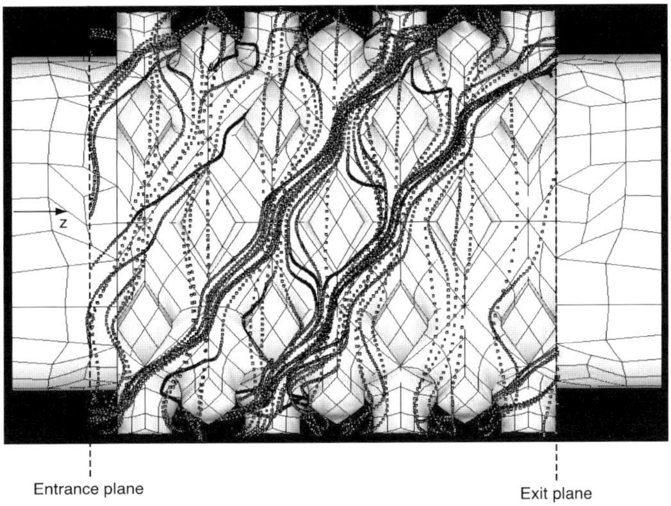

Figure 10.34: Flow patterns inside a rhomboidal distributive mixing head.

where N is the total number of points being tracked and γ_i is the total strain a particle i will undergo, given by

$$\gamma_i = \int_0^{t_{total}} \dot{\gamma}(t)dt \tag{10.115}$$

As an example of the particle tracking procedure, Figure 10.34 presents the particle paths through the -1.6D1.6D rhomboidal mixing section. Figure 10.35 presents the cumulative residence time distribution for the -1.6D1.6D, 1D3D and 1D6D rhomboidal mixing heads. The picture shows that the neutral -1.6D1.6D (pineapple mixer) by far outperforms the other mixing heads. Table 10.2 presents the WATS for the the three rhomboidal sections. Here, too, it is clear that the pineapple mixer applies the largest amount of deformation on the melt.

Rios *et al.* [55] performed an experimental study with the above rhomboidal mixing section configurations using a 45 mm diameter single screw extruder. The mixing sections

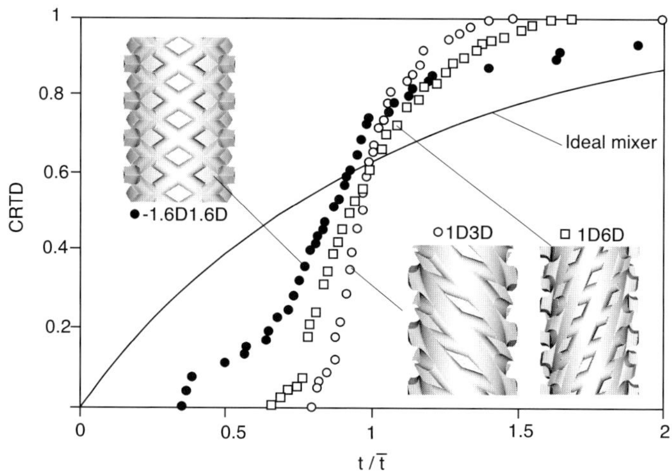

Figure 10.35: Cumulative residence time distribution inside three different rhomboidal mixing heads.

Table 10.2: Weighted Average Total Strain (WATS) for Three Rhomboidal Mixing Heads

Mixing head	WATS
-1.6D1.6D	178
1D6D	109
1D3D	81

are shown in Fig. 10.36. Their experimental study was performed with a PE-HD[2] They introduced a yellow masterbatch pigment in the hopper of the extruder and micrographs were taken of the extrudate for each mixing sections. Figure 10.37 presents the 3 micrographs pertaining to the 3 mixing sections. The micrographs show the superiority of the -1.6D1.6D pineapple mixing section. Also, the most visible striations are seen in the micrograph pertaining to the 1D3D rhomboidal mixing section. Clearly the numerical predictions are qualitatively in agreement with the experimental results.

Similarly, Osswald and Schiffer [59] studied the mixing, deformation of drops, and residence time in single screw extruder rhomboidal mixing sections. Using a boundary element simulation, they were able to modify and optimize an existing mixing section, eliminating the long tail in the residence time distribution. Figure 10.38 presents a drop and its flow line as it travels through the elements of a rhomboidal mixing section. The drop surface was represented with points that were individually tracked as they moved inside the mixer.

Krawinkel *et al.* [37] used the boundary element technique to simulate the flow in co-rotating, double flighted, self cleaning twin screw extruders. Figure 10.39 presents the boundary discretization of the screws along with the pressure distribution on the screw surfaces. Once the surface pressures were solved for, the necessary information for particle tracking was at hand. Figure 10.40 presents several flow lines generated from particle tracking. In order to allow better presentation of these flow lines, they were plotted using

[2]Fitting a power law model to rheological measurements done on the HDPE resulted in a power law index, n, of 0.41 and a consistency index, m, of 16624 Pa-s^n.

Figure 10.36: Rhomboidal mixing sections studied experimentally.

Figure 10.37: Extrudate micrographs ($\times$ 50) for the three different mixing sections.

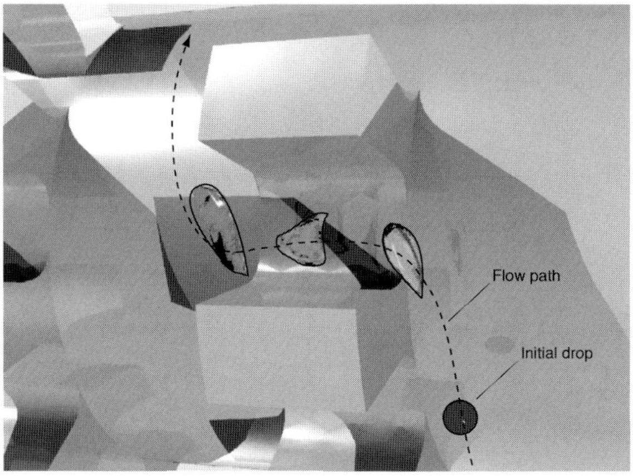

Figure 10.38: Flow line and droplet deformation inside a rhomboidal mixing section.

a moving coordinate system. This moving coordinate system is equivalent to the apparent movement of the flights as they rotate. Using the boundary, it is possible to compute the internal values of velocity, rate of deformation and flow number, to better assess the quality of mixing. Figure 10.41 presents the axial velocity distribution at an arbitrary cross-section along the z-axis. The figure clearly reveals that most of the material conveying, determined by the material transport in the z-direction, occurs near the apex between the screws, in agreement with experimental observation from several other researchers.

Problems

10.1 Determine the velocity profile and traction profiles in a pressure driven slit flow of a Newtonian fluid. Use Δp =1000 Pa, μ =1000 Pa-s, h = 1 mm and a distance from entrance to exit of 1000 mm. Solve the problem using isoparametric 2D quadratic elements and different gauss points, compare your solutions with the analytical solution for slit flow.

10.2 Write a boundary element program that will predict the pressure and velocity fields for the 1 cm thick L-shaped charge depicted in Fig. 10.42. Assume a Newtonian viscosity of 500 Pa-s. Note that for this compression molding problem the volume integral must be included in the analysis.

- **a)** Plot the pressure distribution.
- **b)** Draw the nodal velocity vectors.
- **c)** Comparing with FDM and FEM solutions of the same problem, what are your thoughts regarding the BEM solution?

10.3 In your university library, find the paper Barone, M.R. and T.A. Osswald, *J. of Non-Newt. Fluid Mech.*, 26, 185-206, (1987), and write a 2D BEM program to simulate the compression molding process using the Barone-Caulk model presented in the paper.

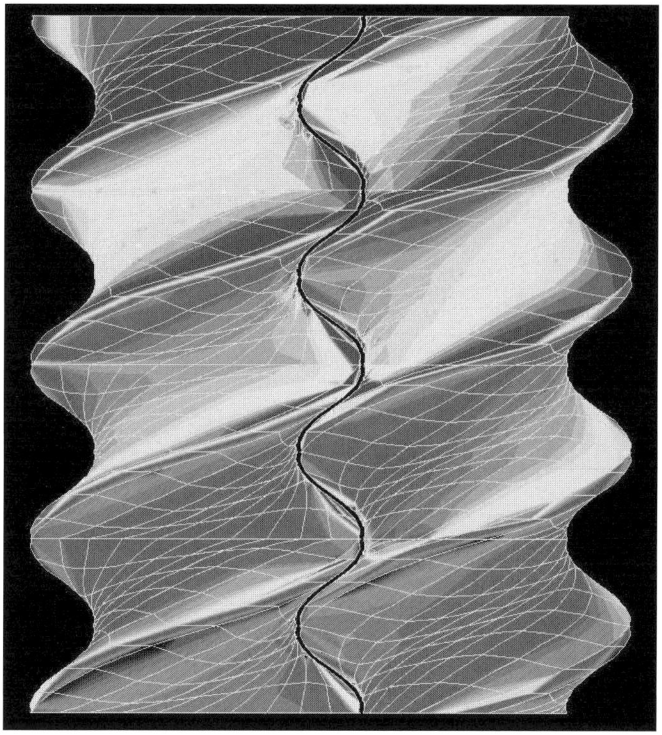

Figure 10.39: Pressure distribution on the screw surfaces of a co-rotating double flighted twin screw extruder.

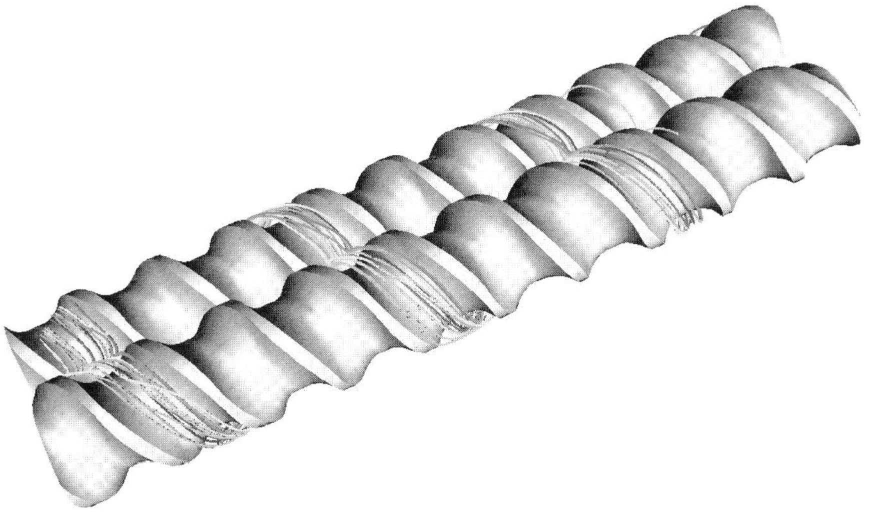

Figure 10.40: Particle tracking inside a co-rotating double flighted twin screw extruder. In order to plot the flow lines, the system is viewed from a coordinate system that moves in the axial direction at a speed of $R\Omega\sin\phi$.

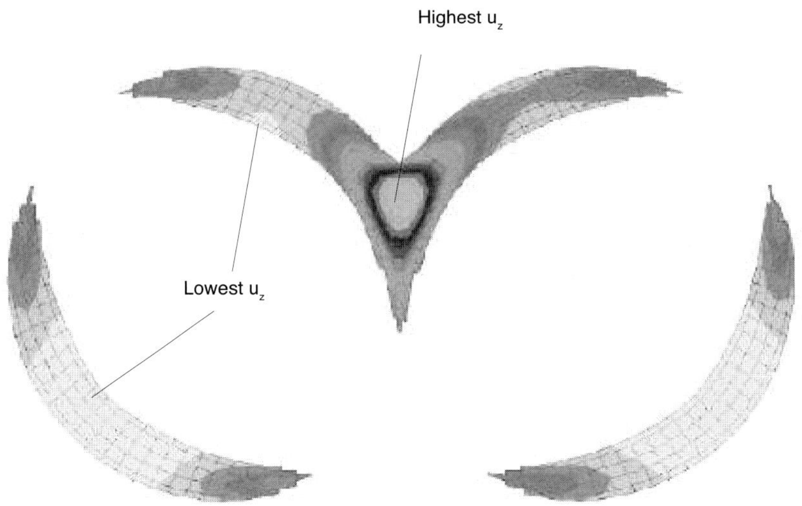

Figure 10.41: Velocity field in the axial direction (u_z) at an arbitrary position along the z-axis.

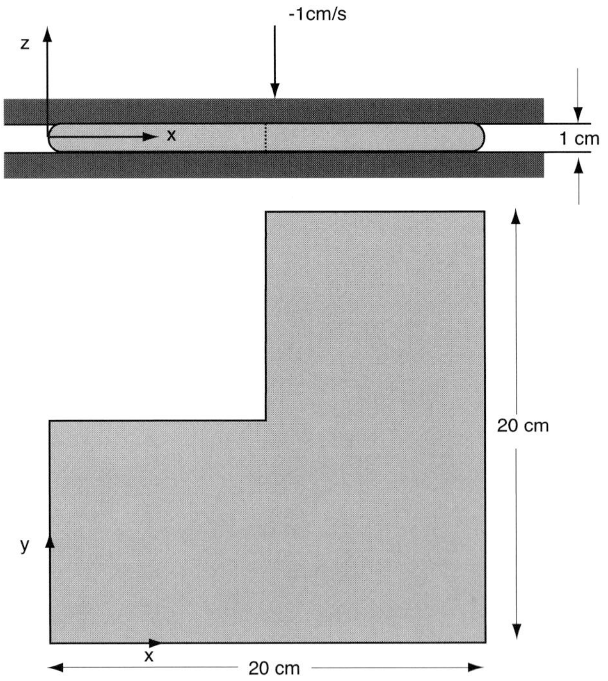

Figure 10.42: Schematic diagram of the compression molding of an L-shaped charge.

Pick one of the geometries and compare your results to the results presented in the paper.

10.4 Consider a drop of a fluid of Newtonian viscosity μ_1 submerged in a Newtonian fluid of viscosity μ_2 with a surface tension, σ.

a) Develop an integral equation for the flow inside the drop (fluid with viscosity μ_1)

b) Develop an integral equation for the flow outside the drop (fluid with viscosity μ_2)

c) Take the limit of these two integral equations when the point approaches the surface of the drop. Note that the velocity field is continuous across the drop surface, but the tractions present a jump given controlled by surface tension. Be careful with the sign and direction of the normal vector. You will obtain a Fredholm integral equation of the second kind for drop deformation.

d) Go to your university library and find the paper Rallison, J.M. and A. Acrivos, *J. Fluid Mech.*, 89, 191, (1978), and compare your equations to those given in the paper.

10.5 Develop the corresponding integral equations for Poisson's equation,

$$\nabla^2 T = b$$

The non-homogeneous term $b(\mathbf{x_o})$ can be expressed as a linear combination of known basis functions $f_R(\mathbf{x_o}, \mathbf{x_i})$ as follows,

$$b(\mathbf{x_o}) = \sum_i^N \alpha_i f_R(\mathbf{x_o}, \mathbf{x_i})$$

With the known functions we can define a set of particular solutions of the following non-homogeneus system,

$$\nabla^2 \hat{T}_i = f_R(\mathbf{x_o}, \mathbf{x_i})$$

a) Substitute the definitions of the particular solutions and the non-homogeneous term and find an equivalent boundary-only integral formulation. This is the commonly known Dual Reciprocity Method.

b) What type of functions are a good selection for f_R?

10.6 Develop a Dual Reciprocity boundary-only integral equation for the Navier-Stokes system of equations.

10.7 Do a literature search and find alternative ways of simulating non-linear equations using BEM. How does the technique compare to other numerical methods.

10.8 What is the indirect BEM formulation? How does it compare to the direct method.

REFERENCES

1. M. Abramowitz and I.A. Stegun. *Handbook of Mathematical Functions.* Dover Publications, New York, 1964.
2. A. Ahmad and P.K. Banerjee. *J. Eng. Mech. ASCE*, 112:682, 1986.

3. S. Ahmad and P.K. Banerjee. Free-vibration analysis by BEM using particular integrals. *J. Eng. Mech.-ASCE*, 112:682, 1986.
4. P.K. Banerjee. *The Boundary Element Methods in Engineering*. Mc-Graw-Hill, London, 1981.
5. G.K. Batchelor. *An introduction to fluid dynamics*. Cambridge University Press, Cambridge, 1967.
6. R.B. Bird, R.C. Armstrong, and O. Hassager. *Dynamics of Polymer Liquids: Fluid Mechanics*, volume 1. John Wiley & Sons, New York, 2nd edition, 1987.
7. C.A. Brebbia and J. Dominguez. *Boundary Elements, an Introductory Course*. Computational Mechanics Publications, Southampton, 1989.
8. C.A. Brebbia, J.C.F. Telles, and L.C. Wrobel. *Boundary Elements Techniques*. Springer-Verlag, New York, 1984.
9. I.N. Bronstein, K.A. Semendjajew, G. Musiol, and H. Muehlig. *Taschenbuch der Mathematik*. Verlag HarriDeutsch, Frankfurt am Main, 2001.
10. J.M. Burgers. On the motion of small particles of elongated form, suspended in a viscous liquid. Report on viscosity and plasticity, Nordemann Publishing, New York, 1938.
11. A. Cheng, O. Lafe, and S. Grilli. *Eng. Anal. Bound. Elements*, 13:303, 1994.
12. B.A. Davis. *Investigation of non-linear flows in polymer mixing using the boundary integral method*. PhD thesis, University of Wisconsin-Madison, Madison, 1995.
13. B.A. Davis, P.J. Gramann, J.C. Maetzig, and T.A. Osswald. The dual-reciprocity method for heat transfer in polymer processing. *Engineering Analysis with Boundary Elements*, 13:249–261, 1994.
14. P.A.M. Dirac. *The principles of quantum mechanics*. Clarendon Press, Oxford, 2nd edition, 1935.
15. A. Einstein. *Ann. Physik*, 19:549, 1906.
16. F.R. Eirich. *Rheology: theory and applications I*. Academic Press, New York, 1956.
17. W.F. Flórez. *Multi-domain dual reciprocity method for the solution of nonlinear flow problems*. PhD thesis, University of Wales, Wessex Institute of Technology, Southampton, 2000.
18. W.F. Flórez. *Nonlinear flow using dual reciprocity*. WIT press, Southampton, 2001.
19. W.F. Flórez and H. Power. *Eng. Anal. Bound. Elements*, 25:57, 2001.
20. W.F. Flórez, H. Power, and F. Chejne. *Num. Meth. Partial Diff. Equations*, 18:469, 2002.
21. M.A. Golberg and C.S. Chen. *Discrete projection methods for integral equations*. Computational Mechanics Publications, Southampton, 1997.
22. P.J. Gramann. PhD thesis, University of Wisconsin-Madison, 1995.
23. P.J. Gramann, L. Stradins, and T. A. Osswald. *Intern. Polymer Processing*, 8:287, 1993.
24. N.M. Gunter. *Potential theory and applications to basic problems of mathematical physics*. Unger, New York, 1967.
25. E. Guth and R. Simha. *Kolloid-Zeitschrift*, 74:266, 1936.
26. J. Happel and H. Brenner. *Low Reynolds Number Hydrodynamics*. Kluwer, Dordrecht, 1991.
27. J.J. Hermans. *Flow properties of disperse systems*. Inter-Science Publishers, New York, 1953.
28. J.P. Hernández-Ortiz. Reciprocidad dual y elementos de frontera para fluidos newtonianos y no-newtonianos en tres dimensiones. Master's thesis, Universidad Pontificia Bolivariana, Medellín, Colombia, 1999.
29. J.P. Hernández-Ortiz. *Boundary integral equations for viscous flows: non-Newtonian behavior and solid inclusions*. PhD thesis, University of Wisconsin-Madison, Madison, 2004.

30. J.P. Hernández-Ortiz, T.A. Osswald, and D.A. Weiss. Simulation of viscous 2d-plannar cylindrical geometry deformation using DR-BEM. *Int. J. Num. Meth. Heat Fluid Flow*, 13:698, 2003.

31. D.J. Jeffery and A. Acrivos. *AIChE J.*, 22:417, 1976.

32. G.B. Jeffery. *Proc. Roy. Soc. Lond.*, A102:161, 1922.

33. G.B. Jeffery. *Proc. Roy. Soc. Lond.*, 111:110, 1926.

34. H. Kardestuncer and D.H. Norrie, editors. *Finite Element Handbook*. McGraw-Hill, New York, 1987.

35. K. Kitawa and C.A. Brebbia. *Eng. Anal. Bound. Elements*, 2:194, 1996.

36. K. Kitawa, C.A. Brebbia, and M. Tanaka. *Topics in Boundary Elements Research*, chapter 5. Springer, New York, 1989.

37. S. Krawinkel, M. Bastian, T.A. Osswald, and H. Potente. Gleichdrall-Doppelschneckenextruder Stroemungssimulation mit BEM. *Plastics Special*, 12:30, 2000.

38. T. Kuroki and K. Onishi. *BEM VII*. Computational Mechanics Publications, Southampton, 1985.

39. P.K. Kythe. *Fundamentals solutions for differential operators and applications*. Birkhaeuser Press, Berlin, 1996.

40. O.A. Ladyzhenskaya. *The mathematical theory of viscous incompressible flow*. Gordon and Beach, New York, 1963.

41. L.D. Landau and E.M. Lifshitz. *Fluid Mechanics*. Butterworth-Heinemann, Oxford, 2nd edition, 1987.

42. D. Nardini and C.A. Brebbia. *App. Math. Model.*, 7:157, 1983.

43. P.W. Partridge, C.A. Brebbia, and L.C. Wrobel. *The dual reciprocity boundary element method*. Computational Mechanics, Southampton, 1991.

44. V. Popov and H. Power. *Bound. Elements Comm.*, 7:1, 1996.

45. V. Popov and H. Power. *Boundary Element Research in Europe*, page 67. Computational Mechanics Publications, 1999.

46. H. Power and R. Mingo. *Eng. Anal. Bound. Elements*, 24:107, 2000.

47. H. Power and R. Mingo. *Eng. Anal. Bound. Elements*, 24:121, 2000.

48. H. Power and P.W. Partridge. *Int. J. Num. Meth. Heat Fluid Flow*, 3:145, 1993.

49. H. Power and P.W. Partridge. *Int. J. Num. Meth. Eng.*, 37:1825, 1994.

50. H. Power and L. C. Wrobel. *Boundary Integral Methods in Fluid Mechanics*. Computational Mechanics Publications, 1995.

51. C. Pozrikidis. *Boundary Integral and Singularity Methods for Linearized Viscous Flow*. Cambridge University, 1992.

52. Y.F. Rashed. Boundary element primer: fundamental solutions. I simple and compund operators. *Bound. Element Comm.*, 13(1):38, 2002.

53. Y.F. Rashed. Boundary element primer: fundamental solutions. II matrix operators. *Bound. Element Comm.*, 13(2):35, 2002.

54. A.C. Rios. *Simulation of mixing in single screw extrusion using the boundary integral method*. PhD thesis, University of Wisconsin-Madison, Madison, 1999.

55. A.C. Rios, P.J. Gramann, T. A. Osswald, M. del P. Noriega, and O.A. Estrada. Experimental and numerical study of rhomboidal mixing sections. *Intern. Polymer Processing*, 15:12–19, 2000.

56. A.C. Rios and T.A. Osswald. Comparative study of rhomboidal mixing sections using the boundary element method. *Engineering Analysis with Boundary Elements*, 24:89–94, 2000.

57. B. Sarler and G. Kuhn. *Eng. Anal. Bound. Elements*, 21:53, 1998.

58. B. Sarler and G. Kuhn. *Eng. Anal. Bound. Elements*, 21:65, 1998.

59. M. Schiffer. Master's thesis, Institut für Kunststoffverarbeitung, Germany, 1999.

60. L. Skerget and M. Hribersek. *Int. J. Num. Meth. Fluids*, 39:115, 1996.

61. L. Skerget and N. Samec. *Eng. Anal. Bound. Elements*, 23:435, 1999.

62. A.E. Taigbenu. *Int. J. Num. Meth. Eng.*, 38:2241, 1995.

63. A.E. Taigbenu and O.O. Onyejekwe. *Appl. Math. Model.*, 19:675, 1995.

64. T.C. Telles. *Int. J. Num. Meth. Eng.*, 24:959, 1987.

65. V. Vand. *J. Phys. Colloid. Chem.*, 52(2):277, 1848.

66. V. Vand. *J. Phys. Colloid. Chem.*, 52(2):300, 1948.

67. O.C. Zienkiewicz. *Finite Element Methods in Stress Analysis*, chapter 13: Iso-parametric and associate elements families for two and three dimensional analysis. Tapir Press, Trondheim, 1969.

68. O.C. Zienkiewicz. *The Finite Element Method*. McGraw-Hill, London, 3rd edition, 1997.

CHAPTER 11

RADIAL FUNCTIONS METHOD

It ain't over till it's over.

—Yogi Berra

Radial functions method (RFM)[1], often referred to as *radial basis functions collocation method* (RBFCM) has gained significant attention in recent years, but compared to FDM, FEM and even BEM, it is a relatively new method. Radial basis functions were originally used by Hardy in 1970 [10, 11] to interpolate topography in maps using sparse and scattered data points. In 1990, Kansa [12, 13] first used the method to solve partial differential equations when studying problems in fluid dynamics. Since then, Kansa and many other researchers have helped promote and advance the technique, making it an accepted tool to solve partial differential equations. Mai-Duy and Tanner [16] used the technique to model non-Newtonian fluid flow of shear thinning and viscoelastic liquids. In their work, they solved flows using a power law shear thinning model, and Ericksen-Filbey and Oldroyd-B viscoelastic models, all with satisfactory results. More recently, Estrada [4], Estrada *et al.* [6, 7] and López and Osswald [5] successfully used the technique to represent coupled energy and momentum balances to model non-Netwonian flows during polymer processing.

The main advantage of the radial functions method (RFM) is that it is a technique that does need neither domain nor boundary meshes as required with FEM and BEM, or

[1]This chapter was written with contributions from O.A. Estrada-Ramírez and I.D. López-Gómez

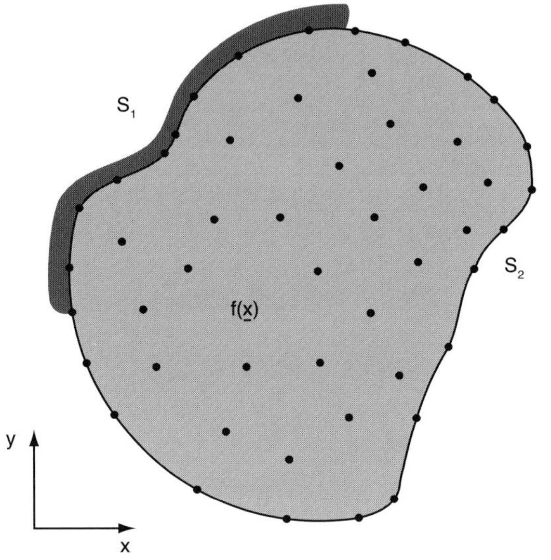

Figure 11.1: Schematic diagram a domain with collocation points.

homogeneous grid points as FDM, to solve partial differential equations. Essentially, it is a meshless technique based on collocation methods. The method has proven to be very accurate compared to other numerical techniques, even for a small number of collocation points [12] and for problems with a large advective components [5], making it an alternative technique to FDM, FEM and BEM. However, one disadvantage of RBFM is that it generates full unsymmetric matrices that require large amounts of storage and computation times. As an alternative, symmetric radial basis functions techniques have been implemented, which reduce storage requirements and computation costs [8]. However, the implementation is more complex, especially for non-linear systems and it presents accuracy problems for nodes that are close to the boundaries [14]. This chapter gives an overview of the radial functions method along with the implementation of the technique, presenting several examples in polymer processing.

11.1 THE KANSA COLLOCATION METHOD

Collocation techniques are based on the fact that a field variable in a continuous space can be approximated with linear interpolation coefficients and basic functions located on discreet points sprinkled on the domain of interest, as schematically presented in Fig. 11.1.

When the value of the field variable is known in some locations, it is possible to determine the field variable at any location in space. A general collocation expression for a two-dimensional space is given by

$$f(x_i, y_i) \approx \sum_{j=1}^{N} g(x_j, y_j)\alpha_j \tag{11.1}$$

where the function $f(x_i, y_i)$ is represented at any collocation point i using N bi-dimensional functions $g(x_j, y_j)$, evaluated at collocation points j, and their corresponding interpolation coefficients α_j that are adjusted to match the field variable of interest. Similar to the general collocation relation, eqn. (11.1), we can represent the field variable $f(x_i, y_i)$ using radial basis function collocation as

$$f(x_i, y_i) \approx \sum_{j=1}^{N} \phi(r_{ij})\alpha_j \tag{11.2}$$

where r_{ij} is the distance between collocation points i and j, and for 2D[2] is given by

$$r_{ij} = \sqrt{(x_i - x_j)^2 + (y_i - y_j)^2} \tag{11.3}$$

If we apply a differential operator $\mathcal{L}[]$ on eqn. (11.2), we can interpolate between operated field variables

$$\mathcal{L}[f(x_i, y_i)] = \sum_{j=1}^{N} \mathcal{L}[\phi(r_{ij})]\alpha_j \tag{11.4}$$

This allows us to represent partial differential equations as found in the balance equations using the collocation method. Equation (11.47) is a solution to a partial differential equation represented by a system of linear algebraic equations, formed by the interpolation coefficients, α_j, and the operated radial functions. The interpolation coefficients are solved for using matrix inversion techniques to approximately satisfy the partial differential equation $\mathcal{L}[f]$.

The sum of the terms in eqn. (11.47) can be regarded as a mathematical series, whose convergence is controlled by the number of terms. Therefore, the accuracy of the solution of an operated system of governing equations is directly linked to the number of collocation points within the domain [2].

There are many choices of radial basis function. Past research has demonstrated that *polyharmhonic thin plate splines* of various order a work best to represent systems governed by the balance equations [4]. A polyharmhonic thin plate spline is given by

$$\phi(r_{ij}) = {r_{ij}}^{2a} \ln(r_{ij}) \tag{11.5}$$

where a can be chosen according to the type of problem being solved. Note that when $r_{ij} \to 0$, the function $\phi(r_{ij})$ as well as its derivatives go to zero. Using the above equation we can define the derivative of the RBF with respect to r as

$$\frac{\partial \phi(r_{ij})}{\partial r} = {r_{ij}}^{2a-1}(2a \ln(r_{ij}) + 1) \tag{11.6}$$

When $\mathcal{L}[]$ is the Laplacian Operator, with d being the dimensionality of the problem, we get

$$\nabla^2 \phi(r_{ij}) = \frac{\partial^2 \phi(r_{ij})}{\partial r^2} + \frac{\partial \phi(r_{ij})}{\partial r} \cdot \frac{d-1}{r_{ij}} \tag{11.7}$$

and

$$\frac{\partial^2 \phi(r_{ij})}{\partial r^2} = {r_{ij}}^{2(a-1)}(4a^2 \ln(r_{ij}) + 4a - 2a \ln(r_{ij}) - 1) \tag{11.8}$$

[2]One- and three-dimensional implementations are similar, where the definition of r_{ij} changes accordingly.

11.2 APPLYING RFM TO BALANCE EQUATIONS IN POLYMER PROCESSING

When solving the balance equations, our primary variables are the temperature, pressure and velocity fields throughout the domain. For example, the temperature at any position (x_i, y_i), T_i, can be represented using

$$T_i = \sum_{j=1}^{N} \phi_T(r_{ij})\alpha_j \tag{11.9}$$

For pressure, the number of nodes can be lower, due to the fact that, when applying velocity Dirichlet boundary conditions, the pressure remains unknown. For pressure we can write

$$p_i = \sum_{j=1}^{Np} \phi_p(r_{ij})\beta_j \tag{11.10}$$

The velocity field in a three-dimensional domain is represented using

$$u_{x_i} = \sum_{j=1}^{N} \phi_u(r_{ij})\lambda_j \tag{11.11}$$

$$u_{y_i} = \sum_{j=1}^{N} \phi_u(r_{ij})\xi_j \tag{11.12}$$

$$u_{z_i} = \sum_{j=1}^{N} \phi_u(r_{ij})\chi_j \tag{11.13}$$

Furthermore, we can also use radial functions to interpolate the magnitude of the rate of deformation tensor using

$$|\dot{\gamma}| = \sum_{j=1}^{N} \varsigma_j \phi_\gamma(r_{ij}) \tag{11.14}$$

11.2.1 Energy Balance

In this section, we implement the radial basis function method in the energy equation and apply the technique to an example problem. We begin with a steady-state energy balance given by

$$\rho C_p \left(u_{x_i} \frac{\partial T_i}{\partial x} + u_{y_i} \frac{\partial T_i}{\partial y} \right) = k\nabla^2 T_i + \eta_i \dot{\gamma}_i^2 \tag{11.15}$$

To approximate this form of the energy balance, we must first define the differential operators applied to the temperature field

$$\frac{\partial T_i}{\partial x} = \sum_{j=1}^{N} \frac{\partial \phi_T(r_{ij})}{\partial r} \frac{\partial r_{ij}}{\partial x} \alpha_j \tag{11.16}$$

$$\frac{\partial T_i}{\partial y} = \sum_{j=1}^{N} \frac{\partial \phi_T(r_{ij})}{\partial r} \frac{\partial r_{ij}}{\partial y} \alpha_j \tag{11.17}$$

and

$$\nabla^2 T_i = \sum_{j=1}^{N} \nabla^2 \phi_T(r_{ij}) \alpha_j \tag{11.18}$$

The resulting RFM form of the steady-state energy balance becomes

$$\sum_{j=1}^{N} \left[\rho C_p \frac{\partial \phi_T(r_{ij})}{\partial r} \left(u_{x_i} \frac{\partial r_{ij}}{\partial x} + u_{y_i} \frac{\partial r_{ij}}{\partial y} \right) - k \nabla^2 \phi_T(r_{ij}) \right] \alpha_j = \eta_i \dot{\gamma}_i^2 \tag{11.19}$$

The boundary conditions must also be written in RFM form. The Dirichlet temperature boundary condition, given by

$$T_i = T_a; i \in \Gamma^D \tag{11.20}$$

is written as

$$\sum_{j=1}^{N} \phi_T(r_{ij}) \alpha_j = T_a \tag{11.21}$$

Furthermore, the Neumann temperature boundary condition

$$-k \left(\frac{\partial T_i}{\partial x} n_{x_i} + \frac{\partial T_i}{\partial y} n_{y_i} \right) = \dot{q}; i \in \Gamma^N \tag{11.22}$$

becomes

$$-k \sum_{j=1}^{N} \left[\frac{\partial \phi_T(r_{ij})}{\partial r} \left(\frac{\partial r_{ij}}{\partial x} n_{x_i} + \frac{\partial r_{ij}}{\partial y} n_{y_i} \right) \right] \alpha_j = \dot{q} \tag{11.23}$$

and the Robin temperature boundary condition

$$-k \left(\frac{\partial T_i}{\partial x} n_{x_i} + \frac{\partial T_i}{\partial y} n_{y_i} \right) = h \left(T_i - T_\infty \right); i \in \Gamma^R \tag{11.24}$$

is written as

$$\sum_{j=1}^{N} \left[k \frac{\partial \phi_T(r_{ij})}{\partial r} \left(\frac{\partial r_{ij}}{\partial x} n_{x_i} + \frac{\partial r_{ij}}{\partial y} n_{y_i} \right) + h \phi_T \left(r_{ij} \right) \right] \alpha_j = h T_\infty \tag{11.25}$$

Similarly to the above derivation, we can also use the technique to predict transient temperature fields. Again, as with finite elements and boundary elements, the time stepping is done using finite difference techniques. For a Crank-Nicholson transient energy equation formulation given by

$$\rho C_p \left[\frac{T_i^l - T_i^{l-1}}{\Delta t} + \frac{1}{2} \left(u_{x_i}^l \frac{\partial T_i^l}{\partial x} + u_{y_i}^l \frac{\partial T_i^l}{\partial y} + u_{x_i}^{l-1} \frac{\partial T_i^{l-1}}{\partial x} + u_{y_i}^{l-1} \frac{\partial T_i^{l-1}}{\partial y} \right) \right] =$$
$$\frac{1}{2} \left[k \nabla^2 T_i^l + \eta_i^l \left(\dot{\gamma}_i^l \right)^2 + k \nabla^2 T_i^{l-1} + \eta_i^{l-1} \left(\dot{\gamma}_i^{l-1} \right)^2 \right]$$

(11.26)

the RFM solution takes the form

$$\sum_{j=1}^{N}\left[\phi_T(r_{ij})+\frac{\Delta t}{2}\left(\rho C_p\frac{\partial\phi_T(r_{ij})}{\partial r}\left(u_{x_i}^l\frac{\partial r_{ij}}{\partial x}+u_{y_i}^l\frac{\partial r_{ij}}{\partial y}\right)-k\nabla^2\phi_T(r_{ij})\right)\right]\alpha_j^l=$$
$$\frac{\Delta t}{2}\left[k\nabla^2 T_i^{l-1}+\eta_i^{l-1}\left(\dot\gamma_i^{l-1}\right)^2+\eta_i^l\left(\dot\gamma_i^l\right)^2-\rho C_p\left(u_{x_i}^{l-1}\frac{\partial T_i^{l-1}}{\partial x}+u_{y_i}^{l-1}\frac{\partial T_i^{l-1}}{\partial y}\right)\right]$$
$$+T_i^{l-1} \tag{11.27}$$

where l denotes the time step number. As will be shown in the next section, material properties that vary in space can also be interpolated throughout the domain with the use of radial functions.

■ EXAMPLE 11.1.

Viscous heating temperature rise due to a combined drag-pressure flow between two parallel plates. Using RFM, Estrada [4] computed the temperature rise of a fluid subjected to a combination of drag and pressure flow between parallel plates and compared the results of an anlytical and a boundary element dual reciprocity (BEM-DRM) solutions presented by Davis *et al.* [3]. Figure 11.2 presents a schematic of the problem with dimensions, physical properties and boundary conditions.

The fluid that is confined between the parallel plates flows due to a drag flow caused by an upper plate velocity, u_0, and a pressure flow caused by a pressure drop in the x-direction of Δp. The combined analytical velocity field is given by

$$u_x=-\frac{1}{2\eta}\left(\frac{\Delta}{L}\right)y^2+u_0\frac{u_0}{h}+\frac{h}{2\eta}\left(\frac{\Delta p}{L}\right)y \tag{11.28}$$

and is presented in Fig. 11.3. Using this velocity profile, the analytical steady-state temperature rise due to viscous heating is given by

$$\Delta T=T-T_0=-\frac{1}{12\eta k}\left(\frac{\Delta p}{L}\right)^2y^4+\frac{1}{6k}\left(\frac{\Delta p}{L}\right)\left[\frac{2u_0}{h}+\frac{h}{\eta}\left(\frac{\Delta p}{L}\right)\right]y^3$$
$$-\frac{1}{2k}\left[\frac{\eta u_0^2}{h^2}+u_0\left(\frac{\Delta p}{L}\right)+\frac{h^2}{4\eta}\left(\frac{\Delta p}{L}\right)^2\right]y^2$$
$$+\left[\frac{h^3}{12\eta k}\left(\frac{\Delta p}{L}\right)^2-\frac{h^2}{6k}\left(\frac{\Delta p}{L}\right)\left(\frac{2u_0}{h}+\frac{h}{\eta}\left(\frac{\Delta p}{L}\right)\right)\right.$$
$$\left.+\frac{h}{2k}\left(\frac{\eta u_0^2}{h^2}+u_0\left(\frac{\Delta p}{L}\right)+\frac{h^2}{4\eta}\left(\frac{\Delta p}{L}\right)^2\right)\right]y \tag{11.29}$$

The RFM solution was compiled using 1029 collocation points (Fig. 11.2) and second order ($a=2$) thin-plate splines. Figure 11.4 presents a comparison between

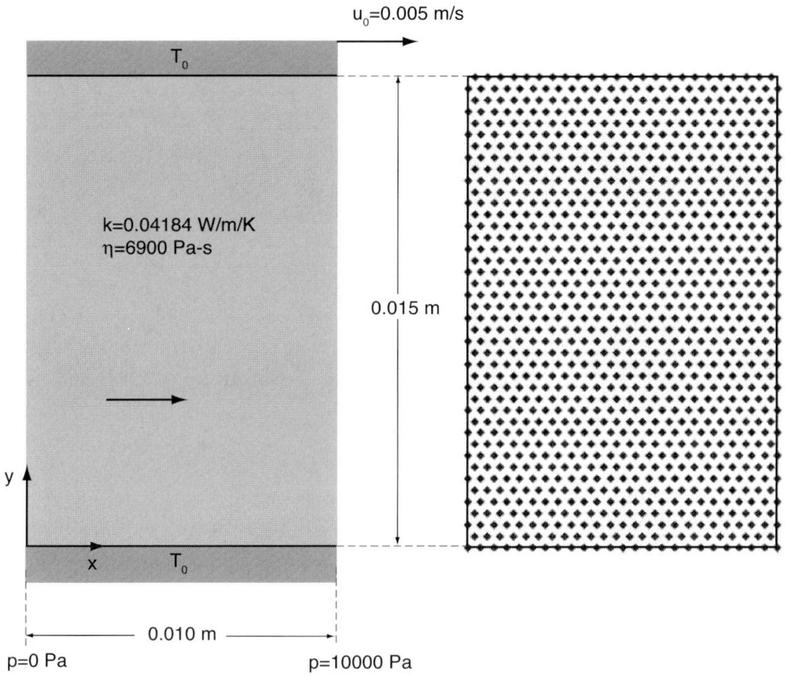

Figure 11.2: Schematic diagram of the viscous dissipation problem with combination drag-pressure flow between parallel plates, and collocation points for the RF method solution.

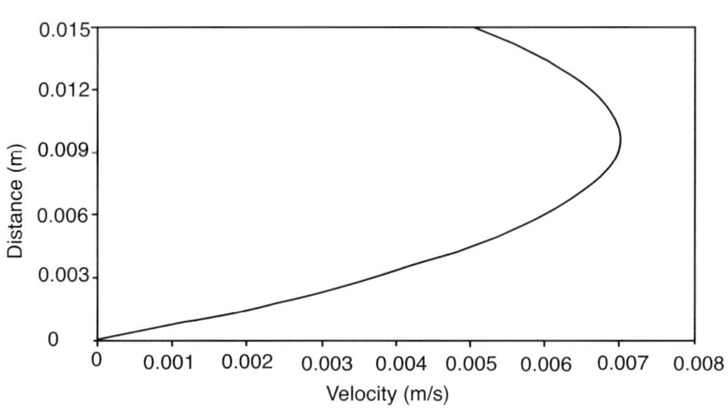

Figure 11.3: Combined drag-pressure velocity field between two parallel plates.

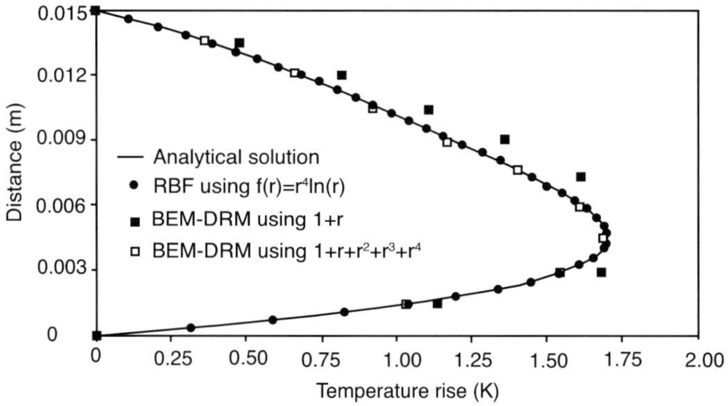

Figure 11.4: Comparison between RFM temperature rise and analytical and BEM-DRM solutions.

the analytical solution and the numerical RFM and BEM-DRM solutions. It is clear from the figure that the RFM solution perfectly captures the viscous dissipation effect in this problem. The average error was 0.296% with a maximum error just under 1%, as expected in the regions of low velocity. The results were comparable with a BEM-DRM solution with fourth order radial polyharmonic polynomic splines [3].

■ **EXAMPLE 11.2.**

One dimensional convection-diffusion problem. As mentioned in Chapter 8, one heat transfer problem that is clearly difficult to solve involves combinations of conduction and convection as the one-dimensional problem illustrated in Fig. 11.5. Here, we have a heat transfer convection-diffusion problem, where the conduction, which results from the temperature gradient, and the flow velocity are both in the x-direction.

As pointed out in Chapter 8, for the case where $D \gg L$ the dimensionless energy balance reduces to

$$Pe\frac{\partial \Theta}{\partial \xi} = \frac{\partial^2 \Theta}{\partial \xi^2} \tag{11.30}$$

where the *Peclet number* is defined by $Pe = \rho C_p u_x L/k$, the dimensionless temperature by $\Theta = (T - T_0)/(T_1 - T_0)$, and $\xi = x/L$. The boundary conditions in dimensionless form are

$$\Theta(0) = 0 \quad \text{and} \quad \Theta(1) = 1 \tag{11.31}$$

This problem was solved using a one-dimensional RFM with 200 collocation points, evenly distributed along the x-axis using Algorithm 16. Figure 11.6 compares the analytical solution with computed RBFM solutions up to Pe=100. As can be seen, even for convection dominated cases, the technique renders excellent results. It is important to point out here that for the radial functions method no up-winding or other special techniques were required.

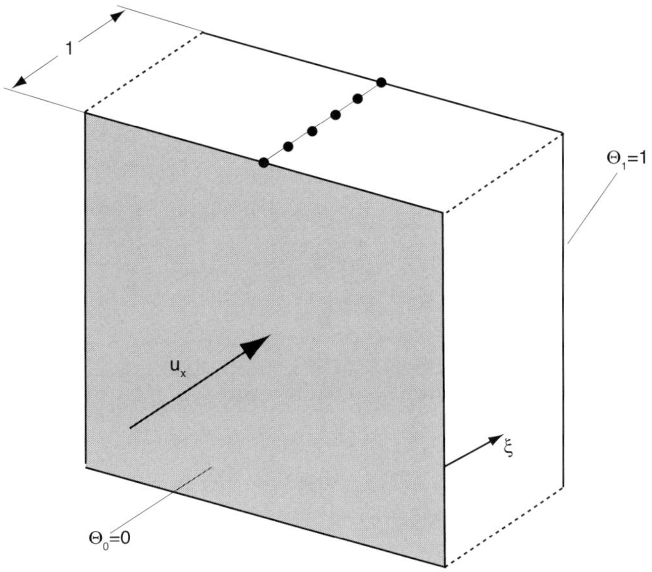

Figure 11.5: Schematic diagram of the convection-conduction problem in dimensionless form.

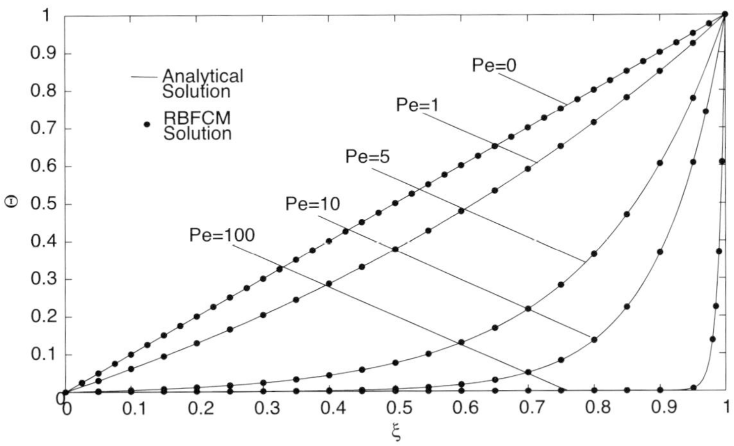

Figure 11.6: Comparison between analytical and RFM solutions for a one-dimensional convection-diffusion heat transfer problem for several Peclet numbers.

Algorithm 16 Convection-diffusion with RFM

```
program RFMconvdiff     RBF and EQ are matrices of NxN and alfa, b and Theta are
vectors of size N
  deltaXi = 1d0/(N-1)
  do i = 1, N
    do j = 1, N
      r = abs((i-1)*deltaXi-(j-1)*deltaXi)
      if(r==0d0) then
        RBF(i,j) = 0d0
      else
        RBF(i,j) = (r**(2*a))*log(r)
      end if
      if(i == 1) then
         b(i)   = 0d0
         Eq(i,j) = RBF(i,j)
      Else if (i == N) then
        b(i)    = 1d0
        Eq(i,j) = RBF(i,j)
      Else
        if(r==0d0) then
          Eq(i,j)=0d0
        Else
          dphi_dr  = r**(2*a-1)*(2*a*dlog(r)+1)
          Lapl_phi = r*(2*(a-1))*(4*a**2*dlog(r)+4*a - 2*a*dlog(r)-1)
          dr_dXi   = ((i-1)*deltaXi-(j-1)*deltaXi)/r
          Eq(i,j)   = Pe*(dphi_dr*dr_dXi) - Lapl_phi
        End If
      b(i) = 0d0
      End If
    End Do
  End do
  call solve-system(Eq,b,alfa)    solve-system is any subroutine to solve linear
systems (AX = b)
  Theta = matmul(RBF,alfa)
end program RFMconvdiff
```

11.2.2 Flow problems

In a similar fashion as with the energy equation, we can also approximate the continuity equation and the equation of motion using radial basis functions. The continuity equation, written as

$$\frac{\partial u_{x_i}}{\partial x} + \frac{\partial u_{y_i}}{\partial y} = 0 \tag{11.32}$$

can be approximated using the RFM by re-writting the above equation as

$$\sum_{j=1}^{N} \left[\frac{\partial \phi_u(r_{ij})}{\partial r} \frac{\partial r_{ij}}{\partial x} \right] \lambda_j + \sum_{j=1}^{N} \left[\frac{\partial \phi_u(r_{ij})}{\partial r} \frac{\partial r_{ij}}{\partial y} \right] \xi_j = 0 \tag{11.33}$$

Similarly, if we write the x-component of the equation of motion for a 2D and 2 1/2D[3] solution as

$$\begin{aligned} \rho \left(u_{x_i} \frac{\partial u_{x_i}}{\partial x} + u_{y_i} \frac{\partial u_{x_i}}{\partial y} \right) = \\ -\frac{\partial p_i}{\partial x} + \eta_i \nabla^2 u_{x_i} + 2 \frac{\partial \eta_i}{\partial x} \frac{\partial u_{x_i}}{\partial x} + \frac{\partial \eta_i}{\partial y} \left(\frac{\partial u_{x_i}}{\partial y} + \frac{\partial u_{y_i}}{\partial x} \right) \end{aligned} \tag{11.34}$$

the RFM approximation is written as

$$\begin{aligned} \sum_{j=1}^{N} \Bigg[\rho \frac{\partial \phi_u(r_{ij})}{\partial r} \left(u_{x_i} \frac{\partial r_{ij}}{\partial x} + u_{y_i} \frac{\partial r_{ij}}{\partial y} \right) - \eta_i \nabla^2 \phi_u(r_{ij}) \\ - \frac{\partial \phi_u(r_{ij})}{\partial r} \left(2 \frac{\partial \eta_i}{\partial x} \frac{\partial r_{ij}}{\partial x} + \frac{\partial \eta_i}{\partial y} \frac{\partial r_{ij}}{\partial y} \right) \Bigg] \lambda_j + \\ \sum_{j=1}^{N} \left[- \frac{\partial \phi_u(r_{ij})}{\partial r} \frac{\partial \eta_i}{\partial y} \frac{\partial r_{ij}}{\partial y} \right] \xi_j + \sum_{j=1}^{N_p} \left[\frac{\partial \phi_p(r_{ij})}{\partial r} \frac{\partial r_{ij}}{\partial x} \right] \beta_j = 0 \end{aligned} \tag{11.35}$$

For the y-component of the equation of motion

$$\begin{aligned} \rho \left(u_{x_i} \frac{\partial u_{y_i}}{\partial x} + u_{y_i} \frac{\partial u_{y_i}}{\partial y} \right) = \\ -\frac{\partial p_i}{\partial y} + \eta_i \nabla^2 u_{y_i} + 2 \frac{\partial \eta_i}{\partial y} \frac{\partial u_{y_i}}{\partial y} + \frac{\partial \eta_i}{\partial x} \left(\frac{\partial u_{x_i}}{\partial y} + \frac{\partial u_{y_i}}{\partial x} \right) \end{aligned} \tag{11.36}$$

we can write

$$\begin{aligned} \sum_{j=1}^{N} \Bigg[\rho \frac{\partial \phi_u(r_{ij})}{\partial r} \left(u_{x_i} \frac{\partial r_{ij}}{\partial x} + u_{y_i} \frac{\partial r_{ij}}{\partial y} \right) - \eta_i \nabla^2 \phi_u(r_{ij}) \\ - \frac{\partial \phi_u(r_{ij})}{\partial r} \left(2 \frac{\partial \eta_i}{\partial y} \frac{\partial r_{ij}}{\partial y} + \frac{\partial \eta_i}{\partial x} \frac{\partial r_{ij}}{\partial x} \right) \Bigg] \xi_j + \\ \sum_{j=1}^{N} \left[- \frac{\partial \phi_u(r_{ij})}{\partial r} \frac{\partial \eta_i}{\partial x} \frac{\partial r_{ij}}{\partial x} \right] \lambda_j + \sum_{j=1}^{N_p} \left[\frac{\partial \phi_p(r_{ij})}{\partial r} \frac{\partial r_{ij}}{\partial y} \right] \beta_j = 0 \end{aligned} \tag{11.37}$$

[3]Within this discussion it is understood that the 2 1/2D solution is a 3D flow where the velocity field does not change in the z-direction. Such a can be flow through a channel such as an unwrapped screw channel of constant cross-section as encountered in the metering section of a single screw extruder.

In a 2 1/2D flow case we must also include the z-component

$$\rho\left(u_{x_i}\frac{\partial u_{z_i}}{\partial x}+u_{y_i}\frac{\partial u_{z_i}}{\partial y}\right)=-\frac{\Delta P}{Z}+\eta_i\nabla^2 u_{z_i}+\frac{\partial \eta_i}{\partial x}\frac{\partial u_{z_i}}{\partial x}+\frac{\partial \eta_i}{\partial y}\frac{\partial u_{z_i}}{\partial y} \tag{11.38}$$

which can be approximated using radial basis functions collocation with

$$\begin{aligned}\sum_{j=1}^{N}\Bigg[\rho\frac{\partial \phi_u(r_{ij})}{\partial r}\left(u_{x_i}\frac{\partial r_{ij}}{\partial x}+u_{y_i}\frac{\partial r_{ij}}{\partial y}\right)-\eta_i\nabla^2\phi_u(r_{ij})-\\ \frac{\partial \phi_u(r_{ij})}{\partial r}\left(\frac{\partial \eta_i}{\partial y}\frac{\partial r_{ij}}{\partial y}+\frac{\partial \eta_i}{\partial x}\frac{\partial r_{ij}}{\partial x}\right)\Bigg]\chi_j=-\frac{\Delta P}{Z}\end{aligned} \tag{11.39}$$

As mentioned earlier, space or field dependent material properties can also be represented with radial functions. This is the case with rate of deformation- or velocity-dependent viscosity. Here, we present two alternatives to represent viscosity and viscosity gradients. The first and simpler form, also referred to a the direct method, simply applies the RFM to the viscosity itself as

$$\eta_i=\sum_{j=1}^{N}\phi_\eta(r_{ij})\omega_j \tag{11.40}$$

where the gradients are represented by

$$\frac{\partial \eta_i}{\partial x}=\sum_{j=1}^{N}\frac{\partial \phi_\eta(r_{ij})}{\partial r}\frac{\partial r_{ij}}{\partial x}\omega_j \tag{11.41}$$

and

$$\frac{\partial \eta_i}{\partial y}=\sum_{j=1}^{N}\frac{\partial \phi_\eta(r_{ij})}{\partial r}\frac{\partial r_{ij}}{\partial y}\omega_j \tag{11.42}$$

The second alternative, or indirect method, applies the RFM to a temperature and rate of deformation dependent function given by

$$\eta_i=f\left(T_i,\dot{\gamma}_i\right) \tag{11.43}$$

where the gradient in the x-direction becomes

$$\frac{\partial \eta_i}{\partial x}=\frac{\partial f\left(T_i,\dot{\gamma}_i\right)}{\partial T}\frac{\partial T_i}{\partial x}+\frac{\partial f\left(T_i,\dot{\gamma}_i\right)}{\partial \dot{\gamma}}\frac{\partial \dot{\gamma}_i}{\partial x} \tag{11.44}$$

This gradient can be approximated using RFM as

$$\frac{\partial \eta_i}{\partial x}=\frac{\partial f\left(T_i,\dot{\gamma}_i\right)}{\partial T}\sum_{j=1}^{N}\frac{\partial \phi_T(r_{ij})}{\partial r}\frac{\partial r_{ij}}{\partial x}\alpha_j+\frac{\partial f\left(T_i,\dot{\gamma}_i\right)}{\partial \dot{\gamma}}\sum_{j=1}^{N}\frac{\partial \phi_\gamma(r_{ij})}{\partial r}\frac{\partial r_{ij}}{\partial x}\varsigma_j \tag{11.45}$$

Similarly, the y-gradient of the viscosity is written as

$$\frac{\partial \eta_i}{\partial y}=\frac{\partial f\left(T_i,\dot{\gamma}_i\right)}{\partial T}\frac{\partial T_i}{\partial y}+\frac{\partial f\left(T_i,\dot{\gamma}_i\right)}{\partial \dot{\gamma}}\frac{\partial \dot{\gamma}_i}{\partial y} \tag{11.46}$$

and can be represented with RFM using

$$\frac{\partial \eta_i}{\partial y} = \frac{\partial f\left(T_i, \dot{\gamma}_i\right)}{\partial T} \sum_{j=1}^{N} \frac{\partial \phi_T(r_{ij})}{\partial r} \frac{\partial r_{ij}}{\partial y} \alpha_j + \frac{\partial f\left(T_i, \dot{\gamma}_i\right)}{\partial \dot{\gamma}} \sum_{j=1}^{N} \frac{\partial \phi_\gamma(r_{ij})}{\partial r} \frac{\partial r_{ij}}{\partial y} \varsigma_j \quad (11.47)$$

Although the indirect method is more difficult to implement, it renders better results because the rate of deformation and temperature fields are smoother and better bounded than the viscosity field. For example, when using the power law shear thinning model, the viscosity goes to infinity when the rate of deformation goes to zero.

The Dirichlet velocity boundary conditions, given by

$$u_{x_i} = u_{a_x}; i \in \Gamma^{Du} \quad (11.48)$$

and

$$u_{y_i} = u_{a_y}; i \in \Gamma^{Du} \quad (11.49)$$

are approximated using

$$\sum_{j=1}^{N} \phi_u(r_{ij})\lambda_j = u_{a_x} \quad (11.50)$$

and

$$\sum_{j=1}^{N} \phi_u(r_{ij})\xi_j = u_{a_y} \quad (11.51)$$

The pressure Dirichlet boundary condition, for a fully developed velocity, can be written as

$$p_i = p_a; i \in \Gamma^{Dp} \quad (11.52)$$

which is terms of RBF is represented using

$$\sum_{j=1}^{N_p} \phi_p(r_{ij})\beta_j = p_a \quad (11.53)$$

$$\frac{\partial u_{x_i}}{\partial x} n_{x_i} + \frac{\partial u_{x_i}}{\partial y} n_{y_i} = 0; i \in \Gamma^{Dp} \quad (11.54)$$

$$\sum_{j=1}^{N} \left[\frac{\partial \phi_u(r_{ij})}{\partial r} \left(\frac{\partial r_{ij}}{\partial x} n_{x_i} + \frac{\partial r_{ij}}{\partial y} n_{y_i} \right) \right] \lambda_j = 0 \quad (11.55)$$

$$\frac{\partial u_{y_i}}{\partial x} n_{x_i} + \frac{\partial u_{y_i}}{\partial y} n_{y_i} = 0; i \in \Gamma^{Dp} \quad (11.56)$$

$$\sum_{j=1}^{N} \left[\frac{\partial \phi_u(r_{ij})}{\partial r} \left(\frac{\partial r_{ij}}{\partial x} n_{x_i} + \frac{\partial r_{ij}}{\partial y} n_{y_i} \right) \right] \xi_j = 0 \quad (11.57)$$

The above approach can be used to model non-Newtonian flows with a relatively high degree of accuracy. For example, Fig. 11.7 presents a comparison between a RFM solution

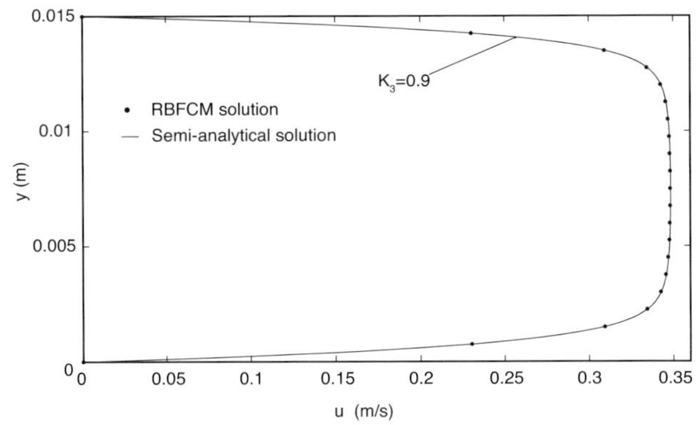

Figure 11.7: Comparison between a semi-analytical and predicted RFM velocity distribution for a pressure flow through a slit for a shear thinning polymer melt.

and a semi-analytical solution for the pressure driven slit flow of a shear thinning melt with a Carreau viscosity model defined by,

$$\eta(\dot{\gamma}, T) = \frac{K_1}{(1 - K_2\dot{\gamma})^{K_3}} \tag{11.58}$$

with constants $K_1 = 24000$ Pa-s, $K_2 = 1.9$ s and a shear thinning constant, K_3, of 0.9[4]. The geometry used to solve the problem was a 0.015 m long slit with a 0.015 m gap and pressure drop of 5000 Pa. As can be seen, the agreement between the solutions is excellent, even for a fluid with relatively high shear thinning behavior. It should be pointed out that the solution of this problem was achieved using a 1D discretization.

■ EXAMPLE 11.3.

Simulation of cavity flow with a Reynolds number of 100. The closed cavity flow is a classical problem used to validate the accuracy of the solution of the equation of motion for a Newtonian fluid flow with inertia effetcs. Here, we present the solution of this problem as presented by Estrada [4]. The geometry and conditions simulated by Estrada are schematically depicted in Fig. 11.8.

Here, the boundary conditions used are all Dirichlet conditions on the walls of the cavity; $u_x = 1$m/s and $u_y = 0$ on the upper wall of the cavity, and the remaining three walls with $u_x = u_y = 0$. Estrada used a viscosity of 1 Pa-s and a density of 100 kg/m^3. Using the above data and dimensions the characteristic Reynolds number for this flow problem is 100. To test the RFM solution, Estrada set up three different geometries with 729 (27×27), 1369 (37×37) and 1849 (43×43) collocation points as depicted in Fig. 11.9 The RFM solutions were compared to FEM solutions. In a similar study, Roldán compared a cavity flow without convective effects to FEM as well as BEM solutions and presented great agreement between the three solutions [17].

Figure 11.10 presents the velocity field inside the cavity for the intermediate size problem with 1369 collocation points. Note that not all collocation points are pre-

[4]A Carreau model constant K_3 of 0.9 is equivalent to a power law index, n, of 0.1.

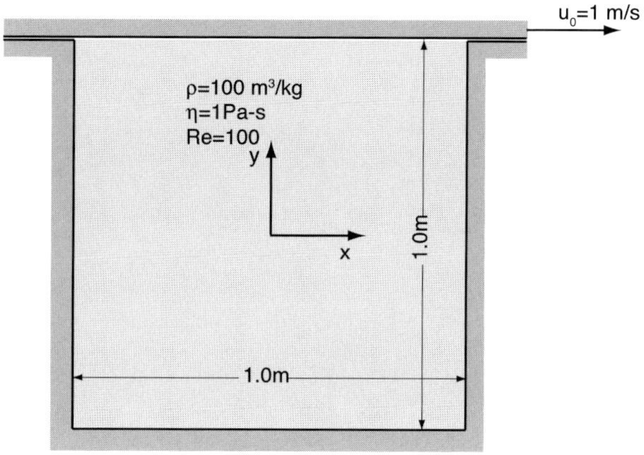

Figure 11.8: Schematic diagram of the cavity flow problem.

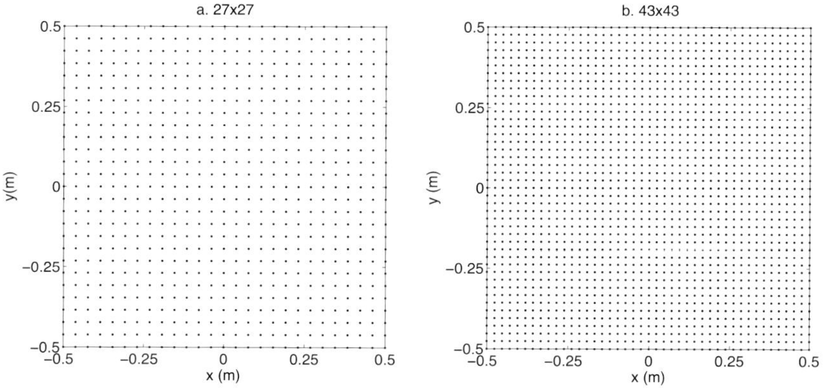

Figure 11.9: Geometry and collocation points for the cavity flow RF method solution.

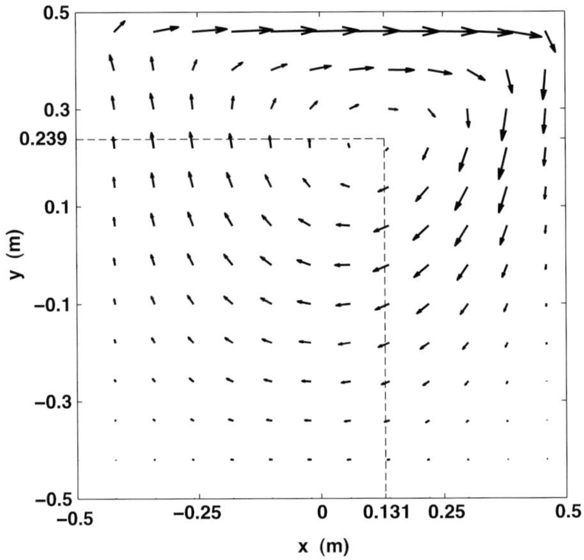

Figure 11.10: Velocity field in the cavity with the RFM prediction of the vortex center location for a Reynolds number of 100.

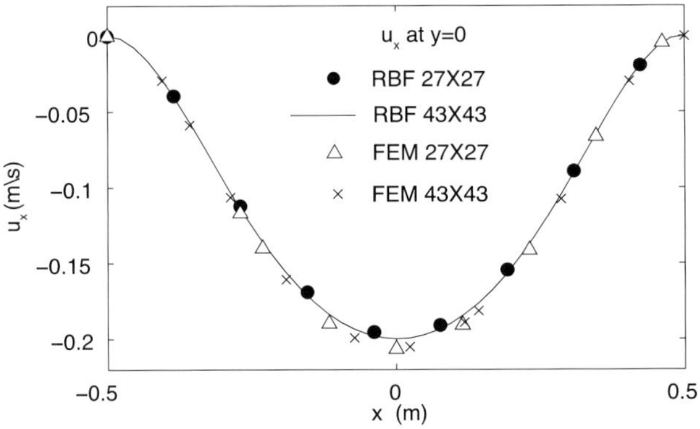

Figure 11.11: Comparison between the predicted velocity distribution in the x-direction at $y = 0$ using RFM and FEM for a cavity flow a Reynolds number of 100.

sented in the graph. The figure also shows the location of the vortex center at the (xy) location of (0.131, 0.239). This location is in agreement with the finite difference solution of the same problem, reported by Liao, of (0.133, 0.248) [15]. The difference in the center location position was only 2 and 4%, in the x- and y-directions, respectively.

Figures 11.11 and 11.12 present the x- and y-velocity components, respectively, along the x-axis ($y = 0$) inside the box, computed using RFM and FEM. The FEM solution is one done with 729 (27×27) and1849 (43×43) nodal points. As can be seen, the solutions are all in good agreement.

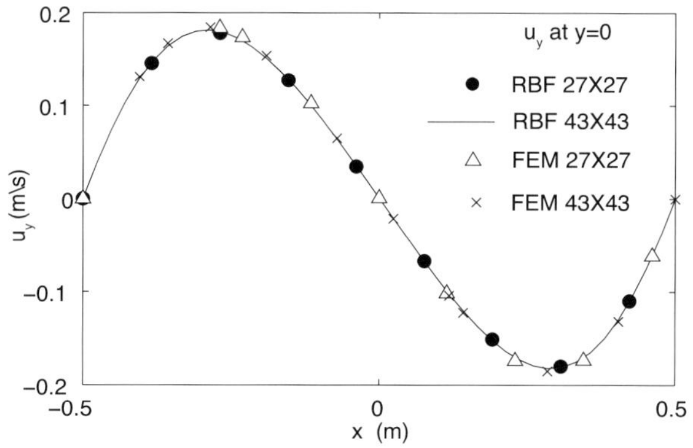

Figure 11.12: Comparison between the predicted velocity in the y-direction at $y = 0$ using RFM and FEM for a cavity flow a Reynolds number of 100.

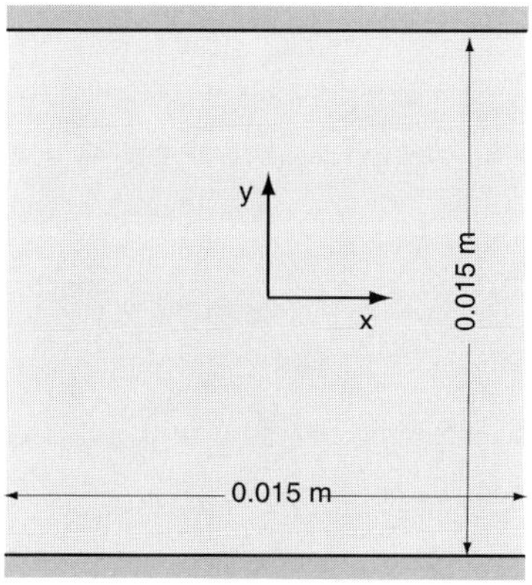

Figure 11.13: Schematic diagram of the flow problem with a fully coupled energy equation and momentum balance.

■ **EXAMPLE 11.4.**

Coupled energy and momentum balances for a pressure flow between parallel plates. In many flows in polymer processing, viscous dissipation is significant ($Br > 1$), with a resulting temperature rise that significantly affects the flow through a temperature dependent viscosity ($Na > 1$). With these types of flows, we have fully coupled energy and momentum equations. In order to test such a flow, Estrada [4] and Lopez and Osswald [5] simulated the pressure flow between two parallel plates for a rate of deformation and temperature dependent polymer melt, $\eta(\dot{\gamma}, T)$. The geometry used is a 0.015 m $\times$0.015 m square, with a fully developed flow, schematically depicted in Fig. 11.13.

The boundary conditions for the momentum balance were $u_x = u_y = 0$ on the upper and lower plates, $p = 105,000$ Pa at the left wall ($x = 0$) and $p = 0$ at the right wall ($x = 0.015$m). The thermal boundary conditions were $T = 200^o$C on the upper and lower plates, and insulated boundary conditions ($\partial T/\partial x$=0) on both

Table 11.1: Thermal Properties for the Coupled Heat Transfer Flow Problem

Parameter	Value
ρ	700 kg/m^3
C_p	2,100 J/kg/K
k	10.0 W/m/K

Table 11.2: Carreau and Arrhenius model constants for the Coupled Heat Transfer Flow Problem

Parameter	Value
K_1	2859.4 Pa-s
K_2	0.077 s
K_3	0.661
E_0	42048.3 J/mol
T_0	200°C

vertical walls. The thermal properties were considered constant and are given in Table 11.1.

The viscosity was modeled using a Carreau model with an Arrhenius temperature dependence given by

$$\eta(\dot{\gamma}, T) = \frac{a_T K_1}{(1 - a_T K_2 \dot{\gamma})^{K_3}} \tag{11.59}$$

where the temperature shift, a_T, is given by

$$a_T = e^{\frac{E_0}{R}\left(\frac{1}{T} - \frac{1}{T_0}\right)} \tag{11.60}$$

The constants for the viscosity model are given in Table 11.2.

In order to solve the coupled equation system, Estrada used RFM with a third order thin plate spline function and 740 collocation points arranged in a grid, while Lopez and Osswald also used 740 collocation points randomly arranged throughout the domain[5] (Fig. 11.14). The RFM solution was compared to an FDM solution. Four different cases were analyzed:

- Slit flow with a constant viscosity, $\eta = \mu$, (Newtonian),
- Slit flow with an Arrhenius temperature dependent viscosity, $\eta(T)$, (Newtonian-Arrhenius),
- Slit flow with a rate of deformation dependent viscosity, $\eta(\dot{\gamma})$, (Carreau), and
- Slit flow with a rate of deformation and Arrhenius temperature dependent viscosity, $\eta(\dot{\gamma}, T)$ (Carreau-Arrhenius).

[5]Lopez and Osswald prescribed a minimum distance of half the lengths found in the grid by Estrada. The minimum distance requirement is necessary to avoid a linear dependence between nodes that are too close to each other, making it difficult to distinguish them from nodes that are far from the clusters, leading to ill-conditioning in the linear set algebraic of equations. In addition, a minimum distance requirement avoids the existence of empty pockets void of collocation points.

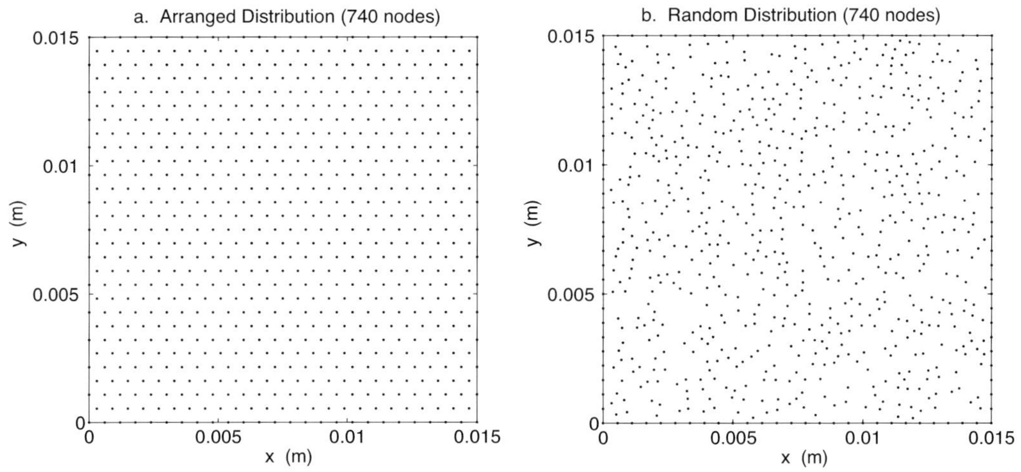

Figure 11.14: Geometry and collocation points for the coupled flow-heat transfer problem RF method solution, with grid-like and randomly arranged collocation points.

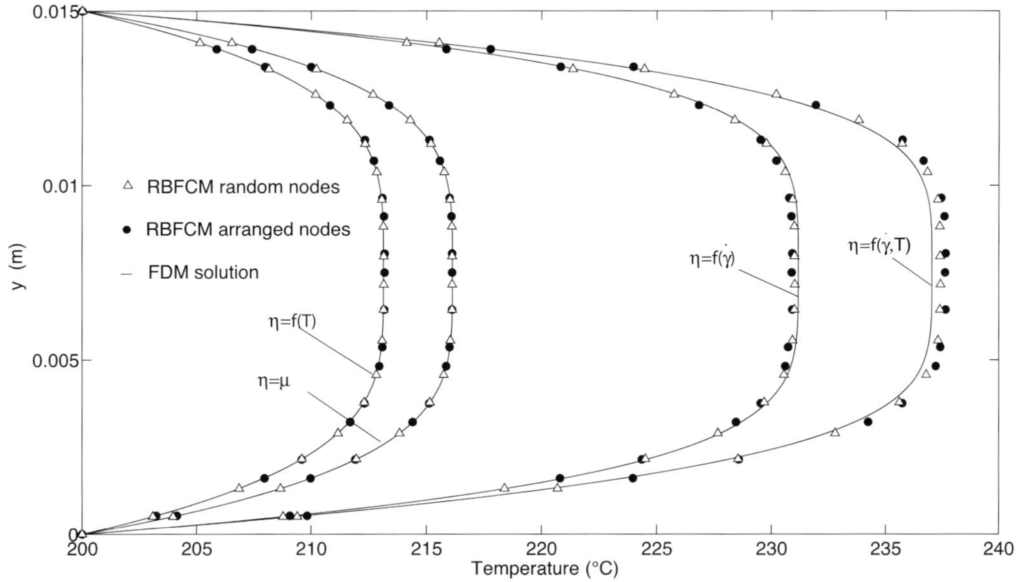

Figure 11.15: Comparison between the temperature field predicted using RFM with arranged and random collocation points and by FDM for the coupled flow-heat transfer problem.

Figures 11.15 and 11.16 compare the temperature and velocity profiles, respectively, for the steady-state, fully developed flow of the coupled flow-heat transfer pressure driven slit flow problem, using RFM and FDM. The agreement between the two solutions is excellent.

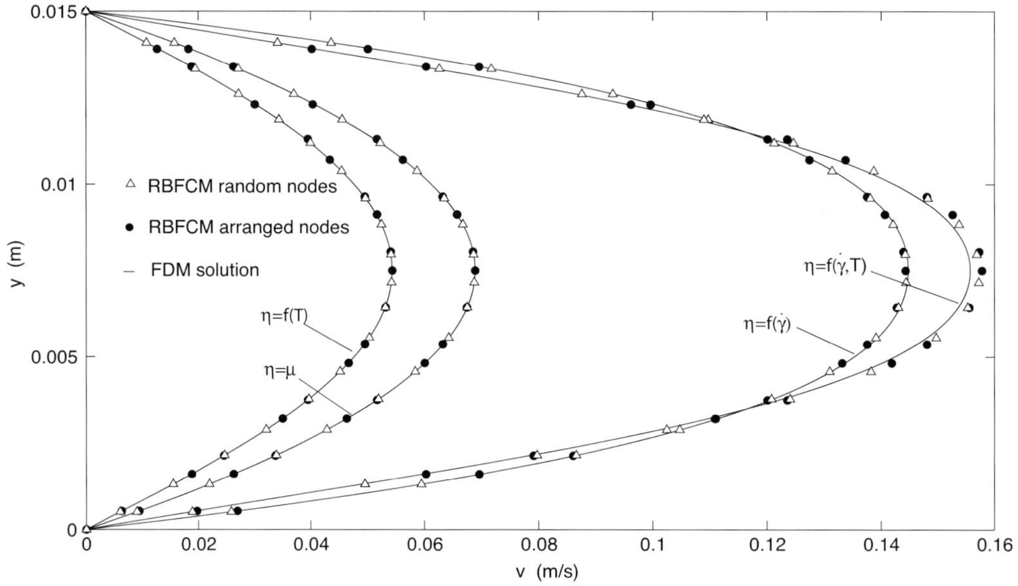

Figure 11.16: Comparison between the velocity distribution predicted by RFM and by FDM for the couple flow-heat transfer problem.

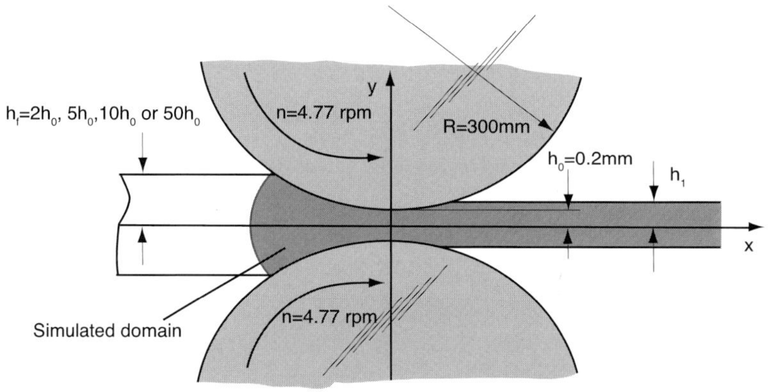

Figure 11.17: Schematic diagram of a calendering process fed with a finite sheet.

■ EXAMPLE 11.5.

Modeling the calendering process for Newtonian and shear thinning polymer melts. Using the RF method, López and Osswald [5] modeled the calendering process for Newtonian and non-Newtonian polymer melts. They used the same dimensions and process conditions used by Agassant *et al.* [1], schematically depicted in Fig. 11.17.

The problem was first solved for a Newtonian fluid (μ = 1000 Pa-s) with bank or fed-sheet-thickness to nip ratios, h_f/h_0, of 2, 5, 10 and 50. The geometry and

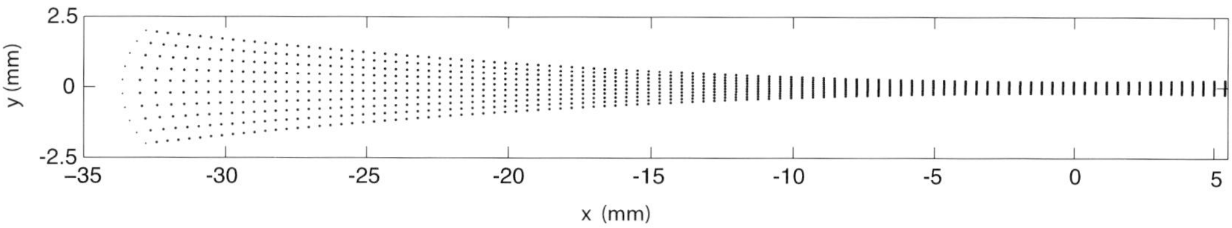

Figure 11.18: Geometry and collocation points for the calendering problem with a bank to nip ratio, h_f/h_0 of 10.

collocation points for $h_f/h_0 = 10$ are shown in Fig. 11.18. The boundary conditions are given by the velocity on the roll surfaces, a zero pressure at the entrance and exit surfaces as well as a zero stress at the entrance surface given by $\partial u_n/\partial \mathbf{n} = 0$. This boundary condition is imposed by setting the velocity of the first two collocation points of each row equal to each other. Furthermore, this problem must be manually iterated, since the final sheet thickness is not known *a priori*. Hence, a sheet separation and thickness is assumed for the first solution. This results in a pressure field with unrealistic oscillations and a point where $p = 0$ that does not coincide with the sheet separation. After the first solution, the separation point is moved to the same x-coordinate where $p = 0$. After a couple of iterations the correct sheet thickness and separation point are achieved, along with a smooth pressure distribution.

Figure 11.19 presents the pressure distribution along the x-axis for a Newtonian solution using several bank-to-nip ratios. The solutions are presented with the analytical predictions using McKelvey's lubrication approximation model presented in Chapter 6. The graph shows that the two solutions are in good agreement. Fig. 11.20 presents a sample velocity field for the Newtonian case with a bank-to-nip ratio of 10. As can be seen, the velocities look plausible and present the recirculation pattern predicted by McKelvey's lubrication approximation model and seen in experimental work done in the past [18].

The same collocation points and geometry for a bank-to-nip ration, h_f/h_0, of 50 were used to solve for the velocity fields and pressure distributions for non-Newtonian shear thinning polymer melts. A power law model with a consistency index, m, of 10^4 Pa-s^n and several power law indices, n, of 0.7, 0.5 and 0.3 were used. The pressure distribution along the x-axis for the shear thinning melt is presented in Fig. 11.21 for various power law indices and in Fig. 11.22 for a power law index of 0.3. Again, the RFM results were compared to analytical solutions presented in Chapter 6. As can be seen, the agreement is excellent. The FEM results presented by Agassant [1] are also in agreement with the RFM. The FEM results predict a slightly higher pressure than the analytical lubrication approximation prediction, whereas the RFM pressure predictions are slightly lower.

■ EXAMPLE 11.6.

Fully developed flow in an unwrapped screw extruder channel. To illustrate the type of problems that can be solved, using a 2 1/2D RFM formulation, López and Osswald [5] modeled the fully developed flow in an unwrapped screw channel as done by Griffith using FDM [9] in 1962. As discussed in Chapters 6 and 8,

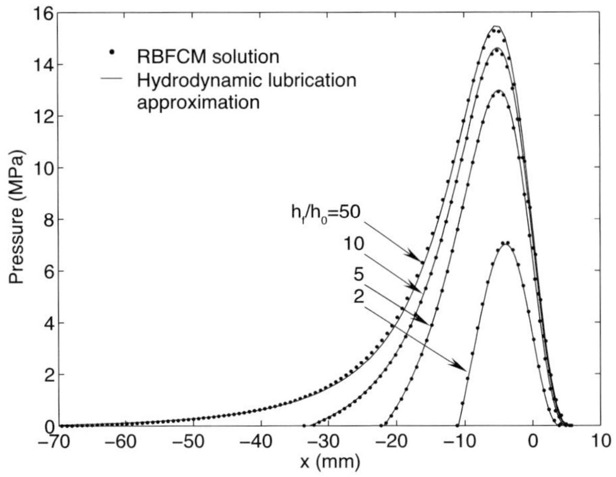

Figure 11.19: Comparison of lubrication approximation solution and RFM solution of the pressure profiles between the rolls for several values of bank, or fed sheet, to nip ratio for Newtonian viscosity model.

the flow in the metering section is a complex three-dimensional flow, that when modeled with a non-Newtonian, shear thinning viscosity, does not have an analytical solution. For their simulation, López and Osswald used the usual unwrapped screw geometry schematically depicted in Fig. 11.25 They used 626 collocation points, evenly distributed on the screw channel cross-section (Fig. 11.26).

A 60 mm$\times$6 mm channel geometry with a barrel velocity, u, of 0.5 m/s was used. The polymer melt was assumed as a shear thinning melt with a power law behavior. A consistency index, m, of 28000 Pa-s^n and a power law index, n, of 0.28 were used. López and Osswald solved for the flow field using different die restriction pressures between open discharge and a pressure high enough (50 MPa/m) that led to a negative volumetric throughput. Figure 11.27 presents the down-channel velocity field for a die restriction pressure gradient, $\partial p/\partial z$, of 20 MPa/m (200 bar/m). The combination of pressure and drag flow can be seen in the figure. Figure 11.28 presents the velocity field caused by the cross-channel component of the flow. Since leakage was neglected in this analysis, as expected, the net cross-flow throughput was zero. Finally, Fig. 11.29 presents a dimensionless throughput versus pressure build-up for the metering section of the screw. The results are compared to Griffith's FDM predictions. As can be seen, the curve computed using RFM with a power law index, n, of 0.28 falls between Griffith's FDM curves for $n = 0.2$ and $n = 0.4$.

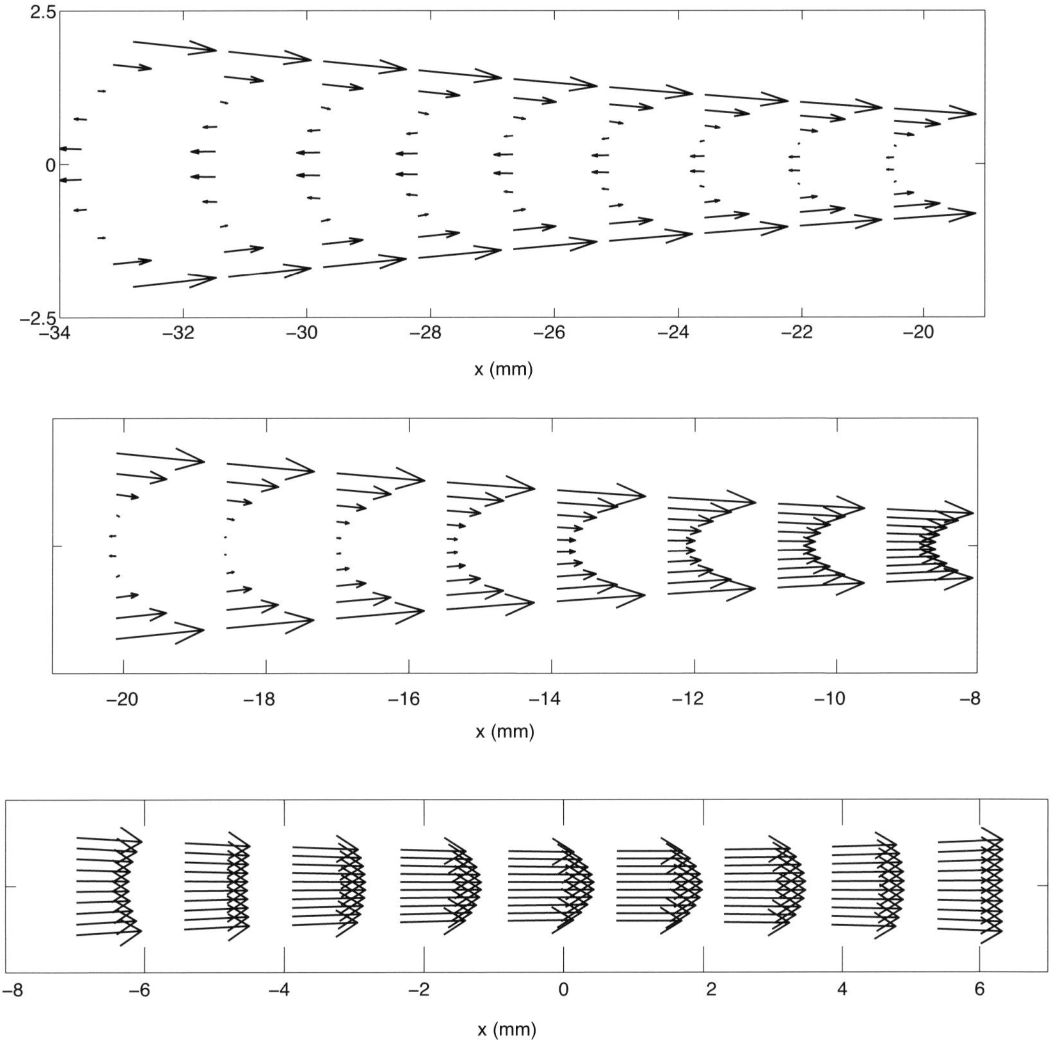

Figure 11.20: RFM solution of the velocity field during calendering of a Newtonian melt for a bank to nip ratio of 10.

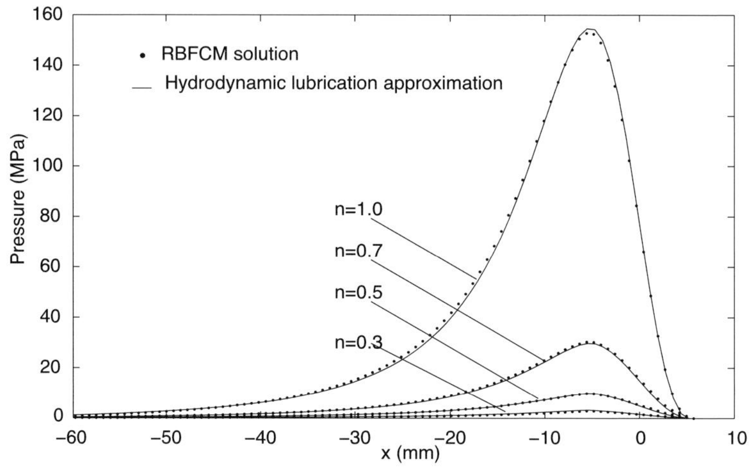

Figure 11.21: Comparison of lubrication approximation solution and RFM solution of the pressure profiles between the rolls for a bank-to-nip ratio of 10, and several power law indices using power law viscosity model.

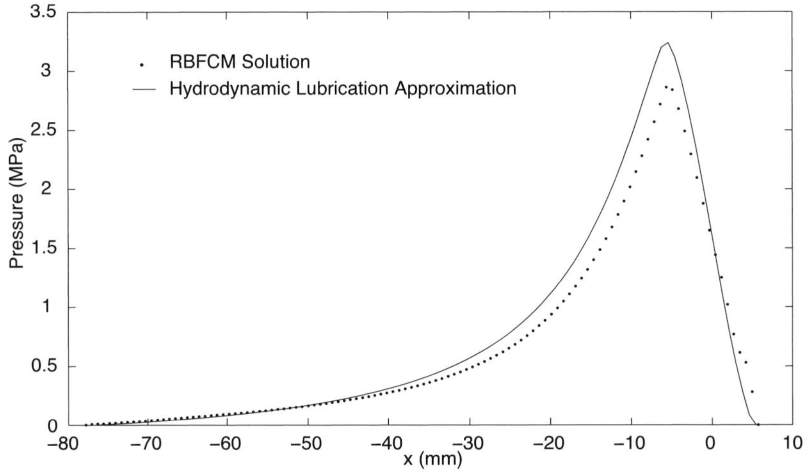

Figure 11.22: Comparison of lubrication approximation solution and RFM solution of the pressure profile between the rolls using a power law viscosity model with a power law index, n, of 0.3.

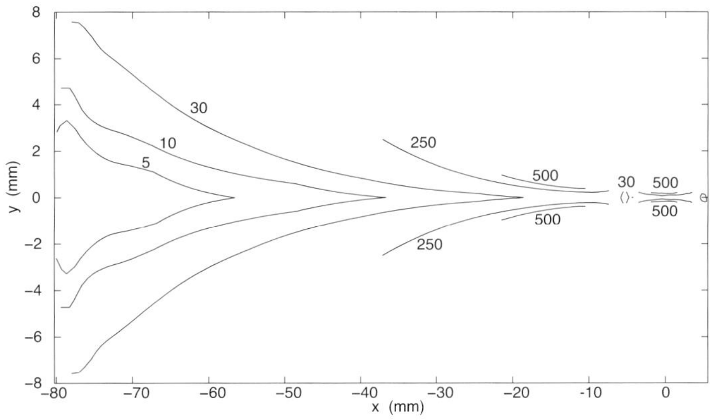

Figure 11.23: RFM rate of deformation profile (1/s units) between the rolls using a power law viscosity model with a power law index, n, of 0.5.

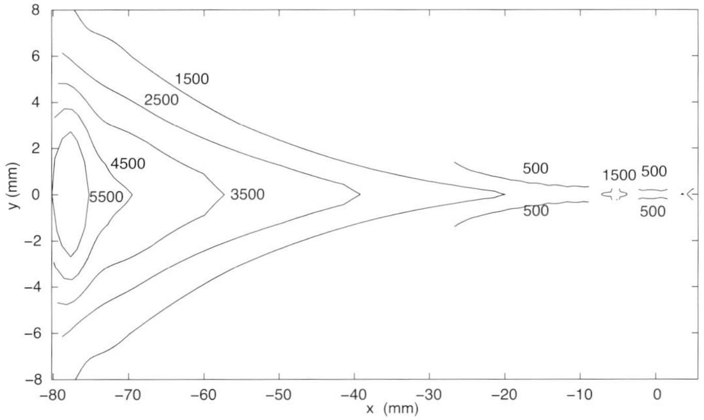

Figure 11.24: RFM viscosity profile (Pa-s units) between the rolls using a power law viscosity model with a power law index, n, of 0.5.

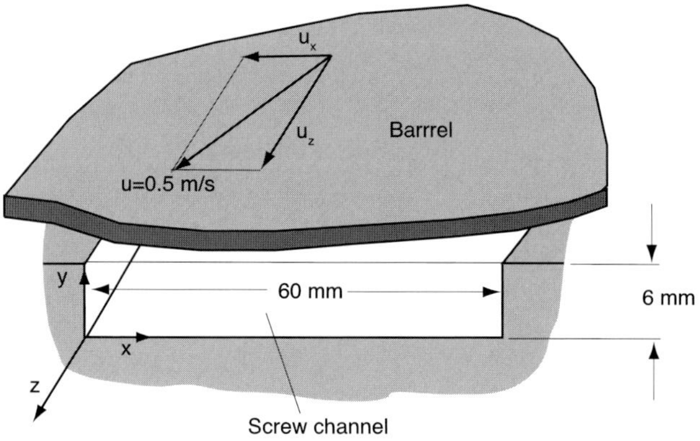

Figure 11.25: Unwrapped screw channel with conditions and dimensions.

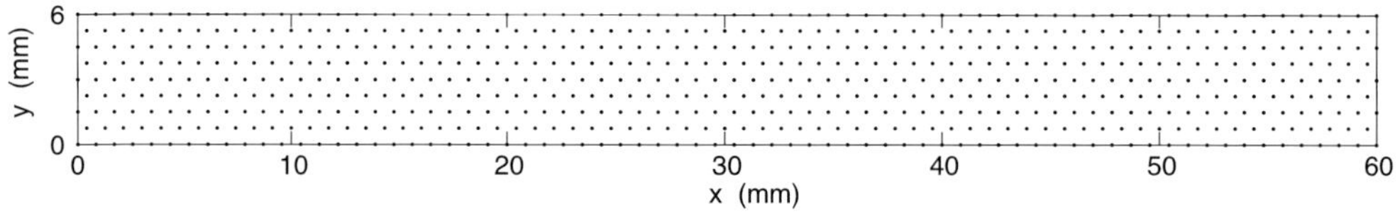

Figure 11.26: Geometry and collocation points to model the unwrapped screw channel.

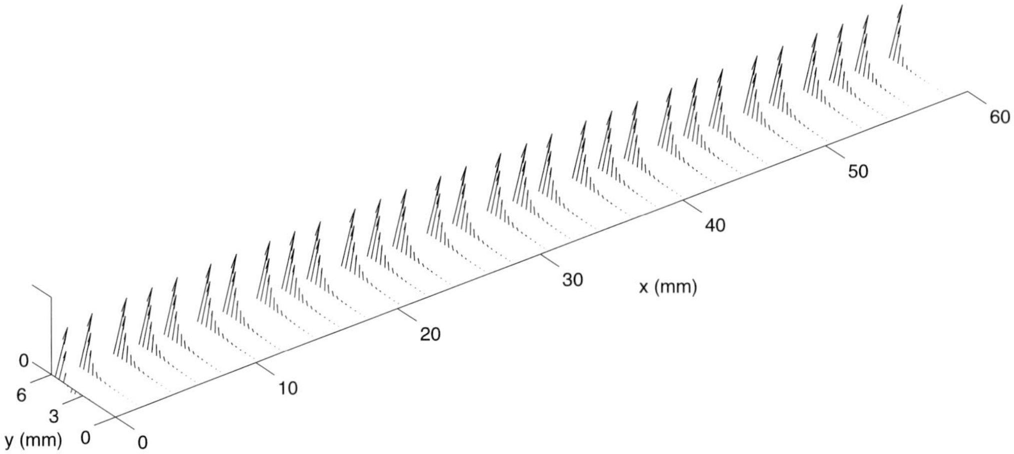

Figure 11.27: Down-channel velocity field for the unwrapped screw channel with a down-channel pressure gradient of 20MPa/m.

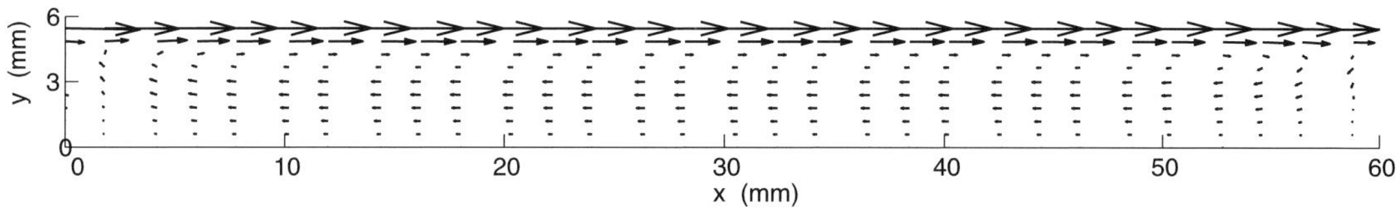

Figure 11.28: Cross-channel velocity field for the unwrapped screw channel with a down-channel pressure gradient of 20MPa/m.

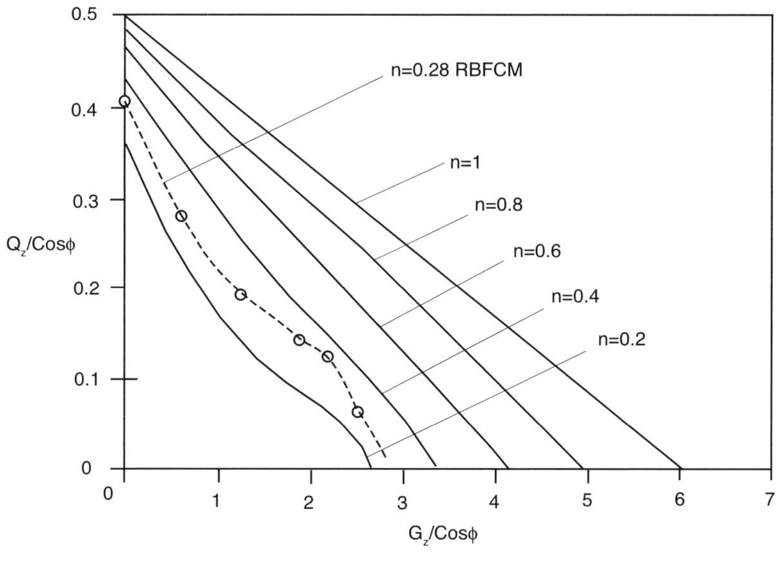

Figure 11.29: Comparison between the dimensionless screw characteristic curve computed using RFM and curves computed using FDM.

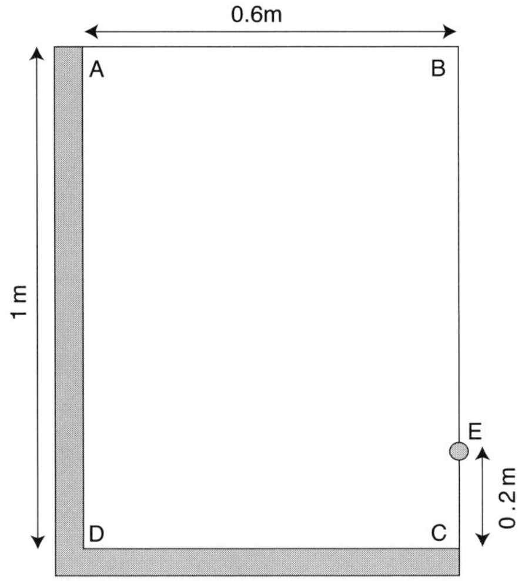

Figure 11.30: Schematic diagram of the NAFEMS benchmark test.

Problems

11.1 Estimate $\frac{\partial f(x,y)}{\partial x}$, $\frac{\partial f(x,y)}{\partial y}$ and $\nabla f(x,y)$ using radial basis functions for the field described by $f(x,y) = 30(y \cdot \sin(3x\pi) + \exp^{(x-2)(y-1)}) + 80$. Use a square domain of size 1×1. Compare the numerical results againts the analytical solution.

a) Use a thin plate spline of order two ($a = 2$), and 200 nodes uniformily arranged.

b) Use thin plate spline of order two (a=2) and a random distribution of 200 nodes following the rule of the minimum distance between points. Use 0.01925, 0.03850 and 0.05775 as the minimun distance values.

11.2 The International Association for the Engineering Analysis Community (NAFEMS) defines a test for the evaluation of the diffusive term of the energy equation using Dirichlet, Neumann and Robin boundary conditions. In this test, the domain defined in the Fig. 11.30 has as boundary conditions, $T = T_a$ in $\bar{CD}$, $-k\frac{\partial T}{\partial x} = 0$ in $\bar{AC}$ and $-k\frac{\partial T}{\partial x} = h(T - T_\infty)$ in $\bar{AB}$ and $\bar{BD}$, where T_a=100°C, k=52 W/mK, h=750 W/m^{2}K and T_∞=0°C. According to the benchmark test, the exact temperature for point E(0.6,0.2) is 18.25375654°C. Write a program using RFM to solve for the temperature field. Use a node distribution that does not include the point (0.6,0.2), and after the simulation, obtain that value using the interpolation with RBFs. Compare the numerical solution with the analytical one.

11.3 The non-isothermal Couette flow between concentric cylinders depicted by Fig. 11.31, has an analytical solution when the viscosity is considered as a constant. The analytical

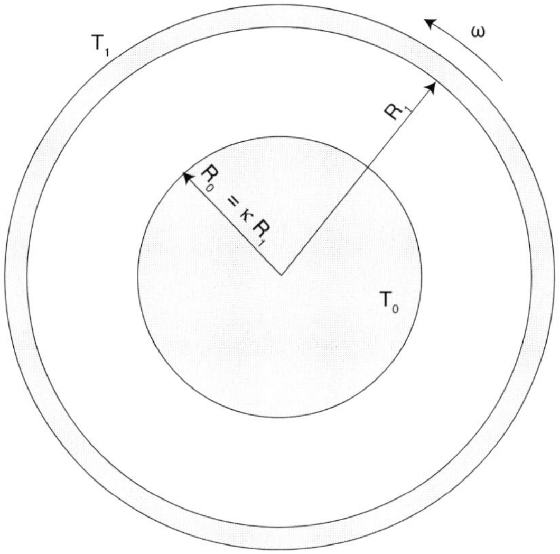

Figure 11.31: Schematic diagram of the Couette flow between concentric cylinders.

solution for the velocity and temperature fields are described by,

$$u_\theta = \omega R_1 \left[\frac{\frac{r}{\kappa R_1} - \frac{\kappa R_1}{r}}{\frac{1}{\kappa} - \kappa} \right]$$

$$T = T_1 + \frac{\mu(\omega R_1)^2}{k(\frac{1}{q^2} - 1)} \left[\frac{\ln(r/R_1)}{\ln(\kappa)} + \frac{R_1^2(\frac{1}{R^2} - \frac{1}{r^2})}{\frac{1}{\kappa^2} - 1} \right] - \frac{T_1 - T_0}{\ln(\kappa)} \ln(r/R_1)$$

Write a program to solve by means of RFM the equation of motion and using the velocity field, calculate viscous dissipation and solve for the energy equation. Neglect inertial and convective effects. Consider T_0=200°C, T_1=150°C, μ=24000 Pa-s, k=0.267 W/mK, R_0=0.1 m, R_1=0.13 m, κ=0.769, ω=0.496 rad/s. Compare the numerical results with the analytical solution. Hint: The couette flow is constant along the angular direction, hence, it is no necessary to use the whole domain.

11.4 Express the x- and y-components of the transient equation of motion using RFM. Consider an explicit formulation. Include the same terms of the steady state formulation presented in this chapter. Consider the viscosity as a function of the rate of deformation.

11.5 For the coupled heat transfer and flow problem (slit flow) presented in this chapter the solution of the energy and equation of motion was obtained considering a two dimensional domain. However, none of the primary variables changes in the x-direction, and the velocity only has a component along the x-axis. Simplify the equations and conditions using a 1D formulation. Recall that the pressure is constant along the y-direction and has a constant drop $(\Delta P/L)$ along the x-axis. Write a program to simulate the same cases as in the example, but using the one-dimensional formulation.

REFERENCES

1. J.F. Agassant, P. Avenas, J.-Ph. Sergent, and P.J. Carreau. *Polymer Processing - Principles and Modeling*. Hanser Publishers, Munich, 1991.
2. M.D. Buhmann. Radial basis functions. *Acta Numerica*, pages 1–38, 2000.
3. B.A. Davis, P.J. Gramann, J.C. Maetzig, and T.A. Osswald. The dual-reciprocity method for heat transfer in polymer processing. *Engineering Analysis with Boundary Elements*, 13:249–261, 1994.
4. O. A. Estrada. Desarrollo de un modelo computacional basado en funciones de interpolación de base radial para la simulación en 2D del flujo no isotérmico de polímeros a través de un cabezal para perfilería. Master's thesis, Universidad EAFIT, 2005.
5. O.A. Estrada, I.D. López-Gómez, and T.A. Osswald. Modeling the non-newtonian calendering process using a coupled flow and heat transfer radial basis functions collocation method. *Journal of Polymer Technology*, 2005.
6. O.A. Estrada, I.D. López-Gómez, C. Roldán, M. del P. Noriega, W.F. Flórez, and T.A. Osswald. Numerical simulation of non-isothermal flow of non-newtonian incompressible fluids, considering viscous dissipation and inertia effects, using radial basis function interpolation. *Numerical Methods for Heat and Fluid Flow*, 2005.
7. Omar Estrada, Iván López, Carlos Roldán, Maria del Pilar Noriega, and Whady Flórez. Solution of steady and transient 2D-energy equation including convection and viscous dissipation effects using radial basis function interpolation. *Journal of Applied Numerical Mathematics*, 2005.
8. G.E. Fasshauer. Solving Partial Differential Equations by Collocation with RBFs., in: *Surface Fitting and Multiresolution Methods*. Vanderbilt Press, 1996.
9. R.M. Griffith. Fully developed flow in screw extruders. *I & EC Fundamentals*, 1(3):180–187, 1962.
10. R.L. Hardy. Multi-quadratic equations of topography and other irregular surfaces. *J. Geophysics Res.*, 176:1905–1915, 1971.
11. R.L. Hardy. Theory and applications of the multi-quadratic-biharmonic method: 20 years of discovery. *Comp. Math. Applic.*, 19(8/9):163–208, 1990.
12. E.J. Kansa. Multiquadratics - a scattered data approximation scheme with applications to computational fluid dynamics: I. surface approximations and partial derivative estimates. *Comput. Math. Appl.*, 19(6-8):127–145, 1990.
13. E.J. Kansa. Multiquadratics- a scattered data approximation scheme with applications to computational fluid dynamics: II. solutions to parabolic, hyperbolic, and elliptic partical differential equations. *Comp. Math. Appl.*, 19(6-8):147–161, 1990.
14. J. Li and C.S. Chen. Some observations on unsymmetric RBF collocation methods for convection-diffusion problems. *Inter. Journal for Numerical Methods in Eng.*, 57:1085–1094, 2003.
15. S.J. Liao. Higher order stream function-vorticity formulation of 2D steady-state Navier-Stokes equation. *International Journal for Numerical Methods in Fluids*, 15:595–612, 1992.
16. N. Mai-Duy and R.I. Tanner. Computing non-Newtonian fluid flow with radial basis function networks. *International Journal for Numerical Methods in Fluids*, 48:1309–1336, 2005.
17. C. Roldan. Implementación computacional usando el método de colocación con funciones de base radial para el modelo de stokes en 2d de fluidos newtonianos con aplicaciones en flujos de polímeros. Diploma thesis, EAFIT University, 2005.
18. W. Unkrüer. *Beitrag zur Ermittlung des Druckverlaufes und der Fließvorgänge im Walzenspalt bei der Kalenderverarbeitung von PVC-hart zu Folien*. PhD thesis, RWTH-Aachen, 1970.

INDEX

S